Lecture Notes in Computer Science   5728

Commenced Publication in 1973
Founding and Former Series Editors:
Gerhard Goos, Juris Hartmanis, and Jan '

Alexander Kurz   Marina Lenisa
Andrzej Tarlecki (Eds.)

Algebra and Coalgebra in Computer Science

Third International Conference, CALCO 2009
Udine, Italy, September 7-10, 2009
Proceedings

 Springer

Volume Editors

Alexander Kurz
University of Leicester
Department of Computer Science
University Road, Leicester, LE1 7RH, UK
E-mail: kurz@mcs.le.ac.uk

Marina Lenisa
Università di Udine
Dipartimento di Matematica e Informatica
via delle Scienze, 206, 33100 Udine, Italy
E-mail: marina.lenisa@dimi.uniud.it

Andrzej Tarlecki
Warsaw University
Faculty of Mathematics, Informatics and Mechanics
ul. Banacha 2, 02-097 Warsaw, Poland
E-mail: tarlecki@mimuw.edu.pl

Library of Congress Control Number: 2009932885

CR Subject Classification (1998): F.4, F.3, F.2, I.1, G.2, G.4, D.2.1

LNCS Sublibrary: SL 1 – Theoretical Computer Science and General Issues

ISSN   0302-9743
ISBN-10   3-642-03740-2 Springer Berlin Heidelberg New York
ISBN-13   978-3-642-03740-5 Springer Berlin Heidelberg New York

springer.com

© Springer-Verlag Berlin Heidelberg 2009
Printed in Germany

Typesetting: Camera-ready by author, data conversion by Scientific Publishing Services, Chennai, India
Printed on acid-free paper   SPIN: 12740369   06/3180   5 4 3 2 1 0

Preface

CALCO, the Conference on Algebra and Coalgebra in Computer Science, is a high-level, bi-annual conference formed by joining CMCS (the International Workshop on Coalgebraic Methods in Computer Science) and WADT (the Workshop on Algebraic Development Techniques). CALCO brings together researchers and practitioners to exchange new results related to foundational aspects and both traditional and emerging uses of algebras and coalgebras in computer science. The study of algebra and coalgebra relates to the data, process and structural aspects of software systems. CALCO is supervised by the Steering Committee, co-chaired by José Luiz Fiadeiro and Jan Rutten.

The first two CALCO conferences took place in 2005 in Swansea, Wales, and in 2007 in Bergen, Norway. CALCO 2009, the third event in the series, took place in Udine, Italy, during September 7–10, 2009.

CALCO 2009 received 42 submissions, out of which 23 were selected for presentation at the conference. In spite of a relatively low number of submissions, their standard was generally very high. The selection process was carried out by the Program Committee, taking into account the originality, quality and relevance of the material presented in each submission, based on the opinions of expert reviewers, four for each submission. The selected and revised papers are included in this volume, together with the contributions from the invited speakers Mai Gehrke, Conor McBride, Prakash Panangaden and Gordon Plotkin.

CALCO 2009 was preceded by two workshops. The CALCO Young Researchers Workshop, CALCO-jnr, was dedicated to presentations by PhD students and by those who completed their doctoral studies within the past few years. CALCO-jnr was organized by Magne Haveraaen, Marina Lenisa, John Power and Monika Seisenberger. The CALCO-tools Workshop, organized by Luigi Liquori and Grigore Roşu, provided presentations of tools, with good time allotted for demonstrations of the running systems. The papers presenting the tools also appear in this volume.

We wish to thank all the authors who submitted their papers to CALCO 2009, the Program Committee for its hard work, and the referees who supported the Program Committee in the evaluation and selection process.

We are grateful to the University of Udine for hosting CALCO 2009 and to the Organizing Committee, chaired by Marina Lenisa, for all the local arrangements. We also thank the Amga spa and the Net spa of Udine, the Municipality of Udine, the International Centre for Mechanical Sciences of Udine, and the Fondazione Crup for their financial support. At Springer, Alfred Hofmann and his team supported the publishing process. We gratefully acknowledge the use of EasyChair, the conference management system by Andrei Voronkov.

June 2009

Alexander Kurz
Andrzej Tarlecki
Marina Lenisa

Organization

CALCO Steering Committee

Jiří Adámek	University of Braunschweig, Germany
Michel Bidoit	INRIA Saclay—Ile-de-France, France
Corina Cîrstea	University of Southampton, UK
José Fiadeiro (Co-chair)	University of Leicester, UK
H. Peter Gumm	Philipps University, Marburg, Germany
Magne Haveraaen	University of Bergen, Norway
Bart Jacobs	University of Nijmegen, The Netherlands
Hans-Jörg Kreowski	University of Bremen, Germany
Ugo Montanari	University of Pisa, Italy
Larry Moss	Indiana University, Bloomington, USA
Till Mossakowski	University of Bremen and DFKI Lab Bremen, Germany
Peter Mosses	University of Wales, Swansea, UK
Fernando Orejas	Polytechnical University Catalonia, Barcelona, Spain
Francesco Parisi-Presicce	Universitá di Roma La Sapienza, Italy
John Power	University of Bath, UK
Horst Reichel	Technical University of Dresden, Germany
Markus Roggenbach	University of Wales Swansea, UK
Jan Rutten (Co-chair)	CWI and Free University, Amsterdam, The Netherlands
Andrzej Tarlecki	Warsaw University, Poland

CALCO 2009 Program Committee

Luca Aceto	Reykjavik University, Iceland
Stephen Bloom	Stevens Institute of Technology, Hoboken, USA
Marcello Bonsangue	Leiden University, The Netherlands
Corina Cîrstea	University of Southampton, UK
Andrea Corradini	University of Pisa, Italy
José Fiadeiro	University of Leicester, UK
Rolf Hennicker	University of Munich, Germany
Furio Honsell	University of Udine, Italy
Bart Jacobs	University of Nijmegen, The Netherlands
Bartek Klin	University of Warsaw, Poland
Alexander Kurz (Co-chair)	University of Leicester, UK
Stefan Milius	University of Braunschweig, Germany

Ugo Montanari	University of Pisa, Italy
Larry Moss	Indiana University, Bloomington, USA
Till Mossakowski	DFKI Lab Bremen and University of Bremen, Germany
Dirk Pattinson	Imperial College London, UK
Dusko Pavlovic	Kestrel Institute, USA
Jean-Eric Pin	CNRS-LIAFA Paris, France
John Power	University of Bath, UK
Grigore Roşu	University of Illinois, Urbana, USA
Jan Rutten	CWI and Free University, Amsterdam, The Netherlands
Davide Sangiorgi	University of Bologna, Italy
Lutz Schröder	DFKI Lab Bremen and University of Bremen, Germany
Eugene Stark	State University of New York, USA
Andrzej Tarlecki (Co-chair)	Warsaw University, Poland
Yde Venema	University of Amsterdam, The Netherlands
James Worrell	University of Oxford, UK

CALCO 2009 Organizing Committee

Fabio Alessi	University of Udine, Italy
Alberto Ciaffaglione	University of Udine, Italy
Pietro Di Gianantonio	University of Udine, Italy
Davide Grohmann	University of Udine, Italy
Furio Honsell	University of Udine, Italy
Marina Lenisa (Chair)	University of Udine, Italy
Marino Miculan	University of Diane, Italy
Ivan Scagnetto	University of Udine, Italy

Table of Contents

Stone Duality

Game Theory

Graph Transformation

Software Development Techniques

CALCO Tools Workshop

Adequacy for Infinitary Algebraic Effects
(Abstract)

Gordon Plotkin

Laboratory for the Foundations of Computer Science,
School of Informatics, University of Edinburgh

Moggi famously proposed a monadic account of computational effects [Mog89, Mog91, BHM02]; this account includes the computational λ-calculus, a core call-by-value functional programming language. Moggi endowed the computational λ-calculus with an effect-oriented denotational semantics by taking the denotations of terms to be morphisms in the Kleisli category of the monad. The question naturally then arises as to whether one can give a correspondingly general treatment of operational semantics.

In the algebraic theory of effects [PP02, PP04, HPP06], a refinement of Moggi's theory, the effects are obtained by appropriate operations, and Moggi's monad is then generated by an equational theory over these operations. In a previous paper with John Power [PP01] we gave a general adequacy theorem for the computational λ-calculus for the case of monads generated by finitary operations. This covers examples such as (probabilistic) nondeterminism and exceptions. The main idea is to evaluate terms symbolically in the absolutely free algebra with the same signature as the equational theory. Without recursion, the evaluated terms are finite; with recursion, they may be infinitely deep.

In general, however, one needs infinitary operations, for example for global state, or for interactive I/O. We show that the previous work can be extended to include such operations by allowing infinitely wide, and possibly also infinitely deep, symbolic evaluation terms. We can also define a general contextual equivalence for any monad and characterise it operationally when the monad is given as a free algebra for a, possibly infinitary, equational theory.

There remain a variety of interesting problems. One can think of the operations of the algebraic theory as *effect constructors*: they create the effects at hand. There are also *effect deconstructors* which react to effects previously created: a typical example is exception-handling. It should be possible to extend the operational treatment to include deconstructors via the homomorphic approach to deconstructors described in [PPr09].

There are also situations where the operational semantics given by the general theory does not immediately correspond to the natural one for the monad at hand. One example is the side-effects monad, where the general theory provides no state component associated to the term being evaluated. Another is that of non-determinism as treated in process calculi, e.g., [Mil80], where the behaviour of a choice between two terms is determined in terms of the behaviours of either of the two terms, rather than as an undetermined move to one of them. The paper [PP08] provides some mathematics intended to help obtain a treatment of side-effects; there the expected pairs of states and terms are treated in terms

A. Kurz, M. Lenisa, and A. Tarlecki (Eds.): CALCO 2009, LNCS 5728, pp. 1–2, 2009.

of a tensor of a coalgebra of states and the algebra of computations. The operational semantics of choice in process-calculus style may be achievable in a general framework by building some of the equational theory—such as that of the choice operator—into the syntax and operational semantics.

References

[BHM02] Benton, N., Hughes, J., Moggi, E.: Monads and effects. In: Barthe, G., Dybjer, P., Pinto, L., Saraiva, J. (eds.) APPSEM 2000. LNCS, vol. 2395, pp. 42–122. Springer, Heidelberg (2002)

[HPP06] Hyland, J.M.E., Plotkin, G.D., Power, A.J.: Combining effects: sum and tensor. In: Artemov, S., Mislove, M. (eds.) Clifford Lectures and the Mathematical Foundations of Programming Semantics. TCS, vol. 357(1-3), pp. 70–99 (2006)

[Mil80] Milner, A.J.R.G.: A Calculus of Communicating Systems. Springer, Heidelberg (1980)

[Mog89] Moggi, E.: Computational lambda-calculus and monads. In: Proc. 3rd. LICS, pp. 14–23. IEEE Press, Los Alamitos (1989)

[Mog91] Moggi, E.: Notions of computation and monads. Inf. & Comp. 93(1), 55–92 (1991)

[PP01] Plotkin, G.D., Power, A.J.: Adequacy for algebraic effects. In: Honsell, F., Miculan, M. (eds.) FOSSACS 2001. LNCS, vol. 2030, pp. 1–24. Springer, Heidelberg (2001)

[PP02] Plotkin, G.D., Power, A.J.: Notions of Computation Determine Monads. In: Nielsen, M., Engberg, U. (eds.) FOSSACS 2002. LNCS, vol. 2303, pp. 342–356. Springer, Heidelberg (2002)

[PP04] Plotkin, G.D., Power, A.J.: Computational effects and operations: an overview. In: Escardó, M., Jung, A. (eds.) Proc. Workshop on Domains VI. ENTCS, vol. 73, pp. 149–163. Elsevier, Amsterdam (2004)

[PP08] Plotkin, G.D., Power, A.J.: Tensors of comodels and models for operational semantics. In: Proc. MFPS XXIV. ENTCS, vol. 218, pp. 295–311. Elsevier, Amsterdam (2008)

[PPr09] Plotkin, G.D., Pretnar, M.: Handlers of algebraic effects. In: Castagna, G. (ed.) ESOP 2009. LNCS, vol. 5502, pp. 80–94. Springer, Heidelberg (2009)

Algebras for Parameterised Monads

Robert Atkey

LFCS, School of Informatics, University of Edinburgh, UK
bob.atkey@ed.ac.uk

Abstract. Parameterised monads have the same relationship to adjunctions with parameters as monads do to adjunctions. In this paper, we investigate algebras for parameterised monads. We identify the Eilenberg-Moore category of algebras for parameterised monads and prove a generalisation of Beck's theorem characterising this category. We demonstrate an application of this theory to the semantics of type and effect systems.

1   Introduction

Monads [7] have a well-known relationship with algebraic theories [10], and have also been used by Moggi [8] to structure denotational semantics for programming languages with effects. Plotkin and Power [9] have used the connection to algebraic theories to generate many of the monads originally identified by Moggi as useful for modelling effectful programming languages. The operations of the algebraic theories are in direct correspondence with the effectful operations of the programming language.

In previous work [1] we have argued that a generalisation of monads, *parameterised monads*, are a useful notion for interpreting programming languages that record information about the effects performed in the typing information [4]. The generalisation is to consider unit and multiplication operations for functors $T : S^{\mathrm{op}} \times S \times C \to C$ for some parameterising category S and some base category C. The two S parameters are used to indicate information about the start and end states of a given computation; thus a morphism $x \to T(s_1, s_2, y)$ represents a computation that, given input of type x, starts in a state described by s_1 and ends in a state described by s_2, with a value of type y.

In this paper, we investigate the connection between parameterised monads and a notion of algebraic theories with parameters. In our previous work [1], we noted that parameterised monads arise from adjunctions with parameters— pairs of functors $F : S \times C \to A$ and $G : S^{\mathrm{op}} \times A \to C$ that, roughly, are adjoint for every $s \in S$. Here, we extend this relationship to show that there are natural notions of Kleisli category and Eilenberg-Moore category for parameterised monads, and that they are respectively initial and final in the category of adjunctions with parameters giving a particular parameterised monad.

We then go on to consider the appropriate notion of algebraic theory for parameterised monads. This turns out to require operations that have an arity, as with normal algebraic theories, and also domain and codomain sorts that are objects of S. We write such operations as $\sigma : s_2 \leftarrow s_1^X$, where X is the arity. These

A. Kurz, M. Lenisa, and A. Tarlecki (Eds.): CALCO 2009, LNCS 5728, pp. 3–17, 2009.

theories are similar to multi-sorted algebraic theories, except that all argument positions in the operation have the same sort. Multi-sorted algebraic theories can be represented as normal monads on presheaf categories over the set of sorts. Our restriction of multi-sorted algebraic theories allows us to expose the sorts in the parameters of the parameterised monad itself.

Algebras for our notion of algebraic theories with parameters are given by functors $a : S^{op} \to C$, with morphisms $\sigma_a : a(s_1)^X \to a(s_2)$ for operations $\sigma : s_2 \leftarrow s_1^X$.

A central result in relating algebraic theories and monads is Beck's theorem [2,7], which characterises the Eilenberg-Moore category of T-algebras for a given monad T in terms of the preservation and creation of coequalisers. In this paper, we prove a generalisation of Beck's theorem for parameterised monads.

As an application of the theory we have developed, we use algebraic theories with parameters to give a semantics to a toy programming language that records information about effects performed.

Overview. In the next section we recall the definition of parameterised monad, and develop the relationship between adjunctions with parameters and parameterised monads, in particular describing the category of algebras for a given monad. In Section 3, we define a notion of algebraic theory with parameters, where the free algebras give rise to parameterised monads. To show that the category of such algebras is isomorphic to the category of algebras for the derived monad, we prove a generalisation of Beck's theorem in Section 4. In Section 5, we consider the case when the parameterising category has structure given by functors and natural transformations. We then apply the results developed to the semantics of type and effect systems in Section 6.

2   Parameterised Monads

Definition. Assume a small category S. An S-parameterised monad on another category C is a 3-tuple $\langle T, \eta, \mu \rangle$, where T is a functor $S^{op} \times S \times C \to C$, the unit η is a family of arrows $\eta_{s,x} : x \to T(s, s, x)$, natural in x and dinatural in s and the multiplication μ is a family of arrows $\mu_{s_1 s_2 s_3 x} : T(s_1, s_2, T(s_2, s_3, x)) \to T(s_1, s_3, x)$, natural in s_1, s_3 and x and dinatural in s_2. These must make the following diagrams commute:

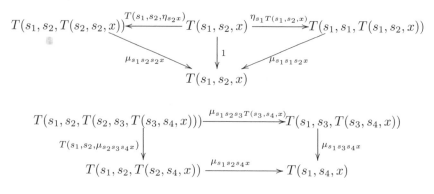

Example 1. Every non-parameterised monad is a parameterised monad for any category of parameters. Set $T(s_1, s_2, x) = Tx$.

Example 2. Our main motivating example for introducing the concept of parameterised monad is for modelling global state in a programming language, where the type of the state may change over time. Take S to be a category of state "types", with a terminal object representing the empty state. Assume that C is cartesian closed, and that there is a functor $\widehat{\cdot} : S \to C$, preserving the terminal object. Take $T(s_1, s_2, x) = (x \times \widehat{s_2})^{\widehat{s_1}}$, with the evident unit and multiplication.

Example 3. It is well-known that (in Set) every monoid M gives a monad $Tx = M \times x$, where the unit of the monad is given using the unit of the monoid, and likewise for multiplication. Analogously, every (small) category S_0 gives an S-parameterised monad, where S is a subcategory of S_0. Set $T(s_1, s_2, x) = S_0(s_1, s_2) \times x$. Monad unit is given using the identities of S_0, and monad multiplication is given using composition.

Adjunctions with Parameters. For categories C and A, an S-parameterised adjunction $\langle F, G, \eta, \epsilon \rangle : C \rightharpoonup A$ consists of functors $F : S \times C \to A$ and $G : S^{\mathrm{op}} \times A \to C$, the unit $\eta_{s,x} : x \to G(s, F(s, x))$, natural in x and dinatural in s, and the counit $\epsilon_{s,a} : F(s, G(s, a)) \to a$, natural in a and dinatural in s, satisfying the following triangular laws:

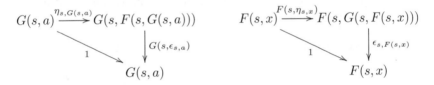

By Theorem §IV.7.3 in Mac Lane [7], if we have a functor $F : S \times C \to A$ such that for every object s, $F(s, -)$ has a right adjoint $G_s : A \to C$, then there is a unique way to make a bifunctor $G : S^{\mathrm{op}} \times A \to C$ such that $G(s, -) = G_s$ and the pair form a parameterised adjunction.

Parameterised adjunctions have the same relationship to parameterised monads as adjunctions have to monads. First, we go from parameterised adjunctions to monads:

Proposition 4. *Given an S-parameterised adjunction $\langle F, G, \eta, \epsilon \rangle : C \rightharpoonup A$, there is an S-parameterised monad on C defined with functor $G(s_1, F(s_2, x))$, unit $\eta_{s,x}$ and multiplication $\mu_{s_1 s_2 s_3 x} = G(s_1, \epsilon_{s_2, F(s_3, x)})$.*

In the opposite direction, from monads to adjunctions, we have the same situation as for non-parameterised monads. There are two canonical adjunctions arising from a parameterised monad; the initial and terminal objects in the category of adjunctions that define the monad.

First, we define the category of adjunctions that we are interested in. Given an S-parameterised monad $\langle T, \eta, \mu \rangle$ on C, the category $\mathrm{PAdj}(T)$ has as objects S-parameterised adjunctions that give the monad T by the construction above;

and arrows $f : (\langle F, G, \eta, \epsilon \rangle : C \rightharpoonup A) \rightarrow (\langle F', G', \eta, \epsilon' \rangle : C \rightharpoonup A')$ are functors $f : A \rightarrow A'$ such that $G = Id \times f; G'$ and $F' = F; f$ and $\epsilon'_{s,fa} = f(\epsilon_{s,a})$. Note that, by the condition that all the adjunctions form the same parameterised monad, all objects of this category have the same unit. The definition of arrow is derived from the standard definition of a transformation of adjoints, extended to parameterised adjunctions and restricted to those that define the same monad.

Kleisli Category. For an S-parameterised monad $\langle T, \eta, \mu \rangle$ on C, the Kleisli category C_T has pairs of objects of S and C as objects and arrows $f : (s_1, x) \rightarrow (s_2, y)$ are arrows $x \rightarrow T(s_1, s_2, y)$ in C. Identities are given by the unit of the monad, and composition is defined using the multiplication.

Proposition 5. *The functors*

$$F_T : S \times C \rightarrow C_T : (s, x) \mapsto (s, x) : (g, f) \mapsto \eta_{s_1, x}; T(s_1, g, f)$$
$$G_T : S^{\mathsf{op}} \times C_T \rightarrow C : (s_1, (s, x)) \mapsto T(s_1, s, x) : (g, f) \mapsto T(g, s_2, f); \mu_{s_1 s_2 s'_2 y}$$

form part of an S-parameterised adjunction between C and C_T. This adjunction is initial in $\mathrm{PAdj}(T)$.

Eilenberg-Moore Category of Algebras. The second canonical parameterised adjunction that arises from a parameterised monad is the parameterised version of the Eilenberg-Moore category of algebras for the monad. For an S-parameterised monad $\langle T, \eta, \mu \rangle$ on C, a T-algebra is a pair $\langle x, h \rangle$ of a functor $x : S^{\mathsf{op}} \rightarrow C$ and an family $h_{s_1 s_2} : T(s_1, s_2, x(s_2)) \rightarrow x(s_1)$, natural in s_1 and dinatural in s_2. These must satisfy the diagrams:

$$
\begin{array}{ccc}
T(s_1, s_2, T(s_2, s_3, x(s_3))) & \xrightarrow{\mu_{s_1 s_2 s_3 x(s_3)}} & T(s_1, s_3, x(s_3)) \\
{\scriptstyle T(s_1, s_2, h_{s_2 s_3})} \downarrow & & \downarrow {\scriptstyle h_{s_1 s_3}} \\
T(s_1, s_2, x(s_2)) & \xrightarrow{h_{s_1 s_2}} & x(s_1)
\end{array}
\qquad
\begin{array}{ccc}
x(s) & \xrightarrow{\eta_{s, x(s)}} & T(s, s, x(s)) \\
& {\scriptstyle 1} \searrow & \downarrow {\scriptstyle h_{s, s}} \\
& & x(s)
\end{array}
$$

A T-algebra map $f : \langle x, h \rangle \rightarrow \langle y, k \rangle$ is a natural transformation $f_s : x(s) \rightarrow y(s)$ such that

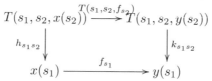

$$
\begin{array}{ccc}
T(s_1, s_2, x(s_2)) & \xrightarrow{T(s_1, s_2, f_{s_2})} & T(s_1, s_2, y(s_2)) \\
{\scriptstyle h_{s_1 s_2}} \downarrow & & \downarrow {\scriptstyle k_{s_1 s_2}} \\
x(s_1) & \xrightarrow{f_{s_1}} & y(s_1)
\end{array}
$$

commutes.

Clearly, T-algebras for a monad and their maps form a category, which we call C^T.

Proposition 6. *Given an S-parameterised monad $\langle T, \eta, \mu \rangle$, the functors*

$$F^T : S \times C \rightarrow C^T : (s, x) \mapsto \langle T(-, s, x), \mu_{s_1 s_2 x} \rangle : (g, f) \mapsto T(-, g, f)$$
$$G^T : S^{\mathsf{op}} \times C^T \rightarrow C : (s, \langle x, h \rangle) \mapsto x(s) : (g, f) \mapsto x(g); f_{s_1}$$

form a parameterised adjunction, with unit η and counit $\epsilon^T_{s, \langle x, h \rangle} = h_{-,s}$. This adjunction is terminal in $\mathrm{PAdj}(T)$.

3   Algebraic Theories with Parameters

Signatures, Terms and Theories. An S-parameterised signature is a collection of operations $\sigma \in \Sigma$, where each operation has an associated arity $ar(\sigma)$, which is a countable set, a domain $dom(\sigma) \in S$ and a co-domain $cod(\sigma) \in S$. As a shorthand, we write an operation σ with $ar(\sigma) = X$, $dom(\sigma) = s_1$ and $cod(\sigma) = s_2$ as $\sigma : s_2 \leftarrow s_1^X$.

Given a signature Σ and a countable set V of variables, we can define the sets $T_\Sigma(s_1, s_2, V)$ of terms as the smallest sets closed under the following two rules:

$$\frac{v \in V \qquad f \in S(s_1, s_2)}{\mathsf{e}(f, v) \in T(s_1, s_2, V)}$$

$$\frac{\sigma : s_2 \leftarrow s_1^X \in \Sigma \qquad \{t_x \in T(s_2, s_2', V)\}_{x \in X} \qquad f \in S(s_1', s_1)}{\mathsf{op}(f, \sigma, \{t_x\}) \in T(s_1', s_2', V)}$$

By this definition, terms consist of trees of operations from the signature, punctuated by morphisms from the parameterising category S and terminated by $\mathsf{e}(f, v)$ terms. The morphisms from S are used to bridge differences between the domains and codomains of each operation.

An S-parameterised algebraic theory is a pair $(\Sigma, \mathcal{E})$ of a signature and a set of equations $t = t' \in \mathcal{E}$ between terms, where for each pair, both t and t' are in $T_\Sigma(s_1, s_2, V)$ for some s_1 and s_2.

Example 7. An example signature is given by the global state parameterised monad from Example 2. Given a category of state types S, with a terminal object 1, we take the underlying category to be Set and the assume that the functor $\hat{\cdot} : S \to$ Set maps every object of S to a countable set. Our signature has two families of operations: $\mathsf{read}_s : s \leftarrow s^s$ and $\mathsf{write}_s(x \in \hat{s}) : 1 \leftarrow s^1$. The required equations are:

$$\mathsf{op}(id, \mathsf{read}_s, \lambda x.\mathsf{e}(id, v)) = v$$
$$\mathsf{op}(id, \mathsf{read}_s, \lambda x.\mathsf{op}(!, \mathsf{write}_s(x), \mathsf{e}(id, v_x))) = \mathsf{op}(id, \mathsf{read}_s, \lambda x.v_x)$$
$$\mathsf{op}(id, \mathsf{read}_s, \lambda x.\mathsf{op}(id, \mathsf{read}_s, \lambda y.\mathsf{e}(id, v_{xy}))) = \mathsf{op}(id, \lambda x.\mathsf{e}(id, v_{xx}))$$
$$\mathsf{op}(id, \mathsf{write}_s(y), \mathsf{op}(id, \mathsf{read}_s(\lambda x.\mathsf{e}(!, v_x)))) = v_y$$

where we use $!$ for the unique terminal morphism in S. These equations state that: (1) reading but not using the result is the same as not reading; (2) reading and then writing back the same value is the same as just reading; (3) reading twice in a row is the same as reading once; and (4) writing, reading and clearing is the same as doing nothing.

Algebras. For a signature Σ and category C with countable products, a Σ-C-algebra consists of a functor $a : S^{\mathsf{op}} \to C$ and for each operation $\sigma : s_2 \leftarrow s_1^X \in \Sigma$ an arrow $\sigma_a : a(s_1)^X \to a(s_2)$. Given two such algebras a homomorphism

$f : (a, \{\sigma_a\}) \to (b, \{\sigma_b\})$ is a natural transformation $f : a \to b$ such that for all $\sigma : s_2 \leftarrow s_1^X \in \Sigma$, the diagram

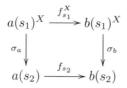

commutes.

Example 8. An algebra for the theory in Example 7 is given by $a(s) = (\widehat{s'} \times x)^s$ for some set x and $s' \in S$. It is easy to see how to give the appropriate implementations of read and write. Elements of $a(s)$ for some s' and x represent reading and writing computations that start with the global state of type s and end with global state of type s' and a result value in x.

Given a Σ-C-algebra a, terms $t \in T(s_1, s_2, V)$ give rise to derived operations $[t] : a(s_2)^V \to a(s_1)$ by induction on the term t:

$$[e(f, v)] = \pi_v; a(f)$$
$$[op(f, \sigma, \{t_i\})] = \langle [t_x] \rangle; \sigma_a; a(f)$$

where π_v is the projection of the countable products of C, and $\langle \cdot \rangle$ is pairing.
   A $(\Sigma, \mathcal{E})$-C-algebra consists of a Σ-C-algebra that satisfies every equation in $\mathcal{E}$: for all $t = t' \in \mathcal{E}(s_1, s_2)$ and valuations $f : 1 \to a(s_2)^V$, then $f; [t] = f; [t']$. The collection of $(\Sigma, \mathcal{E})$-C-algebras and homomorphisms forms a category $(\Sigma, \mathcal{E})$-$Alg(C)$.

Free Algebras. A $(\Sigma, \mathcal{E})$-C-algebra $\langle a, \{\sigma_a\} \rangle$ is free for an S-object s and C-object x if there is an arrow $\eta : x \to a(s)$ such that for any other algebra $\langle b, \{\sigma_b\} \rangle$, there is a unique homomorphism $h : a \to b$ such that $\eta; h_a = f$.

Proposition 9. *Given a functor $F : S \times C \to (\Sigma, \mathcal{E})$-$Alg(C)$ such that for all $s \in S$ and $x \in C$, $F(s, x)$ is free for s and x, then $T(s_1, s_2, x) = F(s_2, x)(s_1)$ can be given the structure of an S-parameterised monad on C.*

Free Algebras in Set. We now construct the free $(\Sigma, \mathcal{E})$-algebra over a set A. We do this in two stages by first constructing the free Σ-algebra and then by quotienting by the equations in $\mathcal{E}$. Fix Σ, $\mathcal{E}$ and A.

Lemma 10. *The sets $T_\Sigma(s_1, s_2, A)$ can be made into a functor $T_\Sigma : S^{op} \times S \times Set \to Set$.*

Lemma 11. *For all s, the functor $T_\Sigma(-, s, A)$ is the carrier of a Σ-algebra.*

Proof. For each operation $\sigma : s_2 \leftarrow s_1^X$, define $\sigma_{T,s,A} : T_\Sigma(s_1, s, A)^X \to T_\Sigma(s_2, s, A)$ as $\sigma_{T_\Sigma}(\langle t_x \rangle) = op(id, \sigma, \langle t_x \rangle)$.

Proposition 12. *For each s, A, the algebra $\langle T_\Sigma(-,s,A), \{\sigma_{T_\Sigma}\}\rangle$ is the free algebra for s and A.*

Proof. (Sketch). Given $f : A \rightarrow B(s)$ for some Σ-algebra $\langle B, \{\sigma_B\}\rangle$, define $h_{s'} : T(s',s,A) \rightarrow B(s')$ by recursion on terms:

$$h_{s'}(\mathsf{e}(g,a)) = B(g)(f(a))$$
$$h_{s'}(\mathsf{op}(g,\sigma,\langle t_i\rangle)) = B(g)(\sigma_B(\langle h_{dom(\sigma)}(t_i)\rangle))$$

The rest of the proof is straightforward, with a little wrinkle in showing that this homomorphism h is unique, which requires a Yoneda-style appeal to the naturality of homomorphisms.

In preparation for quotienting the above free algebra by a set of equations, we define congruences for parameterised algebras and show that quotienting by a congruence gives a parameterised algebra. Given a Σ-algebra $\langle a, \{\sigma_a\}\rangle$ in Set, a family of equivalence relations $\equiv_s$ on a is a congruence if, for all $\sigma : s_1^X \rightarrow s_2$, $\forall x \in X.y_x \equiv_{s_1} z_x$ implies $\sigma_A(\langle y_x\rangle) \equiv_{s_2} \sigma_A(\langle z_x\rangle)$ and for all $g : s \rightarrow s'$, $x \equiv_{s'} y$ implies $a(g)x \equiv_s a(g)y$.

Lemma 13. *If $\langle A, \{\sigma_A\}\rangle$ is a Σ-algebra and $\equiv$ is a congruence, then $\langle A/\equiv, \{\sigma_A\}\rangle$ is also a Σ-algebra.*

Using our set of equations $\mathcal{E}$, for each $s \in S$, define relations $t \equiv_{s'} t'$ on $T_\Sigma(s',s,A)$ as the smallest congruences satisfying the rule:

$$\frac{(t,t') \in \mathcal{E} \qquad \langle t_x, t'_x \in T_\Sigma(s_2,s,A)\rangle \qquad \forall x.t_x \equiv_{s_2} t'_x}{[t]t_x \equiv_{s_1} [t']t'_x}$$

Set $T_{(\Sigma,\mathcal{E})}(-,s,A) = T_\Sigma(-,s,A)/\equiv$.

Proposition 14. *$T_{\Sigma,\mathcal{E}}(-,s,A)$ is the free $(\Sigma,\mathcal{E})$-algebra over s and A.*

4   Beck's Theorem

In this section we prove Beck's theorem [2] characterising the Eilenberg-Moore category of algebras for a parameterised monad. We base our proof on the proof of Beck's theorem for non-parameterised monads given by Mac Lane [7]. Recall that Beck's theorem for non-parameterised monads characterises the category C^T by the use of coequalisers that are preserved and created by the right adjoint. Due to the additional parameterisation, we cannot use coequalisers for parameterised monads, so we use families of pairs of morphisms with common domain that we call S-spans, and consider universal and absolute versions of these.

We will be dealing with functors that have pairs of contra- and co-variant arguments such as $x : S^{\mathsf{op}} \times S \rightarrow A$, where each pair is applied to the same object. We abbreviate such applications like so: $x^s = x(s,s)$, and similarly for multiple pairs of contra-/co-variant arguments.

An S-span in a category A is a pair of functors $x : S^{op} \times S \times S^{op} \times S \to A$ and $y : S^{op} \times S \to A$ and a pair of dinatural transformations $d^{0s_1s_2} : x^{s_1s_2} \to y^{s_1}$ and $d^{1s_1s_2} : x^{s_1s_2} \to y^{s_2}$. A *fill-in* for an S-span is an object z and dinatural transformations $e^s : y^s \to z$ such that, for all s_1, s_2, $d^{0s_1s_2}; e^{s_1} = d^{1s_1s_2}; e^{s_2}$. A *universal* fill-in (e, z) for an S-span d^0, d^1, such that for any other fill-in (f, c) there is a unique arrow $h : z \to c$ such that, for all s, $e^s; h = f^s$. An *absolute* fill-in is a fill-in such that for any functor $T : A \to A'$, the resulting fill-in is universal.

We are interested in S-spans because they arise from T-algebras. Given an S-parameterised monad (T, η, μ) on C, for each T-algebra $\langle x, h \rangle$ we have a fill-in square in $C^{S^{op}}$:

$$\begin{array}{ccc}
T(-, s_1, T(s_1, s_2, x(s_2))) & \xrightarrow{\mu_{-s_1 s_2 x(s_2)}} & T(-, s_2, x(s_2)) \\
{\scriptstyle T(-, s_1, h_{s_1 s_2})}\downarrow & & \downarrow{\scriptstyle h_{-s_2}} \\
T(-, s_1, x(s_1)) & \xrightarrow{\quad h_{-s_1} \quad} & x(-)
\end{array}$$

which is just the associativity law for the T-algebra.

Lemma 15. *This fill-in square is universal and absolute.*

The next lemma will also be useful.

Lemma 16. *Universal fill-ins have the following cancellation property. For two arrows $f, g : z \to c$, if, for all s, $e^s; f = e^s; g$, then $f = g$.*

Theorem 17 (Beck). *Let $\langle F, G, \eta, \epsilon \rangle : C \rightharpoonup A$ be an S-parameterised adjunction and $\langle T, \eta, \mu \rangle$ the derived S-parameterised monad. Then A is isomorphic to C^T iff for each S-span d^0, d^1 in A such that the S-span $G(-, d^0), G(-, d^1)$ has an absolute fill-in $(e^-_, z(-))$, there is a unique universal fill-in (e^-, z) of d^0, d^1 such that $G(s, z) = z(s)$ and $G(s, e^{s'}) = e^{s'}_s$.*

Proof. First, we show the forward direction. We know that A and C^T are isomorphic, so it suffices to show that given S-spans

$$\begin{array}{ccc}
\langle x^{s_1 s_2}, h^{s_1 s_2} \rangle & \xrightarrow{d^{0s_1 s_2}} & \langle y^{s_1}, k^{s_1} \rangle \\
{\scriptstyle d^{1s_1 s_2}}\downarrow & & \\
\langle y^{s_2}, k^{s_2} \rangle & &
\end{array}$$

of T-algebras for which the corresponding S-spans in $C^{S^{op}}$ via G^T have absolute universal fill-ins $(e^s_-, z(-))$

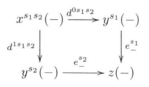

then there is a unique universal fill-in $(e, \langle z, m \rangle)$ in C^T such that $G^T(s, e^{s'}) = e_s^{s'}$ and $G^T(s, \langle z, m \rangle) = z(s)$.

We already know that $z(-)$ is a functor $S^{op} \to C$ and that the e^s are natural, so we use z as the functor part of our T-algebra. We must find a T-algebra structure $m_{s_1, s_2} : T(s_1, s_2, z(s_2)) \to z(s_1)$. This is induced from the universal fill-in:

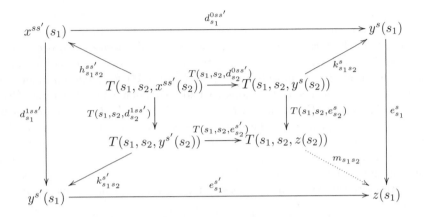

The inner and outer squares are the fill-in properties for $T(s_1, s_2, e_{s_2}^-)$ and $e_{s_1}^-$ respectively. The top and left regions commute because d^0 and d^1 are homomorphisms of T-algebras. Thus, $k_{s_1 s_2}^- ; e_{s_1}^-$ is a fill-in for the inner square. Since e^s is an absolute universal fill-in there is a unique arrow $m_{s_1 s_2} : T(s_1, s_2, z(s_2)) \to z(s_1)$ such that $T(s_1, s_2, e_{s_2}^s) ; m_{s_1 s_2} = k_{s_1 s_2}^s ; e_{s_1}^s$. To complete this direction of the proof we must show that the family $m_{s_1 s_2}$ is natural in s_1 and dinatural in s_2, that it is the structure map for a T-algebra, and that e^s is a universal fill-in of d^0, d^1. These are easily checked by construction of the appropriate diagrams, and use of Lemma 16.

In the reverse direction, we use the fact that for each object $a \in A$, the adjunction $\langle F, G, \eta, \epsilon \rangle : C \rightharpoonup A$ provides a fill-in

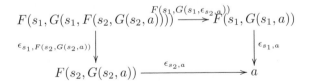

in A. This is the "canonical presentation" of a. This definition, and Lemma 15, are used to prove that there is a unique morphism in $\mathrm{PAdj}(T)$ from any other A' to A in a manner similar to Mac Lane's proof. This then implies the result we desire.

Corollary 18. *For any S-theory $(\Sigma, \mathcal{E})$, $(\Sigma, \mathcal{E})$-Alg is parameterised-monadic over Set.*

5 Structured Parameterisation

In this section we develop a small amount of theory to deal with the case when we have structure on the parameterising category S, expressed using functors, that we wish to have on the parameterised monad itself.

Motivation. Consider the use of parameterised monads to model effectful computation, where the kinds of effects we may perform are regulated by the parameters of the monad. In the type system of the next section, we have typing judgements of the form $\Gamma; a \vdash M : A; b$, where a and b represent objects of our parameterising category, dictating the start and end states of the computation. This is interpreted as a morphism $[\![\Gamma]\!] \to T(a, b, [\![A]\!])$ for some parameterised monad. The system has a "let" construct for sequencing two programs. Given two typed programs $\Gamma; a \vdash M : A; b$ and $\Gamma, x : A; b \vdash N : B; c$ then an obvious way to type the sequencing of these programs is $\Gamma; a \vdash \mathtt{let}\ x \Leftarrow M\ \mathtt{in}\ N : B; c$, matching the bs, and to use the multiplication of the monad for the semantics. However, it is likely that the program N requires some knowledge of its start state beyond that given to it by M. If c denotes the extra knowledge required by N, we write the combination as $b \bullet c$ and the full sequencing rule is given by

$$\frac{\Gamma; a \vdash M : A; b \qquad \Gamma, x : A; b \bullet c \vdash N : B; d}{\Gamma; a \bullet c \vdash \mathtt{let}\ x \Leftarrow M\ \mathtt{in}\ N : B; d.}$$

The operation $- \bullet c$ can be interpreted as a functor on the parameterising category. To give a semantics to the "let" construct, we require some morphism $(- \bullet s)^\dagger : T(s_1, s_2, x) \to T(s_1 \bullet s, s_2 \bullet s, x)$. We call this a *lifting* of the functor $- \bullet s$.

Definition. Given an strong S-parameterised monad (T, η, μ, τ), and a functor $F : S \to S$, a *lifting* of F to T is a natural transformation $F^\dagger_{s_1 s_2 x} : T(s_1, s_2, x) \to T(Fs_1, Fs_2, x)$ that commutes with the unit, multiplication and strength of the monad:

$$F^\dagger; T(Fs_1, Fs_2, F^\dagger); \mu = \mu; F^\dagger \qquad \eta; F^\dagger = \eta_F \qquad \tau; F^\dagger = x \times F^\dagger; \tau$$

This definition is from [1].

Example 19. For the state category, we take the free symmetric monoidal category on the category S of state types used in Example 2, which we label S'. We extend the functor $\widehat{}$ to be $S' \to C$ by mapping objects $s_1 \otimes s_2$ to $\widehat{s_1} \times \widehat{s_2}$, hence $\widehat{}$ becomes a strict symmetric monoidal functor. A lifting of $s \otimes -$ can be defined as: $(s \otimes -)^\dagger = c \mapsto \lambda(s, s_1).\mathtt{let}\ (s_2, a) = c(s_1)\ \mathtt{in}\ ((s, s_2), a)$ and similarly for $(- \otimes s)^\dagger$.

Liftings for Algebraic Theories. For the S-parameterised monads generated from algebraic theories as in Section 3, we can define liftings for endo-functors F on S, provided there is enough structure on the operations of the algebras. Given an S-theory $(\Sigma, \mathcal{E})$, if, for every operation $\sigma : s_1 \leftarrow s_2^X \in \Sigma$ there is an operation $F\sigma : Fs_1 \leftarrow (Fs_2)^X$, and all the equations generated by $\mathcal{E}$ are preserved by lifting all operations, then we can define the following operations on the sets $T_\Sigma(s_1, s_2, x)$:

$$F^\dagger(e(g, x)) = e(Fg, x) \qquad F^\dagger(\mathsf{op}(g, \sigma, \{t_x\})) = \mathsf{op}(Fg, F\sigma, \{F^\dagger(t_x)\})$$

Proposition 20. *This definition defines a lifting of F to $T_{(\Sigma, \mathcal{E})}$.*

6   Semantics for Type and Effect Systems

We now give an application of the theory developed above by using it to give a semantics to a toy language with explicitly typed effectful operations.

Effect Systems. For our language we use a partially-ordered set $\mathcal{A}$ of *permissions* that has an order-preserving binary associative operation $\bullet : \mathcal{A} \times \mathcal{A} \to \mathcal{A}$. Elements of $\mathcal{A}$ are used to represent permissions that permit a program to carry out certain effects. We write the ordering of $\mathcal{A}$ as $a \Rightarrow b$, intended to be reminiscent of logical implication. We use $a, b, c, \ldots$ to range over elements of $\mathcal{A}$.

The types of our toy language are given by:

$$A, B ::= \mathtt{int} \mid \mathtt{bool} \mid \mathtt{unit} \mid A \times B \mid (A; a) \to (B; b)$$

The types are standard except for the function type $(A; a) \to (B; b)$. This is the type of functions that take As and return Bs, but may also perform effects allowed by permissions a and bestow permissions b. We refer to types that do not contain function types as *ground*.

An *effect system* consists of a permission algebra $\mathcal{A}$ and a set Ω of operations $\mathsf{op} : (A; a) \to (B; b)$, where A and B are ground.

Example 21. The traditional effect system recording when a program reads or writes global variables can be expressed in our system. Given a set of locations Loc, we take $\mathcal{A}$ to be the power set of $\{r_l, w_l \mid l \in Loc\}$, ordered by reverse inclusion. The binary operation is defined as $\varepsilon \bullet \varepsilon' = \varepsilon \cup \varepsilon'$. We read $\{r_{l_1}, r_{l_2}, w_{l_2}\}$ as the permission to read the locations l_1 and l_2, the permission to write to location l_2. We have two families of operations:

$$\mathsf{read}_l : (\mathtt{unit}, \{r_l\}) \to (\mathtt{int}, \{r_l\})$$
$$\mathsf{write}_l : (\mathtt{int}, \{w_l\}) \to (\mathtt{unit}, \{w_l\})$$

The read operation requires that we have the permission to perform a read on the required location and it bestows the permission to still read that location afterwards; likewise for writing.

Example 22. This example enforces an ordering on effects performed, in a similar manner to history effects [11]. The permission algebra is given by:

$$H ::= \alpha \mid H_1 + H_2 \mid H_1.H_2 \mid \epsilon$$

$$\overline{\Gamma, x : A \vdash x : A} \qquad \overline{\Gamma \vdash n : \texttt{int}} \qquad \overline{\Gamma \vdash b : \texttt{bool}} \qquad \overline{\Gamma \vdash () : \texttt{unit}}$$

$$\frac{\Gamma \vdash V_1 : \texttt{int} \quad \Gamma \vdash V_2 : \texttt{int}}{\Gamma \vdash V_1 < V_2 : \texttt{bool}} \qquad \frac{\Gamma \vdash V_1 : \texttt{int} \quad \Gamma \vdash V_2 : \texttt{int}}{\Gamma \vdash V_1 + V_2 : \texttt{int}}$$

$$\frac{\Gamma \vdash V_1 : A \quad \Gamma \vdash V_2 : B}{\Gamma \vdash (V_1, V_2) : A \times B} \qquad \frac{\Gamma \vdash V : A_1 \times A_2}{\Gamma \vdash \pi_i V : A_i}$$

$$\frac{\Gamma, x : A; a \vdash M : B; b}{\Gamma \vdash \lambda(x : A; a).M : (A; a) \to (B; b)} \qquad \frac{\Gamma \vdash V : A \quad A \sqsubseteq B}{\Gamma \vdash V : B}$$

$$\frac{\Gamma \vdash V : A}{\Gamma; a \vdash \texttt{val}_a \, V : A; a} \qquad \frac{\Gamma; a \vdash M : A; b \quad \Gamma, x : A; b \bullet c \vdash N : B; d}{\Gamma; a \bullet c \vdash \texttt{let } x \Leftarrow M \texttt{ in } N : B; d}$$

$$\frac{\Gamma \vdash V_1 : (A; a) \to (B; b) \quad \Gamma \vdash V_2 : A}{\Gamma; a \vdash V_1 \, V_2 : B; b}$$

$$\frac{\Gamma \vdash V : \texttt{bool} \quad \Gamma; a \vdash M : A; b \quad \Gamma; a \vdash N : A; b}{\Gamma; a \vdash \texttt{if } V \texttt{ then } M \texttt{ else } N : A; b}$$

$$\frac{\Gamma \vdash V : A \quad \texttt{op} : (A; a) \to (B; b)}{\Gamma; a \vdash \texttt{op } V : B; b} \qquad \frac{a \Rightarrow a' \quad \Gamma; a' \vdash M : A; b' \quad b' \Rightarrow b}{\Gamma; a \vdash M : A; b}$$

$$\frac{\Gamma; a \vdash M : A; b \quad A \sqsubseteq B}{\Gamma; a \vdash M : B; b}$$

Fig. 1. Typing Rules

over some set of basic permissions α. The ordering treats $H_1.H_2$ as an associative binary operator with unit ϵ, and $H_1 + H_2$ as a meet. Define $H_1 \bullet H_2$ as $H_1.H_2$ Operations can now be defined that require some basic permission $\alpha \in \{\texttt{in}, \texttt{out}\}$ to complete:

$$\texttt{input} : (\texttt{unit}, \texttt{in}) \to (\texttt{int}, \epsilon) \qquad \texttt{output} : (\texttt{int}, \texttt{out}) \to (\texttt{unit}, \epsilon)$$

The lack of commutativity in $H_1.H_2$ ensures that operations must be carried out in the predefined order prescribed by the starting set of permissions. The lifting operation ensures that a sub-program need not have complete knowledge of the rest of the program's effects.

Typing Rules. The typing rules for our language are presented in Figure 1. We split the language into effect-free values V and possibly side-effecting programs M, based on the fine-grain call-by-value calculus of Levy *et al* [6].

$$V ::= x \mid n \mid b \mid () \mid V_1 < V_2 \mid V_1 + V_2 \mid (V_1, V_2) \mid \pi_i V \mid \lambda(x : A; a).M$$
$$M ::= \texttt{val}_a \, V \mid \texttt{let } x \Leftarrow M \texttt{ in } N \mid V_1 \, V_2 \mid \texttt{if } V \texttt{ then } M \texttt{ else } N \mid \texttt{op } V$$

Values are typed using a judgement of the form $\Gamma \vdash V : A$, where Γ is a list of variable : type pairs with no duplicated variable names and A is a type. The value typing rules are standard for the given constructs apart from the λ-abstraction rule. This rule's premise is a typing judgement on a program M. Such judgements have the form $\Gamma; a \vdash M : A; b$, where a and b are effect assertions. This judgement states that the program M, when its free variables are typed as in Γ, has effects afforded by a and returns a value of type A, allowing further effects afforded by b.

The language has the usual value and sequencing constructs for a monadic language, extended with permissions. Values are incorporated into programs by the $\mathtt{val}_a\ V$ construct. This is the $\mathtt{return}$ of normal monadic programming extended with the fact that the program starts in a state described by a and does nothing to that state. Two programs are sequenced using the $\mathtt{let}\ x \Leftarrow M\ \mathtt{in}\ N$ construct we introduced in the previous section. There are two subtyping rules for values and computations. The subtyping relation is defined as:

$$\frac{}{\mathtt{int} \sqsubseteq \mathtt{int}} \qquad \frac{}{\mathtt{bool} \sqsubseteq \mathtt{bool}} \qquad \frac{A \sqsubseteq A' \qquad B \sqsubseteq B'}{A \times A' \sqsubseteq B \times B'}$$

$$\frac{A' \sqsubseteq A \qquad a' \Rightarrow a \qquad B \sqsubseteq B' \qquad b \Rightarrow b'}{(A; a) \to (B; b) \sqsubseteq (A'; a') \to (B'; b')}$$

Semantics Using a Parameterised Monad. For an effect system $(\mathcal{A}, \Omega)$, the semantics of the language is defined using an $\mathcal{A}$-parameterised monad (T, η, μ) on Set with a lifting for $\bullet$. This gives the following interpretation of the types:

$$[\![\mathtt{int}]\!] = \mathbb{Z} \qquad [\![\mathtt{bool}]\!] = \mathbb{B} \qquad [\![\mathtt{unit}]\!] = \{\star\} \qquad [\![A \times B]\!] = [\![A]\!] \times [\![B]\!]$$

$$[\![(A; a) \to (B; b)]\!] = [\![A]\!] \to T(a, b, [\![B]\!]).$$

The interpretation of types extends to typing contexts in the standard way. We interpret value judgements $\Gamma \vdash V : A$ as functions $[\![\Gamma]\!] \to [\![A]\!]$, and computation judgements $\Gamma; a \vdash M : B; b$ as functions $[\![\Gamma]\!] \to T(a, b, [\![B]\!])$. For each operation $\mathtt{op} : (A; a) \to (B; b)$ in Ω we require a function $[\![\mathtt{op}]\!] : [\![A]\!] \to T(a, b, [\![B]\!])$.

The interpretation of terms is now relatively straightforward. Due to the consequence and subtyping rules, we must give the interpretation over typing derivations and not the structure of the terms. To resolve this coherence issue we relate this semantics to a semantics in which these rules are no-ops below.

To define the semantics we use the *bind* operator of type $T(a, b, x) \times (x \to T(b, c, y)) \to T(a, c, y)$, derived from the multiplication of the monad. We give the cases for the parameterised monad structure (where the first three are understood to actually apply to the typing rule with the given term constructor):

$$[\![\mathtt{val}_a\ V]\!]\rho = \eta_a([\![V]\!]\rho)$$
$$[\![\mathtt{let}\ x \Leftarrow M\ \mathtt{in}\ N]\!]\rho = bind_{(a\bullet r)(b\bullet r)d}([\![M]\!]\rho) \bullet^\dagger r, \lambda x.[\![N]\!](\rho, x)$$

$$\llbracket \text{op } V \rrbracket \rho = \llbracket \text{op} \rrbracket (\llbracket V \rrbracket \rho)$$

$$\left\llbracket \frac{a \Rightarrow a' \quad \Gamma; a' \vdash M : A; b' \quad b' \Rightarrow b}{\Gamma; a \vdash M : A; b} \right\rrbracket \rho = T(a \Rightarrow a', b' \Rightarrow b, \llbracket M \rrbracket \rho)$$

In the interpretation of `let`, note the use of the lifting operator $(\bullet^\dagger r)$ to lift the interpretation of M so that it can be sequenced with N.

The subtyping relation $A \sqsubseteq B$ is interpreted as an function $\llbracket A \rrbracket \to \llbracket B \rrbracket$ over the derivation, using the functor action of T on $\mathcal{A}$. Subtyping rules are interpreted by composition with this interpretation.

Generating Parameterised Monads for Effect Systems. We now construct an $\mathcal{A}$-parameterised monad suitable for the previous section by using the free algebras in Sections 3 and 5. Given an effect system $(\mathcal{A}, \Omega)$ we define the corresponding $\mathcal{A}$-signature Σ_Ω as the union of sets of operations deriving from each $\text{op} : (A; a) \to (B; b) \in \Omega$:

$$\{\text{op}_x : a \leftarrow b^X \mid x \in \llbracket A \rrbracket\} \cup \{\text{op}_x \bullet c : (a \bullet c) \leftarrow (b \bullet c)^{\llbracket B \rrbracket} \mid x \in \llbracket A \rrbracket, c \in \mathcal{A}\}$$

So for each operation `op` we have a family of operations for each possible input value in the interpretation of A and each possible permission context in which the operation may be used. The duplication of operations over permission contexts will be used to define the lifting operations. The restriction to ground types in operations means that we are not using the definition of the monad that we are currently constructing. By Propositions 9 and 14 we obtain an $\mathcal{A}$-parameterised monad suitable for interpreting our toy language.

Relating Parameterised and Unparameterised Semantics. Finally in this section, we relate our semantics with typed algebraic operations to a standard semantics for a given effect system. The basic idea is to assume that, for an effect system $(\mathcal{A}, \Omega)$ and some set of equations $\mathcal{E}$, there is a non-parameterised monad M that gives an adequate semantics for the chosen operational semantics of this language. If the free algebra for this monad supports the theory $(\Sigma_\Omega, \mathcal{E})$ that we have used to prove any equivalences, then we can define an erasure function $\text{erase} : T(a, b, A) \to MA$ that replays each effect in the typed interpretation in the untyped one. By a logical relations argument using Katsumata's notion of $\top\top$-lifting [5], it is possible to show that, for closed programs M of ground type, this means that $\text{erase}(\llbracket M \rrbracket^t) = \llbracket M \rrbracket^u$, where $\llbracket - \rrbracket^t$ and $\llbracket - \rrbracket^u$ are the semantics in T and M respectively.

Proposition 23. *If the untyped semantics in M is adequate, meaning that if $\llbracket M \rrbracket^u = \llbracket N \rrbracket^u$ then M and N are contextually equivalent, then $\llbracket M \rrbracket^t = \llbracket N \rrbracket^t$ implies that M and N are equivalent for all type derivation contexts $-; a \vdash C[-] : \texttt{bool}; b$.*

Using this proposition it is possible to prove the equivalences given by Benton *et al* [3], using Plotkin and Power's equations for global state [9] and induction on the terms of the free algebra.

7   Conclusions

We have extended the relationship between parameterised monads and adjunctions to include the Eilenberg-Moore category of algebras. We have also given a description of algebraic theories with parameters, and used a generalisation of Beck's theorem to show that the Eilenberg-Moore category and the category of algebras for a given theory coincide. We then used this theory to give a semantics for a toy programming language with effect information recorded in the types.

In future work we wish to develop the theory of algebraic theories with parameters and parameterised monads further:

- We wish to understand more about how structure on the parameterising category S may be carried over to the parameterised monad itself. For every functor $F : S \to S$, with a lifting $F^\dagger$ it is possible to define functors $F^T : C^T \to C^T$ and $F_T : C_T \to C_T$. However, an attempt to define a category of adjunctions with such functors fails because the operation $F^\dagger$ factors across the adjunctions in two different ways for C^T and C_T.
- We wish to construct free algebras with parameters in other categories, especially in some category of domains suitable for interpreting recursion.
- We have not characterised the parameterised monads that arise from algebraic theories. We expect that this will simply be parameterised monads that preserve the right kind of filtered colimits, as for normal monads [10].

References

1. Atkey, R.: Parameterised notions of computation. Journal of Functional Programming 19 (2009)
2. Beck, J.M.: Triples, algebras and cohomology. Reprints in Theory and Applications of Categories 2, 1–59 (2003)
3. Benton, N., Kennedy, A., Hofmann, M., Beringer, L.: Reading, writing and relations. In: Kobayashi, N. (ed.) APLAS 2006. LNCS, vol. 4279, pp. 114–130. Springer, Heidelberg (2006)
4. Gifford, D.K., Lucassen, J.M.: Integrating functional and imperative programming. In: ACM Conference on LISP and Functional Programming (1986)
5. Katsumata, S.-y.: A Semantic Formulation of ⊤⊤-Lifting and Logical Predicates for Computational Metalanguage. In: Ong, L. (ed.) CSL 2005. LNCS, vol. 3634, pp. 87–102. Springer, Heidelberg (2005)
6. Levy, P.B., Power, J., Thielecke, H.: Modelling environments in call-by-value programming languages. Inf. and Comp. 185, 182–210 (2003)
7. Mac Lane, S.: Categories for the Working Mathematician, 2nd edn. Graduate Texts in Mathematics, vol. 5. Springer, Heidelberg (1998)
8. Moggi, E.: Notions of computation and monads. Information and Computation 93(1), 55–92 (1991)
9. Plotkin, G., Power, J.: Notions of computation determine monads. In: Nielsen, M., Engberg, U. (eds.) FOSSACS 2002. LNCS, vol. 2303, pp. 342–356. Springer, Heidelberg (2002)
10. Robinson, E.: Variations on algebra: Monadicity and generalisations of equational theories. Formal Asp. Comput. 13(3–5), 308–326 (2002)
11. Skalka, C., Smith, S.: History Effects and Verification. In: Chin, W.-N. (ed.) APLAS 2004. LNCS, vol. 3302, pp. 107–128. Springer, Heidelberg (2004)

Kleene Monads:
Handling Iteration in a Framework of Generic Effects

Sergey Goncharov[1], Lutz Schröder[1,2], and Till Mossakowski[1,2,*]

[1] Department of Computer Science, University of Bremen
[2] DFKI Laboratory, Bremen

Abstract. Monads are a well-established tool for modelling various computational effects. They form the semantic basis of Moggi's computational metalanguage, the *metalanguage of effects* for short, which made its way into modern functional programming in the shape of Haskell's do-notation. Standard computational idioms call for specific classes of monads that support additional control operations. Here, we introduce Kleene monads, which additionally feature nondeterministic choice and Kleene star, i.e. nondeterministic iteration, and we provide a metalanguage and a sound calculus for Kleene monads, the *metalanguage of control and effects*, which is the natural joint extension of Kleene algebra and the metalanguage of effects. This provides a framework for studying abstract program equality focussing on iteration and effects. These aspects are known to have decidable equational theories when studied in isolation. However, it is well known that decidability breaks easily; e.g. the Horn theory of continuous Kleene algebras fails to be recursively enumerable. Here, we prove several negative results for the metalanguage of control and effects; in particular, already the equational theory of the unrestricted metalanguage of control and effects over continuous Kleene monads fails to be recursively enumerable. We proceed to identify a fragment of this language which still contains both Kleene algebra and the metalanguage of effects and for which the natural axiomatisation is complete, and indeed the equational theory is decidable.

1 Introduction

Program verification by equational reasoning is an attractive concept, as equational reasoning is conceptually simple and in many cases easy to automate. At the same time, equational reasoning is necessarily insufficient for dealing with equality of programs in full Turing-complete programming languages, as the latter will e.g. require induction over data domains; moreover, it is clear that the (observable) equational theory of a Turing-complete programming language fails to be recursively enumerable (it is undecidable by Rice's theorem, and its complement is easily seen to be r.e.) and hence cannot be completely recursively axiomatised. A standard approach is therefore to separate concerns by introducing abstract programming languages that capture only selected aspects of the structure of programs. In such approaches, the abstract level can often be

* Research supported by the DFG project *A construction kit for program logics* (Lu 707/9-1) and by the German Federal Ministry of Education and Research (Project 01 IW 07002 FormalSafe).

A. Kurz, M. Lenisa, and A. Tarlecki (Eds.): CALCO 2009, LNCS 5728, pp. 18–33, 2009.

handled purely equationally, while reasoning in more complex logics, e.g. first or higher order predicate logics, is encapsulated at lower levels, typically the data level. Two approaches of this kind are Kleene algebra (used here in the version of [9]) and Moggi's monad-based computational metalanguage [12], to which we refer for both brevity and distinctness as the *metalanguage of effects*.

Kleene algebra is essentially the equational logic of regular expressions. Seen as a programming language, Kleene algebra features sequential composition, nondeterministic choice, and iteration in the shape of the Kleene star. When extended with tests, Kleene algebra allows encoding complex control structures including e.g. loops with breaks [11]. Thus, the focus of Kleene algebra as an abstract programming language is the modelling of *nondeterministic control*.

The metalanguage of effects [12] is based on the observation that the bind and return operations constituting a *monad* correspond computationally to sequential composition and promotion of values to computations, respectively, which together with explicit computation types and product types form the basic features of the metalanguage of effects. Besides the fact that the language is higher-order w.r.t. effects in that it allows passing around computations as first-class objects, the chief distinctive feature here is the innocuous-looking return operator, which affords a separation between effectful computations and effectfree values. Thus, the focus of the abstraction is indeed on *effects* in this case, while most control structures including in particular any form of loops are absent in the base language. (Of course, Kleene algebra is also about effectful programs, but has no distinction between effectful and pure functions.)

The metalanguage of effects is sound for a wide range of effects, including e.g. state, exceptions, I/O, resumptions, backtracking, and continuations. Consequently, monads are being used in programming language semantics and in functional programming both to encapsulate side effects and to achieve genericity over side effects. E.g. monads have appeared in an abstract modelling of the Java semantics [6] and in region analysis [13, 3], and they form the basis of functional-imperative programming in Haskell [15]; indeed Haskell's do-notation is essentially the metalanguage of effects.

Here, we study the natural combination of Kleene algebra and the metalanguage of effects, which we call the *metalanguage of control and effects* (MCE). The resulting language is rather expressive; e.g. it supports the following slightly abusive implementation of list reverse. Let is_empty, push, and pop be the usual stack operations, where is_empty blocks if the state is nonempty, and otherwise does nothing. Then one can define the reverse operation in the MCE as

$$\text{do } q \leftarrow (\text{init } p \leftarrow \text{ret is\_empty in } (\text{do } x \leftarrow \text{pop}; \text{ret } (\text{do } p; \text{push } x))^*); q$$

The init expression initialises the iteration variable p, and the starred expression is nondeterministically iterated. The program is equivalent to a non-deterministic choice (for all n) of

$$\text{do } q \leftarrow (\text{do } x_1 \leftarrow \text{pop}; \dots; x_n \leftarrow \text{pop}; \text{ret}(\text{do is\_empty}; \text{push } x_1; \dots; \text{push } x_n)); q$$

Hence, this program reads the entire stack into a suspended sequence of push operations and then executes this computation, thus pushing the elements back onto the stack in reversed order. Since the suspended sequence starts with a test for emptiness, it can only be executed if the stack actually is empty — this will discard all sequences of pops not having the length of the stack.

The MCE is interpreted over a class of monads with sufficient additional structure to support iteration, which we call *Kleene monads*; in the same way as in the usual treatment of Kleene algebra, we moreover distinguish *continuous* Kleene monads which interpret the Kleene star as a supremum. We provide a sound calculus for the MCE, which allows for the verification of abstract programs with loops and generic effects. We then proceed to explore the prospects of automatic verification, i.e. we investigate completeness and decidability issues. While Kleene algebra and the metalanguage of effects are equationally complete over the respective natural classes of models [9, 12] and have decidable equational theories (Kleene algebra is even in PSPACE [18]), it is unsurprising in view of our introductory remarks that these properties are sensitive to small extensions of the language; e.g. the Horn theory of continuous Kleene algebras fails to be r.e. [10]. Specifically, we establish the following negative results.

– The equational theory of the MCE over Kleene monads is (r.e. but) undecidable.
– The equational theory of the MCE over continuous Kleene monads is non-r.e. (more precisely at least co-r.e. hard), and hence fails to be finitely axiomatisable.

On the positive side, we show that a natural *regular* fragment of the MCE is complete and decidable. This fragment contains both Kleene algebra and the metalanguage of effects, and hence may be regarded as a suitable abstract framework for combined reasoning about control and effects. The situation is partly summarised in the diagram below.

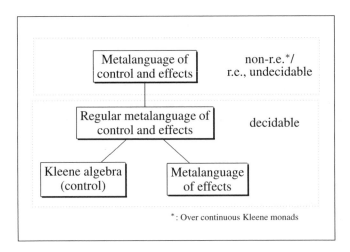

We illustrate the calculus with an extended example, the proof of a property of a further, more natural implementation of list reverse.

2 Preliminaries: Monads and Generic Imperative Programs

Intuitively, a monad associates to each type A a type TA of computations with results in A; a function with side effects that takes inputs of type A and returns values of type B is, then, just a function of type $A \to TB$. This approach abstracts from particular notions of computation, and a surprisingly large amount of reasoning can be carried out without commitment to a particular type of side effect.

Formally, a *monad* on a category $\mathbf{C}$ can be represented as a *Kleisli triple* $\mathbb{T} = (T, \eta, \_^\dagger)$, where $T : \mathrm{Ob}\,\mathbf{C} \to \mathrm{Ob}\,\mathbf{C}$ is a function, the *unit* η is a family of morphisms $\eta_A : A \to TA$, and $\_^\dagger$ assigns to each morphism $f : A \to TB$ a morphism $f^\dagger : TA \to TB$ such that

$$\eta_A^\dagger = id_{TA}, \quad f^\dagger \eta_A = f, \quad \text{and} \quad g^\dagger f^\dagger = (g^\dagger f)^\dagger.$$

The *Kleisli category* $\mathbf{C}_{\mathbb{T}}$ of $\mathbb{T}$ has the same objects as $\mathbf{C}$, and $\mathbf{C}$-morphisms $A \to TB$ as morphisms $A \to B$, with *Kleisli composition* $\diamond$ defined by $f \diamond g = f^\dagger \circ g$.

When $\mathbf{C}$ is *Cartesian*, i.e. has finite products, then $\mathbb{T}$ is *strong* if it is equipped with a natural transformation $t_{A,B} : A \times TB \to T(A \times B)$ called *strength*, subject to certain coherence conditions (see e.g. [12]).

Example 1. [12] Computationally relevant strong monads on the category of sets (and, mutatis mutandis, on other categories with sufficient structure) include the following.

1. *Stateful computations with nontermination:* $TA = S \to (1 + A \times S)$, where S is a fixed set of states.
2. *Nondeterminism:* $TA = \mathcal{P}(A)$, where $\mathcal{P}$ is the covariant power set functor.
3. *Exceptions:* $TA = A + E$, where E is a fixed set of exceptions.
4. *Interactive input:* TA is the smallest fixed point of $\gamma \mapsto A + (U \to \gamma)$, where U is a set of input values.
5. *Interactive output:* TA is the smallest fixed point of $\gamma \mapsto A + (U \times \gamma)$, where U is a set of output values.
6. *Nondeterministic stateful computation:* $TA = S \to \mathcal{P}(A \times S)$.

As originally observed by Moggi [12], strong monads support a *computational metalanguage*, i.e. essentially a generic imperative programming language, which we shall refer to as the *metalanguage of effects*. (We consider the first-order version of this language here, which forms the core of [12]; the study of iteration in the computational λ-calculus is the subject of further work. The negative results presented here are in fact made stronger by using a more economic language.) This language is parametrised over a countable signature Σ including a set of atomic types W, from which the type system is generated by the grammar

$$\mathcal{T} ::= W \mid 1 \mid A \times A \mid TA.$$

Moreover, Σ includes basic programs $f : A \to B$ with given profiles, where A and B are types. For purposes of this work, we require that the source type A for f is T-*free*, i.e. does not mention T.

Remark 2. The (trivial) completeness of the MCE over Kleene monads (see below) holds also for arbitrary signatures, but the proof of completeness and decidability of the regular fragment over continuous Kleene monads does depend on the restriction to T-free arguments; this is unsurprising, as basic programs with occurrences of T in their argument types are essentially user-defined control structures. The negative results presented below are actually made stronger by restricting the language.

The terms of the language and their types are then determined by the term formation rules shown in Fig. 1; judgements $\Gamma \triangleright t : A$ read 'term t has type A in context Γ', where a *context* is a list $\Gamma = (x_1 : A_1, \ldots, x_n : A_n)$ of typed variables.

$$
\textbf{(var)} \ \frac{x : A \in \Gamma}{\Gamma \triangleright x : A} \qquad
\textbf{(app)} \ \frac{f : A \to B \in \Sigma \ \ \Gamma \triangleright t : A}{\Gamma \triangleright f(t) : B} \qquad
\textbf{(1)} \ \frac{}{\Gamma \triangleright \langle\rangle : 1}
$$

$$
\textbf{(pair)} \ \frac{\Gamma \triangleright t : A \ \ \Gamma \triangleright u : B}{\Gamma \triangleright \langle t, u \rangle : A \times B} \qquad
\textbf{(fst)} \ \frac{\Gamma \triangleright t : A \times B}{\Gamma \triangleright \mathrm{fst}(t) : A} \qquad
\textbf{(snd)} \ \frac{\Gamma \triangleright t : A \times B}{\Gamma \triangleright \mathrm{snd}(t) : A}
$$

$$
\textbf{(do)} \ \frac{\Gamma \triangleright p : TA \ \ \Gamma, x : A \triangleright q : TB}{\Gamma \triangleright \mathrm{do}\ x \leftarrow p; q : TB} \qquad
\textbf{(ret)} \ \frac{\Gamma \triangleright t : A}{\Gamma \triangleright \mathrm{ret}\ t : TA}
$$

Fig. 1. Term formation in the metalanguage of effects

Given an interpretation of atomic types W as objects $[W]$ in a Cartesian category $\mathbf{C}$ equipped with a strong monad as above, we obtain obvious interpretations $[A]$, $[\Gamma]$ of types A and contexts Γ as objects in C. Then an interpretation of basic programs $f : A \to B$ as morphisms $[f] : [A] \to [B]$ induces an interpretation $[t] : [\Gamma] \to [A]$ of terms $\Gamma \triangleright t : A$, given by the clauses

$$
\begin{aligned}
&[x_1 : A_1, \ldots, x_n : A_n \triangleright x_i : A_i] = \pi_i^n \\
&[\Gamma \triangleright f(t) : B] = [f] \circ [\Gamma \triangleright t : A], \ f : A \to B \in \Sigma \\
&[\Gamma \triangleright \langle\rangle : 1] = !_{[\Gamma]} \\
&[\Gamma \triangleright \langle t, u \rangle : A \times B] = \langle [\Gamma \triangleright t : A], [\Gamma \triangleright u : B] \rangle \\
&[\Gamma \triangleright \mathrm{fst}(t) : A] = \pi_1 \circ [\Gamma \triangleright t : A \times B] \\
&[\Gamma \triangleright \mathrm{snd}(t) : A] = \pi_2 \circ [\Gamma \triangleright t : A \times B] \\
&[\Gamma \triangleright \mathrm{do}\ x \leftarrow p; q : TB] = [\Gamma, x : A \triangleright q : TB] \diamond t_{[\Gamma],[A]} \circ \langle id, [\Gamma \triangleright p : TA] \rangle \\
&[\Gamma \triangleright \mathrm{ret}\ t : TA] = \eta_A \circ [\Gamma \triangleright t : A]
\end{aligned}
$$

where the π_i denote projections, $!_A : A \to 1$ is the unique morphism into the terminal object, and $\langle \_, \_ \rangle$ denotes pairing of morphisms. We say that an *equation* $\Gamma \triangleright t = s$, where $\Gamma \triangleright t : A$ and $\Gamma \triangleright s : A$, is *satisfied* by the interpretation over $\mathbb{T}$, and shortly write $\mathbb{T} \models \Gamma \triangleright t = s$, if $[\Gamma \triangleright t : A] = [\Gamma \triangleright s : A]$. It is shown in [12] that equality in the metalanguage of effects is completely axiomatised by the standard rules of many-sorted equational logic plus the equations

$$\text{do } x \leftarrow (\text{do } y \leftarrow p; q); r = \text{do } x \leftarrow p; y \leftarrow q; r$$
$$\text{do } x \leftarrow \text{ret } a; p = p[a/x]$$
$$\text{do } x \leftarrow p; \text{ret } x = p$$

By results of [1], it is moreover immediate that the metalanguage of effects is decidable.

3   Kleene Monads

The metalanguage of effects recalled in the previous section is essentially Haskell's do-notation [15]. However, while Haskell's monad libraries include recursively defined loop constructs such as while without further ado, such loops are not provided with reasoning support by the metalanguage of effects: the latter is exclusively a language for *linear sequential programs*. It is the main object of this work to provide an extended generic framework that does support loops. As in classical dynamic logic, we approach this problem from the point of view of regular expressions, i.e. we introduce a nondeterministic iteration construct. We begin with the model for nondeterminism:

Definition 3. An *additive monad* is a monad $\mathbb{T}$ as above, equipped with a factorisation of the hom-functor $\mathsf{Hom}(\_,\_) : \mathbf{C}_\mathbb{T}^{op} \times \mathbf{C}_\mathbb{T} \to \mathbf{Set}$ of its Kleisli category through idempotent commutative monoids. If the underlying category $\mathbf{C}$ is Cartesian, this structure is induced by operations $\oplus : TA \times TA \to TA$ and $0 : 1 \to TA$ which satisfy the idempotent commutative monoid laws and distribute over Kleisli composition; we denote the induced structure on the hom-sets in the Kleisli category by $\oplus$ and 0 as well. A *strong additive monad* on a Cartesian category is an additive monad with a tensorial strength t satisfying two extra conditions:

$$t_{A,B} \circ \langle id, 0 \rangle = 0$$
$$t_{A,B} \circ \langle id, f \oplus g \rangle = t_{A,B} \circ \langle id, f \rangle \oplus t_{A,B} \circ \langle id, g \rangle.$$

The operations $\oplus$ and 0 are analogous to the operations mplus and mzero of Haskell's MonadPlus class. Additive monads typically involve powerset. E.g. among the monads listed in Example 1, the powerset monad and the nondeterministic state monad are additive monads. Similarly, nondeterminism may be introduced in other monads; note however that the nondeterministic exception monad $\mathcal{P}(X + E)$ fails to be additive, as it violates distribution of 0 over Kleisli composition (do $x \leftarrow p; 0$ may terminate if p raises an exception) – one way of dealing with this phenomenon is to treat exceptions separately as an outermost layer, in this case on top of the additive monad $\mathcal{P}$ [16]. Alternative axiomatic formulations can be found in [4, 8]. The computational model behind additive monads is *angelic* nondeterminism, i.e. like in nondeterministic Turing machines, nonterminating computation paths are ignored.

We can compare elements of TA by means of the natural ordering: $p \leq q \Leftrightarrow p \oplus q = q$. This turns every $\mathsf{Hom}(X, TA)$ into a join semilattice with least element 0. This leads to the central concept supporting iteration semantically:

Definition 4. An additive monad $\mathbb{T}$ is a *Kleene monad* if the operators

$$f \mapsto p \oplus f \diamond r \quad \text{and} \quad f \mapsto p \oplus r \diamond f$$

(where we agree that $\diamond$ binds more strongly than $\oplus$) both have least fixed points w.r.t. $\langle \mathrm{Hom}(X, TA), \leq \rangle$, and one of two equivalent conditions holds:

$$\mu f.(p \oplus f \diamond r) \diamond q = p \diamond \mu f.(q \oplus r \diamond f) \tag{1}$$

or

$$\mu f.(p \diamond q \oplus f \diamond r) = p \diamond \mu f.(q \oplus f \diamond r) \tag{2}$$
$$\text{and } \mu f.(p \diamond q \oplus r \diamond f) = \mu f.(p \oplus r \diamond f) \diamond q. \tag{3}$$

Intuitively, (1) means that (p followed by a number of r's) followed by q is the same as p followed by (a number of r's followed by q).

A *strong Kleene monad* is a Kleene monad which is strong and whose strength satisfies an extra *strength continuity* condition:

$$\mu f.(t_{A,B} \circ (id \times p) \oplus t_{A,B} \circ (id \times q) \diamond f) = t_{A,B} \circ (id \times \mu f.(p \oplus q \diamond f)),$$

for $p : A \to TB$ and $q : A \to TB$.

Let us show that indeed (1) is equivalent to the conjunction of (2) and (3). Assume (1). Then $\mu f.(p \diamond q \oplus f \diamond r) = p \diamond q \diamond \mu f.(\eta \oplus r \diamond f) = p \diamond \mu f.(q \oplus f \diamond r)$, hence (1) $\Rightarrow$ (2). By a symmetric argument, also (1) $\Rightarrow$ (3).

Now assume (2) and (3). It suffices to prove that $\mu f.(\eta \oplus f \diamond r) = \mu f.(\eta \oplus r \diamond f)$. As $\eta \oplus r \diamond \mu f.(\eta \oplus f \diamond r) = \eta \oplus r \oplus r \diamond \mu f.(\eta \oplus f \diamond r) \diamond r = \eta \oplus (\eta \oplus r \diamond \mu f.(\eta \oplus f \diamond r)) \diamond r$, $\eta \oplus r \diamond \mu f.(\eta \oplus f \diamond r)$ is a fixed point of the map $f \mapsto \eta \oplus f \diamond r$, hence

$$\eta \oplus r \diamond \mu f.(\eta \oplus f \diamond r) \geq \mu f.(\eta \oplus f \diamond r). \tag{4}$$

On the other hand $\mu f.(\eta \oplus f \diamond r) \diamond r = r \oplus \mu f.(\eta \oplus f \diamond r) \diamond r \diamond r$, which means that $\mu f.(\eta \oplus f \diamond r) \diamond r$ is a fixed point of $f \mapsto r \oplus f \diamond r$, hence $\mu f.(\eta \oplus f \diamond r) \diamond r \geq \mu f.(r \oplus f \diamond r)$. Add η to both sides and obtain

$$\mu f.(\eta \oplus f \diamond r) \geq \eta \oplus r \diamond \mu f.(\eta \oplus f \diamond r). \tag{5}$$

From (4) and (5), we have, $\mu f.(\eta \oplus f \diamond r) = \eta \oplus r \diamond \mu f.(\eta \oplus f \diamond r)$, which means that $\mu f.(\eta \oplus f \diamond r)$ is a fixed point of $f \mapsto \eta \oplus r \diamond f$, hence $\mu f.(\eta \oplus f \diamond r) \geq \mu f.(\eta \oplus r \diamond f)$. By symmetric argument $\mu f.(\eta \oplus r \diamond f) \geq \mu f.(\eta \oplus f \diamond r)$ and we are done. $\square$

Equation (1) is a minimal requirement that allows for defining the Kleene star operator correctly. But typically, one deals with Kleene monads satisfying the more restrictive (but natural) condition that $\mu f.(p \oplus f \diamond r) \diamond q = p \diamond \mu f.(q \oplus r \diamond f)$ is the least upper bound of the family of morphisms of the form $p \diamond r \diamond \ldots \diamond r \diamond q$. We call such Kleene monads *ω-continuous*.

Example 5. A sufficient condition for a strong monad to be an ω-continuous Kleene monad is that its Kleisli category be enriched over cpos with finite joins (i.e., countably complete lattices). This holds, e.g., for strong monads over the category of cpos with finite joins, but also for many other examples including all additive monads mentioned above, in particular the nondeterministic state monad in any topos.

On the logical side, we extend the metalanguage of effects to obtain the *metalanguage of control and effects (MCE)* by adding term formation rules

$$\frac{\Gamma \rhd p : TA \quad \Gamma, x : A \rhd q : TA}{\Gamma \rhd \operatorname{init} x \leftarrow p \operatorname{in} q^* : TA} \qquad \frac{\Gamma \rhd p : TA \quad \Gamma \rhd q : TA}{\Gamma \rhd p + q : TA} \qquad \frac{}{\Gamma \rhd \emptyset : TA}$$

While $\emptyset, +$ just provide the expected syntax for nondeterminism (deadlock and choice), the $\operatorname{init} x \leftarrow p \operatorname{in} q^*$ construct denotes a nondeterministically chosen number of iterated executions of q, initialised by $x \leftarrow p$, with the result x of the computation fed through the loop. Formally, the new language constructs are interpreted over a strong Kleene monad as follows.

- $\llbracket \Gamma \rhd \emptyset : TA \rrbracket = 0$,
- $\llbracket \Gamma \rhd p + q : TA \rrbracket = \llbracket \Gamma \rhd p : TA \rrbracket \oplus \llbracket \Gamma \rhd q : TA \rrbracket$,
- $\llbracket \Gamma \rhd \operatorname{init} x \leftarrow p \operatorname{in} q^* : TA \rrbracket =$
  $T\pi_2 \circ \mu f.(t_{\llbracket \Gamma \rrbracket, \llbracket A \rrbracket} \circ \langle id, \llbracket \Gamma \rhd p : TA \rrbracket \rangle \oplus t_{\llbracket \Gamma \rrbracket, \llbracket A \rrbracket} \circ \langle \pi_1, \llbracket \Gamma, x : A \rhd q : TA \rrbracket \rangle \diamond f)$.

Terms of the Kleene metalanguage will be also referred as *programs*. Programs not involving any monadic constructs (including $\emptyset$) will be called *atomic*. Essentially, atomic programs are combinations of symbols of Σ by means of composition and Cartesian primitives. In Fig. 2 we present an equational axiomatisation MCE, extending the axiomatisation given in the previous section. We leave variable contexts implicit. The order $\leq$ appearing in **(ind$_1$)** and **(ind$_2$)** is the one defined above. The equations **(assoc)**, **(comm)**, and **(idem)** for nondeterministic choice are called the ACI-laws. We refer to the sublanguage of the MCE without $*$ and the corresponding parts of the MCE calculus as the *theory of additive monads*.

Theorem 6 (Soundness). MCE is sound for strong Kleene monads.

Theorem 7 (Completeness). MCE is strongly complete over strong Kleene monads.

We next require some auxiliary machinery for additive monads, which as a side product induces a simple normalisation-based algorithm for deciding equality over additive monads. Consider the following rewriting system, inspired by [2].

$$
\begin{aligned}
(p : 1^n) &\longmapsto \langle\rangle^n \\
\langle \operatorname{fst}(p), \langle\rangle^n \rangle &\longmapsto p \\
\langle \langle\rangle^n, \operatorname{snd}(p) \rangle &\longmapsto p \\
\langle \operatorname{fst}(p), \operatorname{snd}(p) \rangle &\longmapsto p \\
\operatorname{do} x \leftarrow (p : T1^n); \operatorname{ret}\langle\rangle^n &\longmapsto p \\
\operatorname{do} x \leftarrow p; \operatorname{ret} x &\longmapsto p
\end{aligned}
$$

$$
\begin{aligned}
\operatorname{fst}\langle p, q \rangle &\longmapsto p \\
\operatorname{snd}\langle p, q \rangle &\longmapsto q \\
\operatorname{do} x \leftarrow \operatorname{ret} p; q &\longmapsto q[p/x] \quad (*) \\
\operatorname{do} x \leftarrow (\operatorname{do} y \leftarrow p; q); r &\longmapsto \\
\operatorname{do} x \leftarrow p; y \leftarrow q; r
\end{aligned}
$$

Basic monad laws:

(**bind**) $\quad$ do $x \leftarrow (\text{do } y \leftarrow p; q); r = \text{do } x \leftarrow p; y \leftarrow q; r$

(**eta$_1$**) $\quad$ do $x \leftarrow \text{ret } a; p = p[a/x]$

(**eta$_2$**) $\quad$ do $x \leftarrow p; \text{ret } x = p$

Extra axioms for nondeterminism:

(**plus∅**) $\quad p + \emptyset = p$ $\qquad$ (**comm**) $\quad p + q = q + p$

(**idem**) $\quad p + p = p$ $\qquad$ (**assoc**) $\quad p + (q + r) = (p + q) + r$

(**bind∅$_1$**) $\quad$ do $x \leftarrow p; \emptyset = \emptyset$ $\qquad$ (**bind∅$_2$**) do $x \leftarrow \emptyset; p = \emptyset$

(**distr$_1$**) $\quad$ do $x \leftarrow p; (q + r) = \text{do } x \leftarrow p; q + \text{do } x \leftarrow p; r$

(**distr$_2$**) $\quad$ do $x \leftarrow (p + q); r = \text{do } x \leftarrow p; r + \text{do } x \leftarrow q; r$

Extra axioms and rules for Kleene star:

(**unf$_1$**) $\quad$ init $x \leftarrow p$ in $q^* = p + \text{do } x \leftarrow (\text{init } x \leftarrow p \text{ in } q^*); q$

(**unf$_2$**) $\quad$ init $x \leftarrow p$ in $q^* = p + \text{init } x \leftarrow (\text{do } x \leftarrow p; q) \text{ in } q^*$

(**init**) $\quad$ init $x \leftarrow (\text{do } y \leftarrow p; q) \text{ in } r^* = \text{do } y \leftarrow p; \text{init } x \leftarrow q \text{ in } r^* \quad (y \notin FV(r))$

(**ind$_1$**) $\quad \dfrac{\text{do } x \leftarrow p; q \leq p}{\text{init } x \leftarrow p \text{ in } q^* \leq p}$ $\qquad$ (**ind$_2$**) $\quad \dfrac{\text{do } x \leftarrow q; r \leq r}{\text{do } x \leftarrow (\text{init } x \leftarrow p \text{ in } q^*); r \leq \text{do } x \leftarrow p; r}$

Fig. 2. MCE: A calculus for Kleene monads

The rules ($*$) capture the usual monad laws. Here n ranges over naturals (excluding 0), 1^n denotes the n-fold product of the terminal type 1 with itself, $\langle\rangle^n$ is a shortcut for the n-tuple $\langle\langle\rangle, \ldots, \langle\rangle\rangle$ for $n > 1$ and is just $\langle\rangle$ for $n = 1$.

As usual, we have omitted contexts, which are easily reconstructed except possibly in the second and third rules on the left. Their uncut versions look as follows.

$$\Gamma \triangleright \langle \text{fst}(p), \langle\rangle^n \rangle : A \times 1^n \rightarrowtail \Gamma \triangleright p : A \times 1^n$$
$$\Gamma \triangleright \langle \langle\rangle^n, \text{snd}(p) \rangle : 1^n \times A \rightarrowtail \Gamma \triangleright p : 1^n \times A$$

Consider the extra rewrite rules, capturing nondeterminism.

$$
\begin{aligned}
p + \emptyset &\rightarrowtail p & \text{do } x \leftarrow p; \emptyset &\rightarrowtail \emptyset \\
\emptyset + p &\rightarrowtail p & \text{do } x \leftarrow \emptyset; p &\rightarrowtail \emptyset \\
\text{do } x \leftarrow p; (q + r) &\rightarrowtail \text{do } x \leftarrow p; q + \text{do } x \leftarrow p; r \\
\text{do } x \leftarrow (p + q); r &\rightarrowtail \text{do } x \leftarrow p; r + \text{do } x \leftarrow q; r
\end{aligned}
\qquad (**)
$$

Let $\rightarrowtail_\lambda$ stand for the reduction relation defined by rules in ($*$), and $\rightarrowtail_\omega$ for the reduction relation corresponding to the rules in ($**$). With a slight abuse of notation use $\rightarrowtail$ to refer to the combined relation $\rightarrowtail_\lambda \cup \rightarrowtail_\omega$.

Lemma 8. *The rewrite relation $\rightarrowtail$ is confluent and strongly normalising.*

Denote the normal form of a term t under $\rightarrowtail$ by $\mathsf{nf}(t)$. Note that by definition, the normal form $\mathsf{nf}(t)$ of a program t under $\rightarrowtail$ must always look like $\sum_{i=1}^{n} \mathsf{do}\ x_1^i \leftarrow t_{k_i}^i; \ldots; x_{k_i}^i \leftarrow t_{k_i}^i; r^i$, where the topmost construct in the t_j^i and the r^i is either ret or init or an atomic program. As a simple corollary of Lemma 8 we obtain

Theorem 9. *Two programs p and q are provably equal in the theory of additive monads iff $\mathsf{nf}(p)$ and $\mathsf{nf}(q)$ are equivalent modulo the ACl-laws.*

Proof. The required property is local commutativity with ACl. That is, once p and q are related by an ACl law and $p \rightarrowtail r$, there must exist s such that $q \rightarrowtail^+ s$, and s is ACl-equivalent to r. This is easily verified by a case analysis. Together with confluence and strong normalisation, this implies by [7] the Church-Rosser property of $\rightarrowtail$ modulo ACl. As the theory of additive monads is precisely ACl plus the unoriented version of rules, defining $\rightarrowtail$, the last property is equivalent to the claim of the lemma. □

Introduce one more (nonterminating) reduction rule for unfolding the Kleene star:

k-rule: $\quad\quad\quad\quad$ init $x \leftarrow p\,\mathsf{in}\,q^* \rightarrowtail p + \mathsf{init}\ x \leftarrow (\mathsf{do}\ x \leftarrow p; q)\,\mathsf{in}\,q^*$

Define an operator z taking every program to a set of programs as follows:

- $\mathsf{z}(p) = \{p\}$ if is atomic,
- $\mathsf{z}(\mathsf{do}\ x \leftarrow p; q) = \{\mathsf{do}\ x \leftarrow s; t \mid s \in \mathsf{z}(p), t \in \mathsf{z}(q)\}$,
- $\mathsf{z}(\mathsf{ret}\,p) = \{\mathsf{ret}\,p\}$,
- $\mathsf{z}(\emptyset) = \{\,\}$,
- $\mathsf{z}(p + q) = \mathsf{z}(p) \cup \mathsf{z}(q)$,
- $\mathsf{z}(\mathsf{init}\ x \leftarrow p\,\mathsf{in}\,q^*) = \{\,\}$.

Informally, z just flattens nondeterministic choice on the top level into a set, not going under ret's. Then for every program p let $\mathsf{NF}(p) = \bigcup\{\mathsf{z}(\mathsf{nf}_{\lambda\omega}(q)) \mid p \rightarrowtail^*_{\lambda\omega k} q\}$. Now for every normal atomic $a : TA$ and a variable $x : A$ define the *derivative* $\delta_{x \leftarrow a}(t) = \{p \mid \mathsf{do}\ x \leftarrow a; p \in \mathsf{NF}(t)\} \cup \{\mathsf{ret}\,x \mid a \in \mathsf{NF}(t)\}$ and the *residue* $\varepsilon(t) = \{p \mid \mathsf{ret}\,p \in \mathsf{NF}(t)\}$. In order to make the choice of the variable irrelevant we identify the derivatives under α-conversion: $\delta_{x \leftarrow a}(p) = (\delta_{y \leftarrow a}(p))[x/y]$.

Lemma 10. *If the equality $p = q$ holds over (ω-continuous) Kleene monads, then for every a, the sets $\delta_{x \leftarrow a}(p)$ and $\delta_{x \leftarrow a}(q)$ are elementwise equal over (ω-continuous) Kleene monads, and so are $\varepsilon(p)$ and $\varepsilon(q)$.*

The proof is by well-founded induction over proofs. For the continuous case this requires the introduction of an extra (infinitary) law:

$$(\omega) \quad \frac{\forall i\ \ \mathsf{do}\ x \leftarrow (\mathsf{init}\ x \leftarrow p\,\mathsf{in}\,q^i); r \leq t}{\mathsf{do}\ x \leftarrow (\mathsf{init}\ x \leftarrow p\,\mathsf{in}\,q^*); r \leq t}$$

Here, $\mathsf{init}\ x \leftarrow p\,\mathsf{in}\,q^0 = p$, and
$\quad\quad\quad\mathsf{init}\ x \leftarrow p\,\mathsf{in}\,q^{n+1} = \mathsf{init}\ x \leftarrow (\mathsf{do}\ x \leftarrow p; q)\,\mathsf{in}\,q^n.$

Theorem 11 (Undecidability). *The theory* MCE *is undecidable.*

Proof. We show the claim by encoding Post's Correspondence Problem (PCP). Let $\Sigma = \{a, b, c\}$, and let $\{\langle p_1, q_1 \rangle, \ldots, \langle p_n, q_n \rangle\}$ be an instance of PCP in the alphabet $\Sigma \backslash \{c\}$. We define a program s that generates all possible strings of the form $l c r^{-1}$ with $l = p_{i_1} \ldots p_{i_k}$ and $r = q_{i_1} \ldots q_{i_k}$ for some sequence of indices $i_1, \ldots, i_k$, as follows. If we treat symbols of Σ as morphisms with the typing $1 \rightarrow T1$, s can be defined by the term

$$s = \text{do } x \leftarrow \left(\text{init } x \leftarrow \sum_{i=1}^{n} \text{ret(do } p_i; c; q_i^{-1}) \text{ in } \sum_{i=1}^{n} \text{ret(do } p_i; x; q_i^{-1})^* \right); x$$

Then, the fact that the PCP instance does have a solution can be expressed by the inequality

$$\text{ret } s \leq \text{do } x \leftarrow r; \text{ret}(s + x) \tag{6}$$

where r is a metaprogram presenting the nondeterministic collection of terms $\text{ret(do } l; c; l^{-1})$, defined by the term

$$r = \text{init } x \leftarrow \text{ret } c \text{ in ret(do } a; x; a) + \text{ret(do } b; x; b)^*.$$

By Lemma 10, if the inequality (6) is provable in MCE, then s is provably equivalent to some $s' \in \varepsilon(\text{do } x \leftarrow r; \text{ret}(s+x)) = \{s + \text{do } l; c; l^{-1} \mid l \in \{a, b\}^*\}$. Hence for some l, do $l; c; l^{-1} \leq s$ must be provable in MCE. By Lemma 10, this is only possible if $l c l^{-1}$ is an element of the set that defines s. On the other hand, if $l c l^{-1}$ is a solution of the PCP, it suffices to apply one of the unfolding axioms finitely many times in order to show that do $l; c; l^{-1} \leq s$. Likewise, one shows in finitely many steps that do $l; c; l^{-1} \leq r$. Therefore, we have do $x \leftarrow r; \text{ret}(s + x) = \text{do } x \leftarrow (r + \text{ret(do } l; c; l^{-1})); \text{ret}(s + x) = (\text{do } x \leftarrow r; \text{ret}(s + x)) + (\text{do } x \leftarrow \text{ret(do } l; c; l^{-1}); \text{ret}(s + x)) \geq \text{do } x \leftarrow \text{ret(do } l; c; l^{-1}); \text{ret}(s + x) = \text{ret}(s + \text{do } l; c; l^{-1}) = \text{ret } s$. This inequality is just (6), hence we are done. □

4   Continuity

It might seem useful to take the notion of ω-continuous Kleene monad as the standard definition. Indeed, in this case Kleene star is the least upper bound of its finite approximations, which perfectly matches the intuition. However, we now show that continuous Kleene monads in general are logically intractable. On the other hand, we identify a natural restriction on programs which brings the continuous case back into the realm of tractability.

Theorem 12. *Equality in the MCE over continuous Kleene monads is non-r.e.*

Proof. We prove this theorem in much the same manner as Theorem 11. But instead we encode the dual of the Post's Correspondence Problem, which is co-r.e. complete. This encoding is inspired by [10]. Again, let $\Sigma = \{a, b, c\}$, and $\{\langle p_1, q_1 \rangle, \ldots, \langle p_n, q_n \rangle\}$ be an instance of PCP in the alphabet $\Sigma \backslash \{c\}$. Besides the term s from Theorem 11, which presents all pairs of strings generated by the PCP instance, we need a term t presenting pairs of distinct strings. The program t can be defined, using the term r from Theorem 11, as

$$t = \text{do } x \leftarrow r; y \leftarrow \sharp; z \leftarrow \sharp;$$
$$u \leftarrow (\text{do } y; a; x; (b; z + \text{ret}\langle\rangle) + \text{do } y; b; x; (a; z + \text{ret}\langle\rangle) +$$
$$\text{do } (y; a + \text{ret}\langle\rangle); x; b; z + \text{do } (y; b + \text{ret}\langle\rangle); x; a; z); u,$$

where $\sharp$ denotes the 'chaos' program

$$\text{do } x \leftarrow (\text{init } x \leftarrow \text{ret ret}\langle\rangle \text{ in}(\text{ret}(\text{do } x; a) + \text{ret}(\text{do } x; b))^*); x.$$

We are done once we show that the inequality $s \leq t$ holds over all ω-continuous Kleene monads if and only if the PCP instance has no solution. Suppose $s \leq t$ over all ω-continuous monads. In particular, it must be true over the power-set monad $\mathcal{P}$. The interpretation of the terms s and t over $\mathcal{P}$ is precisely as intended, hence the PCP instance cannot have a solution. Now suppose that the PCP instance has no solutions. This means that any string generated by s can also be generated by t. As a consequence of ω-continuity, the interpretations of s and t are precisely the sets of strings they generate. For instance, $[\![s]\!] = \sup_k [\![\text{do } x \leftarrow (\text{init } x \leftarrow \sum_{i=1}^n \text{ret}(\text{do } p_i; c; q_i^{-1}) \text{ in} \sum_{i=1}^n \text{ret}(\text{do } p_i; x; q_i^{-1})^k); x]\!] = \sup_k \sum_{i_1,\dots,i_k} [\![\text{do } p_{i_1}; \dots; p_{i_k}; c; q_{i_k}^{-1}; \dots; q_{i_1}^{-1}]\!]$. As the inequality is stable under suprema, s must be greater then t. $\square$

As an immediate consequence of Theorem 12, the calculus **MCE** is incomplete over ω-continuous Kleene monads. Furthermore, it cannot be completed in any finitary way. Although our encoding of PCP involves only a small part of the expressive power of the language, one may still hope to impose syntactic restrictions on programs in order to regain completeness. One such restriction is to control the use of ret operator, as follows.

Definition 13. The set of *regular programs* is inductively defined by the following rules.

- Any ret-free program is regular,
- All term formation rules, except the rule for Kleene star, generate regular programs from regular premises.

Theorem 14. **MCE** is complete for equality of regular programs over ω-continuous Kleene monads, and equality of regular programs is decidable.

Proof. Let the equality $p = q$ be valid over all ω-continuous Kleene monads, with p and q regular. We prove the claim by induction over the nesting depth of the ret operator. If it is greater then zero we can decrease it by application of the following routine:

1. In both p and q replace (once) every subterm init $x \leftarrow s$ in t^*, innermost first, by the equivalent term $(s + \text{do } x \leftarrow s; \text{init } x \leftarrow t \text{ in } t^*)$, and normalise (by $\rightarrowtail$).
2. In the topmost nondeterministic sum of p and q, replace every atomic t by the equivalent $(\text{do } x \leftarrow t; \text{ret } x)$.

3. At this point, p and q must take the form $\sum_i$ do $x \leftarrow p_1^i; p_2^i + \sum_j$ ret p_3^i and $\sum_k$ do $x \leftarrow q_1^k; q_2^k + \sum_l$ ret q_3^l, respectively, with p_1^i and q_1^k atomic. Replace the original equation by the formula

$$\left(\bigwedge_a \sum_{p_1^i = a} p_2^i = \sum_{q_1^k = a} q_2^k\right) \wedge \left(\bigwedge_j \bigvee_l p_3^j = q_3^l\right) \wedge \left(\bigwedge_l \bigvee_j p_3^j = q_3^l\right) \tag{7}$$

(where a conjunction means that all conjuncts are valid, and a disjunction that one of the disjuncts is valid (!)).

By construction, the depth of the deepest occurrence of ret in the target equations has been decreased by one in comparison to the original equation, if originally it was greater than one. If it was equal to one, because of (2) the whole routine might have to be repeated once in order to decrease it to zero.

Soundness of steps (1) and (2) is clear, because they only appeal to the rules of MCE. Let us now show that step (3) is also legitimate. Indeed, by Lemma 10, for any a, $\delta_{x \leftarrow a}(p) = \delta_{x \leftarrow a}(q)$. The easily verifiable identity $t = \sup \delta_{x \leftarrow a}(\text{do } x \leftarrow a; t)$ shows that the first conjunct in (7) holds, because it represents equalities of suprema of equal sets. Then note that $\varepsilon(p)$ and $\varepsilon(q)$ are finite. By Lemma 10, they must be elementwise equal, which is expressed by the second and the third conjunct of (7).

Now suppose that p and q are both ret-free. W.l.o.g. assume that all the variables in p and q are bound (free ones can be seen as constants). Assign to every normal atomic program a a new functional symbol $\hat{a}$. Let $\widehat{\Sigma}$ be the set of all such symbols. The latter can be seen as an extension of Σ, because any functional symbol from Σ is a special case of an atomic program. Define a map ι, taking every ret-free program to a term of Kleene algebra over $\widehat{\Sigma}$, as follows:

- $\iota(a) = \hat{a}$, for atomic a,
- $\iota(\emptyset) = \emptyset$,
- $\iota(p + q) = \iota(p) + \iota(q)$,
- $\iota(\text{do } x \leftarrow p; q) = \iota(q) \cdot \iota(p)$,
- $\iota(\text{init } x \leftarrow p \text{ in } q^*) = \iota(q)^* \cdot \iota(p)$.

By differentiating the equation $p = q$ appropriately many times, one can show that for any atomic $a_1, \ldots, a_n$, if do $a_1; \ldots; a_n \leq p$, then do $a_1; \ldots; a_n \leq q$ and vice versa. This means precisely that $\iota(p) = \iota(q)$ holds over the algebra of regular events. Hence, by the completeness result of [9], the last equation must be provable in the corresponding calculus for Kleene monads. We are done if we show how to turn this proof into a proof in MCE. To this end, define an operator κ_X taking any Kleene algebra term to a term of the MCE of type $X \to TX$ over the signature $\widehat{\Sigma}$ by

- $\kappa_X(\hat{a}) = \hat{a}(z), \hat{a} \in \widehat{\Sigma}$
- $\kappa_X(\emptyset) = \emptyset$,
- $\kappa_X(p + q) = \kappa_X(p) + \kappa_X(q)$,
- $\kappa_X(p \cdot q) = \text{do } z \leftarrow \kappa_X(q); \kappa_X(p)$,
- $\kappa_X(p^*) = \text{ret } z + \text{init } z \leftarrow \kappa_X(p) \text{ in } \kappa_X(p)^*$.

If we apply κ_X to the proof at hand, we obtain a valid proof in MCE of the equality $\kappa_X(\iota(p)) = \kappa_X(\iota(q))$ – this follows from the obvious similarity between MCE and the calculus for Kleene algebras. The idea is to pick out an interpretation of a symbol from $\widehat{\Sigma}$ in such a way that the last equation turns into $p = q$. W.l.o.g. assume that every variable of p and q is bound only once. Let $\Delta = (x_1 : TA_1, \ldots, x_n : TA_n)$ be the context of all variables occurring in p and q. Put $X = TA_1 \times \ldots \times TA_n \times TA$, where TA is the common type of p and q. Recall that every $\hat{a} \in \widehat{\Sigma}$ occurring in $\iota(p)$, $\iota(q)$ corresponds to an atomic program $(x_{k_1} : TA_{k_1}, \ldots, x_{k_m} : TA_{k_m}) \rhd a : TB$ occurring in p, q, whose return value is either bound to some variable x_k or propagated to the top of the term. In the latter case, we just assume $k = n + 1$. Now put

$$\hat{a}(z) = \text{do } x \leftarrow a\langle \pi_{k_1} z, \ldots, \pi_{k_n} z\rangle; \text{ret}\langle \pi_1 z, \ldots, \pi_{k-1} z, x, \pi_{k+1}, \ldots, \pi_{n+1} z\rangle.$$

Having defined the interpretation of all the symbols $\hat{a}$ in this manner, we obtain an equation over the original signature Σ. It can be shown by induction that the original equation $p = q$ can by obtained from it by application of the operator $\lambda t.\,\text{do } z \leftarrow t; \text{ret } \pi_{n+1} z$. Decidability follows from the algorithmic character of the reduction steps (1) – (3) and the fact that Kleene algebra is decidable.   □

The definition of regularity might seem overly restrictive. It is easy to find examples of programs that do not satisfy regularity directly, but are still equivalent to a regular program. A typical example of this kind is $\text{init } x \leftarrow p\,\text{in}(\text{do } x \leftarrow q; \text{ret } a)^*$, which is equivalent to $p + \text{do } x \leftarrow (\text{init } x \leftarrow q\,\text{in } q[a/x]^*); \text{ret } a$. But in fact, it is not even semi-decidable to check whether a program admits a regular implementation.

Theorem 15. *The problem of checking whether a program in the MCE is semantically equivalent to a regular program over ω-continuous Kleene monads is non-r.e.*

5   Worked Example: Stack Reverse in the MCE

In spite of the negative result proved in Theorem 12 we believe the calculus MCE to be a reasonable framework for proving program equivalence. In fact, we have encoded the calculus in the Isabelle/HOL prover. We now present an example that we have axiomatised as an Isabelle theory and successfully verified[1]: the double-reverse theorem stating that reversing a list twice yields the original list. More specifically, we have proved this theorem for two implementations of reverse, the single-stack example from the introduction and a further implementation using two stacks. For the latter, we declare operations

$$\text{pop}_i : TA, \qquad \text{push}_i : A \to T1, \qquad \text{is\_empty}_i : T1.$$

$(i \in \{1, 2\})$. Some of the axioms imposed on these are

$$\text{do is\_empty}_i; \text{pop}_i = \emptyset \qquad\qquad \text{do } x \leftarrow \text{pop}_i; \text{push}_i\, x \le \text{ret}\langle\rangle$$
$$\text{do push}_i\, x; \text{is\_empty}_i = \emptyset \qquad\qquad \text{do push}_i\, x; \text{pop}_i = \text{ret } x$$
$$\text{do is\_empty}_i; \text{is\_empty}_i = \text{is\_empty}_i \qquad\qquad \text{is\_empty}_i \le \text{ret}\langle\rangle$$

---

[1] The theory files are available under `http://www.informatik.uni-bremen.de/~sergey/StackReverse.tar`

Define $\text{append}_{ij} = \text{do init ret}\langle\rangle \text{ in}(\text{do } x \leftarrow \text{pop}_i; \text{push}_j\, x)^*; \text{is\_empty}_i$, which pops elements from the i-th stack and pushes them into j-th stack in reverse order until the i-th stack is empty. Now the double reverse theorem for two stacks can be encoded as

$$\text{do is\_empty}_2; \text{append}_{12}; \text{append}_{21} \leq \text{is\_empty}_2$$

(where the inequality arises because there is no axiom that guarantees that the stack bottom is reachable, and expresses primarily that the left hand program does not change the state). The proofs are reasonably straightforward but require some degree of manual intervention; the possibility of better proof automation is under investigation.

6   Conclusion and Future Work

We have combined Moggi's computational metalanguage for monads and Kozen's axiomatisation of Kleene algebra to obtain a metalanguage of control and effects (MCE) that can be interpreted over a suitable notion of Kleene monad. While the combined equational language cannot be recursively axiomatised, a quite expressive sublanguage of so-called regular programs does have a complete axiomatisation. This axiomatisation has been formalised in the theorem prover Isabelle/HOL, and some sample verifications have been performed on programs involving monadic do-notation, Kleene star, and nondeterminism, as well as the use of programs as values in a higher-order style. The MCE forms part of an evolving system of formalisms (see e.g. [17, 16, 14]) aimed at the development of a verification framework for functional-imperative programs in the style of Haskell. The further development of this framework will include the integration of Kleene monads with monad-based dynamic logic [17, 14]; an extension of the MCE with side effect free tests in analogy to Kleene algebra with tests [10], which will in particular accommodate the Fischer-Ladner encoding of `if` and `while`; and parametrisation of the MCE over an underlying equational theory of data. A further open point is how our approach to the combination of iteration and effects can be transferred to the related framework of algebraic effects, in particular w.r.t. the combination of effects [5].

Acknowledgement. We thank Erwin R. Catesbeiana for finite discussions about infinite iterations.

References

[1] Benton, P.N., Bierman, G.M., de Paiva, V.: Computational types from a logical perspective. J. Funct. Prog. 8(2), 177–193 (1998)
[2] Curien, P.-L., di Cosmo, R.: A confluent reduction for the lambda-calculus with surjective pairing and terminal object. J. Funct. Program. 6(2), 299–327 (1996)
[3] Fluet, M., Morrisett, G., Ahmed, A.J.: Linear regions are all you need. In: Sestoft, P. (ed.) ESOP 2006. LNCS, vol. 3924, pp. 7–21. Springer, Heidelberg (2006)
[4] Hinze, R.: Deriving backtracking monad transformers. ACM SIGPLAN Notices 35(9), 186–197 (2000)

[5] Hyland, M., Plotkin, G.D., Power, J.: Combining effects: Sum and tensor. Theoret. Comput. Sci. 357, 70–99 (2006)

[6] Jacobs, B., Poll, E.: Coalgebras and Monads in the Semantics of Java. Theoret. Comput. Sci. 291, 329–349 (2003)

[7] Jouannaud, J.-P., Munoz, M.: Termination of a set of rules modulo a set of equations. In: Shostak, R.E. (ed.) CADE 1984. LNCS, vol. 170, pp. 175–193. Springer, Heidelberg (1984)

[8] Kiselyov, O., Shan, C., Friedman, D., Sabry, A.: Backtracking, interleaving, and terminating monad transformers. In: Functional Programming, ICFP 2005, pp. 192–203. ACM Press, New York (2005)

[9] Kozen, D.: A completeness theorem for Kleene algebras and the algebra of regular events. Inf. Comput. 110, 366–390 (1994)

[10] Kozen, D.: Kleene algebra with tests and commutativity conditions. In: Margaria, T., Steffen, B. (eds.) TACAS 1996. LNCS, vol. 1055, pp. 14–33. Springer, Heidelberg (1996)

[11] Kozen, D.: Nonlocal flow of control and Kleene algebra with tests. In: Logic in Computer Science, LICS 2008, pp. 105–117. IEEE Computer Society Press, Los Alamitos (2008)

[12] Moggi, E.: Notions of computation and monads. Inf. Comput. 93, 55–92 (1991)

[13] Moggi, E., Sabry, A.: Monadic encapsulation of effects: A revised approach (extended version). J. Funct. Prog. 11, 591–627 (2001)

[14] Mossakowski, T., Schröder, L., Goncharov, S.: A generic complete dynamic logic for reasoning about purity and effects (Extended version to appear in *Formal Aspects of Computing*). In: Fiadeiro, J.L., Inverardi, P. (eds.) FASE 2008. LNCS, vol. 4961, pp. 199–214. Springer, Heidelberg (2008)

[15] Peyton Jones, S. (ed.): Haskell 98 Language and Libraries — The Revised Report. Cambridge University Press (2003); J. Funct. Prog. 13 (2003)

[16] Schröder, L., Mossakowski, T.: Generic exception handling and the Java monad. In: Rattray, C., Maharaj, S., Shankland, C. (eds.) AMAST 2004. LNCS, vol. 3116, pp. 443–459. Springer, Heidelberg (2004)

[17] Schröder, L., Mossakowski, T.: Monad-independent dynamic logic in HASCASL. J. Logic Comput. 14, 571–619 (2004)

[18] Stockmeyer, L.J., Meyer, A.R.: Word problems requiring exponential time: Preliminary report. In: Symposium on Theory of Computing, STOC 1973, pp. 1–9. ACM Press, New York (1973)

Complete Iterativity for Algebras with Effects

Stefan Milius, Thorsten Palm, and Daniel Schwencke*

Technische Universität Braunschweig,
Institut für Theoretische Informatik,
Mühlenpfordtstraße 22–23, D-38106 Braunschweig, Germany
s.milius@tu-bs.de, palm@iti.cs.tu-bs.de,
schwencke@iti.cs.tu-bs.de

Abstract. Completely iterative algebras (cias) are those algebras in which recursive equations have unique solutions. In this paper we study complete iterativity for algebras with computational effects (described by a monad). First, we prove that for every analytic endofunctor on **Set** there exists a canonical distributive law over any commutative monad M, hence a lifting of that endofunctor to the Kleisli category of M. Then, for an arbitrary distributive law λ of an endofunctor H on **Set** over a monad M we introduce λ-cias. The cia for the corresponding lifting of H (called Kleisli-cias) form a full subcategory of the category of λ-cias. For various monads of interest we prove that free Kleisli-cias coincide with free λ-cias, and these free algebras are given by free algebras for H. Finally, for three concrete examples of monads we prove that Kleisli-cias and λ-cias coincide and give a characterisation of those algebras.

Keywords: iterative algebra, monad, distributive law, initial algebra, terminal coalgebra.

1 Introduction

Iterative theories [3] and iterative algebras [15, 16] were introduced to study the semantics of recursive equations at a purely algebraic level without the need to use extra structure like order or metric. Iterative algebras are algebras for a signature Σ in which every guarded system of recursive equations

$$x_1 \approx t_1(x_1, \ldots, x_n, a_1 \ldots, a_k)$$
$$\vdots \qquad\qquad\qquad\qquad (1.1)$$
$$x_n \approx t_n(x_1, \ldots, x_n, a_1 \ldots, a_k)$$

where $X = \{x_1, \ldots, x_n\}$ is a set of variables, $a_1, \ldots, a_k$ are elements of the algebra (called parameters) and the t_i are Σ-terms that are not single variables $x \in X$, has a unique solution. For example, the algebra $T_\Sigma Y$ of all (not necessarily finite) Σ-trees on the set Y is iterative.

---

* The support by the German Research Foundation (DFG) under the project AD 207/1-1 is acknowledged.

A. Kurz, M. Lenisa, and A. Tarlecki (Eds.): CALCO 2009, LNCS 5728, pp. 34–48, 2009.

The notion of iterative algebra was extended and generalised in [1]; there, iterative algebras are studied for finitary endofunctors on **Set** (or, more generally, on a locally finitely presentable category). And in [11] completely iterative algebras (cias) for an endofunctor were introduced and studied—the completeness refers to the fact that the set of variables in a system (1.1) may be infinite.

It is the aim of this paper to investigate iterativity of algebras in which the operations have side-effects (e. g. partial operations, non-deterministic ones or composite algebras). Such effects are captured by working in the Kleisli category of a monad, see [14]. More precisely, let M be a monad on **Set**. The Kleisli category of M has all sets as objects and morphisms from X to Y are maps $f : X \to MY$, which are understood as functions with a side-effect captured by M. To study algebras with effects we will consider set functors H having a lifting to the Kleisli category of the monad M. An algebra with effect is then an algebra for the lifting of H. It is well known that to have such a lifting is the same as to have a distributive law $\lambda : HM \Rightarrow MH$ of the functor H over the monad.

It has been proved by Hasuo, Jacobs and Sokolova [4] that every polynomial endofunctor on **Set** yields a canonical distributive law over any commutative monad on **Set**. Here we extend this result by showing that every analytic functor, see [6, 7], has a canonical distributive law over every commutative monad (Section 2). We then study two different notions of algebras with "effectful" operations for an endofunctor H: Kleisli-cias are just the completely iterative algebras for a lifting $\bar{H}$ of H, and we introduce the notion of completely λ-iterative algebras. Whereas in Kleisli-cias recursive equations may have side-effects, in λ-cias recursion is effect-free, i. e., only parameters of recursive equations are allowed to have side-effects. We show that every Kleisli-cia is a λ-cia, but the converse does not hold in general (Section 3).

Our next result concerns free iterative algebras with effects: it often turns out that a free H-algebra on a set X is at the same time a free $\bar{H}$-algebra, a free λ-cia and a free Kleisli-cia on X. We prove this in the case where the Kleisli category is suitably cpo-enriched, see Assumption 4.1 for details. This is the setting as studied in [4], and our result is a consequence of the fact proved in loc. cit. that the initial H-algebra is a terminal $\bar{H}$-coalgebra (Section 4).

Finally, we prove that for polynomial functors and the concrete monads $(-) + 1$ (the maybe monad) and $\mathbb{P}$ (the powerset monad), as well as for every endofunctor and the monad $(-)^E$ (the environment monad) Kleisli-cias and λ-cias coincide. We also give in all three cases a characterisation of λ-cias (or, equivalently, of Kleisli-cias): in the first two cases an $\bar{H}$-algebra is a λ-cia iff its carrier has a well-founded order such that each operation is "strictly increasing" (see Theorems 5.1 and 5.2 for a precise statement) and for the environment monad an $\bar{H}$-algebra is a λ-cia iff each component is a cia for H. This also implies that for the environment monad the free λ-cias (or Kleisli-cias) on the set X are given by the terminal coalgebra TX for $H(-) + X$, if it exists (Section 5).

In this version of the paper the longer proofs are omitted. They can be found in the full version [13].

2   Background—Distributive Laws for **Set**-Functors

Assumption 2.1. Throughout the paper we let M be a monad on **Set**, i. e., M : **Set** $\to$ **Set** is an endofunctor equipped with two natural transformations $\eta :$ Id $\Rightarrow M$ (the unit) and $\mu : MM \Rightarrow M$ (the multiplication) subject to the axioms $\mu \cdot \eta M = \mu \cdot M\eta = \mathrm{id}$ and $\mu \cdot M\mu = \mu \cdot \mu M$. We shall also consider **Set** as a symmetric monoidal category with cartesian product as the monoidal product. We denote by $c_{X,Y} : X \times Y \to Y \times X$ the symmetry isomorphisms.

Throughout the paper we denote by $\mathrm{inl} : A \to A + B \leftarrow B : \mathrm{inr}$ the injections of any coproduct $A + B$.

The purpose of the monad M is to represent the type of side-effect for recursive equations, for their solutions and for operations in algebras.

Examples 2.2. (1) The *maybe monad* is given by $MX = X + 1$ with the unit $\mathrm{inl} :$ $X \to X + 1$ and the multiplication $\mathrm{id}_X + \nabla_1 : X + 1 + 1 \to X + 1$ (where $\nabla_1 = [\mathrm{id}, \mathrm{id}] : 1 + 1 \to 1$ is the codiagonal).
(2) The *powerset monad* $M = \mathbb{P}$ has as unit the singleton map $\eta_X : X \to \mathbb{P}X$ and as multiplication the union map $\mu_X : \mathbb{PP}X \to \mathbb{P}X$.
(3) The *subdistribution monad* $M = \mathbb{D}$ assigns to a set X the set $\mathbb{D}X$ of functions $d : X \to [0, 1]$ with finite support and with $\sum_{x \in X} d(x) \le 1$.
(4) The *environment monad* is given by $MX = X^E$, where E is a fixed set. Its unit $\eta_X : X \to X^E$ assigns to an element $x \in X$ the constant map on x, and the multiplication μ_X assigns to an element of $(X^E)^E$ (i.e., an $|E| \times |E|$ matrix with values in X) its diagonal (considered as a map from E to X).
(5) The *finite-list monad* $MX = X^*$ has the unit given by singleton lists and the multiplication by flattening a list of lists.
(6) The *finite-multiset monad* $M = \mathbb{M}$ assigns to a set X the free commutative monoid on X or, equivalently, the set of all finite multisets on X.

In this paper we are interested in endofunctors on **Set** that have a lifting to the Kleisli category of the monad M. In fact, we shall establish that every analytic endofunctor (see [6, 7]) has a canonical lifting to the Kleisli category of every commutative monad (see [8]). This extends a previous result from [4] where this was proved for polynomial endofunctors on **Set**.

Remark 2.3. We shall not present the formal definition of a commutative monad. It is well known that giving the structure of a commutative monad is equivalent to giving the structure of a symmetric monoidal monad, see [8, 9]. So every commutative monad (M, η, μ) on **Set** comes equipped with a *double strength*, i.e., a family of maps $m_{X,Y} :$ $MX \times MY \to M(X \times Y)$ natural in X and Y such that the following axioms hold:

$$m_{X,Y} \cdot (\eta_X \times \eta_Y) = \eta_{X \times Y}, \quad \mu_{X \times Y} \cdot Mm_{X,Y} \cdot m_{MX,MY} = m_{X,Y} \cdot (\mu_X \times \mu_Y)$$

$$\text{and} \quad m_{X \times Y, Z} \cdot (m_{X,Y} \times \mathrm{id}_{MZ}) = m_{X, Y \times Z} \cdot (\mathrm{id}_{MX} \times m_{Y,Z}).$$

Moreover, the monad is symmetric monoidal, i.e., we have $m_{Y,X} \cdot c_{MX,MY} = Mc_{X,Y} \cdot m_{X,Y}$.

Examples 2.4. All monads in Example 2.2, except for the list monad, are commutative monads with a (unique) canonical double strength: see [4] for the first three examples; for the environment monad the double strength is the canonical isomorphism

$X^E \times Y^E \cong (X \times Y)^E$; and the finite-multiset monad has $\mathsf{m}_{X,Y} : \mathbb{M}X \times \mathbb{M}Y \to \mathbb{M}(X \times Y)$ given by $\mathsf{m}_{X,Y}(\langle x_1, \ldots, x_n \rangle, \langle y_1, \ldots, y_m \rangle) = \langle (x_i, y_j) \mid i = 1, \ldots, n; j = 1, \ldots, m \rangle$, where the angular brackets denote multisets.

Next we recall the notion of an analytic functor on **Set** and a characterisation that we shall subsequently use.

Definition 2.5. (A. Joyal [5, 6]) An endofunctor H on **Set** is called *analytic* provided that it is the left Kan extension of a functor from the category $\mathcal{B}$ of natural numbers and bijections to **Set** along the inclusion.

Remark 2.6. (1) In fact, Joyal defined analytic functors by explicitly stating what these Kan extensions are. Let $\mathcal{S}_n$ be the symmetric group of all permutations of n. For every subgroup G of $\mathcal{S}_n$ the *symmetrised representable functor* sends each set X to the set X^n/G of orbits under the action of G on X^n by coordinate interchange, i. e., X^n/G is the quotient of X^n modulo the equivalence $\sim_G$ with $(x_1, \ldots, x_n) \sim_G (y_1, \ldots, y_n)$ iff $(x_{p(1)}, \ldots, x_{p(n)}) = (y_1, \ldots, y_n)$ for some $p \in G$. It is straightforward to work out that an endofunctor on **Set** is analytic iff it is a coproduct of symmetrised representables. So every analytic functor H can be written in the form

$$HX = \coprod_{n \in \mathbb{N}, G \leq \mathcal{S}_n} A_{n,G} \times X^n/G. \tag{2.1}$$

(2) Notice that by (2.1) an analytic functor is a quotient of the corresponding polynomial functor P with $PX = \coprod_{n,G} A_{n,G} \times X^n$.

(3) Clearly every analytic functor is finitary. Joyal proved in [5, 6] that a finitary endofunctor on **Set** is analytic iff it weakly preserves wide pullbacks.

Examples 2.7. (1) Let Σ be a signature of operation symbols with prescribed arity. The associated *signature functor* is the polynomial endofunctor given by $H_\Sigma X = \coprod_{n \in \mathbb{N}} \Sigma_n \times X^n$; the elements of $H_\Sigma X$ are written suggestively as flat terms $\sigma(x_1, \ldots, x_n)$. Clearly H_Σ is analytic (take $A_{n,G} = \Sigma_n$ for the trivial subgroup $G = \{\mathrm{id}\} \leq \mathcal{S}_n$ and $A_{n,G} = 0$ else).

(2) The functor H assigning to a set X the set of finite multisets over X is analytic, since it arises from putting $HX = \coprod_{n \in \mathbb{N}} X^n/\mathcal{S}_n$.

(3) The functor H assigning to a set X the set of trees (always taken to be rooted and ordered) with nodes labelled in X is analytic. In fact, H is the left Kan extension of $t : \mathcal{B} \to \mathbf{Set}$ assigning to the natural number n the set $t(n)$ of trees with $\{0, \ldots, n-1\}$ as the set of nodes.

(4) The finitary powerset functor $\mathbb{P}_{fin}$ is not analytic as it does not preserve weak wide pullbacks.

Recall that liftings of an endofunctor H to the Kleisli category of the monad M are in bijective correspondence with distributive laws of H over M:

Definition 2.8. A *distributive law* of an endofunctor H over a monad M is a natural transformation $\lambda : HM \Rightarrow MH$ such that

$$\lambda \cdot H\eta = \eta H \quad \text{and} \quad \lambda \cdot H\mu = \mu H \cdot M\lambda \cdot \lambda M.$$

Theorem 2.9. *Let H be an analytic functor on* **Set** *and let M be a commutative monad. Then there exists a canonical distributive law λ of H over M.*

Remark 2.10. Let M be a commutative monad and let H be an analytic endofunctor. Write H as a quotient of the polynomial functor P, see Remark 2.6(2), and denote by $\epsilon : P \Rightarrow H$ the natural transformation formed by the canonical surjections. It follows from the proof of Theorem 2.9 (see the full version [13] of this paper) that for the canonical distributive laws $\tilde{\lambda} : PM \Rightarrow MP$ and $\lambda : HM \Rightarrow MH$ we have

$$M\epsilon \cdot \tilde{\lambda} = \lambda \cdot \epsilon M : PM \Rightarrow MH . \tag{2.2}$$

Examples 2.11. We make the distributive law λ of Theorem 2.9 explicit for some combinations of M and H of interest. In the first three items we consider the signature functor $H = H_\Sigma$.

(1) For the maybe monad $MX = X + 1$, $\lambda_X : H_\Sigma(X + 1) \to H_\Sigma X + 1$ maps $\sigma(x_1, \ldots, x_n)$ to itself if all x_i are in X and to the unique element of 1 otherwise.

(2) For $M = \mathbb{P}$, $\lambda_X : H_\Sigma \mathbb{P} X \to \mathbb{P} H_\Sigma X$ acts as follows: $\lambda_X(\sigma(X_1, \ldots, X_n)) = \{\sigma(x_1, \ldots, x_n) \mid x_i \in X_i, i = 1, \ldots, n\}$ for $\sigma \in \Sigma_n$ and $X_i \subseteq X, i = 1, \ldots, n$.

(3) For the environment monad $MX = X^E$, the distributive law $\lambda_X : H_\Sigma X^E \to (H_\Sigma X)^E$ acts as follows: $\lambda_X(\sigma(v_1, \ldots, v_n)) = i \mapsto \sigma(v_1(i), \ldots, v_n(i))$ for $\sigma \in \Sigma_n$ and $v_1, \ldots, v_n : E \to X$ and $i \in E$. More generally, there exists a canonical distributive law of every endofunctor H over M as follows: observe that $X^E \cong \prod_{i \in E} X$ with projections $\pi_i^X : X^E \to X$ for each $i \in E$. Define $\lambda_X : H(X^E) \to (HX)^E$ as the unique morphism such that $\pi_i^{HX} \cdot \lambda_X = H\pi_i^X$ for every $i \in E$. It is easy to prove that λ is a distributive law of H over M.

(4) For $H = \mathrm{Id}$ we obtain from Theorem 2.9 the identity transformation $\lambda = \mathrm{id} : M \Rightarrow M$ (which, in fact, is a distributive law for any monad).

(5) Let H be the finite-multiset functor of Example 2.7(2). Its canonical distributive law λ over the powerset monad is given by

$$\lambda_X(\langle X_1, \ldots, X_n \rangle) = \{\langle x_1, \ldots, x_n \rangle \mid x_i \in X_i, i = 1, \ldots, n\}$$

for $X_j \subseteq X, j = 1, \ldots, n$. In fact, this follows from (2.2) and (2) above, applied to the polynomial functor $PX = \coprod_{n \in \mathbb{N}} X^n$.

3 Iterative Algebras with Effects

Assumption 3.1. In this section we assume that H is an endofunctor on **Set** and that λ is a distributive law of H over the monad (M, η, μ).

Notation 3.2. (1) We denote morphisms in the Kleisli category **Set**$_M$ by the symbol $\longrightarrow\!\!\!\circ\!\!\!\rightarrow$, i.e., $X \longrightarrow\!\!\!\circ\!\!\!\rightarrow Y$ is a map from X to MY. Moreover, $J : \mathbf{Set} \to \mathbf{Set}_M$ denotes the identity-on-objects functor with $Jf = \eta_Y \cdot f$ for any map $f : X \to Y$. Recall that J has a right adjoint V assigning to every $f : X \longrightarrow\!\!\!\circ\!\!\!\rightarrow Y$ the map $\mu_Y \cdot Mf : MX \to MY$. The counit ε of this adjunction is given by the identity maps on MA considered as morphisms $\varepsilon_A : MA \longrightarrow\!\!\!\circ\!\!\!\rightarrow A$ in **Set**$_M$.

(2) The lifting of H introduced by λ is denoted by $\bar{H} : \mathbf{Set}_M \to \mathbf{Set}_M$. It takes a morphism $f : X \to MY$ in $\mathbf{Set}_M$ to $\bar{H}f = \lambda_Y \cdot Hf$.

Remark 3.3. (1) Recall that an algebra for the endofunctor H on $\mathbf{Set}$ is a set A with an algebra structure $\alpha : HA \to A$, and that an H-algebra homomorphism from (A, α) to (B, β) is a map $f : A \to B$ such that $f \cdot \alpha = \beta \cdot Hf$.

(2) Observe that an algebra for the lifting $\bar{H}$ is an algebra whose operations have side-effects described by the monad M. In fact, an $\bar{H}$-algebra is a set A equipped with an algebra structure $\alpha : HA \to MA$. So, for example, for a signature functor H_Σ, an $\bar{H}_\Sigma$-algebra is an algebra for the signature Σ where the operations are partial (in case $MX = X + 1$ is the maybe monad), nondeterministic (in case $M = \mathbb{P}$ is the powerset monad), or are families of operations indexed by elements of E (in case $MX = X^E$ is the environment monad).

Notation 3.4. Let H_Σ be a signature functor, and let $\alpha : H_\Sigma A \to MA$ be an $\bar{H}_\Sigma$-algebra. We denote by $\sigma^A : A^n \to MA$ the component of α corresponding to the n-ary operation symbol σ of Σ.

In the remainder of this paper we shall study algebras with effects in which one can uniquely solve recursive equations. Before we give the formal definition let us discuss a motivating example (without effects first). Let Σ be a signature, and let A be a Σ-algebra. We are interested in unique solutions in A of recursive equation systems (1.1). In fact, it suffices to consider the right-hand side terms of the form $t_i = \sigma(x_{i_1}, \ldots, x_{i_k})$, $\sigma \in \Sigma_k$, or $t_i = a \in A$—every system can be turned into one with right-hand sides of this form by introducing new variables. A solution to a system (1.1) consists of elements $s_1, \ldots, s_n \in A$ turning the formal equations into identities $s_i = t_i^A(s_1, \ldots, s_n, a_1, \ldots, a_p)$ in A. For example, let Σ be the signature with one binary operation symbol $*$ and with the constant symbol c. In the Σ-algebra $A = T_\Sigma Y$ of all Σ-trees on Y, i.e., (rooted and ordered) trees so that nodes with $n > 0$ children are labelled in Σ_n and leaves are labelled in $\Sigma_0 + Y$, the following system

$$x_1 \approx x_2 * c \qquad x_2 \approx x_1 * y$$

has as its unique solution the trees

This motivates the following

Definition 3.5. ([11]) Let $\mathcal{A}$ be a category with finite coproducts and let $H : \mathcal{A} \to \mathcal{A}$. A *flat equation morphism* in an object A (of parameters) is a morphism $e : X \to HX + A$. An H-algebra $\alpha : HA \to A$ is called *completely iterative* (or a *cia*, for short) if every flat equation morphism in A has a unique solution, i.e., for every $e : X \to HX + A$ there exists a unique morphism $e^\dagger : X \to A$ such that

$$e^\dagger = (X \xrightarrow{e} HX + A \xrightarrow{He^\dagger + \mathrm{id}_A} HA + A \xrightarrow{[\alpha, \mathrm{id}_A]} A) . \tag{3.1}$$

Examples 3.6. We recall some examples from previous work.

(1) Let TX denote a terminal coalgebra for $H(-) + X$. Its structure map is an isomorphism by Lambek's Lemma [10], and so its inverse yields (by composing with the coproduct injections) an H-algebra $\tau_X : HTX \to TX$ and a morphism $\eta_X : X \to TX$. Then (TX, τ_X) is a free cia on X with the universal arrow η_X, see [11], Theorems 2.8 and 2.10.

(2) Let H_Σ be a signature functor. The terminal coalgebra for $H_\Sigma(-) + X$ is carried by the set $T_\Sigma X$ of all Σ-trees on X. According to the previous item, this is a free cia for H_Σ on X.

(3) The algebra of addition on $\bar{\mathbb{N}} = \{1, 2, 3, \dots\} \cup \{\infty\}$ is a cia for $HX = X \times X$, see [1].

(4) Let $\mathcal{A} = \mathbf{CMS}$ be the category of complete metric spaces and let H be a contracting endofunctor on $\mathbf{CMS}$, see e. g. [2]. Then any non-empty algebra for H is a cia, see [11] for details. For example, let A be the set of non-empty compact subsets of the unit interval $[0, 1]$ equipped with the Hausdorff metric. This complete metric space can be turned into a cia such that the Cantor set arises as the unique solution of a flat equation morphism, see [12], Example 3.3(v).

(5) Unary algebras of $\mathbf{Set}$. Here we take $\mathcal{A} = \mathbf{Set}$ and $H = \mathrm{Id}$. An algebra $\alpha : A \to A$ is a cia iff α has a fixed point a_0 and there is no infinite sequence $a_1, a_2, a_3, \dots$ with $a_i = \alpha(a_{i+1})$, $i = 1, 2, 3, \dots$, except for the one all of whose members are a_0. The second part of this condition can be put more vividly as follows: the graph with node set $A \setminus \{a_0\}$ and with an edge from $\alpha(a) \neq a_0$ to a for all a is well-founded. Since any well-founded graph induces a well-founded (strict) order on its node set, we have yet another formulation: there is a well-founded order on $A \setminus \{a_0\}$ for which α is strictly increasing in the sense that $\alpha(a) \neq a_0$ implies $a < \alpha(a)$ for all $a \in A$.

(6) Classical algebras are seldom cias. For example, a group or a semilattice is a cia (for $HX = X \times X$) iff they contain one element only (consider the unique solution of $x \approx x \cdot 1$ or $x \approx x \vee x$, respectively).

In this paper we consider cias in the two categories $\mathbf{Set}$ and $\mathbf{Set}_M$. To distinguish ordinary cias for an endofunctor H on $\mathbf{Set}$ from those for a lifting $\bar{H}$ we have the following

Definition 3.7. We call a cia for a lifting $\bar{H} : \mathbf{Set}_M \to \mathbf{Set}_M$ a *Kleisli-cia*.

Remark 3.8. If we spell out the definition of a cia in $\mathbf{Set}_M$, we see that a flat equation morphism is a map $e : X \to M(HX + A)$, and a solution of e in the algebra $\alpha : HA \to MA$ is a map $e^\dagger : X \to MA$ such that

$$e^\dagger = (X \xrightarrow{\ e\ } M(HX + A) \xrightarrow{M(\lambda \cdot He^\dagger + \eta_A)} M(MHA + MA)$$

$$\xrightarrow{\mu_{X+A} \cdot M[M\mathrm{inl}, M\mathrm{inr}]} M(HA + A) \xrightarrow{\mu_A \cdot [\alpha, \eta_A]} MA)$$

This means that the algebra (A, α) as well as the recursive equation e and its solution are "effectful". For example, for $M = \mathbb{P}$ the effect is non-determinism. If $H = H_\Sigma$

for a signature Σ with two binary operation symbols $+$ and $*$, the non-deterministic equation

$$x \approx \{x + x, x * x, a\} \quad \text{where } x \in X \text{ and } a \in A \tag{3.2}$$

gives rise to a flat equation morphism. As we shall see in Example 5.4(2), the set $A = \{2, 3, \ldots, 100\} \subset \mathbb{N}$ together with the operations $+, * : A \times A \to \mathbb{P}A$ returning the sum and product, respectively, as a singleton set if this is less than or equal to 100 and $\emptyset$ otherwise, is a Kleisli-cia. If we choose $a = 10$ in the above equation, the unique solution of x is $\{10, 20, 30, 40, 50, 60, 70, 80, 90, 100\} \subseteq A$.

Observe that, in general, the notion of a Kleisli-cia automatically connects effectful operations of an algebra with effectful recursive equations. One might want to consider these two separately. To this end we propose λ-cias as a notion of algebras with effectful operations where effects in recursive equations are allowed in the parameters only.

Definition 3.9. For the monad M a (flat) M-*equation morphism* in an object A is a morphism $e : X \to HX + MA$. A solution of e in the $\bar{H}$-algebra $\alpha : HA \dashrightarrow A$ is a morphism $e^\dagger : X \dashrightarrow A$ such that

$$e^\dagger = (X \overset{Je}{\dashrightarrow} HX + MA \overset{\bar{H}e^\dagger + \varepsilon_A}{\dashrightarrow} HA + A \overset{[\alpha, \mathrm{id}_A]}{\dashrightarrow} A) \tag{3.3}$$

holds in $\mathbf{Set}_M$. The $\bar{H}$-algebra (A, α) is called *completely λ-iterative* (or λ-*cia*, for short) if every M-equation morphism in A has a unique solution.

Remark 3.10. Observe that an $\bar{H}$-algebra (A, α) is a λ-cia iff $\mu_A \cdot M\alpha \cdot \lambda_A : HMA \to MA$ is an (ordinary) cia for the endofunctor H on **Set**. In fact, this is trivial to see by writing Equation (3.3) in **Set**:

$$e^\dagger = (X \overset{e}{\to} HX + MA \overset{\lambda_A \cdot He^\dagger + \mathrm{id}_A}{\longrightarrow} MHA + MA \overset{[\mu_A \cdot M\alpha, \mathrm{id}_A]}{\longrightarrow} MA).$$

Also notice that in e the monad M is applied only to the second component of the coproduct in the codomain, whereas in a flat equation morphism with respect to a Kleisli-cia M is applied to the whole coproduct, cf. Remark 3.8. Continuing our concrete example for $M = \mathbb{P}$ from Remark 3.8, we see that the formal equation in (3.2) does not give rise to an M-equation morphism in the algebra A, but the system

$$x \approx y * z \qquad y \approx \{2, 3\} \qquad z \approx \{4, 5\}$$

does. Its unique solution assigns to x the set $\{8, 10, 12, 15\}$. In fact, as we shall see in Proposition 3.14, A also is a λ-cia.

Remark 3.11. We analyse the meaning of Equation (3.3) for a signature functor and the distributive laws of Examples 2.11, (1)–(3). Notice that in this case an M-equation morphism e corresponds to a system (1.1) where the right-hand sides are of the form $\sigma(x_{i_1}, \ldots, x_{i_k})$, $\sigma \in \Sigma_k$, or are elements of MA. Further notice that we always have $e^\dagger(x) = e(x)$ if $e(x) \in MA$. We describe the meaning of a solution of an equation $x \approx \sigma(x_1, \ldots, x_k)$.

(1) For the maybe monad $MX = X + \{\bot\}$, we have $e^\dagger(x) = \bot$ if one of the $e^\dagger(x_i)$ is $\bot$, otherwise $e^\dagger(x) = \sigma^A(e^\dagger(x_1), \ldots, e^\dagger(x_k)) \in A + \{\bot\}$.

(2) For the powerset monad $M = \mathbb{P}$, we have

$$e^{\dagger}(x) = \bigcup \{\sigma^A(a_1, \dots, a_k) \mid a_i \in e^{\dagger}(x_i)\} \,.$$

(3) For the environment monad $MX = X^E$, we have

$$e^{\dagger}(x)(i) = \sigma^A(e^{\dagger}(x_1)(i), \dots, e^{\dagger}(x_k)(i))(i), \quad i \in E \,.$$

Examples 3.12. (1) Consider the maybe monad $MX = X + \{\bot\}$ and the distributive law λ of a signature functor H_Σ over M of Example 2.11(1). Let $F_\Sigma Y$ be the algebra of finite Σ-trees on Y. Consider its structure map as partial, so that it becomes an $\bar{H}$-algebra. As such, it is a λ-cia—in fact, the unique solution of an M-equation morphism $e : X \rightarrow H_\Sigma X + F_\Sigma Y + \{\bot\}$ gives its operational semantics, i.e., each variable in X is mapped to the tree unfolding of its recursive definition if this unfolding is finite, and to $\bot$ otherwise.

(2) Analogously, $F_\Sigma Y$ becomes a λ-cia for the distributive law λ of H_Σ over $\mathbb{P}$ of Example 2.11(2). The unique solution of $e : X \rightarrow H_\Sigma X + \mathbb{P} F_\Sigma Y$ assigns to a variable x the set of all possible tree unfoldings (taking into account that $e(x') \subseteq F_\Sigma Y$ for some variables x') of the recursive definition of x if all these unfoldings are finite and $\emptyset$ else. For example, for the signature with one binary operation symbol $*$ the system

$$x \approx x_1 * x_2 \qquad x' \approx x' * x_2 \qquad x_1 \approx \{ \; \overset{*}{\underset{y_1 \quad y_2}{\diagup\diagdown}} \; , y_3\} \qquad x_2 \approx \{ \; \overset{*}{\underset{y_3 \quad y_4}{\diagup\diagdown}} \; \}$$

has the unique solution with $e^{\dagger}(x)$ given by the set of trees with elements

and with $e^{\dagger}(x') = \emptyset$.

(3) Let H be an analytic functor and let λ be the distributive law over one of the monads $MX = X + 1$, $\mathbb{P}X$ or $\mathbb{D}X$ according to Theorem 2.9. Then the initial H-algebra $\phi : H\Phi \rightarrow \Phi$ exists (since H is finitary) and $J\phi : H\Phi \rightarrow M\Phi$ is a λ-cia. In fact, we prove in Section 4 that Φ is the initial λ-cia.

Examples 3.13. Unary λ-cias. Here we consider $H = \mathrm{Id}$ and $\lambda = \mathrm{id} : M \Rightarrow M$. From Remark 3.10 we see that $\alpha : A \rightarrow MA$ is a λ-cia iff $\mu_A \cdot M\alpha$ is a cia for H, i.e., iff $\mu_A \cdot M\alpha$ has a unique fixed point a_0 and $MA \setminus \{a_0\}$ has a well-founded order for which $\mu_A \cdot M\alpha$ is strictly increasing (see Example 3.6(5)).

(1) $\eta_A : A \rightarrow MA$ is a λ-cia iff MA is a singleton set since $\mu_A \cdot M\eta_A = \mathrm{id}_A$.

(2) For the maybe monad $MX = X + \{\bot\}$, the fixed point of MA must be $\bot$. Thus, $\alpha : A \rightarrow A + \{\bot\}$ is a λ-cia iff A has a well-founded order for which α is strictly increasing, i.e., $\alpha(a) \neq \bot$ implies $a < \alpha(a)$.

(3) For $M = \mathbb{P}$, an $\bar{H}$-algebra $\alpha : A \rightarrow \mathbb{P}A$ can be considered as a directed graph with node set A (i.e., a binary relation on A) where there is an edge from v to w iff $v \in \alpha(w)$, and vice versa. Then (A, α) is a λ-cia iff this graph is well-founded, see Corollary 5.3. We could add two equivalent formulations; cf. Example 3.6(5).

(4) For the environment monad $MX = X^E$, $\alpha : A \to A^E$ is a λ-cia iff for each $i \in E$ the map $\pi_i \cdot \alpha : A \to A$ is a unary cia. In fact, this is easy to see by extending (the appropriately simplified version of) Equation (3.3) by each π_i, see also Theorem 5.5.

(5) Let $M = F_\Sigma$ be the monad assigning to a set X the set of all finite Σ-trees on X. If Σ consists of one constant symbol, then M is the maybe monad. In all other cases there are no unary λ-cias. In order to see this, assume that $\alpha : A \to MA$ is a λ-cia. Then $\mu_A \cdot M\alpha$ has a unique fixed point t. Observe that the action of $\mu_A \cdot M\alpha$ is that of replacing in all trees of MA each leaf labelled by $a \in A$ with $\alpha(a)$. This implies that for every leaf of the tree t labelled by an element $a \in A$ we have that $\alpha(a)$ is the single-node tree labelled by a. But then every Σ-tree whose leaves have labels that also appear as leaf labels of t is a fixed point of $\mu_A \cdot M\alpha$. Hence, t does not contain a leaf labelled in A. On the other hand, each tree with no such leaves is a fixed point of $\mu_A \cdot M\alpha$. Thus there must be a unique constant in Σ and no other operation symbols, and so M is the maybe monad.

Proposition 3.14. *Every Kleisli-cia is a λ-cia.*

Proof. Let $\alpha : HA \to MA$ be a Kleisli-cia and let $e : X \to HX + MA$ be an M-equation morphism. We form a flat equation morphism $\bar{e} : X \dashrightarrow \bar{H}X + A$ as follows:

$$\bar{e} = (X \xrightarrow{Je} \bar{H}X + MA \xrightarrow{\mathrm{id}_{\bar{H}X} + \varepsilon_A} \bar{H}X + A) .$$

It is clear from the definitions of solutions (see Definitions 3.5 and 3.9) that a morphism $s : X \dashrightarrow A$ is a solution of $\bar{e}$ in the Kleisli-cia A iff it is a solution of e. Since the former exists uniquely, so does the latter; thus (A, α) is a λ-cia. $\square$

Example 3.15. This example demonstrates that the converse of Proposition 3.14 does not hold in general. Let $H = \mathrm{Id}$, let $MX = X^*$ be the monad of finite lists on X, and let $\lambda = \mathrm{id} : M \Rightarrow M$. Consider the algebra $A = \{0, 1\}$ with the structure $\alpha : A \to A^*$ given by $\alpha(0) = [1]$ and $\alpha(1) = [1, 1]$, where the square brackets denote lists. Then (A, α) is a λ-cia; in fact, $\mu_A \cdot M\alpha$ has as its unique fixed point the empty list, every list starting with 0 is mapped to a list starting with 1, and every list starting with 1 is mapped to a longer list. So an appropriate well-founded order on A^* such that $\mu_A \cdot M\alpha$ is strictly increasing on non-empty lists is given by putting $v < w$ if either v is shorter than w or the lengths agree and v goes before w lexicographically (this is even a well-order on A^*).

To see that (A, α) is not a Kleisli-cia, consider the formal equation $x \approx [x, 1]$ as a flat equation morphism $e : X \to (X + A)^*$. It is not difficult to check that for a solution $e^\dagger(x) = [a_1, \ldots, a_n]$ we have $[a_1, \ldots, a_n] = \alpha(a_1) \cdot \ldots \cdot \alpha(a_n) \cdot [1]$ where $\cdot$ denotes concatenation of lists. Thus $a_n = 1$, and therefore, since $\alpha(a_n) = [1, 1]$ we have $a_{n-1} = 1$ etc., so that we have $a_i = 1$ for all $i = 1, \ldots, n$. But then the two sides of the above equation are lists of different length, a contradiction. So there is no solution of e in A, and A is no Kleisli-cia.

The λ-cias for the endofunctor H together with the usual $\bar{H}$-algebra homomorphisms form a category, which we denote by λ-**Cia**$_H$; and Kleisli-cias and $\bar{H}$-algebra homomorphisms form the category **Cia**$_{\bar{H}}$. From Proposition 3.14 we see that **Cia**$_{\bar{H}}$ is a full

subcategory of λ-$\mathbf{Cia}_H$ which in turn is a full subcategory of the category $\mathbf{Alg}_{\bar{H}}$ of all
algebras for $\bar{H}$:

$$\mathbf{Cia}_{\bar{H}} \hookrightarrow \lambda\text{-}\mathbf{Cia}_H \hookrightarrow \mathbf{Alg}_{\bar{H}} .$$

Our next result is that our choice of morphisms for λ-cias is appropriate. A similar result
holds for cias, see [11], Proposition 2.3.

Notation 3.16. For any M-equation morphism $e : X \to HX + MA$ and any morphism
$f : A \to MB$ we denote by $f \bullet e$ the M-equation morphism $(\mathrm{id}_{HX} + Vf) \cdot e : X \to$
$HX + MB$ (see Notation 3.2(1)), where the parameters in e are "renamed by f".

Proposition 3.17. *Let* $f : A \dashrightarrow B$ *be a morphism, and let* $\alpha : \bar{H}A \dashrightarrow A$ *and*
$\beta : \bar{H}B \dashrightarrow B$ *be* λ-*cias. Then the following statements are equivalent:*

(1) f *is an* $\bar{H}$-*algebra homomorphism from* (A, α) *to* (B, β).
(2) f *preserves solutions, i. e., for all* M-*equation morphisms* $e : X \to HX + MA$ *we*
have $(f \bullet e)^{\dagger} = f \cdot e^{\dagger}$ *(here* $\cdot$ *is composition in* $\mathbf{Set}_M$*).*

4 Free λ-cias

We have already stated in Example 3.12(3) that in a number of concrete cases an initial
algebra for the functor H is an initial λ-cia. In this section we shall establish that free
H-algebras yield free λ-cias whenever $\mathbf{Set}_M$ is suitably cpo-enriched and the lifting $\bar{H}$
is locally monotone. Our results here are an application of the work of Hasuo, Jacobs
and Sokolova [4] and of the work in [11].

Recall that a cpo is a partially ordered set with a least element and with joins of ω-
chains. A category $\mathcal{A}$ is *cpo-enriched* if each hom-set carries the structure of a cpo such
that composition is continuous (i. e., preserves joins of ω-chains). Furthermore, recall
that an endofunctor H on the cpo-enriched category $\mathcal{A}$ is *locally monotone* (*locally continuous*) if each derived function $\mathcal{A}(X, Y) \to \mathcal{A}(HX, HY)$ is monotone (continuous).

Assumption 4.1. Throughout this section we assume that

(1) H is a finitary endofunctor on $\mathbf{Set}$,
(2) λ is a distributive law of H over the monad M (equivalently, $\bar{H}$ is a lifting of H to
    $\mathbf{Set}_M$),
(3) $\mathbf{Set}_M$ is cpo-enriched and the composition in $\mathbf{Set}_M$ is left-strict, i. e., for each
    $f : X \dashrightarrow Y$ the maps $- \cdot f$ preserve the least element,
(4) $\bar{H}$ is locally monotone.

Remark 4.2. (1) Notice that the coproducts of $\mathbf{Set}_M$ are cpo-enriched. In fact, it is
easy to see that the copairing map

$$[-, -] : \mathbf{Set}_M(X, Z) \times \mathbf{Set}_M(Y, Z) \to \mathbf{Set}_M(X + Y, Z)$$

is continuous (and hence monotone). Clearly, for every object X we have the dis-
tributive law $[M\mathrm{inl}, M\mathrm{inr}] \cdot (\lambda + \eta) : H_X M \Rightarrow M H_X$, where $H_X = H(-) + X$.
It follows that $\overline{H}_X = \bar{H}_X = \bar{H}(-) + X$ is locally monotone (locally continuous)
whenever $\bar{H}$ itself is.

(2) Since the functor H is finitary, it has free algebras. We denote by $\phi_X : HFX \to FX$ the structure of a free H-algebra on X and by $u_X : X \to FX$ the universal map. Notice that a free H-algebra on X is the same as an initial algebra for the functor H_X. So by Lambek's Lemma [10], $[\phi_X, u_X]$ is an isomorphism. Finally, for the initial H-algebra we use the notation $\phi : H\Phi \to \Phi$.

Examples 4.3. (1) In this example we let M be one of the monads $(-) + 1$, $\mathbb{P}$ or $\mathbb{D}$. Assume that P is a polynomial functor; then, as shown in [4], Assumption 4.1 is satisfied. In fact, $\mathbf{Set}_M$ is cpo-enriched with a left-strict composition since in all three cases we have a cpo structure $\sqsubseteq$ on each MY: for $(-) + 1$ take the flat cpo structure (i.e., $x \sqsubseteq y$ iff $x = \bot$ or $x = y$), for $\mathbb{P}$ take inclusion, and for $\mathbb{D}$ take the pointwise order ($d \sqsubseteq d'$ iff $d(x) \leq d'(x)$ for all $x \in Y$). This yields a cpo structure $\sqsubseteq$ on $\mathbf{Set}_M(X, Y)$ in a pointwise fashion: $f \sqsubseteq g$ iff for all $x \in X$ we have $f(x) \sqsubseteq g(x)$. It is proved in [4] that there is a canonical distributive law $\tilde{\lambda} : PM \Rightarrow MP$ and that the lifting $\bar{P}$ is locally continuous.

(2) The lifting of every analytic functor H is locally continuous. We know from Theorem 2.9 that there is a canonical distributive law $\lambda : HM \Rightarrow MH$ since the three monads of interest are commutative. Furthermore, recall that H is the quotient of a polynomial functor P via some $\epsilon : P \Rightarrow H$ such that Equation (2.2) holds. Then it is easy to see that $\bar{H}$ is locally continuous: in fact, due to the naturality of ϵ and Equation (2.2) we have for every $f : X \dashrightarrow Y$ in $\mathbf{Set}_M$ a commutative square (written in $\mathbf{Set}$)

$$\begin{array}{ccccc}
PX & \xrightarrow{Pf} & PMY & \xrightarrow{\tilde{\lambda}_Y} & MPY \\
\epsilon_X \downarrow & & \epsilon_{MY} \downarrow & & \downarrow M\epsilon_Y \\
HX & \xrightarrow{Hf} & HMY & \xrightarrow{\lambda_Y} & MHY
\end{array} \tag{4.1}$$

showing that ϵ amounts to a natural transformation $\bar{\epsilon} : \bar{P} \Rightarrow \bar{H}$ with the components $\bar{\epsilon}_X = J\epsilon_X : \bar{P}X \to \bar{H}X$. Suppose that $(f_n)_{n \in \mathbb{N}}$ is an ω-chain in $\mathbf{Set}_M(X, Y)$. Then we have

$$\begin{aligned}
\bar{H}(\bigsqcup f_n) \cdot \bar{\epsilon}_X &= \bar{\epsilon}_Y \cdot \bar{P}(\bigsqcup f_n) && \text{(naturality of } \bar{\epsilon}) \\
&= \bar{\epsilon}_Y \cdot \bigsqcup \bar{P} f_n && \text{(local continuity of } \bar{P}) \\
&= \bigsqcup (\bar{\epsilon}_Y \cdot \bar{P} f_n) && \text{(continuity of } \cdot) \\
&= \bigsqcup (\bar{H} f_n \cdot \bar{\epsilon}_X) && \text{(naturality of } \bar{\epsilon}) \\
&= \bigsqcup \bar{H} f_n \cdot \bar{\epsilon}_X && \text{(continuity of } \cdot).
\end{aligned}$$

Since ϵ_X is a surjective map and the left adjoint J preserves epimorphisms, we can cancel $\bar{\epsilon}_X$ to obtain $\bar{H}(\bigsqcup f_n) = \bigsqcup \bar{H} f_n$, as desired.

Theorem 4.4. *([4]) Under Assumption 4.1, the initial H-algebra $\phi : H\Phi \to \Phi$ yields a terminal $\bar{H}$-coalgebra $J(\phi^{-1}) : \Phi \dashrightarrow \bar{H}\Phi$ and an initial $\bar{H}$-algebra $J\phi : \bar{H}\Phi \dashrightarrow \Phi$.*

Theorem 4.5. *Under Assumption 4.1, the free H-algebra $\phi_X : HFX \to FX$ on X with universal map $u_X : X \to FX$ yields a free $\bar{H}$-algebra $J\phi_X : \bar{H}FX \dashrightarrow FX$ on X with universal map $Ju_X : X \dashrightarrow FX$, and this is a Kleisli-cia.*

Proof. By Remark 4.2(2), $[\phi_X, u_X] : HFX + X \to FX$ is an initial algebra for $H_X = H(-) + X$. Since $\bar{H}_X = \bar{H}(-) + X$ is locally continuous (see Remark 4.2(1)), hence locally monotone, we can apply Theorem 4.4 to see that $J[\phi_X, u_X] : \bar{H}FX + X \dashrightarrow FX$ is an initial algebra and $J([\phi_X, u_X]^{-1})$ a terminal coalgebra for $\bar{H}_X$ on $\mathbf{Set}_M$. Then from Example 3.6(1) we see that $(FX, J\phi_X)$ is a (free) Kleisli-cia on X with universal map Ju_X. □

Corollary 4.6. *Under Assumption 4.1, a free H-algebra yields a free Kleisli-cia and a free λ-cia.*

Example 4.7. For a signature functor H_Σ and $M \in \{(-)+1, \mathbb{P}, \mathbb{D}\}$, the algebra $F_\Sigma X$ of all finite Σ-trees on X yields a free $\bar{H}_\Sigma$-algebra on X and also a free Kleisli- and a free λ-cia on X.

5   Characterisation of λ-cias

In this section we consider three concrete examples of monads: the maybe monad $MX = X + 1$, the powerset monad $M = \mathbb{P}$ and the environment monad $MX = X^E$. We show that for these monads the Kleisli-cias and the λ-cias for a signature functor H_Σ coincide. In fact, in the case of the environment monad we allow arbitrary endofunctors H on $\mathbf{Set}$. In each case we give a characterisation of the λ- or Kleisli-cias.

Theorem 5.1. *Let $MX = X + 1$ be the maybe monad, let H_Σ be a signature functor, and let $\lambda : H_\Sigma M \Rightarrow MH_\Sigma$ be the distributive law of Example 2.11(1). Then the following three conditions are equivalent:*

(1) (A, α) is λ-cia,
(2) (A, α) is Kleisli-cia,
(3) (A, α) is an $\bar{H}_\Sigma$-algebra, and there exists a well-founded order $>$ on A such that every n-ary algebra operation σ^A is strictly increasing in the sense that for all $a_1, \ldots, a_n \in A$ with $\sigma^A(a_1, \ldots, a_n) \neq \bot$ we have $\sigma^A(a_1, \ldots, a_n) > a_i$ for $i = 1, \ldots, n$.

Theorem 5.2. *Let $M = \mathbb{P}$ be the powerset monad, let H_Σ be a signature functor, and let $\lambda : H_\Sigma M \Rightarrow MH_\Sigma$ be the distributive law of Example 2.11(2). Then the following three conditions are equivalent:*

(1) (A, α) is λ-cia,
(2) (A, α) is Kleisli-cia,
(3) (A, α) is an $\bar{H}_\Sigma$-algebra, and there exists a well-founded order $>$ on A such that every n-ary algebra operation σ^A is strictly increasing in the sense that for all $a_1, \ldots, a_n \in A$ with $\sigma^A(a_1, \ldots, a_n) = B$ we have $b > a_i$ for all $b \in B$ and $i = 1, \ldots, n$.

Corollary 5.3. *For $M = \mathbb{P}$, an $\overline{\mathrm{Id}}$-algebra is a λ-cia iff the corresponding graph (cf. Example 3.13(3)) is well-founded.*

In fact, if a graph is well-founded, then the induced (strict) order is, and conversely, a subgraph of a well-founded graph is well-founded.

Notice that this result also shows that even though Kleisli- and λ-cias coincide, and free λ-cias (or Kleisli-cias) coincide with free $\bar{H}$-algebras, not every $\bar{H}$-algebra needs to be a λ-cia; in fact, every directed graph yields an $\overline{\mathrm{Id}}$-algebra.

Examples 5.4. (1) Let $A = \{a_0, a_1, a_2\}$, and define a binary operation $*$ by

$$
\begin{array}{lll}
a_0 * a_0 = \{a_1, a_2\} & a_1 * a_0 = \emptyset & a_2 * a_0 = \emptyset \\
a_0 * a_1 = \{a_2\} & a_1 * a_1 = \{a_2\} & a_2 * a_1 = \emptyset \\
a_0 * a_2 = \emptyset & a_1 * a_2 = \emptyset & a_2 * a_2 = \emptyset
\end{array}
$$

Here we have the well-founded order $a_2 > a_1 > a_0$ for which the operation $*$ is strictly increasing.

(2) Let $A = \{2, 3, \ldots, 100\} \subset \mathbb{N}$ be the $\bar{H}$-algebra of Remark 3.8. For the usual order on A the operations are clearly strictly increasing, and well-foundedness follows from finiteness of A, i.e., A satisfies condition (3) in Theorem 5.2. Observe that for $\mathbb{P}$-equation morphisms $e : X \to HX + \mathbb{P}A$ an *infinitely unfolding variable* x, i.e., one for which there exists an infinite sequence $x_0, x_1, x_2, \ldots$ with $x_0 = x$ and each x_{i+1} appearing in a flat term $e(x_i) \in HX$, is always solved to $\emptyset$. However, for flat equation morphisms $e' : X \dashrightarrow HX + A$ infinitely unfolding variables may be solved to non-empty sets, see Remark 3.8.

Theorem 5.5. *Let $MX = X^E$ be the environment monad, let H be an endofunctor on* **Set***, and let $\lambda : HM \Rightarrow MH$ be the distributive law of Example 2.11(3). Then the following three conditions are equivalent:*

(1) (A, α) is a λ-cia,
(2) (A, α) is a Kleisli-cia,
(3) (A, α) is an $\bar{H}$-algebra such that for every $i \in E$ the H-algebra $(A, \pi_i \cdot \alpha)$ is a cia in **Set***.*

Corollary 5.6. *Let M, H and λ be as in Theorem 5.5, and let TX be a terminal coalgebra for $H(-)+X$. Then TX yields a free λ-cia (and Kleisli-cia) on X. More precisely, the inverse of the coalgebra structure yields an H-algebra structure $\tau_X : HTX \to TX$ and a map $u_X : X \to TX$, and $(TX, J\tau_X)$ is a free λ-cia with universal arrow Ju_X.*

This result follows from the facts that, firstly, TX is a free cia on X for H (see [11]), and that, secondly, $\bar{H}$-algebras are families $(\alpha_i : HA \to A)_{i \in E}$ of H-algebras with the same carrier A, and similarly, $\bar{H}$-algebra homomorphisms are E-indexed families of H-algebra homomorphisms.

6 Conclusions

We have proved that every analytic endofunctor on **Set** admits a canonical distributive law over every commutative monad M. This extends previous work by Hasuo, Jacobs and Sokolova [4]. We then applied this result in our study of algebras whose operations have computational effects (described by the monad M) and that admit unique solutions of recursive equations: we introduced λ-cias and proved that every Kleisli-cia is a λ-cia. We also proved that for the maybe and powerset monads and the respective canonical distributive laws of a signature functor over these monads Kleisli- and λ-cias coincide, and that for the environment monad this is true even for an arbitrary endofunctor on **Set**. It should be interesting to see whether this coincidence can be extended to non-polynomial functors for the maybe and powerset monads, and it would be desirable to have a uniform proof of Theorems 5.1 and 5.2. We leave this for future work.

Concerning free λ-cias we have the following result: for a monad M such that $\mathbf{Set}_M$ is suitably cpo-enriched and for a finitary endofunctor H with a distributive law λ : $HM \Rightarrow MH$ such that the associated lifting $\bar{H}$ is locally monotone we proved that free H-algebras (which always exist for a finitary functor) yield free $\bar{H}$-algebras, which at the same time are free λ-cias and free Kleisli-cias.

References

[1] Adámek, J., Milius, S., Velebil, J.: Iterative algebras at work. Math. Structures Comput. Sci. 16, 1085–1131 (2006)
[2] America, P., Rutten, J.J.M.M.: Solving reflexive domain equations in a category of complete metric spaces. J. Comput. System Sci. 39(3), 343–375 (1989)
[3] Elgot, C.: Monadic computation and iterative algebraic theories. In: Rose, H.E., Shepherdson, J.C. (eds.) Logic Colloquium 1973, pp. 175–230. North-Holland, Amsterdam (1975)
[4] Hasuo, I., Jacobs, B., Sokolova, A.: Generic trace semantics via coinduction. Log. Methods Comput. Sci. 3(4:11), 1–36 (2007)
[5] Joyal, A.: Une théorie combinatoire des séries formelles. Adv. Math. 42, 1–82 (1981)
[6] Joyal, A.: Foncteurs analytiques et espèces de structures. In: Labelle, G., Leroux, P. (eds.) Combinatoire énumérative. Lecture Notes in Math., vol. 1234, pp. 126–159 (1986)
[7] Joyal, A., Street, R.: Braided tensor categories. Adv. Math. 102, 20–78 (1993)
[8] Kock, A.: Monads on symmetric monoidal closed categories. Arch. Math. (Basel) 21, 1–10 (1970)
[9] Kock, A.: Strong functors and monoidal monads. Arch. Math. (Basel) 23, 113–120 (1972)
[10] Lambek, J.: A fixpoint theorem for complete categories. Math. Z. 103(2), 151–161 (1968)
[11] Milius, S.: Completely iterative algebras and completely iterative monads. Inform. and Comput. 196, 1–41 (2005)
[12] Milius, S., Moss, L.S.: The category-theoretic solution of recursive program schemes. Theoret. Comput. Sci. 366, 3–59 (2006)
[13] Milius, S., Palm, T., Schwencke, D.: Complete iterativity for algebras with effects, http://www.stefan-milius.eu
[14] Moggi, E.: Notions of computation and monads. Inform. and Comput. 93(1), 55–92 (1991)
[15] Nelson, E.: Iterative algebras. Theoret. Comput. Sci. 25, 67–94 (1983)
[16] Tiuryn, J.: Unique fixed points vs. least fixed points. Theoret. Comput. Sci. 12, 229–254 (1980)

Semantics of Higher-Order Recursion Schemes

Jiří Adámek[1,*], Stefan Milius[1,*], and Jiří Velebil[2,**]

[1] Institut für Theoretische Informatik, Technische Universität Braunschweig,
Germany
[2] Faculty of Electrical Engineering, Czech Technical University of Prague,
Czech Republic

Abstract. Higher-order recursion schemes are equations defining recursively
new operations from given ones called "terminals". Every such recursion
scheme is proved to have a least interpreted semantics in every Scott's model
of λ-calculus in which the terminals are interpreted as continuous operations.
For the uninterpreted semantics based on infinite λ-terms we follow the idea of
Fiore, Plotkin and Turi and work in the category of sets in context, which are
presheaves on the category of finite sets. Whereas Fiore *et al* proved that the
presheaf F_λ of λ-terms is an initial H_λ-monoid, we work with the presheaf R_λ
of rational infinite λ-terms and prove that this is an initial iterative H_λ-monoid.
We conclude that every guarded higher-order recursion scheme has a unique
uninterpreted solution in R_λ.

Keywords: Higher-order recursion schemes, infinite λ-terms, sets in context,
rational tree.

1 Introduction

Recursion is a fundamental tool for constructing new programs: given a collection Σ of
existing programs of given types (that is, a many-sorted signature of "terminals") one
defines new typed programs $p_1, \ldots, p_n$ (a many-sorted signature of "nonterminals")
using symbols from Σ and $\{p_i\}$. If the recursion only concerns application, we can
formalize this procedure as a collection of equations

$$p_i = f_i \qquad (i = 1, \ldots, n) \tag{1.1}$$

whose right-hand sides f_i are terms in the signature of all terminals and all nonterminals. Such collections are called (first-order) *recursion schemes* and were studied in 1970's by various authors, e.g. B. Courcelle, M. Nivat and I. Guessarian (see
the monograph [8] and references there) or S. J. Garland and D. C. Luckham [7]. Recently, a categorical approach to semantics of first-order recursion schemes was treated
by S. Milius and L. Moss [13]. In the present paper we take a first step in an analogous
approach to the semantics of *higher-order recursion schemes* in which λ-abstraction is

* Supported by the German Research Foundation (DFG) under the project "Coalgebraic Specification".
** Supported by the grant MSM 6840770014 of the Ministry of Education of the Czech Republic.

A. Kurz, M. Lenisa, and A. Tarlecki (Eds.): CALCO 2009, LNCS 5728, pp. 49–63, 2009.

also used as one of the operations. That is, a higher-order recursion scheme, as introduced by W. Damm [4] (see also recent contributions [3] and [14]) is a collection of equations $p_i = f_i$ where f_i are terms using application and λ-abstraction on symbols of Σ and of $\{p_1, \ldots, p_n\}$. As in [13], we first study the uninterpreted semantics, where the given system, is regarded as a purely syntactic construct. At this stage the operation symbols in Σ as well as λ-abstraction and application have no interpretation on actual data. So the semantics is provided by formal (infinite) terms. These terms can be represented by rational trees, i.e., infinite trees having finitely many subtrees. Thus the uninterpreted solution assigns to each of the recursive variables p_i in (1.1) a rational tree $p_i^\dagger$ such that the formal equations become identities if we substitute $p_i^\dagger$ for p_i ($i = 1, \ldots, n$). We assume α-conversion (renaming of bound variables) but no other rules in the uninterpreted semantics. We next turn to an interpreted semantics. Here a recursion schemes is given together with an interpretation of all symbols from Σ as well as λ-abstraction and application. We shall consider as an interpretation a Scott model of λ-calculus as a CPO, D. The symbols of Σ are interpreted as continuous operations on D and formal λ-abstraction and application are the actual λ-abstraction and application in the model D. An interpreted solution in D the assigns to each p_i in the context Γ of all free variables in (1.1), an element of $\mathbf{CPO}(D^\Gamma, D)$ (continuously giving to each assignment of free variables in D^Γ an element of D) such that the formal equations in the recursion scheme become identities in D when the right-hand sides are interpreted in D.

Example 1.1. The fixed-point operator Y is specified by

$$Y = \lambda f. f(Yf)$$

and the uninterpreted semantics is the rational tree

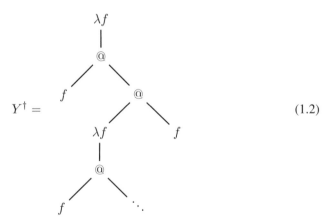

$$Y^\dagger = \tag{1.2}$$

(The symbol @ makes application explicit.) The interpreted solution in D is the least fixed point operator (considered as an element of D).

The above example is untyped, and indeed we are treating the untyped case only in the present paper since its uninterpreted semantics is technically simpler than the typed case; however, the basic ideas are similar. In contrast, the interpreted semantics (based on a specified model of λ-calculus with "terminal" symbols interpreted as operations) is more subtle in the untyped case.

Our main result is that every guarded higher-order recursion scheme has a unique uninterpreted solution, and a least interpreted one. This demonstrates that the methods for iteration in locally finitely presentable categories developed earlier [1] can serve not only for first-order iteration, when applied to endofunctors of **Set**, but also for higher-order iteration: it is possible to apply these methods to other categories, here the category of sets in context.

2 Presheaves as Algebras

Notation 2.1. (1) Throughout the paper a given countably infinite set Var of variables is assumed. Finite subsets $\Gamma \subseteq$ Var are called *contexts* and form a full subcategory $\mathscr{F}$ of **Set**.

When speaking about formulas in context Γ we mean those that have all free variables in Γ. For example, $\lambda x.yx$ is a formula in context $\Gamma = \{y, y'\}$.

(2) We assume that a one-sorted signature Σ is given.

(3) The category **Set**$^{\mathscr{F}}$ of "covariant presheaves" on $\mathscr{F}$ is well known to be equivalent to the category of finitary endofunctors of **Set** (but the word endofunctor is later reserved for endofunctors of **Set**$^{\mathscr{F}}$ throughout our paper). In fact, every endofunctor X yields the presheaf $X/\mathscr{F}$, and conversely, every presheaf X in **Set**$^{\mathscr{F}}$ has a left Kan extension to a finitary endofunctor of **Set**: for every set M we have

$$X(M) = \bigcup X i_\Gamma [X(\Gamma)]$$

where the union ranges over embeddings $i_\Gamma \colon \Gamma \hookrightarrow M$ of contexts Γ into M, and $X i_\Gamma[X(\Gamma)]$ denotes the image of $X i_\Gamma$.

Example 2.2. (i) The *presheaf of variables*, V, is our name for the embedding $\mathscr{F} \hookrightarrow$ **Set**: $V(\Gamma) = \Gamma$.

(ii) *Free presheaf* on one generator of context Γ is our name for the representable presheaf

$$\mathscr{F}(\Gamma, -).$$

In fact, the Yoneda lemma states that this presheaf is freely generated by the element id_Γ of context Γ: for every presheaf X and every $x \in X(\Gamma)$ there exists a unique morphism $f \colon \mathscr{F}(\Gamma, -) \to X$ with $f_\Gamma(\mathrm{id}_\Gamma) = x$.

(iii) The *presheaf F_λ of (finite) λ-terms* is defined via a quotient since we want to treat λ-terms always modulo α-conversion. As explained in [6], the following approach is equivalent to defining λ-terms up to α-equivalence by de Bruijn levels: We first define the set $F'_\lambda(\Gamma)$ of all finite λ-trees τ for every context $\Gamma = \{x_1, \ldots, x_n\}$ as

$$\tau ::= x_i \mid \tau @ \tau \mid \lambda y.\tau \qquad (y \in \mathrm{Var} \setminus \Gamma). \tag{2.1}$$

In the graphic form:

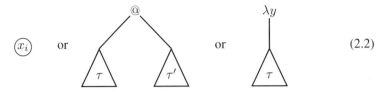

$$(2.2)$$

We then define the presheaf F_λ in every context Γ by

$$F_\lambda(\Gamma) = F'_\lambda(\Gamma)/{\sim_\alpha}$$

where $\sim_\alpha$ represents the α-conversion: this is the least equivalence with $\lambda x_i.\tau \sim_\alpha \lambda x_j.\tau[x_j/x_i]$.

We call the congruence classes of finite λ-trees *finite λ-terms* (more precisely, we should say finite λ-terms modulo α-conversion). Whereas finite λ-trees do not form a presheaf, due to possible crashes of bound and free variables, finite λ-terms do: define F_λ on morphisms $\gamma : \Gamma \to \Gamma'$ by choosing a term $t \in F_\lambda(\Gamma)$, relabelling all bound variables so that they do not lie in Γ', and denoting by $F\gamma(t)$ the tree obtained by relabelling every free variable $x \in \Gamma$ to $\gamma(x) \in \Gamma'$.

(iv) The *presheaf $F_{\lambda,\Sigma}$ of finite λ-Σ-terms* is defined analogously: in (2.1) we just add the term $\sigma(\tau, \ldots, \tau)$ for every $\sigma \in \Sigma_n$, and in (2.2) the corresponding tree.

(v) The presheaf T_λ of all (finite and infinite) λ-terms is defined analogously to F_λ. We first denote by $T'_\lambda(\Gamma)$ the set of all trees (2.2) dropping the assumption of finiteness. Then we use α-conversion: for infinite trees t and t' we write $t \sim_\alpha t'$ if their (finite) cuttings at level k (with label $\bot$ for all leaves at level k) are α-equivalent in the above sense for all $k \in \mathbb{N}$. (Put $\Sigma = \{\bot\}$ with $\bot$ a constant symbol). The presheaf T_λ is defined on objects Γ by $T_\lambda(\Gamma) = T'_\lambda(\Gamma)/{\sim_\alpha}$ and on morphisms $\gamma : \Gamma \to \Gamma'$ by relabelling of variables as in (iii). Observe that since $\mathrm{Var}\setminus\Gamma$ is infinite, the relabelling of bound variables needed here causes no problem.

(v) The presheaf R_λ of *rational λ-terms* is also defined analogously. Recall that a tree is called rational if it has up to isomorphism only finitely many subtrees. We denote by $R'_\lambda(\Gamma)$ the set of all rational trees in $T'_\lambda(\Gamma)$ and define a presheaf R_λ by $R_\lambda(\Gamma) = R'_\lambda(\Gamma)/{\sim_\alpha}$ on objects, and by relabellings of variables (as in (iii)) on morphisms.

(vi) The presheaves $T_{\lambda,\Sigma}$ (of λ-Σ-terms) and $R_{\lambda,\Sigma}$ (of rational λ-Σ-terms) are obvious modifications of (iv) and (v): one adds to (2.2) the case $\sigma(\tau_1, \ldots, \tau_n)$ for all n-ary symbols $\sigma \in \Sigma$ and all (rational) λ-Σ-terms $\tau_1, \ldots, \tau_n$. Observe that, by definition, every rational λ-Σ-term t is represented by a rational λ-Σ-tree. However, t can also be represented by non-rational λ-Σ-trees—for example, if it contains infinitely many λ's, the α-conversion can introduce an infinite number of bound variables.

Notation 2.3. We denote by $\delta\colon \mathbf{Set}^{\mathscr{F}} \to \mathbf{Set}^{\mathscr{F}}$ the endofunctor defined by

$$\delta X(\Gamma) = X(\Gamma + 1).$$

Observe that δ preserves limits and colimits.

Note that an algebra for δ is a presheaf Y together with an operation $Y(\Gamma + 1) \to Y(\Gamma)$ for all contexts Γ—this is precisely the form of λ-abstraction, where to a formula

f in $Y(\Gamma + \{y\})$ we assign $\lambda y.f$ in $Y(\Gamma)$. The other λ-operation, application, is simply a presheaf morphism $X \times X \to X$, that is, a binary operation on X. We put these two together:

Notation 2.4. Let H_λ denote the endofunctor of $\mathbf{Set}^{\mathscr{F}}$ given by

$$H_\lambda X = X \times X + \delta X.$$

Thus, an algebra for H_λ is a presheaf X together with specified operations of application $X(\Gamma) \times X(\Gamma) \to X(\Gamma)$ and abstraction $X(\Gamma + 1) \to X(\Gamma)$ for all contexts Γ; these operations are compatible with renaming of free variables.

Example 2.5. F_λ, T_λ and R_λ are algebras for H_λ in the obvious sense.

Remark 2.6. (i) The slice category $V/\mathbf{Set}^{\mathscr{F}}$ of presheaves X together with a morphism $i \colon V \to X$ is called the category of *pointed presheaves*. For example F_λ is a pointed presheaf in a canonical sense: $i^F \colon V \to F_\lambda$ takes a variable x to the term x. Analogously $i^T \colon V \to T_\lambda$ and $i^R \colon V \to R_\lambda$ are pointed presheaves, and so are $F_{\lambda,\Sigma}$, $R_{\lambda,\Sigma}$ and $T_{\lambda,\Sigma}$.

(ii) Recall that the category $\mathbf{Alg}\,H_\lambda$ of algebras for H_λ has as morphisms the usual H_λ-homomorphisms, i.e., a morphism from $a \colon H_\lambda X \to X$ to $b \colon H_\lambda Y \to Y$ is a presheaf morphism $f \colon X \to Y$ such that $f \cdot a = b \cdot H_\lambda f$. Then $\mathbf{Alg}\,H_\lambda$ is a concrete category over $\mathbf{Set}^{\mathscr{F}}$ with the forgetful functor $(H_\lambda X \to X) \mapsto X$.

Theorem 2.7 (see [6]). *The presheaf F_λ of finite λ-terms is the free H_λ-algebra on V.*

Definition 2.8 (see [1]). *Given an endofunctor H, an algebra $a \colon HA \to A$ is called **completely iterative** if for every object X (of variables) and every (flat equation) morphism $e \colon X \to HX + A$ there exists a unique **solution** which means a unique morphism $e^\dagger \colon X \to A$ such that the square below commutes*

$$
\begin{array}{ccc}
X & \xrightarrow{\;e^\dagger\;} & A \\
{\scriptstyle e}\big\downarrow & & \big\uparrow{\scriptstyle [a,\mathrm{id}]} \\
HX + A & \xrightarrow[H e^\dagger + \mathrm{id}]{} & HA + A
\end{array}
\tag{2.3}
$$

Theorem 2.9. *The presheaf T_λ of infinite λ-terms is the free completely iterative H_λ-algebra on V.*

Proof. As proved in [12], Corollary 6.3, the free completely iterative algebra for H_λ on V is precisely the terminal coalgebra for $H_\lambda(-) + V$. The latter functor clearly preserves limits of ω^{op}-chains. Consequently, its terminal coalgebra is a limit of the chain W with $W_0 = 1$ (the terminal presheaf) and $W_{n+1} = H_{\lambda,\Sigma} W_n + V$, where the connecting maps are the unique $w_0 \colon W_1 \to W_0$ and $w_{n+1} = H_{\lambda,\Sigma} w_n + \mathrm{id}_V$.

For every context Γ identify $W_0(\Gamma)$ with the set $\{\bot\}$ where $\bot \notin \mathrm{Var}$. Then $W_n(\Gamma)$ can be identified with the set of all λ-terms of depth at most n having all leaves of depth n labelled by $\bot$. And w_{n+1} cuts away the level $n + 1$ in trees of $W_{n+1}(\Gamma)$, relabelling level-n-leaves by $\bot$. With this identification we obtain T_λ as a limit of W_n where the limit maps $T_\lambda \to W_n$ cut trees in $T_\lambda(\Gamma)$ at level n and relabel level-n-leaves by $\bot$. $\qquad\square$

Remark 2.10. We are going to characterize the presheaf R_λ as a free iterative algebra for H_λ. This concept differs from 2.8 by admitting only objects X of variables that are *finitely presentable*. This means that the hom-functor $\mathbf{Set}^{\mathscr{F}}(X, -)$ preserves filtered colimits. Recall from [2] that a presheaf X is called *super-finitary* provided that each $X(\Gamma)$ is finite and there exists a nonempty context Γ_0 *generating* X in the sense that for every nonempty context Γ we have

$$X(\Gamma) = \bigcup_{\gamma:\ \Gamma_0 \to \Gamma} X\gamma[X(\Gamma_0)]. \tag{2.4}$$

Example 2.11. A signature Σ defines the polynomial presheaf X_Σ by $X_\Sigma(\Gamma) = \coprod_{\sigma \in \Sigma} \Gamma^{\mathrm{ar}(\sigma)}$. This is a super-finitary presheaf iff Σ is a finite set. Other super-finitary presheaves are precisely the quotients of X_Σ with Σ finite.

Theorem 2.12. *A presheaf in* $\mathbf{Set}^{\mathscr{F}}$ *is finitely presentable iff it is super-finitary.*

Sketch of proof. Every finitely presentable presheaf X is proved to be a directed colimit of presheaves generated by Γ_0 for all possible contexts Γ_0. From finite presentability it then follows that one Γ_0 generates all of X.

Conversely, for every super-finitary presheaf X a finite diagram of representable presheaves is constructed with colimit X, this implies that X is finitely presentable. □

Definition 2.13 (see [1]). *An algebra A for H is called* **iterative** *if every equation morphism $e\colon X \to HX + A$ with X finitely presentable has a unique solution.*

Theorem 2.14. *The presheaf R_λ of rational λ-terms is the free iterative H_λ-algebra on V.*

Sketch of proof. The iterativity of R_λ follows from that of T_λ and the fact that if for an equation morphism $e\colon X \to H_\lambda X + R_\lambda$ the object X is super-finitary, all terms one obtains as solutions in $e^\dagger$ are always rational.

The universal property of R_λ is proved on the basis of the construction presented in [1]: all equation morphisms $e\colon X \to H_\lambda X + V$ with X finitely presentable form a filtered diagram whose colimit is the free iterative algebra on V. We now prove that R_λ is indeed this colimit: for every e we form (using i^R in 2.6(i)) the equation morphism $\tilde{e} = (\mathrm{id}_{H_\lambda X} + i^R) \cdot e\colon X \to H_\lambda X + R_\lambda$, and obtain the unique solution $\tilde{e}^\dagger\colon X \to R_\lambda$; we then verify that the morphisms $\tilde{e}^\dagger$ form the desired colimit cocone. □

Remark 2.15. As mentioned in the Introduction we want to combine application and abstraction with other operations. Suppose $\Sigma = (\Sigma)_{n\in\mathbb{N}}$ is a signature (of "terminals"). Then we can form the endofunctor $H_{\lambda,\Sigma}$ of $\mathbf{Set}^{\mathscr{F}}$ on objects by

$$H_{\lambda,\Sigma}X = X \times X + \delta X + \coprod_{n\in\mathbb{N}} \Sigma_n \bullet X^n \tag{2.5}$$

where $\Sigma_n \bullet X^n$ is the coproduct (that is: disjoint union in every context) of Σ_n copies of the n-th Cartesian power of X. For this endofunctor an algebra is an H_λ-algebra A together with an n-ary operation on $A(\Gamma)$ for every $\sigma \in \Sigma_n$ and every context Γ.

Theorem 2.16. *For every signature* Σ

(i) $F_{\lambda,\Sigma}$ *is the free* $H_{\lambda,\Sigma}$-*algebra on* V,
(ii) $R_{\lambda,\Sigma}$ *is the free iterative* $H_{\lambda,\Sigma}$-*algebra on* V, *and*
(iii) $T_{\lambda,\Sigma}$ *is the free completely iterative* $H_{\lambda,\Sigma}$-*algebra on* V.

In fact, (i) was proved in [6], and the proofs of (ii) and (iii) are completely analogous to the proofs of Theorems 2.14 and 2.9.

3   Presheaves as Monoids

So far we have not treated one of the basic features of λ-calculus: substitution of subterms. For the presheaf $F_{\lambda,\Sigma}$ of finite λ-Σ-terms this was elegantly performed by Fiore *et al* [6] based on the monoidal structure of the category **Set**$^{\mathscr{F}}$. As mentioned in 2.1(2), we can work with the equivalent category Fin(**Set**, **Set**) of all finitary endofunctors of **Set**. Composition of functors makes this a (strict, non-symmetric) monoidal category with unit $\mathrm{Id}_{\mathbf{Set}}$. This monoidal structure, as shown in [6], corresponds to simultaneous substitution. In fact, let X and Y be objects of Fin(**Set**, **Set**). Then the "formulas of the composite presheaf $X \cdot Y$" in context Γ are the elements of

$$X \cdot Y(\Gamma) = X\big(Y(\Gamma)\big) = \bigcup_{u : \bar{\Gamma} \hookrightarrow Y(\Gamma)} X(\bar{\Gamma}), \tag{3.1}$$

where $u : \bar{\Gamma} \hookrightarrow Y(\Gamma)$ ranges over finite subobjects of $Y(\Gamma)$. In fact, X preserves the filtered colimit $Y(\Gamma) = \mathrm{colim}\,\bar{\Gamma}$.

That is, in order to specify an $X \cdot Y$-formula t in context Γ we need (a) an X-formula s in some new context $\bar{\Gamma}$ and (b) for every variable $x \in \bar{\Gamma}$ a Y-formula of context Γ, say, r_x. We can then think of t as the formula $s(r_x/x)$ obtained from s by simultaneous substitution.

Remark 3.1. (i) The monoidal structure on **Set**$^{\mathscr{F}}$ corresponding to composition in Fin(**Set**, **Set**) will be denoted by $\otimes$, with the unit V, see Notation 2.2(i). Observe that every endofunctor $- \otimes X$ preserves colimits, e.g., $(A + B) \otimes X \cong (A \otimes X) + (B \otimes X)$.

(ii) Explicitly, the monoidal structure can be described by the coend

$$(X \otimes Y)(\Gamma) = \int^{\bar{\Gamma}} \mathbf{Set}(\bar{\Gamma}, Y(\Gamma)) \bullet X(\bar{\Gamma}). \tag{3.2}$$

(iii) Recall that monoids in the monoidal category Fin(**Set**, **Set**) are precisely the finitary monads on **Set**.

(iv) The presheaf $F_{\lambda,\Sigma}$ is endowed with the usual simultaneous substitution of λ-terms which defines a morphism $m^F : F_{\lambda,\Sigma} \otimes F_{\lambda,\Sigma} \to F_{\lambda,\Sigma}$. Together with the canonical pointing $i^F : V \to F_{\lambda,\Sigma}$ above this constitutes a monoid.

Analogously the simultaneous substitution of infinite terms defines a monoid $m^T : T_{\lambda,\Sigma} \otimes T_{\lambda,\Sigma} \to T_{\lambda,\Sigma}$. It is easy to see that given a rational term, every simultaneous substitution of rational terms for variables yields again a rational term. Thus, we have a submonoid $m^R : R_{\lambda,\Sigma} \otimes R_{\lambda,\Sigma} \to R_{\lambda,\Sigma}$ of $T_{\lambda,\Sigma}$.

(v) The monoidal operation of $F_{\lambda,\Sigma}$ is well connected to the structure of $H_{\lambda,\Sigma}$-algebra. This was expressed in [6] by the concept of an $H_{\lambda,\Sigma}$-*monoid*.

In order to recall this concept, we need the notion of point-strength introduced in [5] under the name $(I/\mathscr{W})$-strength; this is a weakening of the classical strength (necessary since $H_{\lambda,\Sigma}$ is unfortunately not strong). Recall that objects of the slice category $I/\mathscr{W}$ are morphisms $x\colon I \to X$ for $X \in \mathrm{obj}\,\mathscr{W}$.

Definition 3.2 (see [5]). *Let $(\mathscr{W}, \otimes, I)$ be a strict monoidal category, and H an endofunctor of $\mathscr{W}$. A* **point-strength** *of H is a collection of morphisms*

$$s_{(X,x)(Y,y)}\colon HX \otimes Y \to H(X \otimes Y)$$

natural in (X,x) and (Y,y) ranging through $I/\mathscr{W}$ such that

(i) $s_{(X,x)(I,\mathrm{id})} = \mathrm{id}_{HX}$, *and*

(ii) $s_{(X,x)(Y\otimes Z,y\otimes z)} = s_{(X\otimes Y,x\otimes y)(Z,z)}\cdot\big(s_{(X,x)(Y,y)} \otimes \mathrm{id}_Z\big)$.

Example 3.3. (i) The endofunctor $X \mapsto X \otimes X$ (which usually fails to be strong) has the point-strength

$$s_{(X,x)(Y,y)} = (X\otimes X)\otimes Y = (X\otimes I\otimes X)\otimes Y \xrightarrow{\mathrm{id}_X\,\otimes y\otimes\mathrm{id}_{X\otimes Y}} (X\otimes Y)\otimes(X\otimes Y).$$

(ii) The endofunctor $X \mapsto X^n$ of $\mathbf{Set}^{\mathscr{F}}$ is clearly (point-)strong for every $n \in \mathbb{N}$.

(iii) The functor δ in 2.3 is point-strong, as observed in [6]. The easiest way to describe its point-strength is by working in $\mathrm{Fin}(\mathbf{Set},\mathbf{Set})$. Given pointed objects (X,x), (Y,y), then the point-strength $s_{(X,x)(Y,y)}\colon (\delta X)\cdot Y \to \delta(X\cdot Y)$ has components

$$X\big(Y(\Gamma)+1\big) \xrightarrow{X(\mathrm{id}+y_1)} X\big(Y(\Gamma)+Y(1)\big) \xrightarrow{X\,\mathrm{can}} X\cdot Y(\Gamma+1),$$

where $\mathrm{can}\colon Y(\Gamma)+Y(1) \to Y(\Gamma+1)$ denotes the canonical morphism.

(iv) A coproduct of point-strong functors is point-strong.

Corollary 3.4. *The endofunctors H_λ and $H_{\lambda,\Sigma}$ are point-strong. Their point-strength is denoted by s^H.*

Definition 3.5 (see [6]). *Let H be a point-strong endofunctor of a monoidal category. By an H-**monoid** is meant an H-algebra (A,a) which is also a monoid $m\colon A\otimes A \to A$ and $i\colon I \to A$ such that the square below commutes:*

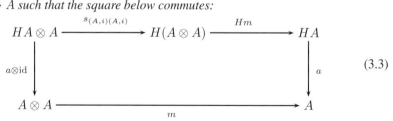

$$(3.3)$$

Theorem 3.6 (see [6]). *The presheaf $F_{\lambda,\Sigma}$ of finite λ-Σ-terms is the initial $H_{\lambda,\Sigma}$-monoid.*

Example 3.7. Although F_λ is the free H_λ-algebra on V, see Theorem 2.7, it is not in general true that $F_\lambda \otimes Z$ is the free H_λ-algebra on a presheaf Z. For example, if $Z = V \times V$, then terms in $(F_\lambda \otimes Z)(\Gamma)$ are precisely the finite λ-terms whose free variables are substituted by pairs in $\Gamma \times \Gamma$. In contrast, the free H_λ-algebra on $V \times V$ contains in context Γ also terms such as $\lambda x.(x, y)$ for $y \in \Gamma$, that is, in variable pairs one member can be bound and one free.

Theorem 3.8 (see [11]). *The presheaf $T_{\lambda, \Sigma}$ of λ-Σ-terms is an $H_{\lambda, \Sigma}$-monoid with simultaneous substitution as monoid structure.*

Although in [11], Example 13, just T_λ is used, the methods of that paper apply to $T_{\lambda, \Sigma}$ immediately.

Theorem 3.9. *The presheaf $T_{\lambda, \Sigma}$ of λ-Σ-terms is the initial completely iterative $H_{\lambda, \Sigma}$-monoid. That is, for every $H_{\lambda, \Sigma}$-monoid A whose $H_{\lambda, \Sigma}$-algebra structure is completely iterative there exists a unique monoid homomorphism from $T_{\lambda, \Sigma}$ to A which is also an $H_{\lambda, \Sigma}$-homomorphism.*

Sketch of proof. Given a completely iterative algebra $a \colon H_{\lambda, \Sigma}A \to A$ which is also a monoid with $m \colon A \otimes A \to A$ and $i \colon V \to A$, and assuming that (3.3) commutes, we have, due to 2.16, a unique homomorphism $h \colon T_{\lambda, \Sigma} \to A$ of $H_{\lambda, \Sigma}$-algebras with $h \cdot i^T = i$. Thus, it remains to prove that h is a monoid homomorphism, that is,

$$h \cdot m^T = m \cdot (h \otimes h) \colon T_{\lambda, \Sigma} \otimes T_{\lambda, \Sigma} \to A. \tag{3.4}$$

To see this we define an equation morphism e with the object $T_{\lambda, \Sigma} \otimes T_{\lambda, \Sigma}$ of variables as follows: we first observe that $T_{\lambda, \Sigma} \cong H_{\lambda, \Sigma}T_{\lambda, \Sigma} + V$ (recall that $T_{\lambda, \Sigma}$ is the terminal coalgebra for $H_{\lambda, \Sigma}(-) + V$, see the proof of Theorem 2.9) and then derive $T_{\lambda, \Sigma} \otimes T_{\lambda, \Sigma} \cong (H_{\lambda, \Sigma}T_{\lambda, \Sigma}) \otimes T_{\lambda, \Sigma} + T_{\lambda, \Sigma}$, see 3.1(i). Put

$$e \equiv (H_{\lambda, \Sigma}T_{\lambda, \Sigma}) \otimes T_{\lambda, \Sigma} + T_{\lambda, \Sigma} \xrightarrow{s^H + h} H_{\lambda, \Sigma}(T_{\lambda, \Sigma} \otimes T_{\lambda, \Sigma}) + A$$

and recall that e has a unique (!) solution $e^\dagger \colon T_{\lambda, \Sigma} \otimes T_{\lambda, \Sigma} \to A$. The verification of (3.4) is performed by proving that both sides are solutions of e. □

Remark 3.10. The presheaf $R_{\lambda, \Sigma}$ is an $H_{\lambda, \Sigma}$-monoid: this is a submonoid of $T_{\lambda, \Sigma}$, and the verification of (3.3) thus follows from that for $T_{\lambda, \Sigma}$.

Theorem 3.11. *The presheaf $R_{\lambda, \Sigma}$ of rational λ-Σ-terms is the initial iterative $H_{\lambda, \Sigma}$-monoid.*

Sketch of proof. For the universal property of being the initial iterative monoid we need to work with the free iterative monad on $H_{\lambda, \Sigma}$ which is the monad $(\mathbb{R}_{\lambda, \Sigma}, \mu^R, \eta^R)$ of free iterative algebras for $H_{\lambda, \Sigma}$. In fact, since $H_{\lambda, \Sigma}$ is clearly a finitary endofunctor of the locally finitely presentable category $\mathbf{Set}^{\mathscr{F}}$, it follows from results of [1] that every presheaf A generates a free iterative algebra $\mathbb{R}_{\lambda, \Sigma}(A)$. We denote its algebra structure by $\varrho_A \colon H_{\lambda, \Sigma}(\mathbb{R}_{\lambda, \Sigma}(A)) \to \mathbb{R}_{\lambda, \Sigma}(A)$ with the universal arrow $\eta_A^R \colon A \to \mathbb{R}_{\lambda, \Sigma}(A)$. In particular, by Theorem 2.14 we have

$$R_{\lambda, \Sigma} = \mathbb{R}_{\lambda, \Sigma}(V). \tag{3.5}$$

We now prove that the endofunctor $\mathbb{R}_{\lambda,\Sigma}$ of $\mathbf{Set}^{\mathscr{F}}$ has a point-strength

$$s^R_{(X,x)(Y,y)}\colon \mathbb{R}_{\lambda,\Sigma}(X) \otimes Y \to \mathbb{R}_{\lambda,\Sigma}(X \otimes Y)$$

such that the monoid structure $m^R\colon \mathbb{R}_{\lambda,\Sigma} \otimes \mathbb{R}_{\lambda,\Sigma} \to \mathbb{R}_{\lambda,\Sigma}$ is simply

$$m^R = \mu^R_V \cdot s_{(V,\mathrm{id})(\mathbb{R}_{\lambda,\Sigma},i^R)}. \tag{3.6}$$

Moreover, we know from [1] that the morphisms ϱ_A and η^R_A form coproduct injections of

$$\mathbb{R}_{\lambda,\Sigma}(A) = H_{\lambda,\Sigma}\big(\mathbb{R}_{\lambda,\Sigma}(A)\big) + A \tag{3.7}$$

and one then proves that this is related to the strength s^R by

$$s^R_{(X,x)(Y,y)} = H_{\lambda,\Sigma}s^R_{(X,x)(Y,y)} + \mathrm{id}_{X \otimes Y}. \tag{3.8}$$

Based on this equation, we prove that for every $H_{\lambda,\Sigma}$-monoid (A, a, m, i) such that the algebra (A, a) is iterative the unique homomorphism $h\colon \mathbb{R}_{\lambda,\Sigma} \to A$ of $H_{\lambda,\Sigma}$-algebras with $h \cdot i^R = i$ (see Theorem 2.16(ii)) preserves monoid multiplication. This proof is based on the construction of $\mathbb{R}_{\lambda,\Sigma}(A)$ in [1] and is rather technical.   □

4 Higher-Order Recursion Schemes

We can reformulate and extend higher-order recursion schemes (1.1) categorically:

Definition 4.1. *A **higher-order recursion scheme** on a signature Σ (of "terminals") is a presheaf morphism*

$$e\colon X \to F_{\lambda,\Sigma} \otimes (X + V) \tag{4.1}$$

where X is a finitely presentable presheaf.

Remark 4.2. (i) The presheaf $F_{\lambda,\Sigma} \otimes (X + V)$ assigns to a context Γ the set $F_{\lambda,\Sigma}(X(\Gamma) + \Gamma)$ of finite λ-terms in contexts $\overline{\Gamma} \subseteq X(\Gamma) + \Gamma$.

(ii) In the introduction we considered, for a given context

$$\Gamma_{nt} = \{p_1, \ldots, p_n\}$$

of "nonterminals", a system of equations $p_i = f_i$ where f_i is a λ-Σ-term in some context $\Gamma_0 = \{x_1, \ldots, x_k\}$. Let X be the free presheaf in n generators $p_1, \ldots, p_n$ of context Γ_0 (a coproduct of n copies of $\mathscr{F}(\Gamma_0, -)$, see Example 2.2(ii)). Then the system of equations defines the unique morphism

$$e\colon X \to F_{\lambda,\Sigma} \otimes (X + V)$$

assigning to every p_i the right-hand side f_i lying in

$$F_{\lambda,\Sigma}(\Gamma_{nt} + \Gamma_0) \subseteq F_{\lambda,\Sigma}(X(\Gamma_0) + \Gamma_0)$$

where we once again consider $F_{\lambda,\Sigma}$ as an object of $\mathrm{Fin}(\mathbf{Set}, \mathbf{Set})$.

(iii) Conversely, every morphism (4.1) yields a system of equations $p_i = f_i$ as follows: let Γ_0 fulfill (2.4) in Remark 2.10, and define $\Gamma_{nt} = X(\Gamma_0)$. The element $f_p = e_{\Gamma_0}(p)$ lies, for every nonterminal $p \in \Gamma_{nt}$, in $F_{\lambda,\Sigma}(\Gamma_{nt} + \Gamma_0)$. We obtain a system of equations $p = f_p$ describing the given morphism e.

(iv) We will use the presheaf $R_{\lambda,\Sigma}$ for our uninterpreted solutions of recursion schemes:

A solution of the system of (formal) equations $p_i = f_i$ are rational terms $p_1^\dagger, \ldots, p_n^\dagger$ making those equations identities in $R_{\lambda,\Sigma}(\Gamma_0)$ when we substitute in f_i the terms $p_j^\dagger$ for the nonterminals p_j ($j = 1, \ldots, n$). This is expressed by the Definition 4.3.

(v) The general case of "equation morphisms" as considered in [1] is (for the endofunctor $H_{\lambda,\Sigma}$) a morphism of type $e : X \rightarrow \mathbb{R}_{\lambda,\Sigma}(X + V)$. Then we see that every higher-order recursion scheme gives an equation morphism via the inclusion $F_{\lambda,\Sigma} \hookrightarrow R_{\lambda,\Sigma}$ and the strength of the monad $\mathbb{R}_{\lambda,\Sigma}$ (but not necessarily conversely). Our solution theorem below is an application of the general result of [1].

Definition 4.3. *A **solution** of a higher-order recursion scheme* $e : X \rightarrow F_{\lambda,\Sigma} \otimes (X + V)$ *is a morphism* $e^\dagger : X \rightarrow R_{\lambda,\Sigma}$ *such that the square below, where* $j : F_{\lambda,\Sigma} \rightarrow R_{\lambda,\Sigma}$ *denotes the embedding, commutes:*

$$
\begin{array}{ccc}
X & \xrightarrow{\quad e^\dagger \quad} & R_{\lambda,\Sigma} \\
\downarrow{\scriptstyle e} & & \uparrow{\scriptstyle m^R} \\
F_{\lambda,\Sigma} \otimes (X + V) & & R_{\lambda,\Sigma} \otimes R_{\lambda,\Sigma} \\
\downarrow{\scriptstyle j \otimes (X+V)} & & \\
R_{\lambda,\Sigma} \otimes (X + V) & \xrightarrow{\quad R_{\lambda,\Sigma} \otimes [e^\dagger, i^R] \quad} & R_{\lambda,\Sigma} \otimes R_{\lambda,\Sigma}
\end{array}
$$

Example 4.4. The fixed-point combinator (see Example 1.1) with $\Sigma = \emptyset$ defines e whose domain is the terminal presheaf 1, that is, $e : 1 \rightarrow F_\lambda \otimes (1 + V)$. The solution $e^\dagger : 1 \rightarrow R_\lambda$ assigns to the unique element of 1 the tree (1.2).

Remark 4.5. Recursion schemes such as $p_1 = p_1$ make no sense—and certainly fail to have a unique solution. In general, we want to avoid right-hand sides of the form p_i. A recursion scheme is called *guarded* if no right-hand side lies in Γ_{nt}. (Theorem 4.7 below shows that no other restrictions are needed.) Guardedness can be formalized as follows: since

$$ R_{\lambda,\Sigma} = H_{\lambda,\Sigma}(R_{\lambda,\Sigma}) + V \qquad \text{with injections } \varrho_V \text{ and } i^R $$

by (3.7), we have (see 3.1(i))

$$ R_{\lambda,\Sigma} \otimes (X + V) \cong H_{\lambda,\Sigma}(R_{\lambda,\Sigma}) \otimes (X + V) + X + V $$

with coproduct injections $\varrho_V \otimes \mathrm{id}_{X+V}$ and $i^R \otimes \mathrm{id}_{X+V}$. Then e is guarded if its extension $(j \otimes (X+V)) \cdot e : X \rightarrow R_{\lambda,\Sigma} \otimes (X+V)$ factorizes through the embedding of the first and third summand of this coproduct:

Definition 4.6. *A higher-order recursion scheme* $e\colon X \to F_{\lambda,\Sigma} \otimes (X + V)$ *is called* **guarded** *if* $(j \otimes (X + V)){\cdot}e$ *factorizes through*

$$\left[\varrho \otimes \mathrm{id}, (i^R \otimes \mathrm{id}){\cdot}\,\mathrm{inr}\right]\colon H_\lambda(R_{\lambda,\Sigma}) \otimes (X + V) + V \to R_{\lambda,\Sigma} \otimes (X + V).$$

Theorem 4.7. *Every guarded higher-order recursion scheme has a unique solution.*

Remark. In Definition 4.1 we restricted high-order recursion schemes to have $F_{\lambda,\Sigma}$ in their codomain. This corresponds well to the classical notion of recursion schemes as explained in Remark 4.2. Moreover, this leads to a simple presentation of the interpreted semantics in Section 5 below. However, Theorem 4.7 remains valid if we replace $F_{\lambda,\Sigma}$ by $R_{\lambda,\Sigma}$ in Definition 4.1 and define solution by $e^\dagger = m^R \cdot R_{\lambda,\Sigma} \otimes [e^\dagger, i^R] \cdot e$. This extends the notion of a higher-order recursion scheme (1.1) to allow the right-hand sides f_i to be rational λ-Σ-terms. We shall prove Theorem 4.7 working with higher-order schemes of the form $e\colon X \to R_{\lambda,\Sigma} \otimes (X + V)$, X finitely presentable. We call e guarded if it factorizes through $[\varrho \otimes \mathrm{id}, (i^R \otimes \mathrm{id}){\cdot}\,\mathrm{inr}]$.

Sketch of proof. Here we work with the free iterative monad $\mathbb{R}_{\lambda,\Sigma}$ generated by the endofunctor $H_{\lambda,\Sigma}$ on $\mathbf{Set}^{\mathscr{F}}$, see 3.11. By (3.5) we have $R_{\lambda,\Sigma} = \mathbb{R}_{\lambda,\Sigma}(V)$. For every guarded recursion program scheme $e\colon X \to \mathbb{R}_{\lambda,\Sigma}(V) \otimes (X + V)$ one constructs a guarded rational equation morphism $\bar{e}\colon X \to \mathbb{R}_{\lambda,\Sigma}(X + V)$ for the monad $\mathbb{R}_{\lambda,\Sigma}$ in the sense of [1]. Since guarded rational equation morphisms have unique solutions for this monad, the proof is finished by verifying that e and $\bar{e}$ have the same solutions.   □

5 Interpreted Solutions

In the uninterpreted semantics of higher-order recursion schemes λ-abstraction and application are only syntactic operations. Therefore terms such as f and $\lambda x.fx$ are unrelated. This is not satisfactory, so we turn to interpreted semantics where the β and η conversions hold. For that we need an interpretation of the λ-calculus plus an interpretation of the terminals in Σ.

We denote by **CPO** the cartesian closed category of posets with directed joins and continuous functions. We assume that a Scott model D of λ-calculus is given, i.e., a CPO with $\perp$ and with an embedding-projection pair

$$\mathsf{fold} : \mathbf{CPO}(D, D) \lhd D : \mathsf{unfold} \tag{5.1}$$

together with continuous operations

$$\sigma^D : D^n \to D \quad \text{for every } n\text{-ary } \sigma \text{ in } \Sigma.$$

We then work with the presheaf $\langle D, D \rangle$ defined by

$$\langle D, D \rangle \Gamma = \mathbf{CPO}(D^\Gamma, D)$$

as our interpretation object. Observe that elements of $\langle D, D \rangle$ can always be interpreted in D: the above function $\mathsf{fold} : \langle D, D \rangle 1 \to D$ yields obvious functions $\mathsf{fold}_\Gamma : \langle D, D \rangle \Gamma \to D$ for all contexts Γ by putting

$$\mathsf{fold}_{\Gamma+1} = \mathsf{fold}_\Gamma \cdot \mathbf{CPO}(D^\Gamma, \mathsf{fold}) \cdot \mathsf{curry},$$

where $\mathsf{curry} : \mathbf{CPO}(D^\Gamma \times D, D) \to \mathbf{CPO}(D^\Gamma, D^D)$ is the currification.

Remark 5.1. The presheaf $\langle D, D \rangle$ is an $H_{\lambda, \Sigma}$-monoid. In fact, application and abstraction are naturally obtained from (5.1), see [6], and the pointing $\iota : V \to \langle D, D \rangle$ assigns to an element $x \in \Gamma$ the x-projection in $\langle D, D \rangle \Gamma = \mathbf{CPO}(D^\Gamma, D)$. The monoid structure

$$m : \langle D, D \rangle \otimes \langle D, D \rangle \to \langle D, D \rangle$$

can be described directly by using the coend formula (3.2) and considering the component of m_Γ corresponding, for an element $f \in \mathbf{Set}(\overline{\Gamma}, \mathbf{CPO}(D^\Gamma, D))$, to the injection

$$\mathrm{in}_f : \mathbf{CPO}(D^{\overline{\Gamma}}, D) \to \int^{\overline{\Gamma}} \mathbf{Set}(\overline{\Gamma}, \mathbf{CPO}(D^\Gamma, D)) \bullet \mathbf{CPO}(D^{\overline{\Gamma}}, D).$$

This component $m_\Gamma \cdot \mathrm{in}_f$ takes $g : D^{\overline{\Gamma}} \to D$ to the function

$$m_\Gamma \cdot \mathrm{in}_f(g) : (x_i) \mapsto g \cdot \langle f(x_i) \rangle \quad \text{for all } (x_i) \text{ in } D^\Gamma. \tag{5.2}$$

There is a much more elegant way of obtaining the monoid structure of $\langle D, D \rangle$. From results of Steve Lack [10] we see that the monoidal category $(\mathbf{Set}^{\mathscr{F}}, \otimes, V)$ has the following monoidal action $*$ on $\mathbf{CPO}$: given X in $\mathbf{Set}^{\mathscr{F}}$ and C in $\mathbf{CPO}$, we put $X * C = \int^\Gamma X(\Gamma) \bullet C^\Gamma$. Moreover, extending the above notation to pairs C, C' of CPO's and defining $\langle C, C' \rangle \Gamma = \mathbf{CPO}(C^\Gamma, C')$ we obtain a presheaf with a natural isomorphism

$$\mathbf{Set}^{\mathscr{F}}(X, \langle C, C' \rangle) \cong \mathbf{CPO}(X * C, C').$$

As observed by George Janelidze and Max Kelly [9] this yields an enriched category whose hom-objects are $\langle C, C' \rangle$. In particular, $\langle D, D \rangle$ receives a monoid structure. It is tedious but not difficult to prove that (a) this monoid structure is given by (5.2) above and (b) it forms an $H_{\lambda, \Sigma}$-monoid (cf. Definition 3.5).

Notation 5.2. We denote by

$$[\![-]\!] : F_{\lambda, \Sigma} \to \langle D, D \rangle$$

the unique $H_{\lambda, \Sigma}$-monoid homomorphism (see Theorem 3.6). For every finite term t in context Γ we thus obtain its interpretation as a continuous function $[\![t]\!]_\Gamma : D^\Gamma \to D$

Remark 5.3. What is our intuition of an interpreted solution of $e : X \to F_{\lambda, \Sigma} \otimes (X + V)$ in the presheaf $\langle D, D \rangle$? This should be an interpretation of X-terms in $\langle D, D \rangle$, that is a natural transformation

$$e^\dagger : X \to \langle D, D \rangle$$

with the following property: Given an X-term x in context Γ then e_Γ assigns to it an element $e_\Gamma(x)$ of $(F_{\lambda, \Sigma} \otimes (X + V))(\Gamma)$ that is a finite term $t \in F_{\lambda, \Sigma}(\overline{\Gamma})$ for some $\overline{\Gamma} \subseteq X(\Gamma) + \Gamma$. We request that the solution assigns to x the same value $e_\Gamma^\dagger(x) : D^\Gamma \to D$ that we obtain from the interpretation $[\![t]\!]$ of the given term by substituting the $\overline{\Gamma}$-variables using $[e^\dagger, \iota] : X + V \to \langle D, D \rangle$. This substitution is given by composing $[\![t]\!] \otimes [e^\dagger, \iota]$ with the monoid structure of $\langle D, D \rangle$. This leads to the following

Definition 5.4. *Given a higher-order recursion scheme $e : X \rightarrow F_{\lambda,\Sigma} \otimes (X + V)$ by an **interpreted solution** is meant a presheaf morphism $e^{\dagger} : X \rightarrow \langle D, D \rangle$ such that the square below commutes:*

$$
\begin{array}{ccc}
X & \xrightarrow{\quad e^{\dagger} \quad} & \langle D, D \rangle \\
{\scriptstyle e}\downarrow & & \uparrow{\scriptstyle m} \\
F_{\lambda,\Sigma} \otimes (X + V) & \xrightarrow[{[\![-]\!] \otimes [e^{\dagger}, \iota]}]{} & \langle D, D \rangle \otimes \langle D, D \rangle
\end{array}
\qquad (5.3)
$$

Theorem 5.5. *Every higher-order recursion scheme has a least interpreted solution in $\langle D, D \rangle$.*

Sketch of proof. Observe that $\mathbf{Set}^{\mathscr{F}}(X, \langle D, D \rangle)$ is a CPO with $\bot$ if the ordering is defined pointwise: for $s, s' : X \rightarrow \langle D, D \rangle$ we put $s \sqsubseteq s'$ if and only if for every context Γ and every $x \in X(\Gamma)$ we have $s_{\Gamma}(x) \sqsubseteq s'_{\Gamma}(x)$ in $\mathbf{CPO}(D^{\Gamma}, D)$. Therefore it is sufficient to prove that the endomap of $\mathbf{Set}^{\mathscr{F}}(X, \langle D, D \rangle)$ given by

$$
s \mapsto m \cdot ([\![-]\!] \otimes [s, \iota]) \cdot e
$$

is continuous, then we can use Kleene Theorem. In fact, from the (obvious) continuity of $s \mapsto [s, \iota]$ it follows (less obviously, but this is not too difficult) that $s \mapsto m \cdot ([\![-]\!] \otimes [s, \iota])$ is continuous, and precomposing with e then also yields a continuous function. $\qquad\square$

6   Conclusions

We proved that guarded higher-order recursion schemes have a unique uninterpreted solution, i.e., a solution as a rational λ-Σ-term. And they also have the least interpreted solution for interpretations based on Scott's models of λ-calculus as CPO's with continuous operations for all "terminal" symbols of the recursion scheme.

Following M. Fiore *et al* [6] we worked in the category $\mathbf{Set}^{\mathscr{F}}$ of sets in context, that is, covariant presheaves on the category $\mathscr{F}$ of finite sets and functions. A presheaf is a set dependent on a context (a finite set of variables). For every signature Σ of "terminal" operation symbols it was proved in [6] that the presheaf $F_{\lambda,\Sigma}$ of all finite λ-Σ-terms is the initial $H_{\lambda,\Sigma}$-monoid. This means that $F_{\lambda,\Sigma}$ has (i) the λ-operations (of abstraction and application) together with operations given by Σ rendering an $H_{\lambda,\Sigma}$-algebra and (ii) the operation expressing simultaneous substitution rendering a monoid in the category of presheaves. And $F_{\lambda,\Sigma}$ is the initial presheaf with such structure. In [11] R. Matthes and T. Uustalu proved that the presheaf $T_{\lambda,\Sigma}$ of finite and infinite λ-Σ-terms is also an $H_{\lambda,\Sigma}$-monoid. Here we proved that this is the initial completely iterative $H_{\lambda,\Sigma}$-monoid. And its subobject $R_{\lambda,\Sigma}$ of all rational λ-Σ-terms is the initial iterative $H_{\lambda,\Sigma}$-monoid. We used that last presheaf in our uninterpreted semantics of recursion schemes.

Our approach was based on untyped λ-calculus. The ideas in the typed version are quite analogous. If T is the set of all types, then we form the full subcategory $\mathscr{F}$ of $\mathbf{Set}^{T}$

of finite T-sorted sets and consider presheaves in $(\mathbf{Set}^T)^{\mathscr{F}}$—the latter category is equivalent to that of finitary endofunctors of the category $\mathbf{Set}^T$ of T-sorted sets. The definition of $H_{\lambda,\Sigma}$ is then completely analogous to the untyped case, and one can form the presheaves $F_{\lambda,\Sigma}$ (free algebra on V), $T_{\lambda,\Sigma}$ (free completely iterative algebra) and $R_{\lambda,\Sigma}$ (free iterative algebra). Each of them is a monoid, in fact, an $H_{\lambda,\Sigma}$-monoid in the sense of [6]. Moreover, every guarded higher-order recursion scheme has a unique solution in $R_{\lambda,\Sigma}$. The interpreted semantics can be built up on a CPO-enriched cartesian closed category (as our model of typed λ-calculus) with additional continuous morphisms for all terminals. The details of the typed version are more involved, and we leave them for future work.

Related results on higher-order substitution can be found e.g. in [11] and [15].

In future work we will, analogously as in [13], investigate the relation of uninterpreted and interpreted solutions.

Acknowledgments. The authors are grateful to an anonymous referee and to Thorsten Palm for several suggestions how to improve the presentation of our paper.

References

1. Adámek, J., Milius, S., Velebil, J.: Iterative algebras at work. Math. Structures Comput. Sci. 16, 1085–1131 (2006)
2. Adámek, J., Trnková, V.: Automata and algebras in a category. Kluwer Academic Publishers, Dordrecht (1990)
3. Aehlig, K.: A finite semantics of simply-typed lambda terms for infinite runs of automata. In: Ésik, Z. (ed.) CSL 2006. LNCS, vol. 4207, pp. 104–118. Springer, Heidelberg (2006)
4. Damm, W.: Higher-order program schemes and their languages. LNCS, vol. 48, pp. 51–72. Springer, Heidelberg (1979)
5. Fiore, M.: Second order dependently sorted abstract syntax. In: Proc. Logic in Computer Science 2008, pp. 57–68. IEEE Press, Los Alamitos (2008)
6. Fiore, M., Plotkin, G., Turi, D.: Abstract syntax and variable binding. In: Proc. Logic in Computer Science 1999, pp. 193–202. IEEE Press, Los Alamitos (1999)
7. Garland, S.J., Luckham, D.C.: Program schemes, recursion schemes and formal languages. J. Comput. Syst. Sci. 7, 119–160 (1973)
8. Guessarian, I.: Algebraic semantics. LNCS, vol. 99. Springer, Heidelberg (1981)
9. Janelidze, G., Kelly, G.M.: A note on actions of a monoidal category. Theory Appl. Categ. 9, 61–91 (2001)
10. Lack, S.: On the monadicity of finitary monads. J. Pure Appl. Algebra 140, 65–73 (1999)
11. Matthes, R., Uustalu, T.: Substitution in non-wellfounded syntax with variable binding. Theoret. Comput. Sci. 327, 155–174 (2004)
12. Milius, S.: Completely iterative algebras and completely iterative monads. Inform. and Comput. 196, 1–41 (2005)
13. Milius, S., Moss, L.: The category theoretic solution of recursive program schemes. Theoret. Comput. Sci. 366, 3–59 (2006); corrigendum 403, 409–415 (2008)
14. Miranda, G.: Structures generated by higher-order grammars and the safety constraint. Ph.D. Thesis, Merton College, Oxford (2006)
15. Power, J.: A unified category theoretical approach to variable binding. In: Proc. MERLIN 2003 (2003)

Coalgebraic Components
in a Many-Sorted Microcosm

Ichiro Hasuo[1,4], Chris Heunen[2], Bart Jacobs[2], and Ana Sokolova[3,*]

[1] RIMS, Kyoto University, Japan
[2] Radboud University Nijmegen, The Netherlands
[3] University of Salzburg, Austria
[4] PRESTO Research Promotion Program, Japan Science and Technology Agency

Abstract. The *microcosm principle*, advocated by Baez and Dolan and formalized for Lawvere theories lately by three of the authors, has been applied to coalgebras in order to describe compositional behavior systematically. Here we further illustrate the usefulness of the approach by extending it to a many-sorted setting. Then we can show that the coalgebraic component calculi of Barbosa are examples, with compositionality of behavior following from microcosm structure. The algebraic structure on these coalgebraic components corresponds to variants of Hughes' notion of *arrow*, introduced to organize computations in functional programming.

1 Introduction

Arguably the most effective countermeasure against today's growing complexity of computer systems is *modularity*: one should be able to derive the behavior of the total system from that of its constituent parts. Parts that were developed and tested in isolation can then safely be composed into bigger systems. Likewise, one would like to be able to prove statements about the compound system based on proofs of substatements about the parts. Therefore, the theoretical models should at the very least be such that their behavior is compositional.

This is easier said than done, especially in the presence of concurrency, that is, when systems can be composed in parallel as well as in sequence. The *microcosm principle* [1,14] brings some order to the situation. Roughly speaking, compositionality means that the behavior of a compound system is the composition of the components' behaviors. The microcosm principle then observes that the very definition of composition of behaviors depends on composition of systems, providing an intrinsic link between the two.

The present article gives a rigorous analysis of compositionality of components as sketched above. Considering models as *coalgebras*, we study Barbosa's calculi of *components* [3, 2] as coalgebras with specified input and output interfaces. Explicitly, a component is a coalgebra for the endofunctor[1]

$$F_{I,J} = (T(J \times \_))^I : \mathbf{Set} \to \mathbf{Set}, \tag{1}$$

---

* Research funded by the Austrian Science Fund (FWF) Project No. V00125.

[1] Note that the functor $F_{I,J}$ also depends on the choice of the parameter T, so could have been denoted e.g. by $F_{I,J}^T$. We shall not do so for simplicity of presentation.

A. Kurz, M. Lenisa, and A. Tarlecki (Eds.): CALCO 2009, LNCS 5728, pp. 64–80, 2009.
© Springer-Verlag Berlin Heidelberg 2009

where I is the set of possible input, and J that of output. The computational effect of the component is modeled by a monad T, as is customary in functional programming [25]. The monad T can capture features such as finite or unbounded non-determinism ($T = \mathcal{P}_\omega, \mathcal{P}$); possible non-termination ($T = 1 + \_$); global states ($T = (S \times \_)^S$); probabilistic branching or combinations of these.

To accommodate component calculi, the surrounding microcosm needs to be *many-sorted*. After all, composing components sequentially requires that the output of the first and the input of the second match up. This is elaborated on in §2. The contribution of the present article is twofold:

- a rigorous development of a many-sorted microcosm principle, in §4;
- an application of the many-sorted microcosm framework to component calculi, in §5.

It turns out that components as $F_{I,J}$-coalgebras carry algebraic structure that is a variant of Hughes' notion of *arrow* [15,20].[2] Arrows, generalizing monads, have been used to model structured computations in semantics of functional programming. In §5 we will prove that components indeed carry such arrow-like structure; however the calculation is overwhelming as it is. To aid the calculation, we exploit the fact that a Kleisli category $\mathcal{K\ell}(T)$—where the calculation takes place—also carries the same arrow-like structure. This allows us to use the axiomatization of the (shared) structure as an "internal language."

2 Leading Example: Sequential Composition

We shall exhibit, using the following example, the kind of phenomena in component calculi that we are interested in.

For simplicity we assume that we have no effect in components (i.e. $T = \mathrm{Id}$, $F_{I,J} = (J \times \_)^I$). Coalgebras for this functor are called *Mealy machines*, see e.g. [8]. A prominent operation in component calculi is *sequential composition*, or *pipeline*. It attaches two components with matching I/O interfaces, one after another:

$$\left(\xrightarrow{I} \boxed{c} \xrightarrow{J} \, , \; \xrightarrow{J} \boxed{d} \xrightarrow{K} \right) \quad \overset{\ggg_{I,J,K}}{\longmapsto} \quad \xrightarrow{I} \boxed{c} \xrightarrow{J} \boxed{d} \xrightarrow{K} \tag{2}$$

Let X and Y be the state spaces of the components c and d, respectively. The resulting component $c \ggg_{I,J,K} d$ has the state space $X \times Y$;[3] given input $i \in I$, first c produces output $j \in J$ that is fed into the input port of d. More precisely, we can define the coalgebra $c \ggg_{I,J,K} d$ to be the adjoint transpose $X \times Y \to (K \times X \times Y)^I$ of the following function.

$$I \times X \times Y \xrightarrow{\hat{c} \times d} J \times X \times (K \times Y)^J \xrightarrow{X \times \mathrm{ev}_J} K \times X \times Y \tag{3}$$

Here $\hat{c} : I \times X \to J \times X$ is the adjoint transpose of the coalgebra c, and $\mathrm{ev}_J : J \times (K \times Y)^J \to K \times Y$ is the obvious evaluation function.

---

[2] Throughout the paper the word "arrow" always refers to Hughes' notion. An "arrow" in a category (as opposed to an object) will be always called a *morphism*.

[3] We will use the infix notation for the operation $\ggg$. The symbol $\ggg$ is taken from that for (Hughes') arrows, whose relevance is explained in §5.

An important ingredient in the theory of coalgebra is "behavior-by-coinduction" [19]: when a state-based system is viewed as an F-coalgebra, then a *final* F-coalgebra (which very often exists) consists of all the "behaviors" of systems of type F. Moreover, the morphism induced by finality is the "behavior map": it carries a state of a system to its behavior. This view is also valid in the current example.

A final $F_{I,J}$-coalgebra—where $F_{I,J} = (J \times \_)^I$—is carried by the set of stream functions $I^\omega \to J^\omega$ which are *causal*, meaning that the n-th letter of the output stream only depends on the first n letters of the input.[4] It conforms to our intuition: the "behavior" of such a component is what we see as an output stream when an input stream is fed. The final coalgebra is concretely as follows.

$$\zeta_{I,J} : \quad \begin{array}{c} Z_{I,J} \\ (\, t : I^\omega \to J^\omega, \text{ causal}\,) \end{array} \xrightarrow{\ \cong\ } \begin{array}{c} F_{I,J}(Z_{I,J}) = (J \times Z_{I,J})^I \\ \lambda i.\left(\, \mathsf{head}(t(i \cdot \vec{i})),\ \lambda \vec{i}.\,\mathsf{tail}(t(i \cdot \vec{i}))\,\right) \end{array}.$$

Here $i \cdot \vec{i}$ is a letter $i \in I$ followed by an arbitrary stream $\vec{i}$; the value of $\mathsf{head}(t(i \cdot \vec{i}))$ does not depend on $\vec{i}$ since t is causal.

Then there naturally arises a "sequential composition" operation that is different from (2): it acts on *behaviors* of components, simply composing two behaviors of matching types.

$$\ggg_{I,J,K} : \quad \begin{array}{c} Z_{I,J} \quad \times \quad Z_{J,K} \\ \left(\, I^\omega \xrightarrow{s} J^\omega \ ,\ J^\omega \xrightarrow{t} K^\omega\,\right) \end{array} \longrightarrow \begin{array}{c} Z_{I,K} \\ I^\omega \xrightarrow{s} J^\omega \xrightarrow{t} K^\omega \end{array} \tag{4}$$

The following observation—regarding the two operations (2) and (4)—is crucial for our behavioral view on component calculi. The "inner" operation (4), although it naturally arises by looking at stream functions, is in fact induced by the "outer" operation (2). Specifically, it arises as the behavior map for the (outer) composition $\zeta_{I,J} \ggg_{I,J,K} \zeta_{J,K}$ of two final coalgebras.

$$\begin{array}{ccc} F_{I,K}(Z_{I,J} \times Z_{J,K}) & \dashrightarrow & F_{I,K}(Z_{I,K}) \\ \zeta_{I,J} \ggg \zeta_{J,K} \uparrow & \text{final}\uparrow \zeta_{I,K} & \\ Z_{I,J} \times Z_{J,K} & \xdashrightarrow{\ \ggg_{I,J,K}\ } & Z_{I,K} \end{array} \quad \text{i.e.} \quad \left(\begin{array}{c} \downarrow I \\ \boxed{\zeta_{I,J}} \\ \downarrow J \\ \boxed{\zeta_{J,K}} \\ \downarrow K \end{array} \right) \xrightarrow[\dashrightarrow]{\ggg_{I,J,K}} \left(\begin{array}{c} \downarrow I \\ \boxed{\zeta_{I,K}} \\ \downarrow K \end{array} \right) \tag{5}$$

Here the coalgebra $\zeta_{I,J} \ggg \zeta_{J,K}$ has a state space $Z_{I,J} \times Z_{J,K}$ due to the definition (3).

As to the two operations (2) and (4), we can ask a further question: are they compatible, in the sense that the the diagram on the right commutes?[5] One can think of this compatibility property as a mathematical formulation of *compositionality*, a fundamental property in the theory of

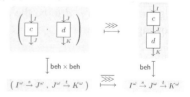

---

[4] This is how they are formalized in [28]. Equivalent formulations are: as string functions $I^* \to J^*$ that are length-preserving and prefix-closed [26]; and as functions $I^+ \to J$ where I^+ is the set of strings of length ≥ 1.

[5] The diagram is simplified for brevity. To be precise, the coalgebras c and d must have their states (say x and y) specified; these states are mapped to the state (x, y) of $c \ggg d$.

processes/components. The characterization of the inner operation by finality (5) is remarkably useful here; finality immediately yields a positive answer.

In fact, the *microcosm principle* is the mathematical structure that has been behind the story. It refers to the phenomenon that the same algebraic structure is carried by a category $\mathbb{C}$ and by an object $X \in \mathbb{C}$, a prototypical example being *a monoid object in a monoidal category* (see e.g. [24, §VII.3]). In [14] we presented another example eminent in the process theory: parallel composition of two coalgebras for the same signature functor, as well as parallel composition of their behaviors. Our story so far is yet another example—taken from component calculi—with its new feature being that the algebraic structure is many-sorted. In the rest of the paper we develop a categorical language for describing the situation, and proving results about it. Among them is the one that ensures compositionality for a wide class of component calculi.

3 FP-Theory

3.1 Presenting Algebraic Structure as a Category

A component calculus—consisting of operations like $\ggg$ and of equations like associativity of $\ggg$—is an instance of *algebraic specification*. So are (the specifications for) monoids, groups, as well as process calculi such as CCS. Component calculi are different from the other examples here, in that they are *many-sorted*. Such a many-sorted algebraic specification consists of

- a set S of *sorts*;
- a set Σ of *operations*. Each operation $\sigma \in \Sigma$ is equipped with its *in-arity* inar(σ) given by a finite sequence of sorts (denoted as a formal product $S_1 \times \cdots \times S_m$), and its *out-arity* outar(σ) that is some sort $S \in S$;
- and a set E of equations.

A straightforward presentation of such is simply as a tuple (S, Σ, E) (see e.g. [18]).

In this paper we prefer a different, categorical presentation of algebraic structure. The idea is that an algebraic specification induces a category $\mathbb{L}$ with:

- all the finite sequences of sorts $S_1 \times \cdots \times S_m$ as its objects;
- operations $\sigma \in \Sigma$ as morphisms inar(σ) $\xrightarrow{\sigma}$ outar(σ). Additionally, projections (such as $\pi_1 : S_1 \times S_2 \to S_1$) and diagonals (such as $\langle \mathrm{id}, \mathrm{id} \rangle : S \to S \times S$) are morphisms. So are (formal) products of two morphisms, equipping the category $\mathbb{L}$ with finite products. Besides we can compose morphisms in the category $\mathbb{L}$; that makes the morphisms in $\mathbb{L}$ induced by the *terms* with operations taken from Σ;
- an equation as a commutative diagram, modulo which we take quotients of terms (as morphisms in $\mathbb{L}$). For example, when $S = \{*\}$ and m is a binary operation, its associativity $\mathrm{m}(x, \mathrm{m}(y, z)) = \mathrm{m}(\mathrm{m}(x, y), z)$ amounts to the diagram on the right.

$$\begin{array}{ccc} 3 & \xrightarrow{\mathrm{m} \times \mathrm{id}} & 2 \\ {\scriptstyle \mathrm{id} \times \mathrm{m}} \downarrow & & \downarrow {\scriptstyle \mathrm{m}} \\ 2 & \xrightarrow{\mathrm{m}} & 1 \end{array} \qquad (6)$$

In fact, what is represented by $\mathbb{L}$ above is not an algebraic specification itself but its *clone* (see e.g. [10]). In categorical terms, the construction is taking a free finite-product category from the objects in S and the arrows in Σ modulo the equations induced by E. See [18, §3.3] for details.

In a one-sorted setting—where arities (objects of $\mathbb{L}$) are identified with natural numbers by taking their length—such a category $\mathbb{L}$ is called a *Lawvere theory* (see e.g. [22, 14, 16]). In a many-sorted setting, such a category $\mathbb{L}$—say a "many-sorted Lawvere theory"—is usually called a *finite-product theory*, or an *FP-theory*, see e.g. [4, 5].

Definition 3.1 (FP-theory). An *FP-theory* is a category with finite products.

The idea of such categorical presentation of algebraic structure originated in [22]. Significant about the approach is that one has a model as a functor.

Definition 3.2 (Set-theoretic model). Let $\mathbb{L}$ be an FP-theory. A *(set-theoretic) model* of $\mathbb{L}$ is a finite-product-preserving (*FP-preserving*) functor $X : \mathbb{L} \to \mathbf{Set}$ into the category $\mathbf{Set}$ of sets and functions.

Later in Def. 4.1 we introduce the notion of *category* with $\mathbb{L}$-structure—this is the kind of models of our interest—based on this standard definition.

To illustrate Def. 3.2 in a one-sorted setting, think about an operation $2 \xrightarrow{\mathrm{m}} 1$ which satisfies associativity (6). Let the image $X(1)$ of $1 \in \mathbb{L}$ be simply denoted by X; then $2 = 1 \times 1 \in \mathbb{L}$ must be mapped to the set X^2 by FP-preservation. By functoriality the morphism m is mapped to a morphism $X(\mathrm{m}) : X^2 \to X$ in $\mathbf{Set}$, which we denote by $[\![\mathrm{m}]\!]_X$. This yields a binary operation on the set X. Moreover, the associativity diagram (6) in $\mathbb{L}$ is carried to a commutative diagram in $\mathbf{Set}$; this expresses associativity of the interpretation $[\![\mathrm{m}]\!]_X$.

When $\mathbb{L}$ arises from a many-sorted algebraic specification, it is not a single set X that carries $\mathbb{L}$-structure; we have a family of sets $\{X(S)\}_{S \in \mathcal{S}}$—one for each sort S—as a carrier. By FP-preservation this extends to interpretation of products of sorts: $X(S_1 \times \cdots \times S_m) \cong X(S_1) \times \cdots \times X(S_m)$. In this way an operation is interpreted with its desired domain and codomain.

Remark 3.3. *Lawvere theories* and *monads* are the two major ways to represent algebraic structure in category theory (see [16]). Both allow straightforward extension to the many-sorted setting: the former to FP-theories and the latter to monads on functor categories (mostly presheaf categories). For the purpose of formalizing the microcosm principle we find the former more useful. Its representation is independent from the domain where the structure is interpreted; hence we can speak of its models in two different domains, as is the case with the microcosm principle. In contrast, a monad on a category $\mathbb{C}$ specifies algebraic structure (i.e. Eilenberg-Moore algebras) that is necessarily carried by an object of $\mathbb{C}$.

3.2   The FP-Theory PLTh

We now present a specific FP-theory; this is our working example. We list its sorts, operations and equations; these altogether induce an FP-theory in the way that we sketched above. We denote the resulting FP-theory by **PLTh**. Later in §5 we will see that this

FP-theory represents Hughes' notion of *arrow*, without its first operation. One can think of **PLTh** as a basic component calculus modeling pipelines (PL for "pipeline").

Assumption 3.4. Throughout the rest of the paper we fix a base category $\mathbb{B}$ to be a cartesian subcategory (i.e. closed under finite products) of **Set**. Its object $I \in \mathbb{B}$ is a set that can play a role of an interface. Its morphism $f : I \to J$—it is a set-theoretic function—represents a "stateless" computation from I to J that can be realized by a component with a single state.

– The sorts $\mathcal{S} = \{(I, J) \mid I, J \in \mathbb{B}\}$. Hence an object of **PLTh** can be written as a formal product $(I_1, J_1) \times \cdots \times (I_m, J_m)$. We denote the nullary product (i.e. the terminal object) by $1 \in$ **PLTh**;
– the operations:

$$\ggg_{I,J,K} : (I, J) \times (J, K) \longrightarrow (I, K) \qquad \textit{sequential composition}$$
$$\text{arr } f : 1 \longrightarrow (I, J) \qquad \textit{pure function}$$

for each object $I, J, K \in \mathbb{B}$ and each morphism $f : I \to J$ in $\mathbb{B}$. Sequential composition is illustrated in (2). The component arr f, intuitively, has a singleton as its state space and realizes "stateless" processing of data stream $\xrightarrow{I}\boxed{\text{arr } f}\xrightarrow{J}$;
– the equations:
  • *associativity:*

$$a : (I, J), b : (J, K), c : (K, L) \vdash a \ggg (b \ggg c) = (a \ggg b) \ggg c$$
$$(\ggg\text{-ASSOC})$$

for each $I, J, K, L \in \mathbb{B}$, omitting the obvious subscripts for $\ggg$, i.e.

$$\begin{array}{ccc}
(I,J) \times (J,K) \times (K,L) & \xrightarrow{\ggg_{I,J,K} \times (K,L)} & (I,K) \times (K,L) \\
{\scriptstyle (I,J) \times \ggg_{J,K,L}}\downarrow & & \downarrow{\scriptstyle \ggg_{I,K,L}} \\
(I,J) \times (J,L) & \xrightarrow[\ggg_{I,J,L}]{} & (I,L)
\end{array}$$

  • *preservation of composition:* for each composable pair of morphisms $f : I \to J$ and $g : J \to K$ in $\mathbb{B}$,

$$\emptyset \vdash \text{arr} (g \circ f) = \text{arr } f \ggg \text{arr } g \qquad (\text{arr-FUNC1})$$

where $\emptyset$ denotes the empty context. On the right is the corresponding diagram.

$$\begin{array}{ccc}
1 & \xrightarrow{\text{arr } f \times \text{arr } g} & (I,J) \times (J,K) \\
& {\scriptstyle \text{arr } (g \circ f)}\searrow & \downarrow{\scriptstyle \ggg_{I,J,K}} \\
& & (I,K)
\end{array}$$

  • *preservation of identities:* for each $I, J \in \mathbb{B}$,

$$a : (I, J) \vdash \text{arr id}_I \ggg_{I,I,J} a = a = a \ggg_{I,J,J} \text{arr id}_J . \qquad (\text{arr-FUNC2})$$

For this FP-theory, the model of our interest is carried by a family of *categories*, namely the category $\mathbf{Coalg}(F_{I,J})$ for each sort (I, J). Formalization of such an *outer model* carried by categories, together with that of an *inner model* carried by final coalgebras, is the main topic of the next section.

4   Microcosm Model of an FP-Theory

In this section we present our formalization of *microcosm models* for an FP-theory $\mathbb{L}$. It is about nested models of $\mathbb{L}$: the outer one ($\mathbb{L}$-*category*) carried by a family $\{\mathbb{C}(S)\}_{S \in \mathcal{S}}$ of categories; the inner one ($\mathbb{L}$-*object*) carried by a family $\{X_S \in \mathbb{C}(S)\}_{S \in \mathcal{S}}$ of objects. We shall use the formalization to prove a compositionality result (Thm. 4.6).

In fact our formalization is essentially the one in our previous work [14]. Due to the space limit we cannot afford sufficient illustration of our seemingly complicated 2-categorical definitions. The reader is suggested to have [14, §3] as her companion; the thesis [13, Chap. 5] of one of the authors has a more detailed account. What is new here, compared to [14], is the following.

- The algebraic structure of our interest is now many-sorted, generalizing $\mathbb{L}$ from a Lawvere theory to an FP-theory.
- Now our framework accommodates categories with "pseudo" algebraic structure, such as (not strict) monoidal categories. We cannot avoid this issue in the current paper, because the concrete model that we deal with is indeed of such a kind with pseudo algebraic structure.

We will depend heavily on 2-categorical notions such as pseudo functors, lax natural transformations, etc. For these notions the reader is referred to [9].

4.1   Outer Model: $\mathbb{L}$-Category

Take the functor $F_{I,J} = (T(J \times \_))^I$, for which coalgebras are components (see §1). We would like that the categories $\{\mathbf{Coalg}(F_{I,J})\}_{I,J \in \mathbb{B}}$ model **PLTh**, the algebraic structure for pipelines in §3.2. That is, we need functors

$$
\begin{aligned}
[\![\ggg_{I,J,K}]\!] &: \mathbf{Coalg}(F_{I,J}) \times \mathbf{Coalg}(F_{J,K}) \to \mathbf{Coalg}(F_{I,K}) \quad \text{for each } I, J, K \in \mathbb{B}, \\
[\![\mathrm{arr}\, f]\!] &: \mathbf{1} \to \mathbf{Coalg}(F_{I,J}) \qquad \text{for each morphism } f : I \to J \text{ in } \mathbb{B},
\end{aligned}
$$

where $\mathbf{1}$ is a (chosen) terminal category. Moreover these functors must satisfy the three classes of equations of **PLTh** in §3.2. One gets pretty close to the desired definition of such "category with $\mathbb{L}$-structure" by replacing "sets" by "categories" in Def. 3.2. That is, by having **CAT**—the 2-category of locally small categories, functors and natural transformations—in place of **Set**. In fact we did so in our previous work [14].

However in our coalgebraic modeling of components we want equations to be satisfied not up-to identity, but up-to (coherent) isomorphisms. For example, consider the functor $[\![\ggg_{I,J,K}]\!]$ above, in particular how it acts on the carrier sets of coalgebras. By the definition in §2 it carries (X, Y) to $X \times Y$; and this operation is associative only up-to isomorphism. This requirement of "satisfaction up-to iso" also applies to monoidal categories; below is how associativity (on the left) is expected to be interpreted, in a monoid and in a monoidal category.

$$
\begin{array}{ccc}
\underline{\text{in } \mathbb{L}} \quad
\begin{array}{ccc}
3 & \xrightarrow{\mathrm{id} \times m} & 2 \\
{\scriptstyle m \times \mathrm{id}} \downarrow & = & \downarrow {\scriptstyle m} \\
2 & \xrightarrow{m} & 1
\end{array}
&
\underline{\text{in } \mathbf{Set}} \quad
\begin{array}{ccc}
X^3 & \xrightarrow{[\![\mathrm{id} \times m]\!]_X} & X^2 \\
{\scriptstyle [\![m \times \mathrm{id}]\!]_X} \downarrow & = & \downarrow {\scriptstyle [\![m]\!]_X} \\
X^2 & \xrightarrow{[\![m]\!]_X} & X
\end{array}
&
\underline{\text{in } \mathbf{CAT}} \quad
\begin{array}{ccc}
\mathbb{C}^3 & \xrightarrow{[\![\mathrm{id} \times m]\!]_\mathbb{C}} & \mathbb{C}^2 \\
{\scriptstyle [\![m \times \mathrm{id}]\!]_\mathbb{C}} \downarrow & \overset{\cong}{\Leftarrow} & \downarrow {\scriptstyle [\![m]\!]_\mathbb{C}} \\
\mathbb{C}^2 & \xrightarrow{[\![m]\!]_\mathbb{C}} & \mathbb{C}
\end{array}
\\[2em]
& x_1 \cdot (x_2 \cdot x_3) = (x_1 \cdot x_2) \cdot x_3 & X_1 \otimes (X_2 \otimes X_3) \overset{\cong}{\to} (X_1 \otimes X_2) \otimes X_3
\end{array}
$$

$$\tag{7}$$

Our approach to such "pseudo algebras" is to relax functorial semantics (Def. 3.2) into *pseudo functorial semantics*, i.e. replacing functor by *pseudo functor*. The idea has been previously mentioned in [14, §3.3] and [13, §5.3.3]; now it has been made rigorous.

Definition 4.1 (𝕃-**category**). An 𝕃-*category* is a pseudo functor $\mathbb{C} : \mathbb{L} \to \mathbf{CAT}$ which is "FP-preserving" in the following sense:[6]

1. the canonical map $\langle \mathbb{C}\pi_1, \mathbb{C}\pi_2 \rangle : \mathbb{C}(A_1 \times A_2) \to \mathbb{C}(A_1) \times \mathbb{C}(A_2)$ is an isomorphism for each $A_1, A_2 \in \mathbb{L}$;
2. the canonical map $\mathbb{C}(1) \to \mathbf{1}$ is an isomorphism;
3. it preserves identities up-to identity: $\mathbb{C}(\mathrm{id}) = \mathrm{id}$;
4. it preserves pre- and post-composition of identities up-to identity: $\mathbb{C}(\mathrm{id} \circ a) = \mathbb{C}(a) = \mathbb{C}(a \circ \mathrm{id})$;
5. it preserves composition of the form $\pi_i \circ a$ up-to identity: $\mathbb{C}(\pi_i \circ a) = \mathbb{C}(\pi_i) \circ \mathbb{C}(a)$. Here $\pi_i : A_1 \times A_2 \to A_i$ is a projection.

We shall often denote $\mathbb{C}$'s action $\mathbb{C}(a)$ on a morphism a by $[\![a]\!]_\mathbb{C}$.

In the definition, what it means exactly to be "FP-preserving" is a delicate issue; for the current purpose of representing pseudo algebras, we found the conditions above to be the right ones. It is justified by the following result.

Proposition 4.2. *Let us denote by* **MonTh** *the Lawvere theory for monoids. The 2-category* **MonCAT** *of monoidal categories, strong monoidal functors and monoidal transformations is equivalent to the 2-category of* **MonTh**-*categories with suitable 1- and 2-cells.*

Proof. The proof involves overwhelming details. It is deferred to [12], where a general result—not only for the specification for monoids but for any algebraic specification—is proved.   □

What the last proposition asserts is stronger than merely establishing a *biequivalence* between the two 2-categories, a claim one would expect from e.g. the coherence result for monoidal categories. See [12] for more details about Def. 4.1 and Prop. 4.2; and see [13, §5.3.3] how pseudo functoriality yields the mediating iso 2-cell in (7).

Remark 4.3. There have been different approaches to formalization of "pseudo algebra." A traditional one (e.g. in [7]) is to find a suitable 2-monad which already takes pseudo satisfaction of equations into account. Another one is by a Lawvere 2-theory, which also includes explicitly the isomorphism up-to which equations are satisfied (see [21]). Neither of them looks quite suitable for the microcosm principle: we want a single representation of algebraic structure interpreted twice; with only the outer one satisfying equations up-to isomorphisms.

The same idea as ours has been pursued by a few other authors. Segal [29] defines pseudo algebras as pseudo functors, for *monoidal* theories (as opposed to *cartesian* theories in our case), with applications to conformal field theory. Fiore's definition [11] is equivalent to ours, but its aspect as a pseudo functor is not emphasized there.

---

[6] To be precise, each of the conditions 3–5 means that the corresponding mediating isomorphism (as part of the definition of a pseudo functor) is actually the identity.

4.2 Inner Model: $\mathbb{L}$-Object

Once we have an outer model $\mathbb{C}$ of $\mathbb{L}$, we can define the notion of *inner model in* $\mathbb{C}$. It is a family of objects $\{X_S \in \mathbb{C}(S)\}_{S \in \mathcal{S}}$ which carries $\mathbb{L}$-structure in the same way as a monoid in a monoidal category does [24, §VII.3]. We have seen in §2 that final coalgebras carry such an inner model and realize composition of behaviors.

Definition 4.4 ($\mathbb{L}$-object). Let $\mathbb{C} : \mathbb{L} \to \mathbf{CAT}$ be an $\mathbb{L}$-category. An $\mathbb{L}$-*object* in $\mathbb{C}$ is a lax natural transformation X as in the diagram above, which is "FP-preserving" in the sense that: it is strictly natural with regard to projections (see [14, Def. 3.4]). Here $1 : \mathbb{L} \to \mathbf{CAT}$ denotes the constant functor into a (chosen) terminal category $\mathbf{1}$.

   An $\mathbb{L}$-object is also called a *microcosm model* of $\mathbb{L}$, emphasizing that it is a model that resides in another model $\mathbb{C}$.

The definition is abstract and it is hard to grasp how it works at a glance. While the reader is referred to [14, 13] for its illustration, we shall point out its highlights.

   An $\mathbb{L}$-object X, as a lax natural transformation, consists of the following data:

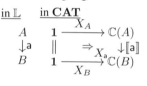

– its components $X_A : \mathbf{1} \to \mathbb{C}(A)$, identified with objects $X_A \in \mathbb{C}(A)$, for each $A \in \mathbb{L}$;
– mediating 2-cells X_{a}, as shown on the right, for each morphism a in $\mathbb{L}$.

Generalizing the illustration in [14, 13] one immediately sees that

– X's components are determined by those $\{X_S\}_{S \in \mathcal{S}}$ on sorts. They extend to an object $S_1 \times \cdots \times S_m$ by:

$$\mathbb{C}(S_1 \times \cdots \times S_m) \ni X_{S_1 \times \cdots \times S_m} \xmapsto{\cong} (X_{S_1}, \ldots, X_{S_m}) \in \mathbb{C}(S_1) \times \cdots \times \mathbb{C}(S_m) \ ;$$

– an operation σ is interpreted on X by means of the mediating 2-cell X_σ;
– equations hold due to the coherence condition on the mediating 2-cells: $X_{\mathsf{b} \circ \mathsf{a}}$ must be given by a suitable composition of X_{a} and X_{b}. In [14, Expl. 3.5] we demonstrate how this coherence condition derives associativity of multiplication for a monoid object (in a monoidal category).

4.3 Categorical Compositionality

Here we shall present a main technical result, namely the compositionality theorem (Thm. 4.6). It is a straightforward many-sorted adaptation of [14, Thm. 3.9]; to which refer for more illustration of the result.

Definition 4.5 ($\mathbb{L}$-functor). Let $\mathbb{C}, \mathbb{D}$ be $\mathbb{L}$-categories. A *lax $\mathbb{L}$-functor* $F : \mathbb{C} \to \mathbb{D}$ is a lax natural transformation $F : \mathbb{C} \Rightarrow \mathbb{D} : \mathbb{L} \to \mathbf{CAT}$ that is FP-preserving in the same sense as in Def. 4.4. Similarly, a *strict $\mathbb{L}$-functor* is a strict natural transformation of the same type.

A lax/strict $\mathbb{L}$-functor determines, as its components, a family of functors $\{F_A : \mathbb{C}(A) \to \mathbb{D}(A)\}_{A \in \mathbb{L}}$. Much like the case for an $\mathbb{L}$-object, it is determined by the components $\{F_S : \mathbb{C}(S) \to \mathbb{D}(S)\}_{S \in \mathcal{S}}$ on sorts.

Theorem 4.6 (Compositionality). *Let $\mathbb{C}$ be an $\mathbb{L}$-category, and $F : \mathbb{C} \to \mathbb{C}$ be a lax $\mathbb{L}$-functor. Assume that there is a final coalgebra $\zeta_A : Z_A \overset{\cong}{\to} F_A(Z_A)$ for each $A \in \mathbb{L}$.*

1. *The family $\{\mathbf{Coalg}(F_A)\}_{A \in \mathbb{L}}$ carries an $\mathbb{L}$-category.*
2. *The family $\{\zeta_A \in \mathbf{Coalg}(F_A)\}_{A \in \mathbb{L}}$ carries a microcosm model of $\mathbb{L}$.*
3. *The family $\{\mathbb{C}(A)/Z_A\}_{A \in \mathbb{L}}$ of slice categories carries an $\mathbb{L}$-category.*
4. *The family of functors $\{\mathrm{beh}_A : \mathbf{Coalg}(F_A) \to \mathbb{C}(A)/Z_A\}_{A \in \mathbb{L}}$, where beh_A is defined by coinduction (see on the right), carries a strict $\mathbb{L}$-functor.*

$$
\begin{array}{ccc}
F_A X & \dashrightarrow & F_A(Z_A) \\
c \uparrow & & \text{final} \uparrow \cong \\
X & \dashrightarrow & Z_A \\
& \mathrm{beh}_A(c) &
\end{array}
$$

Proof. The proof is an adaptation of that of [14, Thm. 3.9]; however it involves additional coherence requirements due to the relaxed definition of $\mathbb{L}$-categories. A detailed proof is found in [12]. $\qquad\square$

An informal reading of the theorem is as follows. To get a "nice" interpretation of a component calculus $\mathbb{L}$ by F-coalgebras, it suffices to check that

- the base category $\mathbb{C}$ models $\mathbb{L}$, and
- the functor F is "lax-compatible" with $\mathbb{L}$.

These data interpret $\mathbb{L}$ on the category of coalgebras, yielding composition of components (Point 1.). Final coalgebras acquire canonical inner $\mathbb{L}$-structure, yielding composition of behaviors (2.). Finally, relating the two interpretations, compositionality is guaranteed (4.).

5   Taxonomy of FP-Theories for Component Calculi

Up to now we have kept an FP-theory $\mathbb{L}$ as a parameter and have developed a uniform framework. Now we turn to: concrete models (components as coalgebras); and three concrete FP-theories **PLTh**, **ArrTh** and **MArrTh** that express basic component calculi. The latter two are equipped with different "parallel composition" operations.

Notably the algebraic structure expressed by **ArrTh** is that of (Hughes') *arrow* [15], equivalently that of *Freyd categories* [23], the notions introduced for modeling structured computations in functional programming.

The main result in this section is that the categories $\{\mathbf{Coalg}(F_{I,J})\}_{I,J}$—modeling components—carry **ArrTh**-structure. If additionally the effect monad T is commutative, then $\{\mathbf{Coalg}(F_{I,J})\}_{I,J}$ a forteriori carries the stronger **MArrTh**-structure. These results parallel classic results in categorical semantics of functional programming, investigating (pre)monoidal structure of a Kleisli category.

5.1 The FP-Theories ArrTh, MArrTh

We shall add, to the FP-theory **PLTh** in §3.2, a suitable "parallel composition" operation and equational axioms to obtain the FP-theory **ArrTh**. By imposing stronger equational axioms we get the FP-theory **MArrTh**.

In **ArrTh** one has additional *sideline* operations

$$\text{first}_{I,J,K} \; : \; (I, J) \longrightarrow (I \times K, J \times K) \; , \quad \text{graphically} \quad \boxed{\begin{smallmatrix} \downarrow I \\ a \\ \downarrow J \end{smallmatrix}} \; \overset{\text{first}_{I,J,K}}{\longmapsto} \; \left(\boxed{\begin{smallmatrix} \downarrow I \\ a \\ \downarrow J \end{smallmatrix}} \; \begin{smallmatrix} K \\ \\ \downarrow K \end{smallmatrix} \right)$$

for each $I, J, K \in \mathbb{B}$. The equations regarding these are:

$$\begin{aligned}
\text{first}_{I,J,1} \, a \ggg \text{arr} \, \pi &= \text{arr} \, \pi \ggg a & (\rho\text{-NAT}) \\
\text{first}_{I,J,K} \, a \ggg \text{arr}(\text{id}_J \times f) &= \text{arr}(\text{id}_I \times f) \ggg \text{first}_{I,J,L} \, a & (\text{arr-CENTR}) \\
(\text{first}_{I,J,K \times L} \, a) \ggg (\text{arr} \, \alpha_{J,K,L}) &= (\text{arr} \, \alpha_{I,K,L}) \ggg \text{first}(\text{first} \, a) & (\alpha\text{-NAT}) \\
\text{first}_{I,J,K} (\text{arr} \, f) &= \text{arr}(f \times \text{id}_K) & (\text{arr-PREMON}) \\
\text{first}_{I,K,L} (a \ggg b) &= (\text{first}_{I,J,L} \, a) \ggg (\text{first}_{J,K,L} \, b) & (\text{first-FUNC})
\end{aligned}$$

In the equations, f denotes a morphism in the base category $\mathbb{B}$. In (ρ-NAT), the π on the left is the projection $\pi : J \times 1 \overset{\cong}{\to} J$ in $\mathbb{B}$. In (α-NAT), α's are associativity isomorphisms like $I \times (J \times K) \overset{\cong}{\to} (I \times J) \times K$ in $\mathbb{B}$.

Remark 5.1. In fact the equation (ρ-NAT) can be derived for any projection $\pi : J \times K \to J$ without requiring $K = 1$. However, the special case above has a clearer role in the corresponding premonoidal structure (see §5.2). Namely, it is the naturality requirement of the right-unit isomorphism $\rho_J = \text{arr} \, \pi_J$ with $\pi_J : J \times 1 \overset{\cong}{\to} J$.

In **MArrTh**, instead of the operations first, one has

$$\|_{I,J,K,L} : \; (I, J) \times (K, L) \longrightarrow (I \times K, J \times L) \qquad \textit{synchronous composition}$$

for each $I, J, K, L \in \mathbb{B}$. The equations are:

$$\begin{aligned}
(a \, \| \, b) \ggg (c \, \| \, d) &= (a \ggg c) \, \| \, (b \ggg d) & (\|\text{-FUNC1}) \\
\text{arr} \, \text{id}_I \, \| \, \text{arr} \, \text{id}_J &= \text{arr} \, \text{id}_{I \times J} & (\|\text{-FUNC2}) \\
a \, \| \, (b \, \| \, c) \ggg \text{arr} \, \alpha &= \text{arr} \, \alpha \ggg (a \, \| \, b) \, \| \, c & (\alpha\text{-NAT}) \\
(a \, \| \, \text{arr} \, \text{id}_1) \ggg \text{arr} \, \pi &= \text{arr} \, \pi \ggg a & (\rho\text{-NAT}) \\
\text{arr}(f \times g) &= \text{arr} \, f \, \| \, \text{arr} \, g & (\text{arr-MON}) \\
(a \, \| \, b) \ggg \text{arr} \, \gamma &= \text{arr} \, \gamma \ggg (b \, \| \, a) & (\gamma\text{-NAT})
\end{aligned}$$

Here α's are associativity isomorphisms, and π's are projections like $J \times 1 \overset{\cong}{\to} J$, as in **ArrTh**. The morphisms γ in (γ-NAT) are symmetry isomorphisms like $J \times I \overset{\cong}{\to} I \times J$ in $\mathbb{B}$. One readily derives, from (ρ-NAT) and (γ-NAT), the equation

$$(\text{arr} \, \text{id}_1 \, \| \, a) \ggg \text{arr} \, \pi' = \text{arr} \, \pi' \ggg a \qquad (\lambda\text{-NAT})$$

where π''s are (second) projections like $1 \times I \overset{\cong}{\to} I$.

The theory **MArrTh** is stronger than **ArrTh**. Indeed, the first operator in **ArrTh** can be defined in **MArrTh**, as: $\text{first}_{I,J,K} \, a := a \, \| \, \text{id}_K$.

The reason that we have the first operation in **ArrTh**, instead of $\|$ as in **MArrTh**, should be noted. The first operation in **ArrTh** yields the operation

$$\text{second}_{I,J,K} \; : \; (I, J) \longrightarrow (K \times I, K \times J) \quad \text{by second} \, a = \text{arr} \, \gamma \ggg \text{first} \, a \ggg \text{arr} \, \gamma,$$

where γ's are symmetry isomorphisms. But the equations in **ArrTh** do not derive $\text{second}\, b \ggg \text{first}\, a =$ $\text{first}\, a \ggg \text{second}\, b$; that is, the two systems on the right should not be identified. There are indeed many situations where they are distinct. Assume that we have the global state monad $T = (S \times \_)^S$ as effect in $F_{I,J}$. One can think of a global state $s \in S$ as residing in the ambience of components, unlike local/internal states that are inside components (i.e. coalgebras). When a component executes, it changes the current global state as well as its internal state; hence the order of execution of a and b does matter. In contrast, when one interprets **MArrTh** in $\{\mathbf{Coalg}(F_{I,J})\}_{I,J}$, the natural axiom ($\|$-FUNC1) requires the two systems above to be equal.

5.2 Set-Theoretic Models: Arrows, Freyd Categories

The FP-theories **PLTh**, **ArrTh** and **MArrTh** and their set-theoretic models are closely related with some notions in semantics of functional programming. Here we elaborate on the relationship; it will be used for calculations later in §5.3.

To start with, **ArrTh** is almost exactly the axiomatization of Hughes' *arrow* [15] (specifically the one in [20]), with the only gap explained in Rem. 5.1. The notion of arrow generalizes that of *monad* (modeling effects, i.e. structured output [25,31]) and that of *comonad* (modeling structured input [30]); the notion of arrow models "structured computations" in general. See e.g. [20, §2.3].

Definition 5.2. An *arrow* is a set-theoretic model (Def. 3.2) of **ArrTh**.

It had been folklore, and was proved in [20], that an arrow is the same thing as a *Freyd category* [27, 23]. A Freyd category is a symmetric premonoidal category $\mathbb{K}$ together with a cartesian category $\mathbb{B}$ embedded via an identity-on-object strict premonoidal functor $\mathbb{B} \to \mathbb{K}$, subject to a condition on center morphisms. In this sense **ArrTh** is an axiomatization of Freyd categories. There is a similar corresponding structure for the stronger FP-theory **MArrTh**, which derives its name (M for *monoidal*).

Definition 5.3. Let $\mathbb{B}$ be a cartesian (hence monoidal) category. A *monoidal Freyd category* on $\mathbb{B}$ is a symmetric monoidal category $\mathbb{K}$ together with a strictly monoidal, identity-on-object functor $\mathbb{B} \to \mathbb{K}$.

Proposition 5.4. *1. A set-theoretic model of* **ArrTh** *is the same as a Freyd category.*
  2. A set-theoretic model of **MArrTh** *is the same thing as a monoidal Freyd category.*

Proof. The first point is simply the correspondence result [20, Thm. 6.1] restated, using Def. 4.4. The proof of the second point goes similar.   □

The notions of (monoidal) Freyd category were introduced in [27], prior to arrows, as axiomatizations of the structure possessed by a Kleisli category $\mathcal{K}\ell(T)$ for a monad T. Here a Kleisli category is the category of types and effectful computations [25]. The results [27, Cor. 4.2 & 4.3]—showing that $\mathcal{K}\ell(T)$ indeed possesses such structure—now read as follows. For the notions of strong/commutative monad, see e.g. [17, §3].

Proposition 5.5. *Let $\mathbb{B}$ be our base category (see Assumption 3.4).*

1. *A monad T on* **Set** *induces a model $\mathcal{K}\ell(T)$:* **ArrTh** $\to$ **Set**. *Specifically, its carrier set $\mathcal{K}\ell(T)(I,J)$ for a sort (I,J) is the homset $\mathrm{Hom}_{\mathcal{K}\ell(T)}(I,J)$; arr is interpreted by the Kleisli inclusion functor; $\ggg$ is by composition in $\mathcal{K}\ell(T)$; and* first *is obtained using T's canonical strength* st. *Note that every monad on* **Set** *is strong.*
2. *Furthermore, if T is commutative then it induces a model $\mathcal{K}\ell(T)$ of* **MArrTh**. *The operation $\|$ is interpreted using T's double strength.* □

Thanks to the proposition we know that all the equations in **ArrTh** or in **MArrTh** hold in $\mathcal{K}\ell(T)$. This will be heavily exploited in the equational reasoning later in §5.3.

For **PLTh**—the parallel-free part of **ArrTh** and **MArrTh**—a set-theoretic model is an arrow without first, or equivalently, a category $\mathbb{K}$ with an identity-on-object functor $\mathbb{B} \to \mathbb{K}$. Yet another characterization of this structure, discovered in [20], is as a monoid object in the monoidal category of bifunctors $[\mathbb{B}^{\mathrm{op}} \times \mathbb{B}, \mathbf{Set}]$. Equivalently it is a monad on $\mathbb{B}$ in the bicategory of profunctors (also called distributors, see e.g. [6,9]).

5.3 PLTh, ArrTh and MArrTh as Component Calculi

We now show that our coalgebraic modeling of components indeed models the calculi **PLTh**, **ArrTh** and **MArrTh**, depending on the monad T. The result parallels Prop. 5.5. Throughout the rest of the section we denote by $\mathbb{L}_{\mathrm{CC}}$ any one of **PLTh**, **ArrTh** and **MArrTh** (CC for "component calculus").

In view of Thm. 4.6, we only need to establish that: 1) the (constant) map $(I,J) \mapsto$ **Set** extends to an $\mathbb{L}_{\mathrm{CC}}$-category; and 2) $\{F_{I,J} : \mathbf{Set} \to \mathbf{Set}\}_{I,J}$ extends to a lax $\mathbb{L}_{\mathrm{CC}}$-functor. Then Thm 4.6 ensures that the components (as coalgebras) and their behaviors (as elements of final coalgebras) carry a microcosm model of $\mathbb{L}_{\mathrm{CC}}$, and that compositionality holds between the two levels.

For 1), we interpret the operations in the following way. The guiding question is: what is the state space of the resulting component.

Definition 5.6. We denote by **Set** the $\mathbb{L}_{\mathrm{CC}}$-category defined as follows. It maps each sort (I,J) to **Set** $\in$ **CAT**; and it interprets operations by

$$1 \overset{[\![\mathrm{arr}\,f]\!]}{\longrightarrow} \mathbf{Set}, \qquad \mathbf{Set} \overset{[\![\mathrm{first}]\!]}{\longrightarrow} \mathbf{Set}, \qquad \mathbf{Set} \times \mathbf{Set} \overset{[\![\ggg]\!]}{\longrightarrow} \mathbf{Set}, \qquad \mathbf{Set} \times \mathbf{Set} \overset{[\![\|]\!]}{\longrightarrow} \mathbf{Set}.$$
$$* \longmapsto 1 \qquad\qquad X \longmapsto X \qquad (X,Y) \longmapsto X \times Y \qquad (X,Y) \longmapsto X \times Y$$

We shall use the same notation **Set** for models of three different FP-theories.

Lemma 5.7. *The data in Def. 5.6 indeed determine an FP-preserving pseudo functor. In particular, all the equations in* **PLTh**, **ArrTh** *and* **MArrTh** *are satisfied up-to coherent isomorphisms.* □

The equations hold only up-to isomorphisms because, for example, the associativity $X \times (Y \times U) \cong (X \times Y) \times U$ in **Set** is only an isomorphism.

The requirement 2)—that $\{F_{I,J} : \mathbf{Set} \to \mathbf{Set}\}_{I,J\in\mathbb{B}}$ extends to a lax $\mathbb{L}_{\mathrm{CC}}$-functor— puts additional demands on a monad T which appears as a parameter in

$F_{I,J} = (T(J \times \_))^I$. This is parallel to Prop. 5.5. Still the actual calculation is overwhelming. We notice that all the operations that appear throughout the calculation can be described as morphisms in the Kleisli category $\mathcal{K}\ell(T)$. Therefore, by Prop. 5.5, they are themselves subject to the equations in $\mathbb{L}_{CC}$. This substantially simplifies the calculation.

Lemma 5.8. *For the endofunctors $F_{I,J} = (T(J \times \_))^I$, the following hold.*

1. *The family $\{F_{I,J} : \mathbf{Set} \to \mathbf{Set}\}_{I,J}$ extends to a lax $\mathbf{ArrTh}$-functor $\mathbf{Set} \to \mathbf{Set}$. Therefore so it does to a lax $\mathbf{PLTh}$-functor.*
2. *If T is commutative, $\{F_{I,J} : \mathbf{Set} \to \mathbf{Set}\}_{I,J}$ extends to a lax $\mathbf{MArrTh}$-functor $\mathbf{Set} \to \mathbf{Set}$.*

Proof. What we need to do is: first to "interpret operations" on $\{F_{I,J}\}_{I,J}$; second to check if these interpreted operations "satisfy equations."

More specifically, to make $\{F_{I,J}\}_{I,J}$ into a lax $\mathbb{L}_{CC}$-functor, we need to have a mediating 2-cell corresponding to each operation (such as $\ggg$). For example:

$$\begin{array}{ccc}
\text{in } \mathbb{L}_{CC} \quad (I,J) \times (J,K) & \text{in } \mathbf{CAT} \quad \mathbf{Set} \times \mathbf{Set} \xrightarrow{F_{I,J} \times F_{J,K}} \mathbf{Set} \times \mathbf{Set} & \\
\downarrow\ggg & [\![\ggg]\!] = \times \downarrow \quad\quad \Downarrow F_{\ggg} \quad\quad \downarrow [\![\ggg]\!] = \times & = \times \quad (8) \\
(I,K) & \mathbf{Set} \xrightarrow{F_{I,K}} \mathbf{Set} &
\end{array}$$

In the diagram, we have denoted the binary product in $\mathbf{Set}$ by the boldface $\times$ to distinguish it from the binary product $\times$ in $\mathbf{CAT}$. The needed 2-cell $F_{\ggg}$ is nothing but a natural transformation

$$F_{\ggg} : F_{I,J}X \times F_{J,K}Y \longrightarrow F_{I,K}(X \times Y) \ . \tag{9}$$

Finding such $F_{\ggg}$ is essentially what we did in the beginning of §2; this time we shall do that for a general monad T (especially involving T's multiplication μ).

After that we have to show that these mediating 2-cells satisfy the coherence condition. In the current setting where the domain category $\mathbb{L}_{CC}$ is syntactically generated from $(\mathcal{S}, \Sigma, E)$, it amounts to checking if the mediating 2-cells "satisfy the equations." Taking $(\ggg\text{-ASSOC})$ as an example, it means that the following 2-cells are equal.

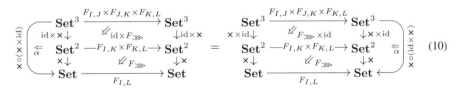

Here α denotes the associativity isomorphism. See [13, Rem. 5.4.1] for more illustration. One readily sees that (10) boils down to commutativity of the following diagram.

$$\begin{array}{c}
F_{I,J}X \times (F_{J,K}Y \times F_{K,L}U) \xrightarrow{\mathrm{id} \times F_{\ggg}} F_{I,J}X \times F_{J,L}(Y \times U) \xrightarrow{F_{\ggg}} F_{I,L}(X \times (Y \times U)) \\
\alpha \downarrow \qquad\qquad\qquad\qquad\qquad\qquad\qquad\qquad\qquad\qquad \downarrow F_{I,L}\alpha \\
(F_{I,J}X \times F_{J,K}Y) \times F_{K,L}U \xrightarrow[F_{\ggg} \times \mathrm{id}]{} F_{I,K}(X \times Y) \times F_{K,L}U \xrightarrow[F_{\ggg}]{} F_{I,L}((X \times Y) \times U)
\end{array} \tag{11}$$

To summarize: we shall introduce a natural transformation F_σ, for each operation σ, like in (9); and check if they satisfy all the equations like in (11). While the first task is straightforward, the second is painfully complicated as it is. We shall only do the aforementioned examples in $\mathbb{L}_{CC}$ for demonstration.

In the sequel, let us denote the set-theoretic model of $\mathbb{L}_{CC}$ induced by the Kleisli category (in Prop. 5.5) by $\mathcal{K}\ell(T)$. In particular, we have $\mathcal{K}\ell(T)(I, J) = (TJ)^I$. Hence the natural transformation $F_{\ggg}$ in (9) is of the type

$$\mathcal{K}\ell(T)(I, J \times X) \times \mathcal{K}\ell(T)(J, K \times Y) \to \mathcal{K}\ell(T)(I, K \times (X \times Y)) .$$

We define it to be the following composition of morphisms in $\mathbb{L}_{CC}$, interpreted in $\mathcal{K}\ell(T)$.

$$
\begin{aligned}
(I, J \times X) \times (J, K \times Y) &\xrightarrow{\text{id} \times \text{first}} (I, J \times X) \times \big(J \times X, (K \times Y) \times X\big) \\
&\overset{\ggg}{\longrightarrow} \big(I, (K \times Y) \times X\big) \xrightarrow{(\_)\ggg(\text{arr}\,\alpha^{-1})} (I, K \times (X \times Y))
\end{aligned}
\tag{12}
$$

One can readily come up with such a morphism in $\mathbb{L}_{CC}$ for each of the other operations.

Let us prove that $F_{\ggg}$ thus defined satisfies (11). The morphisms in the diagram (11) can be also written as morphisms in $\mathbb{L}_{CC}$, interpreted in $\mathcal{K}\ell(T)$. Hence we can use the equational axioms of $\mathbb{L}_{CC}$ (§5.1); in fact they are enough to show commutativity of (11). In the calculation below, we denote the interpretation $[\![\ggg]\!]_{\mathcal{K}\ell(T)}$ simply by $\ggg$; the same for the other operations first and arr. To reduce the number of parentheses, the terms are presented modulo the associativity ($\ggg$-ASSOC).

$$
\begin{aligned}
&\big(F_{\ggg} \circ (F_{\ggg} \times \text{id}) \circ \alpha\big)(c, (d, e)) \\
&= F_{\ggg}\big(F_{\ggg}(c, d), e\big) \\
&= c \ggg \text{first}\, d \ggg \text{arr}\,\alpha^{-1} \ggg \text{first}\, e \ggg \text{arr}\,\alpha^{-1} \\
&= c \ggg \text{first}\, d \ggg \text{first first}\, e \ggg \text{arr}\,\alpha^{-1} \ggg \text{arr}\,\alpha^{-1} &(\alpha\text{-NAT}) \\
&= c \ggg \text{first}\, d \ggg \text{first first}\, e \ggg \text{arr}\,(\alpha^{-1} \circ \alpha^{-1}) &(\text{arr-FUNC1}) \\
&= c \ggg \text{first}\, d \ggg \text{first first}\, e \ggg \text{arr}\,\big((L \times \alpha) \circ \alpha^{-1} \circ (\alpha^{-1} \times U)\big) &(\dagger) \\
&= c \ggg \text{first}\, d \ggg \text{first first}\, e \ggg \text{arr}\,(\alpha^{-1} \times U) \ggg \text{arr}\,\alpha^{-1} \ggg \text{arr}\,(L \times \alpha) &(\text{arr-FUNC1}) \\
&= c \ggg \text{first}\, d \ggg \text{first first}\, e \ggg \text{first}\,(\text{arr}\,\alpha^{-1}) \ggg \text{arr}\,\alpha^{-1} \ggg \text{arr}\,(L \times \alpha) &(\text{first-PREMON}) \\
&= c \ggg \text{first}\big((d \ggg \text{first}\, e) \ggg \text{arr}\,\alpha^{-1}\big) \ggg \text{arr}\,\alpha^{-1} \ggg \text{arr}(L \times \alpha) &(\text{first-FUNC}) \\
&= (F\alpha \circ F_{\ggg})\big(c, F_{\ggg}(d, e)\big) \\
&= \big(F\alpha \circ F_{\ggg} \circ (\text{id} \times F_{\ggg})\big)(c, (d, e))
\end{aligned}
$$

Here the equality ($\dagger$) is because of the "pentagon" coherence for α in **Set**. Naturality of $F_{\ggg}$ in X, Y, as well as satisfaction of the other equations, can be derived by similar calculation.   □

Let us summarize our approach. In the categorical model $\{\mathbf{Coalg}(F_{I,J})\}$ of $\mathbb{L}_{CC}$ that we are interested in, operations are interpreted essentially by $\{F_{I,J}\}$'s lax compatibility with $\mathbb{L}_{CC}$ (Thm. 4.6). Furthermore we notice that the latter can be described using $\mathbb{L}_{CC}$'s specific set-theoretic model $\mathcal{K}\ell(T)$, which we exploit (the above proof). Note that we can employ $\mathcal{K}\ell(T)$ specifically because of the shape $(T(J \times \_))^I$ of $F_{I,J}$.

For obtaining a microcosm model we need final coalgebras. This depends on the "size" of the monad T (see e.g. [26]); all the examples of T listed in §1, except for probabilistic branching and unbounded non-determinism, satisfy this size requirement.

Theorem 5.9. *Assume that, for each $I, J \in \mathbb{B}$, we have a final coalgebra $\zeta_{I,J} : Z_{I,J} \xrightarrow{\cong} F_{I,J}(Z_{I,J})$.*

1. *The family $\{\mathbf{Coalg}(F_{I,J})\}_{I,J}$ carries an $\mathbf{ArrTh}$-category, so a $\mathbf{PLTh}$-category.*
2. *The family $\{Z_{I,J} \in \mathbf{Set}\}_{I,J}$ carries an $\mathbf{ArrTh}$-object, hence a $\mathbf{PLTh}$-object. Compositionality holds in the sense of Thm. 4.6.4.*
3. *If T is commutative, then the above two are also an $\mathbf{MArrTh}$-category and an $\mathbf{MArrTh}$-object, respectively.*

Proof. By Thm. 4.6, Lem. 5.7 and Lem. 5.8.   □

6   Conclusions and Future Work

We have extended our previous formalization of the microcosm principle [14] to a many-sorted setting. This allowed to include Barbosa's component calculi [2] as examples. We studied three concrete calculi that are variants of the axiomatization of arrows, demonstrating similarity between components and structured computations.

As future work, we are interested in further extensions of the component calculi that allow modeling of further interesting examples like wiring and merging components, queues, stacks, and folders of stacks with feedback, etc., as presented in [2]. The proof methods that we derived from the microcosm framework will be useful in its course.

On the more abstract side, it is interesting to elevate the arguments in §5 further, to the bicategory $\mathbf{DIST}$ of distributors, with $\mathbf{CAT}$ embedded in it via the Yoneda embedding. Such a higher level view on the matter might reveal further microcosm instances in our proof methods.

Acknowledgments. Thanks are due to Kazuyuki Asada, Masahito Hasegawa, Paul-André Melliès, John Power and the anonymous referees for helpful discussions and comments.

References

1. Baez, J.C., Dolan, J.: Higher dimensional algebra III: n-categories and the algebra of opetopes. Adv. Math. 135, 145–206 (1998)
2. Barbosa, L.S.: Towards a calculus of state-based software components. Journ. of Universal Comp. Sci. 9(8), 891–909 (2003)
3. Barbosa, L.: Components as Coalgebras. PhD thesis, Univ. Minho (2001)
4. Barr, M., Wells, C.: Toposes, Triples and Theories. Springer, Berlin (1985)
5. Barr, M., Wells, C.: Category Theory for Computing Science, 3rd edn., Centre de recherches mathématiques, Université de Montréal (1999)
6. Bénabou, J.: Distributors at work. Lecture notes by Thomas Streicher (2000), www.mathematik.tu-darmstadt.de/~streicher/FIBR/DiWo.pdf.gz
7. Blackwell, R., Kelly, G., Power, A.: Two-dimensional monad theory. Journ. of Pure & Appl. Algebra 59, 1–41 (1989)
8. Bonsangue, M.M., Rutten, J., Silva, A.: Coalgebraic logic and synthesis of Mealy machines. In: Amadio, R.M. (ed.) FOSSACS 2008. LNCS, vol. 4962, pp. 231–245. Springer, Heidelberg (2008)

9. Borceux, F.: Handbook of Categorical Algebra. Encyclopedia of Mathematics, vol. 50, 51 and 52. Cambridge University Press, Cambridge (1994)
10. Denecke, K., Wismath, S.L.: Universal Algebra and Applications in Theoretical Computer Science. Chapman and Hall, Boca Raton (2002)
11. Fiore, T.M.: Pseudo limits, biadjoints, and pseudo algebras: Categorical foundations of conformal field theory. Memoirs of the AMS 182 (2006)
12. Hasuo, I.: Pseudo functorial semantics (preprint),
    www.kurims.kyoto-u.ac.jp/~ichiro
13. Hasuo, I.: Tracing Anonymity with Coalgebras. PhD thesis, Radboud University Nijmegen (2008)
14. Hasuo, I., Jacobs, B., Sokolova, A.: The microcosm principle and concurrency in coalgebra. In: Amadio, R.M. (ed.) FOSSACS 2008. LNCS, vol. 4962, pp. 246–260. Springer, Heidelberg (2008)
15. Hughes, J.: Generalising monads to arrows. Science of Comput. Progr. 37(1–3), 67–111 (2000)
16. Hyland, M., Power, J.: The category theoretic understanding of universal algebra: Lawvere theories and monads. Elect. Notes in Theor. Comp. Sci. 172, 437–458 (2007)
17. Jacobs, B.: Semantics of weakening and contraction. Ann. Pure & Appl. Logic 69(1), 73–106 (1994)
18. Jacobs, B.: Categorical Logic and Type Theory. North-Holland, Amsterdam (1999)
19. Jacobs, B., Rutten, J.J.M.M.: A tutorial on (co)algebras and (co)induction. EATCS Bulletin 62, 222–259 (1997)
20. Jacobs, B., Heunen, C., Hasuo, I.: Categorical semantics for arrows. Journ. Funct. Progr. (to appear, 2009)
21. Lack, S., Power, J.: Lawvere 2-theories. Presented at CT 2007 (2007),
    www.mat.uc.pt/~categ/ct2007/slides/lack.pdf
22. Lawvere, F.W.: Functorial Semantics of Algebraic Theories and Some Algebraic Problems in the Context of Functorial Semantics of Algebraic Theories. PhD thesis, Columbia University, 1–121 (1963); Reprints in Theory and Applications of Categories, 5 (2004)
23. Levy, P.B., Power, A.J., Thielecke, H.: Modelling environments in call-by-value programming languages. Inf. & Comp. 185(2), 182–210 (2003)
24. Mac Lane, S.: Categories for the Working Mathematician, 2nd edn. Springer, Berlin (1998)
25. Moggi, E.: Notions of computation and monads. Inf. & Comp. 93(1), 55–92 (1991)
26. Pattinson, D.: An introduction to the theory of coalgebras. Course notes for NASSLLI (2003),
    www.indiana.edu/~nasslli
27. Power, J., Robinson, E.: Premonoidal categories and notions of computation. Math. Struct. in Comp. Sci. 7(5), 453–468 (1997)
28. Rutten, J.J.M.M.: Algebraic specification and coalgebraic synthesis of Mealy automata. Elect. Notes in Theor. Comp. Sci. 160, 305–319 (2006)
29. Segal, G.: The definition of conformal field theory. In: Tillmann, U. (ed.) Topology, Geometry and Quantum Field Theory. London Math. Soc. Lect. Note Series, vol. 308, pp. 423–577. Cambridge University Press, Cambridge (2004)
30. Uustalu, T., Vene, V.: Comonadic notions of computation. Elect. Notes in Theor. Comp. Sci. 203(5), 263–284 (2008)
31. Wadler, P.: Monads for functional programming. In: Marktoberdorf Summer School on Program Design Calculi. Springer, Heidelberg (1992)

Complementation of Coalgebra Automata

Christian Kissig[1,*] and Yde Venema[2,**]

[1] Department of Computer Science, University of Leicester, University Road,
Leicester LE1 7RH, United Kingdom
`christian@mcs.le.ac.uk`
[2] Institute for Logic, Language, and Computation, Universiteit van Amsterdam,
Plantage Muidergracht 24, 1018 TV Amsterdam, Netherlands
`Y.Venema@uva.nl`

Abstract. Coalgebra automata, introduced by the second author, generalize the well-known automata that operate on infinite words/streams, trees, graphs or transition systems. This coalgebraic perspective on automata lays foundation to a universal theory of automata operating on infinite models of computation.

In this paper we prove a complementation lemma for coalgebra automata. More specifically, we provide a construction that transforms a given coalgebra automaton with parity acceptance condition into a device of similar type, which accepts exactly those pointed coalgebras that are rejected by the original automaton. Our construction works for automata operating on coalgebras for an arbitrary standard set functor which preserves weak pullbacks and restricts to finite sets.

Our proof is coalgebraic in flavour in that we introduce and use a notion of game bisimilarity between certain kinds of parity games.

1 Introduction

Through their close connection with modal fixpoint logics, automata operating on infinite objects such as words/streams, trees, graphs or transition systems, provide an invaluable tool for the specification and verification of the ongoing behavior of infinite systems. Coalgebra automata, introduced by the second author [9, 10], generalize the well-known automata operating on possibly infinite systems of a specific type. The motivation underlying the introduction of coalgebra automata is to gain a deeper understanding of this branch of automata theory by studying properties of automata in a uniform manner, parametric in the type of the recognized models, that is, the coalgebra functor. The aim is thus to contribute to Universal Coalgebra [8] as a mathematical theory of state-based evolving systems.

* This author would like to thank the Institute for Logic, Language, and Computation, and especially Yde Venema and his Algebra|Coalgebra group for their hospitality during his visit, which this work ensued.
** The research of this author has been made possible by VICI grant 639.073.501 of the Netherlands Organization for Scientific Research (NWO).

A. Kurz, M. Lenisa, and A. Tarlecki (Eds.): CALCO 2009, LNCS 5728, pp. 81–96, 2009.
© Springer-Verlag Berlin Heidelberg 2009

Operationally, coalgebra automata are devices for scanning pointed coalgebras. Structurally the automata rather closely resemble the coalgebras on which they operate; for instance, in the nondeterministic variant, every state in a coalgebra automaton may be seen as capturing various 'realizations' as a T-coalgebra state. The resemblance between coalgebra automata and coalgebras is reflected by the *acceptance game* that is used to determine whether a pointed coalgebra is eventually accepted or rejected by the automaton: This acceptance game, a two-play parity graph game, is a variant of the bisimilarity game [1] that can be played in order to see whether two pointed coalgebras are bisimilar or not.

Earlier work by Kupke and the second author [5] revealed that in fact a large part of the theory of parity automata can be lifted to a coalgebraic level of generality, and thus indeed belongs to the theory of Universal Coalgebra. More specifically, the main result of [5] is a construction tranforming a given alternating coalgebra automaton into an equivalent nondeterministic one; this shows that the nondeterministic model is just as powerful as the alternating one. In addition, coalgebra automata satisfy various *closure properties*: under some conditions on the coalgebra type functor, the collection of recognizable languages (that is, classes of pointed coalgebras that are recognized by some coalgebra automaton) are closed under taking unions, intersections, and existential projections. These results have many applications in the theory of coalgebraic fixpoint logics.

The question whether coalgebra automata admit a *Complementation Lemma* was left as on open problem in [5]. Closure under complementation does not obviously hold, since even for alternating automata the role of the two players in the acceptance game is *not* symmetric. Nevertheless, in this paper we provide a positive answer to the complementation problem, under the same conditions[1] on the functor as in [5]. More precisely, we will prove the following theorem.

Theorem 1 (Complementation Lemma). *Let T be a standard set functor that restricts to finite sets and preserves weak pullbacks. Then the class of recognizable T-languages is closed under complementation.*

Our proof of this Complementation Lemma will be based on an explicit construction which transforms a T-automaton $\mathbb{A}$ into a T-automaton $\mathbb{A}^c$ which accepts exactly those pointed T-coalgebras that are rejected by $\mathbb{A}$. In order to define and use this construction it will be necessary to move from the nondeterministic format of our automata to a wider setting. First of all, since we want to apply the dualization and complementation method of Muller & Schupp [6], we work with automata that are *alternating*, meaning that we increase the role of Abélard in the acceptance game, and *logical* in the sense that the possible unfoldings of an automaton state are expressed as a logical formula rather than as a set of options. Concretely, we would like to move to a setting where the unfolding of an automaton state is a lattice term over nabla-formulas over automaton states,

---

[1] The condition that T should restrict to finite sets is not mentioned in [5], but it is needed there as well, in order to guarantee the correctness of the construction of an equivalent nondeterministic automaton for a given alternating one.

with conjunctions denoting choices for Abélard, disjunctions denoting choices for Éloise, and ∇ denoting Moss' coalgebraic modality. However, it will turn out that by the nature of our construction we also need conjunctions and disjunctions *under* (that is, within the scope of) the modality. Because of this enriched transition structure, the resulting automata will be called *transalternating*.

Concerning these transalternating automata we will prove two results. First, we establish a complementation lemma, by providing a construction which transforms a transalternating automaton $\mathbb{A}$ into a transalternating automaton $\mathbb{A}^c$ which behaves as the complement of $\mathbb{A}$. Note that this construction is linear in size: in fact, $\mathbb{A}^c$ is based on the *same* carrier set as $\mathbb{A}$. And second, we give a construction transforming an arbitrary transalternating automaton into an equivalent alternating one (of size exponential in the size of the original automaton).

At the heart of our construction of the complement automaton lies a new result in coalgebraic modal logic that we call the *One-step Complementation*[2] *Lemma*. This Lemma states that the *Boolean dual* Δ of Moss' modality ∇ can be expressed using disjunctions, Moss' modality itself, and conjunctions. Here 'Boolean dual' refers to the fact that we want for a given formula $\nabla\alpha$ that

$$\Delta\alpha \equiv \neg\nabla(T\neg)(\alpha), \tag{1}$$

For instance, in the case of the power set functor $\mathcal{P}$, defining, for a nonempty α,

$$\Delta\alpha := \nabla\varnothing \vee \bigvee\{\nabla\{a\} \mid a \in \alpha\} \vee \nabla\{\bigwedge\alpha, \top\}$$

we indeed obtain that

$$\Delta\{a_1, \ldots, a_n\} \equiv \neg\nabla\{\neg a_1, \ldots, \neg a_n\}.$$

In the general case, we will see that for each formula $\nabla\alpha$, we can find a set $Q(\alpha)$ (which only uses finite conjunctions over the ingredient formulas from α), such that the following definition

$$\Delta\alpha := \bigvee\{\nabla\beta \mid \beta \in Q(\alpha)\}$$

indeed provides a Boolean dual Δ for ∇, i.e., for this Δ we may prove (1). Note that in order for this definition to give a proper (finitary!) lattice term, we need the set $Q(\alpha)$ to be finite; it is for this reason that we need the functor T to restrict to finite sets.

Applying the methodology of Muller & Schupp [6] to this coalgebraic dualization, we obtain a very simple definition of the complement automaton. Roughly speaking, we obtain the complement of a transalternating automaton by *dualizing* its transition map, and performing a *role switch* on its priority map.

---

[2] Perhaps 'One-step Dualization Lemma' might have been a more adequate name — we chose against this because in the context of coalgebraic logic the word 'dual' has strong connotations towards Stone-type dualities.

Then, in order to *prove* that $\mathbb{A}^c$ is the complement of $\mathbb{A}$, we compare, for an arbitrary pointed coalgebra $(\mathbb{S}, s)$, the acceptance games $\mathcal{G}(\mathbb{A}, \mathbb{S})$ and $\mathcal{G}(\mathbb{A}^c, \mathbb{S})$. We will show that

$$(a_I, s) \in Win_{\exists}(\mathcal{G}(\mathbb{A}^c, \mathbb{S})) \text{ iff } (a_I, s) \in Win_{\forall}(\mathcal{G}(\mathbb{A}, \mathbb{S})).$$

In order to streamline the proof of this result, we introduce a notion of equivalence between positions in parity games. We base this definition on well-known ideas from game theory (for instance, van Benthem's notion of power of players to achieve certain outcomes in finite games [2], and Pauly's bisimulation for coalition logics [7]), adding some features taking care of the acceptance condition. With the resulting notion of 'basic game bisimilarity', we may exploit some of the coalgebraic intuitions we have of parity graph games.

2   Preliminaries

For background information on coalgebra automata, we refer to [10,5]. We will also use notation and terminology from that paper, with the following minor deviations and additions.

Categorical notions. Throughout the paper we asssume that T is a set functor, that is, an endofunctor on the category Set of sets as objects and functions as arrows. This functor is supposed to be (i) standard (that is to preserve embeddings), (ii) to preserve weak pullbacks, and (iii) to restrict to finite sets. We let T_ω denote the finitary version of T, given on objects by $T_\omega X := \bigcup \{T X_0 \mid X_0 \subseteq_\omega X\}$, while on arrows $T_\omega f := Tf$. Given an object $\alpha \in T_\omega X$, we let $Base(\alpha)$ denote the smallest subset $X_0 \subseteq_\omega X$ such that $\alpha \in T_\omega(X_0)$.

Preserving weak pullbacks, T extends to a unique relator $\overline{T}$, which is monotone when T is standard. For a given relation $Z \subseteq X \times Y$, we denote the lifted relation as $\overline{T}Z \subseteq TX \times TY$. Without warning we will assume that $\overline{T}$ is monotone, that it is a relator (that is, it preserves the diagonal relation and distributes over relation composition), and that it commutes with taking relation converse.

Games. Throughout the paper, when discussing parity games, the players will be $0 = \exists$ and $1 = \forall$. We will denote an arbitrary player as Π, and use Σ to denote $\Pi's$ opponent.

Coalgebra automata. As mentioned we will work with coalgebra automata in logical format [10, section 4.5]. It will be convenient to make explicit reference to Moss' modality.

Definition 1. *The functor* $\mathcal{L}$: Set → Set *takes a set* Q *to the carrier* $\mathcal{L}(Q)$ *of the free lattice term algebra over* Q. *Lattice terms* t *are respectively of the form*

$$a ::= q \in Q \mid \bigwedge \tau \mid \bigvee \tau,$$

Table 1. Acceptance game for alternating automata

Position	Player	Admissible Moves	$\Omega_{\mathcal{G}}$
$(q,s) \in Q \times S$	-	$\{(\theta(q), s)\}$	$\Omega(q)$
$(\bigwedge \tau, s) \in \mathcal{L}T^{\nabla}Q \times S$	$\forall$	$\{(a, s) \mid a \in \tau\}$	0
$(\bigvee \tau, s) \in \mathcal{L}T^{\nabla}Q \times S$	$\exists$	$\{(a, s) \mid a \in \tau\}$	0
$(\nabla \alpha, s) \in T^{\nabla}Q \times S$	$\exists$	$\{Z \subseteq Q \times S \mid (\alpha, \sigma(s)) \in \overline{T}Z\}$	0
$Z \subseteq Q \times S$	$\forall$	Z	0

where τ denotes a finite set of lattice terms. Given a set X, let $T^{\nabla}_{\omega}X$ denote the set $\{\nabla \alpha \mid \alpha \in T_{\omega}X\}$, and define $\mathcal{L}_1 Q := \mathcal{L}T^{\nabla}_{\omega}\mathcal{L}(Q)$. Elements of $\mathcal{L}Q$ will be called depth-zero formulas and elements of $\mathcal{L}_1 Q$ depth-one formulas over Q.

Definition 2 (Alternating Automata). An alternating T-automaton is a structure $\mathbb{A} = \langle Q, \theta, q_I, \Omega \rangle$ consisting of a finite set Q of states, a transition function $\theta : Q \to \mathcal{L}T^{\nabla}_{\omega}Q$, a state $q_I \in Q$ distinguished as initial, and a priority function $\Omega : Q \to \mathbb{N}$.

Definition 3 (Acceptance Game). The notion of an alternating T-automaton accepting a pointed T-coalgebra $(\mathbb{S}, s)$ is defined in terms of the parity graph game $\mathcal{G}(\mathbb{A}, \mathbb{S})$ given by Table 1: $\mathbb{A}$ accepts a pointed coalgebra $(\mathbb{S}, s)$ iff $\exists$ has a winning strategy in the acceptance game $\mathcal{G}(\mathbb{A}, \mathbb{S})$ from the initial position (q_I, s). Positions of the form $(q, s) \in Q \times S$ are called basic.

Remark 1. In a basic position (q, s), there is exactly one admissible move, namely, to position $(\theta(q), s)$. As a consequence, technically it does not matter to which player the position is assigned. In Table 1 we have not assigned a player to these positions, since this has some conceptual advantages further on.

One-step semantics. It will be convenient to think of elements of $\mathcal{L}_1 Q$ as formulas indeed, with the following semantics.

Definition 4. Given sets Q and S, a Q-valuation on S is a map $V : Q \to \mathcal{P}(S)$. We define relations $\Vdash^V_0 \subseteq S \times \mathcal{L}Q$ and $\Vdash^V_1 \subseteq TS \times \mathcal{L}_1 Q$, as follows. For $\Vdash^V_0$, we define $s \Vdash^V_0 q$ if $s \in V(q)$, $s \Vdash^V_0 \bigwedge \tau$ ($\bigvee \tau$, respectively) if $s \Vdash^V_0 a$ for all $a \in \tau$ (some $a \in \tau$, respectively); and we define a relation $\Vdash^V_1$ such that $\sigma \Vdash^V_1 \nabla \alpha$ if $(\sigma, \alpha) \in \overline{T}(\Vdash^V_0)$, while for $\bigwedge$ and $\bigvee$ the same clauses apply as for $\Vdash^V_0$.

For clarity of notation we also write $V, s \Vdash_0 a$ for $s \Vdash^V_0 a$, and $V, s \Vdash_1 a$ for $s \Vdash^V_1 a$.

3 One-Step Complementation

The notion of the (Boolean) *dual* of a connective makes a frequent appearance in various branches of logic. For instance, $\wedge$ & $\vee$, and $\diamond$ & $\square$, are well-known pairs of dual operators in propositional and in modal logic, respectively. In this section we introduce the dual, Δ, of Moss' coalgebraic modality, ∇. Building on this we define the dual of a depth-one formula.

Definition 5. *Let T be a standard set functor. Then for each set Q, and each $\alpha \in T_\omega Q$, we define the set $D(\alpha) \subseteq T_\omega \mathcal{P}Q$ as follows:*

$$D(\alpha) := \Big\{ \Phi \in T_\omega \mathcal{P}_\omega Base(\alpha) \mid (\alpha, \Phi) \notin (\overline{T} \not\in) \Big\}$$

In case T restricts to finite sets, define $\Delta\alpha$ as the following formula:

$$\Delta\alpha := \bigvee \Big\{ \nabla T{\bigwedge}(\Phi) \mid \Phi \in D(\alpha) \Big\}.$$

Here we see the connective $\bigwedge$ as a map $\bigwedge : \mathcal{P}_\omega Q \to Q$, so that $T\bigwedge : T_\omega \mathcal{P}_\omega Q \to T_\omega Q$.

We claim that Δ is the Boolean dual of ∇. In case we are dealing with a full Boolean language, we can simply express this fact as in (1). In a setting where negation is not available as a connective, we can formulate the notion of Boolean duality as follows.

Definition 6. *Given a valuation $V : Q \to \mathcal{P}(S)$, we define the* complementary *valuation V^c by putting $V^c(q) := S \setminus V(q)$. Two formulas a, b in $\mathcal{L}Q$ (in $\mathcal{L}_1 Q$, respectively) are* (Boolean) duals *if for all sets S and all Q-valuations on S, $S \setminus \widehat{V^c}(a) = \widehat{V}(b)$ (respectively, if $TS \setminus \widetilde{V^c}(a) = \widetilde{V}(b)$).*

Putting it differently, two depth-zero formulas a and b are Boolean duals iff for all sets S, all valuations V, and all $s \in S$: $V^c, s \Vdash_0 a$ iff $V, s \Vdash_0 b$. Likewise, two depth-one formulas a and b are Boolean duals iff for all sets S, all valuations V, and all $\sigma \in TS$: $V^c, \sigma \nVdash_1 a$ iff $V, \sigma \Vdash_1 b$.

Theorem 2 (One-Step Complementation Lemma). *Let T be a standard set functor preserving weak pullbacks, and let $\alpha \in T_\omega Q$ for some set Q. Then $\nabla\alpha$ and $\Delta\alpha$ are Boolean duals.*

Proof. Fix an arbitrary set S, an arbitrary Q-valuation V on S, and an arbitrary element σ of TS.

First assume that $V, \sigma \Vdash_1 \Delta\alpha$, that is, $V, \sigma \Vdash_1 \nabla(T\bigwedge)(\Phi)$ for some $\Phi \in D(\alpha)$. Then there is some relation $Y \subseteq \mathcal{P}Q \times S$ such that $Y \subseteq (Gr \bigwedge) \circ (\Vdash_0^V)^\smile$ and $(\Phi, \sigma) \in \overline{T}Y$. In order to show that $V^c, \sigma \nVdash_1 \nabla\alpha$, suppose for contradiction that there is some relation Z such that $(\sigma, \alpha) \in \overline{T}Z$ and $V^c, t \Vdash_0 q$ for all pairs $(t, q) \in Z$. It follows that $(\sigma, \alpha) \in \overline{T}Z$ and $Z \cap \Vdash_0^V = \varnothing$.

Now consider the relation $R := Y \circ Z \subseteq \mathcal{P}Q \times Q$, then clearly $(\Phi, \alpha) \in \overline{T}R = \overline{T}Y \circ \overline{T}Z$. On the other hand, it follows from the definition of R that $R \subseteq \not\ni$, because for any $(\phi, q) \in R$ there is an $s \in S$ such that (i) $(\phi, s) \in Y$ implying $V, s \Vdash_0 p$ for all $p \in \phi$, and (ii) $(s, q) \in Z$ meaning that $V, s \nVdash_0 q$. But this gives the desired contradiction since $\Phi \in D(\alpha)$.

Conversely, assume that $V^c, \sigma \nVdash_1 \nabla\alpha$. In order to show that $V, \sigma \Vdash_1 \Delta\alpha$ we need to find some $\Phi \in D(\alpha)$ such that $V, \sigma \Vdash_1 \nabla(T\bigwedge)(\Phi)$. For this purpose, define a map $\phi : S \to \mathcal{P}Base(\alpha)$ by putting, for any $s \in S$, $\phi_s := \{q \in Base(\alpha) \mid V, s \Vdash_0 q\}$.

We claim that $\Phi := T\phi(\sigma)$ has the required properties. First of all, it follows by construction that $Gr(\bigwedge \circ \phi) \subseteq {\Vdash}_0^V$, so that $Gr(T\phi) \circ Gr(T\bigwedge) \subseteq \overline{T}{\Vdash}_0^V$. From this it is immediate that $V, \sigma \Vdash_1 \nabla T \bigwedge(\Phi)$. It remains to show that $\Phi \in D(\alpha)$. For that purpose, consider the relation $Z := Gr(\phi) \circ \not\ni \subseteq S \times Q$. It is easily verified that $V^c, s \Vdash_0 q$ for all $(s, q) \in Z$. Hence, we may derive from the assumption $V^c, \sigma \not\Vdash_1 \nabla\alpha$ that $(\sigma, \alpha) \notin \overline{T}Z = Gr(T\phi) \circ (\overline{T}\not\ni)$. But then it follows from $(\sigma, \Phi) \in Gr(T\phi)$ that $(\Phi, \alpha) \notin (\overline{T}\not\ni)$, as required.

It is an almost immediate consequence of this result, that negation can be defined as an abbreviated connective in the finitary version of Moss' language (with finite conjunctions, disjunctions, and ∇). As we will see, the main result of this paper, viz., the complementation lemma for coalgebra automata, is also a direct corollary of the one-step complementation lemma. As a first step towards proving that statement, let us define the dual of a one-step formula.

Definition 7. *Given a set Q we define the* base dualization *map $\delta_0 : \mathcal{L}Q \to \mathcal{L}Q$ and the* one-step dualization *map $\delta_1 : \mathcal{L}_1 Q \to \mathcal{L}_1 Q$ as follows:*

$$\begin{aligned} \delta_0(q) &:= q & \delta_1(\nabla\alpha) &:= \Delta(T\delta_0)\alpha \\ \delta_0(\bigwedge \phi) &:= \bigvee \delta_0[\phi] & \delta_1(\bigwedge \phi) &:= \bigvee \delta_1[\phi] \\ \delta_0(\bigvee \phi) &:= \bigwedge \delta_0[\phi] & \delta_1(\bigvee \phi) &:= \bigwedge \delta_1[\phi] \end{aligned}$$

Example 1. With $T = \mathcal{P}$ and $\alpha = \{q_1 \vee q_2, q_3 \wedge q_4\}$, we may calculate that

$$\begin{aligned} \delta_1(\nabla\alpha) &= \Delta\{q_1 \wedge q_2, q_3 \vee q_4\} \\ &= \nabla\varnothing \vee \nabla\{q_1 \wedge q_2\} \vee \nabla\{q_3 \vee q_4\} \vee \nabla\{(q_1 \wedge q_2) \wedge (q_3 \vee q_4), \top\}. \end{aligned}$$

The following corollary of the One-Step Complementation lemma states that these dualization operations indeed send formulas to their one-step Boolean duals. The proof of this result is left for the reader.

Corollary 1. *The formulas a and $\delta_1(a)$ are Boolean duals, for any $a \in \mathcal{L}_1 Q$.*

4 Game Bisimulations

In the preliminaries we introduced acceptance games of coalgebra automata as specific parity graph games. In acceptance games we distinguish some positions as basic, which allows us to partition plays into rounds, each delimited by basic positions. From this observation we derive the following definition of basic sets.

Definition 8 (Basic Sets of Positions). *Given a parity graph game $\mathcal{G} = \langle V_0, V_1, E, v_I, \Omega \rangle$, call a set $B \subseteq V_0 \cup V_1$ of positions* basic *if*

1. *the initial position of $\mathcal{G}$ belongs to B;*
2. *any full play starting at some $b \in B$ either ends in a terminal position or it passes through another position in B; and*
3. *$\Omega(v) = 0$ iff $v \notin B$.*

Another way of expressing condition 2 is to say that there are no infinite paths of which the first position is the only basic one. The third condition is there to ensure that who wins an infinite match is determined by the sequence of basic positions induced by the match. It should be clear that the collection of basic positions in acceptance games of coalgebra automata indeed qualifies as a basic set in the sense of Definition 8. By a slight abuse of language, we refer to elements of any basic set as basic positions.

The point behind the introduction of basic positions is that, just as in the special case of acceptance games, we may think of parity games as proceeding in rounds that start and finish at a basic position. Formally we define these rounds, that correspond to subgraphs of the arena, by their unravelling as follows.

Definition 9 (Local Game Trees). *Let B be a basic set related to some parity game $\mathcal{G} = \langle V_0, V_1, E, v_I, \Omega \rangle$. The* local game tree *associated with a basic position $b \in B$ is defined as the following bipartite tree $T^b = \langle V_0^b, V_1^b, E^b, v_I^b \rangle$. Let V^b be the set of those finite paths β starting at b, of which the only basic positions are $first(\beta) = b$ and possibly $last(\beta)$. The bipartition of V^b is given through the bipartition of V, that is $V_\Pi^b := \{\beta \in V^b \mid last(\beta) \in V_\Pi\}$ for both players $\Pi \in \{0, 1\}$. The root v_I^b is $\langle b \rangle$, the path beginning and ending in b. The edge relation is defined as $E^b := \{(\beta, \beta v) \mid v \in E(last(\beta))\}$.*

A node $\beta \in V^b$ is a leaf of T^b if $|\beta| > 1$ and $last(\beta) \in B$; we let $Leaves(T^b)$ denote the set of leaves of T^b, and put $N(b) := \{last(\beta) \mid \beta \in Leaves(T^b)\}$.

Intuitively, T^b can be seen as the tree representation of one round of the game $\mathcal{G}$, starting at b. $N(b) \subseteq V$ denotes the set of positions one may encounter in a match of $\mathcal{G}$ as the next basic position after b — this intuition will be made more precise in Lemma 1 below. It follows from the definition that all paths in T^b are finite, and hence we may use (bottom-up) induction on the immediate successor relation of the tree, or, as we shall say, on the *height* of nodes.

Once we have established the dissection of a parity graph game through basic positions, we may think of the game being in an iterative strategic normal form, in the sense that in each round players make only one choice, determining their complete strategies for that round right at the beginning. We may formalize this using the game-theoretic notion of *power*, which describes the terminal positions in a finitary graph game which a player can force, for instance by deploying a strategy; we refer to van Benthem [2] for an extensive discussion and pointers to the literature. In the following we define the notion of power of a player Π at a basic position b as a collection of sets of basic nodes. Intuitively, when we say that U is in Π's power at basic position b, we mean that at position b, Π has a local strategy ensuring that the next basic position belongs to the set U. Hence, the collection of sets that are in Π's power at a certain basic position is closed under taking supersets. In the definition below we define a collection $P_\Pi(b) \subseteq \mathcal{P}B$. Further on we will see that a set $U \subseteq B$ is in Π's power at b, in the sense described above, iff $U \supseteq V$ for some $V \in P_\Pi(b)$. It will be convenient to use the following notation.

Definition 10. *Given a set $\mathcal{H} \subseteq \mathcal{P}B$, define*

$$\Uparrow(\mathcal{H}) := \{V \subseteq B \mid V \supseteq U \text{ for some } U \in \mathcal{H}\}.$$

Definition 11 (Powers). *Let B be a basic set related to some parity game $\mathcal{G} = \langle V_0, V_1, E, v_I, \Omega \rangle$, and let b be a basic position. By induction on the height of a node $\beta \in V^b$, we define, for each player Π, the* power of Π at β *as a collection $P_\Pi(\beta)$ of subsets of $N(b)$:*

– *If $\beta \in Leaves(T^b)$, we put, for each player,*

$$P_\Pi(\beta) := \Big\{ \{last(\beta)\} \Big\}.$$

– *If $\beta \notin Leaves(T^b)$, we put*

$$P_\Pi(\beta) := \begin{cases} \bigcup \{ P_\Pi(\gamma) \mid \gamma \in E^b(\beta) \} & \text{if } \beta \in V_\Pi^b, \\ \left\{ \bigcup_{\gamma \in E^b(\beta)} Y_\gamma \mid Y_\gamma \in P_\Pi(\gamma), \text{ all } \gamma \right\} & \text{if } \beta \in V_\Sigma^b. \end{cases}$$

where Σ denotes the opponent of Π.

Finally, we define the power of Π at b *as the set $P_\Pi(b) := P_\Pi(\langle b \rangle)$, where $\langle b \rangle$ is the path beginning and ending at b.*

Perhaps some special attention should be devoted to the paths β in T^b such that $E^b(\beta) = \varnothing$. If such a β is a leaf of T^b, then the definition above gives $P_\Pi(\beta) = P_\Sigma(\beta) = \Big\{ \{last(\beta)\} \Big\}$. But if β is not a leaf of T^b, then we obtain, by the inductive clause of the definition:

$$P_\Pi(\beta) := \begin{cases} \varnothing & \text{if } \beta \in V_\Pi^b, \\ \{\varnothing\} & \text{if } \beta \in V_\Sigma^b. \end{cases}$$

This indeed confirms our intuition that at such a position, the player who is to move gets stuck and loses the match.

The following definitions make the notion of a local game more precise.

Definition 12 (Local Games). *Let $\mathcal{G} = \langle V_0, V_1, E, v_I, \Omega \rangle$ be a parity graph game with basic set $B \subseteq V$. We define a game $\mathcal{G}^b$* local *to a basic position $b \in B$ in $\mathcal{G}$ as the (finite length) graph game played by 0 and 1 on the local game tree $T^b = \langle V^b, E^b \rangle$. Matches of this game are won by a player Π if their opponent Σ gets stuck, and end in a tie if the last position of the match is a leaf of T^b.*

Definition 13 (Local Strategies). *A* local strategy *of a player Π in a local game $\mathcal{G}^b = \langle V_0^b, V_1^b, E^b, v_I^b \rangle$ is a partial function $f : V_\Pi^b \to V^b$ defined on such β iff $E^b(\beta)$ is non-empty, then $f(\beta) \in E^b(\beta)$. Such a local strategy for player Π is* surviving *if it guarantees that Π will not get stuck, and thus does not lose the local game; and a local strategy is* winning *for Π if it guarantees that her opponent Σ gets stuck.*

Consider the match of the local game $\mathcal{G}^b$ in which 0 and 1 play local strategies f_0 and f_1, respectively. If this match ends in a leaf β of the local game tree, we let $Res(f_0, f_1)$ denote the basic position $last(\beta)$; if one of the players gets stuck in this match, we leave $Res(f_0, f_1)$ undefined. Given a local strategy f_0 for player 0, we define

$$X_{f_0} := \{\, Res(f_0, f_1) \mid f_1 \text{ a local strategy for player 1}\,\},$$

and similarly we define X_{f_1} for a strategy f_1 for player 1.

Local strategies for Π in $\mathcal{G}^b$ can be linked to (fragments of) strategies for Π in $\mathcal{G}$. Since these links are generally obvious, we will refrain from introducing notation and terminology here.

The following lemma makes precise the links between players' power and their local strategies.

Proposition 1. *Let $\mathcal{G}$, B and b as in Definition 11, let $\Pi \in \{0,1\}$ be a player, and let W be a subset of $N(b)$. Then the following are equivalent:*

1. *$W \in P_\Pi(b)$;*
2. *Π has a surviving strategy f in $\mathcal{G}^b$ such that $W = X_f$.*

In the next section we will need the following lemma, which states a determinacy property of local games.

Proposition 2. *Let $\mathcal{G}$, B and b as in Definition 11, and let $\Pi \in \{0,1\}$ be a player. For any subset $U \subseteq N(b)$, if $U \notin {\Uparrow}(P_\Pi(b))$ then there is a $V \in P_\Sigma(b)$ such that $U \cap V = \varnothing$.*

The partitioning of matches of parity games into rounds between basic positions, and the normalization of the players' moves within each round, lay the foundations to the introduction of a structural equivalence between parity games that we refer to as basic game bisimulation. This equivalence combines (i) a structural part that can be seen as an instantiation of Pauly's bisimulation between extensives strategic games [7], or van Benthem's *power bisimulation* [2], with a combinatorial part that takes care of the parity acceptance condition.

Definition 14 (Basic Game Bisimulation). *Let $\mathcal{G} = \langle V_0, V_1, E, \Omega \rangle$ and $\mathcal{G}' = \langle V_0', V_1', E', \Omega' \rangle$ be parity graph games with basic sets B and B', respectively, and let Π and Π' be (not necessarily distinct) players in $\mathcal{G}$ and $\mathcal{G}'$, respectively.*

A Π, Π'-game bisimulation is a binary relation $Z \subseteq B \times B'$ satisfying for all $v \in V$ and $v' \in V'$ with vZv' the structural conditions

- *(Π,forth) $\forall W \in P_\Pi^{\mathcal{G}}(v).\exists W' \in P_{\Pi'}^{\mathcal{G}'}(v').\forall w' \in W'.\exists w \in W.\ (w, w') \in Z$,*
- *(Σ,forth) $\forall W \in P_\Sigma^{\mathcal{G}}(v).\exists W' \in P_{\Sigma'}^{\mathcal{G}'}(v').\ \forall w' \in W'.\exists w \in W.(w, w') \in Z$,*
- *(Π,back) $\forall W' \in P_\Pi^{\mathcal{G}'}(v').\exists W \in P_{\Pi'}^{\mathcal{G}}(v).\forall w \in W.\exists w' \in W.\ (w, w') \in Z$,*
- *(Σ,back) $\forall W' \in P_\Sigma^{\mathcal{G}'}(v').\exists W \in P_{\Sigma'}^{\mathcal{G}}(v).\forall w \in W.\exists w' \in W.\ (w, w') \in Z$,*

and the priority conditions

- *(parity)* $\Omega(v)$ *mod* $2 = \Pi$ *iff* $\Omega'(v')$ *mod* $2 = \Pi'$,
- *(contraction) for all* $v, w \in V$ *and* $v', w' \in V'$ *with* vZv' *and* wZw', $\Omega(v) \leq$ $\Omega(w)$ *iff* $\Omega(v') \leq \Omega(w')$.

In Condition (parity) above we identify players with their characteristic parity. Note that in fact there are only two kinds of game bisimulations: the $(0,0)$-bisimulations coincide with the $(1,1)$-bisimulations, and the $(0,1)$-bisimulations coincide with the $(1,0)$-bisimulations.

The following theorem bears witness to the fact that game bisimulation is indeed a good notion of structural equivalence between parity games.

Theorem 3. *Let* $\mathcal{G} = \langle V_0, V_1, E, \Omega \rangle$ *and* $\mathcal{G}' = \langle V_0', V_1', E', \Omega' \rangle$ *be parity graph games with basic sets* B *and* B', *respectively, and let* Π *and* Π' *be (not necessarily distinct) players in* $\mathcal{G}$ *and* $\mathcal{G}'$, *respectively. Whenever* v *and* v' *are related through a* Π,Π'*-bisimulation* $Z \subseteq B \times B'$, *we have*

$$v \in Win_\Pi(\mathcal{G}) \text{ iff } v' \in Win_{\Pi'}(\mathcal{G}'). \tag{2}$$

5 Complementation of Alternating Coalgebra Automata

This section is devoted to our main result:

Theorem 4 (Complementation Lemma for Coalgebra Automata). *The class of alternating coalgebra automata is closed under taking complements.*

As mentioned already, based on the approach by Muller & Schupp [6], we will obtain the complement of an alternating automaton by *dualizing* its transition map, and performing a *role switch* on its priority map. More in detail, consider an alternating automaton $\mathbb{A} = \langle Q, q_I, \theta, \Omega \rangle$, with $\theta : Q \to \mathcal{L}T^\nabla Q$. Its complement $\mathbb{A}^c$ will be based on the same carrier set Q and will have the same initial state q_I, while the role switch operation on the priority map can be implemented very simply by putting $\Omega^c(q) := 1 + \Omega(q)$. With *dualizing* the transition map we mean that for each state $q \in Q$, we will define $\theta^c(q)$ as the *dual term* of $\theta(q)$ (as given by Definition 7). However, already when dualizing a simple term $\nabla\alpha$ with $\alpha \in TQ$ (rather than $\alpha \in T\mathcal{L}Q$), we see conjunctions popping up *under* the modal operator ∇. Hence, if we are after a class of automata that is closed under the proposed complementation construction, we need to admit devices with a slightly *richer transition structure*. It follows from the results in section 3 that the set $\mathcal{L}_1 Q$ is the smallest set of formulas containing the set $\mathcal{L}T^\nabla Q$ which is closed under the dualization map δ_1, and hence the 'richer transition structure' that we will propose comprises maps of type $\theta : Q \to \mathcal{L}_1 Q$.

We will call the resulting automata *transalternating* to indicate that there is alternation both under and over the modality ∇. For these *transalternating* automata we devise a simple algorithm to compute complements. Finally we show that the richer transition structure of transalternating automata does not really increase the recognizing power, by providing an explicit construction transforming a transalternating automaton into an equivalent alternating one.

Table 2. Acceptance Games for Transalternating Automata

Position	Player	Admissible Moves	$\Omega_{\mathcal{G}}$
$(q, s) \in Q \times S$	-	$\{(\theta(q), s)\}$	$\Omega(q)$
$(\bigwedge_{i \in I} a_i, s) \in \mathcal{L}_1 Q \times S$	$\forall$	$\{(a_i, s) \mid i \in I\}$	0
$(\bigvee_{i \in I} a_i, s) \in \mathcal{L}_1 Q \times S$	$\exists$	$\{(a_i, s) \mid i \in I\}$	0
$(\nabla \alpha, s) \in T^\nabla \mathcal{L} Q \times S$	$\exists$	$\{Z \subseteq \mathcal{L} Q \times S \mid (\alpha, \sigma(s)) \in \overline{T}Z\}$	0
$Z \subseteq \mathcal{L} Q \times S$	$\forall$	Z	0
$(\bigwedge_{i \in I} a_i, s) \in \mathcal{L} Q \times S$	$\forall$	$\{(a_i, s) \mid i \in I\}$	0
$(\bigvee_{i \in I} a_i, s) \in \mathcal{L} Q \times S$	$\exists$	$\{(a_i, s) \mid i \in I\}$	0

5.1   Transalternating Automata

Formally, transalternating automata are defined as follows.

Definition 15 (Transalternating Automata). *A transalternating T-automaton $\mathbb{A} = \langle Q, \theta, q_I, \Omega \rangle$ consists of a finite set Q of states, a transition function $\theta : Q \to \mathcal{L}_1 Q$, a state $q_I \in Q$ distinguished as initial, and a priority function $\Omega : Q \to \mathbb{N}$.*

Just as for alternating automata, acceptance of a pointed coalgebra $(\mathbb{S}, s)$ is defined in terms of a parity game.

Definition 16 (Acceptance Games for Transalternating Automata). *Let $\mathbb{A} = \langle Q, q_I, \theta, \Omega \rangle$ be a transalternating automaton and let $\mathbb{S} = \langle S, \sigma \rangle$ be a T-coalgebra. The acceptance game $\mathcal{G}(\mathbb{A}, \mathbb{S})$ is the parity graph game $\mathcal{G}(\mathbb{A}, \mathbb{S}) = \langle V_\exists, V_\forall, E, v_I, \Omega_{\mathcal{G}} \rangle$ with $v_I = (q_I, s_I)$ and $V_\exists, V_\forall, E$ and $\Omega_{\mathcal{G}}$ given by Table 2. Positions from $Q \times S$ will be called basic.*

In the next subsection we will define the complementation construction for transalternating automata and show the correctness of this transformation. Our proof will be based on a game bisimulation between two acceptance games, and for that purpose we need to understand the powers of the two players in a single round of such an acceptance game. The next Lemma shows that these powers can be conveniently expressed and understood in terms of the one-step semantics defined in section 3. In order to make sense of this remark, recall that the *basic* positions in an acceptance game are of the form $(q, s) \in Q \times S$. Also, note that this terminology is justified because the set $Q \times S$ is indeed basic in the sense of Definition 8. Hence, we may apply the notions defined in section 4, and in particular, given a basic position (q, s) we may consider the question, what it means for a set $Z \subseteq N(q, s)$ to belong to the set $P_\Pi(q, s)$ for one of the players $\Pi \in \{\exists, \forall\}$. First note that potential elements of $P_\Pi(q, s)$ are sets of basic positions, and as such, binary relations between Q and S. A key observation is that such a binary relation may be identified with a Q-*valuation* on S.

Definition 17. *Given a relation $Z \subseteq Q \times S$, we define the associated Q-valuation on S as the map $V_Z : Q \to \mathcal{P}(S)$ given by*

$$V_Z(q) := \{s \in S \mid (q, s) \in Z\}.$$

Using this perspective on relations between Q and S, we may give the following logical characterization of the players' powers in single rounds of the acceptance game.

Proposition 3. *Let $\mathbb{A}$ be a transalternating T-automaton, and $\mathbb{S}$ a T-coalgebra. Given a basic position (q, s) in $\mathcal{G}(\mathbb{A}, \mathbb{S})$, and a relation $Z \subseteq Q \times S$, we have:*

$$Z \in \Uparrow(P_\exists(q, s)) \iff V_Z, \sigma(s) \Vdash_1 \theta(q)$$
$$Z \in \Uparrow(P_\forall(q, s)) \iff V_Z, \sigma(s) \Vdash_1 \delta_1(\theta(q))$$

5.2   Complementation of Transalternating Automata

As announced, we define the complement of a transalternating automaton as follows.

Definition 18 (Complements of Transalternating Automata). *The complement of a transalternating T-coalgebra automaton $\mathbb{A} = \langle Q, \theta, q_I, \Omega \rangle$ is the transalternating automaton $\mathbb{A}^c = \langle Q, \theta^c, q_I, \Omega^c \rangle$ defined with $\theta^c(q) := \delta_1(\theta(q))$ and $\Omega^c(q) := \Omega(q) + 1$, for all $q \in Q$.*

$\mathbb{A}^c$ is indeed the complement of $\mathbb{A}$.

Proposition 4. *For every transalternating T-coalgebra automaton $\mathbb{A}$, the automaton $\mathbb{A}^c$ accepts precisely those pointed T-coalgebras that are rejected by $\mathbb{A}$.*

Proof. Clearly it suffices to prove, for a given T-coalgebra $\mathbb{S}$, state q of $\mathbb{A}$, and point s in $\mathbb{S}$:

$$(q, s) \in Win_\exists(\mathcal{G}(\mathbb{A}^c, \mathbb{S})) \text{ iff } (q, s) \in Win_\forall(\mathcal{G}(\mathbb{A}, \mathbb{S})).$$

In order to prove this claim we will use a basic game bisimulation.

First we note that $Q \times S$ is a basic set in both acceptance games, $\mathcal{G}(\mathbb{A}, \mathbb{S})$ and $\mathcal{G}(\mathbb{A}^c, \mathbb{S})$. The main observation is that the diagonal relation $Id_{Q \times S} := \{((q, s), (q, s)) \mid q \in Q, s \in S\}$ is an $\forall, \exists$-game bisimulation between $\mathcal{G}(\mathbb{A}, \mathbb{S})$ and $\mathcal{G}(\mathbb{A}^c, \mathbb{S})$. Since it is immediate from the definitions that the diagonal relation satisfies the *priority conditions*, it is left to check the structural conditions.

Leaving the other three conditions for the reader, we establish the condition ($\exists$,forth) as an immediate consequence of the following claim:

$$\text{for all } Z \in P_\exists^\mathcal{G}(q, s) \text{ there is a } Z_\forall \subseteq Z \text{ such that } Z_\forall \in P_\forall^{\mathcal{G}^c}(q, s), \tag{3}$$

which can be proved via the following chain of implications:

$$Z \in P_\exists^{\mathcal{G}}(q,s)$$
$$\Rightarrow V_Z, \sigma(s) \Vdash_1 \theta(q) \qquad\qquad\text{(Proposition 3)}$$
$$\Rightarrow (V_Z)^c, \sigma(s) \nVdash_1 \delta_1(\theta(q)) \qquad\qquad\text{(Corollary 1)}$$
$$\Rightarrow (V_Z)^c, \sigma(s) \nVdash_1 \theta^c(q) \qquad\qquad\text{(definition of }\theta^c)$$
$$\Rightarrow V_{Z^c}, \sigma(s) \nVdash_1 \theta^c(q) \qquad\qquad (\dagger)$$
$$\Rightarrow Z^c \notin \Uparrow(P_\exists^{\mathcal{G}^c}(q,s)) \qquad\qquad\text{(Proposition 3)}$$
$$\Rightarrow \text{there is a } Z_\forall \in P_\forall^{\mathcal{G}^c}(q,s) \text{ with } Z^c \cap Z_\forall = \varnothing \qquad\text{(Proposition 2)}$$
$$\Rightarrow \text{there is a } Z_\forall \subseteq Z \text{ with } Z_\forall \in P_\forall^{\mathcal{G}^c}(q,s) \qquad\text{(elementary set theory)}$$

Here we let Z^c denote the set $(Q \times S) \setminus Z$, and in the implication marked (†) we use the easily verified fact that $(V_Z)^c = V_{Z^c}$.

5.3   Transalternating and Alternating Automata

In the remainder of this section we prove that the enriched structure of transalternating automata does not add expressivity compared to alternating automata. Since there is an obvious way to see an alternating automaton as a transalternating one, this shows that the transalternating automata have exactly the same recognizing power as the alternating ones. Consequently, the complementation lemma for transalternating automata also yields a complementation lemma for alternating ones.

The transformation of a given transalternating automaton $\mathbb{A}$ into an equivalent alternating automaton will be taken care of in two successive steps. The intuitive idea is to first remove the disjunctions, and then the conjunctions, from under the modal operators. In the intermediate stage we are dealing with automata that only allow (certain) conjunctions under the nabla operator.

Definition 19. *Given a set Q we let $\mathcal{S}Q$ denote the set of conjunctions of the form $\bigwedge Q'$, with $Q' \subseteq Q$. We let $\leq \; \subseteq \mathcal{S}Q \times Q$ denote the relation given by $\bigwedge Q' \leq q$ iff $q \in Q'$.*

Definition 20. *A semi-transalternating T-automaton is a automaton of the form $\mathbb{A} = \langle Q, \theta, q_I, \Omega \rangle$ with $\theta : Q \to \mathcal{L}T^\nabla \mathcal{S}Q$.*

We omit the (obvious) definition of the acceptance game associated with this type of automaton.

Proposition 5. *There is an effective algorithm transforming a given transalternating automaton into an equivalent semi-transalternating one based on the same carrier.*

The *proof* of this proposition, which we omit for reasons of space limitations, is based on the fact that every formula in $\mathcal{L}_1 Q$ can be rewritten into an equivalent formula in $\mathcal{L}T^\nabla \mathcal{S}Q$, see [3] for the details.

It is left to prove that every semi-transalternating automaton can be transformed into an equivalent alternating one. Intuitively speaking, one has to carry over the (universal) non-determinism after the ∇-translation to the branching for the next automaton state.

Definition 21. *Let* $\mathbb{A} = \langle Q, q_I, \theta, \Omega \rangle$ *be a semi-transalternating automaton. We define the alternating automaton* $\mathbb{A}^\circ = \langle Q', q'_I, \theta', \Omega' \rangle$ *by putting* $Q' := Q \times SQ$, $q'_I := (q_I, q_I)$, $\Omega'(a, b) := \Omega(a)$, *while* $\theta' : (Q \times SQ) \to \mathcal{L}T^\nabla(Q \times SQ)$ *is given by*

$$\theta'(q, a) := \bigwedge_{p \geq a} \mathcal{L}T^\nabla \iota_p(\theta(p)).$$

Here $\iota_q : SQ \to (Q \times SQ)$, *for* $q \in Q$, *is the map given by* $\iota_q(a) := (q, a)$.

For reasons of space limitations we omit the proof of the equivalence of $\mathbb{A}^\circ$ and $\mathbb{A}$, confining ourselves to an intuitive explanation of this definition. Note that in principle we would like to base $\mathbb{A}^\circ$ on the carrier SQ, defining a transition map $\theta''(a) := \bigwedge_{p \geq a} \theta(p)$. (Note that this gives a well-defined map $\theta'' : SQ \to \mathcal{L}T^\nabla SQ$.) Unfortunately, this set-up is too simple to give a proper account of the acceptance condition. The problem is that in the acceptance game, when moving from a state $(a, s) \in SQ \times Q$ to $(\theta''(a), s)$ and then on to $(\theta(p), s)$ for some $p \geq a$, one would 'bypass' the position (p, s), and thus miss the crucial contribution of the priority of the state p to the information determining the winner of the match. A solution to this problem is to tag p to the new states of $\mathbb{A}^\circ$ occurring in $\theta(p)$. This is exactly the role of the function ι_p in the definition of θ' above: note that $\mathcal{L}T^\nabla \iota_p(\theta(p))$ is nothing but the term $\theta(p)$, seen as an element of $\mathcal{L}T^\nabla SQ$, but with every formula $b \in SQ$ under the ∇ replaced by $(p, b) \in Q \times SQ$.

Thus, intuitively, a $\mathbb{A}^\circ$-state of the form (p, b) represents the conjunction of the states above b (where q is above $b = \bigwedge Q'$ iff it belongs to Q'). However, the *priority* of the state (p, b) is that of p — thus the state (p, b) encodes ('remembers') an earlier visit to p.

Proposition 6. $\mathbb{A}$ *and* $\mathbb{A}^\circ$ *are equivalent, for any semi-transalternating automaton* $\mathbb{A}$.

The latter two propositions establish that there is an effective procedure transforming a transalternating automaton into an equivalent alternating one.

5.4   Size Matters

We conclude this section with some remarks about the size of the automaton resulting from the complementation algorithm presented. By the *size* of an automaton we understand the number of its states.

First of all, since every alternating T-automaton can itself be seen as a transalternating T-automaton, there is no size issue here. Also, our complement of a transalternating T-automaton is of the same size as the original. It is the translation from transalternating to alternating T-automata that introduces exponentially many new states: more specifically, with $\mathbb{A}$ a transalternating automaton of size $|Q| = n$, the equivalent alternating automaton of Definition 21 has $|\mathcal{S}Q \times Q| = 2^n * n$ states. From these observations the following is immediate.

Theorem 5. *For any alternating T-automaton with n states there is a complementing alternating automaton with at most $n * 2^n$ states.*

In case we are dealing with a functor T such that the set $\mathcal{L}T^\nabla Q$ is closed under taking duals, we do not need the concept of transalternation, and we can obtain a much better upper bound. The following result applies for instance to alternating tree automata, see [4] for the details.

Theorem 6. *Let T be such that for any $\alpha \in T_\omega Q$, the formula $\nabla \alpha$ has a dual $\Delta \alpha \in \mathcal{L}T^\nabla Q$. Then for any alternating T-automaton of n states there is a complementing alternating automaton with at most $n + c$ states, for some constant c.*

Acknowledgement. We are grateful to Alessandra Palmigiano and one of the anonymous reviewers for suggesting a simplification of a definition, and to the anonymous reviewers in general for their comments.

References

1. Baltag, A.: A logic for coalgebraic simulation. In: Reichel, H. (ed.) Coalgebraic Methods in Computer Science (CMCS 2000). Electronic Notes in Theoretical Computer Science, vol. 33, pp. 41–60 (2000)
2. van Benthem, J.: Extensive games as process models. Journal of Logic, Language and Information 11(3), 289–313 (2002)
3. Kurz, A., Kupke, C., Venema, Y.: Completeness of the finitary Moss logic. In: Areces, C., Goldblatt, R. (eds.) Advances in Modal Logic, vol. 7, pp. 193–217. College Publications (2008)
4. Kissig, C.: Satisfiability of S2S. Master's thesis, Institute of Logic, Language and Computation, Amsterdam (2007)
5. Kupke, C., Venema, Y.: Coalgebraic automata theory: basic results. Logical Methods in Computer Science 4, 1–43 (2008)
6. Muller, D., Schupp, P.: Alternating automata on infinite trees. Theoretical Computer Science 54, 267–276 (1987)
7. Pauly, M.: Logic for Social Software. PhD thesis, Institute of Logic, Language and Computation, Amsterdam (2001)
8. Rutten, J.: Universal coalgebra: A theory of systems. Theoretical Computer Science 249, 3–80 (2000)
9. Venema, Y.: Automata and fixed point logics for coalgebras. Electronic Notes in Theoretical Computer Science 106, 355–375 (2004)
10. Venema, Y.: Automata and fixed point logic: a coalgebraic perspective. Information and Computation 204, 637–678 (2006)

Characterising Behavioural Equivalence: Three Sides of One Coin

Clemens Kupke[1] and Raul Andres Leal[2,*]

[1] Imperial College London, London, United Kingdom
ckupke@doc.ic.ac.uk
[2] Universiteit van Amsterdam, Amsterdam, The Netherlands
R.A.LealRodriguez@uva.nl

Abstract. We relate three different, but equivalent, ways to characterise behavioural equivalence for set coalgebras. These are: using final coalgebras, using coalgebraic languages that have the Hennessy- Milner property and using coalgebraic languages that have "logical congruences". On the technical side the main result of our paper is a straightforward construction of the final T-coalgebra of a set functor using a given logical language that has the Hennessy-Milner property with respect to the class of T-coalgebras.

1 Introduction

Characterising behavioral equivalence between coalgebras is an important issue of universal coalgebra and coalgebraic logic. Rutten [15] shows that behavioral equivalence can be *structurally characterized* by final coalgebras. Moss [12] and Pattinson [13] provided two different ways for generalizing modal logic to arbitrary coalgebras. These generalizations are called coalgebraic (modal) logics. Moss showed in [12] that his language provides a *logical characterization* of behavioural equivalence: two pointed coalgebras are behaviourally equivalent iff they satisfy the same formulas of the logic. In modal logic terminology this logical characterisation of behavioural equivalence is usually called the "Hennessy-Milner property" ([4]). For the language of coalgebraic modal logic from [13] Lutz Schröder [16] shows how the Hennessy-Milner property can be obtained using certain congruences of coalgebras, which we will call *logical congruences*.

The main contribution of this paper is to introduce a systematic study of the relationship between these three methods: the structural and logical characterisation of behavioural equivalence and the characterisation that uses logical congruences. We work in a general framework that covers all known logics for set coalgebras and easily generalizes to base categories different from Set. Our main theorem (Theorem 10) can be stated as follows: Given a set functor T, a final T-coalgebra exists iff there exists a language for T-coalgebras with the

---

* The research of this author has been made possible by VICI grant 639.073.501 of the Netherlands Organization for Scientific Research (NWO).

A. Kurz, M. Lenisa, and A. Tarlecki (Eds.): CALCO 2009, LNCS 5728, pp. 97–112, 2009.

Hennessy-Milner property iff there exists a language for T-coalgebras that has logical congruences.

The first equivalence was proven by Goldblatt in [6], the second equivalence was discussed by Schröder in [16] for the case of coalgebraic logics. We provide relatively simple proofs for these equivalences in order to obtain our main theorem. In particular, 1) we simplify Goldblatt's proof, 2) generalize Schröder's argument, and in addition to that, 3) we use our framework to construct canonical models and characterize simple coalgebras. Furthermore we demonstrate that our proofs allow for straightforward generalizations to base categories that are different from Set.

The structure of the paper is as follows: In the Section 2, we introduce some preliminaries and mention the relation between final coalgebras and behavioral equivalence. In Section 3 we introduce abstract coalgebraic languages. Using these we present the connection between behavioral equivalence and the Hennessy-Milner property and we construct canonical models. In Section 4 we discuss the connection between coalgebraic congruences and languages with the Hennessy-Milner property and characterize simple coalgebras. Finally in Section 5 we explore generalizations of our work to other categories. For this last part a bit more of knowledge of category theory is assumed.

2   Behaviour and Final Coalgebras (Preliminaries)

Before we start let us stress the fact that one major feature of our work is its simplicity. Therefore we do not require much knowledge of category theory. When possible, we avoid the use of general categorical notions and use set theoretic terminology. The basic notion of this paper is that of coalgebras for a set endofunctor. However, we introduce the notion of coalgebra for an endofunctor on any category, and not just Set, as we will discuss generalisations of our results.

Definition 1. *The category of coalgebras* Coalg(T) *for an endofunctor on a category* $\mathbb{C}$ *has as objects arrows* $\xi : X \longrightarrow TX$ *and morphisms* $f : (X, \xi) \longrightarrow (Y, \gamma)$ *are arrows* $f : X \longrightarrow Y$ *such that* $T(f)\xi = \gamma f$. *The object* X *is called the states of* ξ. *If* $\mathbb{C} =$ Set *we define a* pointed coalgebra *as a pair* (ξ, x) *with* x *an element of the (set of) states of* ξ. *We often write* $f : \xi \longrightarrow \gamma$ *for a morphism* $f : (X, \xi) \longrightarrow (Y, \gamma)$. *We call the arrow* ξ, *in a coalgebra* (X, ξ), *the* structural map *of the coalgebra. We reserve the letter* ξ *for coalgebras with a carrier object* X, *the letter* γ *for coalgebras over an object* Y, *and the letter* ζ *for coalgebras based on an object* Z.

The crucial notion that we want to investigate is *behavioural equivalence*. For now, we only consider set coalgebras. One of the important issues towards a generalisation of the results presented in this paper is to find a notion of behavioural equivalence that can be interpreted in other categories different from Set.

Definition 2. *Let* T : Set $\longrightarrow$ Set *be a functor. Two states* x_1 *and* x_2 *on* T-*coalgebras* ξ_1 *and* ξ_2, *respectively, are* behavioural equivalent, *written* $x_1 \sim x_2$, *if there exists a coalgebra* γ *and morphisms* $g_i : \xi_i \longrightarrow \gamma$ *such that* $g_1(x_1) = g_2(x_2)$.

It was noted by Rutten [15] that behavioural equivalence could be characterised using final systems, also called final coalgebras.

Definition 3. *A* final coalgebra *for an endofunctor T is a terminal object in* $\mathsf{Coalg}(T)$. *Explicitly, a final coalgebra is a coalgebra $\zeta : Z \longrightarrow TZ$ such that for any coalgebra $\xi : X \longrightarrow TX$ there exists a unique morphism $f_\xi : \xi \longrightarrow \zeta$. This morphism is called the* final map *of ξ.*

Final coalgebras are to coalgebra what initial algebras or term algebras are to algebra (cf. e.g. [9]). In this paper we are interested in final coalgebras mainly because they can be used to characterise behavioural equivalence between states.

Proposition 1 ([15]). *If a final coalgebra for a set functor T exists, two states x_i in coalgebras ξ_i are behavioural equivalent if and only if they are mapped into the same state of the final coalgebra, i.e., if $f_{\xi_1}(x_1) = f_{\xi_2}(x_2)$.*

3   The Hennessy-Milner Property and Behaviour

In this section we introduce languages to describe coalgebras. We will show how languages with the Hennessy-Milner property relate to final coalgebras; we will illustrate this interaction constructing canonical models for those languages.

3.1   Abstract Coalgebraic Languages

We begin by showing that the existence of a final coalgebra is equivalent to the existence of a language with the Hennessy-Milner property. Let us first clarify what is meant by an *abstract coalgebraic language*. In the sequel, unless stated otherwise, T will always denote an arbitrary functor $T : \mathsf{Set} \longrightarrow \mathsf{Set}$.

Definition 4. *An* abstract coalgebraic language *for T, or simply a* language *for T-coalgebras, is a set $\mathcal{L}$ together with a function $\Phi_\xi : X \longrightarrow \mathcal{PL}$ for each T-coalgebra $\xi : X \longrightarrow TX$. The function Φ_ξ will be called the* theory map *of ξ, elements of $\mathcal{PL}$ will be called ($\mathcal{L}$-)theories.*

An abstract coalgebraic language is precisely what Goldblatt calls a "small logic" for T ([6]). Examples for such languages are the well-known modal languages for Kripke models.

Example 1. Let Prop be a set of propositional variables and let $\mathsf{K} = \mathcal{P}\mathsf{Prop} \times \mathcal{P}(-)$ where $\mathcal{P}$ denotes the covariant power set functor. Coalgebras for the functor K are *Kripke models*, the K-coalgebra morphisms are known as *bounded morphisms*. A number of languages have been proposed in order to reason about K-coalgebras, eg. the languages of basic modal logic and its variations such as PDL ([4]), temporal languages such as LTL and CTL and the modal μ-calculus ([17]). It can be easily shown that all those languages are abstract coalgebraic languages for K in the sense of Definition 4.

Basic modal logic is a particular instance of a more general class of coalgebraic languages, namely coalgebraic modal logics with predicate liftings, or *languages of predicate liftings* (see [13,16] for details).

The fact that abstract coalgebraic languages are just sets and do not carry by definition any further structure could be seen as a weakness. However, this has the advantage that we are covering any logical language for Set-coalgebras that one can imagine[1]. This means that the results presented in this section hold for any language. For example, for languages with fixpoint operators.

As mentioned before, we are interested in describing behavioural equivalence. To do this we have two requirements on the language, which together lead to what sometimes is called expressivity: 1) *Adequacy:* formulas must be invariant under coalgebra morphisms. 2) *Hennessy-Milner property:* formulas must distinguish states that are not behavioural equivalent.

Definition 5. *An abstract coalgebraic language $\mathcal{L}$ is said to be* adequate *if for every pair of pointed T-coalgebras (ξ_1, x_1) and (ξ_2, x_2),*

$$x_1 \sim x_2 \text{ implies } \Phi_{\xi_1}(x_1) = \Phi_{\xi_2}(x_2).$$

The language $\mathcal{L}$ is said to have the Hennessy-Milner *property if for every pair of pointed T-coalgebras (ξ_1, x_1) and (ξ_2, x_2),*

$$\Phi_{\xi_1}(x_1) = \Phi_{\xi_2}(x_2) \text{ implies } x_1 \sim x_2.$$

Other names used for adequacy and the Hennessy-Milner property are *soundness* and *completeness*, respectively. Using this terminology we can talk of languages that are sound and complete with respect to behavioural equivalence.

Example 2. The languages for the functor K from Example 1 are adequate, but all of them lack the Hennessy-Milner property. It is a consequence of Theorem 10 below that no adequate abstract coalgebraic language can have the Hennessy-Milner property with respect to K-coalgebras because the final K-coalgebra does not exist. The basic modal language has the Hennessy-Milner property with respect to $K_\omega = \mathcal{P}\mathsf{Prop} \times \mathcal{P}_\omega(-)$-coalgebras (image-finite Kripke models). All languages of predicate liftings are adequate. However, not all of them have the Hennessy-Milner property. Sufficient conditions for this can be found in [16].

The reader may worry that the definition of adequacy was not presented as "formulas are invariant under morphisms". The definition above comes from a Modal logic tradition where bismiliar states should satisfy the same formulas, in the case of sets both presentations are equivalent.

Proposition 2. *Let T be a set endofunctor and $\mathcal{L}$ an abstract coalgebraic language for T coalgebras. The language $\mathcal{L}$ is adequate iff formulas are invariant under coalgebra morphisms, i.e. if $f : \xi_1 \longrightarrow \xi_2$ is a morphism of coalgebras, the following diagram commutes:*

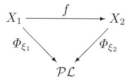

Remark 1. The previous proposition deserves some comments that are particularly relevant for the reader interested into categorical generalizations. First notice that the previous proposition provides a pointless definition of adequacy, ie., this definition can be used in any base category $\mathbb{C}$. With this in mind, the categorically minded reader will recognise a natural transformation $\Phi : U \longrightarrow \Delta_{\mathcal{PL}}$, or equivalently a functor $\Phi : Coalg(T) \longrightarrow (U \downarrow \mathcal{PL})$, where the codomain of Φ is the appropriate comma category. Another issue that deserves attention is that the previous proposition depends on the existence of coproducts.

3.2   An Elementary Construction of Final Coalgebras

One advantage of using abstract coalgebraic languages is that we can easily show how a final coalgebra induces an adequate language that has the Hennessy-Milner property.

Theorem 1. *For any functor $T : Set \longrightarrow Set$, if there exists a final coalgebra then there exists an adequate language for T coalgebras with the Hennessy-Milner property.*

Proof. Let (Z, ζ) be a final coalgebra, and let f_ξ be the final map for each coalgebra ζ. Take $\mathcal{L} = Z$ and for each coalgebra (X, ξ) define $\Phi_\xi(x) = \{f_\xi(x)\}$. Since Z is a final coalgebra this language together with the maps Φ_ξ is adequate and has the Hennessy-Milner property. This concludes the proof.

At first glance it might seem that the previous construction is too abstract and that the language we obtain from a final coalgebra is not interesting. However, this is not the case. We consider the following example from [9].

Example 3. Consider the set endofunctor $T = 1 + (-)$. Coalgebras for this functor can be considered as a black-box machine with one (external) button and one light. The machine performs a certain action only if the button is pressed. And the light goes on only if the machine stops operating. A final coalgebra for T is given by the set $\overline{\mathbb{N}} = \mathbb{N} \cup \{\infty\}$ together with a function $p : \overline{\mathbb{N}} \longrightarrow 1 + \overline{\mathbb{N}}$ defined as follows $p(0) = *$; $p(n+1) = n$; $p(\infty) = \infty$, where $*$ is the only element of 1. This presentation of the final coalgebra for T contains all the information about the observable behaviour of a state in a T-coalgebra as a state can only either lead the machine to stop after n-steps or let the machine run forever.

Other examples demonstrating that final coalgebras are useful for describing the coalgebraic behaviour can be found in [14,2]. Now we will illustrate how to construct a final T-coalgebra from an adequate language which has the Hennessy-Milner property. Our exposition is a slight generalisation and simplification of the construction introduced in [6].

Theorem 2. *For any functor $T : Set \longrightarrow Set$, if there exists an adequate language for T-coalgebras with the Hennessy-Milner property then there exists a final coalgebra.*

We will provide a proof of this theorem after we have made some observations. Our construction has three main points: The first key idea is to notice that if we have a language for T-coalgebras we can identify a concrete set (object) Z which is a natural candidate for the carrier of a final coalgebra. The second observation is that for each coalgebra (X, ξ) there is a natural map $X \longrightarrow TZ$; should the language be adequate and have the Hennessy-Milner property then we can combine these functions into a function $\zeta : Z \longrightarrow TZ$ which endows Z with the structure of a final T-coalgebra. Moreover, using this approach we can show that the function ζ exists if and only if the language has the Hennessy-Milner property. A natural candidate for the carrier of a final coalgebra is the set of satisfiable theories of the language.

Definition 6. *Given a functor T : Set $\longrightarrow$ Set and an abstract coalgebraic language $\mathcal{L}$ for T-coalgebras, the set $Z_{\mathcal{L}}$ of $\mathcal{L}$-satisfiable theories is the set $Z_{\mathcal{L}} := \{\Phi \subseteq \mathcal{L} \,|\, (\exists \xi)(\exists x \in \xi)(\Phi_{\xi}(x) = \Phi)\}$. We often drop the subindex $\mathcal{L}$.*

Remark 2. The reader might worry that in the definition of $Z_{\mathcal{L}}$ we quantify over all coalgebras and all states on them and then we might not be defining a set but a proper class. This is not an issue as we required the language $\mathcal{L}$ to be a set and obviously $Z_{\mathcal{L}} \subseteq \mathcal{L}$.

By definition of $Z_{\mathcal{L}}$ it is clear that for each coalgebra $\xi : X \longrightarrow TX$ there is a canonical map $f_{\xi} : X \longrightarrow Z_{\mathcal{L}}$ that is obtained from $\Phi_{\xi} : X \longrightarrow \mathcal{PL}$ by restricting the codomain. This restriction is possible as the range of Φ_{ξ} is clearly contained in $Z_{\mathcal{L}}$. Using the functions f_{ξ} we can see that for each coalgebra (X, ξ) there is a natural function from X to TZ, namely the lower edge of the following square

$$
\begin{array}{ccc}
X & \xrightarrow{\;\;f_{\xi}\;\;} & Z \\
{\scriptstyle \xi}\big\downarrow & & \\
TX & \xrightarrow[\;T(f_{\xi})\;]{} & TZ
\end{array}
\tag{1}
$$

This square suggests the following assignment ζ: a theory $f_{\xi}(x) = \Phi_{\xi}(x) \in Z$ is assigned to

$$
\zeta(\Phi_{\xi}(x)) = T(f_{\xi})\xi(x).
\tag{2}
$$

Since in general we will have $\Phi = \Phi_{\xi_1}(x_1) = \Phi_{\xi_2}(x_2)$ for different pointed coalgebras (ξ_1, x_1) and (ξ_2, x_2) it is not clear that equation (2) defines a function. We are now going to show that equation (2) defines a function if the language is adequate and has the Hennessy-Milner property. We will prove this in two steps which illustrate that both conditions are really needed.

Lemma 1. *Let $\mathcal{L}$ be an adequate language for T-coalgebras. For any morphism $f : \xi \longrightarrow \gamma$ we have: $(Tf_{\xi})\xi = (Tf_{\gamma})\gamma f$, where f_{ξ} and f_{γ} are obtained from the respective theory maps by restricting the codomain.*

If we assume the language $\mathcal{L}$ to be adequate and we have a morphism $f : \xi \longrightarrow \gamma$, then $\Phi_{\xi}(x) = \Phi_{\gamma}(f(x))$. The previous lemma implies $\zeta(\Phi_{\xi}(x)) = \zeta(\Phi_{\gamma})(f(x))$.

We can show that if in addition to adequacy $\mathcal{L}$ has the Hennessy-Milner property, equation (2) defines a function ζ. In fact these two conditions are equivalent.

Theorem 3. *Let T be a set functor, let $\mathcal{L}$ be an adequate language for T-coalgebras, let $Z_{\mathcal{L}}$ be the set of $\mathcal{L}$-satisfiable theories, and let f_ξ be the function obtained from a theory map Φ_ξ by restricting the domain; the following are equivalent: 1) The language $\mathcal{L}$ has the Hennessy-Milner property. 2) The assignment ζ from (2) which takes an $\mathcal{L}$-theory $\Phi = \Phi_\xi(x) \in Z_{\mathcal{L}}$ to $(Tf_\xi)\xi(x)$ does not depend on the choice of (ξ, x), i.e., (2) defines a function $\zeta : Z \longrightarrow TZ$.*

It is almost immediate by definition of ζ that for each coalgebra ξ the function $f_\xi : X \longrightarrow Z_{\mathcal{L}}$ is a morphism between the coalgebras ξ and ζ. We make this explicit as we will use it in the proof of Theorem 2.

Corollary 1. *Under the conditions of the previous theorem; for any coalgebra ξ, the function $f_\xi : \xi \longrightarrow \zeta$ is a morphism of coalgebras.*

This previous result already implies that a final coalgebra exists but we can do better by showing that (Z, ζ) is already a final object. The next lemma will be useful in several occasions, in particular in the proof of Theorem 2 and in our application to canonical models.

Lemma 2. *For a functor $T : \mathsf{Set} \longrightarrow \mathsf{Set}$ and a language $\mathcal{L}$ that is adequate and has the Hennessy-Milner property, the theory map $\Phi_\zeta : Z \longrightarrow \mathcal{PL}$ is the inclusion, where Z is the set of satisfiable $\mathcal{L}$-theories and ζ is defined as in equation (2).*

Now we have all the material to prove Theorem 2.

Proof (of Theorem 2). Let Z be the set of $\mathcal{L}$-satisfiable theories, let f_ξ be the function obtained from a theory map Φ_ξ by restricting the codomain; Theorem 3 implies that the assignment ζ which takes a theory $\Phi_\xi(x) \in Z$ to $T(f_\xi)\xi(x)$ does not depend on the choice of (ξ, x), i.e., it is a function $\zeta : Z \longrightarrow TZ$. Corollary 1 implies that for each coalgebra ξ the function $f_\xi : \xi \longrightarrow \zeta$ is a morphism of coalgebras. It is only left to prove that $f_\xi : \xi \longrightarrow \zeta$ is the only morphism of coalgebras. Since the language is adequate, this will follow because any morphism of coalgebras $f : \xi \longrightarrow \zeta$ makes the following diagram

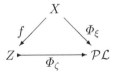

commute and Lemma 2 tells us that the function Φ_ζ is injective. QED.

Gathering Theorem 1 and Theorem 2 we have:

Proposition 3. *The following are equivalent: 1) There exists a final T-coalgebra. 2) There exists an adequate language for T-coalgebras with the Hennessy-Milner property.*

The contrapositive of previous result clearly shows the power of our abstract approach as it tells us that if a functor T fails to have a final coalgebra there is no way to completely describe the behavior of T-coalgebras using an abstract coalgebraic language this is particularly relevant for applications. Also notice that the proof of Theorem 2 tells us a bit more about the relation of final coalgebras and abstract coalgebraic languages; we can improve Theorem 3 into an equivalence as follows.

Theorem 4. *Let $\mathcal{L}$ be a language for T-coalgebras, let Z be the set of $\mathcal{L}$-satisfiable theories and let f_ξ be the function obtained from a theory map Φ_ξ by restricting the codomain; the following are equivalent: 1) The language $\mathcal{L}$ is adequate and has the Hennessy-Milner property. 2) There exists a function $\zeta : Z \longrightarrow TZ$ which furnishes Z with a final coalgebra structure in such a way that for each coalgebra (X, ξ) the function $f_\xi : X \longrightarrow Z$ is the final map.*

An application: Canonical Models. Until here we have illustrated that for set endofunctors, there exists a final coalgebra iff there exists an adequate abstract coalgebraic language with the Hennessy-Milner property. As mentioned before (Example 3), the work in [15,9] presents interesting examples of the implication from left to right. Another non trivial use of the states of a final coalgebra as a language for coalgebras is the work on coequational logic, see e.g. [2], where coequations are in fact the states of an appropriate final coalgebra. Here we illustrate our assembly of final coalgebras from languages with a construction of canonical models. In Lemma 2 we showed that the theory map of (Z, ζ) is the inclusion. Since the states of (Z, ζ) are the satisfiable theories of $\mathcal{L}$ we can rewrite Lemma 2 into a well known theorem of Modal logic.

Lemma 3 (Truth Lemma). *Let $\mathcal{L}$ be an adequate language for T-coalgebras with the Hennessy-Milner property. Let Z be the set of $\mathcal{L}$-satisfiable theories (Definition 6) and let $\zeta : Z \longrightarrow TZ$ be defined as in Equation (2). For any $\Phi \in Z$ and any $\varphi \in \mathcal{L}$ we have $\Phi \models_\zeta \varphi$ iff $\varphi \in \Phi$, where $\Phi \models_\zeta \varphi$ means $\varphi \in \Phi_\zeta(\Phi)$.*

The previous lemma illustrates that our construction is similar to the canonical model construction from modal logic (see [4]). A natural step is to investigate completeness results for different abstract coalgebraic languages. Assuming that $\mathcal{L}$ has some notion of consistency we can ask: are maximally consistent sets satisfiable? We do not peruse this question in this paper, but notice the following result:

Proposition 4. *Let $\mathcal{L}$ be an adequate language for T-coalgebras, let Z be the set of $\mathcal{L}$-satisfiable theories. The set Z is the largest subset of $\mathcal{PL}$ for which we can 1) define a T-coalgebra structure ζ such that the Truth Lemma is satisfied, i.e. the theory map is the inclusion, and 2) such that for each coalgebra the codomain restiricion of the theory maps are morphisms of T-coalgebras.*

As corollary, see [4], we can see that for the canonical model for the logic K; there exists a Kripke frame (X, ξ) for which the modal theory map $\Phi_\xi : X \longrightarrow M$ is not a bounded morphism; also that the canonical model is not a final object.

4   Behaviour and Congruences

The Hennessy-Milner property states that if two states are logically equivalent then they are identified in some coalgebra. However, this coalgebra is not made explicit. The work in the previous section provides a canonical coalgebra where logically equivalent states are identified, namely the final coalgebra. In this section we investigate another construction in order to identify logically equivalent states: taking *logical congruences*. Let us first recall the notion of a coalgebraic congruence and its equivalent characterisations.

Definition 7. *Let (X, ξ) be a T-coalgebra. An equivalence relation θ on the set X is a* congruence of T-coalgebras *iff there exists a coalgebraic structure $\xi_\theta : X/\theta \longrightarrow T(X/\theta)$ such that the following diagram*

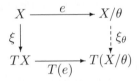

commutes. Here e is the canonical quotient map.

In the category Set, it can be shown that the notion of a congruence for coalgebra can be characterised as the kernel of coalgebra morphisms. In other words, it behaves like the notion of a congruence in universal algebra [8,15]. Congruences for coalgebras where first introduced by Aczel and Mendler in [1].

Fact 5. *Let (X, ξ) be a T-coalgebra for a functor $T : \mathsf{Set} \longrightarrow \mathsf{Set}$, For an equivalence relation θ, on X, the following conditions are equivalent: 1) θ is a congruence of coalgebras. 2) $\theta \subseteq \mathrm{Ker}(T(e)\xi)$. 3) θ is the kernel of some morphism of T-coalgebras with domain ξ.*

We remark that the previous characterisation of congruences depends on the fact that set functors preserve monomorphisms with non-empty domain [8], something that is not generally true in categories different from Set.

4.1   Simple Coalgebras

Before describing how behavioural equivalence relates to congruences, we discuss the notion of a *simple coalgebra*. As it is the case in algebra, the set of all congruences on a coalgebra (X, ξ) is a complete lattice under the partial ordering of set inclusion. In particular, there is a smallest congruence Δ_X (the identity relation on X) and a largest congruence. However, unlike the universal algebra case, the largest congruence may be smaller than the universal relation. Simple coalgebras are defined as coalgebras with only one congruence.

Definition 8. *A coalgebra $\xi : X \longrightarrow TX$ is* simple *if its largest (and hence only) congruence is the identity relation Δ_X on X.*

Example 4. The following Kripke frame is a simple $\mathcal{P}$-coalgebra. In order to see this we have to use that, in this case, Kripke bisimilarity ([4]) coincides with the largest congruence.

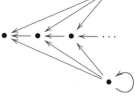

Using coalgebraic languages we can give a more concrete characterisation of simple coalgebras; a first step is given by the following result.

Proposition 5. *Let T be a set functor and let $\mathcal{L}$ be an adequate language for T-coalgebras. A T-coalgebra ξ is simple if the theory map Φ_ξ is injective.*

The converse of the previous proposition is not true in general; the Kripke frame in Example 4 is a counterexample. As we will later see, if the language $\mathcal{L}$ also has the Henessy-Milner property then we obtain an equivalence. The trick for such characterization of simple coalgebras is to make a "logical" representation of each coalgebra; we do this using logical congruences, a concept that we now introduce.

4.2 Logical Congruences

Logical congruences are congruences obtained using logical equivalence of states.

Definition 9. *Given an abstract coalgebraic language $\mathcal{L}$, we say that two pointed coalgebras (ξ_i, x_i) are logically equivalent, written $(\xi_1, x_1) \rightsquigarrow_{\mathcal{L}} (\xi_2, x_2)$, iff $\Phi_{\xi_1}(x_1) = \Phi_{\xi_2}(x_2)$. We call $\rightsquigarrow_{\mathcal{L}}$ the logical equivalence relation of states. Given a coalgebra ξ, we write $\rightsquigarrow_{\mathcal{L}}^{\xi}$ for the relation $\rightsquigarrow_{\mathcal{L}}$ restricted to the states of ξ.*

Our interest into these equivalence relations has two main reasons. The first one was an attempt to make Proposition 5 into an equivalence and then obtain a concrete characterisation of simple coalgebras. The second and most important motivation was to generalise Proposition 3 and Theorem 4 to arbitrary categories. To our surprise logical congruences proved to be remarkably useful to simplify our constructions. In our opinion, logical congruences provide the appropriate categorical generalisation of the Hennessy-Milner property.

Definition 10. *Let $\mathcal{L}$ be an abstract coalgebraic language for T. For each T-coalgebra (X, ξ) we identify the set of satisfiable theories in ξ, $X/\rightsquigarrow_{\mathcal{L}}^{\xi}$, with the set $Z_\xi := \{\Phi \subseteq \mathcal{L} \mid (\exists x \in X)(\Phi_\xi(x) = \Phi)\}$. We use $e_\xi : X \longrightarrow Z_\xi$ for the canonical (quotient) map.*

If our language happens to be adequate and have the Hennessy-Milner property we can show that logical equivalence of states is a congruence of coalgebras.

Lemma 4. *Let $\mathcal{L}$ be a language for T-coalgebras. If $\mathcal{L}$ is adequate and has the Hennessy-Milner property, then for each coalgebra (X, ξ) the relation $\leftrightsquigarrow_{\mathcal{L}}^{\xi}$ is a congruence of coalgebras. Moreover, $\leftrightsquigarrow_{\mathcal{L}}^{\xi}$ is the largest congruence on (X, ξ).*

The main idea is to follow the proof of Theorem 2, on page 103, relativized to the set Z_ξ. Explicitly this is: we define a function $\zeta_\xi : Z_\xi \rightarrow T(Z_\xi)$ such that the canonical map e_ξ is a morphism of coalgebras. See the appendix for more details. Now we can easily make Proposition 5 into an equivalence.

Theorem 6. *Let T be a set endofunctor, and let $\mathcal{L}$ be an adequate language for T-coalgebras with the Hennessy-Milner property. A T-coalgebra ξ is simple iff the theory map Φ_ξ is injective.*

Notice that the construction used in the proof of Lemma 4 generalises the construction of final coalgebras of the previous section. This leads us to the following definition.

Definition 11. *An abstract coalgebraic language $\mathcal{L}$ for T-coalgebras is said to have logical congruences iff for each coalgebra ξ the equivalence relation $\leftrightsquigarrow_{\mathcal{L}}^{\xi}$, is a congruence of T-coalgebras. The quotient of ξ using $\leftrightsquigarrow_{\mathcal{L}}^{\xi}$ is called the logical quotient of ξ and we write (Z_ξ, ζ_ξ) for this coalgebra.*

In [16] Lutz Schröder noticed that languages of predicate liftings that have logical congruences have the Hennessy-Milner property. We turn his observation into a general theorem for abstract coalgebraic languages.

Theorem 7. *If a language $\mathcal{L}$ for T-coalgebras is adequate, the following are equivalent: 1) $\mathcal{L}$ has the Hennessy-Milner property. 2) $\mathcal{L}$ has logical congruences.*

An application: A Concrete Characterization of Simple Coalgebras
As mentioned before, in the work of Schröder [16] we have non trivial use of logical congruences to establish the Hennessy-Milner property for a language. Theorem 7 tells us that in fact the two properties are equivalent. In this section, we illustrate the construction of logical congruences (cf. proof of Lemma 4) giving a concrete characterization of simple coalgebras. We first make a remark concerning the theory maps of logical quotients (Definition 11).

Proposition 6. *Let $\mathcal{L}$ be an adequate language for T-coalgebras that has logical congruences and let $\xi : X \rightarrow TX$ be a T-coalgebra. The theory map $\Phi_{\zeta_\xi} : Z_\xi \rightarrow \mathcal{PL}$ of the logical quotient of ξ (Definition 11) is equal to the inclusion.*

Now we can use logical congruences in order to characterize simple coalgebras as logical quotients, i.e. quotients using the relations $\leftrightsquigarrow_{\mathcal{L}}^{\xi}$.

Theorem 8. *Let $\mathcal{L}$ be an adequate language for T-coalgebras that has logical congruences. Any logical quotient (Z_ξ, ζ_ξ) is simple and any simple T-coalgebra γ is isomorphic to the logical quotient (Z_ξ, ζ_ξ) of some coalgebra ξ.*

Using this characterization of simple coalgebras we can easily prove that truth-preserving functions with simple codomain must be coalgebra morphisms. This was a key result used by Goldblatt in [6] to construct final coalgebras.

Corollary 2. *Let $\mathcal{L}$ be an adequate language for T-coalgebras with the Hennessy-Milner property. Let $f : X \longrightarrow Y$ be a function with $(X, \xi), (Y, \zeta) \in \mathsf{Coalg}(T)$ and ζ simple such that $\Phi_\xi(x) = \Phi_\zeta(f(x))$ for all $x \in X$. Then $f : \xi \longrightarrow \zeta$ is a coalgebra morphism.*

4.3   Logical Congruences and Weak Finality

The results of the previous sections already imply that the existence of logical congruences is equivalent to existence of final coalgebras. Nevertheless, we will do a direct proof of this fact because we can provide a categorical proof that can be reused in several other examples. Our main categorical tool to produce final coalgebras is Freyd's existence Theorem of a final object [11]:

Theorem 9. *A cocomplete category $\mathbb{C}$ has a final object iff it has a small set of objects S which is weakly final, i.e. for every object $c \in \mathbb{C}$ there exists a $s \in S$ and arrow $c \longrightarrow s$; a final object is a colimit of the diagram induced by S^2*

The set S, mentioned in the previous theorem, is called a **solution set**. Freyd's Theorem is strongly related to the Adjoint Functor theorem. Recall that in the category of sets every object only has a set of subobjects (subsets). In Proposition 5 we proved that if a language $\mathcal{L}$ is adequate and has logical congruences each coalgebra (X, ξ) can be mapped to some coalgebra of the form (Z_ξ, ζ_ξ) (Definition 10). This tells us that the coalgebras that are based on subsets of $\mathcal{PL}$ form a solution set, which by Freyd's theorem implies the existence of a final object. Therefore the following holds true.

Proposition 7. *If a language $\mathcal{L}$ is adequate and has logical congruences for T-coalgebras then there exists a final T-coalgebra which is obtained as a colimit of the diagram induced by the T-coalgebras (Z_ξ, ζ_ξ) (Definitions 10 & 11).*

This proposition supplies us with another description of the final coalgebra. Moreover, following the path: Hennessy-Milner $\Rightarrow$ logical congruences $\Rightarrow$ final coalgebras we have another proof of Goldblatt's Theorem. This alternative proof is not as simple as the construction presented in Theorem 2 but illustrates the importance of adequacy. This can be restated saying that the Hennessy-Milner property is a solution set condition to obtain final coalgebras.

4.4   Different Faces of the Hennessy-Milner Property

In summary, in the category of sets and functions, under the assumption of adequacy we have the following equivalents of the Hennessy-Milner property.

---

[2] Recall that the diagram induced by a set of objects is the inclusion functor of the full subcategory of $\mathbb{C}$ generated by S.

Theorem 10. *Let $\mathcal{L}$ be an adequate language for T-coalgebras. The following conditions are equivalent:*

1. *$\mathcal{L}$ has the Hennessy-Milner property.*
2. *The function $\zeta : Z_{\mathcal{L}} \longrightarrow TZ_{\mathcal{L}}$ from equation (2), page 102, on the set of satisfiable $\mathcal{L}$-theories is well-defined and endows $Z_{\mathcal{L}}$ with the structure of a final coalgebra.*
3. *The set $Z_{\mathcal{L}}$ admits a coalgebraic structure, for T, such that for each coalgebra ξ the function $f_{\xi} : X \longrightarrow Z_{\mathcal{L}}$, i.e. the restriction of the codomain of the theory map Φ_{ξ}, is a morphism of coalgebras.*
4. *For each coalgebra ξ the relation $\leftrightsquigarrow_{\mathcal{L}}^{\xi}$ is a congruence of coalgebras.*
5. *For each coalgebra ξ the set of satisfiable theories in ξ admits a coalgebraic structure $\zeta_{\xi} : Z_{\xi} \longrightarrow TZ_{\xi}$, such that the function $e_{\xi} : X \longrightarrow Z_{\xi}$ mapping a state $x \in \xi$ to its $\mathcal{L}$-theory $e_{\xi}(x) = \Phi \in Z_{\xi}$ is a morphism of coalgebras.*
6. *Let (X_1, ξ_1) and (X_2, ξ_2) be T-coalgebras. If the diagram on the left is a pullback (in Set), there exists a coalgebra (Y, γ) and morphisms $f_1 : \xi_1 \longrightarrow \gamma; f_2 : \xi_2 \longrightarrow \gamma$ such that*

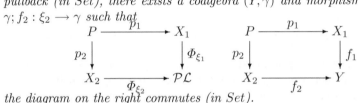

the diagram on the right commutes (in Set).

Proof. The equivalence between 1) and 2) is the content of Theorem 3. The implication from 2) to 3) is obvious and the converse direction is a consequence of the definition of ζ: any map $\zeta' : Z_{\mathcal{L}} \longrightarrow TZ_{\mathcal{L}}$ that turns the theory maps f_{ξ} into coalgebra morphisms must be obviously equal to ζ. The equivalence between 1) and 4) follows from Theorem 7. Item 4) is equivalent to item 5) because $\leftrightsquigarrow_{\mathcal{L}}^{\xi} = \mathrm{Ker}(\Phi_{\xi}) = \mathrm{Ker}(f_{\xi})$. Finally the equivalence between 6) and 1) is immediate from the canonical characterisation of pullbacks in Set.

5   Generalization to Other Categories

In this section we show how the result for coalgebras on Set can be generalised to coalgebras over other base categories. The first part of the section discusses how to generalise the notion of a language to a functor $T : \mathbb{C} \longrightarrow \mathbb{C}$ on an arbitrary category $\mathbb{C}$. After that we focus on a special class of categories, those that are regularly algebraic over Set, and show that the results from the previous section generalise smoothly to these categories. Due to space limitations we cannot provide the necessary categorical definitions. Instead we refer the reader to [3] where all our terminology is explained.

When generalising the notion of an abstract coalgebraic language to categories other than Set we face the problem that we do not know much about the structure of the given base category $\mathbb{C}$. In particular, unlike in the case $\mathbb{C} = \mathsf{Set}$, we do not know how to move freely from an object $\mathcal{L}$ representing the formulas to an object $\mathcal{PL}$ that represents the theories of a given language. This leads us to the following definition of an adequate object for T-coalgebras.

Definition 12. *Let T be a functor $T : \mathbb{C} \to \mathbb{C}$. An object $\mathcal{L}$, in $\mathbb{C}$ is an* adequate object *for T-coalgebras if there exists a natural transformation $\Phi : U \to \Delta_{\mathcal{L}}$, where $U : \mathsf{Coalg}(T) \to \mathbb{C}$ is the forgetful functor and $\Delta_{\mathcal{L}} : \mathsf{Coalg}(T) \to \mathbb{C}$ is the constant functor with value $\mathcal{L}$. We call the components of Φ* theory morphisms.

At first sight it is not completely clear why our definition of an adequate object for T-coalgebras is a good generalisation of a language for T-coalgebras. Under the additional assumption, that we are looking at a category $\mathbb{C}$ that is dual to some category $\mathbb{A}$, our notion seems to be quite natural.

Example 5.   – For $\mathbb{C} = \mathsf{Set}$ we have $\mathcal{L}$ is an adequate abstract coalgebraic language (Definition 4) for T with theory maps $\{\Phi_\xi\}_{\xi \in \mathsf{Coalg}(T)}$ iff $\mathcal{PL}$ together with $\{\Phi_\xi\}_{\xi \in \mathsf{Coalg}(T)}$ is an adequate object for T-coalgebras.
  – Let $\mathbb{C} = \mathsf{Stone}$ the category of Stone spaces and continuous functions and let $T : \mathsf{Stone} \to \mathsf{Stone}$ be a functor. We can use the duality between Stone and the category BA of Boolean algebras to see that an adequate object $\mathcal{L}$ for T-coalgebras corresponds to some Boolean algebra $A_{\mathcal{L}}$. Hence $\mathcal{L}$ is, again by duality, isomorphic to the collection of ultrafilters (=theories) over $A_{\mathcal{L}}$.

In order to arrive at a generalisation of a language for T-coalgebras which has the Hennessy-Milner property, we use the results that we obtained in Section 4. Theorem 10 shows that there are at least three ways to obtain this generalization. Intuitively these are: 1) Our "language" contains somehow the carrier of a final coalgebra. 2) if states are identified by theory morphism then they are identified by some coalgebra morphisms. And finally 3) the image of a theory morphism carries a coalgebraic structure, equivalently, logical equivalence of states is a congruence; this is related to $(RegEpi, Mono)$-factorizations and more generally to factorization structures. The intuition behind $(RegEpi, Mono)$-structured categories is that the image of a morphism is an object in the category, examples of such factorizations are the isomorphism theorems of algebra.

Definition 13. *Let $\mathcal{L}$ be an adequate object for T-coalgebras for a functor $T : \mathbb{C} \to \mathbb{C}$. We say $\mathcal{L}$ is* almost final *if $\mathcal{L}$ has a subobject $m : Z \to \mathcal{L}$ that can be uniquely lifted to a final T-coalgebra (Z, ζ) such that $m = \Phi_\zeta$. If the base category $\mathbb{C}$ has pullbacks we say $\mathcal{L}$ has the* Hennessy-Milner property *if every pullback (P, p_1, p_2) (in $\mathbb{C}$) of theory morphisms Φ_{ξ_1} and Φ_{ξ_2} can be factored (in $\mathbb{C}$) using a pair of coalgebra morphisms. Finally, if the base category $\mathbb{C}$ is $(RegEpi, Mono)$-structured, we say $\mathcal{L}$ has* logical congruences *if for each theory morphism Φ_ξ and each $(RegEpi, Mono)$-factorization (e, Z_ξ, m) of Φ_ξ, there exists a coalgebraic structure $\zeta_\xi : Z_\xi \to T(Z_\xi)$ such that e is a coalgebra morphism from ξ to ζ_ξ.*

As we proved in Theorem 10, all of the three notions from the previous definition are equivalent if our base category $\mathbb{C}$ is Set. How do they relate in other categories?

Proposition 8. *Let $\mathbb{C}$ be a cocomplete and $(RegEpi, Mono)$-structured category with pullbacks. Let T be an endofunctor on $\mathbb{C}$ and let $\mathcal{L}$ be an adequate object for T-coalgebras which is wellpowered. We have: $\mathcal{L}$ has logical congruences $\Rightarrow$ $\mathcal{L}$ is almost final $\Rightarrow$ $\mathcal{L}$ has the Hennessy-Milner property. Furthermore, if T*

preserves monomorphisms, the converse implications are true as well and thus all three notions are equivalent.

In particular, the previous proposition demonstrates that under mild assumptions on our base category we can establish the existence of a final coalgebra for a functor T by proving that there exists some adequate object for T-coalgebras that has logical congruences. We use this fact in order to prove the following

Theorem 11. *Let $\mathbb{A}$ be a category that is regularly algebraic over Set with forgetful functor $V : \mathbb{A} \longrightarrow$ Set and let $T : \mathbb{A} \longrightarrow \mathbb{A}$ be a functor. The functor T has a final coalgebra iff there exists an adequate object $\mathcal{L}$ for T-coalgebras that has the Hennessy-Milner property.*

Our results apply to any category of algebras such as the category BA of Boolean algebras and the category DL of distributive lattices, but also to categories like the category Stone of Stone spaces. We hope to be able to extend the scope of Theorem 11 to categories that are topological over Set such as the category Meas of measurable spaces.

6   Conclusions

In this paper, we have studied three ways to express behavioural equivalence of coalgebra states: using final coalgebras, using coalgebraic languages that have the Hennessy-Milner property and using coalgebraic languages that have logical congruences. We provided a simple proof for the fact that these three different methods are equivalent when used to express behavioural equivalence between set coalgebras. As by-products of our proof we obtained a straightforward construction of final coalgebras as canonical models of coalgebraic logics and a concrete characterisation of simple coalgebras as logical quotients.

A main topic for further research is that of abstract coalgebraic languages for functors on categories different from Set. Section 5 illustrates how abstract coalgebraic languages can be generalised to arbitrary categories. The main result of this section states that for a functor on any category that is regularly algebraic over Set an adequate object with the Hennessy-Milner property exists iff there exists a final coalgebra (Theorem 11). The proof of this theorem demonstrates that logical congruences are useful in order to prove the existence of a final coalgebra. A crucial ingredient for the proof is Freyd's Theorem (Theorem 9). Our hope is that the scope of Theorem 11 can be extended to a larger class of categories that satisfy the conditions of Freyd's theorem.

Another gain of using logical congruences is that they revealed that the Hennessy-Milner property is related to the description of a particular factorisation structure (cf. Def. 13, Prop. 8). In our paper we considered ($RegEpi$, $Mono$)-structured categories, but it is quite natural to generalise the results here to other factorisation structures. We believe that a study of these factorisations will lead to a coalgebraic understanding of non standard bisimulations as, for example, discussed in [7]. A step in this direction has already been made in [10]. There it was shown that using logically invariant morphisms between coalgebras, that are not necessarily coalgebra morphisms, canonical models for abstract coalgebraic languages can be presented as final objects.

Nevertheless, many unsolved issues remain in the case of Set. For instance, it would be interesting to know what conditions on the functor are needed in order to allow to restrict Theorem 10 to countable coalgebraic languages. It is easy to see that this is possible if the functor is finitary and preserves finite sets. Furthermore the theorems presented in our paper can be formulated for covarieties, in fact for classes of coalgebras closed under coproducts, quotients and isomorphisms. The question whether the restriction to those classes of coalgebras yields interesting results is left as future work. Finally, we would like to clarify the seemingly close connection between our work and the construction of the so-called final algebra which can be found in work on hidden algebra ([5]).

References

1. Aczel, P., Mendler, N.P.: A final coalgebra theorem. Category Theory and Computer Science, 357–365 (1989)
2. Adámek, J.: A logic of coequations. In: Ong, L. (ed.) CSL 2005. LNCS, vol. 3634, pp. 70–86. Springer, Heidelberg (2005)
3. Adámek, J., Herrlich, H., Strecker, G.E.: Abstract and Concrete Categories: The Joy of Cats. John Wiley & Sons, Chichester (1990)
4. Blackburn, P., de Rijke, M., Venema, Y.: Modal logic. Cambridge Tracts in Theoretical Computer Science, vol. 53. Cambridge University Press, Cambridge (2002)
5. Goguen, J., Malcolm, G.: A hidden agenda. Theoretical Computer Science 245(1), 55–101 (2000)
6. Goldblatt, R.: Final coalgebras and the Hennessy-Milner property. Annals of Pure and Applied Logic 138(1-3), 77–93 (2006)
7. Goranko, V., Otto, M.: Model Theory of Modal Logic. In: Handbook of Modal Logic, pp. 255–325. Elsevier, Amsterdam (2006)
8. Gumm, H.P.: Elements of the General Theory of Coalgebras, Tech. report, Rand Africaans University, Johannesburg, South Africa (1999) (preliminary version)
9. Jacobs, B., Rutten, J.: A tutorial on (co)algebras and (co)induction. EATCS Bulletin 62, 62–222 (1997)
10. Kurz, A., Pattinson, D.: Coalgebraic Modal Logic of Finite Rank. Mathematical Structures in Computer Science 15(3), 453–473 (2005)
11. MacLane, S.: Categories for the Working Mathematician. Graduate Texts in Mathematics, vol. 5. Springer, New York (1971)
12. Moss, L.S.: Coalgebraic Logic. Annals of Pure and Applied Logic 96(1-3), 277–317 (1999); Erratum published Ann. P. Appl. Log 99, 241–259 (1999)
13. Pattinson, D.: Coalgebraic Modal Logic: Soundness, Completeness and Decidability of Local Consequence. Theoretical Computer Science 309(1-3), 177–193 (2003)
14. Rutten, J.J.M.M.: Automata and coinduction (an exercise in coalgebra). In: Sangiorgi, D., de Simone, R. (eds.) CONCUR 1998. LNCS, vol. 1466, pp. 194–218. Springer, Heidelberg (1998)
15. Rutten, J.J.M.M.: Universal coalgebra: a theory of systems. Theoretical Computer Science 249(1), 3–80 (2000)
16. Schröder, L.: Expressivity of Coalgebraic Modal Logic: The Limits and Beyond. Theoretical Computer Science 390, 230–247 (2008)
17. Stirling, C.: Modal and temporal properties of processes. Springer, Heidelberg (2001)

Let's See How Things Unfold: Reconciling the Infinite with the Intensional
(Extended Abstract)

Conor McBride

University of Strathclyde

1 Introduction

Coinductive types model infinite structures unfolded on demand, like politicians' excuses: for each attack, there is a defence but no likelihood of resolution. Representing such evolving processes coinductively is often more attractive than representing them as functions from a set of permitted observations, such as projections or finite approximants, as it can be tricky to ensure that observations are meaningful and consistent. As programmers and reasoners, we need coinductive definitions in our toolbox, equipped with appropriate computational and logical machinery.

Lazy functional languages like HASKELL [18] exploit call-by-need computation to over-approximate the programming toolkit for coinductive data: in a sense, all data is coinductive and delivered on demand, or not at all if the programmer has failed to ensure the *productivity* of a program.

Tatsuya Hagino pioneered a more precise approach, separating initial data from final codata [10]. The corresponding discipline of 'coprogramming' is given expression in Cockett's work on CHARITY [5,6] and in the work of Turner and Telford on 'Elementary Strong Functional Programming' [22,21,23]. Crucially, all distinguish recursion (structurally decreasing on input) from *corecursion* (structurally increasing in output). As a total programmer, I am often asked 'how do I implement a *server* as a program in your terminating language?', and I reply that I do not: a server is a *coprogram* in a language guaranteeing liveness.

To combine programming and reasoning, or just to program with greater precision, we might look to the proof assistants and functional languages based on intensional type theories, which are now the workhorses of formalized mathematics and metatheory, and the mainspring of innovation in typed programming [16,4,14]. But we are in for a nasty shock if we do. Coinduction in COQ is *broken*: computation does not preserve type. Coinduction in AGDA is *weak*: dependent observations are disallowed, so whilst we can unfold a process, we cannot *see* that it yields its unfolding.

At the heart of the problem is *equality*. Intensional type theories distinguish two notions of equality: the typing rules identify types and values according to an equality *judgment*, decided mechanically during typechecking; meanwhile, we can express equational *propositions* as types whose inhabitants (if we can find them) justify the substitution of like for like.

A. Kurz, M. Lenisa, and A. Tarlecki (Eds.): CALCO 2009, LNCS 5728, pp. 113–126, 2009.
© Springer-Verlag Berlin Heidelberg 2009

In neither CoQ nor AGDA is a coprogram judgmentally equal to its unfolding, hence the failure in the former. That is not just bad luck: in this presentation, I check that it is impossible for any decidable equality to admit unfolding.

Moreover, neither system admits a substitutive propositional equality which identifies bisimilar processes, without losing the basic computational necessity that closed expressions compute to canonical values [13]. That is just bad luck: in this presentation, I show how to construct such a notion of equality, following earlier joint work with Altenkirch on observational equality for functions [2].

The key technical ingredient is the notion of 'interaction structure' due to Hancock and Setzer [11] — a generic treatment of indexed coinductive datatypes, which I show here to be closed under its own notion of bisimulation. This treatment is ready to be implemented in a new version of the EPIGRAM system.

Equipped with a substitutive propositional equality that includes bisimulation, we can rederive CoQ's dependent observation for codata from AGDA's simpler coalgebraic presentation, whilst ensuring that what types we have, we hold. Let's see how things unfold.

2   The Problem

Eduardo Giménez pioneered CoQ's treatment of coinduction [7]. It was a great step forward in its time, giving Coq access to many new application domains. Giménez was aware of the problem with type preservation, giving a counterexample in his doctoral thesis [8]. The problem did not become particularly widely known until recently, when Nicolas Oury broke an overenthusiastic early version of coinduction in AGDA, then backported his toxic program to CoQ, resulting in a flurry of activity on mailing lists which has not yet entirely subsided.

Presented with a categorical flavour, CoQ's treatment is essentially thus: for any given strictly positive functor $F : \mathsf{Set} \to \mathsf{Set}$, we acquire a coinductive set νF equipped with a coconstructor and a coiterator (or 'unfold', or 'anamorphism') which grows νF values on demand by successive applications of a coalgebra to a 'seed' of arbitrary type. Keeping polymorphism implicit, we obtain:

$$\nu F : \mathsf{Set}$$
$$\mathsf{in}_F : F\,(\nu F) \to \nu F$$
$$\mathsf{coit}_F : (S \to F\,S) \to S \to \nu F$$

Of course, CoQ actually provides a much richer notation for coprograms than just coit_F, but a streamlined presentation will help to expose the problem.

For the standard example of coinductive lists, this specializes (modulo high school algebra) to the traditional pair of coconstructors and unfold.

$$\mathsf{CoList}_X : \mathsf{Set}$$
$$\mathsf{nil}_X : \mathsf{CoList}_X$$
$$\mathsf{cons}_X : X \to \mathsf{CoList}_X \to \mathsf{CoList}_X$$
$$\mathsf{unfold}_X : (S \to 1 + X \times S) \to S \to \mathsf{CoList}_X$$

For a given seed $s : S$, the unfold_X coalgebra may deliver a value $\mathsf{inl}\,()$ from the left summand to construct nil_X, or some $\mathsf{inr}\,(x, s')$ to construct a cons_X with head x and a tail grown on demand from seed s'. For a standard example, construct the infinite sequence of natural numbers, starting from a given counter:

$$\mathsf{natsFrom}\ n\ \mapsto\ \mathsf{unfold}_\mathbb{N}\,(\lambda n \mapsto \mathsf{inr}\,(n, n+1))\,n$$

The corresponding elimination behaviour is *dependent* case analysis, not merely branching according to a coconstructor choice, but *seeing* every value as an unfolding to a coconstructor. I write $(x : S) \to T$ for dependent function spaces with explicit application. In general and particular, we acquire:

$$\mathsf{case}_F\ :\ (P : \nu F \to \mathsf{Set})\ \to\ ((t : F\,(\nu F)) \to P\,(\mathsf{in}_F\,t))\ \to\ (x : \nu F) \to P\,x$$

$$\begin{aligned}
\mathsf{caseCoList}_X\ :\ &(P : \mathsf{CoList}_X \to \mathsf{Set}) \to \\
&P\,\mathsf{nil}_X \to \\
&((x : X) \to (xs : \mathsf{CoList}_X) \to P\,(\mathsf{cons}_X\,x\,xs)) \to \\
&(xs : \mathsf{CoList}_X) \to P\,xs
\end{aligned}$$

We may readily recover the destructor — the terminal coalgebra — as a degenerate case analysis:

$$\mathsf{out}_F : \nu F \to F\,(\nu F)$$
$$\mathsf{out}_F = \mathsf{case}_F\,(\lambda_- \mapsto \nu F)\,(\lambda t \mapsto t)$$

Writing $F\,-$ for the functorial action on functions, we can readily see that $\mathsf{coit}_F\,(F\,\mathsf{out}_F)$ does the same job as in_F, but the latter is not redundant, as it is needed to form the type of case_F.

Intensional type theories are computational — if we test a CoList for emptiness, we should certainly receive true or false. It is thus vital to animate case analysis with operational behaviour. Moreover, whatever we choose will play a key role in the type system. Judgmental equality, which I write $\equiv$, is given in CoQ as the congruence and equivalence closure of reduction, $\leadsto$. Decidable typechecking thus rests on terminating reduction. We take care to coiterate only on demand from case analysis, hence finitely often. CoQ computes as follows (but what is the type of the term in the box?)

$$\mathsf{case}_F\ P\ p\ (\mathsf{in}_F\,t)\qquad \leadsto p\,t$$
$$\mathsf{case}_F\ P\ p\ (\mathsf{coit}_F\,f\,s) \leadsto \boxed{p\,(F\,(\mathsf{coit}_F\,f)\,(f\,s))}$$

In the special case of CoList, computing the application of coalgebra to seed delivers a value for $p\,s$ which decides whether the resulting CoList is a nil_X to be replaced by n or a cons_X to be replaced by c. (Again, typecheck the boxes.)

$$\mathsf{caseCoList}_X\ P\ n\ c\ \mathsf{nil}_X\qquad \leadsto n$$
$$\mathsf{caseCoList}_X\ P\ n\ c\ (\mathsf{cons}_X\,x\,xs) \leadsto c\,x\,xs$$
$$\mathsf{caseCoList}_X\ P\ n\ c\ (\mathsf{unfold}_X\,p\,s) \leadsto
\begin{cases}
\boxed{n} & \text{if } p\,s \leadsto^* \mathsf{inl}\,() \\
\boxed{c\,x\,(\mathsf{unfold}\,p\,s')} & \text{if } p\,s \leadsto^* \mathsf{inr}\,(x, s')
\end{cases}$$

Here inl and inr are injections to the sum $1+X{\times}S$ directing coconstructor choice.

Coiteration on demand rules out spontaneous unfolding

$$\mathsf{coit}_F\ f\ s\ \not\leadsto\ \mathsf{in}_F\ (F\ (\mathsf{coit}_F\ f)\ (f\ s))$$

$$\mathsf{unfold}_X\ p\ s\ \not\leadsto\ \begin{cases} \mathsf{nil}_X & \text{if } p\ s \leadsto^* \mathsf{inl}\ () \\ \mathsf{cons}_X\ x\ (\mathsf{unfold}_X\ p\ s') & \text{if } p\ s \leadsto^* \mathsf{inr}\ (x, s') \end{cases}$$

preventing infinite regress. *Ipso facto*, the corresponding judgmental equations do not hold: the reducts shown in boxes above conspicuously fail to share the types of the redices from which they spring. Let us check:

$$\mathsf{case}_F\ P\ p\ (\mathsf{coit}_F\ f\ s)\ \leadsto\ p\ (F\ (\mathsf{coit}_F\ f)\ (f\ s))$$
$$\vdots \qquad\qquad \vdots$$
$$P\ (\mathsf{coit}_F\ f\ s)\ \not\equiv\ P\ (\mathsf{in}_F\ (F\ (\mathsf{coit}_F\ f)\ (f\ s)))$$

Preservation of typing fails precisely because we do not dare compute with laws which we should prefer somehow to hold.

Where does this problem bite? In fact, we can construct a bogus *proof* of exactly the missing equation, via the standard intensional propositional equality:

$$-=-\ :\ X \to X \to \mathsf{Prop}$$
$$\mathsf{refl}\ \ :\ (x{:}X) \to x = x$$

According to the types of our components, we may check that

$$\mathsf{case}_F\ (\lambda x \mapsto x = \mathsf{in}_F\ (F\ (\mathsf{coit}_F\ f)\ (f\ s)))$$
$$(\lambda t \mapsto \mathsf{refl}\ (\mathsf{in}_F\ t))$$
$$(\mathsf{coit}_F\ f\ s)$$
$$:\ \mathsf{coit}_F\ f\ s = \mathsf{in}_F\ (F\ (\mathsf{coit}_F\ f)\ (f\ s))$$

but this proof computes in one step to

$$\mathsf{refl}\ (\mathsf{in}_F\ (F\ (\mathsf{coit}_F\ f)\ (f\ s)))\ \not\leadsto\ \mathsf{coit}_F\ f\ s = \mathsf{in}_F\ (F\ (\mathsf{coit}_F\ f)\ (f\ s))$$

Indeed, this propositional equality has the canonicity property, so in the empty context, all proofs must compute to some $\mathsf{refl}\ t$, making propositional equality correspond exactly to judgmental equality and thus *intensional*. Indeed, this intensionality allows us to define predicates which invalidate dependent case analysis for codata — there are predicates which hold for coconstructed codata but do not hold universally. It should not be possible to prove that a coiteration is intensionally equal to the corresponding coconstruction, just as one cannot prove intensional equality of these implementations of the identity on $\mathbb{N}$:

$$\lambda n \mapsto n \qquad \begin{array}{l} \lambda\ 0 \quad\ \mapsto 0 \\ |\ n+1 \mapsto n+1 \end{array}$$

As so often in type theory, the problem is the struggle for equality.

3   Type Equality versus Reduction

The trade in labour between the judgments for equality and type inhabitation is a crucial aspect of intensional type theories. We expect $\Gamma \vdash - : T$ to be checkable only for specific candidate terms — inhabitation requires evidence. However, we keep $\Gamma \vdash s \equiv t : T$ decidable, and we take advantage by admitting the rule

$$\frac{\Gamma \vdash s : S \quad \Gamma \vdash S \equiv T : \mathsf{Set}}{\Gamma \vdash s : T}$$

If T is a complex type which simplifies mechanically to some trivial type S, then a trivial inhabitant will suffice. As we are free to program in types, we are in a position to shift work from humans to machines, making developments like Gonthier's proof of the Four Colour Theorem tractable [9].

Within the bounds of decidability, and hopefully also good taste, we are free to question whether $\equiv$ should be exactly the equivalence and congruence closure of $\rightsquigarrow$, or whether it might be a little more generous. With care, a *type-directed* test for $\equiv$ can be made to admit rules like

$$\frac{\Gamma ; x{:}S \vdash f\, x \equiv g\, x : T[x]}{\Gamma \vdash f \equiv g : (x{:}S) \to T} \qquad \frac{\Gamma \vdash u : 1 \quad \Gamma \vdash v : 1}{\Gamma \vdash u \equiv v : 1}$$

which are rather more difficult to capture by rewriting alone.

We now face a temptation. Even though coit_F must expand only on demand,

$$\mathsf{coit}_f\, f\, s \not\rightsquigarrow \mathsf{in}_F\, (F\, (\mathsf{coit}_F\, f)\, (f\, s))$$

might we not consider an equality test to be such a demand? When testing an equation, we could unfold coiteration on one side while we have coconstructors on the other, effectively adding the rule

$$\frac{\Gamma \vdash F\, (\mathsf{coit}_F\, f)\, (f\, s) \equiv t : F\, (\nu F)}{\Gamma \vdash \mathsf{coit}_F\, f\, s \equiv \mathsf{in}_F\, t : \nu F}$$

We can safely delay such unfoldings until we are comparing normal forms, as computation will unfold coiteration in redices anyway. Following this strategy, we arrive at a terminating test which is sound with respect to our extended $\equiv$ and which fixes our type error. It looks like we have won!

I used to advocate this approach, but I record it now only to spare others the embarrassment of falling for its charms. No such test can be complete if $\equiv$ is transitive. To see this, consider an automaton (a Turing machine perhaps) specified by a transition function of type $a : S \to 1 + S$ indicating whether the automaton halts or evolves to a new state. We may construct a coalgebra $\mathsf{tick}\, a$ for bitstreams in $\nu(2\times)$ with carrier $1 + S$ as follows

$$\begin{aligned}\mathsf{tick}\, a\; :\quad & 1{+}S \;\to 2 \times (1{+}S) \\ \mathsf{tick}\, a \mapsto\; & \lambda\, (\mathsf{inl}\, ()) \mapsto (\mathsf{true}, \mathsf{inl}\, ()) \\ & \mid (\mathsf{inr}\, s) \;\mapsto (\mathsf{true}, a\, s)\end{aligned}$$

Now ask whether an automaton which has not yet halted ticks the same as one which has halted already:

$$\mathsf{coit}_{2\times} (\mathsf{tick}\ a)\ (\mathsf{inr}\ s) \ \equiv\ \mathsf{coit}_{2\times} (\mathsf{tick}\ a)\ (\mathsf{inl}\ ())\qquad ?$$

If the former ever halts, starting from s, then it is testably equal to a sufficiently large unfolding of form

$$\mathsf{in}_{2\times} (\mathsf{true}, \mathsf{in}_{2\times} (\mathsf{true}, \ldots \mathsf{in}_{2\times} (\mathsf{true}, \mathsf{coit}_{2\times} (\mathsf{tick}\ a)\ (\mathsf{inl}\ ())) \ldots))$$

which is clearly also an unfolding of the latter. By transitivity, a complete decision procedure for $\equiv$ must find the intermediate value for itself, effectively solving the halting problem. Hence it does not exist. We cannot hope to repair CoQ's problem by augmenting the judgmental equality.

4   Weakly Final Coalgebras: A Principled Retreat

Inasmuch as dependent case analysis is not really justified by the intensional behaviour of codata — an uncomputed coiteration is not given by a coconstructor, even if there is no means to observe the difference within the theory — the principled thing to do is to throw out dependent case analysis, retaining only the nondependent observation.

We arrive at the following cut-down presentation, long advocated by Anton Setzer as closer to the destructor-based, coalgebraic intuition behind coinductive sets, and corresponding quite closely to the position implemented in AGDA [15,19,19,20]. For suitable strictly positive functors F, we obtain

$$\nu F : \mathsf{Set}$$
$$\mathsf{out}_F : \nu F \to F\ (\nu F)$$
$$\mathsf{coit}_F : \forall S.(S \to F\ S) \to S \to \nu F$$

$$\mathsf{out}_F\ (\mathsf{coit}_F\ f\ s)\ \rightsquigarrow\ F\ (\mathsf{coit}_F\ f)\ (f\ s)$$

We can restore something like the coconstructor, taking $\mathsf{in}'_F \mapsto \mathsf{coit}_F\ (F\ \mathsf{out}_F)$ for which computation and definitional expansion give us, judgmentally, that $\mathsf{out}_F\ (\mathsf{in}'_F\ t) \equiv F\ \mathsf{in}'_F\ (F\ \mathsf{out}_F\ t)$ which is perhaps less than desirable, but at least believable. Pragmatically, AGDA adds in_F directly, with the more immediate decomposition behaviour

$$\mathsf{out}_F\ (\mathsf{in}_F\ t)\ \rightsquigarrow\ t$$

but this does not mitigate the main problem with this presentation: out_F is only a *weakly* final coalgebra. We are guaranteed that $\mathsf{coit}_F\ f$ is a solution of

$$\mathsf{out}_F\ (g\ s) = F\ g\ (f\ s)$$

but it is not *unique*. Consider $g \mapsto \mathsf{in}'_F \cdot F\ (\mathsf{coit}_F\ f) \cdot f$ and observe that, even with only the pointwise propositional reasoning that intensional type theories admit,

$$\mathsf{out}_F (g\ s)$$

$=$ \hfill {definition}

$$\mathsf{out}_F ((\mathsf{in}'_F \cdot F\ (\mathsf{coit}_F\ f) \cdot f)\ s)$$

$=$ \hfill {applied composition}

$$\mathsf{out}_F\ (\mathsf{in}'_F\ (F\ (\mathsf{coit}_F\ f)\ (f\ s)))$$

$=$ \hfill {computation on demand}

$$F\ \mathsf{in}'_F\ (F\ \mathsf{out}_F\ (F\ (\mathsf{coit}_F\ f)\ (f\ s)))$$

$=$ \hfill {F functorial}

$$F\ (\mathsf{in}'_F \cdot \mathsf{out}_F \cdot \mathsf{coit}_F\ f)\ (f\ s)$$

$=$ \hfill {computation on demand}

$$F\ (\mathsf{in}'_F \cdot F\ (\mathsf{coit}_F\ f) \cdot f)\ (f\ s)$$

$=$ \hfill {definition}

$$F\ g\ (f\ s)$$

However, we cannot prove that $g = \mathsf{coit}_F\ f$, or even that they are pointwise equal. The best we can do is to define an *ad hoc* notion of bisimulation and show that both functions are pointwise bisimilar. We cannot treat bisimulation as a congruence and presume that all predicates respect it for the very good reason that they do not — the intensional equality is again a key counterexample.

This approach is in keeping with the standard properties and difficulties of intensional type theories. On the one hand, we have restored the fact that computation preserves type. On the other hand, we are forced to reason at an intensional level, just as we already are with *functions*. Functions, too, are equipped with only non-dependent elimination behaviour: application allows us to use functions but not to inspect or standardize their construction. Just as we are free to reason dependently about the result of a function application, so we may reason about the result of out_F by whatever means F supports, but the construction of elements of νF remains inscrutable.

5   O for an Observational Propositional Equality!

Earlier, we noted that an unjustified dependent case analysis principle gave rise to a bogus intensional equation between codata which were merely bisimilar. As it happens, we turn that fact on its head and derive dependent case analysis from a bisimulation.

Let us start from the safe, non-dependent presentation of codata given in the previous section. Suppose we could find a notion of equality, $\sim$, say, reflexive and substitutive at each set X

$$\mathsf{refl}_X : (x : X) \to x \sim_X x$$
$$\mathsf{subst}_X : \forall x, y.\ x \sim_X y \to (P : X \to \mathsf{Set}) \to P\ x \to P\ y$$

for which we were able to show

$$\mathsf{io}_F\ :\ (x : \nu F) \to \mathsf{in}_F\ (\mathsf{out}_F\ x) \sim_{\nu F} x$$

We might then construct a general dependent case analysis operator from the destructor out_F.

$$\mathsf{case}_F \;:\; (P\!:\!\nu F \to \mathsf{Set}) \;\to\; ((t\!:\!F\,(\nu F)) \to P\,(\mathsf{in}_F\,t)) \;\to\; (x\!:\!\nu F) \to P\,x$$
$$\mathsf{case}_F\,P\,p\,x \;\mapsto\; \mathsf{subst}_{\nu F}\,(\mathsf{in}_F\,(\mathsf{out}_F\,x))\,x\,(\mathsf{io}_F\,x)\,P\,(p\,(\mathsf{out}_F\,x))$$

If, moreover, we are lucky enough to have that for any $q : x \sim_X x$

$$\mathsf{subst}_X\,q\,P\,p \;\equiv\; p$$

then computing out_F gives us that the following hold judgmentally:

$$\mathsf{case}_F\,P\,p\,(\mathsf{in}_F\,t) \qquad \equiv p\,t$$
$$\mathsf{case}_F\,P\,p\,(\mathsf{coit}_F\,f\,s) \equiv \mathsf{subst}_{\nu F}\,(\mathsf{io}_F\,(\mathsf{coit}_F\,f\,s))\,P\,(p\,(F\,(\mathsf{coit}_F\,f)\,(f\,s)))$$

These equations correspond almost exactly to the computation rules in the CoQ presentation, but the type error in the coiteration case is repaired by explicit appeal to $\mathsf{subst}_{\nu F}$. If only we had an *observational* propositional equality, substitutive for bisimulation, then we should recover the uniqueness of $\mathsf{coit}_F\,f$ propositionally, although it cannot be unique with respect to the judgmental equality $\equiv$. But is this a tall order? I claim not.

6   Observational Equality

Let me sketch how to construct a propositional equality with the desired properties, extending earlier joint work with Altenkirch and Swierstra [2]. The machinery is quite intricate, but most of the choices are forced by the setup.

Start from a type theory equipped at least with 0, 1, dependent function types $(x\!:\!S) \to T$, and dependent pair types $(x\!:\!S) \times T$, but perhaps including sums and inductive types. Now introduce a propositional subuniverse of Set, with a 'decoder' mapping each proposition to the set of its proofs.

$$\mathsf{Prop} \;:\; \mathsf{Set} \qquad [\![-]\!] \;:\; \mathsf{Prop} \to \mathsf{Set}$$

closed under falsity, truth, conjunction, and universal quantification over sets.

$$
\begin{aligned}
[\![\bot]\!] &\;\leadsto\; 0 \\
[\![\top]\!] &\;\leadsto\; 1 \\
[\![P \wedge Q]\!] &\;\leadsto\; [\![P]\!] \times [\![Q]\!] \\
[\![\forall x\!:\!S.\,P]\!] &\;\leadsto\; (x\!:\!S) \to [\![P]\!]
\end{aligned}
$$

We may define implication, $P \Rightarrow Q$, as a degenerate universal quantification, $\forall_{\_}\!:\![\![P]\!].\,Q$, recovering $[\![P \Rightarrow Q]\!] \leadsto [\![P]\!] \to [\![Q]\!]$.

Next, we may introduce equality for sets and values, not as inductive definitions, but rather computing *recursively* over the structure of types and values.

$$\frac{S,T \;:\; \mathsf{Set}}{S \leftrightarrow T \;:\; \mathsf{Prop}} \qquad \frac{s \;:\; S \quad t \;:\; T}{(s\!:\!S) \simeq (t\!:\!T) \;:\; \mathsf{Prop}}$$

Intuitively, $S \leftrightarrow T$ means that there are *structural* coercions between S and T. The heterogeneous value equality is a little more subtle: $(s\!:\!S) \simeq (t\!:\!T)$ indicates that s and t are interchangeable *if* $S \leftrightarrow T$. The intended elimination behaviour helps to motivate the construction: every equation on sets induces a *coherent coercion* between them

$$\mathsf{coe} : (S, T\!:\!\mathsf{Set}) \to S \leftrightarrow T \to S \to T$$
$$\mathsf{coh} : (S, T\!:\!\mathsf{Set}) \to (Q\!:\!S \leftrightarrow T) \to (s\!:\!S) \to (s\!:\!S) \simeq (\mathsf{coe}\ S\ T\ Q\ s\!:\!T)$$

We can transport values between equal sets, and the result remains equal, even though its type may have changed! Correspondingly, set equality must be constructed to ensure that coe can be computed. Meanwhile, value equality should be constructed consistently to reflect the construction of data and the observations on codata.

Set equality follows the structure of sets, yielding $\bot$ for distinct set-formers, and componentwise equations when the set-formers coincide.

$$
\begin{array}{lll}
0 \leftrightarrow 0 & \leadsto \top \\
1 \leftrightarrow 1 & \leadsto \top \\
(x\!:\!S) \to T \leftrightarrow (x'\!:\!S') \to T' & \leadsto S' \leftrightarrow S \ \land\ \forall x'\!:\!S'.\ \forall x\!:\!S.\ (x'\!:\!S') \simeq (x\!:\!S) \Rightarrow T \leftrightarrow T' \\
(x\!:\!S) \times T \leftrightarrow (x'\!:\!S') \times T' & \leadsto S \leftrightarrow S' \ \land\ \forall x\!:\!S.\ \forall x'\!:\!S'.\ (x\!:\!S) \simeq (x'\!:\!S') \Rightarrow T \leftrightarrow T' \\
\quad\vdots & \quad\vdots
\end{array}
$$

The orientation of the equality for the domain component of function and pair types reflects their contra- and co-variance, respectively. The definition of coe follows the same structure, exploiting 0-elimination when the types do not match, and componentwise coercion in the appropriate direction when they do.

$$
\begin{array}{l}
\mathsf{coe}\quad 0 \qquad\qquad 0 \qquad Q\ x \leadsto x \\
\mathsf{coe}\quad 1 \qquad\qquad 1 \qquad Q\ x \leadsto x \\
\mathsf{coe}\ ((x\!:\!S) \to T)\ ((x'\!:\!S') \to T')\ Q\ f \leadsto \lambda x'.\,\mathsf{let}\ x\ \mapsto \mathsf{coe}\ S'\ S\ (\mathsf{fst}\ Q)\ x' \\
\qquad\qquad\qquad\qquad\qquad\qquad\qquad\quad\ \ q\ \mapsto \mathsf{coh}\ S'\ S\ (\mathsf{fst}\ Q)\ x' \\
\qquad\qquad\qquad\qquad\qquad\qquad\qquad\quad\ Q' \mapsto \mathsf{snd}\ Q\ x'\ x\ q \\
\qquad\qquad\qquad\qquad\qquad\qquad\quad\ \mathsf{in}\ \ \mathsf{coe}\ T\ T'\ Q'\ (f\ x) \\
\mathsf{coe}\ ((x\!:\!S) \times T)\ ((x'\!:\!S') \times T')\ Q\ p \leadsto \mathsf{let}\ x\ \mapsto \mathsf{fst}\ p \\
\qquad\qquad\qquad\qquad\qquad\qquad\qquad\quad x' \mapsto \mathsf{coe}\ S\ S'\ (\mathsf{fst}\ Q)\ x \\
\qquad\qquad\qquad\qquad\qquad\qquad\qquad\quad\ q\ \mapsto \mathsf{coh}\ S\ S'\ (\mathsf{fst}\ Q)\ x \\
\qquad\qquad\qquad\qquad\qquad\qquad\qquad\quad Q' \mapsto \mathsf{snd}\ Q\ x\ x'\ q \\
\qquad\qquad\qquad\qquad\qquad\qquad\ \ \mathsf{in}\ (x', \mathsf{coe}\ T\ T'\ Q'\ (\mathsf{snd}\ p)) \\
\quad\vdots \qquad\qquad\qquad\qquad\qquad\quad\vdots
\end{array}
$$

Note how the coherence guarantee from the domain coercion is always exactly what we need to establish the equality of the codomains. To extend coercion to datatypes, exploit that fact that provably equal datatypes have the same constructors and provably equal component types to build a function recursively mapping each constructor to itself.

Crucially, coe computes *lazily* in the proof Q unless the equation is conspicuously absurd. It is this laziness which buys us our freedom to design the value equality observationally: as long as we ensure that $[\![\bot]\!]$ remains uninhabited, coe must always make progress transporting canonical values between canonical types. Moreover, we are free to add whatever propositional axioms we like, provided we retain consistency. Correspondingly, we may take coh as axiomatic, along with

refl $: \forall X. [\![\forall x\!:\!X. (x\!:\!X) \simeq (x\!:\!X)]\!]$
Resp $: \forall X. (P\!:\!X \to \mathsf{Set}) \to [\![\forall x\!:\!X. \forall y\!:\!X. (x\!:\!X) \simeq (y\!:\!X) \Rightarrow P\,x \leftrightarrow P\,y]\!]$

The latter allows us to recover subst from coe.

We can take value equality to be anything we like for types with distinct constructors — it is meaningful only when the types match. For like types, we must be more careful:

$$
\begin{aligned}
(x\!:\!0) &\simeq (y\!:\!0) & &\leadsto \top \\
(x\!:\!1) &\simeq (y\!:\!1) & &\leadsto \top \\
(f\!:\!(x\!:\!S) \to T) &\simeq (f'\!:\!(x'\!:\!S') \to T') & &\leadsto \forall x\!:\!S. \forall x'\!:\!S'. \\
& & & \quad (x\!:\!S) \simeq (x'\!:\!S') \Rightarrow (f\,x\!:\!T) \simeq (f'\,x'\!:\!T') \\
(p\!:\!(x\!:\!S) \times T) &\simeq (p'\!:\!(x'\!:\!S') \times T') & &\leadsto \text{let } x \mapsto \mathsf{fst}\,p \; ; \; y \mapsto \mathsf{snd}\,p \\
& & & \quad\quad x' \mapsto \mathsf{fst}\,p' \; ; \; y' \mapsto \mathsf{snd}\,p' \\
& & & \text{in } (x\!:\!S) \simeq (x'\!:\!S') \wedge (y\!:\!T) \simeq (y'\!:\!T')
\end{aligned}
$$

As you can see, equality for pairs is componentwise and equality for functions is extensional. The coup de grace is that with some care, we can ensure that

$$\mathsf{coe}\,X\,X\,Q\,x \equiv x$$

holds, not by adding a nonlinear reduction rule, but just by extending the way in which normal forms are compared after evaluation — a detailed treatment is given in our paper [2]. This summarizes the basic machinery for observational equality. Let us now add codata, and show that we may indeed take the $x \sim_X y$ we seek to be the homogeneous special case of our value equality $(x\!:\!X) \simeq (y\!:\!X)$.

7   Interaction Structures Closed under Bisimulation

Peter Hancock and Anton Setzer define a notion of *interaction structure* corresponding to potentially infinite trees of traces in a state-dependent command-response dialogue [11,12]. We may introduce a new type constructor for such structures in general, parametrized by a protocol as follows:

$$
\begin{aligned}
\mathsf{IO}\,(S : \mathsf{Set}) & \quad && \text{states of the system} \\
(C : S \to \mathsf{Set}) & && \text{commands for each state} \\
(R : (s\!:\!S) \to C\,s \to \mathsf{Set}) & && \text{responses for each command} \\
(n : (s\!:\!S) \to (c\!:\!C\,s) \to R\,s\,c \to S) & && \text{new state from each response} \\
: S \to \mathsf{Set} & && \text{traces from each state}
\end{aligned}
$$

Interaction structures are the coinductive counterpart of Petersson-Synek trees [17]. The parameters C, R, and n characterize the strictly positive endofunctors on the category $S \to \mathsf{Set}$ of S-indexed sets with index-preserving maps. These are exactly the *indexed containers* [1,3]

$$[S, C, R, n] \; : \; (S \to \mathsf{Set}) \to (S \to \mathsf{Set})$$
$$[S, C, R, n] \, X \, s \; \mapsto \; (c : C \, s) \times (r : R \, c \, s) \to X \, (n \, s \, c \, r)$$

with action on morphisms defined by

$$[S, C, R, n] \, f \; \mapsto \; \lambda s. \, \lambda u. \, (\mathsf{fst} \, u, f \, s \cdot \mathsf{snd} \, u)$$

Now, $(\mathsf{IO} \, S \, C \, R \, n, \mathsf{out})$ is the terminal $[S, C, r, n]$-coalgebra. That is, we take

$$\mathsf{out}_{S,C,R,n} \; : \; (s : S) \to \mathsf{IO} \, S \, C \, R \, n \, s \to [S, C, R, n] \, (\mathsf{IO} \, S \, C \, R \, n) \, s$$
$$\mathsf{coit}_{S,C,R,n} \; : \; \forall X. \, ((s : S) \to X \, s \to [S, C, R, n] \, X \, s) \to$$
$$((s : S) \to X \, s \to \mathsf{IO} \, S \, C \, R \, n \, s)$$

$$\mathsf{out}_{S,C,R,n} \, (\mathsf{coit}_{S,C,R,n} \, f \, s \, x) \; \leadsto \; [S, C, R, n] \, f \, s \, (f \, s \, x)$$

and we may add

$$\mathsf{in}_{S,C,R,n} \; : \; (s : S) \to [S, C, R, n] \, (\mathsf{IO} \, S \, C \, R \, n) \, s \to \mathsf{IO} \, S \, C \, R \, n \, s$$

$$\mathsf{out}_{S,C,R,n} \, (\mathsf{in}_{S,C,R,n} \, s \, t) \; \leadsto \; t$$

Markus Michelbrink and Anton Setzer have shown that $[S, C, R, n]$ has a final coalgebra [15]. For our purposes, however, we must now choose answers to three additional questions:

1. *When are two* IO *sets provably equal?*   When their parameters are provably pointwise equal.
2. *How can we transport codata between provably equal* IO *structures?*   By a 'device driver' coiteration which translates commands one way and responses the other, exploiting the equality of command and response sets.
3. *When are* IO *values provably equal?*   When they are bisimilar.

Fleshing out the first two answers is largely mechanical, although some care must be taken with the precise computational details. For the third, we need to ensure that our universe Prop is capable of expressing bisimulations. Correspondingly, let us extend Prop with coinductive *predicates*, generally.

$\mathsf{HAP} \, (S \; : \mathsf{Set})$	states of the system
$(H : S \to \mathsf{Prop})$	happiness in each state
$(R : (s : S) \to [\![H \, s]\!] \to \mathsf{Set})$	responses to happiness
$(n \; : (s : S) \to (h : [\![H \, s]\!]) \to R \, s \, h \to S)$	new state from each response
$: \; S \to \mathsf{Prop}$	happiness from each state on

The predicate $\mathsf{HAP} \, S \, H \, R \, n$ is parametrized by a predicate H which indicates happiness with a current state drawn from S and a pair R, n indicating how states may evolve when we are happy: $\mathsf{HAP}\ S\ H\ R\ n\ s$ then holds for any state s whenceforth eternal happiness is assured. It is easy to interpret these propositions as coinductive sets of proofs:

$$[\![\mathsf{HAP}\ S\ H\ R\ n\ s]\!] \ \leadsto\ \mathsf{IO}\ S\ (\lambda s.\ [\![H\ s]\!])\ R\ n\ s$$

We may now define value equality for IO processes as a kind of happiness corresponding to bisimilarity — I precede each component with its motivation:

$(p\!:\!\mathsf{IO}\ S\ C\ R\ n\ s) \simeq (p'\!:\!\mathsf{IO}\ S'\ C'\ R'\ n'\ s') \ \leadsto$
    HAP — we consider two processes, each in its own state
        $(((s\!:\!S)\times\mathsf{IO}\ S\ C\ R\ n\ s)\times((s'\!:\!S')\times\mathsf{IO}\ S'\ C'\ R'\ n'\ s'))$
        — we are happy *now* if they issue equal commands
        $(\lambda((s,p),(s',p')).$
            let $c \mapsto \mathsf{fst}\ (\mathsf{out}\ p)$; $\ c' \mapsto \mathsf{fst}\ (\mathsf{out}\ p')$
          in $(c\!:\!C\ s) \simeq (c'\!:\!C'\ s'))$
            — we need to stay happy whenever they receive equal responses
        $(\lambda((s,p),(s',p')).\ \lambda\_.$
            let $c \mapsto \mathsf{fst}\ (\mathsf{out}\ p)$; $\ c' \mapsto \mathsf{fst}\ (\mathsf{out}\ p')$
          in $(r\!:\!R\ s\ c)\times(r'\!:\!R'\ s'\ c')\times[\![(r\!:\!R\ s\ c)\simeq(r'\!:\!R'\ s'\ c')]\!])$
            — each process evolves according to its own protocol
        $(\lambda((s,p),(s',p')).\ \lambda\_.\ \lambda(r,r',\_).$
            let $c \mapsto \mathsf{fst}\ (\mathsf{out}\ p)$; $\ c' \mapsto \mathsf{fst}\ (\mathsf{out}\ p')$
               $u \mapsto \mathsf{snd}\ (\mathsf{out}\ p)$; $\ u' \mapsto \mathsf{snd}\ (\mathsf{out}\ p')$
          in $((n\ s\ c\ r, u\ r),(n'\ s'\ c'\ r', u'\ r')))$
            — and we start from the processes at hand
        $((s,p),(s',p'))$

We may now establish that

$$(\mathsf{in}_{S,C,R,n}\ s\ (\mathsf{out}_{S,C,R,n}\ s\ p)\!:\!\mathsf{IO}\ S\ C\ R\ n\ s)\simeq(p\!:\!\mathsf{IO}\ S\ C\ R\ n\ s)$$

by a coiterative proof of bisimilarity. We may similarly establish that $\mathsf{coit}_{S,C,R,n}f$ is the unique solution of its defining equation, up to $\simeq$, again by coiteration. By generalising our treatment of codata to the indexed case, we effectively *closed* coinductive types under observational equality, re-using the same IO structures with a suitably enriched notion of state. This pleasing uniformity is rather to be expected when starting from such an expressive class of definitions.

8 Epilogue

If we want to reason about coprograms in intuitive way, we need a substitutive equality which includes bisimilarity. Equivalently, we need a dependent case analysis to see how things unfold. We have seen why the latter leads to the failure of type preservation in CoQ and that there is no chance to fix this problem by strengthening the judgmental equality. We have also seen how to construct an observational propositional equality which amounts to extensionality for functions and bisimilarity for coinductive values, giving rise to a structural coercion between equal sets. By explicit appeal to this observational equality, we can restore the type safety of dependent case analysis for codata.

As proof of concept, I have constructed in AGDA a small universe of sets and propositions containing interaction structures and observational equality, equipped with coercion as described above. This development will inform a new treatment of codata in the next version of EPIGRAM, corresponding directly to the categorical notion of terminal coalgebra, and with bisimulation substutive in all contexts.

A curious consequence of this choice is that predicates of intensional stamp, like inductively defined equality, must be ruthlessly excised from the type theory. If a predicate holds for one implementation but not another, it can hardly respect observational equality. As we reason, so we must define — up to observation. It as almost as if a greater fidelity to abstract, extensional mathematics is being forced upon us, whether we like it or not. Not only *can* we express coalgebraic and algebraic structure precisely in dependent type theories, but increasingly, we *must*. 'Category theory working, for the programmer' is becoming a viable prospect and a vital prospectus.

Acknowledgements

I should like to thank Nicolas Oury for showing me this problem, Alexander Kurz for his patience with my laziness, and Peter Hancock for provocation.

References

1. Abbott, M., Altenkirch, T., Ghani, N.: Containers - constructing strictly positive types. Theoretical Computer Science 342, 3–27 (2005); Applied Semantics: Selected Topics
2. Altenkirch, T., McBride, C., Swierstra, W.: Observational equality, now! In: Stump, A., Xi, H. (eds.) PLPV, pp. 57–68. ACM, New York (2007)
3. Altenkirch, T., Morris, P.: Indexed Containers. In: Proceedings of LICS (2009)
4. Bertot, Y., Castéran, P.: Interactive Theorem Proving And Program Development: Coq'Art: the Calculus of Inductive Constructions. Springer, Heidelberg (2004)
5. Cockett, R., Fukushima, T.: About Charity. Yellow Series Report No. 92/480/18, Department of Computer Science, The University of Calgary (June 1992)
6. Cockett, R., Spencer, D.: Strong categorical datatypes I. In: Seely, R.A.G. (ed.) International Meeting on Category Theory 1991, Canadian Mathematical Society Proceedings, AMS (1992)
7. Giménez, E.: Codifying guarded definitions with recursive schemes. In: Smith, J., Dybjer, P., Nordström, B. (eds.) TYPES 1994. LNCS, vol. 996, pp. 39–59. Springer, Heidelberg (1995)
8. Giménez, E.: Un Calcul de Constructions Infinies et son application à la vérification de systèmes communicants. PhD thesis, Ecole Normale Supérieure de Lyon (1996)
9. Gonthier, G.: A computer-checked proof of the Four-Colour theorem. Technical report, Microsoft Research (2005)

10. Hagino, T.: A Categorical Programming Language. PhD thesis, Laboratory for Foundations of Computer Science, University of Edinburgh (1987)
11. Hancock, P., Setzer, A.: Interactive programs in dependent type theory. In: Clote, P.G., Schwichtenberg, H. (eds.) CSL 2000. LNCS, vol. 1862, pp. 317–331. Springer, Heidelberg (2000)
12. Hancock, P., Setzer, A.: Interactive programs and weakly final coalgebras in dependent type theory. In: Crosilla, L., Schuster, P. (eds.) From Sets and Types to Topology and Analysis. Towards Practicable Foundations for Constructive Mathematics, Oxford, pp. 115–134. Clarendon Press (2005)
13. Hofmann, M.: Extensional concepts in intensional type theory. PhD thesis, Laboratory for Foundations of Computer Science, University of Edinburgh (1995), http://www.lfcs.informatics.ed.ac.uk/reports/95/ECS-LFCS-95-327/
14. McBride, C., McKinna, J.: The view from the left. Journal of Functional Programming 14(1), 69–111 (2004)
15. Michelbrink, M., Setzer, A.: State dependent IO-monads in type theory. Electronic Notes in Theoretical Computer Science 122, 127–146 (2005)
16. Norell, U.: Towards a Practical Programming Language based on Dependent Type Theory. PhD thesis, Chalmers University of Technology (2007)
17. Petersson, K., Synek, D.: A set constructor for inductive sets in martin-löf's type theory. In: Dybjer, P., Pitts, A.M., Pitt, D.H., Poigné, A., Rydeheard, D.E. (eds.) Category Theory and Computer Science. LNCS, vol. 389, pp. 128–140. Springer, Heidelberg (1989)
18. Jones, S.P. (ed.): Haskell 98 Language and Libraries: The Revised Report. Cambridge University Press, Cambridge (2003)
19. Setzer, A.: Guarded recursion in dependent type theory. In: Talk at Agda Implementors Meeting, Göteborg, Sweden, May 2007, vol. 6 (2007)
20. Setzer, A.: Coalgebras in dependent type theory. In: Talk at Agda Intensive Meeting, Sendai, Japan, November 2008, vol. 9 (2008)
21. Telford, A., Turner, D.: Ensuring streams flow. In: Johnson, M. (ed.) AMAST 1997. LNCS, vol. 1349, pp. 509–523. Springer, Heidelberg (1997)
22. Turner, D.A.: Elementary strong functional programming. In: Hartel, P.H., Plasmeijer, R. (eds.) FPLE 1995. LNCS, vol. 1022, pp. 1–13. Springer, Heidelberg (1995)
23. Turner, D.A.: Total functional programming. J. UCS 10(7), 751–768 (2004)

Circular Coinduction:
A Proof Theoretical Foundation

Grigore Roşu[1] and Dorel Lucanu[2]

[1] Department of Computer Science
University of Illinois at Urbana-Champaign, USA
grosu@illinois.edu
[2] Faculty of Computer Science
Alexandru Ioan Cuza University, Iaşi, Romania
dlucanu@info.uaic.ro

Abstract. Several algorithmic variants of circular coinduction have been proposed and implemented during the last decade, but a proof theoretical foundation of circular coinduction in its full generality is still missing. This paper gives a three-rule proof system that can be used to formally derive circular coinductive proofs. This three-rule system is proved behaviorally sound and is exemplified by proving several properties of infinite streams. Algorithmic variants of circular coinduction now become heuristics to search for proof derivations using the three rules.

1 Introduction

Circular coinduction at a glance. Circular coinduction is a generic name associated to a series of algorithmic techniques to prove behavioral equivalence mechanically. A variant of circular coinduction was published for the first time in [15] in March 2000, and then other more general or more operational variants not too long afterwards first in [13] and then in [5,6]; the system BOBJ implemented an earlier variant of circular coinduction back in 1998. The name "circular coinduction" was inspired from how it operates: it systematically searches for circular behaviors of the terms to prove equivalent. More specifically, it derives the behavioral equational task until one obtains, on every path, either a truth or a cycle.

Let us intuitively discuss how circular coinduction operates, by means of an example; rigorous definitions and proofs are given later in the paper. The usual *zeros* and *ones* infinite streams containing only 0 and 1 bits, respectively, as well as a *blink* stream of alternating 0 and 1 bits and the *zip* binary operation on streams, can be defined equationally as in Fig. 1. Lazy evaluation languages like Haskell support such stream definitions. One can prove $zip(zeros, ones) = blink$ by circular coinduction as follows: (1) Check that the two streams have the same head, namely 0; (2) Take the tail of the two streams and generate new goal $zip(ones, zeros) = 1:blink$; this becomes the next task;

$$zeros = 0 : zeros$$
$$ones = 1 : ones$$
$$blink = 0 : 1 : blink$$
$$zip(B : S, S') = B : zip(S', S)$$

Fig. 1. Stream definitions

A. Kurz, M. Lenisa, and A. Tarlecki (Eds.): CALCO 2009, LNCS 5728, pp. 127–144, 2009.

(3) Check that the two new streams have the same head, 1 in this case; (4) Take the tail of the two new streams; after simplification one gets the new goal $zip(zeros, ones) = blink$, which is nothing but the original proof task; (5) Conclude that $zip(zeros, ones) = blink$ holds. This is also shown in Fig. 2.

The intuition for the above "proof" is that the two streams have been exhaustively tried to be distinguished by iteratively deriving them, that is, checking their heads and taking their tails. Ending up in circles (we obtained the same new proof task as the original one) means that the two streams are indistinguishable with experiments involving stream heads and tails, so they are equal. In fact, the correctness of circular coinduction can be informally argued in several ways; we found that colleagues with different backgrounds prefer

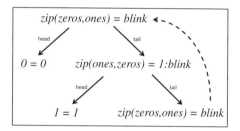

Fig. 2. Intuitive proof by circular coinduction of $zip(zeros, ones) = blink$

different arguments: (1) Since each derived path ends up in a cycle, it means that there is no way to show the two original terms behaviorally different, so they must be equal; (2) The obtained circular graph structure can be used as a backbone to "consume" any possible experiment applied on the two original terms, rolling each experiment until it becomes a visible equality, so the graph is a witness that the two terms are "indistinguishable under experiments"; (3) The equalities that appear as nodes in the obtained graph can be regarded as lemmas inferred in order to prove the original task, and the graph itself as a dependence relation among these lemmas; one can take all these and produce a parallel proof by context induction [9] for all of them; (4) When/if it stabilizes, it "discovers" a binary relation which is compatible with the derivatives and is the identity on data, so the stabilized set of equations, which includes the original task, is included in the behavioral equivalence (because the latter is the largest with that property); (5) It incrementally completes a given equality into a bisimulation relation on terms, so the two original terms must be behaviorally equivalent (bisimilar).

At our knowledge, variants of circular coinduction have been implemented in three systems so far: in a behavioral extension of OBJ called BOBJ [13] (not maintained anymore), in Isabelle/HOL for CoCasl [8], and in CIRC [11].

Novel proof system. The definition and proof of correctness of the original circular coinduction variant in [15,13] were rather complex. Moreover, the setting in those and other papers on circular coinduction was model theoretical, with hidden algebras as models, making one wonder whether the applicability of circular coinduction was limited to hidden algebra models or not. Here we adopt a general proof theoretical approach, without fixing any particular models (and implicitly any particular model-based notions of behavioral equivalence) and any particular base deductive system. Instead, our approach here is parametric in a given base *entailment relation* $\vdash$, which one can use to discard "obvious" proof tasks, and in a set of *derivatives* Δ, which are special terms with a hole (i.e., contexts) allowing us to define a syntactic notion of *behavioral equivalence* on terms without referring to any model but only to $\vdash$. This syntactic notion of behavioral equivalence is so natural that it is straightforwardly sound in all models that we are aware of. More importantly, it allows for more succinct and clean definitions and proofs for circular coinduction, independent of the underlying models.

Fig. 3 shows our novel formulation of circular coinduction as a three-rule proof system deriving pairs of the form $\mathcal{H} \Vdash^{\circlearrowleft} \mathcal{G}$, where $\mathcal{H}$ (*hypotheses*) and $\mathcal{G}$ (*goals*) are sets of equations. Some equations can be *frozen* and written in a box (e.g., $\boxed{e}$), with the intuition that those equations cannot be used in contextual reasoning (i.e., the congruence rule of equational logic cannot be applied on them), but only at the top; this is easily achieved by defining the box as a wrapper

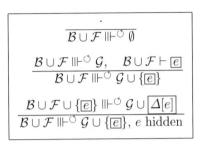

Fig. 3. Circular coinduction

operator of a fresh sort result, say *Frozen*. $\mathcal{G}$ contains only frozen equations, while $\mathcal{H}$ may contain both frozen (denoted by $\mathcal{F}$) and normal (denoted by $\mathcal{B}$), or "unfrozen" equations; the unfrozen equations are those in the original behavioral specification $\mathcal{B}$. The first rule says that we are done when there are no goals left. The second rule discards those goals which can be proved with the base entailment system. The third rule is the distinguished rule of circular coinduction that enables circular reasoning and it essentially says the following, where the involved equations are all frozen:

To prove behavioral property e, assume it and prove its derivatives $\Delta[e]$; or, metaphorically, "if one cannot show it wrong then it is right".

Let us discuss the circular coinductive proof system for infinite streams, whose derivatives Δ are the head and tail operations. The third rule in Fig. 3 says that in order to prove an equality of two infinite streams s_1 and s_2 (let us informally write it $s_1 \sim s_2$ for the time being), just assume $s_1 \sim s_2$ as a hypothesis and then prove $head(s_1) = head(s_2)$ and $tail(s_1) \sim tail(s_2)$. To prove the latter, one may need to apply the third rule again, assuming $tail(s_1) \sim tail(s_2)$ and then proving $head(tail(s_1)) = head(tail(s_2))$ and $tail(tail(s_1)) \sim tail(tail(s_2))$, and so on and so forth. The proof system therefore attempts to incrementally perform a bisimulation, or behavioral closure of the property to prove. Practice and experience tell us that naive attempts to perform such a closure, without an appropriate reasoning infrastructure, will easily lead to nontermination in many cases. The strength of circular coinduction comes from the fact that it uses the available knowledge, i.e. both the equations in $\mathcal{B}$ and the accumulated hypotheses in $\mathcal{F}$, during this closure process; however, using the available knowledge soundly and effectively is a much more complex task than it may appear.

There are several reasoning aspects encapsulated in the second condition of the second rule in Fig. 3, namely in $\mathcal{B} \cup \mathcal{F} \vdash \boxed{e}$, that are key to termination and therefore core to circular coinduction. For example, can one freely use the original equations in $\mathcal{B}$, including those between infinite streams, to simplify the terms s_1 and s_2 in a new proof task $s_1 \sim s_2$? The answer is *yes*. Also, can one freely use the accumulated hypotheses in $\mathcal{F}$? This time the answer is *no*, because otherwise the new tasks $\Delta[e]$ added by the third rule in Fig. 3 would hold vacuously by the congruence rule of equational reasoning. Freezing inhibits the applications of congruence, so one needs to find other means to eliminate the derivative operations. It turns out that the other rules of equational deduction are sound in combination with the accumulated hypotheses in $\mathcal{F}$, including substitution and transitivity, but why that is the case is not immediate.

A way to understand freezing by means of context induction [9] intuitions is to regard it as the application of an arbitrary but fixed context C to an equation e; with this, the third rule in Fig. 3 says that in order to prove e, assume $C[e]$ for an arbitrary but fixed context C and prove $C[\delta[e]]$ for any $\delta \in \Delta$. This view does not explain the need for freezing e in $\mathcal{G} \cup \{\boxed{e}\}$ in the third rule in Fig. 3, but it gives an intuitive argument why congruence is not sound in combination with the accumulated hypotheses. This way of applying context induction is reminiscent to the inductive proof techniques in [2], but there C is a variable of sort *context* and one needs to do an explicit induction on it. The use of freezing relieves the user of our proof system from performing explicit induction on contexts; in fact, the user of our proof system needs not be aware of any contexts at all (except for the derivatives), nor of induction on contexts. Nevertheless, the soundness of our proof system is complex and makes use of a well-founded induction on contexts (in Theorem 2), but the proof system itself is simple and easy to use.

The circular coinductive proof system in Fig. 3 is the result of many years of experimentation with BOBJ [13], a behavioral variant of OBJ implementing an early version of circular coinduction, and with CIRC [11], a behavioral extension of Maude [3] optimized for automated inductive and coinductive proving. The operational simplicity of our proof system may, however, be deceiving. For example, a crucial detail may have passed unnoticed to the hasty reader: the third rule in Fig. 3 does *not* split the proof of $\mathcal{G} \cup \{\boxed{e}\}$ into a proof of $\mathcal{G}$ and a proof of $\Delta[\boxed{e}]$. Instead, it generates a proof task for $\mathcal{G} \cup \Delta[\boxed{e}]$. This allows one to move some equations in the set of goals $\mathcal{G}$ as hypotheses in $\mathcal{F}$ and use them to prove other, completely unrelated equations in $\mathcal{G}$! This allows us, in particular, to prove a set of mutually related behavioral equalities in parallel (an example is given later in the paper). The soundness of our proof system (Theorem 3, which makes use of Theorem 2) is a monolithic and difficult result, depending upon the existence of a *complete* derivation proof for $\mathcal{B} \Vdash^{\circ} \boxed{G}$. In other words, its soundness does not follow from the soundness of each derivation rule, as it is commonly done in soundness proofs; in fact, the third rule is not sound in isolation. We are not aware of any other similar proof systems in the literature.

A question is whether the proposed proof system can prove all behavioral equalities. Unfortunately, the answer is negative even for the particular case of streams. Indeed, [14] shows that stream behavioral equivalence is a Π_2^0 complete problem, which means in particular that there is no complete proof system for behavioral equivalence. Thus, the best we can do is to algorithmically approximate behavioral equivalence, aiming at covering more practical situations, but there will always be room for better proof systems. In fact, the latest version of CIRC [10] already implements an extended and more complex variant of the discussed proof system. Nevertheless, the circular coinductive proof system presented in this paper is simple, powerful and relatively efficiently implementable.

Implementation in CIRC. The proposed circular coinductive reasoning system has been implemented and extensively evaluated in the latest version of CIRC. Even though versions of CIRC have been available for download and on-line experimentation at http : //fsl.cs.uiuc.edu/circ for more than 3 years, this is the first paper about its foundations from a proof theoretical rather than algorithmic perspective, explaining how its automated coinductive engine works and why it is sound. Since CIRC can automatically prove using the discussed technique all the behavioral properties in this paper (see its website above) and since its syntax minimally extends that of Maude, which we assume known, from here on we give our running specification, that of streams, using CIRC:

```
op hd : Stream -> Bit .          derivative hd(*:Stream) .
op tl : Stream -> Stream .       derivative tl(*:Stream) .
```

Declaring `hd` (head) and `tl` (tail) as *derivatives* tells CIRC that one can use them to derive (or observe) the task to prove. One can define several operations on streams. For example, `not(S)` reverses each bit in stream `S` and is trivial to define; `odd`, `even` and `zip` are standard and can be defined as follows (we use now the definitional style of CIRC, explicitly using derivatives in equations):

```
ops odd even : Stream -> Stream .  op zip : Stream Stream -> Stream .
eq hd(odd(S)) = hd(S) .            eq hd(zip(S,S')) = hd(S) .
eq tl(odd(S)) = even(tl(S)) .      eq tl(zip(S,S')) = zip(S',tl(S)) .
eq even(S) = odd(tl(S)) .
```

One can now ask CIRC to prove $\texttt{zip(odd(S), even(S))} = \texttt{S}$. Using the circular coinductive proof system, CIRC proves it automatically as follows: (1) assume $\texttt{zip(odd(S), even(S))} = \texttt{S}$ as frozen hypothesis; (2) add its derivatives $\texttt{hd(zip(odd(S), even(S)))} = \texttt{hd(S)}$ and $\texttt{tl(zip(odd(S), even(S)))} = \texttt{tl(S)}$ as new proof tasks; (3) prove first task using the stream equations; (4) prove second task by first reducing it to $\texttt{zip(odd(tl(S)), even(tl(S)))} = \texttt{tl(S)}$ using stream equations and then using frozen hypothesis (1) at its top with substitution $\texttt{S} \mapsto \texttt{tl(S)}$.

Sometimes it is easier to prove more tasks at the same time than each task separately, because they can help each other. Here is an example. The famous Thue-Morse sequence (see, e.g., [1]) has several interesting properties and can be defined as a stream in several ways (we give one definition below). Interestingly, `morse` is a fixed point of the operation `f` also defined below:

```
--- Thue-Morse sequence:          --- morse is a fixed point of f
op morse : -> Stream .            op f : Stream -> Stream .
eq hd(morse) = 0 . eq hd(tl(morse)) = 1 .  eq hd(f(S)) = hd(S) .
eq tl(tl(morse))                  eq hd(tl(f(S))) = not(hd(S)) .
   = zip(tl(morse), not(tl(morse))) .  eq tl(tl(f(S))) = f(tl(S)) .
```

If we try to prove $f(\text{morse}) = \text{morse}$ (either manually or using CIRC), then we see that larger and larger frozen equations are derived and the process does not terminate. Analyzing the new equations (as humans), we see that the source of problems is the lack of an equation $f(S) = \text{zip}(S, \text{not}(S))$. We can then either prove it first as a lemma, or simply add it to the set of goals and then restart the coinductive proving process with the goal $\{f(\text{morse}) = \text{morse}, f(S) = \text{zip}(S, \text{not}(S))\}$, this time succeeding in ten steps (see Example 4 in Section 5).

Paper structure. Section 2 introduces basic notions and notations. A key notion is that of a Δ-contextual entailment system; one such system is assumed given and acts as a parameter of our generic techniques developed in the rest of the paper. Section 3 recasts behavioral satisfaction in our generic setting. Section 4 is dedicated to defining behavioral equivalence and coinduction in our generic setting, showing that behavioral equivalence remains the largest relation closed under derivatives. Our novel circular coinductive proof system is presented in Section 5, together with its behavioral soundness result.

2   Preliminaries

We assume the reader familiar with basics of many sorted algebraic specifications and only briefly recall our notation. An algebraic specification, or simply a *specification*, is a triple (S, Σ, E), where S is a set of *sorts*, Σ is a $(S^* \times S)$-*signature* and E is a set of Σ-*equations* of the form $(\forall X) t = t'$ if $\wedge_{i \in I} u_i = v_i$ with t, t', u_i, and v_i Σ-terms with variables in X, $i = 0, \ldots, n$; the two terms appearing in any equality in an equation, that is the terms t, t' and each pair u_i, v_i for each $i \in I$, have, respectively, the same sort. If the sort of t and t' is s we may say that the sort of the equation is also s. When $i = 0$ we call the equation unconditional and omit the condition (i.e., write it $(\forall X) t = t'$). When $X = \emptyset$ we drop the quantifier and call the equation *ground*.

If Σ is a many sorted signature, then a Σ-*context* C is a Σ-term which has one occurrence of a distinguished variable $*:s$ of sort s; to make this precise, we may write $C[*:s]$ instead of just C. When Σ is understood, a Σ-context may be called just a *context*. When the sort s of $*$ is important, we may call $C[*:s]$ a *context for sort* s; also, when the sort of a context C (regarded as a term), say s', is important, C may be called a *context of sort* s'. If $C[*:s]$ is a context for sort s of sort s' and t is a term of sort s, then $C[t]$ is the term of sort s' obtained by replacing t for $*{:}s$ in C. A Σ-context $C[*{:}s]$ induces a partially defined *equation transformer* $e \mapsto C[e]$: if e is an equation $(\forall X)\, t = t'$ if c of sort s, then $C[e]$ is the equation $(\forall X \cup Y)\, C[t] = C[t']$ if c, where Y is the set of non-star variables occurring in $C[*{:}s]$. Moreover, if $\mathbf{C}$ is a set of contexts and E a set of equations, then $\mathbf{C}[e] = \{C[e] \mid C \in \mathbf{C}\}$, $C[E] = \{C[e] \mid e \in E\}$ and $\mathbf{C}[E] = \bigcup_{e \in E} \mathbf{C}[e]$.

The theoretical results in this paper will be parametric in a given entailment relation $\vdash$ on many sorted equational specifications, which may, but is not enforced to, be the usual equational deduction relation [7]. For instance, it can also be the "rewriting" entailment relation ($E \vdash t = t'$ iff t and t' rewrite to the same term using E as a rewrite system), or some behavioral entailment system, etc. We need though some properties of $\vdash$, which we axiomatize here by adapting to our context the general definition of entailments system as given in [12]. Fix a signature Σ (in a broader setting, $\vdash$ could be regarded as a set $\{\vdash_\Sigma \mid \Sigma \text{ signature}\}$ indexed by the signature; however, since we work with only one signature in this paper and that signature is understood, for simplicity we fix the signature).

Definition 1. *If Δ is a set of Σ-contexts, then a Δ-**contextual entailment system** is an (infix) relation $\vdash$ between sets of equations and equations, with: (reflexivity) $\{e\} \vdash e$; (monotonicity) If $E_1 \supseteq E_2$ and $E_2 \vdash e$ then $E_1 \vdash e$; (transitivity) If $E_1 \vdash E_2$ and $E_2 \vdash e$ then $E_1 \vdash e$; (Δ-congruence) If $E \vdash e$ then $E \vdash \Delta[e]$. Above, E, E_1, E_2 range over sets of equations and e over equations; we tacitly extended $\vdash$ to sets of equations as expected: $E_1 \vdash E_2$ iff $E_1 \vdash e$ for any $e \in E_2$. We let $E^\bullet$ denote the set of equations $\{e \mid E \vdash e\}$.*

One can use the above to prove many properties of $\vdash$ on sets of equations. Here are some of them used later in the paper (their proofs are simple exercises): $E \vdash \emptyset$, $E \vdash E$, if $E_1 \vdash E_2$ and $E_2 \vdash E_3$ then $E_1 \vdash E_3$, if $E_1 \vdash E_2$ then $E \cup E_1 \vdash E \cup E_2$, if $E_1 \vdash E_2$ then $E_1 \vdash \Delta[E_2]$, if $E_1 \vdash E_2$ then $E_1 \vdash E_1 \cup E_2$, if $E_1 \supseteq E_2$ then $E_1 \vdash E_2$, if $E \vdash E_1$ and $E \vdash E_2$ then $E \vdash E_1 \cup E_2$.

We take the liberty to slightly abuse the syntax of entailment and allow one to write a specification instead of a set of equations, with the obvious meaning: if $\mathcal{B} = (S, \Sigma, E)$ is a specification and e is a Σ-equation, then $\mathcal{B} \vdash e$ iff $E \vdash e$. Also, if $\mathcal{B} = (S, \Sigma, E)$ then we may write $\mathcal{B}^\bullet$ instead of $E^\bullet$.

3   Behavioral Specification and Satisfaction

Our approach in this paper is purely proof theoretical, that is, there are no models involved. We prefer this approach because we think that it is the most general one for our purpose. As pointed out in [13], there is a plethora of model theoretical variants of behavioral logics, some with models with fixed data others with loose data, some which are co-algebraic (i.e., isomorphic to a category of coalgebras) others which are not (e.g., some do not even admit final models), some which come with a behavioral equality others whose behavioral equality can be computed, etc. However, the proof theoretical behavioral entailment relation defined in this section is sound for all these different variants of behavioral logics. On the other hand, as shown in [13,14], neither of the behavioral logics is complete. Therefore, one looses nothing by working with a proof theoretical instead of a model theoretical notion of behavioral entailment. On the contrary, this allows us to define behavioral equivalence and coinduction in a more general way, independent upon the particular (adhoc) choice for models.

A *behavioral specification* is a pair $(\mathcal{B}, \Delta)$, where $\mathcal{B} = (S, \Sigma, E)$ is a many sorted algebraic specification and Δ is a set of Σ-contexts, called *derivatives*. We let Δ_s denote all the derivatives of sort s in Δ. If $\delta[*{:}h] \in \Delta$ is a derivative, then the sort h is called a *hidden sort*; we let $H \subseteq S$ denote the set of all hidden sorts of $\mathcal{B}$. Remaining sorts are called *data, or visible, sorts* and we let $V = S - H$ denote their set. A *data operator* is an operator in Σ taking and returning only visible sorts; a *data term* is a term built with only data operators and variables of data sorts; a *data equation* is an equation built with only data terms.

Sorts are therefore split into hidden and visible, so that one can derive terms of hidden sort until they possibly become visible. Formally, a Δ-*experiment* is a Δ-context of visible sort, that is: (1) each $\delta[*{:}h] \in \Delta_v$ with $v \in V$ is an experiment, and (2) if $C[*{:}h']$ is an experiment and $\delta[*{:}h] \in \Delta_{h'}$, then so is $C[\delta[*{:}h]]$. An equation $(\forall X)\, t = t'$ **if** c is called a *hidden equation* iff the common sort of t and t' is hidden, and it is called a *data, or visible, equation* iff the common sort of t and t' is visible. In this paper we consider only equations whose conditions are conjunctions of visible equalities. If Δ is understood, then we may write experiment for Δ-experiment and context for Δ-context; also, we may write only $\mathcal{B}$ instead of $(\mathcal{B}, \Delta)$ and, via previous conventions, identify $\mathcal{B}$ with its set of equations. If $\mathcal{G}$ is a set of Σ-equations, we let *visible*$(\mathcal{G})$ and *hidden*$(\mathcal{G})$ denote the sets of $\mathcal{G}$'s visible and hidden equations, respectively.

From here on in the paper we fix a signature Σ, a set of Σ-contexts Δ, and a generic Δ-contextual entailment system, $\vdash$. For the sake of concreteness, one may think of $\vdash$ as the ordinary equational entailment relation; however, the proofs in the sequel rely only on the axioms in Definition 1. Unless otherwise specified, we also assume a fixed behavioral specification $(\mathcal{B}, \Delta)$.

Example 1. Let STREAM be the behavioral specification of streams discussed in Section 1, with the conventional equational reasoning as basic entailment relation; in other words, STREAM $\vdash$ e iff e is derivable using equational reasoning from the equations in STREAM. The set Δ of derivatives is $\{\mathtt{hd}(*{:}\mathtt{Stream}), \mathtt{tl}(*{:}\mathtt{Stream})\}$ and the experiments one can perform on streams are therefore contexts of the form $\mathtt{hd}(\mathtt{tl}^i(*{:}\mathtt{Stream}))$, where $\mathtt{i} \geq 0$.

Definition 2. $\mathcal{B}$ **behaviorally satisfies** *equation* e, *written* $\mathcal{B} \Vdash e$, *iff:* $\mathcal{B} \vdash e$ *if* e *is visible, and* $\mathcal{B} \vdash C[e]$ *for each appropriate experiment* C *if* e *is hidden.*

Example 2. In the case of streams, we have that STREAM $\Vdash$ str $=$ str$'$ iff STREAM $\vdash$ hd(tl^i(str)) $=$ hd(tl^i(str$'$)) for all i $\geq$ 0. For example, one can directly show that STREAM $\Vdash$ zip(odd(S), even(S)) $=$ S by showing by induction on i that hd(tl^i(zip(odd(S), even(S)))) $=$ hd(tl^i(S)). Also, one can show that STREAM $\Vdash$ f(not(S)) $=$ not(f(S)), by showing that STREAM $\vdash$ hd(tl^i(f(not(S)))) $=$ g(i, hd(S)) and STREAM $\vdash$ hd(tl^i(not(f(S)))) $=$ g(i, hd(S)) for all i $\geq$ 0, where g(i, B) $=$ not(B) if i is even, and g(i, B) $=$ B if i is odd. Such direct proofs based on direct exhaustive analysis of experiments are essentially proofs by context induction [9], known to be problematic in practice due to the large number of lemmas needed (see [4] for an early reference to this aspect) and diffi-cult to mechanize. We are going to give more elegant behavioral proofs of these properties, first by coinduction and then by circular coinduction.

Proposition 1. $\Vdash$ *extends* $\vdash$, *that is, if* $\mathcal{B} \vdash e$ *then* $\mathcal{B} \Vdash e$.

Proof. If e is visible then it holds by Definition 2. If e is hidden, then by the Δ-congruence property of $\vdash$ in Definition 1, it follows that $\mathcal{B} \vdash C[e]$ for any appropriate Δ-context C, in particular for any Δ-experiment. Hence, $\mathcal{B} \Vdash e$. □

4   Behavioral Equivalence and Coinduction

Definition 3. *The **behavioral equivalence** of B is the set* $\equiv_\mathcal{B} \overset{def}{=} \{e \mid \mathcal{B} \Vdash e\}$.

Thus, the behavioral equivalence of $\mathcal{B}$ is the set of equations behaviorally satisfied by $\mathcal{B}$. When $\mathcal{B}$ is clear, we may write $\equiv$ instead of $\equiv_\mathcal{B}$. Note that $\equiv$ may contain also conditional equations with visible conditions. However, the visible conditions can be regarded as data side conditions, so we took the liberty to make the slight abuse of terminology above and called $\equiv_\mathcal{B}$ an equivalence.

A major result of any behavioral logic, making proofs by coinduction sound, is that behavioral equivalence is the largest relation that is consistent with the data and is compatible with the behavioral operations, or the observers. In what follows we capture this same result in our proof theoretical setting. To do that, we first need to formally define what is meant here by "consistent with data and compatible with behavioral operators". Recall that $\mathcal{B}^\bullet$ uses the original entailment relation, that is, $\mathcal{B}^\bullet = \{e \mid \mathcal{B} \vdash e\}$.

Definition 4. *A set of equations $\mathcal{G}$ is **behaviorally closed** iff $\mathcal{B} \vdash$ visible($\mathcal{G}$) and $\Delta(\mathcal{G} - \mathcal{B}^\bullet) \subseteq \mathcal{G}$.*

In other words, a set of equations $\mathcal{G}$ is behaviorally closed iff for any $e \in \mathcal{G}$, we have that $\mathcal{B} \vdash e$ or otherwise e is hidden and $\Delta[e] \subseteq \mathcal{G}$. Therefore, $\mathcal{G}$ can contain as many equations entailed by $\mathcal{B}$ as needed, both visible and hidden, even all of them. The first condition says that if $\mathcal{G}$ contains any visible equation at all, then that equation must be entailed by $\mathcal{B}$. The second condition says that if a hidden equation $e \in \mathcal{G}$ is not entailed by $\mathcal{B}$ then it must be the case that its derivatives are also in $\mathcal{G}$. A degenerate case is when $\mathcal{G} = \mathcal{B}$, in which case the second condition is superfluous. In other words, $\mathcal{G}$ is closed under, or compatible with, the derivatives and the only way to "escape" this derivative closing process is by a direct proof using $\mathcal{B}$ and the original entailment system.

Theorem 1. *(Coinduction) For any behavioral specification, the behavioral equivalence $\equiv$ is the largest behaviorally closed set of equations.*

One can prove Theorem 1 directly by induction on contexts; however, we are going to prove a more general result later in the paper, the coinductive circularity principle (Theorem 2), so we refrain from giving a direct proof here.

Theorem 1 justifies the correctness of behavioral proofs by coinduction. Indeed, in order to prove that $\mathcal{B}$ behaviorally satisfies an equation e, all we need to do is to find some behaviorally closed set of equations $\mathcal{G}$ such that $e \in \mathcal{G}$.

Experience with many coinductive proofs suggests that the following simple coinductive proving technique works in most practical cases of interest:

(Coinductive proving technique) Aim is to prove $\mathcal{B} \Vdash e$.

1. *If one can show $\mathcal{B} \vdash e$ then stop with* **success**, *otherwise continue;*
2. *If e is visible then stop with* **failure**;
3. *Find some (typically finite) set of hidden equations G with $e \in G$;*
4. *Let $\mathcal{G}$ be the closure of $G \cup \mathcal{B}^{\bullet}$ under substitution, symmetry, and transitivity;*
5. *Show $\Delta[G] \subseteq \mathcal{G}$;*

The most difficult part is to find the right G at step 3. If G is too small, in particular if one picks $G = \{e\}$, then $\mathcal{G}$ may be too small and thus one may not be able to show step 5 because it may be the case for some $f \in G$ and appropriate $\delta \in \Delta$ that $\delta[f] \notin \mathcal{G}$ and one cannot show $\mathcal{B} \vdash \delta[f]$. If G is too large, then the number of proof tasks at step 5 may be prohibitive. Finding a good G at step 3 is the most human-intensive part (the rest is, to a large extent, mechanizable). Steps 4 and 5 above are combined in practice, by showing that one can prove $\Delta[G]$ using unrestricted equational reasoning with the equations in $\mathcal{B}$ and equational reasoning without the congruence rule with the equations in G. The correctness of the technique above follows from the following proposition:

Proposition 2. *Let G be a set of hidden equations such that $\Delta[G] \subseteq \mathcal{G}$, where $\mathcal{G}$ is the closure of $G \cup \mathcal{B}^{\bullet}$ under substitution, symmetry, and transitivity. Then $\Delta[\mathcal{G}] \subseteq \mathcal{G}$ and $\mathcal{G}$ is behaviorally closed.*

Proof. If $\mathcal{E}$ is a set of equations, let $\overline{\mathcal{E}}$ be the closure of $\mathcal{E}$ under substitution, symmetry, and transitivity, and note that $\Delta[\overline{\mathcal{E}}] = \overline{\Delta[\mathcal{E}]}$. Taking $\mathcal{E} = G \cup \mathcal{B}^{\bullet}$, we get $\Delta[\mathcal{G}] = \Delta[\overline{G \cup \mathcal{B}^{\bullet}}] = \overline{\Delta[G \cup \mathcal{B}^{\bullet}]} = \overline{\Delta[G] \cup \Delta[\mathcal{B}^{\bullet}]} \subseteq \overline{\mathcal{G} \cup \mathcal{B}^{\bullet}} = \overline{\mathcal{G}} = \mathcal{G}$. Since G contains only hidden equations and since substitution, symmetry and transitivity are sort preserving, we conclude that $visible(\mathcal{G}) = visible(\mathcal{B}^{\bullet})$. It is now clear that $\mathcal{G}$ is behaviorally closed (both conditions in Definition 4 hold). □

Proposition 2 therefore implies that the $\mathcal{G}$ in the coinductive proving technique above is behaviorally closed. Since $e \in G \subseteq \mathcal{G}$, Theorem 1 implies that $\mathcal{B} \Vdash e$, so our coinductive proving technique is indeed sound.

Example 3. Let us prove by coinduction the two stream properties in Example 2, STREAM $\Vdash$ zip(odd(S), even(S)) = S and STREAM $\Vdash$ f(not(S)) = not(f(S)).

For the first property, let G be the set $\{$zip(odd(S), even(S)) = S$\}$ containing only the equation to prove, and let $\mathcal{G}$ be the closure of $G \cup$ STREAM$^{\bullet}$ under substitution, symmetry and transitivity. Since

$$hd(zip(odd(S), even(S))) = hd(S)$$
$$tl(zip(odd(S), even(S))) = zip(odd(tl(S)), even(tl(S)))$$

are in STREAM$^{\bullet}$ (and hence in $\mathcal{G}$), the following equations are in $\mathcal{G}$ as well:

$$zip(odd(tl(S)), even(tl(S))) = tl(S) \qquad \text{(substitution)}$$
$$tl(zip(odd(S), even(S))) = tl(S) \qquad \text{(transitivity)}$$

Therefore $\Delta(G) \subseteq \mathcal{G}$, so we are done.

For the second property, f(not(S)) = not(f(S)), let G be the two equation set $\{$f(not(S)) = not(f(S)), tl(f(not(S))) = tl(not(f(S)))$\}$, which includes the equation to prove, and let $\mathcal{G}$ the closure under substitution, symmetry and transitivity of $G \cup$ STREAM$^{\bullet}$. Since the following equations

$$hd(f(not(S))) = not(hd(S)) \qquad\qquad hd(not(f(S))) = not(hd(S))$$
$$\qquad\qquad\qquad\qquad\qquad\qquad\qquad tl(not(f(S))) = not(tl(f(S)))$$
$$hd(tl(f(not(S)))) = hd(S) \qquad\qquad hd(tl(not(f(S)))) = hd(S)$$
$$tl(tl(f(not(S)))) = f(not(tl(S))) \qquad tl(tl((not(f(S)))) = not(f(tl(S)))$$

are all in STREAM$^{\bullet}$ (and hence in $\mathcal{G}$), the following equations are in $\mathcal{G}$ as well:

$$hd(f(not(S))) = hd(not(f(S))) \qquad \text{(symm. \& trans.)}$$
$$hd(tl(f(not(S)))) = hd(tl(not(f(S)))) \qquad \text{(symm. \& trans.)}$$
$$tl(tl(f(not(S)))) = tl(tl(not(f(S)))) \qquad \text{(subst. \& symm. \& trans.)}$$

Therefore $\Delta(G) \subseteq \mathcal{G}$, so we are done with the second proof as well.

The set G was straightforward to choose in the first proof in the example above, but it was not that easy in the second one. If we take G in the second proof to consist only of the initial goal like in the first proof, then we fail to prove that tl(f(not(S))) = tl(not(f(S))) is in $\mathcal{G}$.

Unfortunately, there is no magic recipe to choose good sets G, which is what makes coinduction hard to automate. In fact, a naive enumeration of all sets G followed by an attempt to prove $\Delta(G) \subseteq \mathcal{G}$ leads to a Π_2^0 algorithm, which matches the worst case complexity of the behavioral satisfaction problem even for streams [14]. In practice, however, we can do better than enumerating all G. For example, after failing to prove tl(f(not(S))) = tl(not(f(S))) in the second example above when one naively chooses $G = \{$f(not(S)) = not(f(S))$\}$, then one can add the failed task to G and resume the task of proving that $\Delta(G) \subseteq \mathcal{G}$, this time successfully. We will see in the next section that this way of searching for a suitable G is the basis on which the circular coinductive reasoning is developed.

5   Circular Coinduction

A key notion in our formalization and even implementation of circular coinduction is that of a "frozen" equation. The motivation underlying frozen equations is that they structurally inhibit their use underneath proper contexts; because of that, they will allow us to capture the informal notion of "circular behavior" elegantly, rigorously, and generally (modulo a restricted form of equational reasoning). Formally, let $(\mathcal{B}, \Delta)$ be a behavioral specification and let us extend its signature Σ with a new sort *Frozen* and a new operation $\boxed{-} : s \to$ *Frozen* for each sort s. If t is a term, then we call $\boxed{t}$ the *frozen (form of)* t. Note that freezing only acts on the original sorts in Σ, so double freezing, e.g., $\boxed{\boxed{t}}$, is not allowed. If e is an equation $(\forall X)\, t = t'$ if c, then we let $\boxed{e}$ be the *frozen equation* $(\forall X)\boxed{t} = \boxed{t'}$ if c; note that the condition c stays unfrozen, but recall that we only assume visible conditions. By analogy, we call the equations over the original signature Σ *unfrozen equations*. If e is an (unfrozen) visible equation then $\boxed{e}$ is called a *frozen visible equation*; similarly when e is hidden. If C is a context for e, then we take the freedom to write $C[\boxed{e}]$ as a shortcut for $\boxed{C[e]}$. It is important to note here that if $E \cup \mathcal{F} \vdash \mathcal{G}$ for some unfrozen equation set E and frozen equation sets $\mathcal{F}$ and $\mathcal{G}$, then it is not necessarily the case that $E \cup \mathcal{F} \vdash C[\mathcal{G}]$ for a context C. Freezing therefore inhibits the free application of the congruence deduction rule of equational reasoning.

Recall that, for generality, we work with an arbitrary entailment system in this paper, which may or may not necessarily be the entailment relation of equational deduction. We next add two more axioms:

Definition 5. *A Δ-contextual entailment system with freezing is a Δ-contextual entailment system such that (below, E ranges over unfrozen equations, e over visible unfrozen equations, and $\mathcal{F}$ and $\mathcal{G}$ over frozen hidden equations):*

(A1) $E \cup \mathcal{F} \vdash \boxed{e}$ *iff* $E \vdash e$;

(A2) $E \cup \mathcal{F} \vdash \mathcal{G}$ *implies* $E \cup \delta[\mathcal{F}] \vdash \delta[\mathcal{G}]$ *for each $\delta \in \Delta$, equivalent to saying that for any Δ-context C, $E \cup \mathcal{F} \vdash \mathcal{G}$ implies $E \cup C[\mathcal{F}] \vdash C[\mathcal{G}]$.*

The first axiom says that frozen hidden equations do not interfere with the entailment of frozen or unfrozen visible equations. The second equation says that if some frozen hidden equations $\mathcal{F}$ can be used to derive some other frozen hidden equations $\mathcal{G}$, then one can also use the frozen equations $\delta[\mathcal{F}]$ to derive $\delta[\mathcal{G}]$. Freezing acts as a protective shell that inhibits applications of the congruence rule on frozen equations but instead allows the application of derivatives both onto the hypothesis and the conclusion of an entailment pair at the same time.

Therefore, our working entailment system $\vdash$ is now defined over both unfrozen and frozen equations. Note, however, that this extension is conservative, in that one cannot infer any new entailments of unfrozen equations that were not possible before; because of that, we take the liberty to use the same symbol, $\vdash$, for the extended entailment system. It is easy to check these additional axioms for concrete entailment relations. Consider, for example, the equational deduction entailment: (A1) follows by first noticing that $E \cup \mathcal{F} \vdash \boxed{e}$ iff $E \vdash \boxed{e}$ (because there is no way to make use of the equations in $\mathcal{F}$ in any derivation of $\boxed{e}$) and then that $E \vdash \boxed{e}$ iff $E \vdash e$ (because E contains no frozen equations); and (A2) holds because any proof π of a frozen hidden equation $\boxed{e_{\mathcal{G}}}$ in $\mathcal{G}$ from $E \cup \mathcal{F}$ can use the frozen equations in $\mathcal{F}$ only "at the top", that is, not underneath operators via an equational congruence deduction step, so one can simply replace any frozen term $\boxed{t}$ in π by $\boxed{\delta[t]}$ and thus get a proof for $\boxed{e_{\mathcal{G}}}$.

Theorem 2. *(coinductive circularity principle) If $\mathcal{B}$ is a behavioral specification and F is a set of hidden equations with $\mathcal{B} \cup \boxed{F} \vdash \boxed{\Delta[F]}$ then $\mathcal{B} \Vdash F$.*

Proof. We first prove by well-founded induction on the depth of C that the hypothesis of the theorem implies $\mathcal{B} \cup \boxed{F} \vdash \boxed{C[F]}$ for any Δ-context C. If C is a degenerated context $*{:}h$ then $C[F]$ is a subset of F (its equations of sort h), so the result follows by the fact that $\vdash$ is an entailment system. The case $C \in \Delta$ follows from the theorem hypothesis. If $C[*{:}h] = C_1[C_2[*{:}h]]$ for some proper contexts C_1 and C_2 strictly smaller in depth than C, then we have by the induction hypothesis that $\mathcal{B} \cup \boxed{F} \vdash \boxed{C_1[F]}$ and $\mathcal{B} \cup \boxed{F} \vdash \boxed{C_2[F]}$. Since $\vdash$ is an entailment system with freezing, by (A2) it follows that $\mathcal{B} \cup \boxed{C_1[F]} \vdash \boxed{C_1[C_2[F]]}$. Now, since $\mathcal{B} \cup \boxed{F} \vdash \boxed{C_1[F]}$ implies $\mathcal{B} \cup \boxed{F} \vdash \mathcal{B} \cup \boxed{C_1[F]}$, by the transitivity of $\vdash$ (both of these properties of entailment systems), we get $\mathcal{B} \cup \boxed{F} \vdash \boxed{C[F]}$.

Therefore, $\mathcal{B} \cup \boxed{F} \vdash \boxed{C[F]}$ for any Δ-context C, in particular $\mathcal{B} \cup \boxed{F} \vdash \boxed{C[F]}$ for any Δ-experiment C. Then by axiom (A1) of entailment systems with freezing we obtain that $\mathcal{B} \vdash C[e]$ for any $e \in F$ and any Δ-experiment C for which $C[e]$ is defined. Therefore, $\mathcal{B} \Vdash F$.   □

Theorem 2 serves as the foundation of circular coinduction, because what circular coinduction essentially does is to iteratively attempt to complete a set of hidden equations that it starts with until it becomes a set F that satisfies the hypothesis of Theorem 2. Interestingly, Theorem 1 becomes now a simple corollary of Theorem 2: given a behaviorally closed $\mathcal{G}$ like in the hypothesis of Theorem 1, take F to be the set of all hidden equations in $\mathcal{G}$; then for an $e \in F$ of sort h and a $\delta[*{:}h] \in \Delta$, $\delta[e]$ either is visible and so $\mathcal{B} \vdash \delta[e]$ or is in F, so the hypothesis of Theorem 2 holds vacuously; therefore, $\mathcal{B} \Vdash \mathcal{G}$. This is not entirely unexpected, because the F in Theorem 2 is stated to be a fixed point w.r.t. Δ-derivability; what may be slightly unexpected is that such fixed points can be safely calculated modulo reasoning into an entailment system with freezing, in particular modulo equational deduction with inhibited congruence.

Fig. 4 defines circular coinduction as a proof system for deriving pairs of the form $\mathcal{H} \Vdash^{\circlearrowleft} \mathcal{G}$, where $\mathcal{H}$, the *hypotheses*, can contain both frozen and unfrozen equations, and where $\mathcal{G}$, the *goals*, contains only frozen equations. Initially, $\mathcal{H}$ is the original behavioral specification $\mathcal{B}$ and $\mathcal{G}$ is the frozen version $\boxed{G}$ of the original goals G to prove. The circular coinductive proving process proceeds as follows. At each moment when $\mathcal{G} \neq \emptyset$, a frozen equation $\boxed{e}$ is picked from $\mathcal{G}$. If $\mathcal{H} \vdash \boxed{e}$ holds, then $\boxed{e}$ may be discarded via a [Reduce] step. The core rule of circular coinduction is [Derive], which allows one to assume $\boxed{e}$ as hypothesis and generate $\boxed{\Delta[e]}$ as new proof obligations. This rule gives the user of our proof system the impression of circular reasoning, because one appears to assume what one wants to prove and go on. The key observation here is that the equation $\boxed{e}$ is assumed as hypothesis in its frozen form, which means that it cannot be freely used to prove its derivatives; otherwise, those would follow immediately by applying the congruence rule of equational deduction. One is done when the set of goals becomes empty. When that happens, one can conclude that $\mathcal{B} \Vdash G$.

The soundness of the circular coinductive proof system in Fig. 4, which is proved below, is a monolithic result depending upon a derivation of the complete proof for a task of the form $\mathcal{B} \Vdash^{\circlearrowleft} \boxed{G}$. We do not know any way to decompose the proof of soundness by proving the soundness of each derivation rule, as it is commonly done in soundness proofs. For example, one may attempt to show that $\mathcal{H} \Vdash^{\circlearrowleft} \mathcal{G}$ derivable implies $\mathcal{H}^{\circ} \Vdash \mathcal{G}^{\circ}$, where

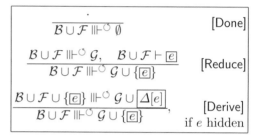

$$\frac{\cdot}{\mathcal{B} \cup \mathcal{F} \Vdash^{\circlearrowleft} \emptyset} \quad \text{[Done]}$$

$$\frac{\mathcal{B} \cup \mathcal{F} \Vdash^{\circlearrowleft} \mathcal{G}, \quad \mathcal{B} \cup \mathcal{F} \vdash \boxed{e}}{\mathcal{B} \cup \mathcal{F} \Vdash^{\circlearrowleft} \mathcal{G} \cup \{\boxed{e}\}} \quad \text{[Reduce]}$$

$$\frac{\mathcal{B} \cup \mathcal{F} \cup \{\boxed{e}\} \Vdash^{\circlearrowleft} \mathcal{G} \cup \boxed{\Delta[e]}}{\mathcal{B} \cup \mathcal{F} \Vdash^{\circlearrowleft} \mathcal{G} \cup \{\boxed{e}\}}, \quad \begin{array}{l} \text{[Derive]} \\ \text{if } e \text{ hidden} \end{array}$$

Fig. 4. Circular coinduction proof system: If $\mathcal{B} \Vdash^{\circlearrowleft} \boxed{G}$ is derivable then $\mathcal{B} \Vdash G$

$\mathcal{H}^{\circ}$ unfreezes all the frozen equations in $\mathcal{H}$ (and similarly for $\mathcal{G}^{\circ}$). Unfortunately, this simplistic approach cannot be used to show the [Derive] rule "sound", as it is *not* sound in this sense: indeed, $\mathcal{H}^{\circ} \cup \{e\} \Vdash \mathcal{G}^{\circ} \cup \Delta[e]$ is equivalent to $\mathcal{H}^{\circ} \cup \{e\} \Vdash \mathcal{G}^{\circ}$ because $\{e\} \Vdash \Delta[e]$, and there is no way to show from these that $\mathcal{H}^{\circ} \Vdash \mathcal{G}^{\circ} \cup \{e\}$. The soundness arguments of circular coinduction can be found in the complete proof, not in each particular derivation rule, which is what makes the theorem below unexpectedly hard to prove (note it also uses Theorem 2).

Theorem 3. *(soundness of circular coinduction) If $\mathcal{B}$ is a behavioral specification and G is a set of equations such that $\mathcal{B} \Vdash^{\circlearrowleft} \boxed{G}$ is derivable using the proof system in Fig. 4, then $\mathcal{B} \Vdash G$.*

Proof. Any derivation of $\mathcal{H} \Vdash^{\circlearrowleft} \mathcal{G}$ using the proof system in Fig. 4 yields a sequence of pairs $\mathcal{H}_0 \Vdash^{\circlearrowleft} \mathcal{G}_0, \mathcal{H}_1 \Vdash^{\circlearrowleft} \mathcal{G}_1, \ldots \mathcal{H}_n \Vdash^{\circlearrowleft} \mathcal{G}_n$, where $\mathcal{H}_0 = \mathcal{B}$, $\mathcal{G}_0 = \boxed{G}$, $\mathcal{G}_n = \emptyset$, and for every $0 \leq i < n$, there is some $\boxed{e} \in \mathcal{G}_i$ such that one of the following holds, each corresponding to one of the rules [Reduce] or [Derive]:

    [Reduce] $\mathcal{H}_i \vdash \boxed{e}$ and $\mathcal{G}_{i+1} = \mathcal{G}_i - \{\boxed{e}\}$ and $\mathcal{H}_{i+1} = \mathcal{H}_i$; or
    [Derive] e is hidden and $\mathcal{G}_{i+1} = (\mathcal{G}_i - \{\boxed{e}\}) \cup \boxed{\Delta[e]}$ and $\mathcal{H}_{i+1} = \mathcal{H}_i \cup \boxed{e}$.

Let $\mathcal{G} = \bigcup_{i=0}^{n} \mathcal{G}_i$, let $\mathcal{G}^{\circ} = \{e \mid \boxed{e} \in \mathcal{G}\}$, and let $F = hidden(\mathcal{G}^{\circ})$. Note that for each $0 \leq i < n$, $\mathcal{H}_i = \mathcal{B} \cup \mathcal{F}_i$ for some set of frozen hidden equations $\mathcal{F}_i$ with

$\mathcal{F}_i \cup \Delta[\mathcal{F}_i] \subseteq \mathcal{G}$: indeed, only frozen hidden equations are added to $\mathcal{H}$ and only by the rule [Derive], which also adds at the same time the derivatives of those equations to $\mathcal{G}$. This implies that if i corresponds to a [Reduce] step with $\mathcal{H}_i \vdash \boxed{e}$ for some $\boxed{e} \in \mathcal{G}$, then either $\mathcal{B} \vdash e$ by (A1) when e is visible, or $\mathcal{B} \cup \Delta[\mathcal{F}_i] \vdash \boxed{\Delta[e]}$ by (A2) when $e \in F$.

If $e \in \mathcal{G}^\circ$ visible, then there must be some $0 \leq i < n$ such that $\mathcal{H}_i \vdash \boxed{e}$, so $\mathcal{B} \vdash e$. Since $\Delta[\mathcal{F}_i] \subseteq \mathcal{G}$, equations $\boxed{f} \in \Delta[\mathcal{F}_i]$ either are visible and so $\mathcal{B} \vdash f$, or are hidden and so $\boxed{f} \in \boxed{F}$; in either case, we conclude that $\mathcal{B} \cup \boxed{F} \vdash \Delta[\mathcal{F}_i]$ for any $0 \leq i < n$, so $\mathcal{B} \cup \boxed{F} \vdash \boxed{\Delta[e]}$ for any $e \in F$ such that $\mathcal{H}_i \vdash \boxed{e}$ in some [Reduce] rule applied at step $0 \leq i < n$. If $e \in F$ such that a $\boxed{\delta[e]}$ for each appropriate $\delta \in \Delta$ is added to $\mathcal{G}$ via a [Derive] rule, then it is either that $\delta[e] \in \mathcal{G}^\circ$ and so $\mathcal{B} \vdash \delta[e]$, or that $\delta[e] \in F$; in either case, for such $e \in F$ we conclude that $\mathcal{B} \cup \boxed{F} \vdash \boxed{\Delta[e]}$. We covered all cases for $e \in F$, so $\mathcal{B} \cup \boxed{F} \vdash \boxed{\Delta[F]}$; then Theorem 2 implies that $\mathcal{B} \Vdash F$. Since F contains all the hidden equations of $\mathcal{G}^\circ$ and since we already proved that $\mathcal{B} \vdash e$, i.e., $\mathcal{B} \Vdash e$, for all $e \in \mathcal{G}^\circ$ visible, we conclude that $\mathcal{B} \Vdash \mathcal{G}^\circ$. Since $G \subseteq \mathcal{G}^\circ$, it follows that $\mathcal{B} \Vdash G$. $\qquad\square$

Example 4. Let us now derive by circular coinduction the two stream properties proved manually by coinduction in Example 3, as well as the fixed point property of the `morse` stream defined in Section 1.

The proof tree below summarizes the derivation steps for the first property. We only show the proof steps using the circular coinduction rules, that is, we omit the proof steps associated to the base entailment system $\vdash$:

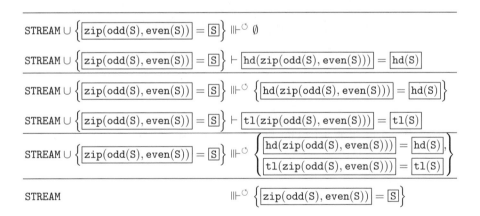

The top horizontal line marks a [Done] step. The second one marks a [Reduce] step which follows by equational reasoning using only the equations of STREAM. The third horizontal line marks also a [Reduce] step but where the frozen hypothesis is used this time: the goal $\mathtt{tl}(\mathtt{zip}(\mathtt{odd}(S), \mathtt{even}(S))) = \mathtt{tl}(S)$ is reduced first to $\mathtt{zip}(\mathtt{odd}(\mathtt{tl}(S)), \mathtt{even}(\mathtt{tl}(S))) = \mathtt{tl}(S)$ using the equations of STREAM and then to $\mathtt{tl}(S) = \mathtt{tl}(S)$ using the frozen hypothesis. The bottom horizontal line marks a [Derive] step: the initial goal is added as a frozen hypothesis and its derivatives become the new goals. In terms of the proof of Theorem 3, we have $\mathcal{B} = \mathtt{STREAM}$, $F_i = \{\mathtt{zip}(\mathtt{odd}(\mathtt{S}), \mathtt{even}(\mathtt{S})) = \mathtt{S}\}$ for $i = 1, 2, 3$, $\mathcal{G}_1^\circ = \{\mathtt{hd}(\mathtt{zip}(\mathtt{odd}(\mathtt{S}), \mathtt{even}(\mathtt{S}))) = \mathtt{hd}(\mathtt{S}), \mathtt{tl}(\mathtt{zip}(\mathtt{odd}(\mathtt{S}), \mathtt{even}(\mathtt{S}))) = \mathtt{tl}(\mathtt{S})\}$, $\mathcal{G}_2^\circ = \{\mathtt{hd}(\mathtt{zip}(\mathtt{odd}(\mathtt{S}), \mathtt{even}(\mathtt{S}))) = \mathtt{hd}(\mathtt{S})\}$, and $\mathcal{G}_3^\circ = \emptyset$. We observe that the circular coinduction proof system can be easily mechanized: the above proof tree can be automatically built in a bottom-up manner. This feature is fully exploited by the CIRC tool [10].

Similarly, next proof tree shows a derivation of $\mathtt{STREAM} \Vdash^\circlearrowright \mathtt{f}(\mathtt{not}(\mathtt{S})) = \mathtt{not}(\mathtt{f}(\mathtt{S}))$. The following notations are used: $F_1 = \{\mathtt{f}(\mathtt{not}(\mathtt{S})) = \mathtt{not}(\mathtt{f}(\mathtt{S}))\}$, $F_2 = F_1 \cup \{\mathtt{tl}(\mathtt{f}(\mathtt{not}(\mathtt{S}))) = \mathtt{tl}(\mathtt{not}(\mathtt{f}(\mathtt{S})))\}$. Note that this time the index of F does not reflect the step number, as it is the case in the proof of Theorem 3.

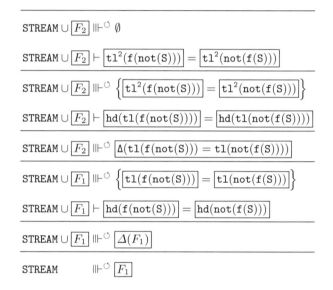

The top-down sequence of the applied proof rules is: [Done], [Reduce], [Reduce], [Derive], [Reduce], [Derive]. The first [Reduce] uses a frozen hypothesis from $\boxed{F_2}$. The goal $\mathtt{tl}^2(\mathtt{f}(\mathtt{not}(\mathtt{S}))) = \mathtt{tl}^2(\mathtt{not}(\mathtt{f}(\mathtt{S})))$ is reduced to $\mathtt{f}(\mathtt{not}(\mathtt{tl}(\mathtt{S}))) = \mathtt{not}(\mathtt{f}(\mathtt{tl}(\mathtt{S})))$ by equational reasoning using only the equations of $\mathtt{STREAM}$, and then to $\mathtt{not}(\mathtt{f}(\mathtt{tl}(\mathtt{S}))) = \mathtt{not}(\mathtt{f}(\mathtt{tl}(\mathtt{S})))$ using the frozen hypothesis from $\boxed{F_1} \subset \boxed{F_2}$. Note that the first [Derive] step in the above sequence exhibits how the circular coinduction has automatically "discovered" the second equation that was necessary to prove the property by plain coinduction in Example 3.

Finally, next proof tree shows a circular coinductive proof that, with the streams defined in Section 1, $\mathtt{morse}$ is a fixed point of $\mathtt{f}$. Note that one cannot prove directly the fixed point property (see [10] for how CIRC fails to prove it); instead, we prove $G = \{\mathtt{f}(\mathtt{morse}) = \mathtt{morse}, (\forall \mathtt{S})\,\mathtt{zip}(\mathtt{S}, \mathtt{not}(\mathtt{S})) = \mathtt{f}(\mathtt{S})\}$.

The description of the proof tree uses the following notations:

$$F_1 = \{\mathtt{f(morse)} = \mathtt{morse}, \mathtt{zip(S, not\ S)} = \mathtt{f(S)}\}$$
$$F_2 = F_1 \cup \{\mathtt{tl(f(morse))} = \mathtt{tl(morse)}\}$$
$$F_3 = F_2 \cup \{\mathtt{tl(zip(S, not\ S))} = \mathtt{tl(f(S))}\}$$

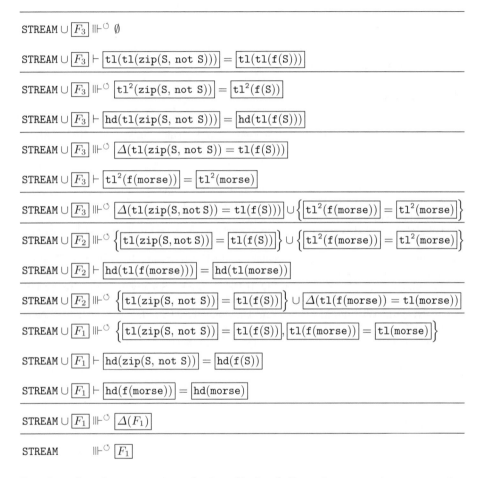

For the sake of presentation, the last [Reduce] discards two goals, representing in fact two proof steps.

6 Conclusion

Previous formalizations of circular coinduction were either algorithmic in nature, or limited. For example, [5] introduces circular coinductive rewriting as an operational technique to extend rewriting with coinductive steps. On the other hand, [6] attempts to capture circular coinduction as a proof rule, but, unfortunately, it only works with properties that need at most one derivation step and it is melted away within a particular entailment system for hidden algebra, making it hard to understand what circular coinduction really is.

This paper presented circular coinduction as a sound, generic, self-contained and easy to understand proof system. We believe that this result will enhance understanding of circular coinduction, will allow it to be applicable to various coalgebraic settings, and will lead to improved implementations and extensions.

Acknowledgment. The paper is supported in part by NSF grants CCF-0448501, CNS-0509321 and CNS-0720512, by NASA contract NNL08AA23C, and by CNCSIS grant PN-II-ID-393.

References

1. Allouche, J.-P., Shallit, J.: The ubiquitous Prouhet-Thue-Morse sequence. In: Ding, T.H.C., Niederreiter, H. (eds.) Sequences and Their applications (Proc. SETA 1998), pp. 1–16. Springer, Heidelberg (1999)
2. Bidoit, M., Hennicker, R.: Constructor-based observational logic. J. Log. Algebr. Program. 67(1-2), 3–51 (2006)
3. Clavel, M., Durán, F., Eker, S., Lincoln, P., Martí-Oliet, N., Meseguer, J., Talcott, C. (eds.): All About Maude - A High-Performance Logical Framework. LNCS, vol. 4350. Springer, Heidelberg (2007)
4. Gaudel, M.-C., Privara, I.: Context induction: an exercise. Technical Report 687, LRI, Université de Paris-Sud (1991)
5. Goguen, J., Lin, K., Roşu, G.: Circular coinductive rewriting. In: ASE 2000: Proceedings of the 15th IEEE international conference on Automated software engineering, Washington, DC, USA, pp. 123–132. IEEE Computer Society, Los Alamitos (2000)
6. Goguen, J., Lin, K., Roşu, G.: Conditional circular coinductive rewriting with case analysis. In: Wirsing, M., Pattinson, D., Hennicker, R. (eds.) WADT 2003. LNCS, vol. 2755, pp. 216–232. Springer, Heidelberg (2003)
7. Goguen, J., Meseguer, J.: Completeness of Many-Sorted Equational Logic. Houston Journal of Mathematics 11(3), 307–334 (1985)
8. Hausmann, D., Mossakowski, T., Schröder, L.: Iterative circular coinduction for CoCasl in Isabelle/HOL. In: Cerioli, M. (ed.) FASE 2005. LNCS, vol. 3442, pp. 341–356. Springer, Heidelberg (2005)
9. Hennicker, R.: Context induction: a proof principle for behavioral abstractions. Formal Aspects of Computing 3(4), 326–345 (1991)
10. Lucanu, D., Goriac, E.-I., Caltais, G., Roşu, G.: CIRC: A behavioral verification tool based on circular coinduction. In: Kurz, A., Lenisa, M., Tarlecki, A. (eds.) Designing Privacy Enhancing Technologies. LNCS, vol. 5728, pp. 433–442. Springer, Heidelberg (2009)
11. Lucanu, D., Roşu, G.: CIRC: A circular coinductive prover. In: Mossakowski, T., Montanari, U., Haveraaen, M. (eds.) CALCO 2007. LNCS, vol. 4624, pp. 372–378. Springer, Heidelberg (2007)
12. Meseguer, J.: General logics. In: Ebbinghaus, H.-D., et al. (eds.) Logic Colloquium 1987, pp. 275–329. North Holland, Amsterdam (1989)
13. Roşu, G.: Hidden Logic. PhD thesis, University of California at San Diego (2000)
14. Roşu, G.: Equality of streams is a Π_2^0-complete problem. In: Proceedings of the 11th ACM SIGPLAN International Conference on Functional Programming (ICFP 2006), ACM Press, New York (2006)
15. Roşu, G., Goguen, J.: Circular coinduction. Short paper at the International Joint Conference on Automated Reasoning, IJCAR 2001 (2001)

Approximating Labelled Markov Processes Again!

Philippe Chaput[1,*], Vincent Danos[2], Prakash Panangaden[1,*],
and Gordon Plotkin[2,**]

[1] School of Computer Science, McGill University
[2] School of Informatics, University of Edinburgh

Abstract. Labelled Markov processes are continuous-state fully probabilistic labelled transition systems. They can be seen as co-algebras of a
suitable monad on the category of measurable space. The theory as developed so far included a treatment of bisimulation, logical characterization
of bisimulation, weak bisimulation, metrics, universal domains for LMPs
and approximations. Much of the theory involved delicate properties of
analytic spaces.

Recently a new kind of averaging procedure was used to construct approximations. Remarkably, this version of the theory uses a dual view of
LMPs and greatly simplifies the theory eliminating the need to consider
aanlytic spaces. In this talk I will survey some of the ideas that led to
this work.

1 Introduction

The study of continuous systems is becoming a more common part of computer
science. Control theory, or systems theory has worked with an abstract notion
of a system; see, for example, the text by Sontag [1]. These notions were general
enough to include the continuous physical systems normally studied by control
engineers as well as systems such as automata which were inherently discrete in
nature. General concepts such as "state", "transformation" and "observation"
are used in both the discrete and the continuous settings. Labelled Markov processes (LMPs) [2,3] were invented as a formalism to study *interacting probabilistic
systems* that may be either discrete or continuous.

The basic idea of LMPs is to consider probabilistic labelled transition systems
with continuous state spaces. There are states and actions or labels. A state and
an action determines a probability distribution over the next states. There is no
other nondeterminism as seen, for example, in probabilistic automata[4]. Nor is
there any attempt to assign probabilities to the actions; the nondeterminism is
captured by the arbitrariness in the choice of actions. This model is very close
to the notion of Markov decision process (MDP) [5] used in optimization theory
and machine learning. The only difference is that we do not have a concept of

---

[*] Research supported by NSERC.
[**] Research supported by a Royal Society – Wolfson Research Merit Award.

A. Kurz, M. Lenisa, and A. Tarlecki (Eds.): CALCO 2009, LNCS 5728, pp. 145–156, 2009.

"reward" or "payoff." The discrete version of this model is due to Larsen and Skou [6] and is called by them the *reactive model*.

The systems that we consider - continuous state space and discrete time - are of interest for two reasons. First, this is a reasonable middle ground before we get to a completely continuous model. Several authors have looked at discrete space, continuous time models. It is clear that fully continuous models will force us to make a significant deepening of the mathematics. One will have to consider stochastic differential equations rather than transition systems to describe the evolution of systems: we are not there yet.

Second, these systems occur in natural examples. In the past one of us was involved with a case study from avionics where the discrete-time-continuous-space character of the system was clearly in evidence. Many control systems have this character. It is possible that this may even be fruitful for biology.

Very similar notions were introduced at around the same time by de Vink and Rutten[7]. The main difference is that they worked with ultrametric spaces; one can argue that these spaces are essentially discrete. For example, they are totally disconnected and are nothing like the spaces that arise in physical systems. Their important contribution was to emphasize the co-algebraic point of view – noted in passing in [3] – and show that bisimulation could be presented as a span of co-algebra homomorphisms. The same idea, span of "zigzags", is used by Desharnais et al. [3]. In the present work we emphasize the importance of co-spans rather than spans.

The notion of bisimulation is central to the study of concurrent systems. While there is a variety of different equivalence relations between processes (two-way simulation, trace equivalence, failures equivalence and many more), bisimulation enjoys some fundamental mathematical properties, most notably its characterization as a fixed-point, which make it the most discussed process equivalence.

One might take the view that any automated analysis or logical reasoning must be inherently discrete in character. In particular, even if one is interested in reasoning about a physical system, one has to first discretize the system. In fact, this point of view actually provides a good argument for retaining the continuous view of the system. A given system may well be described in continuous terms. Without formalizing the continuous system - and having a notion of equivalence between discrete and continuous systems - how does one argue that the discretized system is a faithful model of the underlying continuous system? Even suppose one is willing to treat a discrete model as given, what if one needs to refine the model? For example, a given discretization may arise from some type of approximation based on a given *tolerance*; how does one refine the tolerance or discretize *adaptively*? Clearly the underlying continuous model has to be retained and used if we want to construct different discrete approximations.

In brief, a labelled Markov process can be described as follows. There is a set of states and a set of labels. The system is in a state at a point in time and moves between states. The state which it moves to is governed by which interaction with the environment is taking place and this is indicated by the labels. The system evolves according to a probabilistic law. If the system interacts with the environment by synchronizing on a label it makes a transition to a new state governed by a transition probability distribution. So far, this is essentially the model developed by Larsen and Skou [6]. They specify the transitions by giving, for each label, a probability for going from one state to another. Bisimulation then amounts to matching the moves; this means that both the labels and the probabilities must be the same.

In the case of a continuous state space, however, one cannot simply specify transition probabilities from one state to another. In most interesting systems all such transition probabilities would be zero! Instead one must work with probability densities. In so doing, one has to confront the major issues that arose when probability theory was first formalized, such as the existence of subsets for which the notion of probability does not make sense. In the present case, we have to introduce a notion of set for which "probabilities make sense" (i.e. a σ-field) and instead of talking about probabilities of going from a state s to another state s', we have to talk about going from a state s to a *set of* states A.

The notion of bisimulation for these systems is a mild modification of the definition of Larsen and Skou. One has to add a measure theoretic condition. In the logical characterization it was shown that two states are bisimilar if and only if they satisfy all the same formulas of a modal logic similar to Hennessy-Milner logic. The striking aspect of this result is that the logic is completely negation free. Even for purely discrete systems this result was new and quite unexpected. It shows that fully probabilistic systems are very close to determinate systems. The nature of the proof is quite different from proofs of other Hennessy-Milner type results.

In recent work we take an entirely new approach, in some ways "dual" to the normal view of probabilistic transition systems. We think of a Markov process as a transformer of functions defined on it rather than as a transformer of the state. Thus, instead of working directly with a Markov kernel $\tau(s, A)$ that takes a state s to a probability distribution over the state space, we think of a Markov process as transforming a function f into a new function $\int f(s')\tau(s, \mathrm{d}s')$ over the state space. This is the probabilistic analogue of working with predicate transformers, a point of view advocated by Kozen [8] in a path-breaking early paper on probabilistic systems and logic.

This new way of looking at things leads to three new results:

1. It is possible to define bisimulation on general spaces – not just on analytic spaces – and show that it is an equivalence relation with easy categorical constructions. The logical characterization of bisimulation can also be done generally, and with no complicated measure theoretic arguments.
2. A new and flexible approach to approximation based on averaging can be given. This vastly generalizes and streamlines the idea of using conditional expectations to compute approximation [9].
3. It is possible to show that there is a minimal bisimulation equivalent to a process obtained as the limit of finite approximants.

2   Labelled Markov Processes

Labelled Markov processes are probabilistic versions of labelled transition systems. Corresponding to each label a Markov process is defined. The transition probability is given by a *stochastic kernel* (Feller's terminology [10]) also commonly called a *Markov kernel*. Thus the indeterminacy has two sources: the "choice" made by the environment - no probabilities are attributed to this at all - and the probabilistic transitions made by the process. This is the "reactive" model of Larsen and Skou [6] who used it in a discrete state-space setting.

A key ingredient in the theory is the stochastic kernel or Markov kernel. We will call it a *transition probability function*.

Definition 1. *A **transition (sub-)probability function** on a measurable space (S, Σ) is a function $\tau : S \times \Sigma \longrightarrow [0, 1]$ such that for each fixed $s \in S$, the set function $\tau(s, \cdot)$ is a (sub-)probability measure, and for each fixed $X \in \Sigma$ the function $\tau(\cdot, X)$ is a measurable function.*

One interprets $\tau(s, X)$ as the probability of the process starting in state s making a transition into one of the states in X. The transition probability is really a *conditional probability*; it gives the probability of the process being in one of the states of the set X after the transition, *given* that it was in the state s before the transition. In general the transition probabilities could depend on time, in the sense that the transition probability could be different at every step, but still independent of past history; we always consider the time-independent case.

We will work with *sub-probability* functions; i.e. with functions where $\tau(s, S) \le 1$ rather than $\tau(s, S) = 1$. The mathematical results go through in this extended case. We view processes where the transition functions are only sub-probabilities as being *partially defined*, opening the way for a notion of approximation.

The stochastic systems studied in the literature are usually only the very special version where $\tau(s, S)$ is either 1 or 0. We call such processes *total* and the general processes are called *partial*. We capture the idea that an action is rejected by setting $\tau(s, S)$ to be 0.

Definition 2. *A **labelled Markov process (LMP)** S with label set $\mathcal{A}$ is a structure $(S, i, \Sigma, \{\tau_a \mid a \in \mathcal{A}\})$, where S is the set of states, i is the initial state, and Σ is the Borel σ-field on S, and*

$$\forall a \in \mathcal{A}, \tau_a : S \times \Sigma \longrightarrow [0, 1]$$

is a transition sub-probability function.

We will fix the label set to be $\mathcal{A}$ once and for all. We will write (S, i, Σ, τ) for labelled Markov processes, instead of the more precise $(S, i, \Sigma, \{\tau_a \mid a \in \mathcal{A}\})$.

We give a simple example to illustrate the ideas.

Example 1. (From [11]) Consider a process with two labels $\{a, b\}$. The state space is upper right quadrant of the real plane, $\mathbb{R}^2$ together with a single extra point. In order to describe this system conveniently we will pretend, at first, that the state space is the entire real plane. When the process makes an a-move from state (x_0, y_0), it jumps to (x, y_0), where the probability distribution for x is given by the density $K_\alpha \exp(-\alpha(x - x_0)^2)$, where $K_\alpha = \sqrt{\alpha/\pi}$ is the normalizing factor. When it makes a b-move it jumps from state (x_0, y_0) to (x_0, y), where the distribution of y is given by the density function $K_\beta \exp(-\beta(y - y_0)^2)$. The meaning of these densities is as follows. The probability of jumping from (x_0, y_0) to a state with x-coordinate in the interval $[s, t]$ under an a-move is $\int_s^t K_\alpha \exp(-\alpha(x - x_0)^2)dx$. All points with $x < 0$ or $y < 0$ are identified as a single absorbing state. Once it is in this state no more transitions are possible. Note that the probability of jumping to any given point is, of course, 0. In this process the interaction with the environment controls whether the jump is along the x-axis or along the y-axis but the actual extent of the jump is governed by a probability distribution. If there were just a single label we would have an ordinary (time-independent) Markov process; in fact it would be a brownian motion with absorbing walls.

3   Bisimulation and Logic

The fundamental process equivalence that we consider is *strong probabilistic bisimulation*. Probabilistic bisimulation means matching the moves and probabilities *exactly* – thus each system must be able to make the same transitions with the same probabilities as the other. Larsen and Skou define a bisimulation relation R as an equivalence relation on the states satisfying the condition that, for each label a, equivalent states have equal probability of making an a-transition to any R-equivalence class of states. In the continuous case, we demand that equivalent states have equal probability of making an a-transition to any union of equivalence classes of states *provided that the union is measurable*.

Instead of talking about sets of equivalence classes we will instead use the notion of R-*closed* sets. Let R be a binary relation on a set S. We say a set $X \subseteq S$ is R-*closed* if $R(X) := \{t | \exists s \in X, sRt\}$ is a subset of X. If R is reflexive, this becomes $R(X) = X$. If R is an equivalence relation, a set is R-closed if and only if it is a union of equivalence classes.

Definition 3. *Let $\mathcal{S} = (S, i, \Sigma, \tau)$ be a labelled Markov process. An equivalence relation R on S is a **bisimulation** if whenever sRs', with $s, s' \in S$, we have that for all $a \in \mathcal{A}$ and every R-closed measurable set $X \in \Sigma$, $\tau_a(s, X) = \tau_a(s', X)$. Two states are bisimilar if they are related by a bisimulation relation.*

Alternately, bisimulation on the states of a labelled Markov process can be viewed as the maximum fixed point of the following (monotone) functional F on the lattice of equivalence relations on $(S \times S, \subseteq)$:

$$s \, F(R) \, t \text{ if for all } a \in \mathcal{A}, \text{ and all } R\text{-closed } C \in \Sigma, \tau_a(s, C) \leq \tau_a(t, C).$$

It is easy to prove that bisimulation is an equivalence relation.

The intuition of this definition is that the relation R relates those states that can be "lumped" together. Bisimulation is the largest such relation. In fact the notion of bisimulation was known in the queuing theory community [12] under the name of "lumpability".

One can define a simple modal logic and prove that two states are bisimilar if and only if they satisfy exactly the same formulas. As before we assume that there is a fixed set of "actions" $\mathcal{A}$. The logic is called $\mathcal{L}$ and has the following syntax:

$$\mathsf{T} \mid \phi_1 \wedge \phi_2 \mid \langle a \rangle_q \phi$$

where a is an action and q is a rational number. This is the basic logic with which we establish the logical characterization.

Given a labelled Markov process $\mathcal{S} = (S, i, \Sigma, \tau)$ we write $s \models \phi$ to mean that the state s satisfies the formula ϕ. The definition of the relation $\models$ is given by induction on formulas. The definition is obvious for the propositional constant T and conjunction. We say $s \models \langle a \rangle_q \phi$ if and only if $\exists X \in \Sigma.(\forall s' \in X.s' \models \phi) \wedge (\tau_a(s, X) > q)$. In other words, the process in state s can make an a-move to a state, that satisfies ϕ, with probability strictly greater than q. We write $\llbracket \phi \rrbracket_\mathcal{S}$ for the set $\{s \in S \mid s \models \phi\}$, and $\mathcal{S} \models \phi$ if $i \models \phi$. We often omit the subscript when no confusion can arise.

The logic that Larsen and Skou used in [6] has more constructs including disjunction and some negative constructs. They show that for systems satisfying a "minimum deviation condition" – a uniform bound on the degree of branching everywhere – two *states* of the same process are bisimilar if and only if they satisfy the same formulas of their logic.

The main theorem relating $\mathcal{L}$ and bisimulation is the following:

Theorem 4. *Let (S, i, Σ, τ) be a labelled Markov process. Two states $s, s' \in S$ are bisimilar if and only if they satisfy the same formulas of $\mathcal{L}$.*

It does not have any kind of minimal deviation condition, nor any negative constructs in the logic. In [13] a logical characterization of simulation was also proved.

4   Approximation by Unfolding

In this section I review the "old" view of approximation. The key idea is the construction of some approximants via an "unfolding" construction. As the approximation is refined there are more and more transitions possible. There are two parameters to the approximation, one is a natural number n, and the other is a positive rational ϵ. The number n gives the number of successive transitions possible from the start state. The number ϵ measures the accuracy with which the probabilities approximate the transition probabilities of the original process.

Given a labelled Markov process $\mathcal{S} = (S, i, \Sigma, \tau)$, an integer n and a rational number $\epsilon > 0$, we define $\mathcal{S}(n, \epsilon)$ to be an n-step unfolding approximation of $\mathcal{S}$.

Its state-space is divided into $n+1$ levels which are numbered $0, 1, \ldots, n$. Bisimulation is a fixed point and that one has - for each n - a level n approximation to bisimulation. At each level, say n, the states of the approximant is a partition of S; these partitions correspond to the equivalence classes corresponding to the level n approximation to bisimulation. The initial state of $\mathcal{S}(n, \epsilon)$ is at level n and transitions only occur between a state of one level to a state of one lower level. Thus, in particular, states of level 0 have no outgoing transitions. In the following we omit the curly brackets around singletons.

Let (S, i, Σ, τ) be a labelled Markov process, $n \in \mathbf{N}$ and ϵ a positive rational. We denote the finite-state approximation by $\mathcal{S}(n, \epsilon) = (P, p_0, \rho)$ where P is a subset of $\Sigma \times \{0, \ldots, n\}$. It is defined as follows, for $n \in \mathbf{N}$ and $\epsilon > 0$. $\mathcal{S}(n, \epsilon)$ has $n+1$ levels. States are defined by induction on their level. Level 0 has one state $(S, 0)$. Now, given the sets from level l, we define states of level $l+1$ as follows. Suppose that there are m states at level l, we partition the interval $[0, 1]$ into intervals of size ϵ/m. Let $(B_j)_{j \in I}$ stand for this partition; i.e. for $\{\{0\}, (0, \epsilon/m], (\epsilon/m, 2\epsilon/m], \ldots\}$. States of level $l+1$ are obtained by the partition of S that is generated by the sets $\tau_a(\cdot, C)^{-1}(B_j)$, for every set C corresponding to state at level l and every label $a \in \{a_1, \ldots, a_n\}$, $i \in I$. Thus, if a set X is in this partition of S, $(X, l+1)$ is a state of level $l+1$. Transitions can happen from a state of level $l+1$ to a state of level l, and the transition probability function is given by

$$\rho_a((X, k), (B, l)) = \begin{cases} \inf_{t \in X} \tau_a(t, B)) & \text{if } k = l+1, \\ 0 & \text{otherwise.} \end{cases}$$

The initial state p_0 of $\mathcal{S}(n, \epsilon)$ is the unique state (X, n) such that X contains i, the initial state of $\mathcal{S}$.

The next theorem is the main result.

Theorem 5. *If a state $s \in S$ satisfies a formula $\phi \in \mathcal{L}_\vee$, then there is some approximation $\mathcal{S}(n, \epsilon)$ such that $(X_s, n) \models \phi$.*

In the original approximation paper [13], a universal domain of LMPs was defined and a simulation order was given. It was shown that the approximants form a directed set and that any LMP could be given as a sup of the approximants. The approximants gave a countable base for the domain which was thus a continuous domain. It is a final co-algebra for a suitable category of LMPs. Independently, van Breugel and Worrell showed that one could define a final co-algebra for LMPs using other techniques.

In a later paper, Desharnais et al. [14] developed a metric and showed that the approximations converge in that metric. Thus this seems to strangthen the idea that this is a robust notion of convergence. The trouble is that the approximants are from below. In other words, if we cluster some of the states to form an approximate state say B, to estimate the transition probability from B to A we take the inf of $\tau(x, A)$ as x ranges over B. It is far more natural to average the states in B. This idea was developed in [9]. The construction that we develop next absorbs this idea but in a completely new way.

5   The Dual View

Definition 6. *An abstract Markov process (AMP) on a probability space X is a ω-continuous linear map $\tau : L_\infty^+(X) \longrightarrow L_\infty^+(X)$ with $\tau(\mathbf{1}_X) \leq_p \mathbf{1}_X$.*

The condition that $\tau(\mathbf{1}_X) \leq_p \mathbf{1}_X$ is equivalent to requiring that the operator norm of τ be less than one, i.e. that $\|\tau(f)\|_\infty \leq \|f\|_\infty$ for all $f \in L_\infty^+(X)$. This is natural, as the function $\tau(\mathbf{1}_X)$, evaluated at a point x, is the probability of jumping from x to X, which is less than one. AMPs are often called Markov operators in the literature, and have been first introduced in [15].

We now formalize the notion of conditional expectation. We work in a subcategory of **Prb**, called **Rad**$_\infty$, where we require the image measure to be bounded by a multiple of the measure in the codomain; that is, measurable maps $\alpha : (X, \Sigma, p) \longrightarrow (Y, \Lambda, q)$ such that $M_\alpha(p) \leq Kq$ for some real number K.

Let us define an operator $\mathbb{E}_\alpha : L_\infty^+(X, p) \longrightarrow L_\infty^+(Y, q)$, as follows: $\mathbb{E}_\alpha(f) = \frac{\mathrm{d}M_\alpha(f \triangleright p)}{\mathrm{d}q}$. As α is in **Rad**$_\infty$, the Radon-Nikodym derivative is defined and is in $L_\infty^+(X, p)$. That is, the following diagram commutes by definition:

$$
\begin{array}{ccc}
L_\infty^+(X, p) & \xrightarrow{\;\triangleright p\;} & \mathcal{M}^{\leq Kp}(X) \\
\Big\downarrow{\scriptstyle \mathbb{E}_\alpha} & & \Big\downarrow{\scriptstyle M_\alpha(-)} \\
L_\infty^+(Y, q) & \xleftarrow{\;\frac{\mathrm{d}}{\mathrm{d}q}\;} & \mathcal{M}^{\leq Kq}(Y)
\end{array}
$$

Note that if (X, Σ, p) is a probability space and $\Lambda \subseteq \Sigma$ is a sub-σ-algebra, then we have the obvious map $\lambda : (X, \Sigma, p) \longrightarrow (X, \Lambda, p)$ which is the identity on the underlying set X. This map is in **Rad**$_\infty$ and it is easy to see that $\mathbb{E}_\lambda$ is precisely the conditional expectation onto Λ. Thus the operator $\mathbb{E}_-$ truly generalizes conditional expectation. It is easy to show that $\mathbb{E}_{\alpha \circ \beta} = \mathbb{E}_\alpha \circ \mathbb{E}_\beta$ and thus $\mathbb{E}_-$ is functorial.

Given an AMP on (X, p) and a map $\alpha : (X, p) \longrightarrow (Y, q)$ in **Rad**$_\infty$, we thus have the following approximation scheme:

$$
\begin{array}{ccc}
L_\infty^+(Y, q) & \xdashrightarrow{\;\alpha(\tau)\;} & L_\infty^+(Y, q) \\
{\scriptstyle (-) \circ \alpha}\Big\downarrow & & \Big\uparrow{\scriptstyle \mathbb{E}_\alpha} \\
L_\infty^+(X, p) & \xrightarrow{\;\tau\;} & L_\infty^+(X, p)
\end{array}
$$

Note that $\|\alpha(\tau)\| \leq \|(-) \circ \alpha\| \cdot \|\tau\| \cdot \|\mathbb{E}_\alpha\| = \|\tau\| \cdot \|d(\alpha)\|_\infty$. Here the norm of $(\cdot) \circ \alpha$ is 1. As an AMP has a norm less than 1, we can only be sure that a map α yields an approximation for every AMP on X if $\|d(\alpha)\|_\infty \leq 1$. We call the AMP $\alpha(\tau)$ the projection of τ on Y.

6   Bisimulation

The notion of probabilistic bisimulation was introduced by Larsen and Skou [6] for discrete spaces and by Desharnais et al. [3] for continuous spaces.

Subsequently a dual notion called event bisimulation or probabilistic co-congruence was defined independently by Danos et al. [16] and by Bartels et al. [17]. The idea of event bisimulation was that one should focus on the measurable sets rather than on the points. This meshes exactly with the view here.

Definition 7. *Given a (usual) Markov process (X, Σ, τ), an event-bisimulation is a sub-σ-algebra Λ of Σ such that (X, Λ, τ) is still a Markov process [16].*

The only additional condition that needs to be respected for this to be true is that the Markov process $\tau(x, A)$ is Λ-measurable for a fixed $A \in \Lambda$. Translating this definition in terms of AMPs, this implies that the AMP τ sends the subspace $L_\infty^+(X, \Lambda, p)$ to itself, and so that the following commutes:

$$
\begin{array}{ccc}
L_\infty^+(X, \Sigma) & \xrightarrow{\ \tau\ } & L_\infty^+(X, \Sigma) \\
\uparrow & & \uparrow \\
L_\infty^+(X, \Lambda) & \xrightarrow[\ \tau\ ]{} & L_\infty^+(X, \Lambda)
\end{array}
$$

A generalization to the above would be a $\mathbf{Rad}_\infty$ map α from (X, Σ, p) to (Y, Λ, q), respectively equipped with AMPs τ and ρ, such that the following commutes:

$$
\begin{array}{ccc}
L_\infty^+(X, p) & \xrightarrow{\ \tau\ } & L_\infty^+(X, p) \\
\uparrow {\scriptstyle(-)\circ\alpha} & & \uparrow {\scriptstyle(-)\circ\alpha} \\
L_\infty^+(Y, q) & \xrightarrow[\ \rho\ ]{} & L_\infty^+(Y, q)
\end{array}
$$

We will call such a map a *zigzag*.

Definition 8. *We say that two objects of $\mathbf{AMP}$, (X, Σ, p, τ) and (Y, Λ, q, ρ), are bisimilar if there is a third object (Z, Γ, r, π) with a pair of zigzags*

$$
\alpha : (X, \Sigma, p, \tau) \longrightarrow (Z, \Gamma, r, \pi)
$$
$$
\beta : (Y, \Lambda, q, \rho) \longrightarrow (Z, \Gamma, r, \pi)
$$

making a cospan diagram

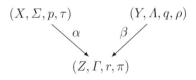

Note that the identity function on an AMP is a zigzag, and thus that any zigzag between two AMPs X and Y implies that they are bisimilar.

The great advantage of cospans is that one needs pushouts to exist rather than pullbacks (or weak pullbacks); pushouts are much easier to construct. The following theorem shows that bisimulation is an equivalence.

Theorem 9. *Let $\alpha : X \longrightarrow Y$ and $\beta : X \longrightarrow Z$ be a span of zigzags. Then the pushout W exists and the pushout maps $\delta : Y \longrightarrow W$ and $\gamma : Z \longrightarrow W$ are zigzags.*

Corollary 10. *Bisimulation is an equivalence relation on the objects of* **AMP**.

It turns out that there is a "smallest" bisimulation. Given an AMP (X, Σ, p, τ), one question one may ask is whether there is a "smallest" object $(\tilde{X}, \tilde{\Sigma}, r, \xi)$ in **AMP** such that, for every zigzag from X to another AMP (Y, Λ, q, ρ), there is a zigzag from (Y, Λ, q, ρ) to $(\tilde{X}, \tilde{\Sigma}, r, \xi)$. It can be shown that such an object exists.

Proposition 11. *Let $\{\alpha_i : (X, \Sigma, p, \tau) \longrightarrow (Y_i, \Lambda_i, q_i, \rho_i)\}$ be the set of all zigzags in* **AMP** *with domain (X, Σ, p, τ). This yields a generalized pushout diagram, and as in Theorem 9, the pushout $(\tilde{X}, \tilde{\Sigma}, r, \xi)$ exists and the pushout maps are zigzags.*

A version of the logical characterization theorem follows immediately from this.

7   Approximation Revisited

In this section, we let the measurable map $i_\Lambda : (X, \Sigma) \longrightarrow (X, \Lambda)$ be the identity on the set X, restricting the σ-field. The resulting AMP morphism is denoted as $i_\Lambda : (X, \Sigma, p, \tau) \longrightarrow (X, \Lambda, p, \Lambda(\tau))$, as p is just restricted on a smaller σ-field, with $\Lambda(\tau)$ being the projection of τ on the smaller σ-field Λ.

Let (X, Σ, p, τ_a) be a labelled AMP. Let $\mathcal{P}$ be a finite set of rationals in $[0, 1]$; we will call it a *rational partition*. We define a family of finite π-systems [18], subsets of Σ, as follows:

$$\Phi_{\mathcal{P},0} = \{X, \emptyset\}$$
$$\Phi_{\mathcal{P},n} = \pi\left(\left\{\tau_a(\mathbf{1}_A)^{-1}(q_i, 1] : q_i \in \mathcal{P}, A \in \Phi_{\mathcal{P},n-1}, a \in \mathcal{A}\right\} \cup \Phi_{\mathcal{P},n-1}\right)$$

where $\pi(\Omega)$ is the π-system generated by the class of sets Ω.

For each pair $(\mathcal{P}, M)$ consisting of a rational partition and a natural number, we define a σ-algebra $\Lambda_{\mathcal{P},M}$ on X as $\Lambda_{\mathcal{P},M} = \sigma(\Phi_{\mathcal{P},M})$, the σ-algebra generated by $\Phi_{\mathcal{P},M}$. We shall call each pair $(\mathcal{P}, M)$ consisting of a rational partition and a natural number an *approximation pair*. These σ-algebras have a very important property:

Proposition 12. *Given any labelled AMP (X, Σ, p, τ_a), the σ-field $\sigma\left(\bigcup \Lambda_{\mathcal{P},M}\right)$, where the union is taken over all approximation pairs, is precisely the σ-field $\sigma[\![\mathcal{L}]\!]$ obtained from the logic.*

In fact one can use a theorem of Choksi's from 1958 [19] to construct projective limits of finite approximants and establish the following fundamental theorem.

Theorem 13. *Given a labelled AMP (X, Σ, p, τ_a), the projective limit of its finite approximants $(\text{proj}\lim \hat{X}, \Gamma, \gamma, \zeta_a)$ is isomorphic to its smallest bisimulation $(\tilde{X}, \tilde{\Sigma}, r, \xi_a)$.*

8   Conclusions

The main contribution of the present work is to show how one can obtain a powerful and general notion of approximation of Markov processes using the dualized view of Markov processes as transformers of random variables (measurable functions). We view Markov processes as "predicate transformers". Our main result is to show that this way of working with Markov processes greatly simplifies the theory: bisimulation, logical characterization and approximation. Working with the functions (properties) one is less troubled by having to deal with things that are defined only "almost everywhere" as happens when one works with states.

A very nice feature of the theory is the ability to show that a minimal bisimulation exists. Furthermore, this minimal object can be constructed as the projective limit of finite approximants.

References

1. Sontag, E.: Mathematical Control Theory. Texts in Applied Mathematics, vol. 6. Springer, Heidelberg (1990)
2. Blute, R., Desharnais, J., Edalat, A., Panangaden, P.: Bisimulation for labelled Markov processes. In: Proceedings of the Twelfth IEEE Symposium On Logic In Computer Science, Warsaw, Poland (1997)
3. Desharnais, J., Edalat, A., Panangaden, P.: Bisimulation for labeled Markov processes. Information and Computation 179(2), 163–193 (2002)
4. Segala, R., Lynch, N.: Probabilistic simulations for probabilistic processes. In: Jonsson, B., Parrow, J. (eds.) CONCUR 1994. LNCS, vol. 836, pp. 481–496. Springer, Heidelberg (1994)
5. Puterman, M.L.: Markov Decision Processes: Discrete Stochastic Dynamic Programming. Wiley, Chichester (1994)
6. Larsen, K.G., Skou, A.: Bisimulation through probablistic testing. Information and Computation 94, 1–28 (1991)
7. de Vink, E., Rutten, J.J.M.M.: Bisimulation for probabilistic transition systems: A coalgebraic approach. In: Proceedings of the 24th International Colloquium On Automata Languages And Programming (1997)
8. Kozen, D.: A probabilistic PDL. Journal of Computer and Systems Sciences 30(2), 162–178 (1985)
9. Danos, V., Desharnais, J., Panangaden, P.: Conditional expectation and the approximation of labelled markov processes. In: Amadio, R.M., Lugiez, D. (eds.) CONCUR 2003. LNCS, vol. 2761, pp. 477–491. Springer, Heidelberg (2003)
10. Feller, W.: An Introduction to Probability Theory and its Applications II, 2nd edn. John Wiley and Sons, Chichester (1971)
11. Desharnais, J., Gupta, V., Jagadeesan, R., Panangaden, P.: The metric analogue of weak bisimulation for labelled Markov processes. In: Proceedings of the Seventeenth Annual IEEE Symposium On Logic In Computer Science, pp. 413–422 (2002)
12. Kemeny, J.G., Snell, J.L.: Finite Markov Chains. Van Nostrand (1960)

13. Desharnais, J., Gupta, V., Jagadeesan, R., Panangaden, P.: Approximating labeled Markov processes. Information and Computation 184(1), 160–200 (2003)
14. Desharnais, J., Gupta, V., Jagadeesan, R., Panangaden, P.: A metric for labelled Markov processes. Theoretical Computer Science 318(3), 323–354 (2004)
15. Hopf, E.: The general temporally discrete Markoff process. J. Rational Math. Mech. Anal. 3, 13–45 (1954)
16. Danos, V., Desharnais, J., Laviolette, F., Panangaden, P.: Bisimulation and cocongruence for probabilistic systems. Information and Computation 204(4), 503–523 (2006)
17. Bartels, F., Sokolova, A., de Vink, E.: A hierarchy of probabilistic system types. Theoretical Computer Science 327, 3–22 (2004)
18. Billingsley, P.: Probability and Measure. Wiley Interscience, Hoboken (1995)
19. Choksi, J.: Inverse limits on measure spaces. Proc. London Math. Soc. 8(3), 321–342 (1958)

Weak Bisimilarity Coalgebraically

Andrei Popescu

University of Illinois at Urbana-Champaign and
Institute of Mathematics Simion Stoilow of the Romanian Academy

Abstract. We argue that weak bisimilarity of processes can be conveniently captured in a semantic domain by a combination of traces and coalgebraic finality, in such a way that important process algebra aspects such as parallel composition and recursion can be represented compositionally. We illustrate the usefulness of our approach by providing a fully-abstract denotational semantics for CCS under weak bisimilarity.

1   Introduction

Weak bisimilarity (henceforth referred to simply as *bisimilarity*) is one of the most interesting equivalences defined on nondeterministic interactive processes. A nondeterministic process p may take one of several actions a_i and transit to a process p_i; among the possible actions, there is the *silent, unobservable* action, τ. Intuitively, a process may take a τ action at any time without being noticed, while any other action is *observable*. Now, the *(behavioral) identity* of two processes p_1 and p_2 can be determined by the following two clauses (and their symmetrics): **(i)** if p_1 may do something internally (i.e., take zero or more τ actions) becoming p_1', then p_2 may as well do something internally becoming p_2', behaviorally identical to p_1'; **(ii)** if p_1 may do something internally, then take an observable action a, then again do something internally eventually becoming p_1', then p_2 may as well go through a similar series of steps (consisting of action a possibly succeeded and/or preceded by some silent steps) eventually becoming p_2', behaviorally identical to p_1'. The *strong bisimilarity* of processes is a useful, but often too crude, approximation of bisimilarity, obtained by assuming the internal action τ observable.

The typical route for introducing processes is through some syntactic constructs building terms whose behavior is given by an SOS-specified labeled transition system. The identity of a process can then be taken to be the bisimilarity class of a term. A more direct, and often more insightful way of establishing the identity of the described processes is to show that process terms really describe processes, in other words, to assign a denotation from a *domain of processes* (or *behaviors*) to each term. The agreement between bisimilarity defined operationally and the kernel of the denotation map is referred to as *full abstraction*.

Bisimilarity, although extensively studied in various operational settings [14,15,20], has not benefited yet, in our opinion, from a suitable compositional denotational-semantics account within process algebra. (Although a certain ad hoc compositionality for CCS and CSP has been achieved in [2] and could be achieved along the lines of [18], and non-compositional denotational models for

CCS, CSP and π under weak bisimilarity have been sketched in [12,13] – see also Section 4.) To the contrary, *strong* bisimilarity, having a simpler structure, has received compositional fully-abstract semantics, both domain-theoretic [1,10,22] and coalgebraic [18,24], based on hypotheses that hold (or merely using techniques that work) for many process calculi in current use. The same holds true for *may/must testing equivalence*, also simpler than bisimilarity but in a different way (namely, in that establishing it does not involve any alternating game as for strong bisimilarity and bisimilarity) [3,6,11].

The difficulty with assigning compositional denotational semantics for bisimilarity seems to emerge from the fundamental divorce between two features:

(I) the traditional one-action-depth, "branching-time" presentation of a process calculus;

(II) the "linear-time" consideration of τ^*-sequences in the definition of bisimilarity.

Considering (I), one is tempted to transform right away the conditional rules from the SOS presentation of the system into corresponding corecursive or fixpoint recursive definitions, leading to problems with (II).

The solution proposed by this paper comes from resisting this temptation, and acknowledging that, due to the possibility of melting observable actions into silent actions via communication, *in order to handle what happens arbitrarily τ-deep into processes, one also needs to deal with what happens arbitrarily deep along any sequence of actions.* This might suggest to abandon coalgebraic semantics altogether and go after some form of pure trace semantics,[1] but this would be a mistake, since the very nature of bisimilarity is coalgebraic – the infinite game that defines bisimilaritiy is a journey into processes with infinitely often occurring stops and restarts. Instead, we shall combine traces with coalgebra, identifying a process with all pairs (traceOfActions,continuationProcess) such that that trace of actions is possible and leads to that continuation.

Technically, we shall regard bisimilarity as strong bisimilarity where the "actions" are now sequences of *observable* actions, and where a chain of zero or more τ actions is represented by the empty sequence ϵ. The monoidal properties of the set of action sequences shall be reflected in modal axioms for the final coalgebraic semantic domain. Of course, defining semantic operations within such an approach requires a preliminary combinatorial study of sequences/traces of actions, that need to be shuffled in consistent ways. As it happens, paying a priori attention to traces has its rewards – operations on processes like parallel composition and iteration, main characters in process algebra, receive concise, uniform and conceptually clear definitions.

After we discuss the above constructions in an abstract semantic domain (without reference to syntax), we illustrate the usefulness of our approach by defining a novel denotational semantics for Milner's Calculus of Communicating Systems (CCS) [14]. The semantics will be both *compositional* and *fully abstract.*

---

[1] In this paper "coalgebraic semantics" will mean: semantics involving a domain which is a final coalgebra of some kind, and thus it is inhabited by items having a relevant *branching structure.* We contrast this with a (pure) *trace semantics*, where the branching structure is ignored.

Here is how the remaining of the paper is structured. The rest of this section is dedicated to general mathematical preliminaries and conventions. Section 2 discusses, in coalgebraic terms, semantic models for weak bisimilarity. Section 3 shows that our approach yields naturally a fully-abstract semantics for CCS under weak bisimilarity. Section 4 draws conclusions and discusses related work.

Preliminaries and conventions. When we introduce a metavariable (such as x) to range over a certain domain, we implicitly assume that this is also the case for its version decorated with accents and subscripts (such as x_1, x'). We use λ-abstractions informally, at the meta-level.

Throughout our exposition, we shall make the standard distinction between classes and sets, and we shall occasionally refer to collections and families of classes – such references could be avoided in well-known ways, or alternatively we could consider a three-level tower of Grothendieck universes, where we could place our sets, classes and collections. Thanks to their standardness w.r.t. our constructions, we shall not worry about foundational issues in this paper, beyond insisting that for a class C, $\mathcal{P}(C)$ denotes the class of *subsets* (and not subclasses) of C.

By "relation" we mean, by default, "binary relation". For a relation $R \subseteq A \times B$, $R^\smile \subseteq B \times A$ is its converse. Given a function $f : A \to B$ and $A' \subseteq A$, $Im(f)$ is the image of f and $f|_{A'}$ is the restriction of f to A'. We usually denote function application by juxtaposition, as in $f\,x$, but sometimes (especially when dealing with *fixed* operators) also employ the parenthesized notation $f(x)$. Given $f : A \to B$, $a \in A$ and $b \in B$, $f[a \leftarrow b] : A \to B$ is defined by $(f[a \leftarrow b])a' = b$ if $a' = a$ and $= f\,a'$ otherwise. A^* is the set of finite sequences/traces of items in A, i.e., the set of words over the alphabet A. $\#$ denotes, in an infixed manner, word concatenation, and ϵ the empty word. (I.e., $(A^*, \#, \epsilon)$ is the monoid freely generated by A.) Thus, for two words $w_1, w_2 \in A^*$ (unless otherwise stated), $w_1 \# w_2$, and *not* the simple juxtaposition $w_1 w_2$, denotes their concatenation. However, given the elements $a_1, \ldots, a_n \in A$, we write $a_1 \ldots a_n$ for the word (sequence) built with these letters (in this order). A^+ denotes $A^* \setminus \{\epsilon\}$. $\mathbb{N}$ is the set of natural numbers. For $m, n \in \mathbb{N}$, $\overline{m, n}$ is $\{i.\, m \leq i \leq n\}$. Unless otherwise stated, m, n, i, j, k range over $\mathbb{N}$. In proofs, "IH" means "the inductive hypothesis".

2   Semantic Domain for Weak Bisimilarity

In this section, we present a semantic domain for weak bisimilarity based on traces of actions, and show its connection with, and its advantage over a more standard domain based on single actions.

Cls denotes the hyper-category of classes and functions between classes. In what follows, we shall employ basic concepts and results about coalgebras [19]. Given a functor $F : Cls \to Cls$, an F-coalgebra is a pair $(D, \delta : D \to FD)$. A *morphism* between two F-coalgebras (D, δ) and (D', δ') is a map $f : D \to D'$ such that $(F\,f) \circ \delta = \delta' \circ f$. A *stable part* of an F-coalgebra (D, δ) is a subclass $X \subseteq D$ such that $\forall x \in X.\, \delta x \in F X$; a stable part X yields a *subcoalgebra* $(X, \delta|_X)$. Given a fixed set Z, the functor $\mathcal{P}(Z \times \_)$ on *Cls* maps each X to

$\mathcal{P}(Z \times X)$ and each $f : X \to Y$ to $\lambda K. \{(z, f\, x). (z, x) \in K\}$. Fix Act, a set of *actions*. We only consider coalgebras for $\mathcal{P}((Act \cup \{\epsilon\}) \times \_)$ and $\mathcal{P}(Act^* \times \_)$, which we call *one-step*, and *multi-step coalgebras*, respectively. (We have these functors act on *classes* rather than sets in order to ensure the existence of their final coalgebras – see below. An alternative would have been to replace $\mathcal{P}(X)$ with $\mathcal{P}_k(X)$, the set of subsets of X of cardinality less than k, for a large enough cardinal k (such as $2^{\aleph_0}$) – we prefer our current solution since it does not commit to any arbitrary choice.)

Domains of processes under strong bisimilarity are typically modeled as final *one-step* coalgebras. However, if the desired process identity is bisimilarity, then (also having in mind that operations like parallel composition need to take a deeper, multi-step look into the argument processes) it is more natural to consider a suitable *multi-step* coalgebra as the domain. But also we would like to keep in sight that bisimilarity is a weakening of strong bisimilarity by internalizing τ, in particular, to be able to infer bisimilarity from strong bisimilarity without any detour, and also to infer bisimilarity as usual, by showing how to simulate single steps only (by multi-steps). These lead to the following constructions.

Given a multi-step coalgebra (D, δ), an element $d \in D$ is said to be:

- *reflexive*, if $(\epsilon, d) \in \delta\, d$;
- *transitive*, if $\forall w, w', d', d''. (w, d') \in \delta\, d \wedge (w', d'') \in \delta\, d' \Rightarrow (w \# w', d'') \in \delta\, d$;
- *prefix-closed*, if $\forall w, w', d''. (w \# w', d'') \in \delta\, d \Rightarrow (\exists d'. (w, d') \in \delta\, d \wedge (w', d'') \in \delta\, d')$;
- *monoidal*, if it is reflexive, transitive and prefix-closed.

(D, δ) is said to be *monoidal* if all elements of D are monoidal.

Let $(Preproc, unf : Preproc \to \mathcal{P}(Act^* \times Preproc))$ be the final multi-step coalgebra. We call the elements of $Preproc$ *preprocesses*, and unf the *unfolding map*. Let $Proc$, the class of *processes*, be the stable part of $(Preproc, unf)$ cogenerated by the class of all monoidal preprocesses. This notion of process encompasses two ideas. First, processes have a *linear-time* structure which is compatible with the monoid of action sequences (via reflexivity and transitivity) and has no discontinuities (via prefix-closeness). Second, processes have the above linear-time properties preserved by the *branching-time*, coalgebraic structure – one should think of a process as an *hereditarily* monoidal preprocess, that is, a preprocess π such that π is monoidal and the preprocesses from all its arbitrarily deep unfoldings are so.

The properties of the linear-time structure, making sense for any transition system labeled with sequences of actions (i.e., for any multi-step coalgebra), are the essential features of weak bisimilarity. Our choice to impose these properties hereditarily deep for elements in the final multi-step coalgebra makes $(Proc, unf : Proc \to \mathcal{P}(Act^* \times Proc))$ (where unf here denotes the restriction of $unf : Preproc \to \mathcal{P}(Act^* \times Preproc)$) the *final monoidal multi-step coalgebra*, featuring the following *corecursive definition* principle: to define a (parameterized) operation $f : Param \times Proc^n \to Proc$, it suffices to organize $Param \times Proc^n$ as a *monoidal* multi-step coalgebra by defining $\delta : Param \times Proc^n \to \mathcal{P}(Act^* \times (Param \times Proc^n))$ (then take f to be the unique morphism between $(Param \times Proc^n, \delta)$ and $(Proc, unf)$).

Moreover, the fact that $(Proc, unf)$ is *simple* (being a subcoalgebra of the (absolutely) final coalgebra), yields standardly a proof principle: Assume $\theta \subseteq Proc \times Proc$ is a strong bisimulation on $Proc$ (regarded as an Act^*-labeled transition system), in that the following hold for all $(\pi, \pi') \in \theta$:

- $\forall (w, \pi'') \in unf(\pi). \exists \pi'''. (w, \pi''') \in unf(\pi') \wedge (\pi'', \pi''') \in \theta;$
- $\forall (w, \pi'') \in unf(\pi'). \exists \pi'''. (w, \pi''') \in unf(\pi) \wedge (\pi''', \pi'') \in \theta.$

Then $\pi = \pi'$ for all $(\pi, \pi') \in \theta$.

Thanks to the affinity between $Proc$ and the monoidal structure of Act^*, we also have a simpler (but equally powerful) proof principle which reflects more closely the traditional way of dealing with weak bisimilarity, by showing how to simulate single actions only:

Let $\theta \subseteq Proc \times Proc$ be such that the following hold for all $(\pi, \pi') \in \theta$:

- $\forall (w, \pi'') \in unf(\pi). |w| \in \{0, 1\} \Rightarrow (\exists \pi'''. (w, \pi''') \in unf(\pi') \wedge (\pi'', \pi''') \in \theta);$
- $\forall (w, \pi'') \in unf(\pi'). |w| \in \{0, 1\} \Rightarrow (\exists \pi'''. (w, \pi''') \in unf(\pi) \wedge (\pi''', \pi'') \in \theta).$

Then $\pi = \pi'$ for all $(\pi, \pi') \in \theta$.

The latter proof principle may appear, at first, as if employing *strong* bisimulation (w.r.t. *single* actions) – but remember that processes absorb the monoidal properties of action sequences; in particular: ϵ is identified with (any sequence in the language) ϵ^*, thus meaning "zero or more silent steps"; the single action a is identified with $\epsilon^* \# a \# \epsilon^*$, meaning "$a$ preceded and succeeded by zero or more silent steps". Thus, *weak* bisimulation is what we really deal with here, strong bisimulation being only a particular case.

The fact that each process is uniquely identified by its behavior w.r.t. sequences of length 0 or 1 (i.e., elements of $Act \cup \{\epsilon\}$) suggests a more compact representation of the domain of processes as a one-step coalgebra. As already mentioned, the choice of the (absolutely) final one-step coalgebra as the semantic domain for *strong* bisimilarity typically already yields the desired properties (in particular, full abstraction [18]). However, here, since we are after bismilarity, we shall require that processes, even in this more compact representation with single steps, retain the affinity with the monoid of action sequences – this means here that zero or more ϵ-steps can always be appended and/or prepended "silently", yielding a property similar to monoidality that we shall call ϵ-monoidality. (Although the constructions below are not needed later in the paper, they are useful for placing our domain in a context that clarifies its connection with the traditional view on bisimilarity and its benefit w.r.t. compositionality.)

Let $(Cpreproc, unfc)$ be the final one-step coalgebra. We call the elements of *Cpreproc compact (representations of) preprocesses*. Given a one-step coalgebra (D, δ), an element $d \in D$ is said to be: ϵ-*reflexive*, if $(\epsilon, d) \in \delta d$; ϵ-*transitive*, if $\forall d', d'' \in D. (\epsilon, d') \in \delta d \wedge (\epsilon, d'') \in \delta d' \Rightarrow (\epsilon, d'') \in \delta d$; ϵ-*loud-transitive*, if $\forall d', d'', d''' \in D. \forall a \in Act. (\epsilon, d') \in \delta d \wedge (a, d'') \in \delta d' \wedge (\epsilon, d''') \in \delta d'' \Rightarrow (a, d''') \in \delta d$; ϵ-*monoidal*, if it is ϵ-reflexive, ϵ-transitive and ϵ-loud-transitive. (D, δ) is said to be ϵ-*monoidal* if all elements of D are ϵ-monoidal. (The ϵ-monoidal one-step coalgebras are essentially Aczel's τ-colagebras [2].) Let *Cproc*, the class of *compact (representations of) processes*, be the stable part of the one-step coalgebra (*Cpreproc*, *unfc*) cogenerated by the class of all ϵ-monoidal compact preprocesses. Then (*Cproc*, *unfc* : *Cproc* $\to$ $\mathcal{P}((Act \cup \{\epsilon\}) \times Cproc))$ is the final ϵ-monoidal one-step coalgebra.

Next, we write π for processes and σ for compact processes. Using the finality of *Cproc* and *Proc*, we define two maps, *pack* : *Proc* $\to$ *Cproc* and *unpack* : *Cproc* $\to$ *Proc*, for moving back and forth between a process and its compact representation:

- $unfc\,(pack(\pi)) = \{(w, pack(\pi')) . (w, \pi') \in \pi \land w \in Act \cup \{\epsilon\}\}$;
- $unf(unpack(\sigma)) = \{(w_1\# \ldots \#w_n, unf(unpack(\sigma'))) . n \in I\!N \land (\forall i \in \overline{1,n}. w_i \in Act \cup \{\epsilon\}) \land (\exists \sigma_1, \ldots, \sigma_{n+1}. \sigma_1 = \sigma \land \sigma_{n+1} = \sigma' \land (\forall i \in \overline{1,n}. (w_i, \sigma_{i+1}) \in \sigma_i))\}$.

Thus, $pack(\pi)$ retains from π (and from all its (arbitrarily deep) continuations) only the one-step continuations, while *unpack* σ expands σ (and all its continuations) along all possible sequences of steps. *pack* and *unpack* are mutually inverse bijections.

Compact processes are coalgebraically only one-step deep, hence correspond more directly to the traditional operational semantics presentation of process calculi. However, in our context of (weak) bisimilarity, these compact one-step representations have a salient disadvantage compared to (multi-step) processes: crucial operations in process calculi, such as parallel composition and iteration, are not definable purely coalgebraically[2] on compact processes, the one-step coalgebra falling short on providing means to describe the composite process.

To illustrate this point, assume that *Act* is endowed with a bijection $^{-}$: *Act* $\to$ *Act* which is involutive (in that $\overline{\overline{a}} = a$ for all $a \in Act$), and say we would like to define CCS-like parallel composition on processes (thus, we assume that a and $\overline{a}$ are to be interpreted as complementary actions, whose synchronization yields a silent action). The main task in front of us is to show how sequences of actions interact, possibly nondeterministically. Knowing how *single actions* interact, and assuming an *interleaving semantics*, we obtain the following recursive definition of the *parallel composition* (or *synchronized shuffle*) $|$: $Act^* \times Act^* \to \mathcal{P}(Act^*)$:

- $\epsilon|\epsilon = \{\epsilon\}$;
- $(a\#w_1)|(b\#w_2) = a\#(w_1|(b\#w_2)) \cup b\#((a\#w_1)|w_2)$, if $\overline{a} \neq b$;
- $(a\#w_1)|(b\#w_2) = a\#(w_1|(b\#w_2)) \cup b\#((a\#w_1)|w_2) \cup w_1|w_2$, if $\overline{a} = b$.

(Above, we overloaded $\#$ in the usual fashion, to denote both $\#$: $Act^* \times Act^* \to Act^*$ and its componentwise extension to $Act^* \times \mathcal{P}(Act^*) \to \mathcal{P}(Act^*)$, given by $w \# S = \{w \# w' . w' \in S\}$.)

Now, parallel composition of processes, $|$: $Proc \times Proc \to Proc$, simply follows coinductively the interaction law prescribed by action sequence composition: $unf(\pi_1|\pi_2) = \{(w, \pi_1'|\pi_2') : \exists w_1, w_2. w \in w_1|w_2 \land (w_1, \pi_1') \in unf(\pi_1) \land (w_2, \pi_2') \in unf(\pi_2)\}$.

---

[2] In the sense that a definition based entirely on coalgebraic finality is not available.

Note the separation of concerns in our definition of |: first we dealt with action sequence combinatorics, so that afterwards the definition of parallel composition of processes was stated *purely coalgebraically*, composing together the continuations of the components to yield the continuations of the composite. On the other hand, if trying to define parallel composition on *compact processes*, one finds the coalgebraic one-step depth of the latter insufficient for describing even the one-step behavior of the composite – this is because a step of the composite $\sigma_1|\sigma_2$ may result from an *arbitrarily long* sequence of interactions between steps taken by continuations of σ_1 and σ_2. In fact, any reasonable definition of parallel composition on *Cproc* would essentially appeal to our multi-step domain *Proc*, unpacking the components σ_1 and σ_2, composing these unpacked versions in *Proc*, and then packing the result (i.e., the operation on *Cproc* would be essentially $pack((unpack\ \sigma_1)|(unpack\ \sigma_2)))$. This suggests that *Proc*, and not *Cproc*, is the fundamental domain for compositional weak bisimilarity.

One may argue that, via the bijections *pack* and *unpack*, *Proc* and *Cproc* are really the same domain, so that considering one or the other are two sides of the same coin. However, *Proc* and *Cproc* take nevertheless two *distinct views* to processes, with the multi-step view brought by *Proc*, as explained above, allowing for a cleaner compositionality and separation of concerns when defining process composition. When the effect of actions becomes more complex, with possible changes of the communication topology, the multi-step view brings an insight into the semantics of a calculus under bisimilarity which is essential for apprehending the "right" semantic operations for full abstraction – this is illustrated in [17], where we define a fully-abstract semantics for the π-calculus under weak bisimilarity. A simpler illustration of our approach is presented in the next section, where we give a novel semantics for CCS which is both fully abstract and compositional.

3   Denotational Semantics for CCS

In this section we use our approach from the previous section to endow CCS under (weak) bisimilarity with a fully abstract compositional denotational semantics. We achieve this in two steps:

– first, we consistently modify the CCS transition system to employ traces of actions rather than single actions;
– second, we define the semantic operators on the domain *Proc* from the previous section and show that they reflect the behavior of trace-based CCS.

The first step, although consisting of a purely syntactic transformation, is the actual realization of our insight concerning the semantics of weak bisimilarity, since it emphasizes behavior along arbitrarily long traces. The second step is in many aspects routine, as it merely "institutionalizes" in a final coalgebra the above trace-based behavior. If it were not for the second-order features given by the fixpoint operators, we could have directly employed the general theory from [18,24] for this second step – however, fixpoints in combination with arbitrary traces did bring additional difficulties which required ad hoc resolutions, not covered yet, as far as we see, by any general theory of denotational semantics for SOS formats.

Concepts and notations concerning transition systems. We shall consider several *transition system specifications*, which are parameterized by classes *State*, of states, and classes *Label*, of labels and consist of descriptions (of some kind) of labeled transition systems. In our cases:

- *State* will be *Term* or *Proc* and *Label* will be $Act \cup \{\tau\}$, $(Act \cup \{\tau\})^*$ or Act^*;
- the description will consist of either a set of SOS rules (such as CCS and CCST below) or the indication of a coalgebra which yields standardly a labeled transition system (such as $(Proc, unf)$).

Given a transition system specification TS, we write $\vdash_{TS} s \xrightarrow{l} s'$ to indicate that $s \xrightarrow{l} s'$ is inferable in TS (i.e., belongs to the described labeled transition system). TS has the following associated notions. A relation $\theta \subseteq State \times State$ is called:

- a *strong TS-simulation*, if the following holds for all $(s, t) \in \theta$: if $\vdash_{TS} s \xrightarrow{l} s'$ for some l and s', then there exists t' such that $\vdash_{TS} t \xrightarrow{l} t'$ and $(s', t') \in \theta$;
- a *strong TS-bisimulation*, if both θ and $\theta^\smile$ are strong TS-simulations.

The *strong TS-bisimilarity* relation, denoted $\sim_{TS}$, is the union of all strong TS-bisimulations, and is also the largest strong TS-bisimulation.

In case $Label = Act \cup \{\tau\}$, we also have *weak* versions of the above concepts,[3] which we describe below. Given a regular language L over $Act \cup \{\tau\}$, we write $\vDash_{TS} s \xrightarrow{L} s'$ to indicate that there exists a word $w = \alpha_1 \dots \alpha_n \in L$ (with the α_i-s in $Act \cup \{\tau\}$) and $s_0, \dots, s_n$ such that $s_0 = s$, $s_n = s'$, and $s_i \xrightarrow{\alpha_{i+1}} s_{i+1}$ for all $i \in \overline{0, n-1}$. (Notice the usage of the symbol $\vDash$, distinct from $\vdash$, to denote a notion of "deduction" which is not primitive in TS, but derived from $\vdash$.) Examples of regular languages L include τ^* and $\tau^* \# a \# \tau^*$ for some $a \in Act$. A relation $\theta \subseteq State \times State$ is called:

- a *TS-simulation*, if the following hold for all $(s, t) \in \theta$:
  - if $\vDash_{TS} s \xrightarrow{\tau^*} s'$ for some s', then there exists t' such that $\vDash_{TS} t \xrightarrow{\tau^*} t'$ and $(s', t') \in \theta$;
  - if $\vDash_{TS} s \xrightarrow{\tau^* \# a \# \tau^*} s'$ for some s' and $a \in Act$, then there exists t' such that $\vDash_{TS} t \xrightarrow{\tau^* \# a \# \tau^*} t'$ and $(s', t') \in \theta$.
- a *TS-bisimulation*, if both θ and $\theta^\smile$ are TS-simulations.

The *TS-bisimilarity* relation, denoted $\approx_{TS}$, is the union of all TS-bisimulations, and is also the largest TS-bisimulation.

The following trace-based characterization of simulation (hence of bisimilarity), as seen in Section 2 crucial for our approach to compositionality, is known already from [14]. Let $del_\tau : (Act \cup \{\tau\})^* \to Act^*$ be the map that deletes all the τ-s. A relation $\theta \subseteq State \times State$ is a TS-simulation iff the following hold for all $(s, t) \in \theta$: if $\vDash_{TS} s \xrightarrow{v} s'$ for some s' and $v \in (Act \cup \{\tau\})^*$, then there exist t'

---

[3] As before, we shall omit the prefix "weak".

and $w \in (Act \cup \{\tau\})^*$ such that $del_\tau(v) = del_\tau(w)$, $\vdash_{TS} t \xrightarrow{w} t'$ and $(s', t') \in \theta$. Thus, according to this characterization, in the bisimilarity game one has to match a trace of actions with a trace "τ-equivalent" to it. Henceforth we take this characterization as the very definition of TS-simulation. (Notice that, in our semantic domain *Proc* from Section 2, τ is absorbed in the monoidal structure, hence "τ-equivalence", that is, "ϵ-equivalence", coincides with *strong* bisimilarity, which in turn coincides, by the internal full abstraction, with equality.)

CCS recalled. For the presentation of CCS, we follow [14] rather closely. We shall not consider the full CCS, but a version restricted in one essential way and in several inessential ways.

The essential restriction consists of allowing only *guarded* sums. For arbitrary sums, (weak) bisimilarity is not a congruence, hence a priori impossible to capture by a compositional denotation. On the other hand, the congruence cogenerated by bisimilarity can receive such a denotation by an adaptation of our approach, but we feel that this would bring technical complications orthogonal to the main ideas of this text. Thus, we take the (rather customary) view that the "real CCS calculus" is that with guarded sums only – already in [14] (at page 113) it is noted: "[weak bisimilarity] is a congruence if we limit the use of Summation to *guarded* sums [...]. In fact, in applications we hardly ever use Summation in any other way." (Also, unguarded sums are excluded already at the syntax level in monographs (of related calculi) such as [20].)

The inessential restrictions consist of excluding the *restriction* and *renaming* operators and dealing with simple (non-mutual) recursion only. These features are of course essential to the calculus, but inessential to the presentation of our semantics, since it is immediate how they can be handled in our framework.

On the other hand, we do not require the fixpoint expressions to be guarded. Guardedness is a convenient/desired feature in the context of strong bisimilarity, since there it guarantees (in the presence of finite summations) finite branching. But since weak bisimilarity is infinitely branching anyway, here unguarded replication does not raise new fundamental problems, and yields perfectly valid behavior – for example, the π-calculus-style replication $!P$ is only definable by an unguarded fixpoint expression, $\mu X. X | P$.

Henceforth, we call "CCS" the indicated restricted version of CCS, presented below in detail. We fix the following:

- A set *Act*, of *loud actions*, ranged over by a; an item $\tau \notin Act$, called the *silent action*; $Act \cup \{\tau\}$ is called the set of *actions* and is ranged over by α;
- A map $\overline{} : Act \to Act$ such that $\overline{\overline{a}} = a$ for all $a \in Act$.
- An infinite set *Var*, of (process) variables, ranged over by X, Y.
- A set *Index*, of *indexes* (for summations), ranged over by i, j; I will range over $\mathcal{P}(Index)$. (*Index* is not required to be finite.)

The set *Term*, of *(process) terms*, ranged over by P, Q, is given by the following grammar:

$$P ::= X \mid \sum_{i \in I} \alpha_i. P_i \mid P|Q \mid \mu X. P.$$

In a term $\mu X. P$, X is bound in P – this notion of binding yields standardly a notion of alpha-equivalence on terms. We identify terms modulo alpha-equivalence. $P[Q/X]$ denotes the term obtained from P by (capture-free) substituting all free variables of P with Q. $P[Q/X]^n$ denotes $P[Q/X]\dots[Q/X]$ (n times).

The CCS transition system infers triples $P \xrightarrow{\alpha} P'$:

$$\frac{\cdot}{\sum_{i\in I} \alpha_i. P_i \xrightarrow{\alpha_j} P_j}\,\text{(Sum)}\,[j \in I] \qquad \frac{P \xrightarrow{\alpha} P'}{P|Q \xrightarrow{\alpha} P'|Q}\text{(ParL)} \qquad \frac{Q \xrightarrow{\alpha} Q'}{P|Q \xrightarrow{\alpha} P|Q'}\text{(ParR)}$$

$$\frac{P \xrightarrow{a} P' \quad Q \xrightarrow{\bar{a}} Q'}{P|Q \xrightarrow{\tau} P'|Q'}\text{(Com)} \qquad \frac{P[(\mu X. P)/X] \xrightarrow{\alpha} P'}{\mu X. P \xrightarrow{\alpha} P'}\text{(Rec)}$$

Trace-based CCS. We first employ our insight concerning arbitrarily long traces for compositional bisimilarity *at the syntactic level*, by defining the following system CCST, a variation of CCS with *traces* instead of single actions as labels. (In the rule (ParT) below, $| : Act^* \times Act^* \to \mathcal{P}(Act^*)$ is the synchronized shuffle defined in Section 2.)

$$\frac{\cdot}{P \xrightarrow{\epsilon} P}\text{(Silent)} \qquad \frac{P_j \xrightarrow{w} P'}{\sum_{i\in I} \alpha_i. P_i \xrightarrow{\alpha_j \# w} P'}\,\text{(SumT)}\,[j \in I]$$

$$\frac{P \xrightarrow{w_1} P' \quad Q \xrightarrow{w_2} Q'}{P|Q \xrightarrow{w} P'|Q'}\,\text{(ParT)}\,[w \in w_1|w_2] \qquad \frac{P[(\mu X. P)/X] \xrightarrow{w} P'}{\mu X. P \xrightarrow{w} P'}\text{(RecT)}$$

The rules of CCST were produced by taking the reflexive-transitive closure of the rules of CCS, i.e., by composing the system CCS with itself *horizontally* an indefinite number of times. In the process, we also made sure that zero or more τ actions were identified with the empty trace ϵ, and in particular we have added the rule (Silent).

Lemma 1. $\vdash_{\text{CCS}}$ *is closed under the rules that define* $\vdash_{\text{CCST}}$ *(modulo the absorbtion of τ for the case of parallel composition), namely:*

(1) $\vdash_{\text{CCS}} P \xrightarrow{\epsilon} P$.

(2) If $j \in I$ and $\vdash_{\text{CCS}} P_j \xrightarrow{w} P'$, then $\vdash_{\text{CCS}} \sum_{i\in I} \alpha_i. P_i \xrightarrow{\alpha_j \# w} P'$.

(3) If $\vdash_{\text{CCS}} P \xrightarrow{w_1} P'$, $\vdash_{\text{CCS}} Q \xrightarrow{w_2} Q'$, and $v \in del_\tau(w_1)|del_\tau(w_2)$, then there exists w such that $del_\tau(w) = v$ and $\vdash_{\text{CCS}} P|Q \xrightarrow{w} P'|Q'$.

(4) If $\vdash_{\text{CCS}} P[(\mu X. P)/X] \xrightarrow{w} P'$, then $\vdash_{\text{CCS}} \mu X. P \xrightarrow{w} P'$.

Proof. (1): Immediate. (2) and (4): By case analysis on whether w is empty. (3): By an easy (but tedious) induction on the sum of the lengths of w_1 and w_2, using the definition of $|$ on traces. □

The next proposition states the fact we naturally expect from CCST – that it indeed produces precisely the traces of actions produced by CCS, *with the τ-actions absorbed.*

Proposition 1. *The following hold for all $P, P' \in Term$, $w \in (Act \cup \{\tau\})^*$, and $v \in Act^*$:*

(1) ⊢$_{\text{CCS}}$ $P \xrightarrow{w} P'$ *implies* ⊢$_{\text{CCST}}$ $P \xrightarrow{del_\tau(w)} P'$.

(2) ⊢$_{\text{CCST}}$ $P \xrightarrow{v} P'$, *then there exists* $w' \in (Act \cup \{\tau\})^*$ *such that* $del_\tau(w') = v$
and ⊢$_{\text{CCS}}$ $P \xrightarrow{w'} P'$.

Proof. (1): Easy induction on w.   (2): Immediate consequence of Lemma 1.   □

As a consequence, we have the identity between CCS-bisimilarity and strong
CCST-bisimilarity:

Corollary 1. *For all* $P, Q \in Term$, $P \approx_{\text{CCS}} Q$ *iff* $P \sim_{\text{CCST}} Q$.

Proof. Immediate from Proposition 1 (using the aforementioned trace-based
characterization of bisimilarity).   □

Semantic domain and operators. The semantic domain is the one advocated in Section 2, namely the carrier set *Proc* of the final monoidal multi-step
coalgebra $(Proc, unf : Proc \rightarrow \mathcal{P}(Proc \times Act^*))$.

As for the semantic operators, the ones corresponding to the summation and
parallel composition syntactic constructs have *almost* necessary definitions induced automatically by the de Simone SOS format of the CCST rules that involve
these constructs:

- For each $I \subseteq Index$, $\mathsf{Sum}_I : (I \rightarrow (Act \cup \{\epsilon\}) \times Proc) \rightarrow Proc$, by
$unf(\mathsf{Sum}_I((\alpha_i, \pi_i)_{i \in I})) = \{(\epsilon, \mathsf{Sum}_I((\alpha_i, \pi_i)_{i \in I}))\} \cup \{(\alpha_i \# w, \pi') . i \in I \wedge (w, \pi') \in unf(\pi_i)\}$;

- $\mathsf{Par} : Proc \times Proc \rightarrow Proc$, by
$unf(\mathsf{Par}(\pi_1, \pi_2)) = \{(w, \mathsf{Par}(\pi_1', \pi_2')). \exists w_1, w_2. (w_1, \pi_1) \in unf(\pi_1') \wedge (w_2, \pi_2) \in unf(\pi_2') \wedge w \in w_1 | w_2\}$ (where again $| : Act^* \times Act^* \rightarrow \mathcal{P}(Act^*)$ is the synchronized shuffle defined in Section 2).

Notice that, in the case of summation, we have explicitly added a self-ϵ-transition
to ensure reflexivity of the result. The other properties, namely, transitivity and
prefix-closeness, can be easily seen to be preserved by Sum_I. Moreover, Par is
easily seen to preserve monoidality. These mean that the definitions of these
operators are *correct corecursive definitions on Proc*.

In order to define a semantic operator corresponding to the μ construct, we
cannot simply transliterate the corresponding rule (RecT) as we did for the other
operators. Indeed, (RecT) is not in de Simone or in other amenable format that
would allow a local semantic definition. However, (RecT) can be replaced with
the following:

$$\frac{P[P/X]^n \xrightarrow{w} Q'}{\mu X. P \xrightarrow{w} Q'[(\mu X. P)/X]} (\mathsf{RecT'})$$

Intuitively, the recursive unfolding of a (sub)term $\mu X. P$ in a derivation needs
to be performed only once, atomically, for a sufficiently large depth n.

In the next lemma (and henceforth) we write ⊢$_{\text{CCST},k}$ $P \xrightarrow{w} Q$ to indicate
the fact that $P \xrightarrow{w} Q$ has a derivation of height *at most* k in CCST.

Lemma 2. *(1) If $\vdash_{CCST,k} P[(\mu X. P)/X] \xrightarrow{w} P'$, then there exist $n \in \mathbb{N}$ and Q' such that $Q'[(\mu X. P)/X] = P'$ and $\vdash_{CCST,k} P[P/X]^n \xrightarrow{w} Q'$.*
(2) If $\vdash_{CCST} P[P/X]^n \xrightarrow{w} P'$, then $\vdash_{CCST} \mu X. P \xrightarrow{w} P'[(\mu X. P)/X]$

Proof. (1): By induction on k.   (2) By induction on the definition of $\vdash_{CCST}$.   □

Now we are ready to define the semantic operator $\mathsf{Fix} : (Proc \to Proc) \to Proc$ (corresponding to μ) corecursively:
$unf(\mathsf{Fix}(F)) = \bigcup_{n \geq 1}\{(w, F'(\mathsf{Fix}(F))). \forall \pi \in Proc. (w, F' \pi) \in unf(F^n \pi)\}$.
(where $F : Proc \to Proc$ and F^n denotes $F \circ \ldots \circ F$ (n times))

Again, one can immediately check that the above is a correct definition. The definition is a rather standard semantic representation of the rule (RecT'), with the opening of the scope of X (regarding (RecT') as being applied backwards) handled semantically via the interplay between functional binding and universal quantification. On the other hand, the reader may legitimately wonder whether a more abstract semantic definition of recursion, namely one employing genuine fixpoints instead of numbers, would be possible. Unfortunately, an operator on processes whose (least) fixed point (according to some ordering) would give our semantic recursive item does not seem to be available, because of the non-continuous behavior of infinitely-branching recursion, which expands the trees *locally*. By contrast, a domain-theoretic setting (not suitable here due to infinite branching) allows itself to take *global* approximating snapshots from the recursively growing trees, enabling the mentioned more elegant treatment of recursion. (In the context of infinite branching, such snapshots would not be enough to characterize the final trees.)

The coalgebra $(Proc, unf)$ yields standardly an Act^*-labeled transition system with processes as states: $\vdash_{Proc} \pi \xrightarrow{w} \pi'$ iff $(w, \pi') \in unf(\pi)$. It is convenient (including more readable) to rephrase the above coinductive definitions in terms of this transition system:

- (C1) $\vdash_{Proc} \mathsf{Sum}_I((\alpha_i, \pi_i)_{i \in I}) \xrightarrow{w} \pi'$ iff $(w = \epsilon \land \pi' = \mathsf{Sum}_I((\alpha_i, \pi_i)_{i \in I})) \lor$
  $(\exists i \in I, w'. w = \alpha_i \# w' \land \vdash_{Proc} \pi_i \xrightarrow{w'} \pi')$.
- (C2) $\vdash_{Proc} \mathsf{Par}(\pi_1, \pi_2) \xrightarrow{w} \pi'$ iff $\exists w_1, w_2, \pi_1', \pi_2'. w \in w_1|w_2 \land \pi' = \mathsf{Par}(\pi_1', \pi_2')$
  $\land \vdash_{Proc} \pi_1 \xrightarrow{w_1} \pi_1' \land \vdash_{Proc} \pi_2 \xrightarrow{w_2} \pi_2'$.
- (C3) $\vdash_{Proc} \mathsf{Fix}(F) \xrightarrow{w} \pi'$ iff $\exists F', n \geq 1. \pi' = F'(\mathsf{Fix}(F)) \land (\forall \pi. F^n \pi \xrightarrow{w} F' \pi)$.

Everything is now almost set in place for the proof of full abstraction, except for one final detail: in (RecT'), we need to make sure that $P[P/X]^n$ is smaller than $\mu X. P$ in some way, so that $\mu X. P$ is simplified by an application of (RecT'). To this end, we define the relation $\succ \subseteq Term \times Term$ as follows: $P \succ Q$ iff one of the following holds:

- $\exists Q'. P = Q|Q' \lor P = Q'|Q$;
- $\exists I, (\alpha_i)_{i \in I}, (P_i)_{i \in I}, j \in I. P = \sum_{i \in I} \alpha_i. P_i \land Q = P_j$;
- $\exists R, X, n. P = \mu X. R \land Q = R[R/X]^n$.

Lemma 3. $\succ$ *is well-founded, in that there exists no infinite sequence $(P_k)_{k \in \mathbb{N}}$ such that $\forall k. P_k \succ P_{k+1}$.*

Proof. Define $\rightsquigarrow \subseteq Term \times Term$ inductively by the following clauses:
- $\mu X. P \rightsquigarrow P[P/X]^n$;
- If $P \rightsquigarrow P'$, then $P|Q \rightsquigarrow P'|Q$ and $Q|P \rightsquigarrow Q|P'$;
- If $j \in I$ and $P \rightsquigarrow P'$, then $\sum_{i \in I} P_i \rightsquigarrow \sum_{i \in I} P_i'$, where $P_j' = P'$ and $P_i' = P_i$ form all $i \neq j$.

Let $WF_{\rightsquigarrow}$ be the set of all terms well-founded w.r.t. $\rightsquigarrow$, i.e., of all terms P for which there exists no infinite sequence $(P_k)_{k \in \mathbb{N}}$ such that $P_0 = P$ and $\forall k. P_k \rightsquigarrow P_{k+1}$.

Claim1: If $P \in WF_{\rightsquigarrow}$ and $\forall X. \sigma X \in WF_{\rightsquigarrow}$, then $P[\sigma] \in WF_{\rightsquigarrow}$ (where σ is any map in $Var \rightarrow Term$ and $P[\sigma]$ is the term obtained from P by applying the substitution σ). (Proof: By induction on P.)

Claim2: If $P \in WF_{\rightsquigarrow}$, then $P[P/X]^n \in WF_{\rightsquigarrow}$. (Proof: By induction on n, using Claim1.)

Claim3: $\rightsquigarrow$ is well-founded. (Proof: By induction on terms, using Claim2 for μ-terms.)

Now, we can prove that $\succ$ is well-founded: Assume an infinite sequence $(P_k)_{k \in \mathbb{N}}$ such that $\forall k. P_k \succ P_{k+1}$. We construct, recursively on k, a sequence $(Q_k)_{k \in \mathbb{N}}$ such that $\forall k. Q_k \rightsquigarrow Q_{k+1}$, thus deriving a contradiction with Claim3. In fact, we define the Q_k's via the contexts (i.e., terms with one hole) $C_k[*]$, by $Q_k = C_k[P]$, where the $C_k[*]$-s are defined as follows:

- $C_0[*] = *$ (the trivial context);
- if $P_{n+1} = R|P_n$, then $C_{n+1}[*] = R | C_n[*]$;
- if $P_{n+1} = P_n|R$, then $C_{n+1}[*] = C_n[*] | R$;
- if $P_{n+1} = \sum_{i \in I} \alpha_i. R_i$, $j \in I$ and $P_n = R_j$, then $C_{n+1}[*] = \sum_{i \in I} \alpha_i. H_i$, where $H_j = C_n[*]$ and $H_j = R_j$ for all $j \neq i$;
- if $P_{n+1} = \mu X. R$ and $P_n = R[R/X]^n$, then $C_{n+1}[*] = C_n[*]$.

It is easy to see that indeed the Q_k-s (i.e., the $C_k[P_k]$-s) form a $\rightsquigarrow$-chain. $\square$

Denotational semantics. The above semantic operators yield standardly an interpretation $\langle \_ \rangle : Term \rightarrow (Var \rightarrow Proc) \rightarrow Proc$ of the CCS syntax in environments (via a higher-order version of the initial algebra semantics):

- $\langle \sum_{i \in I} \alpha_i. P_i \rangle \, \rho = \mathsf{Sum}_I((\alpha_i, \langle P_i \rangle \, \rho)_{i \in I})$;
- $\langle P|Q \rangle \, \rho = \mathsf{Par}(\langle P \rangle \, \rho, \langle Q \rangle \, \rho)$;
- $\langle \mu X. P \rangle \, \rho = \mathsf{Fix}(\lambda \pi. \langle P \rangle (\rho[X \leftarrow \pi]))$.

Lemma 4. *The following hold for all $X \in Var$, $P, Q \in Term$, $\pi \in Proc$, ρ, ρ' : $Var \rightarrow Proc$ and $n \in \mathbb{N}$:*
(1) If $\rho Y = \rho' Y$ for all $Y \in FV(P)$, then $\langle P \rangle \, \rho = \langle P \rangle \, \rho'$.
(2) $\langle P[Q/X] \rangle \, \rho = \langle P \rangle (\rho[X \leftarrow \langle Q \rangle \, \rho])$.
(3) $\langle P[P/X]^n \rangle (\rho[X \leftarrow \pi]) = F^{n+1} \pi$, where $F = \lambda \pi. \langle P \rangle (\rho[X \leftarrow \pi])$.

Proof. Points (1) and (2) state well-known facts holding for any standard interpretation of syntax with static bindings in a domain (as is the case here). We prove point (3) by induction on n. The base case, $n = 0$, follows from point (2). For the inductive case, we have the following chain of equalities:

$\langle P[P/X]^{n+1}\rangle(\rho[X \leftarrow \pi]) = \langle P[P/X]^n[P/X]\rangle(\rho[X \leftarrow \pi]) =$ (by point (2)) $=$ $\langle P[P/X]^n\rangle(\rho[X \leftarrow \pi][X \leftarrow \langle P\rangle(\rho[X \leftarrow \pi])]) = \langle P[P/X]^n\rangle(\rho[X \leftarrow \langle P\rangle(\rho[X \leftarrow \pi])]) =$ (by IH) $= F^{n+1}(\langle P\rangle(\rho[X \leftarrow \pi])) = F^{n+1}(F\,\pi) = F^{n+2}\,\pi.$ □

We are finally ready to state the connection between CCST and *Proc* (which leads to full abstraction):

Proposition 2. *The following hold for all* $P, P' \in Term$, $\pi' \in Proc$, $w \in Act^*$, *and* $\rho : Var \to Proc$:
(1) If $\vdash_{CCST} P \xrightarrow{w} P'$, *then* $\vdash_{Proc} \langle P\rangle \rho \xrightarrow{w} \langle P'\rangle \rho.$
(2) If $\vdash_{Proc} \langle P\rangle \rho \xrightarrow{w} \pi'$, *then there exists* P'' *such that* $\vdash_{CCST} P \xrightarrow{w} P''$ *and* $\pi' = \langle P''\rangle \rho.$
(In other words, for every ρ, *the map* $P \mapsto \langle P\rangle \rho$ *is a morphism between the multi-step coalgebra induced by CCST and* $(Proc, unf).)$

Proof. (1): We prove by induction on k that $\vdash_{CCST,k} P \xrightarrow{w} P'$ implies $\forall \rho.\ \vdash_{Proc} \langle P\rangle \rho \xrightarrow{w} \langle P'\rangle \rho.$ The cases where the last applied rule was (SumT) or (ParT) are immediate. We only treat the case of (RecT).

Assume $\vdash_{CCST,k} P[(\mu X.\,P)/X] \xrightarrow{w} P'$. Fix ρ. We need to show $\vdash_{Proc} \langle \mu X.\,P\rangle \rho \xrightarrow{w} \langle P'\rangle \rho$. Let $F = \lambda \pi.\langle P\rangle(\rho[X \leftarrow \pi])$.

Fix π. By Lemma 2.(1), there exist n and Q' such that $P' = Q'[(\mu X.\,P)/X]$ and $\vdash_{CCST,k} P[P/X]^n \xrightarrow{w} Q'$. Let $F' = \lambda \pi.\langle Q'\rangle(\rho[X \leftarrow \pi])$. With IH, we have $\vdash_{Proc} \langle P[P/X]^n\rangle(\rho[X \leftarrow \pi]) \xrightarrow{w} \langle Q'\rangle(\rho[X \leftarrow \pi])$, i.e., $\vdash_{Proc} \langle P[P/X]^n\rangle(\rho[X \leftarrow \pi]) \xrightarrow{w} F'\,\pi$, hence, using Lemma 4.(3), $\vdash_{Proc} F^{n+1}\pi \xrightarrow{w} F'\,\pi.$

We thus proved $\forall \pi.\ \vdash_{Proc} F^{n+1}\pi \xrightarrow{w} F'\,\pi$, hence, with (C3), $\vdash_{Proc} \mathsf{Fix}(F) \xrightarrow{w} F'(\mathsf{Fix}(F))$. Moreover, by Lemma 4.(2), we have $F'(\mathsf{Fix}(F)) = \langle Q'\rangle(\rho[X \leftarrow \mathsf{Fix}(F)]) = \langle Q'\rangle(\rho[X \leftarrow \mu X.\,P\ \rho]) = \langle Q'[(\mu X.P)/X]\rangle(\rho) = \langle P'\rangle \rho.$ Ultimately, we obtain $\vdash_{Proc} \langle \mu X.\,P\rangle \rho \xrightarrow{w} \langle P'\rangle \rho$, as desired.

(2): By well-founded induction on P (the lefthand side of the hypothesis) w.r.t. $\succ$ (since $\succ$ is well-founded according to Lemma 3). Again, the cases where P is a summation or a parallel composition are easy. We only consider the case of recursion.

Assume $\langle \mu X.\,P\rangle \rho \xrightarrow{w} \pi'$. Let $F = \lambda \pi.\langle P\rangle(\rho[X \leftarrow \pi])$. Then $\langle \mu X.\,P\rangle \rho = \mathsf{Fix}(F)$, hence   Fact0: $\mathsf{Fix}(F) \xrightarrow{w} \pi'$. By Lemma 4.(3), we have
Fact1: $\langle P[P/X]^n\rangle\,(\rho[X \leftarrow \mathsf{Fix}(F)]) = F^n(\mathsf{Fix}(F))$,
Moreover, from Fact0 and (C3), we obtain $n \geq 1$ and F' such that $\pi' = F'(\mathsf{Fix}(F))$ and $\forall \pi.\ \vdash_{Proc} F^n\,\pi \xrightarrow{w} F'\,\pi$. In particular, $\vdash_{Proc} F^n(\mathsf{Fix}(F)) \xrightarrow{w} F'(\mathsf{Fix}(F))$, i.e., $\vdash_{Proc} F^n(\mathsf{Fix}(F)) \xrightarrow{w} \pi'$, hence, with Fact1, $\vdash_{Proc} \langle P[P/X]^n\rangle\,(\rho[X \leftarrow \mathsf{Fix}(F)]) \xrightarrow{w} \pi'.$

With IH, we find P' such that $\langle P'\rangle\,(\rho[X \leftarrow \mathsf{Fix}(F)]) = \pi'$ and $\vdash_{CCST} P[P/X]^n \xrightarrow{w} P'$. By Lemma 2.(2), we have $\vdash_{CCST} \mu X.\,P \xrightarrow{w} P'[(\mu X.\,P)/X].$

Moreover, by Lemma 4.(2), we have $\langle P'[(\mu X.\,P)/X]\rangle \rho = \langle P'\rangle\,(\rho[X \leftarrow \langle \mu X.\,P\rangle \rho]) = F'(\mathsf{Fix}(F)) = \pi'$, making $P'[(\mu X.\,P)/X]$ the desired term.   □

Corollary 2. *The following are equivalent for all $P, Q \in$ Term: $P \sim_{CCST} Q$ iff $\langle P \rangle = \langle Q \rangle$.*

Proof. Immediately from Proposition 2 (since coalgebra morphisms preserve and reflect strong bisimilarity, and since strong bisimilarity in *Proc* is equality). □

Now, Corollaries 1 and 2 immediately yield full abstraction:

Theorem 1. *The following are equivalent for all $P, Q \in$ Term: $P \approx_{CCS} Q$ iff $\langle P \rangle = \langle Q \rangle$.*

4   Conclusions and Related Work

We described an approach to (weak) bisimilarity and illustrated it by providing a novel denotational semantics for CCS under bisimilarity which is both *fully abstract* and *compositional*. Previous approaches to the denotational semantics of bisimilarity CCS [2], as well as instances of more generic approaches [18] that can in principle cover this case, essentially consider final *one-step* coalgebras for the semantic domain. Other works on the denotation of related calculi either do not feature compositionality [12,13] or cover only strong bisimilarity and related strong equivalencies [10,22,4,23].

Models for process calculi which are *intensional* (in that the identity of the semantic items does not coincide with bisimilarity or other targeted notion of operational equivalence) were proposed in [8,16,9,7,21] (among other places). The framework of [9] (extending the one of [16]) offers facilities to define and reason "syntax freely" about weak bisimilarity in models which are already fully abstract w.r.t. strong bisimilarity, via a suitable hiding functor. The relationship between the abstract categorical machinery from there aimed at hiding the τ-actions and our packing and unpacking bijections between processes and their compact representations deserves future research.

Our technique of combining traces with coalgebra seems able to capture bisimilarity of a wide range of process calculi. In particular, our transformation of CCS into CCST could be soundly performed on systems in an SOS format of a quite general kind, e.g., of the kind singled out in [5,25] to ensure that weak bisimilarity is a congruence. We have not worked out the details yet though.

With a bit of extra effort, we could have finetuned our definitions into domain-theoretic ones using Abramsky powerdomains instead of powersets, along the lines of [10,22]. However, our semantics would then have not captured bisimilarity, but the weaker relation of having all finite subtrees bisimilar [1].

References

1. Abramsky, S.: A domain equation for bisimulation. Inf. Comput. 92(2), 161–218 (1991)
2. Aczel, P.: Final universes of processes. In: Main, M.G., Melton, A.C., Mislove, M.W., Schmidt, D., Brookes, S.D. (eds.) MFPS 1993. LNCS, vol. 802, pp. 1–28. Springer, Heidelberg (1994)

3. Baldamus, M., Parrow, J., Victor, B.: A fully abstract encoding of the *pi*-calculus with data terms. In: Caires, L., Italiano, G.F., Monteiro, L., Palamidessi, C., Yung, M. (eds.) ICALP 2005. LNCS, vol. 3580, pp. 1202–1213. Springer, Heidelberg (2005)
4. Baldamus, M., Stauner, T.: Modifying Esterel concepts to model hybrid systems. Electr. Notes Theor. Comput. Sci. 65(5) (2002)
5. Bloom, B.: Structural operational semantics for weak bisimulations. Theor. Comput. Sci., 146(1&2), 25–68 (1995)
6. Boreale, M., Gadducci, F.: Denotational testing semantics in coinductive form. In: Rovan, B., Vojtáš, P. (eds.) MFCS 2003. LNCS, vol. 2747, pp. 279–289. Springer, Heidelberg (2003)
7. Buscemi, M.G., Montanari, U.: A first order coalgebraic model of π-calculus early observational equivalence. In: Brim, L., Jančar, P., Křetínský, M., Kucera, A. (eds.) CONCUR 2002. LNCS, vol. 2421, pp. 449–465. Springer, Heidelberg (2002)
8. Cattani, G.L., Sewell, P.: Models for name-passing processes: interleaving and causal. In: LICS 2000, pp. 322–333 (2000)
9. Fiore, M., Cattani, G.L., Winskel, G.: Weak bisimulation and open maps. In: LICS 1999, pp. 67–76 (1999)
10. Fiore, M.P., Moggi, E., Sangiorgi, D.: A fully-abstract model for the π-calculus. In: LICS 1996, pp. 43–54 (1996)
11. Hennessy, M.: A fully abstract denotational semantics for the π-calculus. Theor. Comput. Sci. 278(1-2), 53–89 (2002)
12. Honsell, F., Lenisa, M., Montanari, U., Pistore, M.: Final semantics for the pi-calculus. In: PROCOMET 1998, pp. 225–243 (1998)
13. Lenisa, M.: Themes in Final Semantics. Dipartimento di Informatica, Universita' di Pisa, TD 6 (1998)
14. Milner, R.: Communication and concurrency. Prentice-Hall, Englewood Cliffs (1989)
15. Milner, R., Parrow, J., Walker, D.: A calculus of mobile processes, parts i and ii. Inf. Comput. 100(1), 1–77 (1992)
16. Nielsen, M., Chang, A.: Observe behaviour categorically. In: FST&TCS 1995, pp. 263–278 (1995)
17. Popescu, A.: A fully abstract coalgebraic semantics for the pi-calculus under weak bisimilarity. Tech. Report UIUCDCS-R-2009-3045. University of Illinois (2009)
18. Rutten, J.J.M.M.: Processes as terms: Non-well-founded models for bisimulation. Math. Struct. Comp. Sci. 2(3), 257–275 (1992)
19. Rutten, J.J.M.M.: Universal coalgebra: a theory of systems. Theor. Comput. Sci. 249(1), 3–80 (2000)
20. Sangiorgi, D., Walker, D.: The π-calculus. A theory of mobile processes. Cambridge (2001)
21. Sokolova, A., de Vink, E.P., Woracek, H.: Weak bisimulation for action-type coalgebras. Electr. Notes Theor. Comput. Sci. 122, 211–228 (2005)
22. Stark, I.: A fully-abstract domain model for the π-calculus. In: LICS 1996, pp. 36–42 (1996)
23. Staton, S.: Name-passing process calculi: operational models and structural operational semantics. PhD thesis, University of Cambridge (2007)
24. Turi, D., Plotkin, G.: Towards a mathematical operational semantics. In: LICS 1997, pp. 280–291 (1997)
25. van Glabbeek, R.J.: On cool congruence formats for weak bisimulations. In: Van Hung, D., Wirsing, M. (eds.) ICTAC 2005. LNCS, vol. 3722, pp. 318–333. Springer, Heidelberg (2005)

Coalgebraic Symbolic Semantics[*]

Filippo Bonchi[1,2] and Ugo Montanari[1]

[1] Dipartimento di Informatica, Università di Pisa
[2] Centrum voor Wiskunde en Informatica (CWI)

Abstract. The operational semantics of interactive systems is usually described by labeled transition systems. Abstract semantics (that is defined in terms of bisimilarity) is characterized by the final morphism in some category of coalgebras. Since the behaviour of interactive systems is for many reasons infinite, *symbolic semantics* were introduced as a mean to define smaller, possibly finite, transition systems, by employing symbolic actions and avoiding some sources of infiniteness. Unfortunately, symbolic bisimilarity has a different "shape" with respect to ordinary bisimilarity, and thus the standard coalgebraic characterization does not work. In this paper, we introduce its coalgebraic models.

1 Introduction

A compositional interactive system is usually defined as a labelled transition system (LTS) where states are equipped with an algebraic structure. Abstract semantics is often defined as bisimilarity. Then a key property is that bisimilarity be a congruence, i.e. that abstract semantics respects the algebraic operations.

Universal Coalgebra [1] provides a categorical framework where the behaviour of dynamical systems can be characterized as *final semantics*. More precisely, if **Coalg$_B$** (i.e., the category of **B**-coalgebras and **B**-cohomomorphisms for a certain endofunctor **B**) has a final object, then the behavior of a **B**-coalgebra is defined as a final morphism. Intuitively, a final object is a universe of abstract behaviors and a final morphism is a function mapping each system in its abstract behavior. Ordinary LTSs can be represented as coalgebras for a suitable functor. Then, two states are bisimilar if and only if they are identified by a final morphism. The image of a certain LTS through a final morphism is its minimal representative (with respect to bisimilarity), which in the finite case can be computed via the list partitioning algorithm [2].

When bisimilarity is not a congruence, the abstract semantics is defined either as the *largest congruence contained into bisimilarity* [3] or as the *largest bisimulation that is also a congruence* [4]. In this paper we focus on the latter and we call it *saturated bisimilarity* ($\sim^S$). Indeed it coincides with ordinary bisimilarity on the *saturated transition system*, that is obtained by the original LTS by adding the transition $p \xrightarrow{c,a} q$, for every context c, whenever $c(p) \xrightarrow{a} q$.

---

[*] This work was carried out during the tenure of an ERCIM "Alain Bensoussan" Fellowship Programme and supported by the IST 2004-16004 SENSORIA.

Many interesting abstract semantics are defined in this way. For example, since late and early bisimilarity of π-calculus [5] are not preserved under substitution (and thus under input prefixes), in [6] Sangiorgi introduces *open bisimilarity* as the largest bisimulation on π-calculus agents which is closed under substitutions. Other noteworthy examples are asynchronous π-calculus [7,8] and mobile ambients calculus [9,10]. The definition of saturated bisimilarity as ordinary bisimulation on the saturated LTS, while in principle operational, often makes the portion of LTS reachable by any nontrivial agent infinite state, and in any case is very inefficient, since it introduces a large number of additional states and transitions. Inspired by [11], Sangiorgi defines in [6] a symbolic transition system and symbolic bisimilarity that efficiently characterizes open bisimilarity. After this, many formalisms have been equipped with a symbolic semantics.

In [12], we have introduced a general model that describes at an abstract level both saturated and symbolic semantics. In this abstract setting, a symbolic transition $p \xrightarrow{c,\alpha}_\beta p'$ means that $c(p) \xrightarrow{\alpha} p'$ and c is a smallest context that allows p to performs such transition. Moreover, a certain *derivation relation* $\vdash$ amongst the transitions of a systems is defined: $p \xrightarrow{c_1,\alpha_1} p_1 \vdash p \xrightarrow{c_2,\alpha_2} p_2$ means that the latter transition is a logical consequence of the former. In this way, if all and only the saturated transitions are logical consequences of symbolic transitions, then saturated bisimilarity can be retrieved via the symbolic LTS.

However, the ordinary bisimilarity over the symbolic transition system differs from saturated bisimilarity. Symbolic bisimilarity is thus defined with an asymmetric shape. In the bisimulation game, when a player proposes a transition, the opponent can answer with a move with a different label. For example in the open π-calculus, a transition $p \xrightarrow{[a=b],\tau} p'$ can be matched by $q \xrightarrow{\tau} q'$. Moreover, the bisimulation game does not restart from p' and q', but from p' and $q'\{b/a\}$.

For this reason, ordinary coalgebras fail to characterize symbolic bisimilarity. Here, we provide coalgebraic models for it by relying on the framework of [12].

Consider the example of open bisimilarity discussed above. The fact that open bisimulation does not relate the arriving states p' and q', but p' and $q'\{b/a\}$, forces us to look for models equipped with an algebraic structure. In [13], *bialgebras* are introduced as a both algebraic and coalgebraic model, while an alternative approach based on *structured coalgebras*, i.e. on coalgebras in categories of algebras, is presented in [14]. In this paper we adopt the latter and we introduce **Coalg$_H$** (Sec. 4), a category of structured coalgebras where the saturated transition system can be naively modeled in such a way that $\sim^S$ coincides with the kernel of a final morphism. Then, we focus only on those **H**-coalgebras whose sets of transitions are closed w.r.t. the derivation relation $\vdash$. These form the category of *saturated coalgebras* **Coalgs$_T$** (Sec. 5.1) that is a covariety of **Coalg$_H$**. Thus, it has a final object and bisimilarity coincides with the one in **Coalg$_H$**.

In order to characterize symbolic bisimilarity, we introduce the notions of *redundant transition* and *semantically redundant transition*. Intuitively, a transition $p \xrightarrow{c_2,\alpha_2} q$ is redundant if there exists another transition $p \xrightarrow{c_1,\alpha_1} p_1$ that logically implies it, that is $p \xrightarrow{c_1,\alpha_1} p_1 \vdash p \xrightarrow{c_2,\alpha_2} q$, while it is semantically redundant, if it is redundant up to bisimilarity, i.e., $p \xrightarrow{c_1,\alpha_1} p_1 \vdash p \xrightarrow{c_2,\alpha_2} p_2$ and

q is bisimilar to p_2. Now, in order to retrieve saturated bisimilarity by disregarding redundant transitions, we have to remove from the saturated transition system not only all the redundant transitions, but also the semantically redundant ones. This is done in the category of *normalized coalgebras* $\mathbf{Coalg_{N_T}}$ (Sec. 5.2). These are defined as coalgebras without redundant transitions. Thus, by definition, a final coalgebra in $\mathbf{Coalg_{N_T}}$ has no semantically redundant transitions. The main peculiarity of $\mathbf{Coalg_{N_T}}$ relies in its morphisms. Indeed, ordinary (cohomo)morphisms between LTSs must preserve and reflect all the transitions ("zig-zag" morphisms), while morphisms between normalized coalgebras must preserve only those transitions that are not semantically redundant.

Moreover, we prove that $\mathbf{Coalg_{S_T}}$ and $\mathbf{Coalg_{N_T}}$ are isomorphic (Sec. 5.3). This means that a final morphism in the latter category still characterizes $\sim^S$, but with two important differences w.r.t. $\mathbf{Coalg_{S_T}}$. First of all, in a final $\mathbf{N_T}$-coalgebra, there are not semantically redundant transitions. Intuitively, a final $\mathbf{N_T}$-coalgebra is a universe of *abstract symbolic behaviours* and a final morphism maps each system in its abstract symbolic behaviour. Secondly, minimization in $\mathbf{Coalg_{N_T}}$ is feasible, while in $\mathbf{Coalg_{S_T}}$ is not, because saturated coalgebras have all the redundant transitions. Minimizing in $\mathbf{Coalg_{N_T}}$ coincides with a *symbolic minimization algorithm* that we have introduced in [15] (Sec. 6). The algorithm shows another peculiarity of normalized coalgebras: minimization relies on the algebraic structure. Since in bialgebras bisimilarity abstracts away from this, we can conclude that our normalized coalgebras are not bialgebras. This is the reason why we work with structured coalgebras.

The background is in Sec. 2 and 3.

2   Saturated and Symbolic Semantics

In this section we recall the general framework for symbolic bisimilarity that we have introduced in [12]. As running example, we will use open Petri nets [16]. However, our theory has as special cases the abstract semantics of many formalisms such as ambients [9], open [6] and asynchronous [7] π-calculus.

2.1   Saturated Semantics

Given a small category $\mathbf{C}$, a $\Gamma(\mathbf{C})$-algebra is an algebra for the algebraic specification in Fig. 1(A) where $|\mathbf{C}|$ denotes the set of objects of $\mathbf{C}$, $||\mathbf{C}||$ the set of arrows of $\mathbf{C}$ and, for all $i, j \in |\mathbf{C}|$, $\mathbf{C}[i, j]$ denotes the set of arrows from i to j.

Thus, a $\Gamma(\mathbf{C})$-algebra $\mathbb{X}$ consists of a $|\mathbf{C}|$-sorted family $X = \{X_i \mid i \in |\mathbf{C}|\}$ of sets and a function $c_{\mathbb{X}} : X_i \to X_j$ for all $c \in \mathbf{C}[i, j]$.[1]

The main definition of the framework presented in [12] is that of *context interactive systems*. In our theory, an interactive system is a state-machine that can interact with the environment (contexts) through an evolving interface.

---

[1] Note that $\Gamma(\mathbf{C})$-algebras coincide with functors from $\mathbf{C}$ to $\mathbf{Set}$ and $\Gamma(\mathbf{C})$-homomorphisms coincide with natural transformations amongst functors. Thus, $\mathbf{Alg_{\Gamma(C)}}$ is isomorphic to $\mathbf{Set^C}$(the category of covariant presheaves over $\mathbf{C}$).

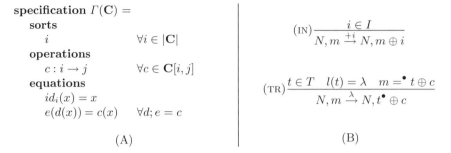

$$
\begin{array}{ll}
\textbf{specification } \Gamma(\mathbf{C}) = & \\
\quad \textbf{sorts} & \\
\qquad i & \forall i \in |\mathbf{C}| \\
\quad \textbf{operations} & \\
\qquad c : i \to j & \forall c \in \mathbf{C}[i, j] \\
\quad \textbf{equations} & \\
\qquad id_i(x) = x & \\
\qquad e(d(x)) = c(x) & \forall d; e = c
\end{array}
$$

$$(\text{IN})\dfrac{i \in I}{N, m \xrightarrow{+i} N, m \oplus i}$$

$$(\text{TR})\dfrac{t \in T \quad l(t) = \lambda \quad m =^{\bullet} t \oplus c}{N, m \xrightarrow{\lambda} N, t^{\bullet} \oplus c}$$

(A) | (B)

Fig. 1. (A) Algebraic specification $\Gamma(\mathbf{C})$. (B) Operational semantics of open nets.

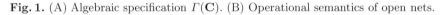

Definition 1 (Context Interactive System). *A context interactive system* $\mathcal{I}$ *is a quadruple* $\langle \mathbf{C}, \mathbb{X}, O, tr \rangle$ *where:*

- $\mathbf{C}$ *is a category,*
- $\mathbb{X}$ *is a* $\Gamma(\mathbf{C})$*-algebra,*
- O *is a set of observations,*
- $tr \subseteq X \times O \times X$ *is a labeled transition relation* $(p \xrightarrow{o} p'$ *means* $(p, o, p') \in tr)$.

Intuitively, objects of $\mathbf{C}$ are interfaces of the system, while arrows are contexts. Every element p of X_i represents a state with interface i and it can be inserted into the context $c \in \mathbf{C}[i, j]$, obtaining a new state $c_{\mathbb{X}}(p)$ that has interface j. Every state can evolve into a new state (possibly with different interface) producing an observation $o \in O$.

The abstract semantics of interactive systems is usually defined through behavioural equivalences. In [12] we proposed a general notion of bisimilarity that generalizes the abstract semantics of a large variety of formalisms [9,7,6,17,18,19]. The idea is that two states of a system are equivalent if they are indistinguishable from an external observer that, in any moment of their execution, can insert them into some environment and then observe some transitions.

Definition 2 (Saturated Bisimilarity). *Let* $\mathcal{I} = \langle \mathbf{C}, \mathbb{X}, O, tr \rangle$ *be a context interactive system. Let* $R = \{R_i \subseteq X_i \times X_i \mid i \in |\mathbf{C}|\}$ *be a* $|\mathbf{C}|$*-sorted family of symmetric relations.* R *is a* saturated bisimulation *iff,* $\forall i, j \in |\mathbf{C}|$, $\forall c \in \mathbf{C}[i, j]$, *whenever* pR_iq:

- $c_{\mathbb{X}}(p)\, R_j\, c_{\mathbb{X}}(q)$,
- *if* $p \xrightarrow{o} p'$, *then* $q \xrightarrow{o} q'$ *and* $p'R_kq'$ *(for some* $k \in |\mathbf{C}|$*).*

We write $p \sim_i^S q$ *iff there is a saturated bisimulation* R *such that* pR_iq.

An alternative but equivalent definition can be given by defining the *saturated transition system* (SATTS) as follows: $p \xrightarrow{c,o}_S q$ if and only if $c(p) \xrightarrow{o} q$. Trivially the ordinary bisimilarity over SATTS coincides with $\sim^S$.

Proposition 1. $\sim^S$ *is the coarsest bisimulation congruence.*

2.2   Running Example: Open Petri Nets

Differently from process calculi, Petri nets have not a widely known interactive behaviour. Indeed they model concurrent systems that are closed, in the sense that they do not interact with the environment. *Open nets* [16] are P/T Petri nets that can interact by exchanging tokens on *input* and *output places*.

Definition 3 (Open Net). *An* open net *is a tuple* $N = (S, T, pre, post, l, I, O)$ *where* S *and* T *are the sets of places and transitions* $(S \cap T = \varnothing)$; $pre, post : T \to S^\oplus$ *are functions mapping each transition to its pre- and post-set;* $l : T \to \Lambda$ *is a labeling function* $(\Lambda$ *is a set of labels) and* $I, O \subseteq S$ *are the sets of input and output places. A* marked open net *is a pair* $\langle N, m \rangle$ *where* N *is an open net and* $m \in S^\oplus$ *is a marking.*

Fig.2 shows five open nets where, as usual, circles represents places and rectangles transitions (labeled with α, β). Arrows from places to transitions represent *pre*, while arrows from transitions to places represent *post*. Input places are denoted by ingoing edges, thus the only input place of N_1 is \$. To make examples easier, hereafter we only consider *open input nets*, i.e., open nets without output places. The operational semantics of marked open nets is expressed by the rules on Fig.1(B), where we use $^\bullet t$ and $t^\bullet$ to denote $pre(t)$ and $post(t)$. The rule (TR) is the standard rule of P/T nets (seen as multisets rewriting). The rule (IN) states that in any moment a token can be inserted inside an input place and, for this reason, the LTS has always an infinite number of states. Fig.2(A) shows part of the infinite transition system of $\langle N_1, a \rangle$. The abstract semantics (denoted by $\sim^N$) is defined in [20] as the ordinary bisimilarity over such an LTS. It is worth noting that $\sim^N$ can be seen as an instance of saturated semantics, where multisets over open places are contexts and transitions are only those generated by the rule (TR).

In the following we formally define $\mathcal{N} = \langle \mathbf{Tok}, \mathbb{N}, \Lambda, tr_\mathcal{N} \rangle$ that is the context interactive system of all open nets (labeled over the set of labels Λ).

The category $\mathbf{Tok}$ is formally defined as follows,

- $|\mathbf{Tok}| = \{I \mid I$ is a set of places$\}$,
- $\forall I \in |\mathbf{Tok}|, \mathbf{Tok}[I, I] = I^\oplus, id_I = \varnothing$ and $\forall i_1, i_2 \in I^\oplus, i_1; i_2 = i_1 \oplus i_2$.

Intuitively objects are sets of input places I, while arrows are multisets of tokens on the input places. We say that a marked open net $\langle N, m \rangle$ has interface I if the set of input places of N is I. For example the marked open net $\langle N_1, a \rangle$ has interface $\{\$\}$. Let us define the $\Gamma(\mathbf{Tok})$-algebra $\mathbb{N}$. For any sort I, the carrier set N_I contains all the marked open nets with interface I. For any operator $i \in \mathbf{Tok}[I, I]$, the function $i_\mathbb{N}$ maps $\langle N, m \rangle$ into $\langle N, m \oplus i \rangle$.

The transition structure $tr_\mathcal{N}$ (denoted by $\to_\mathcal{N}$) associates to a state $\langle N, m \rangle$ the transitions obtained by using the rule (TR) of Fig.1(B). In [12], it is proved that saturated bisimilarity for $\mathcal{N}$ coincides with $\sim^N$. In the remainder of the paper we will use as running example the open nets in Fig.2. Since all the places have different names (with the exception of \$), in order to make lighter the

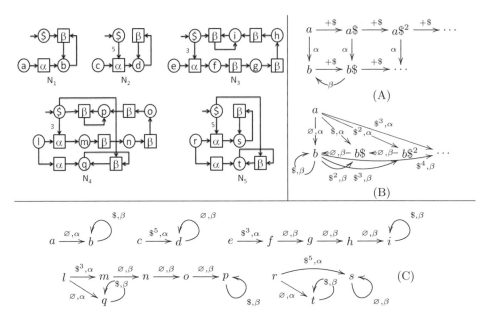

Fig. 2. The open nets N_1, N_2, N_3, N_4 and N_5.(A)Part of the infinite transition system of $\langle N_1, a \rangle$. (B)Part of the infinite saturated transition system of $\langle N_1, a \rangle$.(C)The symbolic transition systems of $\langle N_1, a \rangle$,$\langle N_2, c \rangle$,$\langle N_3, e \rangle$,$\langle N_4, l \rangle$ and $\langle N_5, r \rangle$.

notation, we write only the marking to mean the corresponding marked net, e.g. $b^2\$$ means the marked net $\langle N_1, b^2\$\rangle$.

The marked net a (i.e., $\langle N_1, a \rangle$) represents a system that provides a service β. After the activation α, it provides β whenever the client pay one \$ (i.e., the environment insert a token into \$). The marked net c instead requires five \$ during the activation, but then provides the service β for free. The marked net e, requires three \$ during the activation. For three times, the service β is performed for free and then it costs one \$. It is easy to see that all these marked nets are not bisimilar. Indeed, a client that has only one \$ can have the service β only with a, while a client with five \$ can have the service β for six times only with c. The marked net r represents a system that offers the behaviour of both a and c, i.e. either the activation α is for free and then the service β costs one, or the activation costs five and then the service is for free. Also this marked net is different from all the others.

Now consider the marked net l. It offers the behaviour of both a and e, but it is equivalent to a, i.e. $l \sim^N a$. Roughly, the behaviour of e is absorbed by the behaviour of a. This is analogous to what happens in asynchronous π-calculus [7] where it holds that $a(x).(\bar{a}x \mid p) + \tau.p \sim \tau.p$.

2.3   Symbolic Semantics

Saturated bisimulation is a good notion of equivalence but it is hard to check, since it involves a quantification over all contexts. In [12], we have introduced a general notion of *symbolic bisimilarity* that coincides with saturated bisimilarity, but it avoids to consider all contexts. The idea is to define a symbolic transition system where transitions are labeled both with the usual observation and also with the minimal context that allows the transition.

Definition 4 (Symbolic Context Transition System). *A symbolic context transition system (SCTS for short) for a system $\mathcal{I} = \langle \mathbf{C}, \mathbb{X}, O, tr \rangle$ is a transition system $\beta \subseteq \mathbb{X} \times ||\mathbf{C}|| \times O \times \mathbb{X}$.*

In [12], we have introduced a SCTS for open nets. Intuitively the symbolic transition $N, m \xrightarrow{i,\lambda}_\eta N, m'$ is possible if and only if $N, m \oplus i \xrightarrow{\lambda}_{\mathcal{N}} N, m'$ and i is the smallest multiset (on input places) allowing such transition. This SCTS is formally defined by the following rule.

$$\frac{t \in T \quad l(t) = \lambda \quad m = (m \cap {}^\bullet t) \oplus n \quad i \subseteq I^\oplus \quad {}^\bullet t = (m \cap {}^\bullet t) \oplus i}{N, m \xrightarrow{i,\lambda}_\eta N, t^\bullet \oplus n}$$

The marking $m \cap {}^\bullet t$ contains all the tokens of m that are needed to perform the transition t. The marking n contains all the tokens of m that are not useful for performing t, while the marking i contains all the tokens that m needs to reach ${}^\bullet t$. Note that i is exactly the *smallest* multiset that is needed to perform the transition t. Indeed if we take i_1 strictly included into i, $m \oplus i_1$ cannot match ${}^\bullet t$. As an example consider the net N_2 in Fig.2 with marking $cd\$^2$ and let t be the only transition labeled with α. We have that $cd\$^2 \cap {}^\bullet t = c\$^2$, $n = d$ and $i = \$^3$. Thus $N_2, cd\$^2 \xrightarrow{\$^3, \alpha}_\eta N_2, dd$. Fig.2(C) shows symbolic transition systems of marked open nets discussed in the previous subsection.

Definition 5 (Inference System). *An inference system T for a context interactive system $\mathcal{I} = \langle \mathbf{C}, \mathbb{X}, O, tr \rangle$ is a set of rules of the following format, where $i, j \in |\mathbf{C}|$, $o, o' \in O$, $c \in \mathbf{C}[i, i']$ and $d \in \mathbf{C}[j, j']$.*

$$\frac{p_i \xrightarrow{o} q_j}{c(p_i) \xrightarrow{o'} d(q_j)}$$

The above rule states that all processes with interface i that perform a transition with observation o going into a state q_j with interface j, when inserted into the context c can perform a transition with the observation o' going into $d(q_j)$.

In the following, we write $c \xrightarrow[o']{o} d$ to mean a rule like the above. The rules $c \xrightarrow[o']{o} c'$ and $d \xrightarrow[o'']{o'} d'$ *derive* the rule $c; d \xrightarrow[o'']{o} c'; d'$ if $c; d$ and $c'; d'$ are defined. Given an inference system T, $\Phi(T)$ is the set of all the rules derivable from T together with the identities rules ($\forall o \in O$ and $\forall i, j \in |\mathbf{C}|$, $id_i \xrightarrow[o]{o} id_j$).

Definition 6 (Derivations, soundness and completeness). *Let $\mathcal{I}$ be a context interactive system, β an SCTS and T an inference system.*

We say that $p \xrightarrow{c_1,o_1} p_1$ derives $p \xrightarrow{c_2,o_2} p_2$ in T (written $p \xrightarrow{c_1,o_1} p_1 \vdash_T p \xrightarrow{c_2,o_2} p_2$) if there exist $d, e \in ||\mathbf{C}||$ such that $d \xrightarrow[o_2]{o_1} e \in \Phi(T)$, $c_1; d = c_2$ and $e_{\mathbb{X}}(p_1) = p_2$. We say that β and T are sound and complete *w.r.t. $\mathcal{I}$ if*

$$p \xrightarrow{c,o}_S q \text{ iff } p \xrightarrow{c',o'}_\beta q' \text{ and } p \xrightarrow{c',o'}_\beta q' \vdash_T p \xrightarrow{c,o}_S q.$$

A sound and complete SCTS could be considerably smaller than the saturated transition system, but still containing all the information needed to recover $\sim^S$. Note that the ordinary bisimilarity over SCTS (hereafter called *syntactical bisimilarity* and denoted by $\sim^W$) is usually stricter than $\sim^S$. As an example consider the marked open nets a and l. These are not syntactically bisimilar, since $l \xrightarrow{\$^3,\alpha}_\eta m$ while a cannot (Fig.2(C)). However, they are saturated bisimilar, as discussed in the previous subsection. In order to recover $\sim^S$ through the symbolic transition system we need a more elaborated definition of bisimulation.

Definition 7 (Symbolic Bisimilarity). *Let $\mathcal{I} = \langle \mathbf{C}, \mathbb{X}, O, tr \rangle$ be an interactive system, T be a set of rules and β be a symbolic transition system. Let $R = \{R_i \subseteq X_i \times X_i \mid i \in |\mathbf{C}|\}$ be an $|\mathbf{C}|$ sorted family of symmetric relations. R is a* symbolic bisimulation *iff $\forall i \in |\mathbf{C}|$, whenever pR_iq:*

- *if $p \xrightarrow{c,o}_\beta p'$, then $q \xrightarrow{c_1,o_1}_\beta q'_1$ and $q \xrightarrow{c_1,o_1}_\beta q'_1 \vdash_T q \xrightarrow{c,o} q'$ and $p'R_kq'$.*

We write $p \sim^{SYM}_i q$ iff there exists a symbolic bisimulation R such that pR_iq.

Theorem 1. *Let $\mathcal{I}$ be a context interactive system, β an SCTS and T an inference system. If β and T are sound and complete w.r.t. $\mathcal{I}$, then $\sim^{SYM}=\sim^S$.*

In the remainder of this section we focus on open Petri nets. The inference system $T_\mathcal{N}$ is defined by the following parametric rule.

$$\frac{N, m \xrightarrow{\lambda}_\mathcal{N} N, m'}{N, m \oplus i \xrightarrow{\lambda}_\mathcal{N} N, m' \oplus i}$$

The intuitive meaning of this rule is that for all possible observations λ and multiset i on input places, if a marked net performs a transition with observation λ, then the addition of i preserves this transition.

Now, consider derivations between transitions of open nets. It is easy to see that $N, m \xrightarrow{i_1,\lambda_1} N, m_1 \vdash_{T_\mathcal{N}} N, m \xrightarrow{i_2,\lambda_2} N, m_2$ if and only if $\lambda_2 = \lambda_1$ and there exists a multiset x on input places such that $i_2 = i_1 \oplus x$ and $m_2 = m_1 \oplus x$. For all the nets N_k of our example, this just means that for all observations λ and for all multisets m, n, we have that $\langle N_k, m \rangle \xrightarrow{\$^i,\lambda}_\eta \langle N_k, n \rangle \vdash_{T_\mathcal{N}} \langle N_k, m \rangle \xrightarrow{\$^{i+j},\lambda} \langle N_k, n\$^j \rangle$.

In [12] we have shown that $T_\mathcal{N}$ and η are sound and complete w.r.t. $\mathcal{N}$.

3   (Structured) Coalgebras

In this section we recall the notions of the theory of coalgebras that will be useful in the following to give coalgebraic models for saturated and symbolic semantics.

Definition 8 (Coalgebra). *Let* $\mathbf{B} : \mathbf{C} \to \mathbf{C}$ *be an endofunctor on a category* $\mathbf{C}$. *A* $\mathbf{B}$*-coalgebra is a pair* $\langle X, \alpha \rangle$ *where* X *is an object of* $\mathbf{C}$ *and* $\alpha : X \to \mathbf{B}(X)$ *is an arrow. A* $\mathbf{B}$*-cohomomorphism* $f : \langle X, \alpha \rangle \to \langle Y, \beta \rangle$ *is an arrow* $f : X \to Y$ *of* $\mathbf{C}$ *such that* $f; \beta = \alpha; \mathbf{B}(f)$. $\mathbf{B}$*-coalgebras and* $\mathbf{B}$*-cohomomorphisms forms the category* $\mathbf{Coalg_B}$.

For example, the functor $\mathbf{P_L} : \mathbf{Set} \to \mathbf{Set}$ defined as $\mathbf{P}(L \times id)$ (for L a set of labels and $\mathbf{P}$ the powerset functor) defines the category $\mathbf{Coalg_{P_L}}$ of L-labeled transition systems and "zig-zag" morphisms.

If $\mathbf{Coalg_B}$ has a final object, one can define the behaviour of a $\mathbf{B}$-coalgebra as the final morphism. Thus behavioural equivalence, hereafter referred as bisimilarity, is defined as the kernel of a final morphism. Moreover, in a final coalgebra all bisimilar states are identified, and thus, the image of a coalgebra through a final morphism is its minimal realization (w.r.t. bisimilarity). In the finite case, this can be computed via a minimization algorithm.

Unfortunately, due to cardinality reasons, $\mathbf{Coalg_{P_L}}$ does not have a final object [1]. One satisfactory solution consists in replacing the powerset functor $\mathbf{P}$ by the *countable* powerset functor $\mathbf{P}_c$, which maps a set to the family of its countable subsets. Then, by defining $\mathbf{P_L^c} = \mathbf{P}_c(L \times id)$, one has that coalgebras for this endofunctor are one-to-one with transition systems with *countable degree*. Unlike functor $\mathbf{P_L}$, functor $\mathbf{P_L^c}$ admits final coalgebras (Ex. 6.8 of [1]).

The coalgebraic representation using functor $\mathbf{P_L^c}$ is not completely satisfactory, because the intrinsic algebraic structure of states is lost. This calls for the introduction of *structured coalgebras* [21], i.e. coalgebras for an endofuctor on a category $\mathbf{Alg_\Gamma}$ of algebras for a specification Γ. Since cohomomorphisms in a category of structured coalgebras are also Γ-homomorphisms, *bisimilarity* (i.e. the kernel of a final morphism) *is a congruence* w.r.t. the operations in Γ.

Moreover, since we would like that the structured coalgebraic model is compatible with the unstructured, set-based one, we are interested in functors $\mathbf{B^\Gamma} : \mathbf{Alg_\Gamma} \to \mathbf{Alg_\Gamma}$ that are *lifting* of some functor $\mathbf{B} : \mathbf{Set} \to \mathbf{Set}$ along the *forgetful* functor $\mathbf{V^\Gamma} : \mathbf{Alg_\Gamma} \to \mathbf{Set}$ (i.e., $\mathbf{B^\Gamma}; \mathbf{V^\Gamma} = \mathbf{V^\Gamma}; \mathbf{B}$).

Proposition 2 (From [21]). *Let* Γ *be an algebraic specification. Let* $\mathbf{V^\Gamma} : \mathbf{Alg_\Gamma} \to \mathbf{Set}$ *be the forgetful functor. If* $\mathbf{B_\Gamma} : \mathbf{Alg_\Gamma} \to \mathbf{Alg_\Gamma}$ *is a lifting of* $\mathbf{P_L^c}$ *along* $\mathbf{V^\Gamma}$, *then* (1) $\mathbf{Coalg_{B_\Gamma}}$ *has a final object,* (2) *bisimilarity is uniquely induced by bisimilarity in* $\mathbf{Coalg_{P_L^c}}$ *and* (3) *bisimilarity is a congruence.*

In [13], *bialgebras* are used as structures combining algebras and coalgebras. Bialgebras are richer than structured coalgebras, in the sense that they can be seen both as coalgebras on algebras and also as algebras on coalgebras. In Section 5.2, we will introduce *normalized coalgebras* that are not bialgebras. This explains why we decided to use structured coalgebras.

4   Coalgebraic Saturated Semantics

In this section we introduce the coalgebraic model for the saturated transition system. First we model it as a coalgebra over $\mathbf{Set}^{|\mathbf{C}|}$, i.e., the category of $|\mathbf{C}|$-sorted families of sets and functions. Therefore in this model, all the algebraic structure is missing. Then we lift it to $\mathbf{Alg}_{\Gamma(\mathbf{C})}$ that is the category of $\Gamma(\mathbf{C})$-algebras and $\Gamma(\mathbf{C})$-homomorphisms.

In the following, we assume that the SATTS has countable degree.

Definition 9. $\mathbf{G} : \mathbf{Set}^{|\mathbf{C}|} \to \mathbf{Set}^{|\mathbf{C}|}$ *is defined for each $|\mathbf{C}|$-sorted family of set X and for each $i \in |\mathbf{C}|$ as $\mathbf{G}(X_i) = \mathbf{P}_c(\sum_{j \in |\mathbf{C}|}(\mathbf{C}[i,j] \times O \times \sum_{k \in |\mathbf{C}|} X_k))$. Analogously for arrows.*

A $\mathbf{G}$-coalgebra is a $\mathbf{C}$-sorted family $\alpha = \{\alpha_i : X_i \to \mathbf{G}(X_i) \mid i \in |\mathbf{C}|\}$ of functions assigning to each $p \in X_i$ a set of transitions (c, o, q) where c is an arrow of $\mathbf{C}$ (context) with source i, o is an observation and q is the arriving state. Note that q can have any possible sort ($q \in \sum_{k \in |\mathbf{C}|} X_k$).

Then, given $\mathcal{I} = \langle \mathbf{C}, \mathbb{X}, O, tr \rangle$, we define the $\mathbf{G}$-coalgebra $\langle X, \alpha_{\mathcal{I}} \rangle$ corresponding to the SATTS, where $\forall i \in |\mathbf{C}|$, $\forall p \in X_i$, $(c, o, q) \in \alpha_{\mathcal{I}}(p)$ iff $(c_{\mathbb{X}}(p), o, q) \in tr$.

Now we want to define an endofunctor $\mathbf{H}$ on $\mathbf{Alg}_{\Gamma(\mathbf{C})}$ that is a lifting of $\mathbf{G}$ and such that $\langle \mathbb{X}, \alpha_{\mathcal{I}} \rangle$ is a $\mathbf{H}$-coalgebra. In order to do that, we must define how $\mathbf{H}$ modifies the operations of $\Gamma(\mathbf{C})$-algebras. Usually, this is done by giving a set of GSOS rules.

In our case, the following (parametric) rule defines $\mathbf{H}$.

$$\frac{p \xrightarrow{c_1, l} q \quad c_1 = d; c_2}{d(p) \xrightarrow{c_2, l} q}$$

Hereafter, in order to make lighter the notation, we will avoid to specify sorts. We will denote a $\Gamma(\mathbf{C})$-algebra $\mathbb{X}$ as $\langle X, d_{\mathbb{X}}^0, d_{\mathbb{X}}^1, \dots \rangle$ where X is the $|\mathbf{C}|$-sorted carrier set of $\mathbb{X}$ and $d_{\mathbb{X}}^i$ is the function corresponding to the operator $d^i \in ||\mathbf{C}||$.

Definition 10. $\mathbf{H} : \mathbf{Alg}_{\Gamma(\mathbf{C})} \to \mathbf{Alg}_{\Gamma(\mathbf{C})}$ *maps each $\mathbb{X} = \langle X, d_{\mathbb{X}}^0, d_{\mathbb{X}}^1, \dots \rangle \in \mathbf{Alg}_{\Gamma(\mathbf{C})}$ into $\langle \mathbf{G}(X), d_{\mathbf{H}(\mathbb{X})}^0, d_{\mathbf{H}(\mathbb{X})}^1, \dots \rangle$ where $\forall d \in \Gamma(\mathbf{C})$, $\forall A \in \mathbf{G}(X)$, $d_{\mathbf{H}(\mathbb{X})} A = \{(c_2, l, x) \text{ s.t. } (c_1, l, x) \in A \text{ and } c_1 = d; c_2\}$. For arrows, it is defined as $\mathbf{G}$.*

It is immediate to see that $\mathbf{H}$ is a lifting of $\mathbf{G}$. Thus, by Prop.2, follows that $\mathbf{Coalg}_\mathbf{H}$ has final object and that bisimilarity is a congruence.[2] In the following, we proved that $\alpha_{\mathcal{I}} : \mathbb{X} \to \mathbf{H}(\mathbb{X})$ is a $\Gamma(\mathbf{C})$-homomorphism.

Theorem 2. $\langle \mathbb{X}, \alpha_{\mathcal{I}} \rangle$ *is a $\mathbf{H}$-coalgebra.*

Now, since a final coalgebra $F_\mathbf{H}$ exists in $\mathbf{Coalg}_\mathbf{H}$ and since $\langle \mathbb{X}, \alpha_{\mathcal{I}} \rangle$ is a $\mathbf{H}$-coalgebra, then the kernel of its final morphism coincides with $\sim^S$.

---

[2] Prop. 2 holds also for many-sorted algebras and many sorted sets [22].

5   Coalgebraic Symbolic Semantics

In Sec. 4 we have characterized saturated bisimilarity as the equivalence induced by a final morphisms from $\langle \mathbb{X}, \alpha_{\mathcal{I}} \rangle$ (i.e., the **H**-coalgebra corresponding to SATTS) to $F_{\mathbf{H}}$. This is theoretically interesting, but pragmatically useless. Indeed SATTS is usually infinite branching (or in any case very inefficient), and so it is the minimal model. In this section we use symbolic bisimilarity in order to give an efficient and coalgebraic characterization of $\sim^S$. We provide a notion of *redundant transitions* and we introduce *normalized coalgebras* as coalgebras without redundant transitions. The category of normalized coalgebras (**Coalg$_{\mathbf{N_T}}$**) is isomorphic to the category of *saturated coalgebras*, i.e., the full subcategory of **Coalg$_{\mathbf{H}}$** that contains only those coalgebras "satisfying" an inference system T. From the isomorphism follows that $\sim^S$ coincides with the kernel of the final morphism in **Coalg$_{\mathbf{N_T}}$**. This provides a characterization of $\sim^S$ really useful: every equivalence class has a canonical model that is smaller than that in **Coalg$_{\mathbf{H}}$** because normalized coalgebras have not redundant transitions. Moreover, minimizing in **Coalg$_{\mathbf{N_T}}$** is feasible since it abstracts away from redundant transitions.

5.1   Saturated Coalgebras

Hereafter we refer to a context interactive system $\mathcal{I} = \langle \mathbf{C}, \mathbb{X}, O, tr \rangle$ and to an inference system T. First, we extend $\vdash_T$ (Def.6) to $\Gamma(\mathbf{C})$-algebras.

Definition 11 (extended derivation). *Let $\mathbb{X}$ be a $\Gamma(\mathbf{C})$-algebra. A transition $p \xrightarrow{c_1, o_1} q_1$ derives a transition $d_{\mathbb{X}}(p) \xrightarrow{c_2, o_2} q_2$ in $\mathbb{X}$ through T (written $(c_1, o_1, q_1) \vdash^d_{T, \mathbb{X}} (c_2, o_2, q_2)$) iff there exists $e \in ||\mathbf{C}||$ such that $c_1; e = d; c_2$ and $\phi : o_1 \xrightarrow[e']{e} o_2 \in \Phi(T)$ and $e'_{\mathbb{X}}(q_1) = q_2$.*

Intuitively, $\vdash^d_{T, \mathbb{X}}$ allows to derive from the set of transitions of a state p some transitions of $d_{\mathbb{X}}(p)$. Consider the symbolic transition $p \xrightarrow{\$^3, \alpha}_\eta m$ of l (Fig.2) in our running example. The derivation $(\$^3, \alpha, m) \vdash^{\$^2}_{T_\mathcal{N}, \mathbb{N}} (\$, \alpha, m) \vdash^{\$^2}_{T_\mathcal{N}, \mathbb{N}} (\$, \alpha, m\$^2)$ means that $l\$^2 \xrightarrow{\$, \alpha} m$ and $l\$^4 \xrightarrow{\$, \alpha} m\$^2$. The latter is not a symbolic transition.

Definition 12 (Sound Inference System). *A inference system T is* sound *w.r.t. a **H**-coalgebra $\langle \mathbb{X}, \alpha \rangle$ (or viceversa, $\langle \mathbb{X}, \alpha \rangle$ satisfies T) iff whenever $(c, o, q) \in \alpha(p)$ and $(c, o, q) \vdash^d_{T, \mathbb{X}} (c', o', q')$ then $(c', o', q') \in \alpha(d_{\mathbb{X}}(p))$.*

For example, $\langle \mathbb{N}, \alpha_\mathcal{N} \rangle$ (i.e., the **H**-coalgebra corresponding to the SATTS of our running example) satisfies $T_\mathcal{N}$, while the coalgebra corresponding to the symbolic transition systems does not. Hereafter we use $\vdash_{T, \mathbb{X}}$ to mean $\vdash^{id}_{T, \mathbb{X}}$.

Definition 13 (Saturated set). *Let $\mathbb{X}$ be a $\Gamma(\mathbf{C})$-algebra. A set $A \in \mathbf{G}(X)$ is* saturated *in T and $\mathbb{X}$ if it is closed w.r.t. $\vdash_{T, \mathbb{X}}$. The set $\mathbf{S^{T\mathbb{Y}}}(X)$ is the subset of $\mathbf{G}(X)$ containing all and only the saturated sets in T and $\mathbb{X}$.*

Definition 14. $\mathbf{S_T} : \mathbf{Alg}_{\Gamma(\mathbf{C})} \rightarrow \mathbf{Alg}_{\Gamma(\mathbf{C})}$ *maps each $\mathbb{X} = \langle X, d^0_{\mathbb{X}}, d^1_{\mathbb{X}}, \dots \rangle \in \mathbf{Alg}_{\Gamma(\mathbf{C})}$ into $\mathbf{S_T}(\mathbb{X}) = \langle \mathbf{S^{T\mathbb{X}}}(X), d^0_{\mathbf{S_T}(X)}, d^1_{\mathbf{S_T}(X)}, \dots \rangle$ where $\forall d \in \Gamma(\mathbf{C}), \forall A \in$*

$\mathbf{G}(X)$, $d_{\mathbf{S_T}(\mathbb{X})}A = \{(c_2, o_2, x_2)\ s.t.\ (c_1, o_1, x_1) \in A\ and\ (c_1, o_1, x_1) \vdash^d_{T,\mathbb{X}} (c_2, o_2, x_2)\}$. *For arrows, it is defined as* $\mathbf{G}$.

There are two differences w.r.t $\mathbf{H}$. First, we require that all the sets of transitions are saturated. Then the operators are defined by using the relation $\vdash^d_{T,\mathbb{X}}$.

Notice that $\mathbf{S_T}$ cannot be regarded as a lifting of any endofunctor over $\mathbf{Set}^{|C|}$. Indeed the definition of $\mathbf{S^{T^{\mathbb{X}}}}(X)$ depends from the algebraic structure $\mathbb{X}$. For this reason we cannot use Prop.2. In order to prove the existence of final object in $\mathbf{Coalgs_T}$, we show that $\mathbf{Coalgs_T}$ is the full subcategory of $\mathbf{Coalg_H}$ containing all and only the coalgebras satisfying T. More precisely, we show that $|\mathbf{Coalgs_T}|$ is a *covariety* of $\mathbf{Coalg_H}$.

Lemma 1. *A* $\mathbf{H}$*-coalgebra is a* $\mathbf{S_T}$*-coalgebra iff satisfies* T.

Proposition 3. $|\mathbf{Coalgs_T}|$ *is a covariety of* $\mathbf{Coalg_H}$.

From this follows that we can construct a final object in $\mathbf{Coalgs_T}$ as the biggest subobject of $F_\mathbf{H}$ satisfying T. Thus, the kernel of final morphism from $\langle \mathbb{X}, \alpha_\mathcal{I} \rangle$ in $\mathbf{Coalgs_T}$ coincides with $\sim^S$.

5.2   Normalized Coalgebras

In this subsection we introduce normalized coalgebras, in order to characterize $\sim^S$ without considering the whole SATTS and by relying on the derivation relation $\vdash_{T,\mathbb{X}}$. The following observation is fundamental to explain our idea.

Lemma 2. *Let* $\mathbb{X}$ *be a* $\Gamma(\mathbf{C})$*-algebra. If* $(c_1, o_1, p_1) \vdash_{T,\mathbb{X}} (c_2, o_2, p_2)$ *then* $p_2 = e_{\mathbb{X}}(p_1)$ *for some* $e \in ||\mathbf{C}||$. *Moreover* $\forall q_1 \in X$, $(c_1, o_1, q_1) \vdash_{T,\mathbb{X}} (c_2, o_2, e_{\mathbb{X}}(q_1))$.

Consider a $\mathbf{H}$-coalgebra $\langle \mathbb{X}, \gamma \rangle$ and the equivalence $\sim^\gamma$ induced by the final morphism. Suppose that $p \xrightarrow{c_1, o_1}_\gamma p_1$ and $p \xrightarrow{c_2, o_2}_\gamma p_2$ such that $(c_1, o_1, p_1) \vdash_{T,\mathbb{X}} (c_2, o_2, e_{\mathbb{X}}(p_1))$. If $\langle \mathbb{X}, \gamma \rangle$ satisfies T (i.e., it is a $\mathbf{S_T}$-coalgebra), we can forget about the latter transition. Indeed, for all $q \in X$, if $q \xrightarrow{c_1, o_1}_\gamma q_1$ then also $q \xrightarrow{c_2, o_2}_\gamma e_{\mathbb{X}}(q_1)$ (since $\langle \mathbb{X}, \gamma \rangle$ satisfies T) and if $p_1 \sim^\gamma q_1$, then also $e_{\mathbb{X}}(p_1) \sim^\gamma e_{\mathbb{X}}(q_1)$ (since $\sim^\gamma$ is a congruence). Thus, when checking bisimilarity, we can avoid to consider those transitions that are derivable from others. We call such transitions *redundant*.

The wrong way to efficiently characterize $\sim^\gamma$ by exploiting $\vdash_{T,\mathbb{X}}$, consists in removing all the redundant transitions from $\langle \mathbb{X}, \gamma \rangle$ obtaining a new coalgebra $\langle \mathbb{X}, \beta \rangle$ and then computing $\sim^\beta$ (i.e., the ordinary bisimilarity on $\langle \mathbb{X}, \beta \rangle$). When considering $\langle \mathbb{X}, \alpha_\mathcal{I} \rangle$ (i.e., the $\mathbf{H}$-coalgebra corresponding to SATTS), this roughly means to build a symbolic transition system and then computing the ordinary bisimilarity over this. But, as we have seen in Sec.2, the resulting bisimilarity (denoted by $\sim^W$) does not coincide with the original one.

Generally, this happens when $p \xrightarrow{c_1, o_1}_\beta p_1$ and $p \xrightarrow{c_2, o_2}_\beta p_2$ with $(c_1, o_1, p_1) \vdash_{T,\mathbb{X}} (c_2, o_2, e_{\mathbb{X}}(p_1))$ and $e_{\mathbb{X}}(p_1) \neq p_2$, but $e_{\mathbb{X}}(p_1) \sim^\gamma p_2$. Notice that the latter transitions is not removed, because it is not considered redundant since $e_{\mathbb{X}}(p_1)$ is different from p_2 (even if semantically equivalent). A transition as the latter is called *semantically redundant* and it causes the mismatch between $\sim^\beta$ and $\sim^\gamma$.

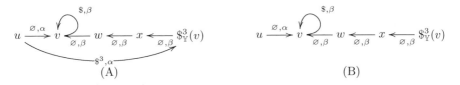

Fig. 3. (A) Part of a not normalized coalgebra $\langle \mathbb{Y}, \gamma \rangle$. (B) Part of a normalized coalgebra $\langle \mathbb{Y}, \gamma; norm_{\mathbb{Y}, T_\mathcal{N}} \rangle$.

Indeed, take a process q performing $q \xrightarrow{c_1, o_1}_\beta q_1$ with $p_1 \sim^\gamma q_1$. Clearly $p \not\sim^\beta q$, but $p \sim^\gamma q$. Indeed $(c_1, o_1, q_1) \vdash_{T, \mathbb{X}} (c_2, o_2, e_\mathbb{X}(q_1))$ and thus $q \xrightarrow{c_2, o_2}_\gamma e_\mathbb{X}(q_1)$ (since $\langle \mathbb{X}, \gamma \rangle$ satisfies T) and $p_2 \sim^\gamma e_\mathbb{X}(p_1) \sim^\gamma e_\mathbb{X}(q_1)$ (since $\sim^\gamma$ is a congruence).

As an example consider the symbolic transition system of l (Fig.2). $l \xrightarrow{\varnothing, \alpha}_\eta q$ and $l \xrightarrow{\$^3, \alpha}_\eta m$. Moreover, $l \xrightarrow{\varnothing, \alpha}_\eta q \vdash_{T_\mathcal{N}, \mathbb{N}} l \xrightarrow{\$^3, \alpha} \$^3_\mathbb{N}(q)$ and $\$^3_\mathbb{N}(q) = q\$^3 \neq m$, but $q\$^3 \sim^S m$. Now consider a. $a \xrightarrow{\varnothing, \alpha}_\eta b$. Clearly $l \not\sim^W a$ but $l \sim^S a$ (Sec. 2).

The above observation tells us that we have to remove not only the redundant transition, i.e., those derivable from $\vdash_{T, \mathbb{X}}$, but also the *semantically redundant* ones. But immediately a problem arises. How can we decide which transitions are semantically redundant, if semantic redundancy itself depends on bisimilarity?

Our solution is the following: we define a category of coalgebras without redundant transitions ($\mathbf{Coalg_{N_T}}$) and, as a result, a final coalgebra contains no semantically redundant transitions.

Definition 15 (Normalized Set and Normalization). *Let* $\mathbb{X}$ *be a* $\Gamma(\mathbf{C})$-*algebra. A transition* (c', o', q') *is equivalent to* (c, o, q) *in* $T, \mathbb{X}$ *(written* $(c', o', q') \equiv_{T, \mathbb{X}} (c, o, q)$*) iff* $(c', o', q') \vdash_{T, \mathbb{X}} (c, o, q)$ *and* $(c, o, q) \vdash_{T, \mathbb{X}} (c', o', q')$. *A transition* (c', o', q') *dominates* (c, o, q) *in* $T, \mathbb{X}$ *(written* $(c', o', q') \prec_{T, \mathbb{X}} (c, o, q)$*) iff* $(c', o', q') \vdash_{T, \mathbb{X}} (c, o, q)$ *and* $(c, o, q) \nvdash_{T, \mathbb{X}} (c', o', q')$. *Let and* $A \in \mathbf{G}(Y)$. *A transition* $(c, o, q) \in A$ *is redundant in* A *w.r.t.* $T, \mathbb{X}$ *if* $\exists (c', o', q') \in A$ *such that* $(c', o', q') \prec_{T, \mathbb{X}} (c, o, q)$. *The set* A *is normalized w.r.t.* $T, \mathbb{X}$ *iff it does not contain redundant transitions and it is closed by equivalent transitions. The set* $\mathbf{N}_\mathbb{X}^\mathbf{T}(Y)$ *is the subset of* $\mathbf{G}(Y)$ *containing all and only the normalized sets w.r.t.* $T, \mathbb{X}$. *The normalization function* $norm_{T, \mathbb{Y}} : \mathbf{G}(Y) \to \mathbf{N}_\mathbb{X}^\mathbf{T}(Y)$ *maps* $A \in \mathbf{G}(Y)$ *into* $\{(c', o', q') s.t. \exists (c, o, q) \in A \ s.t. \ (c', o', q') \equiv_{T, \mathbb{X}} (c, o, q)$ *and* (c, o, q) *not redundant in* A *w.r.t.* $T, \mathbb{X}\}$.

Recall $\mathcal{N} = \langle \mathbf{Tok}, \mathbb{N}, \Lambda, tr_\mathcal{N} \rangle$ and $T_\mathcal{N}$. Consider the coalgebra $\langle \mathbb{Y}, \gamma \rangle$ partially depicted in Fig.3(A). Here we have that $(\varnothing, \alpha, v) \vdash_{T_\mathcal{N}, \mathbb{Y}} (\$^3, \alpha, \$^3_\mathbb{Y}(v))$ but $(\$^3, \alpha, \$^3_\mathbb{Y}(v)) \nvdash_{T_\mathcal{N}, \mathbb{Y}} (\varnothing, \alpha, v)$. Thus the set $\gamma(u)$, i.e., the set of transitions of u, is not normalized (w.r.t $T_\mathcal{N}, \mathbb{Y}$) since the transition $(\$^3, \alpha, \$^3_\mathbb{Y}(v))$ is redundant in $\gamma(u)$ (it is dominated by $(\varnothing, \alpha, v)$). By applying $norm_{\mathbb{Y}, T_\mathcal{N}}$ to $\gamma(u)$, we get the normalized set of transitions $\{(\varnothing, \alpha, v)\}$ (Fig.3(B)). It is worth noting that in open petri nets, two transitions are equivalent iff they are the same transition.

Definition 16. $\mathbf{N_T} : \mathbf{Alg}_{\Gamma(C)} \to \mathbf{Alg}_{\Gamma(C)}$ *maps each* $\mathbb{X} = \langle X, d_{\mathbb{X}}^1, d_{\mathbb{X}}^2, \dots \rangle \in$
$\mathbf{Alg}_{\Gamma(C)}$ *into* $\mathbf{N_T}(\mathbb{X}) = \langle \mathbf{N}_{\mathbb{X}}^{\mathbf{T}}(X), d_{\mathbf{S_T}(\mathbb{X})}^1; norm_{T,\mathbb{X}}, d_{\mathbf{S_T}(\mathbb{X})}^2; norm_{T,\mathbb{X}}, \dots \rangle$. *For all*
$h : \mathbb{X} \to \mathbb{Y}$, $\mathbf{N_T}(h) = \mathbf{H}(h); norm_{T,\mathbb{Y}}$.

Thus, the coalgebra $\langle \mathbb{Y}, \gamma \rangle$ partially represented in Fig.3(A) is not normalized, while $\langle \mathbb{Y}, \gamma; norm_{T,\mathbb{Y}} \rangle$ in Fig.3(B) is. In order to get a normalized coalgebra for our running example, we can normalize its saturated coalgebra $\langle \mathbb{N}, \alpha_{\mathcal{N}} \rangle$ obtaining $\langle \mathbb{N}, \alpha_{\mathcal{N}}; norm_{T_{\mathcal{N}}, \mathbb{N}} \rangle$. For the nets shown in Fig.2, this coincides with the SCTS η.

The most important idea behind normalized coalgebra is in the definition of $\mathbf{N_T}(h)$: we first apply $\mathbf{H}(h)$ and then the normalization $norm_{T,\mathbb{Y}}$. Thus $\mathbf{N_T}$-cohomomorphisms must preserve not all the transitions of the source coalgebras, but only those that are not redundant when mapped into the target.

As an example consider the coalgebras $\langle \mathbb{N}, \alpha_{\mathcal{N}}; norm_{T_{\mathcal{N}}, \mathbb{N}} \rangle$. For the state l, it coincides with the SCTS η (Fig. 2(C)). Consider the coalgebra $\langle \mathbb{Y}, \gamma; norm_{T_{\mathcal{N}}, \mathbb{Y}} \rangle$ (partially represented in Fig.3(B)) and the $\Gamma(\mathbf{Tok})$-homomorphism $h : \mathbb{N} \to \mathbb{Y}$ that maps l, m, n, o into $u, \$_{\mathbb{Y}}^3(v), x, w$ (respectively) and p, q into v. Note that the transition $l \xrightarrow{\$^3, \alpha}_{\eta} m$ is not preserved (i.e., $u \xcancel{\xrightarrow{\$^3, \alpha}}_{\gamma} h(m)$), but h is however a $\mathbf{N_T}$-cohomomorphism, because the transition $(\$^3, \alpha, h(m)) \in \mathbf{H}(h)(\eta(l))$ is removed by $norm_{T_{\mathcal{N}}, \mathbb{Y}}$. Indeed $h(m) = \$_{\mathbb{Y}}^3(v)$ and $(\varnothing, \alpha, v) \vdash_{T_{\mathcal{N}}, \mathbb{Y}} (\$^3, \alpha, \$_{\mathbb{Y}}^3(v))$. Thus, we forget about $l \xrightarrow{\$^3, \alpha}_{\eta} m$ that is, indeed, semantically redundant.

5.3   Isomorphism Theorem

Now we prove that $\mathbf{Coalg}_{\mathbf{N_T}}$ is isomorphic to $\mathbf{Coalg}_{\mathbf{S_T}}$.

Definition 17 (Saturation). *The function* $sat_{T,\mathbb{X}} : \mathbf{H}(X) \to \mathbf{S}^{T\mathbb{X}}(X)$ *maps all* $A \in \mathbf{H}(X)$ *in* $\{(c', o', x') \text{ s.t. } (c, o, x) \in A \text{ and } (c, o, x) \vdash_{T,\mathbb{X}} (c', o', x')\}$.

Saturation is intuitively the opposite of normalization. Indeed saturation adds to a set all the redundant transitions, while normalization junks all of them. Thus, if we take a saturated set of transitions, we first normalize it and then we saturate it, we obtain the original set. Analogously for a normalized set.

However, in order to get such correspondence, we must add a constraint to our theory. Indeed, according to the actual definitions, there could exist a $\mathbf{S_T}$-coalgebra $\langle \mathbb{X}, \gamma \rangle$ and an infinite descending chain like: $\cdots \prec_{T,\mathbb{X}} p \xrightarrow{c_2, o_2}_{\gamma} p_2 \prec_{T,\mathbb{X}} p \xrightarrow{c_1, o_1}_{\gamma} p_1$. In this chain, all the transitions are redundant and thus if we normalize it, we obtain an empty set of transitions.

Definition 18 (Normalizable System). *A context interactive system* $\mathcal{I} = \langle \mathbf{C}, \mathbb{X}, O, tr \rangle$ *is normalizable w.r.t.* T *iff* $\forall \mathbb{X} \in \mathbf{Alg}_{\Gamma(C)}$, $\prec_{T,\mathbb{X}}$ *is well founded.*

Lemma 3. *Let* $\mathcal{I}$ *be a normalizable system w.r.t.* T. *Let* $\mathbb{X}$ *be* $\Gamma(\mathbf{C})$-algebra and $A \in \mathbf{G}(Y)$. *Then* $\forall (d, o, x) \in A$, $\exists (d', o', x') \in norm_{T,\mathbb{Y}}(A)$, *such that* $(d', o', x') \prec_{T,\mathbb{X}} (d, o, x)$.

All the examples of [12] are normalizable [23]. Hereafter, we always refer to normalizable systems.

Proposition 4. *Let $norm_T$ and sat_T be respectively the families of morphisms* $\{norm_{T,\mathbb{X}} : \mathbf{S_T}(\mathbb{X}) \to \mathbf{N_T}(\mathbb{X}), \ \forall \mathbb{X} \in |\mathbf{Alg}_{\Gamma(\mathbf{C})}|\}$ *and* $\{sat_{T,\mathbb{X}} : \mathbf{N_T}(\mathbb{X}) \to \mathbf{S_T}(\mathbb{X}), \ \forall \mathbb{X} \in |\mathbf{Alg}_{\Gamma(\mathbf{C})}|\}$. *Then* $norm_T : \mathbf{S_T} \Rightarrow \mathbf{N_T}$ *and* $sat_T : \mathbf{N_T} \Rightarrow \mathbf{S_T}$ *are natural transformations. More precisely, they are natural isomorphisms, one the inverse of the other.*

It is well-known that any natural transformation between endofunctors induces a functor between the corresponding categories of coalgebras [1]. In our case, $norm_T : \mathbf{S_T} \Rightarrow \mathbf{N_T}$ induces the functor $\mathbf{NORM_T} : \mathbf{Coalgs_{S_T}} \to \mathbf{Coalgn_{N_T}}$ that maps every coalgebra $\langle \mathbb{X}, \alpha \rangle$ in $\langle \mathbb{X}, \alpha; norm_{T,\mathbb{X}} \rangle$ and every cohomomorphism h in itself. Analogously for sat_T.

Theorem 3. $\mathbf{Coalgs_{S_T}}$ *and* $\mathbf{Coalgn_{N_T}}$ *are isomorphic.*

Thus $\mathbf{Coalgn_{N_T}}$ has a final coalgebra $F_{\mathbf{N_T}}$ and the final morphisms from $\langle \mathbb{X}, \alpha_{\mathcal{I}}; norm_{T,\mathbb{X}} \rangle$ (that is $\mathbf{NORM_T}\langle \mathbb{X}, \alpha_{\mathcal{I}} \rangle$)) still characterizes $\sim^S$. This is theoretically very interesting, since the minimal canonical representatives of $\sim^S$ in $\mathbf{Coalgn_{N_T}}$ do not contain any (semantically) redundant transitions and thus they are much smaller than the (possibly infinite) minimal representatives in $\mathbf{Coalgs_{S_T}}$. Pragmatically, it allows for an effective procedure for minimizing that we will discuss in the next section. Notice that minimization is usually unfeasible in $\mathbf{Coalgs_{S_T}}$.

6   From Normalized Coalgebras to Symbolic Minimization

In [15], we have introduced a partition refinement algorithm for symbolic bisimilarity. First, it creates a partition P_0 equating all the states (with the same interface) of a symbolic transition system β and then, iteratively, refines this partition by splitting non equivalent states. The algorithm terminates whenever two subsequent partitions are equivalent. It computes the partition P_{n+1} as follows: p and q are equivalent in P_{n+1} iff whenever $p \xrightarrow{c,o}_\beta p_1$ is *not-redundant in* P_n, then $q \xrightarrow{c,o}_\beta q_1$ is *not-redundant in* P_n and p_1, q_1 are equivalent in P_n (and viceversa). By "not-redundant in P_n", we mean that no transition $p \xrightarrow{c',o'}_\beta p_1'$ exists such that $(c', o', p_1') \vdash_{T,\mathbb{X}} (c, o, p_2')$ and p_2', p_1' are equivalent in P_n.

Fig. 4 shows the partitions computed by the algorithm for the SCTS η of the marked nets a and l. Notice that a and l are equivalent in the partition P_1, because the transition $l \xrightarrow{\$^3, \alpha}_\eta m$ is redundant in P_0. Indeed, $l \xrightarrow{\varnothing, \alpha}_\eta q$, $(\varnothing, \alpha, q) \vdash_{T_\mathcal{N}, \mathbb{N}} (\$^3, \alpha, q\$^3)$ and m is equivalent to $q\$^3$ in P_0.

Hereafter, we show that this algorithm corresponds to computing *the approximations* of a final morphism from $\langle \mathbb{X}, \alpha_{\mathcal{I}}; norm_{T,\mathbb{X}} \rangle$ to $F_{\mathbf{N_T}}$, as induced by the terminal sequence $1 \leftarrow \mathbf{N_T}(1) \leftarrow \mathbf{N_T^2}(1) \leftarrow \ldots$ where 1 is a final $\Gamma(\mathbf{C})$-algebra.

The $n + 1$ approximation $!_{n+1}$ is defined as $\alpha_{\mathcal{I}}; norm_{T,\mathbb{X}}; \mathbf{N_T}(!_n)$. It is worth noting that $\forall n$ the kernel of $!_n$ coincides with the partition P_n computed by the algorithm. Indeed, we can safely replace $\alpha_{\mathcal{I}}; norm_{T,\mathbb{X}}$ with the SCTS β.

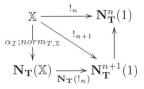

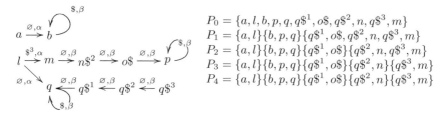

$$P_0 = \{a, l, b, p, q, q\$^1, o\$, q\$^2, n, q\$^3, m\}$$
$$P_1 = \{a, l\}\{b, p, q\}\{q\$^1, o\$, q\$^2, n, q\$^3, m\}$$
$$P_2 = \{a, l\}\{b, p, q\}\{q\$^1, o\$\}\{q\$^2, n, q\$^3, m\}$$
$$P_3 = \{a, l\}\{b, p, q\}\{q\$^1, o\$\}\{q\$^2, n\}\{q\$^3, m\}$$
$$P_4 = \{a, l\}\{b, p, q\}\{q\$^1, o\$\}\{q\$^2, n\}\{q\$^3, m\}$$

Fig. 4. The partitions computed for the marked nets a and l

Then, $!_{n+1} = \beta; \mathbf{N_T}(!_n)$. By the peculiar definition of $\mathbf{N_T}$ on arrows, $\mathbf{N_T}(!_n) = \mathbf{H}(!_n); norm_{T,\mathbf{N_T^n}(1)}$ and the normalization $norm_{T,\mathbf{N_T^n}(1)}$ exactly removes all the transitions that are redundant in P_n.

We end up this section by showing why we have used structured coalgebras instead of bialgebras. Bialgebras abstract away from the algebraic structure, while this is employed by the minimization in $\mathbf{Coalg_{N_T}}$. Indeed, in Fig.4, in order to compute the partitions of l, the algorithm needs the state $q\$^3$ that is not reachable from l. This happens because the algorithm must check if $l \xrightarrow{\$^3,\alpha}_\eta m$ is redundant. In [15], we have shown that we do not need the whole algebra but just a part of it, that can be computed in the initialization of the algorithm.

7   Conclusions and Related Works

The paper introduces two coalgebraic models for context interactive systems [12]. In the first one, the saturated transition system is an ordinary structured coalgebra $\langle \mathbb{X}, \alpha_{\mathcal{I}} \rangle$ and its final morphism induces $\sim^S$. The second model is the normalized coalgebra $\langle \mathbb{X}, \alpha_{\mathcal{I}}; norm_{T,\mathbb{X}} \rangle$ that is obtained by pruning all the redundant transitions from the first one. The equivalence induced by its final morphism is still $\sim^S$, but this characterization is much more convenient. Indeed, in the final normalized coalgebra all the (semantically) redundant transitions are eliminated. Moreover, minimization is feasible with normalized coalgebras and coincides with the symbolic minimization algorithm introduced in [15].

In [24], we have used normalized coalgebras for *Leifer and Milner's reactive systems* [25]. These are an instance of our contexts interactive systems (as shown in [12]) and thus the normalized coalgebras of [24] are a just special case of the one presented in this paper. More precisely, the coalgebras in [24] are defined for a special inference system, while those presented here are parametric w.r.t. it. This provides a flexible theory that gives coalgebraic semantics to many formalisms, such as mobile ambients [9], open [6] and asynchronous [8] π-calculus.

A coalgebraic model for mobile ambients has been proposed in [26]. However it characterizes *action bisimilarity* that is strictly included into *reduction barbed congruence* [10]. In [27], the authors show a context interactive system for mobile ambients, where the symbolic bisimilarity coincides with [10].

For asynchronous and open π-calculus, a minimization algorithm has been proposed in [28] and [29], respectively. In [15], we showed that these are special cases of our algorithm.

References

1. Rutten, J.: Universal coalgebra: a theory of systems. TCS 249(1), 3–80 (2000)
2. Kanellakis, P.C., Smolka, S.A.: Ccs expressions, finite state processes, and three problems of equivalence. Information and Computation 86(1), 43–68 (1990)
3. Milner, R.: Communicating and Mobile Systems: the π-Calculus. Cambridge University Press, Cambridge (1999)
4. Montanari, U., Sassone, V.: Dynamic congruence vs. progressing bisimulation for ccs. Fundam. Inform. 16(1), 171–199 (1992)
5. Milner, R., Parrow, J., Walker, D.: A calculus of mobile processes, i and ii. Information and Computation 100(1), 1–40, 41–77 (1992)
6. Sangiorgi, D.: A theory of bisimulation for the pi-calculus. Acta Inf. 33(1), 69–97 (1996)
7. Amadio, R.M., Castellani, I., Sangiorgi, D.: On bisimulations for the asynchronous pi-calculus. In: Sassone, V., Montanari, U. (eds.) CONCUR 1996. LNCS, vol. 1119, pp. 147–162. Springer, Heidelberg (1996)
8. Honda, K., Tokoro, M.: An object calculus for asynchronous communication. In: America, P. (ed.) ECOOP 1991. LNCS, vol. 512, pp. 133–147. Springer, Heidelberg (1991)
9. Cardelli, L., Gordon, A.D.: Mobile ambients. TCS 240(1), 177–213 (2000)
10. Merro, M., Nardelli, F.Z.: Bisimulation proof methods for mobile ambients. In: Baeten, J.C.M., Lenstra, J.K., Parrow, J., Woeginger, G.J. (eds.) ICALP 2003. LNCS, vol. 2719, pp. 584–598. Springer, Heidelberg (2003)
11. Hennessy, M., Lin, H.: Symbolic bisimulations. TCS 138(2), 353–389 (1995)
12. Bonchi, F., Montanari, U.: Symbolic semantics revisited. In: Amadio, R.M. (ed.) FOSSACS 2008. LNCS, vol. 4962, pp. 395–412. Springer, Heidelberg (2008)
13. Turi, D., Plotkin, G.D.: Towards a mathematical operational semantics. In: Proc. of LICS, pp. 280–291. IEEE, Los Alamitos (1997)
14. Corradini, A., Große-Rhode, M., Heckel, R.: Structured transition systems as lax coalgebras. ENTCS 11 (1998)
15. Bonchi, F., Montanari, U.: Minimization algorithm for symbolic bisimilarity. In: Proc. of ESOP. LNCS, vol. 5502, pp. 267–284. Springer, Heidelberg (2009)
16. Kindler, E.: A compositional partial order semantics for petri net components. In: Azéma, P., Balbo, G. (eds.) ICATPN 1997. LNCS, vol. 1248, pp. 235–252. Springer, Heidelberg (1997)
17. Parrow, J., Victor, B.: The fusion calculus: Expressiveness and symmetry in mobile processes. In: Proc. of LICS, pp. 176–185 (1998)
18. Wischik, L., Gardner, P.: Explicit fusions. TCS 340(3), 606–630 (2005)
19. Buscemi, M., Montanari, U.: Cc-pi: A constraint-based language for specifying service level agreements. In: De Nicola, R. (ed.) ESOP 2007. LNCS, vol. 4421, pp. 18–32. Springer, Heidelberg (2007)
20. Baldan, P., Corradini, A., Ehrig, H., Heckel, R., König, B.: Bisimilarity and behaviour-preserving reconfiguration of open petri nets. In: Mossakowski, T., Montanari, U., Haveraaen, M. (eds.) CALCO 2007. LNCS, vol. 4624, pp. 126–142. Springer, Heidelberg (2007)

21. Corradini, A., Große-Rhode, M., Heckel, R.: A coalgebraic presentation of structured transition systems. TCS 260, 27–55 (2001)
22. Corradini, A., Heckel, R., Montanari, U.: Tile transition systems as structured coalgebras. In: Proc. of FCT, pp. 13–38 (1999)
23. Bonchi, F.: Abstract Semantics by Observable Contexts. PhD thesis (2008)
24. Bonchi, F., Montanari, U.: Coalgebraic models for reactive systems. In: Caires, L., Vasconcelos, V.T. (eds.) CONCUR 2007. LNCS, vol. 4703, pp. 364–379. Springer, Heidelberg (2007)
25. Leifer, J.J., Milner, R.: Deriving bisimulation congruences for reactive systems. In: Palamidessi, C. (ed.) CONCUR 2000. LNCS, vol. 1877, pp. 243–258. Springer, Heidelberg (2000)
26. Hausmann, D., Mossakowski, T., Schröder, L.: A coalgebraic approach to the semantics of the ambient calculus. TCS 366(1-2), 121–143 (2006)
27. Bonchi, F., Gadducci, F., Monreale, G.V.: Reactive systems, barbed semantics and the mobile ambients. In: de Alfaro, L. (ed.) FOSSACS 2009. LNCS, vol. 5504, pp. 272–287. Springer, Heidelberg (2009)
28. Montanari, U., Pistore, M.: Finite state verification for the asynchronous pi-calculus. In: Cleaveland, W.R. (ed.) TACAS 1999. LNCS, vol. 1579, pp. 255–269. Springer, Heidelberg (1999)
29. Pistore, M., Sangiorgi, D.: A partition refinement algorithm for the pi-calculus. Information and Computation 164(2), 264–321 (2001)

Relating Coalgebraic Notions of Bisimulation
with Applications to Name-Passing Process Calculi
(Extended Abstract)

Sam Staton

Computer Laboratory, University of Cambridge

Abstract. A labelled transition system can be understood as a coalgebra for a particular endofunctor on the category of sets. Generalizing, we are led to consider coalgebras for arbitrary endofunctors on arbitrary categories.

Bisimulation is a crucial notion in the theory of labelled transition systems. We identify four definitions of bisimulation on general coalgebras. The definitions all specialize to the same notion for the special case of labelled transition systems. We investigate general conditions under which the four notions coincide.

As an extended example, we consider the semantics of name-passing process calculi (such as the pi-calculus), and present a new coalgebraic model for name-passing calculi.

1 Introduction

Notions of bisimulation play a central role in the theory of transition systems. As different kinds of system are encountered, different notions of bisimulation arise, but the same questions are posed: Is there a fixed-point characterization for the maximal bisimulation, bisimilarity? Is there a minimal system, where bisimilar states are equated? And is there a procedure for constructing a minimal system, or for verifying bisimilarity?

The theory of coalgebras provides a setting in which different notions of transition system can be understood at a general level. In this paper we investigate notions of bisimulation at this general level, and determine how and when these questions can be answered.

To explain the generalization from transition systems to coalgebras, we begin with the traditional model of a labelled transition system,

$$(X, \ (\to_X) \subseteq X \times L \times X)$$

(for some set L of labels). A labelled transition system can be considered coalgebraically as a set X of states equipped with a 'next-state' function $X \to \mathcal{P}(L \times X)$. (Here, $\mathcal{P}$ is the powerset operator.) Generalizing, we are led to consider an arbitrary category $\mathcal{C}$ and an endofunctor B on it; then a coalgebra is an object X in $\mathcal{C}$ of 'states', and a 'next-state' morphism $X \to \mathrm{B}(X)$.

A. Kurz, M. Lenisa, and A. Tarlecki (Eds.): CALCO 2009, LNCS 5728, pp. 191–205, 2009.

Coalgebras in different categories. The generalization to different endofunctors on different categories has proved appropriate in various settings: concepts from modal logic have been studied in terms of coalgebras over Stone spaces [2,6,29]; basic process calculi with recursion can be described using coalgebras over categories of domains [28,35]; stochastic transition systems have been studied in terms of coalgebras over metric and measurable spaces [8,12,42,43].

In this paper we revisit work [14,16,20,40] on models of name-passing calculi, such as the π-calculus, where it is appropriate to work in a sheaf topos. Endofunctors that describe transition-system-like behaviour often decompose as B = $P \circ$ B′, where B′ is an endofunctor of a particularly simple form, and P is a powerset functor for a class of small maps, in the sense of algebraic set theory. A contribution of the present work is the introduction of a powerset that is appropriate for name-passing. It arises by combining the theory of semilattices with a theory of name-equality testing.

Notions of bisimulation. Once coalgebras are understood as generalized transition systems, we can consider bisimulation relations for these systems. Recall that, for labelled transition systems $(X, \to_X)$ and $(Y, \to_Y)$, a relation $R \subseteq X \times Y$ is a (strong) *bisimulation* if, whenever $x\,R\,y$, then for all $l \in L$:

- For $x \in X$, if $x \xrightarrow{l}_X x'$ then there is $y' \in Y$ such that $y \xrightarrow{l}_Y y'$ and $x'\,R\,y'$;
- For $y \in Y$, if $y \xrightarrow{l}_Y y'$ then there is $x' \in X$ such that $x \xrightarrow{l}_X x'$ and $x'\,R\,y'$.

In this article, we identify four notions of bisimulation that have been proposed in the context of coalgebras for endofunctors on arbitrary categories. Aczel and Mendler [3] introduced two definitions; we also consider a definition due to Hermida and Jacobs [22]; and lastly a definition that gives an immediate connection with 'final coalgebra' semantics.

The four notions coincide for the particular case of labelled transition systems. We investigate conditions under which the notions are related in the more general setting of coalgebras.

Relationship with the terminal sequence. Various authors have constructed terminal coalgebras as a limit of a transfinite sequence, beginning

$$1 \xleftarrow{!} B(1) \xleftarrow{B(!)} B(B(1)) \xleftarrow{B(B(!))} B(B(B(1))) \leftarrow \cdots \leftarrow \cdots \quad .$$

On the other hand, bisimulations can often be characterized as post-fixed points of a monotone operator Φ on a lattice of relations, so that a maximal bisimulation ('bisimilarity') arises as a limit of a transfinite sequence, beginning

$$X \times Y \supseteq \Phi(X \times Y) \supseteq \Phi(\Phi(X \times Y)) \supseteq \Phi(\Phi(\Phi(X \times Y))) \supseteq \cdots \supseteq \cdots \quad .$$

These sequences suggest algorithms for minimizing and computing bisimilarity for arbitrary coalgebras (see e.g. [13]). We investigate conditions under which the steps of the terminal coalgebra sequence are precisely related with the steps of this relation refinement sequence.

Other approaches not considered. In this article we are concerned with (internal) relations between the carrier objects of two fixed coalgebras: a relation is an object of the base category. Some authors (e.g. [12]) instead work with an equivalence relation on the class of all coalgebras, setting two coalgebras as 'bisimilar' if there is a span of surjective homomorphisms between them. Others work with relations as bimodules [36,8]. We will not discuss these approaches here.

Outline. This paper is structured as follows. In Section 2 we recall some examples of coalgebras for endofunctors. We recall the four notions of bisimulation in Section 3. In Section 4 we investigate how the different notions of bisimulation are related. In Section 5 we investigate the connection between the terminal sequence and the relation refinement sequence. Finally, in Section 6, we provide a novel analysis of models of name-passing calculi.

Many of the results in Sections 4 and 5 are well-known where $C = \mathbf{Set}$. In other cases, some results are probably folklore; I have tried to ascribe credit where it is due.

2   Coalgebras: Definition and Examples

Recall the definition of a coalgebra for an endofunctor:

Definition 1. *Consider an endofunctor* B *on a category* C. *A* B*-coalgebra is given by an object* X *of* C *together with morphism* $X \to B(X)$ *in* C. *A homomorphism of* B*-coalgebras, from* (X, h) *to* (Y, k), *is a morphism* $f : X \to Y$ *that respects the coalgebra structure, i.e. such that* $Bf \circ h = k \circ f$.

We collect some examples of structures that arise as coalgebras for endofunctors. For further motivation, see [4,23,37].

Streams and coinductive datatypes. Let A be a set, and consider the endofunctor $(A \times (-) + 1)$ on the category **Set** of sets. Coalgebras for this endofunctor are a form of stream, of elements of A. Other coinductive datatypes arise from other 'polynomial' endofunctors, built out of constants, sums and products (see e.g. [37, Sec. 10]).

Labelled transition systems. In the introduction, we discussed how labelled transition systems correspond to coalgebras for the endofunctor $\mathcal{P}(L \times (-))$. Here, $\mathcal{P}$ is the powerset functor, that acts by direct image. For finite non-determinism, and image-finite transition systems, one can instead consider the endofunctor $\mathcal{P}_{\mathrm{f}}(L \times (-))$ where $\mathcal{P}_{\mathrm{f}}$ is the finite powerset functor.

Powersets and small maps. A general treatment of powersets is suggested by algebraic set theory [26]. A model of algebraic set theory is a category C together with a class of 'small' maps $\mathcal{S}$ in C, all subject to certain conditions — see Figure 1. In such a situation, an $\mathcal{S}$-*relation* is a jointly-monic span $(X \leftarrow R \to Y)$ of which the projection $R \to X$ is in $\mathcal{S}$. Axiom (P1) entails that there is an endofunctor $\mathcal{P}_{\mathcal{S}}$ on C that classifies $\mathcal{S}$-relations.

Axioms for a class $\mathcal{S}$ of small maps in a regular category:

(A1) $\mathcal{S}$ is closed under composition, and all identity morphisms are in $\mathcal{S}$.

(A2) $\mathcal{S}$ is stable under pullback: in dgm. 1, if $f \in \mathcal{S}$, then $f' \in \mathcal{S}$.

$$\begin{array}{ccc} A' & \longrightarrow & A \\ f' \downarrow & \lrcorner & \downarrow f \\ B' & \underset{g}{\longrightarrow} & B \end{array} \qquad (1)$$

(A3) In dgm. 1, if $f' \in \mathcal{S}$ and g is a cover (aka strong epi), then $f \in \mathcal{S}$.

(A6) In the following triangle, if $f \in \mathcal{S}$ and e is a cover, then $g \in \mathcal{S}$.

$$\begin{array}{ccc} A & \overset{e}{\dashrightarrow} & A' \\ & f \searrow & \downarrow g \\ & & B \end{array}$$

(P1) For every object A of $\mathcal{C}$ there is an $\mathcal{S}$-relation $(\mathcal{P}_\mathcal{S}(A) \leftarrow \ni_B \rightarrow A)$ such that for every $\mathcal{S}$-relation $(I \leftarrow R \rightarrow A)$ there is a unique morphism $I \rightarrow P(A)$ making the following diagram a pullback.

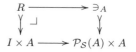
$$\begin{array}{ccc} R & \longrightarrow & \ni_A \\ \downarrow & \lrcorner & \downarrow \\ I \times A & \longrightarrow & \mathcal{P}_\mathcal{S}(A) \times A \end{array}$$

Additional axioms for an extensive regular category:

(A4) Maps $0 \rightarrow 1$ and $1+1 \rightarrow 1$ are in $\mathcal{S}$.

(A5) If $A \rightarrow A'$ and $B \rightarrow B'$ are in $\mathcal{S}$, then so is $(A + B) \rightarrow (A' + B')$.

Additional axioms (optional):

(A7) (*Collection*, assuming (P1)) The endofunctor $\mathcal{P}_\mathcal{S}$ preserves covers.

(M) All monos in $\mathcal{C}$ are in $\mathcal{S}$.

Fig. 1. Axioms for small maps (see [26]; we do not need representability here.) Recall that a category is regular if it has finite limits and stable cover-image factorizations (e.g. [7, Vol. 2, Ch. 2], [25, A1.3]), and that it is extensive if it has finite sums that are stable under pullback (e.g. [10]). An extensive regular category is sometimes called positive (e.g. [25, A1.4]).

For instance, in the category of sets, we can say that a function $f : X \rightarrow Y$ is small if for every $y \in Y$ the set $\{x \in X \mid f(x) = y\}$ is finite. An $\mathcal{S}$-relation is precisely an image-finite one, and the $\mathcal{S}$-powerset is the finite powerset, $\mathcal{P}_f$.

For more complex examples, see Section 6 and [5].

Continuous-state systems. For systems with continuous state spaces it is appropriate to work in a category of topological spaces. For a first example, consider a Stone space X, and let $K(X)$ be the space of compact subsets of X, with the finite (aka Vietoris) topology. The endofunctor K on the category of Stone spaces has attracted interest because of a duality with modal logics (e.g.[2,29]).

Continuous phenomena also arise in other application areas. For recursively defined systems, it is reasonable to investigate coalgebras for powerdomain constructions on a category of domains (see e.g. [28,35,1]). For continuous stochastic systems, researchers have investigated coalgebras for probability distribution functors on categories of metric or measurable spaces (see e.g. [8,12,42,43]).

3   Four Definitions of Bisimulation

We now recall four coalgebraic notions of bisimulation from the literature.

Context. Throughout this section, we consider an endofunctor B on a category $\mathcal{C}$. We assume that $\mathcal{C}$ has finite limits and images (i.e. (strong-epi/mono) factorizations; see e.g. [7, Vol. 2, Ch. 2], [25, A1.3]). We fix two B-coalgebras, $h : X \rightarrow B(X)$ and $k : Y \rightarrow B(Y)$. We write $\mathrm{Rel}(X,Y)$ for the preorder of relations, viz. jointly-monic spans $(X \leftarrow R \rightarrow Y)$.

3.1   The Lifting-Span Bisimulation of Aczel and Mendler

Definition 2 (following [3]). *A relation R in $\mathrm{Rel}(X,Y)$ is an* AM-bisimulation *if there is a B-coalgebra structure $R \rightarrow B(R)$ making the following diagram commute.*

$$
\begin{array}{ccccc}
X & \longleftarrow & R & \longrightarrow & Y \\
{\scriptstyle h}\downarrow & & \downarrow & & \downarrow{\scriptstyle k} \\
B(X) & \longleftarrow & B(R) & \longrightarrow & B(Y)
\end{array}
$$

3.2   The Relation-Lifting Bisimulation of Hermida and Jacobs

For any relation R in $\mathrm{Rel}(X,Y)$, the 'relation lifting' $\bar{B}(R)$ in $\mathrm{Rel}(B(X),B(Y))$ is the image of the composite morphism $B(R) \rightarrow B(X \times Y) \rightarrow B(X) \times B(Y)$.

Definition 3 (following [22]). *A relation R in $\mathrm{Rel}(X,Y)$ is an* HJ-bisimulation *if there is a morphism $R \rightarrow \bar{B}(R)$ making the left-hand diagram below commute.*

$$
\begin{array}{ccccc}
X & \longleftarrow & R & \longrightarrow & Y \\
{\scriptstyle h}\downarrow & & \downarrow & & \downarrow{\scriptstyle k} \\
B(X) & \longleftarrow & \bar{B}(R) & \longrightarrow & B(Y)
\end{array}
\qquad\qquad
\begin{array}{ccc}
\Phi^{\mathrm{HJ}}(R) & \longrightarrow & \bar{B}(R) \\
\downarrow & \lrcorner & \downarrow \\
X \times Y & \underset{h \times k}{\longrightarrow} & B(X) \times B(Y)
\end{array}
$$

Equivalently: let the right-hand diagram be a pullback; R is an HJ-bisimulation iff $R \leq \Phi^{\mathrm{HJ}}(R)$ in $\mathrm{Rel}(X,Y)$.

Proposition 4. *The operator Φ^{HJ} on $\mathrm{Rel}(X,Y)$ is monotone.*

For an example, return to the situation where $\mathcal{C} = \mathbf{Set}$ and $B = \mathcal{P}(L \times -)$. For any relation R in $\mathrm{Rel}(X,Y)$, the refined relation $\Phi^{\mathrm{HJ}}(R)$ is the set of all pairs $(x,y) \in X \times Y$ for which

(i) $\forall (l, x') \in h(x).\ \exists y' \in Y.\ (l, y') \in k(y)$ and $(x', y') \in R$;

(ii) $\forall (l, y') \in k(y).\ \exists x' \in X.\ (l, x') \in h(x)$ and $(x', y') \in R$.

Thus Φ^{HJ} is the construction $\mathcal{F}$ considered by Milner in [33, Sec. 4].

3.3   The Congruences of Aczel and Mendler

Definition 5 (following [3]). *A relation R in $\mathrm{Rel}(X,Y)$ is an* AM-precongruence *if for every cospan $(X \xrightarrow{i} Z \xleftarrow{j} Y)$,*

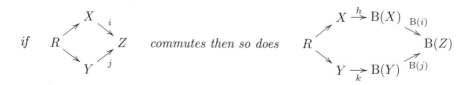

if $R \overset{X}{\underset{Y}{\vphantom{X}}} \overset{i}{\underset{j}{\vphantom{i}}} Z$ *commutes then so does* $R \overset{X \overset{h}{\to} B(X)}{\underset{Y \underset{k}{\to} B(Y)}{\vphantom{X}}} \overset{B(i)}{\underset{B(j)}{\vphantom{B(i)}}} B(Z)$

When $\mathcal{C}$ has pushouts, then we let $\Phi^{\mathrm{AM}}(R)$ be the pullback of the cospan $(X \overset{h}{\to} \mathrm{B}X \overset{i}{\to} \mathrm{B}(X +_R Y) \overset{j}{\leftarrow} \mathrm{B}Y \overset{k}{\leftarrow} Y)$. By definition, a relation R is an AM-precongruence iff $R \le \Phi^{\mathrm{AM}}(R)$.

This definition differs from that of [3] in that we consider relations between different coalgebras. If $(X, h) = (Y, k)$, then an equivalence relation is an AM-precongruence exactly when it is a congruence in the sense of [3].

Proposition 6. *The operator Φ^{AM} on $\mathrm{Rel}(X, Y)$ is monotone.*

(N.B. Φ^{AM} is different from Φ^{HJ}, even when B is the identity functor on **Set**.)

3.4   Terminal Coalgebras and Kernel-Bisimulations

Suppose for a moment that there is a terminal B-coalgebra, $(Z, z\colon Z \to \mathrm{B}(Z))$. This induces a relation in $\mathrm{Rel}(X, Y)$ as the pullback of the unique terminal morphisms $(X \to Z \leftarrow Y)$. Many authors have argued that this relation is the right notion of bisimilarity. We can formulate a related notion of bisimulation without assuming terminal coalgebras, as follows.

Definition 7. *A relation R is a* kernel-bisimulation *if there is a cospan of B-coalgebras, $(X, h) \to (Z, z) \leftarrow (Y, k)$, and R is the pullback of $(X \to Z \leftarrow Y)$.*

(The term 'cocongruence' is sometimes used to refer directly to the cospan involved, e.g. [30].)

4   Relating the Notions of Bisimulation

In this section we establish when the notions of bisimulation, introduced in the previous section, are related. As is well known, the four definitions coincide for the case of labelled transition systems.

4.1   Conditions on Endofunctors

We begin by recalling five conditions that might be assumed of our endofunctor B.

1. We say that B *preserves relations*, if a jointly-monic span is mapped to a jointly-monic span.

To introduce the remaining conditions, we consider a cospan $(A_1 \to Z \leftarrow A_2)$ in $\mathcal{C}$, and in particular the mediating morphism m from the image of the pullback to the pullback of the image:

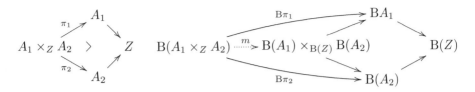

Here are four conditions on B, listed in order of decreasing strength:

2. B *preserves pullbacks* if m is always an isomorphism;
3. B *preserves weak pullbacks*, if m is always split epi (see e.g. [21]);
4. B *covers pullbacks* if m is always a cover (=strong epi);
5. B *preserves pullbacks along monos* if m is an iso when $A_1 \to Z$ is monic.

Tying up with item (1), note that B preserves pullbacks if and only if it preserves relations and covers pullbacks (e.g. [9, Sec. 4.3]).

4.2   Relevance of the Conditions on Endofunctors

We now discuss which of the conditions (1–5) are relevant for the endofunctors considered in Section 2. First, note that polynomial endofunctors on extensive categories preserve all pullbacks.

Regarding powerset functors, we have the following general result.

Proposition 8. *Let C be a regular category with a class S of small maps. Let $\mathcal{P}_S$ be an S-powerset.*

1. *The functor $\mathcal{P}_S$ preserves pullbacks along monomorphisms.*
2. *If S contains all monomorphisms (M), then $\mathcal{P}_S$ preserves weak pullbacks.*
3. *Let C also be extensive, and let S satisfy Axioms (A1–7), but not necessarily (M). The functor $\mathcal{P}_S$ preserves covers and covers pullbacks.*

Moving to the setting of continuous state spaces, the compact-subspace endofunctor K does not preserve weak pullbacks, but it does cover pullbacks on the category of Stone spaces [6]; this seems to be an instance of Prop. 8(3). More sophisticated continuous settings are problematic. Plotkin [35] discusses problems with coalgebraic bisimulation in categories of domains. The convex powerdomain does not even preserve monomorphisms. Counterexamples to the weak-pullback-preservation of probability distributions on measurable spaces are discussed in [42].

4.3   Relating Notions of Bisimulation

Theorem 9. *Let B be an endofunctor on a category C with finite limits and images.*

1. *Every AM-bisimulation is an HJ-bisimulation.*
2. *Every HJ-bisimulation is an AM-precongruence.*
3. *Every AM-precongruence is contained in a kernel bisimulation that is an AM-precongruence, provided C has pushouts.*

4. *Every kernel bisimulation is an AM-bisimulation, provided* B *preserves weak pullbacks.*
5. *Every kernel bisimulation is an HJ-bisimulation, provided* B *covers pullbacks.*
6. *Every kernel bisimulation is an AM-precongruence, provided* B *preserves pullbacks along monos.*
7. *Every HJ-bisimulation is an AM-bisimulation, provided either* B *preserves relations, or every epi in* C *is split.*

In summary:

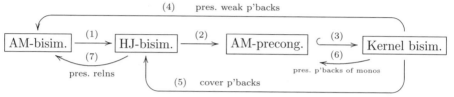

The second part of Theorem 9(7) accounts for the following well-known fact: when $C = \mathbf{Set}$, assuming the axiom of choice, HJ-bisimulation is the same thing as AM-bisimulation. We can achieve close connections in more constructive settings:

Theorem 10. *Let* C *be a regular category with a class* S *of small maps, with an* S-*powerset,* $\mathcal{P}_S$. *Suppose that* $\mathrm{B}(-) \cong \mathcal{P}_S(\mathrm{B}'(-))$, *for some endofunctor* B' *that preserves relations. If all monomorphisms are in* S *(axiom (M)), then every HJ-bisimulation is an AM-bisimulation.*

5   Bisimilarity through Transfinite Constructions

In this section we consider a procedure for constructing the maximal bisimulation. We relate it with the terminal sequence, which is used for finding final coalgebras.

Context. In this section we assume that the ambient category C is complete with images and pushouts. We fix an endofunctor B on C, and fix two B-coalgebras, $h : X \to \mathrm{B}(X)$ and $k : Y \to \mathrm{B}(Y)$.

Relation refinement sequences. HJ-bisimulations and AM-precongruences can be understood as postfixed points of operators Φ^{HJ} and Φ^{AM} respectively. When C is well-powered, greatest fixed points can be obtained as limits. For the case of HJ-bisimulations, we define an ordinal-indexed cochain $(r_{\beta,\alpha} : R_\beta^{\mathrm{HJ}} \rightarrowtail R_\alpha^{\mathrm{HJ}})_{\alpha \leq \beta}$ in $\mathrm{Rel}(X,Y)$, in the usual way (see e.g. [11]):

- *Limiting case:* If λ is limiting, $R_\lambda^{\mathrm{HJ}} = \bigcap_{\alpha < \lambda} R_\alpha^{\mathrm{HJ}}$; e.g. $R_0^{\mathrm{HJ}} = X \times Y$.
- *Inductive case:* $R_{\alpha+1}^{\mathrm{HJ}} = \Phi^{\mathrm{HJ}}(R_\alpha^{\mathrm{HJ}})$.

If this sequence is eventually stationary then it achieves the maximal post-fixed point of Φ^{HJ}, the greatest HJ-bisimulation.

Similarly, we consider a cochain $(R_\beta^{\mathrm{AM}} \rightarrowtail R_\alpha^{\mathrm{AM}})_{\alpha \leq \beta}$ for the operator Φ^{AM}.

For the case of the endofunctor $\mathcal{P}(L \times (-))$ on **Set**, the relation refinement sequences $(R_\alpha^{\mathrm{HJ}})_\alpha$ and $(R_\alpha^{\mathrm{AM}})_\alpha$ coincide, giving a transfinite extension of Milner's sequence $\sim_0 \supseteq \sim_1 \supseteq \cdots \supseteq \sim$ (c.f. [32, Sec 5.7]).

The terminal sequence. The terminal sequence is an ordinal-indexed cochain $(z_{\beta,\alpha} : Z_\beta \to Z_\alpha)_{\alpha \leq \beta}$ that can be used to construct a final coalgebra for an endofunctor (see e.g. [44]). The cochain commutes and satisfies the following conditions:

- *Limiting case:* If λ is limiting, $Z_\lambda = \lim\{z_{\beta,\alpha} : Z_\beta \to Z_\alpha \mid \alpha \leq \beta < \lambda\}$, and the cone $\{z_{\lambda,\alpha} : Z_\lambda \to Z_\alpha \mid \alpha < \lambda\}$ is the limiting one; e.g. $Z_0 = 1$.
- *Inductive case:* $Z_{\alpha+1} = \mathrm{B}(Z_\alpha)$; and $z_{\beta+1,\alpha+1} = \mathrm{B}(z_{\beta,\alpha}) : Z_{\beta+1} \to Z_{\alpha+1}$.

5.1   Relating the Relation and Terminal Sequences

The coalgebra (X, h), (Y, k) determine two cones over the terminal sequence: $(x_\alpha : X \to Z_\alpha)_\alpha$ and $(y_\alpha : X \to Z_\alpha)_\alpha$. The cone $(x_\alpha)_\alpha$ is given as follows.

- *Limiting case:* If λ is a limit ordinal then the morphisms $x_\alpha : X \to Z_\alpha$ for $\alpha < \lambda$ form a cocone over the cochain $(z_{\beta,\alpha} : Z_\beta \to Z_\alpha)_{\alpha \leq \beta < \lambda}$, with apex X. We let $x_\lambda : X \to Z_\lambda$ be the unique mediating morphism. For instance, when $\lambda = 0$, then $x_\lambda : X \to Z_\lambda$ is the terminal map $X \to 1$.

- *Inductive case:* Let $x_{\alpha+1}$ be the composite $X \xrightarrow{h} \mathrm{B}(X) \xrightarrow{\mathrm{B}x_\alpha} \mathrm{B}(Z_\alpha) = Z_{\alpha+1}$.

The other cone, $(y_\alpha : X \to Z_\alpha)_\alpha$, is defined similarly.

Proposition 11. *Consider an ordinal α, and consider the pullback $X \times_{Z_\alpha} Y$ of the cospan $(X \xrightarrow{x_\alpha} Z_\alpha \xleftarrow{y_\alpha} Y)$.*

1. *If B preserves pullbacks along monos then $X \times_{Z_\alpha} Y = R_\alpha^{\mathrm{AM}}$.*
2. *If B covers pullbacks, then $X \times_{Z_\alpha} Y = R_\alpha^{\mathrm{HJ}}$ $(= R_\alpha^{\mathrm{AM}})$.*

The relation refinement sequence may converge before the terminal sequence:

Corollary 12. *Suppose that B covers pullbacks. If, for some ordinal α, the morphism $z_{\alpha+1,\alpha} : Z_{\alpha+1} \to Z_\alpha$ is monic, then the relation refinement sequence converges at α.*

If $\mathcal{C}$ is **Set** and B preserves filtered colimits, then the terminal sequence does not converge until $(\omega + \omega)$, but it becomes monic at ω [44].

6   Models of Name-Passing Process Calculi

As a case study, we now investigate models for name-passing process calculi. A fragment of the π-calculus is given in Figure 2. To simplify the presentation, we omit restriction (name-generation) for now; we return to this issue in Section 6.4.

It is unreasonable to model name-passing in the category of sets. One alternative is the category of nominal substitutions [15,17], as we now investigate. We describe an endofunctor, for which coalgebras capture the model-theoretic properties of the π-calculus, and for which bisimulation is open bisimulation [38].

Throughout this section we fix an infinite set $\mathbb{A}$ of channel names.

Terms (T_π): $p ::= a(b).p \mid \bar{a}b.p \mid p|p \mid \mathbf{0}$
Labels (L_π): $l ::= ab \mid \bar{a}b \mid \tau$

Structural operational semantics:

$$\overline{a(b).p \xrightarrow{ac} \{{}^c\!/_b\}p} \qquad \overline{\bar{a}b.p \xrightarrow{\bar{a}b} p} \qquad \frac{p \xrightarrow{l} p'}{p|q \xrightarrow{l} p'|q} \qquad \frac{p \xrightarrow{\bar{a}b} p' \quad q \xrightarrow{ab} q'}{p|q \xrightarrow{\tau} p'|q'}$$

Fig. 2. A fragment of the π-calculus [34]. Symmetric versions of rules are elided.

6.1   Labelled Transition Systems in Nominal Substitutions

Definition 13. *A nominal substitution is a set X together with a function* $\mathsf{sub} : \mathbb{A} \times \mathbb{A} \times X \to X$, *written* $\{{}^b\!/_a\}x = \mathsf{sub}(b, a, x)$, *such that*

1. $\{{}^a\!/_a\}x = x$
2. $\{{}^c\!/_b\}\{{}^b\!/_a\}x = \{{}^c\!/_b\}\{{}^c\!/_a\}x$
3. *If $c \neq b \neq a \neq d$ then $\{{}^d\!/_b\}\{{}^c\!/_a\}x = \{{}^c\!/_a\}\{{}^d\!/_b\}x$.*
4. *If $a \neq b$ then $\{{}^c\!/_a\}\{{}^b\!/_a\}x = \{{}^b\!/_a\}x$.*
5. *For each $x \in X$, the set* $\mathrm{supp}(x) = \{a \in \mathbb{A} \mid \exists b \in \mathbb{A}. \{{}^b\!/_a\}x \neq x\}$ *is finite.*[1]

A homomorphism of nominal substitutions, $f : (X, \mathsf{sub}) \to (Y, \mathsf{sub})$, is a function that preserves the structure: $f\{{}^b\!/_a\}x = \{{}^b\!/_a\}(f(x))$. We write **NomSub** *for the resulting category.*

The set $\mathbb{A}$ is itself a nominal substitution. Other important examples are set T_π of π-calculus terms up-to α-equivalence, and the set L_π of labels, both with the evident (capture-avoiding) substitution structure.

   The transition relation for the π-calculus, as given in Figure 2, is a subset $(\longrightarrow) \subseteq T_\pi \times L_\pi \times T_\pi$. Moreover, the relation is a relation in **NomSub**; that is to say, $(\longrightarrow)$ has a nominal substitution structure, and $(\longrightarrow) \rightarrowtail T_\pi \times L_\pi \times T_\pi$ preserves the structure. That is, if $t \xrightarrow{l} u$, then $\{{}^b\!/_a\}t \xrightarrow{\{{}^b\!/_a\}l} \{{}^b\!/_a\}u$. (see e.g. [41]).

6.2   Coalgebras and Notions of Non-determinism

We now investigate how the transition relation for the π-calculus can be viewed in a coalgebraic way. As a starting point, we note that **NomSub** is a topos, and hence to give an substitution-closed transition relation is to give a coalgebra for $\mathcal{P}(L_\pi \times (-))$, i.e. a homomorphism $X \to \mathcal{P}(L_\pi \times X)$ (writing $\mathcal{P}$ for the power object of **NomSub**). This approach is not without drawbacks, however: (i) an explicit description of $\mathcal{P}$ is cumbersome; (ii) the functor $\mathcal{P}$ is not finitary, and there is no final coalgebra for cardinality reasons. To remedy this, we now consider an explicit description of a finitary subfunctor of $\mathcal{P}$.

A theory of equality testing and non-determinism. In Figure 3, we present a theory of 'conditional-semilattices'. An conditional-semilattice has enough structure to describe both nondeterminism and name-equality testing. Conditional-semilattices form a category $\mathcal{CSL}$, with the evident morphisms, and the

---

[1] I am grateful to Andy Pitts for pointing out this simplified condition.

forgetful functor $\mathcal{CSL} \to$ **NomSub** is monadic. The free conditional-semilattice, $\mathcal{P}_{\text{csl}}(X)$, on a nominal substitution X, can be constructed by considering the set of all well-formed terms built from $+$, $\mathbf{0}$, if, and elements of X, quotiented by the equations in Figure 3. Alternatively, Figure 3 can be understood as a presentation of an enriched algebraic theory in **NomSub**, following [27], and a finitary monad $\mathcal{P}_{\text{csl}}$ arises from the general results there.

Proposition 14. *The monad* $\mathcal{P}_{\text{csl}}$ *classifies homomorphisms* $f\colon Y \to X$ *in* **NomSub** *for which* $\{y \in Y \mid f(y) = x\}$ *is finite, for each* $x \in X$. *That is: the following data are equivalent, for all nominal substitutions* X *and* Y.

1. *A homomorphism* $X \to \mathcal{P}_{\text{csl}}(Y)$.
2. *A subset* $R \subseteq X \times Y$ *such that*
   (a) *If* $x\,R\,y$ *then* $\{^b/_a\}x\,R\,\{^b/_a\}y$ *("R is substitution closed"); and*
   (b) *For all* $x \in X$, *the image* $\{y \in Y \mid x\,R\,y\}$ *is finite.*

We write $\mathcal{S}_{\text{csl}}$ for the class of morphisms that $\mathcal{P}_{\text{csl}}$ classifies. This class satisfies all the axioms of Figure 1.

For the curious topos theorist, we record that $\mathcal{P}_{\text{csl}}$ is the free semilattice generated by the partial map classifier.

Coalgebraic models of name-passing calculi. Coalgebras for the endofunctor $\mathcal{P}_{\text{csl}}(L_\pi \times (-))$ are the same thing as substitution closed, image-finite relations. Unfortunately, the transition system for the π-calculus is not image-finite — the process $a(b).\mathbf{0}$ can input an infinite number of different names. The π-calculus behaviour is essentially finite, though, in a sense that we now make precise.

For each nominal substitution X, we define a set $X^{\mathbb{A}} = (\mathbb{A} \times X)/_{=_\alpha}$, where $=_\alpha$ is the equivalence relation generated by setting $(a, x) =_\alpha (b, \{^b/_a\}x)$, for any name b not in $\text{supp}(x)$. This construction extends to an endofunctor on nominal substitutions. We thus have an appropriate endofunctor for name-passing behaviour: let $B_\pi = \mathcal{P}_{\text{csl}}(\mathbb{A} \times (-)^{\mathbb{A}} + \mathbb{A} \times \mathbb{A} \times (-) + (-))$.

An *conditional-semilattice* is a nominal substitution X together with three homomorphisms

$$+\colon X \times X \to X \qquad \mathbf{0}\colon 1 \to X \qquad \text{if}\colon \mathbb{A} \times \mathbb{A} \times X \to X$$

that satisfy the following equations. (We abbreviate $\text{if}(a, b, x)$ by $[a = b]\,x$.)

$$x + x = x$$
$$x + y = y + x$$
$$x + (y + z) = (x + y) + z$$
$$x + \mathbf{0} = x$$

$$[a = b]\,\mathbf{0} = \mathbf{0}$$
$$[a = b]\,(x + y) = ([a = b]\,x) + ([a = b]\,y)$$
$$[a = b]\,x + x = x$$

$$[a = b]\,[a = b]\,x = [a = b]\,x$$
$$[a = b]\,[c = d]\,x = [c = d]\,[a = b]\,x$$
$$[a = b]\,[b = c]\,x = [a = b]\,[a = c]\,x$$
$$[a = a]\,x = x$$
$$[a = b]\,x = [b = a]\,x$$

$$[a = b]\,x = [a = b]\,(\{^a/_b\}x)$$

Fig. 3. Theory of conditional-semilattices (c.f. [38, Sec. 4.1])

Theorem 15. *The following data are equivalent.*

1. *A B_π-coalgebra.*
2. *An substitution-closed relation $(\longrightarrow) \subseteq X \times L_\pi \times X$ for which*
    (a) *for every $x \in X$, the sets $\{(a,b,x') \mid x \xrightarrow{\bar{a}b} x'\}$, $\{x' \mid x \xrightarrow{\tau} x'\}$, are both finite;*
    (b) *for all $x \in X$ there is a name c and a finite subset $\{(a_1,y_1),\ldots,(a_n,y_n)\}$ of $\mathbb{A} \times X$ such that if $x \xrightarrow{ab} x'$ then, for some $i \le n$, we have $a = a_i$ and $x' = \{^b/_c\}y_i$;*
    (c) *if $x \xrightarrow{ab} x'$ then there is a name $c \notin \mathrm{supp}(x)$ and $x'' \in X$ such that $x \xrightarrow{ac} x''$ and $\{^b/_c\}x'' = x'$.*

The transition relation for the π-calculus, given in Figure 2, satisfies the three conditions of Theorem 15(2), and indeed these conditions essentially exhaust the model-theoretic properties of this fragment of the π-calculus (see [39, Sec. 1.4.2]).

For instance, in the 'next state' coalgebra $T_\pi \to B_\pi(T_\pi)$ for the π-calculus, the process $(a(c).x \mid \bar{b}d.y)$ maps to the equivalence class in $B_\pi(T_\pi)$ that represents the following element. We specify the input and output transitions, and, although the process cannot perform a τ action, we must record that it could, if a and b were equal.

$$\mathsf{inj}_1(a, [c, (x \mid \bar{b}d.y)]_{=_\alpha}) \ + \ \mathsf{inj}_2(b, d, (a(c).x \mid y)) \ + \ \mathsf{if}(a, b, \mathsf{inj}_3(((\{^d/_c\}x) \mid y)). \quad (2)$$

Free semilattices aren't enough. In (2), we see that the operator 'if' in the theory of conditional-semilattices is essential for modelling name-passing. To emphasise this, consider the free semilattice monad $\mathcal{P}_{sl}$ on **NomSub**: $\mathcal{P}_{sl}(X)$ is the set of finite subsets S of X, with the pointwise substitution structure. To understand why this is inadequate for name-passing, consider the following result:

Proposition 16. *Consider nominal substitutions X and Y. The following data are equivalent.*

1. *A homomorphism $X \to \mathcal{P}_{sl}(Y)$.*
2. *A substitution-closed subset $R \subseteq X \times Y$ such that for all $x \in X$, the image $\{y \in Y \mid x \mathrel{R} y\}$ is finite, and such that if $\{^b/_a\}x \mathrel{R} y$ then there is $y' \in Y$ such that $\{^b/_a\}y' = y$ and $x \mathrel{R} y'$.*

6.3   Bisimulation for Name-Passing Calculi

HJ-bisimulations for B_π are essentially open bisimulations [38, Sec. 3.3]:

Proposition 17. *Let (X,h) and (Y,k) be B_π-coalgebras, corresponding to transition systems $(X, \to_X)$ and $(Y, \to_Y)$. A substitution-closed relation $R \subseteq X \times Y$ is an HJ-bisimulation if and only if it is a bisimulation between the transition systems in the classical sense.*

The class $\mathcal{S}_{csl}$ contains all monomorphisms, so, by the results in Section 4, an HJ-bisimulation is the same thing as an AM-bisimulation, and the greatest such bisimulation is the greatest AM-precongruence and the kernel of the final map.

6.4 Remarks on Functor Categories and Other Models of Name-Passing

The category of nominal substitutions is equivalent to the category $[\mathbf{F}_{ne}, \mathbf{Set}]$ of set-valued functors, where $\mathbf{F}_{ne}$ is the category of non-empty finite subsets of $\mathbb{A}$ and all functions between them[2]. Via Prop. 14 and the Yoneda lemma, we have an explicit description of $\mathcal{P}_{csl}$.

In some circumstances it is appropriate to work in the category $[\mathbf{I}, \mathbf{Set}]$, where $\mathbf{I}$ is the category of all finite subsets of $\mathbb{A}$ and injections between them. The free semilattice monad $\mathcal{P}_{sl}$ on $[\mathbf{I}, \mathbf{Set}]$ has a simple description: for F in $[\mathbf{I}, \mathbf{Set}]$ and $A \subseteq_f \mathbb{A}$, the set $(\mathcal{P}_{sl}F)(A)$ is the set of finite subsets of $F(A)$. Unlike the situation in **NomSub**, the monad $\mathcal{P}_{sl}$ on $[\mathbf{I}, \mathbf{Set}]$ is an appropriate monad for non-determinism in the semantics of name-passing languages [14,16].

In any topos, the free semilattice monad $\mathcal{P}_{sl}$ classifies the morphisms that are Kuratowski-finite. By Prop. 8, $\mathcal{P}_{sl}$ covers pullbacks, and if the topos is Boolean, then it also preserves weak pullbacks: note that monos are Kuratowski-finite if and only if they are complemented (see also [24]). Indeed, the example of [24] can be adapted to show that $\mathcal{P}_{sl}$ does *not* preserve weak pullbacks in $[\mathbf{I}, \mathbf{Set}]$ (correcting oversights in [16,20,40]). In spite of this, HJ-bisimulations can still be related with AM-bisimulations:

Proposition 18 (c.f. [40, Sec. 3.3.3]). *Let B' be an endofunctor on $[\mathbf{I}, \mathbf{Set}]$ that preserves complemented relations, and consider the composite endofunctor $\mathcal{P}_{sl} \circ B'$. The $\neg\neg$-closure of an HJ-bisimulation is an AM-bisimulation.*

The situation simplifies if we restrict attention to the category of $\neg\neg$-sheaves. The sheaves can be understood as 'nominal sets' [18,40] or as 'named-sets with symmetry', giving a connection with 'history dependent automata' [13,14,19].

Modelling restriction. To simplify the presentation, we have so-far ignored the restriction operator in the π-calculus, which introduces the facility of name generation. To properly handle name generation, the category **NomSub** is inadequate. Instead, following [20,31], one can work in the category $[\mathbf{D}, \mathbf{Set}]$ of set-valued functors, where $\mathbf{D}$ is the category of finite irreflexive undirected graphs and homomorphisms between them. The idea is to keep track of which names are known to be different, described by an edge in a graph.

The free semilattice monad on $[\mathbf{D}, \mathbf{Set}]$ is not a sufficient notion of non-determinism, for the reasons discussed above (correcting an oversight in [20]). Instead, one can work with the theory in Figure 3, with an additional axiom saying that when names a and b are distinct then $[a = b]x = \mathbf{0}$ (c.f. [38, Sec. 6, Ax. P]). The resulting monad corresponds to a class of small maps, but not all monos are small. It would be interesting to find a generalization of Prop. 18 that applies to this situation.

---

[2] I am grateful to Alexander Kurz for pointing this out.

Acknowledgements. It has been helpful to discuss the material in this article with various people, particularly Marcelo Fiore. Benno van den Berg gave some advice on algebraic set theory. Referees' comments were helpful, and I am sorry not to have been able to accommodate all their suggestions at this point. I acknowledge support from EPSRC grants GR/T22049/01 and EP/E042414/1.

References

1. Abramsky, S.: A domain equation for bisimulation. Inform. and Comput. 92, 161–218 (1991)
2. Abramsky, S.: A cook's tour of the finitary non-well-founded sets. In: We Will Show Them! Essays in Honour of Dov Gabbay, vol. 1, pp. 1–18 (2005)
3. Aczel, P., Mendler, N.: A final coalgebra theorem. In: Proc. of CTCS 1989, pp. 357–365 (1989)
4. Adámek, J.: Introduction to coalgebra. Theory Appl. of Categ. 14(8), 157–199 (2005)
5. van den Berg, B., Marchi, F.D.: Models of non-well-founded sets via an indexed final coalgebra theorem. J. Symbolic Logic 72(3), 767–791 (2007)
6. Bezhanishvili, N., Fontaine, G., Venema, Y.: Vietoris bisimulations. J. Logic Comput. (2008) (to appear)
7. Borceux, F.: Handbook of Categorical Algebra. Cambridge University Press, Cambridge (1994)
8. van Breugel, F., Hermida, C., Makkai, M., Worrell, J.B.: An accessible approach to behavioural pseudometrics. In: Caires, L., Italiano, G.F., Monteiro, L., Palamidessi, C., Yung, M. (eds.) ICALP 2005. LNCS, vol. 3580, pp. 1018–1030. Springer, Heidelberg (2005)
9. Carboni, A., Kelly, G.M., Wood, R.J.: A 2-categorical approach to geometric morphisms I. Cah. Topol. Géom. Différ. Catég. XXXII(1), 47–95 (1991)
10. Carboni, A., Lack, S., Walters, R.F.C.: Introduction to extensive and distributive categories. J. Pure Appl. Algebra 83, 145–158 (1993)
11. Cousot, P., Cousot, R.: Constructive versions of Tarski's fixed point theorems. Pacific J. Math. 82(1), 43–57 (1979)
12. Danos, V., Desharnais, J., Laviolette, F., Panangaden, P.: Bisimulation and cocongruence for probabilistic systems. Inform. and Comput. 204(4), 503–523 (2006)
13. Ferrari, G., Montanari, U., Pistore, M.: Minimizing transition systems for name passing calculi: A co-algebraic formulation. In: Nielsen, M., Engberg, U. (eds.) FOSSACS 2002. LNCS, vol. 2303, pp. 129–158. Springer, Heidelberg (2002)
14. Fiore, M.P., Staton, S.: Comparing operational models of name-passing process calculi. Inform. and Comput. 204(4), 435–678 (2006)
15. Fiore, M.P., Staton, S.: A congruence rule format for name-passing process calculi from mathematical structural operational semantics. In: Proc. of LICS 2006, pp. 49–58 (2006)
16. Fiore, M.P., Turi, D.: Semantics of name and value passing (extended abstract). In: Proc. of LICS 2001, pp. 93–104 (2001)
17. Gabbay, M.J., Hofmann, M.: Nominal renaming sets. In: Cervesato, I., Veith, H., Voronkov, A. (eds.) LPAR 2008. LNCS (LNAI), vol. 5330, pp. 158–173. Springer, Heidelberg (2008)
18. Gabbay, M.J., Pitts, A.M.: A new approach to abstract syntax with variable binding. Formal Aspects of Computing 13, 341–363 (2001)

19. Gadducci, F., Miculan, M., Montanari, U.: About permutation algebras (pre)sheaves and named sets. Higher-Order Symb. Comput. 19(2–3), 283–304 (2006)
20. Ghani, N., Yemane, K., Victor, B.: Relationally staged computations in calculi of mobile processes. In: Proc. of CMCS 2004, pp. 105–120 (2004)
21. Gumm, H.P.: Functors for coalgebras. Algebra Univers 45, 135–147 (2001)
22. Hermida, C., Jacobs, B.: Structural induction and coinduction in a fibrational setting. Inform. and Comput. 145(2), 107–152 (1998)
23. Jacobs, B., Rutten, J.: A tutorial on coalgebras and coinduction. EATCS Bulletin 62, 222–259 (1997)
24. Johnstone, P., Power, J., Tsujishita, T., Watanabe, H., Worrell, J.: On the structure of categories of coalgebras. Theoret. Comput. Sci. 260(1-2), 87–117 (2001)
25. Johnstone, P.T.: Sketches of an Elephant: A Topos Theory Compendium. Oxford University Press, Oxford (2002)
26. Joyal, A., Moerdijk, I.: Algebraic Set Theory. Cambridge University Press, Cambridge (1995)
27. Kelly, G.M., Power, A.J.: Adjunctions whose counits are coequalizers, and presentations of finitary enriched monads. J. Pure Appl. Algebra 89(1–2), 163–179 (1993)
28. Klin, B.: Adding recursive constructs to bialgebraic semantics. J. Log. Algebr. Program. 61, 259–286 (2004)
29. Kupke, C., Kurz, A., Venema, Y.: Stone coalgebras. Theoret. Comput. Sci. 327(1-2), 109–134 (2004)
30. Kurz, A.: Logics for Coalgebras and Applications to Computer Science. PhD thesis, Ludwig-Maximilians-Universität München (2000)
31. Miculan, M., Yemane, K.: A unifying model of variables and names. In: Sassone, V. (ed.) FOSSACS 2005. LNCS, vol. 3441, pp. 170–186. Springer, Heidelberg (2005)
32. Milner, R.: A Calculus of Communication Systems. LNCS, vol. 92. Springer, Heidelberg (1980)
33. Milner, R.: Calculi for synchrony and asynchrony. Theoret. Comput. Sci. 25, 267–310 (1983)
34. Milner, R., Parrow, J., Walker, D.: A calculus of mobile processes, I and II. Inform. and Comput. 100(1), 1–77 (1992)
35. Plotkin, G.D.: Bialgebraic semantics and recursion (extended abstract). In: Proc. of CMCS 2001. Electron. Notes Theor. Comput. Sci, vol. 44(1), pp. 1–4 (2001)
36. Rutten, J.J.M.M.: Relators and metric bisimulations. In: Proc. of CMCS 1998. Electron. Notes Theor. Comput. Sci, vol. 11 (1998)
37. Rutten, J.J.M.M.: Universal coalgebra: a theory of systems. Theoret. Comput. Sci. 249(1), 3–80 (2000)
38. Sangiorgi, D.: A theory of bisimulation for the π-calculus. Acta Inform 33(1), 69–97 (1996)
39. Sangiorgi, D., Walker, D.: The π-calculus: a theory of mobile processes. Cambridge University Press, Cambridge (2001)
40. Staton, S.: Name-passing process calculi: operational models and structural operational semantics. PhD thesis, University of Cambridge (2007)
41. Staton, S.: General structural operational semantics through categorical logic. In: Proc. of LICS 2008, pp. 166–177 (2008)
42. Viglizzo, I.D.: Final sequences and final coalgebras for measurable spaces. In: Fiadeiro, J.L., Harman, N.A., Roggenbach, M., Rutten, J. (eds.) CALCO 2005. LNCS, vol. 3629, pp. 395–407. Springer, Heidelberg (2005)
43. de Vink, E.P., Rutten, J.J.M.M.: Bisimulation for probabilistic transition systems: a coalgebraic approach. Theoret. Comput. Sci. 221, 271–293 (1999)
44. Worrell, J.: On the final sequence of a finitary set functor. Theoret. Comput. Sci. 338, 184–199 (2005)

Traces, Executions and Schedulers, Coalgebraically

Bart Jacobs[1] and Ana Sokolova[2,*]

[1] Radboud University Nijmegen, The Netherlands
[2] University of Salzburg, Austria

Abstract. A theory of traces of computations has emerged within the field of coalgebra, via finality in Kleisli categories. In concurrency theory, traces are traditionally obtained from executions, by projecting away states. These traces and executions are sequences and will be called "thin". The coalgebraic approach gives rise to both "thin" and "fat" traces/executions, where in the "fat" case the structure of computations is preserved. This distinction between thin and fat will be introduced first. It is needed for a theory of schedulers in a coalgebraic setting, of which we only present the very basic definitions and results.

1 Introduction

This paper is about traces and executions, in the general setting of coalgebra. It introduces what we call "thin" and "fat" style semantics, both for traces and executions. Roughly speaking, "thin" semantics is what is traditionally considered for traces and executions, especially for labelled transition systems (LTSs) [6]. It involves sequences/lists of observable actions (for traces) or lists of actions and intermediate states (for executions). The "fat" approach emerged from more recent work on traces in a coalgebraic setting [9,10]. It applies to systems as coalgebras $c: X \to TF(X)$ of type TF where T is a monad for computational effect or branching type, and F is a functor that determines the transition type, subject to a set of conditions. The semantics is described as a map $X \to I$ in the Kleisli category of the monad T, where I is the initial algebra of F. Elements of this initial algebra incorporate the "fat", tree-like structure of computations of type F.

Here we describe how to understand the thin and fat approaches in a common framework. Figure 1 gives an overview. It will be explained in the course of this article. At this stage we can already see that thin semantics (of both traces and executions) involves lists, via the Kleene star $(-)^*$—which can of course be described via the μ fixed point (initial algebra) operator. The fat semantics involves initial algebras of the functors F and $F(X \times -)$, where the latter involves the state space X, in order to accommodate states in executions. These initial algebras may have much more (tree) structure than lists. It should be noted however that they need not always exist.

* Research funded by the Austrian Science Fund (FWF) Project Nr. V00125.

A. Kurz, M. Lenisa, and A. Tarlecki (Eds.): CALCO 2009, LNCS 5728, pp. 206–220, 2009.

Initial algebras wrt. $X \to TF(X)$ with $F = F(0) + F_\bullet$	**Thin** (non-determinism only, with $T = \mathcal{P}$)	**Fat** (for general monads T)
Traces	$\begin{aligned} &F_\bullet(1)^\star \times F(0) \\ &= \mu Y.\, F(0) + F_\bullet(1) \times Y \end{aligned}$	$\mu Y.\, F(Y)$
Executions	$\begin{aligned} &(F_\bullet(X) \times X)^\star \times F(0) \\ &= \mu Y.\, F(0) + (F_\bullet(X) \times X) \times Y \end{aligned}$	$\mu Y.\, F(X \times Y)$

Fig. 1. Traces and executions

The first part of this paper concentrates on this table. It is needed in order to properly capture schedulers. Schedulers are often used to resolve non-determinism, by making particular choices. They are used for instance in the semantics of programming languages [16], (probabilistic) verification [5,14,15] or security [2,4]. They are not always described in the mathematically most rigorous manner. Hence a precise understanding is valuable. It was the original focus of the paper. But the "preliminary" work on thin and fat traces and executions turned out to be more involved than expected, so that in the end only the last part of the paper (Section 8) is left for schedulers. It does not do much more than setting the scene, by introducing some basic definitions and a "soundness" theorem. It ends with a definition of "completeness" of scheduler semantics, as a cliffhanger. It will be further developed and illustrated in subsequent work.

2   Preliminaries

We assume the reader is reasonably familiar with categorical notation and terminology and with the theory of coalgebras. We shall briefly review our notation. Cartesian projections will be written as $\pi_i \colon X_1 \times X_2 \to X_i$ with $\langle f, g \rangle$ for tupling. By δ we denote the diagonal, $\delta = \langle \mathrm{id}, \mathrm{id} \rangle \colon X \to X \times X$. Dually, coprojections are written as $\kappa_i \colon X_i \to X_1 + X_2$ with cotupling $[f, g]$. In **Sets** coprojections are disjoint, meaning that the pullback of κ_1 and κ_2 is empty. Coproducts are also universal: given $f \colon Y \to X_1 + X_2$ we can split up $Y_1 + Y_2 \overset{\cong}{\to} Y$ via the two pullbacks $Y_i \to Y$ of the coprojections along f, see *e.g.* [3].

We recall that every functor $F \colon \mathbf{Sets} \to \mathbf{Sets}$ is strong, via a "strength" map $\mathsf{st} \colon X \times F(Y) \to F(X \times Y)$ given by $\mathsf{st}(x, v) = F(\lambda y.\, \langle x, y \rangle)(v)$. If F happens to be a monad, then it is strong, meaning that its unit η and multiplication μ commute appropriately with this strength. A map of monads $\sigma \colon S \Rightarrow T$ is called strong if it commutes with strength. Examples of monads that occur in this setting are powerset $\mathcal{P}$ for non-determinism, lift $1 + (-)$ for partiality, or (sub)distribution $\mathcal{D}$ for probabilism. The Kleisli category of a monad T will be written as $\mathcal{K}\ell(T)$.

3   A Motivating Example: Binary Trees with Output

We start with a simple example of a transition type functor generating binary trees with output, namely $F(X) = A + (B \times X^2)$ for constant sets A and B.

In a state $x \in X$ a transition in $F(X)$ of such a binary tree functor either produces an output in A and terminates, or makes a step in $B \times X^2$, consisting of an observable output element in B together with a pair of children states which will (both!) be active in the next step. In this section we shall concretely describe both "thin" and "fat" executions and traces for this transition type functor F. The general construction of these executions and traces via finality is described later, namely in Sections 5 and 7.

Thin traces and executions. As illustration we consider a coalgebra $c \colon X \to \mathcal{P}(FX)$, where $FX = A + B \times X^2$ as above. Starting from a state $x_0 \in X$ we can consider the "thin" executions starting in x_0. They are (finite) sequences of the form:

$$b_0, x_1, b_1, x_2, \ldots, b_n, x_n, a \tag{1}$$

where:

$$c(x_0) \ni \begin{cases} (b_0, x_1, x_1') \text{ for some } x_1' \in X, \text{ or} \\ (b_0, x_1', x_1) \text{ for some } x_1' \in X \end{cases}$$

$$c(x_1) \ni \begin{cases} (b_1, x_2, x_2') \text{ for some } x_2' \in X, \text{ or} \\ (b_1, x_2', x_2) \text{ for some } x_2' \in X \end{cases}$$

$$\vdots$$

$$c(x_n) \ni a.$$

These executions thus capture possible computation paths, involving a specific choice of left or right successor state.

We shall write $\mathsf{texc}(x_0)$ for the set of all such "thin" executions; hence $\mathsf{texc}(x_0) \in \mathcal{P}((B \times X)^\star \times A)$. It will be described later as a map $\mathsf{texc} \colon X \to (B \times X)^\star \times A$, in the Kleisli category $\mathcal{K}\ell(\mathcal{P})$, obtained by finality using the result of Section 4 below.

Roughly, a trace is an execution with the states removed. So if we remove states x_i from (1) we are left with a "thin" trace, as element of $B^\star \times A$. The trace is a Kleisli map $\mathsf{ttr} \colon X \to \mathcal{P}(B^\star \times A)$. As we shall see, it can also be obtained by finality.

Fat traces and executions. Thin executions and traces describe computation *paths*. What we call fat executions or traces does not involve paths but trees that retain the structure of the transition type. Hence, examples of fat executions from a state $x_0 \in X$ are:

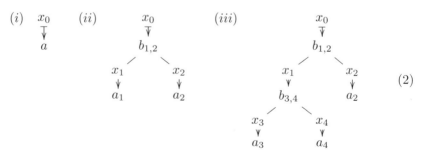

$$\tag{2}$$

A fat trace can be understood as what remains when states are removed from such trees. But there is a more direct way of understanding such traces, namely as elements of the initial algebra of the functor F. As usual, this initial algebra I is obtained as a colimit $I = \operatorname{colim}_{i \in \mathbb{N}} F^i(\emptyset)$ of the initial sequence, where:

$$F(\emptyset) = A$$
$$F^2(\emptyset) = A + (B \times A^2)$$
$$F^3(\emptyset) = A + B \times \left(A + (B \times A^2)\right)^2$$
$$= A + (B \times A^2) + (B \times A \times B \times A^2)$$
$$+ (B \times B \times A^2 \times A) + (B \times B \times A^2 \times B \times A^2)$$
$$F^4(\emptyset) = \cdots$$

Given a coalgebra $c \colon X \to \mathcal{P}(FX)$, coalgebraic trace theory provides us with a trace map $\mathsf{ftr}_c \colon X \to I$ in the Kleisli category $\mathcal{K}\ell(\mathcal{P})$, obtained by finality. For a state $x \in X$ the set $\mathsf{ftr}_c(x) \in \mathcal{P}(I)$ contains:

$$a \in \mathsf{ftr}_c(x) \iff a \in c(x)$$
$$\langle b, \alpha_1, \alpha_2 \rangle \in \mathsf{ftr}_c(x) \iff \exists x_1, x_2 \in X. \, \langle b, x_1, x_2 \rangle \in c(x) \text{ and}$$
$$\alpha_1 \in \mathsf{ftr}_c(x_1) \text{ and } \alpha_2 \in \mathsf{ftr}_c(x_2).$$

How to understand "fat" executions? It is not hard to see that the trees in (2) are elements of the initial algebra of the functor $F(X \times -)$. Indeed, this initial algebra is obtained as colimit of the chain $(F(X \times -))^i(\emptyset)$, which starts with:

$$(F(X \times -))^1(\emptyset) = F(X \times \emptyset)$$
$$= A$$
$$(F(X \times -))^2(\emptyset) = F(X \times F(X \times \emptyset))$$
$$= F(X \times A)$$
$$= A + (B \times (X \times A)^2)$$
$$(F(X \times -))^3(\emptyset) = \cdots$$

Let us write I_X for this initial algebra. The fat execution map then appears as a map $\mathsf{fexc}_c \colon X \to I_X$ in the Kleisli category $\mathcal{K}\ell(\mathcal{P})$, that is, as function $\mathsf{fexc}_c \colon X \to \mathcal{P}(I_X)$ in **Sets**. It can be obtained by finality. Its key properties are:

$$a \in \mathsf{fexc}_c(x) \iff a \in c(x)$$
$$\langle b, x_1, \alpha_1, x_2, \alpha_2 \rangle \in \mathsf{fexc}_c(x) \iff \langle b, x_1, x_2 \rangle \in c(x) \text{ and}$$
$$\alpha_1 \in \mathsf{fexc}_c(x_1) \text{ and } \alpha_2 \in \mathsf{fexc}_c(x_2).$$

An analogous definition of fat traces and executions, as well as a connection between them, can be made for an arbitrary monad T and an arbitrary functor F satisfying the requirements of the coalgebraic trace theorem. These fat traces and executions will be presented below, in Section 5.

4   Final Coalgebras in Kleisli Categories

Coalgebraic trace semantics has been developed for coalgebras of the form $X \to TF(X)$ where T is a suitable monad, see [9,10]. It can be formulated for arbitrary categories, but here we shall restrict ourselves to **Sets**. There are some technical requirements.

- There must be a distributive law $\lambda \colon FT \Rightarrow TF$; it induces a lifting of F to $\overline{F} \colon \mathcal{K}\ell(T) \to \mathcal{K}\ell(T)$, which commutes with the canonical functor $J \colon \mathbf{Sets} \to \mathcal{K}\ell(T)$;
- The Kleisli category $\mathcal{K}\ell(T)$ must be suitably order-enriched, with order $\sqsubseteq$ on Kleisli homsets, bottom element $\bot$ and suprema $\bigvee$ of directed subsets;
- The lifting $\overline{F} \colon \mathcal{K}\ell(T) \to \mathcal{K}\ell(T)$ must be locally monotone.

The requirements are discussed in detail in [10]. Examples of monads T and functors F that satisfy these requirements are: the powerset monad $\mathcal{P}$, the sub-distribution monad $\mathcal{D}$ (with probability distributions with sum less than or equal to 1), the lift monad $1+(-)$, and the list monad $(-)^*$, together with all "shapely" functors. We shall not concentrate on these requirements, and assume that they simply hold for the monads/functors that we use in this paper. Of crucial importance is the following result, describing how final coalgebras arise in Kleisli categories from initial algebras in the underlying category. As will be amply illustrated, it will be used throughout to obtain traces and executions, of various forms.

Theorem 1. *Let F and T be a functor and a monad on* **Sets** *satisfying the above requirements. If there is an initial algebra $\alpha \colon F(I) \xrightarrow{\cong} I$ in* **Sets***, then the associated coalgebra $J(\alpha^{-1}) \colon I \xrightarrow{\cong} F(I)$ is final in the category of coalgebras of the lifting $\overline{F}$ to $\mathcal{K}\ell(T)$.*   □

Coalgebraic trace semantics shows that linear-time semantics fits into the paradigm of final coalgebra semantics, and can thus benefit from the associated machinery, for instance in showing compositionality / congruence of bisimilarity and trace equivalence for various coalgebras [11]. In the second part of this paper we shall restrict ourselves to the special case $T = \mathcal{P}$ of the powerset monad. Recall that the associated Kleisli category $\mathcal{K}\ell(\mathcal{P})$ is the category of sets and relations between them. The distributive law then follows from preservation of weak pullbacks of F. It means that there is a "relation lifting" operation $\mathrm{Rel}(F) \colon \mathcal{P}(X \times Y) \longrightarrow \mathcal{P}(FX \times FY)$, which induces the "power" distributive law $\lambda \colon F\mathcal{P} \Rightarrow \mathcal{P}F$, namely as $\lambda_X(u) = \{v \in FX \mid \mathrm{Rel}(F)(\in)(u, v)\}$, see [10, Lemma 2.3] (going back to [12]) for details. This λ commutes with the monad operations $\eta = \{-\}$ and $\mu = \bigcup$ of the powerset monad. The order on the Kleisli homsets is the pointwise inclusion order, and will be denoted by $\subseteq$.

For reasoning about schedulers (in Section 8) we borrow some results from Hasuo [7,8] on oplax morphisms in Kleisli categories—for the special case $T = \mathcal{P}$. To start, an *oplax morphism* in $\mathcal{K}\ell(\mathcal{P})$ from a coalgebra $c \colon X \to FX$ to a coalgebra $d \colon Y \to FY$ is a Kleisli map $f \colon X \to Y$ such that:

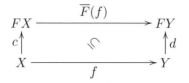

Proposition 1 ([7,8]). *For an arbitrary coalgebra $c: X \to TF(X)$ we write $\mathsf{tr}_c: X \to T(I)$ for the $\overline{F}$-coalgebra morphism in $\mathcal{K}\ell(T)$ obtained by finality. This map tr_c is the smallest one among the oplax morphisms from $c: X \to FX$ to the final coalgebra $I \overset{\cong}{\to} FI$ in $\mathcal{K}\ell(\mathcal{P})$.* □

Corollary 1. *If $f: X \to Y$ is an oplax morphism from $c: X \to FX$ to $d: Y \to FY$, then $\mathsf{tr}_d \circ f$ is also an oplax morphism, so that $\mathsf{tr}_c \subseteq \mathsf{tr}_d \circ f$ in $\mathcal{K}\ell(\mathcal{P})$.* □

5 Fat Traces and Executions

Fat traces. We assume that the functor F has an initial algebra I, in addition to the assumptions from Section 4, with map $\alpha: F(I) \overset{\cong}{\to} I$. It yields by the trace Theorem 1 a final coalgebra $J(\alpha^{-1}): I \to F(I)$ in the Kleisli category $\mathcal{K}\ell(T)$. Each coalgebra $c: X \to TF(X)$ gives rise to a "fat" trace map $\mathsf{ftr}_c: X \to I$ in $\mathcal{K}\ell(T)$ by finality, as in:

$$
\begin{array}{ccc}
F(X) & \overset{\overline{F}(\mathsf{ftr}_c)}{\dashrightarrow} & F(I) \\
{\scriptstyle c}\uparrow & & \cong\uparrow{\scriptstyle J(\alpha^{-1})} \\
X & \underset{\mathsf{ftr}_c}{\dashrightarrow} & I
\end{array}
$$

Fat executions. We fix a particular coalgebra $c: X \to TF(X)$, and assume that the functor $F(X \times -)$ has an initial algebra I_X, with map $\alpha_X: F(X \times I_X) \overset{\cong}{\to} I_X$. We obtain a "fat" execution map, again by finality in $\mathcal{K}\ell(T)$:

$$
\begin{array}{ccc}
F(X \times X) & \overset{\overline{F}(\mathsf{id} \times \mathsf{fexc}_c)}{\dashrightarrow} & F(X \times I_X) \\
{\scriptstyle \overline{F}J(\delta) \circ c}\uparrow & & \cong\uparrow{\scriptstyle J(\alpha_X^{-1})} \\
X & \underset{\mathsf{fexc}_c}{\dashrightarrow} & I_X
\end{array}
$$

The notation is a bit sloppy. The map on top involves the lifting $\overline{F(\mathsf{id} \times -)}$, which exists via strength.

Relating fat executions and traces. We first construct a map $\pi_X: I_X \to I$ that projects away states by initiality in **Sets**, as in:

$$
\begin{array}{ccc}
F(X \times I_X) & \overset{F(\mathsf{id} \times \pi_X)}{\dashrightarrow} & F(X \times I) \\
{\scriptstyle \alpha_X}\downarrow\cong & & \downarrow{\scriptstyle \alpha \circ F(\pi_2)} \\
I_X & \underset{\pi_X}{\dashrightarrow} & I
\end{array}
\tag{3}
$$

Then we obtain the basic execution-trace equation in $\mathcal{K}\ell(T)$:

$$\mathsf{ftr}_c = J(\pi_X) \circ \mathsf{fexc}_c. \tag{4}$$

It is proven by a uniqueness argument in $\mathcal{K}\ell(T)$:

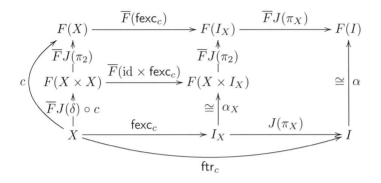

The square on the right is essentially $J\colon \mathbf{Sets} \to \mathcal{K}\ell(T)$ applied to (3). The lower-left rectangle commutes by definition of fexc_c and the upper-left one commutes because $\overline{F}J(\pi_2)$ is a natural transformation $\overline{F}(\mathrm{id} \times -) \Rightarrow \overline{F}$ between liftings.

6 Splitting Up Functors

The problem that we address in this section can best be illustrated with an example. Consider the functor $F(X) = A + (B \times X)$. It consists of two (sum) components, one containing a successor state, namely $B \times X$, and one without, namely A. If we wish to consider non-terminating executions of an F-coalgebra only the part of the functor containing states is relevant. For traces, on the other hand, one looks at sequences/trees of observables. Then the "state-less" part is also relevant, at the end of a such a sequence. In this section we shall see how to get a handle on these different parts of a functor (with and without states), via a form of linearisation of a functor[1].

In the remainder of this paper we shall work with what we shall call subpower functors F, as "transition type" functors, in coalgebras $X \to \mathcal{P}(FX)$.

Definition 1. *A functor $F\colon \mathbf{Sets} \to \mathbf{Sets}$ will be called a "subpower" functor if it preserves weak pullbacks and comes with a natural transformation $\rho\colon F \Rightarrow \mathcal{P}$ with the special property that $u \in F(\rho_X(u))$ for each $u \in F(X)$.*

We recall that weak pullback preserving functors preserve monos/injections. Hence the inclusion $\rho_X(u) \hookrightarrow X$, for $u \in F(X)$, yields an injection $F(\rho_X(u)) \rightarrowtail$

---

[1] This linearisation does not keep track of "positions" or "holes" like in derivatives of functors [13,1].

$F(X)$ so that the requirement $u \in F(\rho_X(u))$ should formally be understood as a (necessarily unique) factorisation:

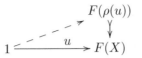

Like in this diagram, we often omit the subscript X in ρ_X.

It is not hard to see that the identity functor is subpower, with ρ as singleton map and that constant functors are subpower via the empty map. Further, subpower functors are closed under coproducts and products. Additionally, certain special functors are subpower, like the probability distribution functor $\mathcal{D}$, via the support map. Notice that taking the greatest subset $\rho_X(u) = X$ does not yield a natural transformation.

We now use that a powerset $\mathcal{P}(X)$ can be written as coproduct $\mathcal{P}(X) = 1 + \mathcal{P}_\bullet(X)$, where $\mathcal{P}_\bullet(X)$ contains the non-empty subsets of X and 1 corresponds to the empty subset. For a subpower functor with $\rho: F \Rightarrow \mathcal{P}$ we can thus form the following two pullbacks.

$$
\begin{array}{ccc}
F_\emptyset(X) \rightarrowtail & F(X) \longleftarrow & F_\bullet(X) \\
\downarrow \quad \lrcorner & \downarrow \rho_X & \llcorner \quad \downarrow \\
1 \rightarrowtail \quad \emptyset \quad \longrightarrow & \mathcal{P}(X) \longleftarrow & \mathcal{P}_\bullet(X)
\end{array} \tag{5}
$$

Since coproducts are universal [3] in **Sets**, the induced cotuple $F_\emptyset(X) + F_\bullet(X) \rightarrow F(X)$ is an isomorphism. In this way we can split up F in two parts, one with output states, and one without them.

Lemma 1. *Consider a subpower functor F.*

1. *Both $F_\emptyset$ and $F_\bullet$ are functors with (coprojection) natural transformations $F_\emptyset \Rightarrow F$ and $F_\bullet \Rightarrow F$.*
2. *$F_\emptyset(X) = F(0)$, for each X, so that $F(X) = F(0) + F_\bullet(X)$—where 0 is the initial object (empty set) in **Sets**.*
3. *The functor $F_\bullet$ is again subpower, via $\rho_\bullet = (F_\bullet \Rightarrow F \overset{\rho}{\Rightarrow} \mathcal{P})$; as a consequence, there is a distributive law $\lambda_\bullet \colon F_\bullet \mathcal{P} \Rightarrow \mathcal{P} F_\bullet$, commuting with λ via $F_\bullet \Rightarrow F$ in the obvious way.*

Proof. For the first point we shall do the proof for $F_\bullet$. Consider for a map $f \colon X \rightarrow Y$ the following diagram:

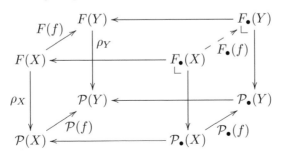

The maps $F_\bullet(X) \to F(X)$ are natural by construction of $F_\bullet$. Similarly, there is a natural transformation $F_\emptyset \Rightarrow F$.

For point (2), note first that the empty set is an isomorphism $\emptyset\colon 1 \xrightarrow{\cong} \mathcal{P}(0)$. Hence for $X = 0$ the map $F_\emptyset(0) \to F(0)$ in (5) is also an isomorphism. Thus $F(0) \cong F_\emptyset(0) \hookrightarrow F_\emptyset(X)$. Conversely, if $u \in F_\emptyset(X)$, then $u \in F(\rho(u)) = F(0)$.

For the third point we use that the natural transformation $F_\bullet \Rightarrow F = F(0) + F_\bullet$ is essentially the second coprojection κ_2. Hence if $u \in F_\bullet(X)$, then $\kappa_2(u) \in F(\rho(\kappa_2(u))) = F(0) + F_\bullet(\rho(\kappa_2(u)))$, so that $u \in F_\bullet(\rho_\bullet(u))$.

The distributive law $\lambda_\bullet\colon F_\bullet\mathcal{P} \Rightarrow \mathcal{P}F_\bullet$ exists because $F_\bullet$ preserves weak pullbacks: assume $p_i\colon P \to X_i$ is the weak pullback of $f_i\colon X_i \to Y$, for $i \in \{1,2\}$. Let $g_i\colon Z \to F_\bullet(X_i)$ satisfy $F_\bullet(f_1) \circ g_1 = F_\bullet(f_2) \circ g_2$. Then by post-composition with $F_\bullet(X_i) \hookrightarrow F(0) + F_\bullet(X_i) = F(X_i)$ we obtain, because F preserves weak pullbacks, a mediating map $h\colon Z \to F(P) = F(0) + F_\bullet(P)$. This h must then factor through $F_\bullet(P)$.

Hence $\rho_\bullet$ makes $F_\bullet$ a subpower functor. The proof of the connection between $\lambda_\bullet$ and λ uses relation lifting. The details are skipped. $\quad\square$

The next map $\mathsf{split}\colon FX \to \mathcal{P}(F(0) + F_\bullet(X) \times X)$ will be important:

$$FX = F(0) + F_\bullet(X) \xrightarrow{\mathrm{id}+\delta} F(0) + F_\bullet(X) \times F_\bullet(X)$$

$$\mathrm{id}+\mathrm{id}\times\rho_\bullet \downarrow$$

$$F(0) + F_\bullet(X) \times \mathcal{P}(X) \xrightarrow{\mathrm{id}+\mathsf{st}} F(0) + \mathcal{P}(F_\bullet(X) \times X) \qquad (6)$$

$$\downarrow [\eta\circ\kappa_1, \mathcal{P}(\kappa_2)]$$

$$\mathcal{P}(F(0) + F_\bullet(X) \times X)$$

It is natural $F \Rightarrow \mathcal{P}(F(0) + F_\bullet \times \mathrm{id})$, in **Sets**.

7 Thin Traces and Executions for Non-determinism

We now restrict ourselves to the powerset monad $\mathcal{P}$ for non-determinism and will assume that the transition type functor is a subpower functor, via $\rho\colon F \Rightarrow \mathcal{P}$. Hence we can write $F(X) = F(0) + F_\bullet(X)$.

Thin traces. We shall write L for the set of lists $L = F_\bullet(1)^\star \times F(0)$, ending with an element in $F(0)$. This L is of course the initial algebra of the functor $Y \mapsto F(0) + F_\bullet(1) \times Y$. The initial algebra structure will be written as $[\mathsf{end}, \mathsf{cons}]\colon F(0) + F_\bullet(1) \times L \xrightarrow{\cong} L$. Therefore, using trace semantics in the Kleisli category $\mathcal{K}\ell(\mathcal{P})$ of the powerset monad, we obtain a "thin" trace map by finality:

$$
\begin{array}{ccc}
F(0) + F_\bullet(1) \times X & \xdashrightarrow{\mathrm{id}\,+\,\mathrm{id}\,\times\,\mathsf{ttr}_c} & F(0) + F_\bullet(1) \times L \\
c_{\mathsf{lt}} \uparrow & & \uparrow \cong \\
X & \xdashrightarrow{\qquad \mathsf{ttr}_c \qquad} & L
\end{array}
$$

where the coalgebra $c_{\mathsf{lt}}\colon X \to \mathcal{P}(F(0) + F_\bullet(1) \times X)$ is defined as composite

$$c_{\mathsf{lt}} = \mathcal{P}(\mathrm{id} + (F_\bullet(!) \times \mathrm{id})) \circ \mu \circ \mathcal{P}(\mathsf{split}) \circ c, \qquad (7)$$

with split defined in (6). We note that for "linear" functors such as $F(X) = A + B \times X$ for which $F(X) = F(0) + F_{\bullet}(1) \times X$ there is no difference between "thin" and "fat" traces (or executions).

Thin executions. We now fix a non-deterministic coalgebra $c: X \to \mathcal{P}(FX)$ in advance and write $L_X = (F_{\bullet}(X) \times X)^{\star} \times F(0)$ for the set of lists of executions. This L_X is the initial algebra of the functor $F(0) + (F_{\bullet}(X) \times X) \times (-)$. The associated thin execution map is obtained in:

$$
\begin{array}{ccc}
F(0) + (F_{\bullet}(X) \times X) \times X & \xrightarrow{\text{id} + \text{id} \times \text{texc}_c} & F(0) + (F_{\bullet}(X) \times X) \times L_X \\
c_{\text{le}} \Big\uparrow & & \Big\uparrow \cong \\
X & \xrightarrow{\quad \text{texc}_c \quad} & L_X
\end{array}
$$

where the coalgebra $c_{\text{le}}: X \to \mathcal{P}(F(0) + (F_{\bullet}(X) \times X) \times X)$ is defined as:

$$
c_{\text{le}} = \mathcal{P}(\text{id} + \langle \text{id}, \pi_2 \rangle) \circ \mu \circ \mathcal{P}(\text{split}) \circ c, \tag{8}
$$

where $\text{split}: F(X) \to \mathcal{P}(F(0) + F_{\bullet}(X) \times X)$ is from (6).

Relating thin executions and traces. The first step in relating thin executions and traces is to get a map $L_X \to L$ between the corresponding sequences. It is of course obtained by initiality (of L_X) in **Sets**, as in:

$$
\begin{array}{ccc}
F(0) + (F_{\bullet}(X) \times X) \times L_X & \xrightarrow{\text{id} + \text{id} \times p_X} & F(0) + (F_{\bullet}(X) \times X) \times L \\
[\text{end}, \text{cons}] \Big\downarrow \cong & & \Big\downarrow [\text{end}, \text{cons}] \circ (\text{id} + (F_{\bullet}(!) \circ \pi_1) \times \text{id}) \\
L_X & \xrightarrow{\quad p_X \quad} & L
\end{array}
$$

As before we obtain the basic execution-trace equation in $\mathcal{K\ell}(\mathcal{P})$, but this time for the thin case:

$$
\text{ttr}_c = J(p_X) \circ \text{texc}_c. \tag{9}
$$

It is proven by uniqueness in:

$$
\begin{array}{ccccc}
F(0) + F_{\bullet}(1) \times X & \xrightarrow{\text{id} + \text{id} \times \text{texc}_c} & F(0) + F_{\bullet}(1) \times L_X & \xrightarrow{\text{id} + \text{id} \times J(p_X)} & F(0) + F_{\bullet}(1) \times L \\
J(\text{id} + (F_{\bullet}(!) \circ \pi_1) \times \text{id}) \Big\uparrow & & J(\text{id} + (F_{\bullet}(!) \circ \pi_1) \times \text{id}) \Big\uparrow & & \Big\uparrow \cong \\
F(0) + (F_{\bullet}(X) \times X) \times X & \xrightarrow{\text{id} + \text{id} \times \text{texc}_c} & F(0) + (F_{\bullet}(X) \times X) \times L_X & & \\
c_{\text{le}} \Big\uparrow & & \Big\uparrow \cong & & \\
X & \xrightarrow{\quad \text{texc}_c \quad} & L_X & \xrightarrow{\quad J(p_X) \quad} & L \\
& & & \text{ttr}_c &
\end{array}
$$

It requires that we prove that the vertical composite on the left equals c_{lt}. This follows from an easy calculation in **Sets**:

$$\mathcal{P}(\text{id} + (F_\bullet(!) \circ \pi_1) \times \text{id}) \circ c_{\text{le}}$$
$$= \mathcal{P}(\text{id} + (F_\bullet(!) \circ \pi_1) \times \text{id}) \circ \mathcal{P}(\text{id} + \langle \text{id}, \pi_2 \rangle) \circ \mu \circ \mathcal{P}(\text{split}) \circ c$$
$$= \mathcal{P}(\text{id} + (F_\bullet(!) \times \text{id})) \circ \mu \circ \mathcal{P}(\text{split}) \circ c$$
$$= c_{\text{lt}}.$$

From fat to thin traces. As we have seen in the previous sections one can define thin and fat traces separately. Here we show that one can also obtain thin traces from fat ones via a special "paths" map between the corresponding initial algebras, as in the following diagram (in $\mathcal{K}\ell(\mathcal{P})$).

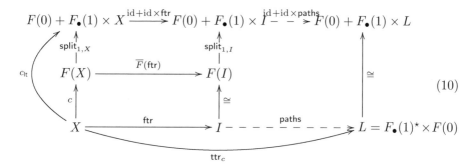

$$(10)$$

where $\text{split}_{1,X} = \mathcal{P}(\text{id} + (F_\bullet(!) \times \text{id})) \circ \text{split}_X$. The Kleisli map $\text{paths}: I \to \mathcal{P}(L)$ that is (implicitly) defined by finality yields the set of paths/sequences in a tree. The upper left square commutes by naturality of $\text{split}_{1,X}$, from $\overline{F}$ to $\overline{F(0) + F_\bullet(1) \times \text{id}}$ in $\mathcal{K}\ell(\mathcal{P})$. This naturality requires the additional assumption that $\lambda: F\mathcal{P} \Rightarrow \mathcal{P}F$ and $\rho: F \Rightarrow \mathcal{P}$ are compatible in the sense that $\mu \circ \mathcal{P}(\rho) \circ \lambda = \mu \circ \rho$. Details are skipped.

Thin and fat executions are similarly related via a paths map between initial algebras I_X to L_X.

8   Scheduling

Scheduling is about resolving non-determinism, by choosing some structure like singletons, lists, multisets or distributions instead of plain, unstructured, subsets. How this non-determinism is resolved will be described generically, at first, in terms of another monad S with a (strong) monad map $\sigma: S \Rightarrow \mathcal{P}$. Possible examples of S are identity Id, lift $1 + (-)$, list $(-)^\star$, multiset $\mathcal{M}$, or distribution $\mathcal{D}$, each with "obvious" mappings σ to the powerset monad. Very roughly, scheduling "of type S" involves a suitable inverse to this mapping $\sigma: S \Rightarrow \mathcal{P}$.

For a set X we shall abbreviate

$$\Xi = X \times (F_\bullet(X) \times X)^\star.$$

It contains all the finite (non-terminating) thin "executions", appended to a starting state, as first component. Of course the elements of Ξ are not really executions since there is no coalgebra involved and no one-step connections between the constituents. There are three obvious maps:

$$X \xrightarrow{\ \text{in}\ } \Xi = X \times (F_\bullet(X) \times X)^\star \underset{\text{last}}{\overset{\text{first}}{\rightrightarrows}} X$$

where $\text{in}(x) = \langle x, \langle\rangle\rangle$. The map first yields the first state of the execution, simply via the first projection, and last yields the last state of the execution defined as:

$$\text{last}(x, \alpha) = \begin{cases} x & \text{if } \alpha \text{ is the empty sequence } \langle\rangle \\ y & \text{if } \alpha = \beta \cdot \langle u, y\rangle. \end{cases}$$

Clearly, first $\circ$ in $=$ id, but also last $\circ$ in $=$ id.

We now come to the crucial notion of scheduler. Informally it chooses a computation of type S for a non-deterministic computation, given a scheduling type $\sigma: S \Rightarrow \mathcal{P}$.

Definition 2. *A scheduler of type $\sigma: S \Rightarrow \mathcal{P}$ for a coalgebra $c: X \to \mathcal{P}F(X)$ is a mapping ξ in:*

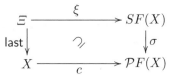

Intuitively, a scheduler chooses a next step starting from the last state of an execution in Ξ. The inclusion presented by the diagram, together with the definition of last, ensures that the chosen next step is contained in the next-step options that the coalgebra c provides for the last state of the execution.

Because such a scheduler ξ takes elements from Ξ as input it may be called *history dependent*: the scheduler may take previous execution steps into account when making the current scheduling decision. Here one may object that the set Ξ has too many elements—not just the proper executions. One way to handle this is to let ξ choose in these non-proper execution cases a "bottom" element.

Example 1. We illustrate several variants of schedulers, for the simplest and most well known example of LTS with termination, namely $FX = \{\checkmark\} + A \times X$, in which case $\Xi = X \times (A \times X)^\star$.

1. *Deterministic schedulers* have type $\eta: Id \Rightarrow \mathcal{P}$. They are maps $\xi: \Xi \to (\{\checkmark\} + A \times X)$ that choose a single possibility out of the last state of an execution.
2. *Non-deterministic schedulers* are schedulers of type $id: \mathcal{P} \Rightarrow \mathcal{P}$. They merely reduce non-determinism by pruning out some of the possible options in the last state of an execution. These are maps $\xi: \Xi \to \mathcal{P}(\{\checkmark\} + A \times X)$.

3. *Randomized schedulers* replace non-deterministic choice by a probability distribution via the scheduling type $\mathsf{supp}\colon \mathcal{D} \Rightarrow \mathcal{P}$. They are maps $\xi\colon \Xi \to \mathcal{D}(\{\checkmark\} + A \times X)$ such that $\xi(x,\alpha)(z) \neq 0$ implies $z \in c(\mathsf{last}(x,\alpha))$. It means that the support of the distribution $\varphi = \xi(x,\alpha)$ that is produced by the scheduler ξ on a non-terminating execution $\langle x,\alpha \rangle$ is a subset of the set of transitions resulting from the last state $\mathsf{last}(x,\alpha)$ of the execution.

Definition 3. *Let ξ be a scheduler for $c\colon X \to \mathcal{P}F(X)$. The coalgebra of executions of c under the scheduler ξ,*

$$\Xi \xrightarrow{\ \ c_\xi\ \ } \mathcal{P}F(\Xi)$$

is the composite of the following pile of maps.

$$
\begin{array}{ll}
\Xi \xrightarrow{\quad \langle id,\xi\rangle \quad} & \Xi \times SF(X) \\[4pt]
\xrightarrow{\quad id \times \sigma \quad} & \Xi \times \mathcal{P}F(X) \\[4pt]
\xrightarrow{\quad \mathsf{st} \quad} & \mathcal{P}(\Xi \times (F(0) + F_\bullet(X))) \\[4pt]
\xrightarrow{\quad \mathcal{P}(\mathsf{dist}) \quad} & \mathcal{P}(\Xi \times F(0) + \Xi \times F_\bullet(X)) \\[4pt]
\xrightarrow{\quad \mathcal{P}(\pi_2 + \langle id,\pi_2\rangle) \quad} & \mathcal{P}(F(0) + \Xi \times F_\bullet(X) \times F_\bullet(X)) \\[4pt]
\xrightarrow{\quad \mathcal{P}(id + \mathsf{st}) \quad} & \mathcal{P}(F(0) + F_\bullet(\Xi \times F_\bullet(X) \times X)) \\[4pt]
\xrightarrow{\quad \mathcal{P}(id + F_\bullet(\mathsf{cons})) \quad} & \mathcal{P}F(\Xi)
\end{array}
$$

where cons extends executions; it satisfies $\mathsf{last} \circ \mathsf{cons} = \pi_3$.

The coalgebra c_ξ yields a fat trace map $\mathsf{ftr}_{c_\xi}\colon \Xi \to I$ in the Kleisli category $\mathcal{K}\ell(\mathcal{P})$, for I being the initial F-algebra. One can also look at the thin trace map of c_ξ but it can be obtained simply via the paths map from (10). The next lemma relates the scheduler-induced coalgebra c_ξ on executions to the original coalgebra c on states.

Lemma 2. *The map $J(\mathsf{last})$ is an oplax morphism from $c_\xi\colon \Xi \to F(\Xi)$ to $c\colon X \to F(X)$ in $\mathcal{K}\ell(\mathcal{P})$, i.e.*

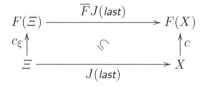

Proof. We first note that $\overline{F}J = JF$, and that $JF(f) \circ c \subseteq d \circ J(f)$ in $\mathcal{K}\ell(\mathcal{P})$ if and only if $\mathcal{P}F(f) \circ c \subseteq d \circ f$ in **Sets**. Therefore, it is enough to show that the following oplax diagram commutes in **Sets**.

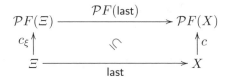

which we get from the following calculation

$$
\begin{aligned}
\mathcal{P}F(\mathsf{last}) \circ c_\xi &= \mathcal{P}F(\mathsf{last}) \circ \mathcal{P}(\mathsf{id} + F_\bullet \mathsf{cons}) \circ \mathcal{P}(\mathsf{id} + \mathsf{st}) \circ \mathcal{P}(\pi_2 + \langle \mathsf{id}, \pi_2 \rangle) \circ \\
&\qquad \mathcal{P}(\mathsf{dist}) \circ \mathsf{st} \circ \langle \mathsf{id}, \sigma \circ \xi \rangle \\
&= \mathcal{P}(\mathsf{id} + F_\bullet \mathsf{last}) \circ \mathcal{P}(\mathsf{id} + F_\bullet \mathsf{cons}) \circ \mathcal{P}(\mathsf{id} + \mathsf{st}) \circ \mathcal{P}(\pi_2 + \langle \mathsf{id}, \pi_2 \rangle) \circ \\
&\qquad \mathcal{P}(\mathsf{dist}) \circ \mathsf{st} \circ \langle \mathsf{id}, \sigma \circ \xi \rangle \\
&= \mathcal{P}(\mathsf{id} + F_\bullet \pi_3) \circ \mathcal{P}(\mathsf{id} + \mathsf{st}) \circ \mathcal{P}(\pi_2 + \langle \mathsf{id}, \pi_2 \rangle) \circ \\
&\qquad \mathcal{P}(\mathsf{dist}) \circ \mathsf{st} \circ \langle \mathsf{id}, \sigma \circ \xi \rangle \\
&= \mathcal{P}(\mathsf{id} + \pi_3) \circ \mathcal{P}(\pi_2 + \langle \mathsf{id}, \pi_2 \rangle) \circ \mathcal{P}(\mathsf{dist}) \circ \mathsf{st} \circ \langle \mathsf{id}, \sigma \circ \xi \rangle \\
&= \mathcal{P}(\pi_2 + \pi_2) \circ \mathcal{P}(\mathsf{dist}) \circ \mathsf{st} \circ \langle \mathsf{id}, \sigma \circ \xi \rangle \\
&= \mathcal{P}(\pi_2) \circ \mathsf{st} \circ \langle \mathsf{id}, \sigma \circ \xi \rangle \\
&= \pi_2 \circ \langle \mathsf{id}, \sigma \circ \xi \rangle \\
&= \sigma \circ \xi \\
&\subseteq c \circ \mathsf{last}
\end{aligned}
$$

where the inclusion holds by the definition of a scheduler. $\square$

Definition 4. *For a coalgebra* $c \colon X \to \mathcal{P}F(X)$ *we define the fat "scheduler" trace map* $\mathit{fstr}_c \colon X \to \mathcal{P}(I)$ *as:*

$$
\mathit{fstr}_c(x) = \bigcup \{ \mathit{ftr}_{c_\xi}(\mathsf{in}(x)) \mid \xi \text{ is scheduler for } c \}.
$$

Theorem 2 (Soundness). *The fat scheduler traces are contained in the fat traces, that is, for any coalgebra* $c \colon X \to \mathcal{P}F(X)$ *it holds that*

$$
\mathit{fstr}_c \subseteq \mathit{ftr}_c.
$$

Proof. By Lemma 2 we have that $J(\mathsf{last})$ is an oplax morphism from $c_\xi \colon \varXi \to F(\varXi)$ to $c \colon X \to F(X)$ in $\mathcal{K}\ell(\mathcal{P})$. Therefore, Corollary 1 yields $\mathit{ftr}_{c_\xi} \subseteq \mathit{ftr}_c \circ J(\mathsf{last})$ in $\mathcal{K}\ell(\mathcal{P})$, or equivalently $\mathit{ftr}_{c_\xi} \subseteq \mathit{ftr}_c \circ \mathsf{last}$, with composition in **Sets**, which further implies that

$$
\mathit{ftr}_{c_\xi} \circ \mathsf{in} \subseteq \mathit{ftr}_c \circ \mathsf{last} \circ \mathsf{in} = \mathit{ftr}_c. \qquad \square
$$

We end with a definition, which introduces many new questions, such as: which types of scheduling are complete for which functors. These questions will be postponed to future work.

Definition 5. *A scheduler type* $\sigma \colon S \Rightarrow \mathcal{P}$ *is fat complete if* $\mathit{fstr}_c = \mathit{ftr}_c$ *for any coalgebra* $c \colon X \to \mathcal{P}F(X)$.

9   Conclusions and Future Work

This paper deepens the study of "traces" in a coalgebraic setting and brings schedulers within scope. Many research issues remain, like completeness of scheduling, for instance for deterministic or probabilistic systems, or scheduling for other monads than powerset. Also, the relevance of derivatives of functors—capturing the idea of a hole where a scheduler should continue—needs further investigation.

References

1. Abbott, M., Altenkirch, T., Ghani, N., McBride, C.: Derivatives of Containers. In: Hofmann, M.O. (ed.) TLCA 2003. LNCS, vol. 2701, pp. 16–30. Springer, Heidelberg (2003)
2. Bhargava, M., Palamidessi, C.: Probabilistic anonymity. In: Abadi, M., de Alfaro, L. (eds.) CONCUR 2005. LNCS, vol. 3653, pp. 171–185. Springer, Heidelberg (2005)
3. Carboni, A., Lack, S., Walters, R.F.C.: Introduction to extensive and distributive categories. JPAA 84(2), 145–158 (1993)
4. Chatzikokolakis, K., Palamidessi, C.: Making random choices invisible to the scheduler. In: Caires, L., Vasconcelos, V.T. (eds.) CONCUR 2007. LNCS, vol. 4703, pp. 42–58. Springer, Heidelberg (2007)
5. Cheung, L.: Reconciling Nondeterministic and Probabilistic Choices. PhD thesis, Radboud University Nijmegen (2006)
6. van Glabbeek, R.J.: The linear time – branching time spectrum (extended abstract). In: Baeten, J.C.M., Klop, J.W. (eds.) CONCUR 1990. LNCS, vol. 458, pp. 278–297. Springer, Heidelberg (1990)
7. Hasuo, I.: Generic forward and backward simulations. In: Baier, C., Hermanns, H. (eds.) CONCUR 2006. LNCS, vol. 4137, pp. 406–420. Springer, Heidelberg (2006)
8. Hasuo, I.: Tracing Anonymity with Coalgebras. PhD thesis, Radboud University Nijmegen (2008)
9. Hasuo, I., Jacobs, B.: Context-free languages via coalgebraic trace semantics. In: Fiadeiro, J.L., Harman, N.A., Roggenbach, M., Rutten, J. (eds.) CALCO 2005. LNCS, vol. 3629, pp. 213–231. Springer, Heidelberg (2005)
10. Hasuo, I., Jacobs, B., Sokolova, A.: Generic trace semantics via coinduction. LMCS, 3(4:11) (2007)
11. Hasuo, I., Jacobs, B., Sokolova, A.: The microcosm principle and concurrency in coalgebra. In: Amadio, R.M. (ed.) FOSSACS 2008. LNCS, vol. 4962, pp. 246–260. Springer, Heidelberg (2008)
12. Jacobs, B.: Trace semantics for coalgebras. In: Adámek, J., Milius, S. (eds.) CMCS 2004. ENTCS, vol. 106 (2004)
13. Joyal, A.: Foncteurs analytiques et espces de structures. In: Labelle, G., Leroux, P. (eds.) Combinatoire Enumerative. LNM, vol. 1234, pp. 126–159. Springer, Berlin (1986)
14. Segala, R.: A compositional trace-based semantics for probabilistic automata. In: Lee, I., Smolka, S.A. (eds.) CONCUR 1995. LNCS, vol. 962, pp. 234–248. Springer, Heidelberg (1995)
15. Segala, R.: Modeling and verification of randomized distributed real-time systems. PhD thesis, MIT (1995)
16. Varacca, D., Winskel, G.: Distributing probability over non-determinism. MSCS 16, 87–113 (2006)

Non-strongly Stable Orders Also Define Interesting Simulation Relations[*]

Ignacio Fábregas, David de Frutos Escrig, and Miguel Palomino

Departamento de Sistemas Informáticos y Computación, UCM
fabregas@fdi.ucm.es, {miguelpt,defrutos}@sip.ucm.es

Abstract. We present a study of the notion of coalgebraic simulation introduced by Hughes and Jacobs. Although in their original paper they allow any functorial order in their definition of coalgebraic simulation, for the simulation relations to have good properties they focus their attention on functors with orders which are strongly stable. This guarantees a so-called "composition-preserving" property from which all the desired good properties follow. We have noticed that the notion of strong stability not only ensures such good properties but also "distinguishes the direction" of the simulation. For example, the classic notion of simulation for labeled transition systems, the relation "p is simulated by q", can be defined as a coalgebraic simulation relation by means of a strongly stable order, whereas the opposite relation, "p simulates q", cannot. Our study was motivated by some interesting classes of simulations that illustrate the application of these results: covariant-contravariant simulations and conformance simulations.

1 Introduction and Presentation of Our New Results

Simulations are a very natural way to compare systems defined by transition systems or other related mechanisms based on the description of systems by means of the actions they can execute at each of their states [11]. They can be enriched in several ways to obtain, in particular, the important ready simulation semantics [2,8], as well as other more elaborated ones such as nested simulations [5]. Quite recently we have studied the general concept of constrained simulation [3], proving that all the simulation relations constrained by an adequate condition have similar properties. The semantics of these constrained simulations is also the basis for our unified presentation of the semantics of processes [4], where all the semantics in the ltbt-spectrum [13] (and other new semantics) are classified in a systematic way.

Hughes and Jacobs [6] have also developed a systematic study of simulation-like relations, this time in a purely coalgebraic context, so that simulations are studied in connection with bisimulations [11], the fundamental concept to define equivalence in the coalgebraic world. Their coalgebraic simulations are defined in terms of an order ⊑ associated to the functor F corresponding to the coalgebra $c : X \longrightarrow FX$ that we want to observe. In this way they obtain a very general notion of coalgebraic simulation, not only because all functors F are considered, including in particular the important

---

[*] Research supported by the Spanish projects DESAFIOS TIN2006-15660-C02-01, WEST TIN2006-15578-C02-01, PROMESAS S-0505/TIC/0407 and UCM-BSCH GR58/08/910606.

A. Kurz, M. Lenisa, and A. Tarlecki (Eds.): CALCO 2009, LNCS 5728, pp. 221–235, 2009.

class of polynomial functors, but also because by changing the family of orders $\sqsubseteq_X$ many different families of simulation relations can be obtained. The general properties of these simulations can be studied in the defined coalgebraic framework, thus avoiding the need of similar proofs for each of the particular classes of simulations.

Certainly, this generic presentation of the notion of coalgebraic simulation has as advantage that it provides a wide and abstract framework where one can try to isolate and take advantage of the main properties of all the simulation-like relations. However, at the same time it can be argued that the proposal fails to capture in a tight manner the spirit of simulation relations because, in addition to the natural notions of simulations, the framework also allows for other less interesting relations. This has as a result that some natural properties of simulations cannot be proved in general, simply due to the fact that they are not satisfied by all of the permitted coalgebraic simulation relations. For instance, the induced similarity relation between systems is not always an order because transitivity is not always satisfied. In order to guarantee transivity, and other related properties of coalgebraic simulations, Jacobs and Hughes introduce in [7] the composition-preserving property to the order $\sqsubseteq$ that induces the simulation relation. In [6] they continue with the study of the topic and present *stability* of orders as a natural categorical property to guarantee that an order is composition-preserving. They also comment that stability is not easy to check and introduce a stronger condition (that we will call right-stability) so that, whenever applicable, the checking of the main properties of coalgebraic simulations becomes much simpler than in the general case.

Roughly speaking, given an order $\sqsubseteq_X$ on FX for each set X, the induced coalgebraic simulations are defined in the same way as bisimulations for F, but allowing a double application of $\sqsubseteq$ on the two sides of the defined relation. More precisely, instead of the functor $\mathrm{Rel}(F)$ defining plain bisimulations, $\mathrm{Rel}_{\sqsubseteq}(F)$ defined as $\sqsubseteq_Y \circ \mathrm{Rel}(F) \circ \sqsubseteq_X$ is used. There are several interesting facts hidden behind the apparent simplicity of this definition. The first one is that, in general, it only defines an order and not an equivalence relation, even if it is based on bisimulations (that always define an equivalence relation, namely, bisimilarity). The reason is that the order $\sqsubseteq$ appears "in the same direction" on both sides of the definition, thus breaking its symmetry. However, we can also define some equivalence relations weaker than bisimilarity by using an equivalence relation $\equiv$ as the order $\sqsubseteq$. Another interesting fact is that whenever we define a coalgebraic simulation by using $\sqsubseteq$, the inverse order $\sqsupseteq$ defines the inverse relation of that defined by $\sqsubseteq$ once we also interchange the roles of the related sets X and Y (so we could say that we are defining in fact the same relation but looking at it from the other side). Stability is also a symmetric condition, so that whenever an order $\sqsubseteq$ on a functor F is stable, the inverse order $\sqsupseteq$ is stable for F, too. This is quite reasonable, since stability is imposed in order to guaratee transitivity of the generated similarity relation and the inverse of a transitive relation is also transitive, so that whenever $\sqsubseteq$ generates an "admissible" similarity relation (meaning that it is an order), the inverse order $\sqsupseteq$ must be also admissible.

It is worth noting that the stronger condition guaranteeing stability is asymmetric. In fact, Hughes and Jacobs prove in [6] that "right-stability" implies that

$$\mathrm{Rel}(F)(R) \circ \sqsubseteq_X \subseteq \sqsubseteq_Y \circ \mathrm{Rel}(F)(R), \tag{1}$$

which in fact motivates our name for the condition.

A second surprise was to notice that, in most cases, right-stability induces a "natural direction" on the orders defining the coalgebraic simulation. For instance, for plain similarity over labeled transition systems, the inclusion order $\subseteq$ induces the classic simulation relation while the reversed inclusion $\supseteq$ induces the opposite "simulated by" relation: the first one is right-stable while the second is not.

All these general results arose when trying to integrate two new simulation-like notions as coalgebraic simulations definable by a stable order, so that we could obtain for free all the good properties that have been proved in [6] for this class of relations.

The first new simulation notion is that of covariant-contravariant simulations, where the alphabet of actions Act is partitioned into three disjoint sets Act^l, Act^r, and Act^{bi}. The intention is for the simulation to treat the actions in Act^l like in the ordinary case, to interchange the role of the related processes for those actions in Act^r, and to impose a symmetric condition like that defining bisimulation for the actions in Act^{bi}.

The second notion, conformance simulations, captures the conformance relations [9,12] that several authors introduced in order to formalize the notion of possible implementations. Like covariant-contravariant simulations, they can be defined as coalgebraic simulations for some stable order which is not right-stable neither left-stable. We show that the good properties of these two classes of orders are preserved in those orders that can be seen as a kind of composition of right-stable and left-stable orders. We use this fact to derive the stability of the orders defining both covariant-contravariant and conformance simulations.

2 Coalgebraic Simulations and Stability

Given a category $\mathbb{C}$ and an endofunctor F in $\mathbb{C}$, an F-coalgebra, or just a coalgebra, consists of an object $X \in \mathbb{C}$ together with a morphism $c : X \longrightarrow FX$. We often call X the state space and c the transition or coalgebra structure.

An arbitrary endofunctor $F : \textbf{Sets} \longrightarrow \textbf{Sets}$ can be lifted to a functor in the category **Rel** over $\textbf{Sets} \times \textbf{Sets}$ of relations, $\text{Rel}(F) : \textbf{Rel} \longrightarrow \textbf{Rel}$. In set-theoretic terms, for a relation $R \subseteq X_1 \times X_2$,

$$\text{Rel}(F)(R) = \{\langle u, v \rangle \in FX_1 \times FX_2 \mid \exists w \in F(R). F(r_1)(w) = u, F(r_2)(w) = v\}.$$

A *bisimulation* for coalgebras $c : X \longrightarrow FX$ and $d : Y \longrightarrow FY$ is a relation $R \subseteq X \times Y$ which is "closed under c and d":

$$\text{if } (x, y) \in R \text{ then } (c(x), d(y)) \in \text{Rel}(F)(R),$$

where the r_i are the projections of R into X and Y. Sometimes we shall use the term F-bisimulation to emphasize the functor we are working with.

Bisimulations can also be characterized by means of spans, using the general categorical definition by Aczel and Mendler [1]:

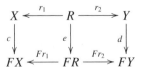

R is a bisimulation iff it is the carrier of some coalgebra e making the above diagram commute. Alternatively, bisimulations can also be defined as the $\text{Rel}(F)$-coalgebras in the category **Rel**.

We will also need the general concept of simulation introduced by Hughes and Jacobs [6] using orders on functors. Let $F : \textbf{Sets} \longrightarrow \textbf{Sets}$ be a functor. *An order on F is defined by means of a functorial collection of preorders* $\sqsubseteq_X \subseteq FX \times FX$ that must be preserved by renaming: for every $f : X \longrightarrow Y$, if $u \sqsubseteq_X u'$ then $Ff(u) \sqsubseteq_Y Ff(u')$.

Given an order $\sqsubseteq$ on F, a $\sqsubseteq$-*simulation* for coalgebras $c : X \longrightarrow FX$ and $d : Y \longrightarrow FY$ is a relation $R \subseteq X \times Y$ such that

$$\text{if } (x, y) \in R \text{ then } (c(x), d(y)) \in \text{Rel}_\sqsubseteq(F)(R),$$

where the lax relation lifting $\text{Rel}_\sqsubseteq(F)(R)$ is $\sqsubseteq_Y \circ \text{Rel}(F)(R) \circ \sqsubseteq_X$, which can be expanded to

$$\text{Rel}_\sqsubseteq(F)(R) = \{(u, v) \mid \exists w \in F(R). \, u \sqsubseteq_X Fr_1(w) \wedge Fr_2(w) \sqsubseteq_Y v\}.$$

Alternatively, $\sqsubseteq$-simulations are just the $\text{Rel}_\sqsubseteq(F)$-coalgebras in **Rel**.

Sometimes, when $f : X \longrightarrow Y$ and $A \subseteq X$ we will simply write $f(A)$ for the image $\coprod_f(A)$.

A functor with order $\sqsubseteq$ is *stable* [6] if the relation lifting $\text{Rel}_\sqsubseteq(F)$ commutes with substitution, that is, if for every $f : X \longrightarrow Z$ and $g : Y \longrightarrow W$, $\text{Rel}_\sqsubseteq(F)((f \times g)^{-1}(R)) = (Ff \times Fg)^{-1}(\text{Rel}_\sqsubseteq(F)(R))$.[1] They also define a stronger condition that we are going to call right-stability.

Definition 1 ([6]). *We will say that a functor F with order $\sqsubseteq$ is **right-stable** if, for every function $f : X \longrightarrow Y$, we have*[2]

$$(id \times Ff)^{-1} \sqsubseteq_Y \; \subseteq \coprod_{Ff \times id} \sqsubseteq_X . \tag{2}$$

According to [6], condition (2) is equivalent to (a) F being stable and (b) for every relation $R \subseteq X \times Y$,

$$\text{Rel}(F)(R) \circ \sqsubseteq_X \; \subseteq \; \sqsubseteq_Y \circ \text{Rel}(F)(R). \tag{3}$$

Right-stability was introduced by arguing that it is easier to check than plain stability, while being satisfied by nearly all orders discussed in that paper. Surprisingly, one cannot find in [6] a clear explanation of the reason why right-stable orders are easier to analyze. In our opinion, the crucial fact is that from (3) we can immediately conclude that

$$\sqsubseteq_Y \circ \text{Rel}(F)(R) \circ \sqsubseteq_X \; = \; \sqsubseteq_Y \circ \text{Rel}(F)(R), \tag{4}$$

so that the coalgebraic simulations for a right-stable order $\sqsubseteq$ can be equivalently defined by means of the asymmetric definition on the right-hand side of equality (4). If the order $\sqsubseteq$ can be used only on one of the sides of the definition, the verification of the

---

[1] In fact, the inclusion $\subseteq$ always holds.

[2] Again, the other inclusion is always true since $\sqsubseteq$ functorial means that $Ff(u) \sqsubseteq_Y Ff(v)$ if $u \sqsubseteq_X v$.

properties of the induced coalgebraic simulations becomes much easier than when using the original definition.

It was quite surprising to discover that the easiest way to prove the properties of the "simulated by" relations which come from symmetric properties such as composition-preserving (that are also satisfied by the corresponding inverse relations "simulates") is to break that symmetry by considering the asymmetric definition of coalgebraic simulations that only use $\sqsubseteq_Y$; certainly, this is only possible when the defining order $\sqsubseteq$ is right-stable.

Stability is used in [6, Lemma 5.3] to prove that lax relation lifting preserves composition of relations, which is needed to prove [6, Lemma 5.4(2)], the crucial fact that the induced similarity relation is transitive; this need not be the case for the simulation notion defined by an arbitrary order $\sqsubseteq$.

3   On Stability of Simulation and Anti-simulation

Plain simulations between labeled transition systems can be defined as coalgebraic simulations considering the functor $F = \mathcal{P}^A$ (G^A denote the functor $X \mapsto (G(X))^A$) with the order $\sqsubseteq$ given by $\alpha \sqsubseteq \beta$ for $\alpha, \beta : A \longrightarrow \mathcal{P}X$ iff for all $a \in A$, $\alpha(a) \subseteq \beta(a)$.

Lemma 1. *The order $\sqsubseteq$ defining plain simulations for labeled transition systems is right-stable.*

Corollary 1. *Plain simulations between labeled transition systems can be defined as the $(\sqsubseteq_Y \circ \mathrm{Rel}(F))$-coalgebras.*

It is worth examining the consequences of the removal of $\sqsubseteq_X$ from the original definition of coalgebraic simulations in this particular case. Both $\sqsubseteq_X$ and $\sqsubseteq_Y$ correspond to the inclusion order, but when applied at the right-hand side it means that we can reduce the set of successors of the simulating process q when simulating the execution of a by p. This means that starting from a set $Y' \subseteq Y$ we can obtain an adequate subset $Y'' \subseteq Y'$. Instead, the application of $\sqsubseteq_X$ at the left-hand side allows to enlarge the set of successors of the simulated process p and this produces a set X'' larger than the given X': one could say that we need to consider "new" information not in X', while going from Y' to Y'' just "removes" some known information.

Another interesting point arises from the fact that every use of $\sqsubseteq_X$ at the left-hand side can be "compensated" by removing at Y the added states and this is why Corollary 1 was correct, because we can always avoid the introduction of new successors in the simulated process by simply removing them at the right-hand side. However, the opposite procedure, to compensate the removal of states by adding them at the simulated process side is not always possible, since in general X could be not big enough.

The anti-simulations can be defined as coalgebraic simulations by taking the reversed inclusion order instead of $\subseteq$. It is interesting to note that it is not right-stable as the following counterexample shows. Let $X = \{x\}$ and $Y = \{y_1, y_2\}$ and let $f : X \longrightarrow Y$ be such that $f(x) = y_1$. With these definitions the pair $(Y, X) \in (id \times \mathcal{P}f)^{-1}(\supseteq)$, since $Y \supseteq \{y_1\} = \mathcal{P}f(X)$, but it is obvious that there is no $A \subseteq X$ such that $Y = f(A)$ because f is not surjective.

However, the order defining anti-simulations is stable as a consequence of the following general result.

Lemma 2. *F with an order $\sqsubseteq$ is stable iff it is stable with the order $\sqsubseteq^{op}$.*

Proof. It is shown in [6, Lemma 4.2(4)] that $\mathrm{Rel}_{\sqsubseteq^{op}}(F)(R) = (\mathrm{Rel}_{\sqsubseteq}(F)(R^{op}))^{op}$. Then, on the one hand,

$$(Ff \times Fg)^{-1}(\mathrm{Rel}_{\sqsubseteq^{op}}(F)(R)) = (Ff \times Fg)^{-1}(\mathrm{Rel}_{\sqsubseteq}(F)(R^{op}))^{op}$$
$$= ((Fg \times Ff)^{-1}\mathrm{Rel}_{\sqsubseteq}(F)(R^{op}))^{op},$$

and on the other hand,

$$\mathrm{Rel}_{\sqsubseteq^{op}}(F)((f \times g)^{-1}(R)) = (\mathrm{Rel}_{\sqsubseteq}(F)((f \times g)^{-1}(R))^{op})^{op}$$
$$= (\mathrm{Rel}_{\sqsubseteq}(F)((g \times f)^{-1}(R^{op})))^{op}.$$

Since $R^{op} \subseteq Y \times X$ is a relation whenever $R \subseteq X \times Y$ is so, and f, g, and R are arbitrary, we have shown that

$$\mathrm{Rel}_{\sqsubseteq}(F)((f \times g)^{-1}(R)) = (Ff \times Fg)^{-1}(\mathrm{Rel}_{\sqsubseteq}(F)(R))$$

if and only if

$$\mathrm{Rel}_{\sqsubseteq^{op}}(F)((f \times g)^{-1}(R)) = (Ff \times Fg)^{-1}(\mathrm{Rel}_{\sqsubseteq^{op}}(F)(R)),$$

and therefore F is stable for $\sqsubseteq$ iff it is stable for $\sqsubseteq^{op}$. $\qquad\qquad\square$

Corollary 2. *The order $\sqsubseteq^{op}$ defining anti-simulations for transition systems as coalgebraic simulations is stable.*

One could conclude from the observation above that there is indeed a natural argument supporting plain similarity as a "right" coalgebraic similarity, definable by a right-stable order. This criterion could be adopted to define right coalgebraic simulations, which plain similarity would satisfy while the opposite relation "is simulated by" would not. However, we immediately noticed that we could define "left-stable" orders by interchanging the roles of Ff and id in the definition of right-stable order, obtaining the inverse inclusion in (1).

Definition 2. *We will say that a functor F with order $\sqsubseteq$ is **left-stable** if, for every function $f : X \longrightarrow Y$, we have*

$$(Ff \times id)^{-1} \sqsubseteq_Y \; \subseteq \coprod_{id \times Ff} \sqsubseteq_X . \tag{5}$$

It is inmediate to check that an order $\sqsubseteq$ is left-stable iff the inverse order $\sqsubseteq^{op}$ is right-stable. Moreover, left-stable orders have the same structural properties that right-stable ones so that, in particular, they are also stable and hence composition-preserving. But in this case it would be the inverse simulations, corresponding to the "is simulated by" notion, that would be natural instead of plain simulations. As a conclusion, we could use right or left-stability as a criterion to choose a natural direction for the simulation order. But the important fact in both cases is that the simplified asymmetric definitions (using either $\sqsubseteq_X$ or $\sqsubseteq_Y$) of coalgebraic simulations are much easier to handle than the symmetric original definition (where both $\sqsubseteq_X$ and $\sqsubseteq_Y$ have to be used).

4 Covariant-Contravariant Simulations and Conformance Simulations

Covariant-contravariant simulations are defined by combining the conditions "to simulate" and "be simulated by", using a partition of the alphabet of actions of the compared labeled transition systems.

Definition 3. *Given* $c : X \longrightarrow \mathcal{P}(X)^{Act}$ *and* $d : Y \longrightarrow \mathcal{P}(Y)^{Act}$ *labeled transition systems for the alphabet Act, and* $\{Act^r, Act^l, Act^{bi}\}$ *a partition of this alphabet, a* (Act^r, Act^l)-*simulation between c and d is a relation* $S \subseteq X \times Y$ *such that for every* $(x, y) \in S$ *we have:*

- *for all* $a \in Act^r \cup Act^{bi}$ *and all* $x \xrightarrow{a} x'$ *there exists* $y \xrightarrow{a} y'$ *with* $(x', y') \in S$.
- *for all* $a \in Act^l \cup Act^{bi}$, *and all* $y \xrightarrow{a} y'$ *there exists* $x \xrightarrow{a} x'$ *with* $(x', y') \in S$.

We write $x \,_{Act^r}S_{Act^l}\, y$, *and say that x is* (Act^r, Act^l)-*simulated by y, if and only if there exists some* (Act^r, Act^l)-*simulation S with* xSy.

A very interesting application of this kind of simulations is related with the definition of adequate simulation notions for input/output (I/O) automata [10]. The classic approach to simulations is based on the definition of semantics for reactive systems, where all the actions of the processes correspond to input actions that the user must trigger. Instead, whenever we have explicit output actions the situation is the opposite: it is the system that produces the actions and the user who is forced to accept the produced output. Then, it is natural to conclude that in the simulation framework we have to dualize the simulation condition when considering output actions, and this is exactly what our anti-simulation relations do.

Covariant-contravariant simulations can be easily obtained as coalgebraic simulations, as the following proposition proves.

Proposition 1. (Act^r, Act^l)-*simulations can be defined as the coalgebraic simulations for the functor* $F = \mathcal{P}^{Act}$ *with functorial order* $_{Act^r}\sqsubseteq_{Act^l}$ *where, for each set X and* $\alpha, \alpha' : Act \longrightarrow \mathcal{P}(X)$, *we have* $\alpha \,_{Act^r}\sqsubseteq_{Act^l}\, \alpha'$ *if:*

- *for all* $a \in Act^r \cup Act^{bi}$, $\alpha(a) \subseteq \alpha'(a)$, *and*
- *for all* $a \in Act^l \cup Act^{bi}$, $\alpha(a) \supseteq \alpha'(a)$.

Note that in particular we have $\alpha(a) = \alpha'(a)$ *for all* $a \in Act^{bi}$.

Proof. Intuitively, using the order $_{Act^r}\sqsubseteq_{Act^l}$ on the left-hand side of $\text{Rel}_\sqsubseteq(F)(R)$ allows us to remove a'-transitions when $a' \in Act^l$, whereas using it on the right-hand side of $\text{Rel}_\sqsubseteq(F)(R)$ allows us to remove a-transitions when $a \in Act^r$.

Let us suppose that we have a classic covariant-contravariant simulation $_{Act^r}S_{Act^l}$ between labeled transition systems $c : P \longrightarrow \mathcal{P}(P)^{Act}$ and $d : Q \longrightarrow \mathcal{P}(Q)^{Act}$ defined by $c(p)(a) = \{p' \mid p \xrightarrow{a} p'\}$ and $d(q)(a) = \{q' \mid q \xrightarrow{a} q'\}$. We must show that if $p \,_{Act^r}S_{Act^l}\, q$ then there exist p^* and q^* such that

$$c(p) \,_{Act^r}\sqsubseteq_{Act^l}\, p^* \text{Rel}(\mathcal{P}^{Act})(_{Act^r}S_{Act^l})q^* \,_{Act^r}\sqsubseteq_{Act^l}\, d(q). \tag{6}$$

We define p^* and q^* as follows:

- p^* has the same transitions as $c(p)$, except for those transitions $p \xrightarrow{a'} p'$ with $a' \in Act^l$ such that there is no q' with $q \xrightarrow{a'} q'$ and $p' \,_{Act^r}S_{Act^l}\, q'$.
- q^* has the same transitions as $d(q)$, except for those transitions $q \xrightarrow{a} q'$ with $a \in Act^r$ such that there is no p' with $p \xrightarrow{a} p'$ and $p' \,_{Act^r}S_{Act^l}\, q'$.

It is immediate from these definitions that $c(p) \,_{Act^r}\sqsubseteq_{Act^l}\, p^*$ and $q^* \,_{Act^r}\sqsubseteq_{Act^l}\, d(q)$, so we are left with checking that $p^*\text{Rel}(\mathcal{P}^{Act})q^*$.

Let $p' \in p^*(a)$ with $a \in Act^r$. By construction of p^*, since we have not dropped any a-transitions from p^*, $p \xrightarrow{a} p'$. Using the fact that $_{Act^r}S_{Act^l}$ is a classic covariant-contravariant simulation, there exists q' such that $q \xrightarrow{a} q'$ with $p' \,_{Act^r}S_{Act^l}\, q'$, and, again by construction, $q' \in q^*(a)$ because there is some $p \xrightarrow{a} p'$ with $p' \,_{Act^r}S_{Act^l}\, q'$. Similarly, if $p' \in p^*(a)$ with $a' \in Act^l$, by construction of p^* there must exist some q' such that $q \xrightarrow{a'} q'$ with $p' \,_{Act^r}S_{Act^l}\, q'$. Again, since we have not removed any a'-transitions from $d(q)$ in q^*, it must be true that $q' \in q^*(a)$. Finally, if $p' \in p^*(a)$ with $a \in Act^{bi}$ we have that $p \xrightarrow{a} p'$ and hence there exists q' such that $q \xrightarrow{a} q'$ with $p' \,_{Act^r}S_{Act^l}\, q'$, but also $q' \in q^*(a)$.

The argument that shows that for every $q' \in q^*(a)$ there exists some $p' \in p^*(a)$ with $p' \,_{Act^r}S_{Act^l}\, q'$ is analogous.

We show now the other implication, that a coalgebraic covariant-contravariant simulation is a classic one. In this case we start from coalgebras c and d that satisfy relation (6) whenever $p \,_{Act^r}S_{Act^l}\, q$.

If $p \xrightarrow{a} p'$ for $a \in Act^r$, then $p' \in p^*(a)$ because $c(p) \,_{Act^r}\sqsubseteq_{Act^l}\, p^*$ and, since $p^*\text{Rel}(\mathcal{P}^{Act})(_{Act^r}S_{Act^l})q^*$, there is some $q' \in q^*(a)$ with $p' \,_{Act^r}S_{Act^l}\, q'$. Again, the definition of $_{Act^r}\sqsubseteq_{Act^l}$ ensures that $q^*(a) \subseteq d(q)(a)$ and hence $q \xrightarrow{a} q'$ as required. Similarly, if $q \xrightarrow{a'} q'$ for $a' \in Act^l$, then $q' \in q^*(a)$ because $q^* \,_{Act^r}\sqsubseteq_{Act^l}\, d(q)$ and thus, as in the previous case, there exists $p' \in p^*(a)$ with $p' \,_{Act^r}S_{Act^l}\, q'$ and $p \xrightarrow{a'} p'$. Finally if $p \xrightarrow{a} p'$ for $a \in Act^{bi}$ (resp. $q \xrightarrow{a} q'$), again by the definition of $_{Act^r}\sqsubseteq_{Act^l}$ we have $p' \in p^*(a)$ (resp. $q' \in q^*(a)$) and, from $p^*\text{Rel}(\mathcal{P}^{Act})(_{Act^r}S_{Act^l})q^*$, it follows that there exists $q' \in q^*(a)$ (resp. $p' \in p^*(a)$) such that $p' \,_{Act^r}S_{Act^l}\, q'$; by the definition of $_{Act^r}\sqsubseteq_{Act^l}$, $q \xrightarrow{a} q'$ (resp. $p \xrightarrow{a} p'$). $\qquad\square$

The other new kind of simulations in which we are interested is that of conformance simulations, where the conformance relation in [9,12] meets the simulation world in a nice way. In the definition below we will write $p \xrightarrow{a}$ if $p \xrightarrow{a} p'$ for some p'.

Definition 4. *Given $c : X \longrightarrow \mathcal{P}(X)^A$ and $d : Y \longrightarrow \mathcal{P}(Y)^A$ two labeled transition systems for the alphabet A, a **conformance simulation** between them is a relation $R \subseteq X \times Y$ such that whenever pRq, then:*

- *For all $a \in A$, if $p \xrightarrow{a}$ we must also have $q \xrightarrow{a}$ (this means, using the usual notation for process algebras, that $I(p) \subseteq I(q)$).*
- *For all $a \in A$ such that $q \xrightarrow{a} q'$ and $p \xrightarrow{a}$, there exists some p' with $p \xrightarrow{a} p'$ and $p'Rq'$.*

Conformance simulations allow the extension of the set of actions offered by a process, so that in particular we will have $a < a + b$, but they also consider that a process can be "improved" by reducing the nondeterminism in it, so that $ap + aq < ap$. In this way we have again a kind of covariant-contravariant simulation, not driven by the alphabet of actions executed by the processes but by their nondeterminism.

Once again, conformance simulations can be defined as coalgebraic simulations taking the adequate order on the functor defining labeled transition systems.

Proposition 2. *Conformance simulations can be obtained as the coalgebraic simulations for the order $\sqsubseteq^{Conf}$ on the functor $\mathcal{P}^A$, where for any set X we have $u \sqsubseteq_X^{Conf} v$ if for every $u, v : A \longrightarrow \mathcal{P}X$ and $a \in A$:*

- *either $u(a) = \emptyset$, or*
- *$u(a) \supseteq v(a)$ and $v(a) \neq \emptyset$.*

Proof. Let us first prove that $\sqsubseteq_X^{Conf}$ is indeed an order. It is clear that the only not immediate property is transitivity. To check it, let us take $u \sqsubseteq_X^{Conf} v \sqsubseteq_Y^{Conf} w$: if $u(a) = \emptyset$ we are done; otherwise, we have $u(a) \supseteq v(a)$ and $v(a) \neq \emptyset$, so that we also have $v(a) \supseteq w(a)$ and $w(a) \neq \emptyset$, obtaining $u(a) \supseteq w(a)$ and $w(a) \neq \emptyset$.

Now, we can interpret that using the order $\sqsubseteq^{Conf}$ on the left-hand side of $\text{Rel}_\sqsubseteq(F)(R)$ allows us to remove all a-transitions except for the last one, whereas using it on the right-hand side allows us to remove all b-transitions for $b \in B$, where B is any set of actions. But again, as in the proof of Proposition 1, we can compensate these additions with the corresponding removals at the other side and the proof follows in an analogous way.   □

Next we check that the order $_{Act^r}\sqsubseteq_{Act^l}$ defining covariant-contravariant simulations is stable.

Lemma 3. *Given a partition $\{Act^r, Act^l, Act^{bi}\}$ of Act the order $_{Act^r}\sqsubseteq_{Act^l}$ for the functor $\mathcal{P}^{Act}$ defining covariant-contravariant simulations for transition systems is stable.*

Proof. It is clear that the order $_{Act^r}\sqsubseteq_{Act^l}$ can be obtained as the product of a family of orders $\sqsubseteq^a$ for the functor $\mathcal{P}$, with $a \in Act$. This is indeed the case taking $\sqsubseteq_X^a = \subseteq_X$ for $a \in Act^r$, $\sqsubseteq_X^a = \supseteq_X$ for $a \in Act^l$ and $\sqsubseteq_X^a = =_X$ for $a \in Act^{bi}$. Then it is easy to see that to obtain that $_{Act^r}\sqsubseteq_{Act^l}$ is stable it is enough to prove that each of the orders $\sqsubseteq^a$ is stable.

This latter requirement is straightforward because, for $a \in Act^r$, $\sqsubseteq^a$ is right-stable; for $a \in Act^l$ the order $\sqsubseteq^a$ is left-stable; and for $a \in Act^{bi}$, $\sqsubseteq^a$ is the equality relation, which is both right and left-stable, for every functor F.   □

Certainly, the order defining covariant-contravariant simulations is not right-stable nor left-stable, but in the proof above we have used the power of these two properties thanks to the fact that the order $_{Act^r}\sqsubseteq_{Act^l}$ can be factorised as the product of a family of orders that are either right-stable or left-stable. Then we can obtain the following sequence of general definitions and results, from which Lemma 3 could be obtained as a simple particular case.[3]

---

[3] Instead of removing the above, we have preferred to maintain the sequence of results in the order in which we got them, starting with our motivating example.

Definition 5. *We say that an order $\sqsubseteq$ on a functor F^A is **action-distributive** if there is a family of orders $\sqsubseteq^a$ on F such that*

$$f \sqsubseteq g \iff f(a) \sqsubseteq^a g(a) \text{ for all } a \in A.$$

Whenever $\sqsubseteq$ can be distributed in this way we will write $\sqsubseteq = \prod_{a \in A} \sqsubseteq^a$.

Definition 6. *We say that an action-distributive order $\sqsubseteq$ on F^A is **side stable** if for the decomposition $\sqsubseteq = \prod_{a \in A} \sqsubseteq^a$ we have that each order $\sqsubseteq^a$ is either right-stable or left-stable.*

By separating the right-stable and the left-stable components we obtain $\sqsubseteq = \sqsubseteq^l \times \sqsubseteq^r$, where A^r (resp. A^l) collects the set of arguments[4] $a \in A$ with $\sqsubseteq^a$ right-stable (resp. left-stable). We extend $\sqsubseteq^l$ and $\sqsubseteq^r$ to obtain a pair of orders on F^A, $\sqsubseteq^{\bar{l}}$ and $\sqsubseteq^{\bar{r}}$, defined by:

- $f \sqsubseteq^{\bar{r}} g$ iff $f(a) \sqsubseteq^a g(a)$ for all $a \in A^r$ and $f(a) = g(a)$ for all $a \in A^l$.
- $f \sqsubseteq^{\bar{l}} g$ iff $f(a) \sqsubseteq^a g(a)$ for all $a \in A^l$ and $f(a) = g(a)$ for all $a \in A^r$.

Proposition 3. *The order $\sqsubseteq^{\bar{l}}$ is left-stable, while $\sqsubseteq^{\bar{r}}$ is right-stable. We have $\sqsubseteq = (\sqsubseteq^{\bar{l}} \circ \sqsubseteq^{\bar{r}}) = (\sqsubseteq^{\bar{r}} \circ \sqsubseteq^{\bar{l}})$, and therefore we also have $\sqsubseteq = (\sqsubseteq^{\bar{l}} \cup \sqsubseteq^{\bar{r}})^*$.*

Proposition 4. *For any side stable order $\sqsubseteq$ on F^A, if we have a decomposition $\sqsubseteq = \sqsubseteq^l \times \sqsubseteq^r$ based on a partition of A into a set of right-stable components A^r and another set of left-stable components A^l, then we can obtain the coalgebraic simulations for $\sqsubseteq$ as the $(\sqsubseteq^{\bar{r}}_Y \circ \mathrm{Rel}(F) \circ \sqsubseteq^{\bar{l}}_X)$-coalgebras.*

Proof. By definition, $\mathrm{Rel}_{\sqsubseteq}(F)(R) = \sqsubseteq_Y \circ \mathrm{Rel}(F)(R) \circ \sqsubseteq_X$. Since $\sqsubseteq = (\sqsubseteq^{\bar{r}} \circ \sqsubseteq^{\bar{l}}) = (\sqsubseteq^{\bar{l}} \circ \sqsubseteq^{\bar{r}})$, we have:

$$
\begin{aligned}
\sqsubseteq_Y \circ \mathrm{Rel}(F)(R) \circ \sqsubseteq_X &= (\sqsubseteq^{\bar{l}}_Y \circ \sqsubseteq^{\bar{r}}_Y) \circ \mathrm{Rel}(F)(R) \circ (\sqsubseteq^{\bar{r}}_X \circ \sqsubseteq^{\bar{l}}_X) \\
&= \sqsubseteq^{\bar{l}}_Y \circ (\sqsubseteq^{\bar{r}}_Y \circ \mathrm{Rel}(F)(R) \circ \sqsubseteq^{\bar{r}}_X) \circ \sqsubseteq^{\bar{l}}_X \\
&= (\sqsubseteq^{\bar{l}}_Y \circ \sqsubseteq^{\bar{r}}_Y) \circ \mathrm{Rel}(F)(R) \circ \sqsubseteq^{\bar{l}}_X \quad \text{(by right-stability of } \sqsubseteq^{\bar{r}}) \\
&= \sqsubseteq^{\bar{r}}_Y \circ (\sqsubseteq^{\bar{l}}_Y \circ \mathrm{Rel}(F)(R) \circ \sqsubseteq^{\bar{l}}_X) \quad \text{(since } \sqsubseteq^{\bar{r}} \text{ and } \sqsubseteq^{\bar{l}} \text{ commute)} \\
&= \sqsubseteq^{\bar{r}}_Y \circ \mathrm{Rel}(F)(R) \circ \sqsubseteq^{\bar{l}}_X \quad \text{(by left-stability of } \sqsubseteq^{\bar{l}})
\end{aligned}
$$

$\square$

The characterization above still requires the use of the order on both sides of the $\mathrm{Rel}(F)(R)$ operator. However, the fact that $\sqsubseteq^{\bar{r}}_Y$ (resp. $\sqsubseteq^{\bar{l}}_X$) is right-stable (resp. left-stable) makes the application of this decomposition as simple as when coping with either a right or left-stable order.

Proposition 5. *If $\sqsubseteq = \prod_{a \in A} \sqsubseteq^a$ and $\sqsubseteq^a$ is stable for all $a \in A$, then $\sqsubseteq$ is stable.*

---

[4] We have assumed here a partition $\{A^l, A^r\}$ of the set A into two sets of right-stable and left-stable components. Obviously, if there were some arguments $a \in A$ on which $\sqsubseteq^a$ is both right-stable and left-stable then the decomposition would not be unique, but the result would be valid for any such decomposition.

Proof. The result follows from the following chain of implications:

$$(u, v) \in (Ff \times Fg)^{-1}\text{Rel}_{\sqsubseteq}(F)(R)$$
$$\Longleftrightarrow \quad Ff(u) \sqsubseteq z'\text{Rel}(F)(R)w' \sqsubseteq Fg(v)$$
$$\Longleftrightarrow \quad Ff(u)(a) \sqsubseteq^a z'(a)\text{Rel}(F^a)(R)w'(a) \sqsubseteq^a Fg(v)(a), \text{ for all } a$$
$$\Longleftrightarrow \quad (u(a), v(a)) \in (Ff \times Fg)^{-1}\text{Rel}_{\sqsubseteq^a}(F)(R), \text{ for all } a$$
$$\Longrightarrow \quad (u(a), v(a)) \in \text{Rel}_{\sqsubseteq^a}(F)((f \times g)^{-1}R), \text{ for all } a$$
$$\Longleftrightarrow \quad u(a) \sqsubseteq^a x'(a)\text{Rel}(F)((f \times g)^{-1}R)y'(a) \sqsubseteq^a v(a), \text{ for all } a$$
$$\Longleftrightarrow \quad (u, v) \in \text{Rel}_{\sqsubseteq}(F)((f \times g)^{-1}R) \qquad \qquad \square$$

Corollary 3. *Any side stable order is stable.*

Corollary 4. *The order* $_{Act'}\sqsubseteq_{Act^l}$ *defining covariant-contravariant simulations is side stable and therefore it is stable too.*

Next we consider the case of conformance simulations, for which we can obtain similar results to those proved for covariant-contravariant simulations.

Lemma 4. *The order* $\sqsubseteq^{Conf}$ *defining conformance simulations for transition systems is stable.*

Proof. Let $R \subseteq Z \times W$ be a relation and $f : X \longrightarrow Z$, $g : Y \longrightarrow W$ arbitrary functions. If $(u, v) \in (\mathcal{P}^A f \times \mathcal{P}^A g)^{-1}(\text{Rel}_{\sqsubseteq^{Conf}}(\mathcal{P}^A)(R))$, then there exist z and w such that

$$\mathcal{P}^A f(u) \sqsubseteq^{Conf} z \, \text{Rel}(\mathcal{P}^A)(R) \, w \sqsubseteq^{Conf} \mathcal{P}^A g(v). \tag{7}$$

We have to show that $(u, v) \in \text{Rel}_{\sqsubseteq^{Conf}}(\mathcal{P}^A)((f \times g)^{-1}(R))$, that is, there exist x and y such that

$$u \sqsubseteq^{Conf} x \, \text{Rel}(\mathcal{P}^A)((f \times g)^{-1}(R)) \, y \sqsubseteq^{Conf} v.$$

Let us define $x : A \longrightarrow \mathcal{P}(X)$ by $x(a) = u(a) \cap f^{-1}(z(a))$ and $y : A \longrightarrow \mathcal{P}(Y)$ by $y(a) = g^{-1}(w(a))$. Then we have:

1. $u \sqsubseteq^{Conf} x$.
   If $u(a) = \emptyset$, there is nothing to prove. Otherwise, since $\mathcal{P}^A f(u) \sqsubseteq^{Conf} z$ and $f(u(a)) \neq \emptyset$, we have $f(u(a)) \supseteq z(a) \neq \emptyset$ and hence $u(a) \supseteq u(a) \cap f^{-1}(z(a)) = x(a) \neq \emptyset$.
2. $y \sqsubseteq^{Conf} v$.
   If $w(a) = \emptyset$, then $y(a) = g^{-1}(w(a)) = \emptyset$. Otherwise, since $w \sqsubseteq^{Conf} \mathcal{P}^A g(v)$, we have $w(a) \supseteq g(v(a)) \neq \emptyset$, so that $v(a) \neq \emptyset$ and $y(a) = g^{-1}(w(a)) \supseteq g^{-1}(g(v(a))) \supseteq v(a)$.
3. $x \, \text{Rel}(\mathcal{P}^A)((f \times g)^{-1}(R)) \, y$.
   For every $a \in A$ we need to show that $x(a) \, \text{Rel}(\mathcal{P})((f \times g)^{-1}(R)) \, y(a)$, which means:
   (a) for every $p \in x(a)$ there exists $q \in y(a)$ such that $p \, (f \times g)^{-1}(R) \, q$, that is, $f(p)Rg(q)$; and
   (b) for every $q \in y(a)$ there exists $p \in x(a)$ such that $p \, (f \times g)^{-1}(R) \, q$, that is, $f(p)Rg(q)$.
   In the first case, let $p \in x(a)$; by definition of x, $f(p) \in z(a)$. Now, from $z \, \text{Rel}(\mathcal{P}^A)(R) \, w$ we obtain that for each $p' \in z(a)$ there exists $q' \in w(a)$ such that $p'Rq'$. Then, for $f(p) \in z(a)$ there exists $q' \in w(a)$ with $f(p)Rq'$; and by definition of y, there exists $q \in y(a)$ with $q' = g(q)$ as required.
   In the second case, let $q \in y(a)$ so that $g(q) \in w(a)$. Again, from $z \, \text{Rel}(\mathcal{P}^A)(R) \, w$ it follows that there is $p' \in z(a)$ with $p'Rg(q)$. Now, $f(u(a)) \supseteq z(a)$ because $u \sqsubseteq^{Conf} z$, so there exists $p \in u(a) \cap f^{-1}(z(a))$ with $f(p) = p'$, as required. $\qquad \square$

As in the case of covariant-contravariant simulations, conformance simulations cannot be defined as coalgebraic simulations using neither a right-stable order nor a left-stable order. But we can find in the arguments above the basis for a decomposition of the involved order $\sqsubseteq^{Conf}$, according to the two cases in its definition. Once again $\sqsubseteq^{Conf}$ is an action-distributive order on $\mathcal{P}^A$, but in order to obtain the adequate decomposition of $\sqsubseteq^{Conf}$ now we also need to decompose the component orders $\sqsubseteq^a$.

Definition 7. We define the conformance orders $\sqsubseteq^{C\neg\emptyset}$, $\sqsubseteq^{C\emptyset}$, and $\sqsubseteq^C$ on the functor $\mathcal{P}$ by:

- $x_1 \sqsubseteq^{C\emptyset} x_2$ if $x_1 = \emptyset$ or $x_1 = x_2$.
- $x_1 \sqsubseteq^{C\neg\emptyset} x_2$ if $x_1 \supseteq x_2$ and $x_2 \neq \emptyset$, or $x_1 = x_2$.
- $x_1 \sqsubseteq^C x_2$ if $x_1 \sqsubseteq^{C\neg\emptyset} x_2$ or $x_1 \sqsubseteq^{C\emptyset} x_2$.

Proposition 6. The two relations $\sqsubseteq^{C\emptyset}$ and $\sqsubseteq^{C\neg\emptyset}$ commute with each other:

$$(\sqsubseteq^{C\emptyset} \circ \sqsubseteq^{C\neg\emptyset}) = (\sqsubseteq^{C\neg\emptyset} \circ \sqsubseteq^{C\emptyset}),$$

from where it follows that $(\sqsubseteq^{C\emptyset} \cup \sqsubseteq^{C\neg\emptyset})^* = (\sqsubseteq^{C\emptyset} \circ \sqsubseteq^{C\neg\emptyset}) = (\sqsubseteq^{C\neg\emptyset} \circ \sqsubseteq^{C\emptyset})$. We also have $\sqsubseteq^C = (\sqsubseteq^{C\emptyset} \circ \sqsubseteq^{C\neg\emptyset})$, from where we conclude that $\sqsubseteq^C$ is indeed an order relation.

Proof. Let $u (\sqsubseteq^{C\emptyset} \circ \sqsubseteq^{C\neg\emptyset}) v$: there is some w such that $u \sqsubseteq^{C\neg\emptyset} w$ and $w \sqsubseteq^{C\emptyset} v$. We need to find w' such that $u \sqsubseteq^{C\emptyset} w'$ and $w' \sqsubseteq^{C\neg\emptyset} v$. If $w = \emptyset$ then it must be $u = \emptyset$ too, and we can take $w' = v$; otherwise, it must be $v = w$ and we can take $w' = u$. The other inclusion is similar. □

Corollary 5. The order $\sqsubseteq^{Conf}$ defining conformance simulations can be decomposed into $\prod_{a\in A} \sqsubseteq^a$ where, for each $a \in A$, we have $\sqsubseteq^a = \sqsubseteq^C$ as defined above. Then, $\sqsubseteq^{Conf} = \prod_{a\in A}(\sqsubseteq^{a,\neg\emptyset} \cup \sqsubseteq^{a,\emptyset})^* = \prod_{a\in A}(\sqsubseteq^{a,\neg\emptyset}) \circ \prod_{a\in A}(\sqsubseteq^{a,\emptyset}) = \prod_{a\in A}(\sqsubseteq^{a,\emptyset}) \circ \prod_{a\in A}(\sqsubseteq^{a,\neg\emptyset})$, so that we obtain $\sqsubseteq^{Conf}$ as the composition of a right-stable order and a left-stable order that commute with each other.

Proposition 7. For any pair of right (resp. left)-stable orders $\sqsubseteq^1$, $\sqsubseteq^2$ on F, their composition also defines a right (resp. left)-stable order on F.

Proof. Given $f : X \longrightarrow Y$ we must show that

$$(id \times Ff)^{-1}(\sqsubseteq^1_Y \circ \sqsubseteq^2_Y) \subseteq \coprod_{(Ff\times id)} (\sqsubseteq^1_X \circ \sqsubseteq^2_X).$$

Let us assume that $(y, x) \in (id \times Ff)^{-1}(\sqsubseteq^1_Y \circ \sqsubseteq^2_Y)$, that is, $y (\sqsubseteq^1 \circ \sqsubseteq^2) y' = Ff(x)$; then, there exists $y'' \in FY$ such that $y \sqsubseteq^2_Y y''$ and $y'' \sqsubseteq^1_Y y'$. Graphically,

$$y \quad \sqsubseteq^2_Y \quad y'' \quad \sqsubseteq^1_Y \quad y' \qquad (8)$$
$$\Big\uparrow Ff$$
$$x$$

Since $\sqsubseteq^1_Y$ is right-stable we have that $(id \times Ff)^{-1} \sqsubseteq^1_Y \subseteq \coprod_{(Ff \times id)} \sqsubseteq^1_X$. Hence, there exists $x'' \in FX$ such that $Ff(x'') = y''$ and $x'' \sqsubseteq^1_X x$, thus turning diagram (8) into the following:

$$y \sqsubseteq^2_Y y'' \qquad\qquad (9)$$

Now, we can apply right-stability of $\sqsubseteq^2$: since we have $(y, x'') \in (id \times Ff)^{-1} \sqsubseteq^2_Y \subseteq \coprod_{(Ff \times id)} \sqsubseteq^2_X$, there exists $x' \in FX$ such that $Ff(x') = y$ and $x' \sqsubseteq^2_X x''$. Thus, diagram (9) becomes

$$y \qquad\qquad (10)$$

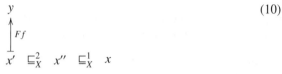

which means that there exist $x', x'' \in FX$ such that $Ff(x') = y$, $x' \sqsubseteq^2_X x''$ and $x'' \sqsubseteq^1_X x$, or equivalently, that $(y, x) \in \coprod_{(Ff \times id)}(\sqsubseteq^1_X \circ \sqsubseteq^2_X)$, as we had to prove.   □

Proposition 8. *If $\sqsubseteq^r$ is a right-stable order on F and $\sqsubseteq^l$ is a left-stable order on F that commute with each other, then their composition defines a stable order on F. Moreover, the coalgebraic simulations for the order $\sqsubseteq = \sqsubseteq^r \circ \sqsubseteq^l$ can be equivalently defined as the $(\sqsubseteq^r \circ \mathrm{Rel}(F)(R) \circ \sqsubseteq^l)$-coalgebras.*

Proof. Let $R \subseteq Z \times W$ be a relation, $f : X \longrightarrow Z$ and $g : Y \longrightarrow W$ arbitrary functions, and $\sqsubseteq = \sqsubseteq^r \circ \sqsubseteq^l$. Let us suppose that $(u, v) \in (Ff \times Fg)^{-1}(\mathrm{Rel}_{\sqsubseteq}(F)(R))$. Then, since $\sqsubseteq^r$ and $\sqsubseteq^l$ commute with each other, using Proposition 4, there exist z', w' such that

$$Ff(u) \sqsubseteq^l_Z z' \ \mathrm{Rel}(F)(R) \ w' \sqsubseteq^r_W Fg(v). \qquad\qquad (11)$$

If we write z for $Ff(u)$ and w for $Fg(v)$, then equation (11) is equivalent to

$$z \sqsubseteq^l_Z z' \ \mathrm{Rel}(F)(R) \ w' \sqsubseteq^r_W w \qquad\qquad (12)$$

and we have to show that $(u, v) \in \mathrm{Rel}_{\sqsubseteq}(F)((f \times g)^{-1}(R))$, that is, that there exist x and y such that

$$u \sqsubseteq^l_X x \ \mathrm{Rel}(F)((f \times g)^{-1}(R)) \ y \sqsubseteq^r_Y v.$$

Using that $\sqsubseteq^r$ is right-stable on the rhs of equation (11), we get $(w', v) \in (id \times Fg)^{-1} \sqsubseteq^r_W \subseteq \coprod_{(Fg \times id)} \sqsubseteq^r_Y$, so that there is some $y \in FY$ such that $Fg(y) = w'$, with $y \sqsubseteq^r_Y v$. Graphically, diagram (12) becomes

$$z \sqsubseteq^l_Z z' \ \mathrm{Rel}(F)(R) \ w' \qquad\qquad (13)$$

Analogously, applying the left-stability of order $\sqsubseteq_Z^l$ we get that there is some $x \in FX$ with $Ff(x) = z'$ such that $u \sqsubseteq_X^l x$. Or graphically,

$$z' \quad \text{Rel}(F)(R) \quad w' \tag{14}$$

$$
\begin{array}{ccc}
z' & \text{Rel}(F)(R) & w' \\
\uparrow & & \uparrow \\
Ff \Big| & & \Big| Fg \\
| & & | \\
u \quad \sqsubseteq_X^l \quad x & & y \quad \sqsubseteq_Y^r \quad v
\end{array}
$$

But diagram (14) is just what we had to prove, since we have found x, y such that $(x, y) \in (Ff \times Fg)^{-1}(\text{Rel}(F)(R)) = \text{Rel}(F)((f \times g)^{-1}(R))$ with $u \sqsubseteq_X^l x$, $y \sqsubseteq_Y^r v$ or, in other words, $(u, v) \in \text{Rel}_{\sqsubseteq}(F)((f \times g)^{-1}(R))$.   □

In particular, for our running example of conformance simulations we obtain the corresponding factorization of the definition of coalgebraic simulations for the order $\sqsubseteq^{Conf}$:

Corollary 6. *Coalgebraic simulations for the conformance order $\sqsubseteq^{Conf}$ can be equivalently defined as the $(\prod_{a \in A}(\sqsubseteq_Y^{a, \neg \emptyset}) \circ \text{Rel}(F)(R) \circ \prod_{a \in A}(\sqsubseteq_X^{a, \emptyset}))$-coalgebras.*

5   Conclusion

We have presented in this paper two new simulation orders induced by two criteria that capture the difference between input and output actions and the implementation notions that are formalized by the conformance relations.

In order to apply the general theory of coalgebraic simulations to them, we identified the corresponding orders on the functor defining labeled transition systems. However, it was not immediate to prove that the obtained orders had the desired good properties since the usual way to do it, namely, by establishing stability as a consequence of a stronger property that we have called right-stability, is not applicable in this case.

Trying to adapt that property to our situation we have discovered several interesting consequences. We highlight the fact that right-stability is an assymetric property which has proved to be very useful for the study of a "reversible" concept such as that of relation, since it is clear that any structural result on the theory of relations should remain true when we reverse the relations, simply "observing" them "from the other side". Two consequences of that assymetric approach followed: first we noticed that we could use it to point the simulation orders in some natural way; secondly we also noticed that by dualizing the right-stability condition we could obtain left-stability.

But the crucial result in order to be able to manage more complicated simulation notions, as proved to be the case for our new covariant-contravariant simulations and the conformance simulations, was the discovery of the fact that both of them could be factorized into the composition of a right-stable and a left-stable component. Exploiting this decomposition we have been able to easily adapt all the techniques that had proved to be very useful for the case of right-stable orders.

We plan to expand our work here in two different directions. The first one is concerned with the two new simulated notions introduced in this paper: once we know that they can be defined as stable coalgebraic simulations and therefore have all the desired basic properties of simulations, we will continue with their study by integrating them into our unified presentation of the semantics for processes [4]. Hence we expect to obtain, in particular, a clear relation between conformance similarity and the classic similarity orders as well as an algebraic characterization for the new semantics. In addition, we plan to continue with our study of stability, which has proved to be a crucial property in order to understand the notion of coalgebraic simulation, thus making it possible to apply the theory to other examples like those studied in this paper.

References

1. Aczel, P., Mendler, N.P.: A final coalgebra theorem. In: Dybjer, P., Pitts, A.M., Pitt, D.H., Poigné, A., Rydeheard, D.E. (eds.) Category Theory and Computer Science. LNCS, vol. 389, pp. 357–365. Springer, Heidelberg (1989)
2. Bloom, B., Istrail, S., Meyer, A.R.: Bisimulation can't be traced. J. ACM 42(1), 232–268 (1995)
3. de Frutos-Escrig, D., Gregorio-Rodríguez, C.: Universal coinductive characterisations of process semantics. In: Ausiello, G., Karhumäki, J., Mauri, G., Ong, C.-H.L. (eds.) IFIP TCS. IFIP, vol. 273, pp. 397–412. Springer, Heidelberg (2008)
4. de Frutos Escrig, D., Gregorio-Rodríguez, C., Palomino, M.: On the unification of semantics for processes: observational semantics. In: Nielsen, M., Kucera, A., Miltersen, P.B., Palamidessi, C., Tuma, P., Valencia, F.D. (eds.) SOFSEM 2009. LNCS, vol. 5404, pp. 279–290. Springer, Heidelberg (2009)
5. Groote, J.F., Vaandrager, F.W.: Structured operational semantics and bisimulation as a congruence. Inf. Comput. 100(2), 202–260 (1992)
6. Hughes, J., Jacobs, B.: Simulations in coalgebra. TCS 327(1-2), 71–108 (2004)
7. Jacobs, B., Hughes, J.: Simulations in coalgebra. In: Gumm, H.P. (ed.) CMCS 2003: 6th International Workshop on Coalgebraic Methods in Computer Science, vol. 82 (2003)
8. Larsen, K.G., Skou, A.: Bisimulation through probabilistic testing. Inf. Comput. 94(1), 1–28 (1991)
9. Leduc, G.: A framework based on implementation relations for implementing LOTOS specifications. Computer Networks and ISDN Systems 25(1), 23–41 (1992)
10. Lynch, N.A., Tuttle, M.R.: Hierarchical correctness proofs for distributed algorithms. In: Sixth Annual ACM Symposium on Principles of Distributed Computing, pp. 137–151 (1987)
11. Park, D.: Concurrency and automata on infinite sequences. In: Deussen, P. (ed.) GI-TCS 1981. LNCS, vol. 104, pp. 167–183. Springer, Heidelberg (1981)
12. Tretmans, J.: Conformance testing with labelled transition systems: Implementation relations and test generation. Computer Networks and ISDN Systems 29(1), 49–79 (1996)
13. van Glabbeek, R.J.: The linear time-branching time spectrum I: The semantics of concrete, sequential processes. In: Bergstra, J.A., Ponse, A., Smolka, S.A. (eds.) Handbook of process algebra, pp. 3–99 (2001)

Stone Duality and the Recognisable Languages over an Algebra

Mai Gehrke

Radboud University Nijmegen, The Netherlands

Abstract. This is a theoretical paper giving the extended Stone duality perspective on the recently discovered connection between duality theory as studied in non-classical logic and theoretical computer science and the algebraic theory of finite state automata. As a bi-product we obtain a general result about profinite completion, namely, that it is the dual under extended Stone duality of the recognisable languages over the original algebra equipped with certain residuation operations.

1   Introduction

In the algebraic theory of finite state automata, finite semigroups are associated with the regular or recognisable languages which are exactly the ones recognised by finite state automata. The highlights of this theory include tools such as syntactic monoids as defined by Rabin and Scott [12], which are powerful algebraic invariants of finite state automata, and the Reiterman theorems that give equational properties for various classes of automata, thus rendering membership in these classes decidable. In [3] new results were reported identifying many of these tools and theorems as special cases of extended Stone duality and thereby vastly generalising them. There the focus was on the impact of these results for automata theory and the form of the underlying duality results was somewhat hidden. Nevertheless, the entire development may be seen from a duality theory perspective. Showing this is the purpose of this paper.

The paper is organised as follows: we start with the necessary preliminaries on duality, then we look at algebras as relational frames and identify the ultrafilter extension and the residuation operations both of which will turn out to play a central role. We then look at the algebra of recognisable subsets of an algebra and show that, with the residuation operations, it is the dual of the profinite completion for any algebra. Finally we *briefly* sketch how these results tie up with the tools of automata theory such as syntactic monoids, profinite terms and Reiterman theorems. For more details and examples on this see [3] and [10].

We work with algebras M with one binary operation which we denote as multiplication. Eventually we will apply the theory in the case of monoids, and finitely generated free monoids in particular, but most of what we will talk about here works for algebras in any signature of operations of finite arity.

A. Kurz, M. Lenisa, and A. Tarlecki (Eds.): CALCO 2009, LNCS 5728, pp. 236–250, 2009.
© Springer-Verlag Berlin Heidelberg 2009

2   Preliminaries on Duality

In this section we collect a few basic and well-known facts about extended duality that we will need. Stone duality lies at the core of many dualities between algebras and coalgebras as it connects Boolean algebras with (topologised) sets. Extended Stone or Priestley duality takes additional structure into account and treats dualities between algebras and coalgebras based on this duality.

One fundamental perspective on duality is that it is about representing abstract algebras in concrete ones, that is, embedding an abstract distributive lattice in a concrete one consisting of subsets with the operations of $\wedge$ and $\vee$ provided by the set-theoretic intersection and union. The embedding perspective is emphasised in the approach introduced by Jónsson and Tarski in their 1951 paper [9] on so-called canonical extensions. This approach has since been generalised to distributive lattices [4] and is particularly useful when additional structure (i.e., algebras and coalgebras) must be treated.

A canonical extension [4] of a distributive lattice D is an embedding $e : D \hookrightarrow D^\sigma$ with the properties:

(1) D^σ is a *complete* lattice;
(2) e is *dense* in the sense that each element of D^σ is a join of meets and a meet of joins of elements from D;
(3) e is *compact* in the sense that for $S, T \subseteq D$ we have $\bigwedge e(S) \leqslant \bigvee e(T)$ implies the existence of finite subsets $S' \subseteq S$ and $T' \subseteq T$ with $\bigwedge S' \leqslant \bigvee T'$.

Each distributive lattice, D, has a canonical extension and it is unique up to isomorphism. In the presence of the Axiom of Choice, it is an equivalent encoding of the dual space of D. Given a distributive lattice D and its canonical extension $D \hookrightarrow D^\sigma$, one may obtain the points of the dual space of D as the set $J^\infty(D^\sigma)$ of completely join irreducible elements of D^σ ordered as a subposet of D^σ and equipped with the topology generated by the basis of subsets of the form $\hat{d} = \{x \in J^\infty(D^\sigma) \mid x \leqslant d\}$ for $d \in D$. Conversely, given the Priestley dual $(X, \leqslant, \tau)$ with the corresponding embedding of D into the clopen downsets of the dual, we obtain the canonical extension by considering the codomain of this embedding to be the collection $\mathcal{D}(X)$ of downsets of $(X, \leqslant)$.

Given a distributive lattice D and its canonical extension $D \hookrightarrow D^\sigma$, we will think of D as a subset of D^σ. We define $K(D^\sigma)$ to be the $\bigwedge$-closure of D and refer to the elements of $K(D^\sigma)$ as closed elements. Dually, $O(D^\sigma)$ is the $\bigvee$-closure of D in D^σ and the elements of $O(D^\sigma)$ are referred to as open elements. The compactness of the embedding $D \hookrightarrow D^\sigma$ is the lattice theoretic encoding of the fact that the dual space $Spec(D)$ of D is topologically compact. It is a consequence of compactness that $K(D^\sigma) \cap O(D^\sigma) = D$ so that the 'clopen' elements of D^σ are exactly the elements of D.

The points of the dual space of D may be recognised in D^σ either as the set $J^\infty(D^\sigma)$ of completely join-prime elements or as the dually defined set $M^\infty(D^\sigma)$ of completely meet-prime elements. These two subposets of D^σ are isomorphic via the isomorphism

$$\kappa : J^\infty(D^\sigma) \to M^\infty(D^\sigma)$$

$$x \mapsto \kappa(x) = \bigvee \{u \in D^\sigma \mid x \not\leqslant u\}.$$

In fact, just as in finite distributive lattices, the completely join-prime and completely meet-prime elements come in pairs $(x, \kappa(x))$ that split the lattice in two disjoint pieces: $D^\sigma = {\uparrow}x \cup {\downarrow}\kappa(x)$. Or, in an alternative formulation which we shall make use of here, we have for $x \in J^\infty(D^\sigma)$

$$\forall u \in D^\sigma \ (x \leqslant u \iff u \not\leqslant \kappa(x)).$$

The elements of $J^\infty(D^\sigma)$ correspond to the prime filters of D (*with the reverse order of inclusion*) via $x \mapsto ({\uparrow}\, x) \cap D$ for $x \in J^\infty(D^\sigma)$ and $p \mapsto \bigwedge p$ where the meet is taken in D^σ for p a prime filter of D. The elements of $M^\infty(D^\sigma)$ correspond to the prime ideals of D, and the isomorphism κ witnesses the correspondence given by set-complementation between prime filters and prime ideals of a distributive lattice. A homomorphism, $h : D \to E$, between bounded distributive lattices extends uniquely to a complete homomorphism $h^\sigma : D^\sigma \to E^\sigma$, and the lower adjoint of such a map sends $X_E = J^\infty(E^\sigma)$ into $X_D = J^\infty(D^\sigma)$. This is the dual map which is order preserving and continuous.

In extended Priestley or Stone duality [7], additional operations on a distributive lattice are captured by additional relational structure on the dual space, see also [5,6]. Here we give a brief description of the relational dual of the additional operations we will be most concerned with. It is easiest to start with an operation $\cdot : D \times D \to D$ preserving $\vee$ in each coordinate. Because the embedding of D in D^σ is dense, D^σ is $\bigvee$-generated by $K(D^\sigma)$. Thus we may extend $\cdot : D \times D \to D$ to an operation on D^σ in two tempi: For $x_1, x_2 \in K(D^\sigma)$ we let

$$x_1 \cdot^\sigma x_2 = \bigwedge \{d_1 \cdot d_2 \mid x_i \leqslant d_i \in D, i = 1, 2\}$$

and for $u_1, u_2 \in D^\sigma$ in general we let

$$u_1 \cdot^\sigma u_2 = \bigvee \{x_1 \cdot^\sigma x_2 \mid u_i \geqslant x_i \in K(D^\sigma), i = 1, 2\}.$$

This is an extension of the original multiplication on D preserving $\bigvee$ in each coordinate, see [4, Theorem 2.21, p. 25]. Thus its action on D^σ is completely captured by its restriction to $X = J^\infty(D^\sigma)$ which is given by:

$$R_{\,.} = \{(x, y, z) \in X^3 \mid x \cdot^\sigma y \geqslant z\}.$$

This yields a relation R that is order compatible in the sense that

$$[\geqslant \times \geqslant] \circ R \circ \geqslant = R$$

if we think of R as a relation from $X \times X$ to X. Topologically, as derived in [7], such a relation R comes from an operation $\cdot : D \times D \to D$ if and only if

(1) For each $x \in X$ the set $R[{\_}, {\_}, x]$ is closed;
(2) For all U, V clopen downsets of X the set $R[U, V, {\_}]$ is clopen;
(Notice that our last coordinate is the first coordinate in [7]).

For operations with other preservation properties, one has to apply some order duality. For this to work it is important that all domain coordinates transform to joins in the codomain or all transform to meets. For example, for an operation $\backslash : D \times D \to D$ that sends joins in the first coordinate to meets and meets in the second coordinate to meets (when the other coordinate remains fixed), we must first extend $\backslash$ to $K(D^\sigma) \times O(D^\sigma)$ by setting

$$x \backslash^\pi y = \bigvee \{d \backslash d' \mid x \leqslant d \in D \text{ and } y \geqslant d' \in D\}$$

for $x \in K(D^\sigma)$ and $y \in O(D^\sigma)$ and then let

$$u \backslash^\pi v = \bigwedge \{x \backslash^\pi y \mid u \geqslant x \in K(D^\sigma) \text{ and } v \leqslant y \in O(D^\sigma)\}$$

for $u, v \in D^\sigma$. This results in an operation on D^σ which sends arbitrary joins in the first coordinate to meets, and arbitrary meets in the second coordinate to meets. Consequently the operation is fully captured by its action on $J^\infty(D^\sigma) \times M^\infty(D^\sigma)$ which is encoded on $X = J^\infty(D^\sigma)$ by

$$S_\backslash = \{(x, y, z) \in X^3 \mid x(\backslash)^\pi \kappa(z) \leqslant \kappa(y)\}.$$

In the sequel we will be applying these results in a situation where we have operations $(\cdot, \backslash, /)$ that form a residuated family on D. That is, for all $a, b, c \in D$ we have

$$a \cdot b \leqslant c \iff b \leqslant a \backslash c \iff a \leqslant c/b.$$

In this case one can prove that the extended operations $(\cdot^\sigma, \backslash^\pi, /^\pi)$ also form a residuated family and consequently that $R_\cdot$, $S_\backslash$ and $S_/$ are identical relations given simultaneously by

$$R(x, y, z) \iff x \cdot^\sigma y \geqslant z \iff x \backslash^\pi \kappa(z) \leqslant \kappa(y) \iff \kappa(z)/^\pi y \leqslant \kappa(x). \quad (1)$$

Given an order-compatible relation R, it always will give rise to a residuated family of maps on the lattice $\mathcal{D}(X)$ of downsets of X given by

$$S \cdot T = R[S, T, \_] = \{n \mid \text{there exist } \ell, m \, [\ell \in S \text{ and } m \in T \text{ and } R(\ell, m, n)]\}$$
$$S \backslash T = (R[S, \_, T^c])^c = \{m \mid \text{for all } \ell, n \, [(\ell \in S \text{ and } R(\ell, m, n)) \implies n \in T]\}$$
$$T/S = (R[\_, S, T^c])^c = \{\ell \mid \text{for all } m, n \, [(m \in S \text{ and } R(\ell, m, n)) \implies n \in T]\}.$$

However, one of these can be a topological dual without the other two being such. This is determined by the topological properties of the relation. As stated above, this multiplication $\cdot$ on $\mathcal{D}(X) \cong D^\sigma$ coming from R is the extension of an operation on D that has R as its dual relation if and only if

(1) For each $x \in X$ the set $R[\_, \_, x]$ is closed;
(2) For all U, V clopen downsets of X the set $R[U, V, \_]$ is clopen.

The properties for $\backslash$ and $/$ to be topological duals are, respectively:

(1) For each $x \in X$ the set $R[\_, x, \_]$ is closed;

(2) For all U clopen downset of X and V clopen upset of X, the set $R[U, \_, V]$ is clopen,

and

(1) For each $x \in X_B$ the set $R[x, \_, \_]$ is closed;

(2) For all U clopen downset of X and V clopen upset of X, the set $R[\_, U, V]$ is clopen.

3   Algebras as Relational Frames

Now we get to the subject proper of this work. One can think of an algebra M as a *relational structure*, playing a role dual to a Boolean algebra with operators. Indeed M is a relational structure as any n-ary operation is an $n+1$-ary relation (which just happens to be functional). We may lift this relation to an algebraic structure over the Boolean algebra, $\mathcal{P}(M)$, of all subsets of M. We work with the case of a single binary operation as it is typical. We will refer to the operation as a multiplication though we do not assume it to be associative. The multiplication lifts to a binary operation on $\mathcal{P}(M)$

$$XY = \{xy \mid x \in X \text{ and } y \in X\}.$$

The complex algebra (after Frobenius' *algebra of complexes* in group theory) of the structure $(M, \cdot)$ is the multimodal algebra $(\mathcal{P}(M), \cap, \cup, (\ )^c, \emptyset, M, \cdot)$ (in the sequel we will suppress the Boolean operations $\cap$, $\cup$, $(\ )^c$, $\emptyset$, M and only specify the additional operations, thus in this case $(\mathcal{P}(M), \cdot)$). Since the multiplication is obtained by relation lifting, it preserves arbitrary unions in each coordinate and is thus residuated. That is, there are operations

$$\backslash, / : \mathcal{P}(M) \times \mathcal{P}(M) \to \mathcal{P}(M)$$

such that, for all $H, K, L \in \mathcal{P}(M)$,

$$HK \subseteq L \iff K \subseteq H\backslash L \iff H \subseteq L/K.$$

More explicitly, the *left* and *right residuals* of L by K are defined by:

$$K\backslash L = \{m \in M \mid Km \subseteq L\} = \{m \in M \mid \text{ for all } u \in K, um \in L\}$$
$$L/K = \{m \in M \mid mK \subseteq L\} = \{m \in M \mid \text{ for all } u \in K, mu \in L\}.$$

We will see in the sequel that in order to understand M as a dual structure the residual operations are a lot more important than the lifted multiplication. This should come as no surprise since, in duality theory, duals of morphisms are given by residuation of the lifted maps. It is thus to be expected that if we think of M as a frame (with the forward direction of multiplication as the direction of the relation $\cdot$), then what we might call the quotient-operation-algebra $(\mathcal{P}(M), /, \backslash)$ (with the Boolean operations) should be its dual complex algebra. However, the algebra $\mathcal{P}(M)$ is closed under all three operations and is thus a residuated Boolean algebra $(\mathcal{P}(M), \cdot, \backslash, /)$.

We will be interested in subalgebras of $\mathcal{P}(M)$ that are not necessarily complete and atomic and thus we need the dual of $\mathcal{P}(M)$ under topological duality rather than under the discrete duality between complex algebras and their atom structures. The extended Stone dual of the residuated Boolean algebra, $(\mathcal{P}(M), \cdot, \backslash, /)$, is the topo-relational space $(\beta(M), \tau, R)$ where $(\beta(M), \tau)$ is the Stone-Čech compactification of the discrete space M (or equivalently the Stone dual of the Boolean algebra $\mathcal{P}(M)$). We have the embedding

$$M \hookrightarrow \beta(M)$$

$$m \mapsto \mu_m = \bigwedge \{L \in \mathcal{P}(M) \mid m \in L\}$$

where the meet is taken in $\mathcal{P}(M)^\sigma$ and the resulting element, μ_m, is the atom corresponding to the principal ultrafilter generated by $m \in M$. The relation R is given by the definition in (1). In the modal logic literature this (or at least the non-topologised version) is known as the ultrafilter extension of the original frame $(M, \cdot)$. It is not hard to see that this is an extension of M. The following result is true about ultrafilter extensions also for non-functional (image finite) frames.

Theorem 1. *Let $(M, \{f_\alpha\}_\alpha)$ be any algebra for a type of finitary operations. The ultrafilter extension of $(M, \{f_\alpha\}_\alpha)$ as a relational frame is the topo-relational structure $(\beta(M), \tau, \{R_\alpha\}_\alpha)$ where $M \hookrightarrow (\beta(M), \tau)$ is the Stone-Čech compactification of M as a discrete space, and R_α is the closure of the graph of f_α for each α. Furthermore, if f_α is n-ary, then the restriction of R_α to domain elements from A^n is functional and equal to f_α.*

If a function on a dense subset of a topological space has a continuous extension to the whole space then the graph of the extension is the closure of the graph on the dense subset, so the relation R above is the graph of this extension if it exists. Since the power set $\mathcal{P}(M)$ is closed under the complexified operations of M as well as each of the residuals, the dual relation satisfies all topological conditions for R being dual to each of these. In the binary case these are the six properties listed at the end of Section 2. Clearly this implies that R is continuous *if* it is functional. However, this is typically not the case. In particular, even for the monoid, $\mathbb{N}$, of non-negative integers under addition, the Stone-Čech compactification, $\beta(\mathbb{N})$, is well known not to admit any simultaneously right and left continuous extension of the addition on $\mathbb{N}$, see [8].

4   The Algebra of Recognisable Subsets of an Algebra

A fundamental notion in automata theory is that of recognition. In an algebraic form: A subset L of M is *recognised* by a surjective homomorphism $\varphi : M \to N$ if there is a $P \subseteq N$ such that $L = \varphi^{-1}(P)$ and L is *recognised by* N if there is a $\varphi : M \to N$ which recognises L. Here we are interested in finite recognition.

Definition 1. *Let M be an algebra. The recognisable subsets of M are those that are recognised by some finite algebra of the same type as M. That is,*

$$\mathrm{Rec}(M) = \{\varphi^{-1}(P) \mid \varphi : M \to F \text{ is an onto morphism, } F \text{ is finite, } F \supseteq P\}.$$

Proposition 1. *Let M be an algebra. Then $\mathrm{Rec}(M)$ is closed under the Boolean operations as well as under the residuals w.r.t. arbitrary subsets of M.*

Proof. If L is recognisable then L^c is recognised by the same morphism and the complementary subset. The intersection of two recognisable subsets is recognised by the image of the product map using the intersection of the product of the two subsets with that image. Finally, for any subsets S of M. If L is recognised by a monoid morphism $\varphi : M \to F$, then $S\backslash L$ and L/S are also recognised by φ as $\varphi^{-1}(\varphi(S)\backslash P) = S\backslash\varphi^{-1}(P)$ and $\varphi^{-1}(P/\varphi(S)) = \varphi^{-1}(P)/S$.   □

In general $\mathrm{Rec}(M)$ is not closed under product, but as we have seen above it is always closed under the residuals. In fact, more is true since it is closed under taking the left or right residual with respect to *any* subset of M. That is, $\mathrm{Rec}(M)$ is a kind of *residuation ideal*. As we will see, this property has essential consequences. For classes of recognisable sets closed under finite intersections, closure under the residuals with arbitrary denominators amounts to the same as closure under the unary residuals which we call quotient operators.

Proposition 2. *Let M be an algebra and C a subset of $\mathrm{Rec}(M)$ closed under finite intersections. Then C is closed under a residual of one of the operations of M for any denominator in $\mathcal{P}(M)$ if and only if C is closed under the residual w.r.t. singleton denominators.*

Proof. Consider a binary operation $\cdot$ and its residual $\backslash$. The result follows easily from the fact that for an arbitrary subset S and a recognisable language L recognised by a homomorphism into a finite algebra, $\varphi : M \to F$, we have $S\backslash L = \varphi^{-1}(\varphi(S)\backslash P)$ where $P = \varphi(L)$, and $\varphi(S)\backslash P = \bigcap_{v\in S}\varphi(\{v\})\backslash P = \bigcap_{v\in S'}\varphi(v)\backslash P$ for some finite subset S' of S since $\varphi(S) \subseteq F$ is finite. That is, $S\backslash L = \varphi^{-1}(\bigcap_{v\in S'}\varphi(v)\backslash P) = \bigcap_{v\in S'}\varphi^{-1}(\varphi(v)\backslash P) = \bigcap_{v\in S'} v\backslash L$.   □

From this result we see that the unary residuals, that is, the quotient operators and thus also the modal algebra $(\mathcal{P}(M), (v^{-1}(\,), (\,)v^{-1})_{v\in M})$ play an important role for recognisable languages.

5   Profinite Completions as Dual Spaces

We will show that the profinite completion of an algebra M is equal to the extended Stone dual of $\mathrm{Rec}(M)$ as a residuation ideal. We start by looking at the underlying Boolean algebra. Since $\mathrm{Rec}(M) \hookrightarrow \mathcal{P}(M)$ is a Boolean embedding, dually, we have a topological quotient map $\beta(M) \twoheadrightarrow \widehat{M}$ where $\widehat{M}$ is the dual space of $\mathrm{Rec}(M)$. In general the composition $M \hookrightarrow \beta(M) \twoheadrightarrow \widehat{M}$ is not injective.

Proposition 3. *Let M be an algebra. The map*

$$M \to \widehat{M}$$

$$m \mapsto \bigwedge \{L \in \mathrm{Rec}(M) \mid m \in L\}$$

is an injection if and only if M is residually finite. In particular, for $M = A^*$, the free monoid over a finite set A, this map is injective.

Proof. The kernel, θ_M, of the above map is easily seen to be $\theta_M = \bigcap \{\theta \in Con(M) \mid M/\theta$ is finite$\}$. Notice that this kernel is always a congruence and that it is trivial if and only if the finite quotients of M separate the points of M. That is, exactly if M is residually finite. The fact that A^* is residually finite is not hard to prove.   □

In fact more can be said than this proposition. The quotient M/θ_M is the residually finite reflection of M and $\widehat{M}$ is homeomorphic to the dual space of the Boolean algebra of recognisable subsets over M/θ_M. We will henceforth assume that M is residually finite.

Example 1. Let A be a finite set, A^* the free monoid generated by A. Since A^* is infinite, there are many points in the remainder of $\widehat{A^*}$, some of which have been used to great advantage in the algebraic theory of automata. For example, one may show that for any $u \in A^*$

$$u^\omega = \bigwedge \{L \in \mathrm{Rec}(A^*) : \exists n \; \forall k \; (n \leqslant k \implies u^{k!} \in L\}$$

is an element of $\widehat{A^*}$. To see this, let $L \in \mathrm{Rec}(A^*)$. Pick a monoid homomorphism $\varphi : A^* \to F$ into a finite monoid that recognises L with the set P. Let n be the order of the monoid F. Then it is an algebraic observation that $e = \varphi(u^{n!})$ must be an idempotent element of F, it follows that $e = \varphi(u^{k!})$ for all $k \geqslant n$ and since this element is in P or in P^c either L or L^c is above u^ω.

Theorem 2. *Let M be any algebra. The topological space underlying the profinite completion of M is the dual space, $\widehat{M}$, of the Boolean algebra $\mathrm{Rec}(M)$.*

Proof. The profinite completion of M is, by definition, the topological algebra obtained as the inverse limit of the finite quotients of M. For each such quotient $\varphi : M \to F$, viewed simply as a set map, the dual via the discrete duality corresponds to the Boolean algebra embedding $\varphi^{-1} : \mathcal{P}(F) \to \mathcal{P}(M)$. In fact, by definition of $\mathrm{Rec}(M)$, this map is into $\mathrm{Rec}(M)$ and $\mathrm{Rec}(M)$ is the direct union of the corresponding sub-Boolean algebras. Since direct limits are carried to inverse limits by the duality, the result follows.   □

Profinite methods have been studied intensely in connection with automata theory [1]. The connection of the theorem was used directly in this context by Pippenger [11]. As we will now see the connection to duality also encompasses the additional operations and this is what really makes it powerful and interesting.

Lemma 1. *Let M be an algebra, $\widehat{M}$ the dual space of $\mathrm{Rec}(M)$. For each opera-tion on M, the relation dual to the residual operations on $\mathrm{Rec}(M)$ is functional.*

Proof. We show it for a binary operation. By definition $R(x, y, z)$ if and only if $x(\backslash)^{\pi}\kappa(z) \leqslant \kappa(y)$ if and only if $y \not\leqslant x(\backslash)^{\pi}\kappa(z)$. Since $x(\backslash)^{\pi}\kappa(z) = \bigwedge\{H\backslash L \mid x \leqslant H \in \mathrm{Rec}(M), z \not\leqslant L \in \mathrm{Rec}(M)\}$, we obtain

$$
\begin{aligned}
R(x, y, z) \quad &\Longleftrightarrow \quad \forall H, L \in \mathrm{Rec}(M) \quad [(x \leqslant H, z \not\leqslant L) \Rightarrow y \not\leqslant H\backslash L] \\
&\Longleftrightarrow \quad \forall H, L \in \mathrm{Rec}(M) \quad [(x \leqslant H, y \leqslant H\backslash L) \Rightarrow z \leqslant L] \\
&\Longleftrightarrow \quad \forall L \in \mathrm{Rec}(M) \quad [\exists H(x \leqslant H, y \leqslant H\backslash L) \Rightarrow z \leqslant L].
\end{aligned}
$$

That is, given $x, y \in \widehat{M}$ we have $R(x, y, z)$ if and only if $z \leqslant \bigwedge \mu$ where $\mu = \{L \in \mathrm{Rec}(M) \mid \exists H(x \leqslant H, y \leqslant H\backslash L)\}$. We show that μ is an ultrafilter of $\mathrm{Rec}(M)$ and thus there is exactly one z with $R(x, y, z)$, namely $z = \bigwedge \mu$. Suppose $L_1, L_2 \in \mathrm{Rec}(M)$ and that there are $H_1, H_2 \in \mathrm{Rec}(M)$ with $x \leqslant H_i$ and $y \leqslant H_i\backslash L_i$ for $i = 1, 2$. Then $x \leqslant H_1 \cap H_2$ and

$$
\begin{aligned}
(H_1 \cap H_2)\backslash(L_1 \cap L_2) &= [(H_1 \cap H_2)\backslash L_1] \cap [(H_1 \cap H_2)\backslash L_2] \\
&\supseteq (H_1\backslash L_1) \cap (H_2\backslash L_2) \geqslant y
\end{aligned}
$$

and μ is closed under intersection. Now let $L \in \mathrm{Rec}(M)$ and let $\varphi : M \to F$ be a homomorphism into a finite algebra and $P \subseteq F$ such that $\varphi^{-1}(P) = L$. First ob-serve that $x \leqslant M = \bigcup_{a \in F} \varphi^{-1}(a)$ implies $x \leqslant \varphi^{-1}(a)$ for some $a \in F$ since each $\varphi^{-1}(a) \in \mathrm{Rec}(M) \subseteq \mathrm{Rec}(M)^{\sigma}$ and x is completely join prime in $\mathrm{Rec}(M)^{\sigma}$. Fur-thermore, a residual operation always satisfies $K\backslash 1 = 1$ and since $1_{\mathrm{Rec}(M)} = M$, we have $y \leqslant M = \varphi^{-1}(a)\backslash M$. Since $\varphi^{-1}(F) = M$ then, as we saw in the proof of Proposition 1, $\varphi^{-1}(a)\backslash M = \varphi^{-1}(\varphi(\varphi^{-1}(a))\backslash F) = \varphi^{-1}(a\backslash F)$. Using the fact that residuals w.r.t. singletons preserve union, we now obtain $y \leqslant M = \varphi^{-1}(a\backslash F) = \varphi^{-1}(a\backslash P)\cup\varphi^{-1}(a\backslash P^c)$ and thus $y \leqslant \varphi^{-1}(a\backslash P)$ or $y \leqslant \varphi^{-1}(a\backslash P^c)$. Finally, using again the rewriting from Proposition 1, we get $\varphi^{-1}(a\backslash P) = \varphi^{-1}(a)\backslash\varphi^{-1}(P) = \varphi^{-1}(a)\backslash L$ and similarly $\varphi^{-1}(a\backslash P^c) = \varphi^{-1}(a)\backslash L^c$ so that $L \in \mu$ or $L^c \in \mu$. $\quad\square$

We now show that the continuity of a functional relation on a dual space is equivalent to the topological conditions guaranteeing that the relation is the topological dual of the residual operations corresponding to the relation.

Proposition 4. *Let X be a Boolean space and let $R \subseteq X^n \times X$ be the graph of a n-ary operation f, that is, $R(x_1, \ldots, x_n, z)$ if and only if $f(x_1, \ldots, x_n) = z$. Then , for each i with $1 \leqslant i \leqslant n$, the following conditions are equivalent:*

(1) *the operation f is continuous;*

(2) *for all $x \in X$, $R[\_\_, x, \_\_]$ (where x is in the ith spot) is closed in X^n and for all clopen sets $U_j, V \subseteq X$ the relational image $R[U_1, \ldots, U_{i-1}, \_, U_{i+1}, \ldots, V^c]$ is clopen.*

If these conditions are satisfied, the operation f is an open map if and only if for all $z \in X$, $R[\_\_, z]$ is closed in X^n and for all clopen sets $U_1, \ldots, U_n \subseteq X$, $R[U, \ldots, U_n, \_]$ is clopen.

Proof. We just prove the result for $i = 1$ to minimise notation. First we as-sume (2) holds and prove that f is continuous. Let $(x_1, \ldots, x_n) \in X^n$ with

$z = f(x_1, \ldots, x_n) \in V$, where V is clopen in X. Since $R[x_1, \_]$ is closed and V is clopen, it follows that both $R[x_1, \_] \cap (X^n \times V)$ and $R[x_1, \_] \cap (X^n \times V^c)$ are closed. Furthermore, since the projection $\pi : X^{n-1} \times X \to X^{n-1}$ is a projection along a compact space, it is a closed map and thus both

$$\pi(R[x_1, \_] \cap (X^{n-1} \times V)) = f_{x_1}^{-1}(V)$$

and

$$\pi(R[x_1, \_] \cap (X^{n-1} \times V^c)) = f_{x_1}^{-1}(V^c)$$

are closed where $f_{x_1} : X^{n-1} \to X$ is given by $\overline{x} \mapsto f(x_1, \overline{x})$. The sets $f_{x_1}^{-1}(V)$ and $f_{x_1}^{-1}(V^c)$ are complementary, and thus they are also open. Since $\overline{x} = (x_2, \ldots, x_n) \in f_{x_1}^{-1}(V)$, there are clopens $U_2, \ldots, U_n$ with $\overline{x} \in U_2 \times \ldots \times U_n \subseteq f_{x_1}^{-1}(V)$. We have $x_1 \notin R[\_, U_2, \ldots, U_n, V^c]$, and thus $x_1 \in (R[\_, U_2, \ldots, U_n, V^c])^c = U_1$ which is clopen by condition (2). That is, $(x_1, \ldots, x_n) \in U_1 \times U_2 \times \ldots \times U_n$ and each U_i is clopen. Furthermore, if $(y_1, \ldots, y_n) \in U_1 \times U_2 \times \ldots \times U_n$ then $y_1 \in U_1 = (R[\_, U_2, \ldots, U_n, V^c])^c$ and thus

$$y_1 \notin R[\_, U_2, \ldots, U_n, V^c]$$
$$= \{y' \mid \exists (y_2', \ldots, y_n') \in U_2 \times \ldots \times U_n \; f(y', y_2', \ldots, y_n') \notin V\}.$$

That is, for all $(y_2', \ldots, y_n') \in U_2 \times \ldots \times U_n$ we have $f(y_1, y_2', \ldots, y_n') \in V$ and in particular $f(y_1, y_2, \ldots, y_n) \in V$. We have shown then that $(x_1, \ldots, x_n) \in U_1 \times U_2 \times \ldots \times U_n \subseteq f^{-1}(V)$ and thus that f is continuous.

On the other hand, if f is continuous, then, for each $x_1 \in X$ we have that the function $f_{x_1} : X^{n-1} \to X$ as defined above is continuous and thus its graph, $R[x_1, \_]$, is closed in X^n. Also, by the definition of continuity, for any clopen set $V \subseteq X$ we have $f^{-1}(V) = R[\_, V^c]$ is clopen. But for clopen sets $U_2, \ldots, U_n \subseteq X$ we have that

$$R[\_, U_2, \ldots, U_n, V^c] = R[\_, V^c] \cap (U_2 \times \ldots \times U_n \times X)$$

which is is clopen.

Finally, assuming f is continuous, we have that $R[\_, z] = f^{-1}(\{z\})$ is closed and thus the first property adds no condition. The second property says that f maps clopen sets to clopen sets. Since forward image preserves unions, it follows that this condition is equivalent to f being an open map. □

Note that since the operations on $\widehat{M}$ are continuous extensions of the ones on M and since M is dense in $\widehat{M}$, it follows that $\widehat{M}$ satisfies exactly the same equations as M. By combining the above proposition and lemma and Theorem 2, we obtain the following result:

Theorem 3. *Let M be an algebra then, $\widehat{M}$, the dual space under the extended Stone duality of $\mathrm{Rec}(M)$ with the residuals of the liftings of the operations of M, is the profinite completion of M as a topological algebra.*

This result identifies profinite completions of algebras as special cases of extended Stone duality. This predicts potential cross-fertilisation between areas (e.g. of logic and theoretical computer science) where extended Stone duality methods are employed and areas (e.g. of algebra, geometry, functional analysis) where profinite completions are used. In fact, Proposition 4 indicates that even for more general topological algebras there is a connection. This theorem also gives a constructive access to profinite completions (including their algebraic operations!): Obtaining the extended dual space $\widehat{M}$ requires the Axiom of Choice in general, but one can always get the Boolean algebra with residuation operations, $\mathrm{Rec}(M)$, and the two carry the same information as they are dual to each other. As we will see in Section 7, the fruitfulness goes beyond this level. Sophisticated tools such as Reiterman theorems are also cases of extended Stone duality.

6   Syntactic Monoids and Profinite Terms

The connection between automata theory and semigroup theory originates in the fact that every finite state automaton has a so-called syntactic monoid. A second fact central to the advanced algebraic theory of automata is the fact that, in the case where the algebra M is a finitely generated free algebra, the elements of $\widehat{M}$ may be viewed as the 'term algebra' for finite algebras of the given type.

Both of these facts are easy to understand from a duality point of view and depend just on the dual understanding of surjective Boolean topological algebra morphisms. We work with monoids from now on as these are the pertinent algebras for automata theory. A result more general than the following is proved in Mirte Dekkers' thesis [2, p.90].

Theorem 4. *Let A be a finite set, A^* the free monoid over A. The Boolean residuation ideals of $\mathrm{Rec}(A^*)$ correspond dually to the Boolean topological semigroup homomorphic images of $\widehat{A^*}$. That is, a Boolean algebra embedding $\varphi : B \hookrightarrow \mathrm{Rec}(A^*)$ embeds $(B, \backslash, /)$ as a residuation ideal if and only if the extended Stone dual of $(B, \backslash, /)$ is a Boolean topological monoid X and the continuous surjection, $f : \widehat{A^*} \to X$, dual to φ is a monoid morphism.*

Note that Boolean topological monoids that are finite are simply finite monoids.

We now outline how semigroups end up being related to automata. Given an automata $\mathcal{A}$ over the alphabet A, the associated language $L_{\mathcal{A}}$ is the set of words recognised by $\mathcal{A}$ (in the automata sense). It is easy to see that, for all $u, v \in A^*$, the language $u \backslash L_{\mathcal{A}}/v$ is recognised by an automaton obtained from $\mathcal{A}$ by moving the initial state and the set of final states around. Since this only results in finitely many automata, it follows that the residuation ideal generated by $\mathcal{A}$, $<L_{\mathcal{A}}>$, is finite. By extended duality we obtain the following result.

Theorem 5. *Let $\mathcal{A}$ be an automaton, $L_{\mathcal{A}}$ the associated language, then the dual of $<L_{\mathcal{A}}>$ is a finite quotient of $(\beta(A^*), R)$ that is a monoid.*

Proof. This follows easily from the fact that $<L_{\mathcal{A}}>$ is finite because A^* is dense in $\beta(A^*)$ so that it has to map onto the finite quotient and R is a monoid operation on A^*.   □

The dual, $S(L_{\mathcal{A}})$, of $<L_{\mathcal{A}}>$ is the syntactic monoid of $L_{\mathcal{A}}$. Note that by the way duality works, the languages in $<L_{\mathcal{A}}>$ are recognised by $S(L_{\mathcal{A}})$ (in the semigroup sense) and thus belong to $\mathrm{Rec}(A^*)$. Conversely, every 'stamp' $\varphi : A^* \to F \supseteq P$ may be seen as an automaton that recognises $\varphi^{-1}(P)$ so that the languages recognised by automata are exactly those in $\mathrm{Rec}(A^*)$.

We now turn to the profinite terms. Let $A = \{a_1, \ldots, a_n\}$ be a finite set, A^* the free monoid freely generated by A, and F be a finite monoid. If we see the elements of A as variables, then an evaluation of these variables in F is a set function $\varphi : a_i \mapsto m_i \in F$. By freeness of A^* over A this uniquely extends to a surjective monoid homomorphism $\varphi^* : A^* \to F'$ with F' a submonoid of F. This map in turn gives an embedding

$$(\varphi^*)^{-1} : \mathcal{P}(F') \hookrightarrow \mathcal{P}(A^*)$$

of $(\mathcal{P}(F'), \backslash, /)$ as a residuation ideal. Now, by the definition of recognisable languages, in fact, $(\varphi^*)^{-1}$ maps into $\mathrm{Rec}(A^*)$ and by the above theorem, such an embedding corresponds to a Boolean topological semigroup quotient

$$\widehat{\varphi} : \widehat{A^*} \to X'_F.$$

But for finite algebras, the topological and the discrete dual coincide, and thus X'_F is just F' and we have interpreted each element of $\widehat{A^*}$ in F. That is, for $x \in \widehat{A^*}$, we have a term function $x^F(a_1, \ldots, a_n) : F^n \to F$ where, for each tuple $(m_1, \ldots, m_n) \in F^n$ we set $x^F(m_1, \ldots, m_n) = \widehat{\varphi}(x)$ where $\widehat{\varphi}$ is obtained as above from the map $\varphi : A \to F$ which maps a_i to m_i.

Example 2. Recall that in Example 1, we introduced, for each word $u \in A^*$ the profinite word $u^\omega = \bigwedge\{L \in \mathrm{Rec}(A^*) : \exists n \; \forall k \; (n \leqslant k \Rightarrow u^{k!} \in L\}$. Let F be a finite monoid and $m \in F$. We now show that $(a^\omega)^F(m) = m^\omega$ is the unique idempotent in the cyclic semigroup generated by m in F. If $A = \{a\}$ then $A^* = \mathbb{N}$, and if $\varphi : a \mapsto m$ then $\varphi^* : n \mapsto m^n \in F$. As mentioned in Example 1, $e = m^{n!}$ where $n = |F|$ is the unique idempotent in the cyclic semigroup generated by m in F. We have $\widehat{\varphi}(a^\omega) \leqslant e$ if and only if $a^\omega \leqslant (\varphi^*)^{-1}(e)$ and $(\varphi^*)^{-1}(e) = \{\ell \mid m^\ell = e\} \supseteq \{k! \mid k \geqslant n\} \geqslant a^\omega$. So indeed, $\widehat{\varphi}(a^\omega) \leqslant e$, and since e is an atom, it follows that $\widehat{\varphi}(a^\omega) = e$.

7 Duality for Subalgebras and Eilenberg-Reiterman Theorems

In this final section we will sketch how Eilenberg-Reiterman theorems essentially are special cases of duality theory for subalgebras. This yields the main theorem of [3]. Let $A \hookrightarrow B$ be an embedding of bounded distributive lattices. Then A gives rise to a quasiorder on X_B given by

$$x \preceq y \iff \forall a \in A \; (y \leqslant a \Rightarrow x \leqslant a)$$

where $\leqslant$ is the order in B^σ. It is easy to verify that $\preceq$ is a quasiorder extending the order on X_B.

Definition 2. *Let B be a bounded distributive lattice, X_B the dual frame of B. A quasiorder $\preceq$ on X_B extending the order $\leqslant$ on X_B is said to be compatible provided it satisfies*

$$\forall x, y \in X_B \quad [x \npreceq y \Rightarrow \exists a \in B \ (y \leqslant a \ and \ x \nleqslant a \ and \ \hat{a} \ is \ a \ \preceq\text{-}downset)].$$

The set of compatible quasiorders on X_B is a poset under set inclusion.

This yields a dual characterisation of bounded sublattices.

Theorem 6. *Let B be a bounded distributive lattice, X_B the dual frame of B. The assignments*

$$E \mapsto A_E = \{a \in B \mid \forall(x,y) \in E \ (y \leqslant a \ \Rightarrow x \leqslant a)\}$$

for $E \subseteq X \times X$ and

$$S \mapsto \preceq_S = \{(x,y) \in X \mid \forall a \in S \ (y \leqslant a \ \Rightarrow x \leqslant a)\}$$

for $S \subseteq B$ establish a Galois connection whose Galois closed sets are the compatible quasiorders and the bounded sublattices, respectively.

Eilenberg theorems relate classes of languages with classes of finite monoids, Reiterman theorems classes of finite monoids and equational theories in profinite terms. The above result, applied to $B = \mathrm{Rec}(A^*)$, relates classes of languages and equational theories, thus yielding an 'Eilenberg-Reiterman theorem'. For $u, v \in \widehat{A^*}$, we write an 'equation' of this most general form as $u \to v$. As above, such an equation holds for a language $L \in \mathrm{Rec}(M)$ if and only if $u \leqslant L$ implies $v \leqslant L$. Note that if $\varphi : A^* \to F$ is the syntactic monoid of L with $L = \varphi^{-1}(P)$, then this means that the 'stamp' $\varphi : A^* \to F \supseteq P$ satisfies $u \to v$, namely $\widehat{\varphi}(u) \in P$ implies $\widehat{\varphi}(v) \in P$. The above theorem specialises to the following result.

Theorem 7. *A set of recognisable languages of A^* is a lattice of languages if and only if it can be defined by a set of equations of the form $u \to v$, where $u, v \in \widehat{A^*}$. Furthermore, given $\mathcal{D} \subseteq \mathcal{P}(A^*)$, the set of all equations $u \to v$ that hold in $\mathcal{D}$ form a compatible quasiorder on the space $\widehat{A^*}$.*

Note that Boolean subalgebras are exactly those for which the corresponding compatible quasiorder is an equivalence relation. Writing $u \leftrightarrow v$ for ($u \to v$ and $v \to u$), we get an equational description of the Boolean algebras of languages.

Corollary 1. *A set of recognisable languages of A^* is a Boolean algebra of languages if and only if it can be defined by a set of equations of the form $u \leftrightarrow v$, where $u, v \in \widehat{A^*}$.*

We say that a lattice of recognisable languages $\mathcal{Q}$ is closed under quotienting provided, for every $L \in \mathcal{Q}$ and $v \in A^*$, $v^{-1}L$ and Lv^{-1} are also in $\mathcal{Q}$ and we call $\mathcal{Q}$ a quotienting subalgebra of $\mathrm{Rec}(A^*)$. As we've seen in Proposition 2, $\mathcal{Q}$ is a quotienting subalgebra if and only if it is a residuation ideal of $\mathrm{Rec}(A^*)$. By

Theorem 4 these correspond to Priestley quotients of $\widehat{A^*}$ that are also monoid quotients. Consequently, if for u and v in $\widehat{A^*}$ we say that L *satisfies the equation* $u \leqslant v$ if, for all $x, y \in \widehat{A^*}$, it satisfies the equation $xvy \rightarrow xuy$ (that is, we generate a monoid quotient) then we obtain the following theorem.

Theorem 8. *A set of recognisable languages of A^* is a lattice of languages closed under quotients if and only if it can be defined by a set of equations of the form $u \leqslant v$, where $u, v \in \widehat{A^*}$.*

For a language L and the quotienting lattice $\mathcal{Q}_L$ generated by L we get the following formulation in terms of ordered syntactic monoids.

Proposition 5. *Let L be a recognisable language of A^*, let $(M, \leqslant_L)$ be its syntactic ordered monoid and let $\varphi : A^* \rightarrow M$ be its syntactic morphism. Then L satisfies the equation $u \leqslant v$ if and only if $\widehat{\varphi}(u) \leqslant_L \widehat{\varphi}(v)$.*

Theorem 8 can be readily extended to Boolean algebras. Let u and v be two profinite words. We say that a recognisable language L *satisfies the equation $u = v$* if it satisfies the equations $u \leqslant v$ and $v \leqslant u$. Proposition 5 now gives immediately:

Corollary 2. *A set of recognisable languages of A^* is a Boolean algebra of languages closed under quotients if and only if it can be defined by a set of equations of the form $u = v$, where $u, v \in \widehat{A^*}$.*

Example 3. A language with zero is a language whose syntactic monoid has a zero. The class of regular languages with zero is closed under Boolean operations and quotients, but *not* under inverse of morphisms so that this example isn't covered by previous results. One can show that a regular language has a zero iff it satisfies the equations $u\rho_A = \rho_A = \rho_A u$ for all $u \in A^*$. Here ρ_A is the limit of the sequence $(v_n)_{n \geqslant 0}$ where $v_0 = u_0$, $v_{n+1} = (v_n u_{n+1} v_n)^{(n+1)!}$ obtained by fixing a total order on A and letting $u_0, u_1, \ldots$ be the ordered sequence of all words of A^* in the induced shortlex order. For more details and further examples, see [3].

So far our 'equations' are 'local', that is, they are not invariant under substitution. The last ingredient of the original Reiterman theorem is this invariance. Substitutions are given by maps $A \rightarrow B^*$ where B is another finite alphabet, or equivalently monoid homomorphisms $f : A^* \rightarrow B^*$. We define a *class of recognisable languages* to be a correspondence $\mathcal{V}$ which associates with each alphabet A a set $\mathcal{V}(A^*)$ of recognisable languages of A^*. Let $\mathcal{C}$ be a class of morphisms between finitely generated free monoids that is closed under composition and contains all length-preserving morphisms. Examples include the classes of all length-preserving morphisms (morphisms for which the image of each letter is a letter), of all length-multiplying morphisms (morphisms such that, for some integer k, the length of the image of a word is k times the length of the word), all non-erasing morphisms (morphisms for which the image of each letter is a non-empty word), all length-decreasing morphisms (morphisms for which the image of each letter is either a letter of the empty word) and all morphisms. We say a class $\mathcal{V}$ is closed under $\varphi \colon A^* \rightarrow B^*$ provided $L \in \mathcal{V}(B^*)$ implies

$\varphi^{-1}(L) \in \mathcal{V}(A^*)$. Since substitutions extend to the profinite terms $\widehat{f} : \widehat{A^*} \to \widehat{B^*}$, a class $\mathcal{V}$ is closed under f if and only if the corresponding class of equations is closed under the substitution $\widehat{f}$. In summary we obtain a fully modular generalised Reiterman theorem that is summed up in the following table.

Closed under	Equations	Definition
$\cup, \cap$	$u \to v$	$\hat{\eta}(u) \in \eta(L) \Rightarrow \hat{\eta}(v) \in \eta(L)$
quotient	$u \leqslant v$	$xuy \to xvy$
complement	$u \leftrightarrow v$	$u \to v$ and $v \to u$
quotient and complement	$u = v$	$xuy \leftrightarrow xvy$
Closed under inverse of morphisms		**Interpretation of variables**
all morphisms		words
nonerasing morphisms		non-empty words
length multiplying morphisms		words of equal length
length preserving morphisms		letters

References

1. Almeida, J.: Finite Semigroups and Universal Algebra. World Scientific, Singapore (1994)
2. Dekkers, M.: Stone duality: An application in the theory of formal languages, Master's thesis, Radboud Universiteit Nijmegen, the Netherlands (December 2008)
3. Gehrke, M., Grigorieff, S., Pin, J.-É.: Duality and equational theory of regular languages. In: Aceto, L., Damgård, I., Goldberg, L.A., Halldórsson, M.M., Ingólfsdóttir, A., Walukiewicz, I. (eds.) ICALP 2008, Part II. LNCS, vol. 5126, pp. 246–257. Springer, Heidelberg (2008)
4. Gehrke, M., Jónsson, B.: Bounded distributive lattices expansions. Mathematica Scandinavica 94(2), 13–45 (2004)
5. Gehrke, M., Priestley, H.A.: Canonical extensions of certain algebras with binary operations, J. Pure Appl. Alg. 209, 269–290 (2007)
6. Gehrke, M., Priestley, H.A.: Duality for certain algebras with binary operations via their canonical extensions. Stud. Log. 86(1), 31–68 (2007)
7. Goldblatt, R.: Varieties of complex algebras. APAL 44, 173–242 (1989)
8. Hindman, N., Strauss, D.: Algebra in the Stone-Čech compactification. de Gruyter Expositions in Mathematics, vol. 27. de Gruyter & Co., Berlin (1998)
9. Jónsson, B., Tarski, A.: Boolean algebras with operators I. Amer. J. Math. 73, 891–939 (1951)
10. Pin, J.-E.: Profinite Methods in Automata Theory. In: Albers, Marion (eds.) STACS 2009, Schloss Dagstuhl. Leibniz International Proceedings in Informatics, vol. 3, pp. 31–50 (2009)
11. Pippenger, N.: Regular Languages and Stone Duality. Theory Comput. Syst. 30(2), 121–134 (1997)
12. Rabin, M.O., Scott, D.: Finite automata and their decision problems. Journal of IBM 3, 114–125 (1959)

Free Heyting Algebras: Revisited

Nick Bezhanishvili[1,*] and Mai Gehrke[2,**]

[1] Department of Computing, Imperial College London, United Kingdom
[2] IMAPP, Radbout Universiteit Nijmegen, the Netherlands

Abstract. We use coalgebraic methods to describe finitely generated free Heyting algebras. Heyting algebras are axiomatized by rank 0-1 axioms. In the process of constructing free Heyting algebras we first apply existing methods to weak Heyting algebras—the rank 1 reducts of Heyting algebras—and then adjust them to the mixed rank 0-1 axioms. On the negative side, our work shows that one cannot use arbitrary axiomatizations in this approach. Also, the adjustments made for the mixed rank axioms are not just purely equational, but rely on properties of implication as a residual. On the other hand, the duality and coalgebra perspectives do allow us, in the case of Heyting algebras, to derive Ghilardi's (Ghilardi, 1992) powerful representation of finitely generated free Heyting algebras in a simple, transparent, and modular way using Birkhoff duality for finite distributive lattices.

1 Introduction

Coalgebraic methods and techniques are becoming increasingly important in investigating non-classical logics [19]. In particular, logics axiomatized by rank 1 axioms allow coalgebraic representation as coalgebras for a functor [14,18]. We recall that an equation is of rank 1 for an operation f if each variable occurring in the equation is under the scope of exactly one occurrence of f. As a result the algebras for these logics become algebras for a functor. Consequently, free algebras in the corresponding varieties are initial algebras in the category of algebras for this functor. This correspondence immediately gives a constructive description of free algebras for rank 1 logics [11,1,5]. Examples of rank 1 logics are the basic modal logic **K**, basic positive modal logic, graded modal logic, probabilistic modal logic, coalition logic and so on [18]. For a coalgebraic approach to the complexity of rank 1 logics we refer to [18]. On the other hand, rank 1 axioms are too simple—very few well-known logics are axiomatized by rank 1 axioms. Therefore, one would, of course, want to extend the existing coalgebraic techniques to non-rank 1 logics. As follows from [15] algebras for these logics cannot be represented as algebras for a functor. Therefore, for these algebras we cannot use the standard construction of free algebras in a straightforward way.

In this paper, which is a facet of a larger joint project with Alexander Kurz [5], we try to take the first steps toward a coalgebraic treatment of modal logics beyond rank 1. We recall that an equation is of rank 0-1 for an operation f if each variable occurring in the equation is under the scope of at most one occurrence of f. With the ultimate

* Partially supported by EPSRC EP/C014014/1 and EP/F032102/1.
** Partially supported by EPSRC EP/E029329/1 and EP/F016662/1.

goal of generalizing a method of constructing free algebras for varieties axiomatized by rank 1 axioms to the case of rank 0-1 axioms, we consider the case of Heyting algebras (intuitionistic logic, which is of rank 0-1 for $f =\rightarrow$). In particular, we construct free Heyting algebras. For an extension of coalgebraic techniques to deal with the finite model property of non-rank 1 logics we refer to [17].

Free Heyting algebras have been the subject of intensive investigation for decades. The one-generated free Heyting algebra was constructed by Rieger and Nishimura in the 50s. In the 70s Urquhart gave an algebraic characterization of finitely generated free Heyting algebras. A very detailed description of finitely generated free Heyting algebras in terms of their dual spaces was obtained in the 80s by Grigolia, Shehtman, Bellissima and Rybakov. This method is based on a description of the points of finite depth of the dual frame of the free Heyting algebra. For the details of this construction we refer to [9, Section 8.7] and [4, Section 3.2] and the references therein. Finally, Ghilardi [10] introduced a different method for describing free Heyting algebras. His technique builds the free Heyting algebra on a distributive lattice step by step by freely adding to the original lattice the implications of degree n, for each $n \in \omega$. Ghilardi [10] used this technique to show that every finitely generated free Heyting algebra is a bi-Heyting algebra. A more detailed account of Ghilardi's construction can be found in [7] and [12]. Ghilardi and Zawadowski [12], based on this method, derive a model-theoretic proof of Pitts' uniform interpolation theorem. In [3] a similar construction is used to describe free linear Heyting algebras over a finite distributive lattice and [16] uses the same method to construct high order cylindric Heyting algebras.

Our contribution is to derive Ghilardi's representation of finitely generated free Heyting algebras in a simple, transparent, and modular way using Birkhoff duality for finite distributive lattices. We split the process into two parts. We first apply the initial algebra construction to weak Heyting algebras—the rank 1 reducts of Heyting algebras. Then we adjust this method to the mixed rank 0-1 axioms. Finally, by using Birkhoff duality we obtain Ghilardi's [10] powerful representation of the finite approximants of the dual of finitely generated free Heyting algebra in a simple and systematic way. On the negative side, our work shows that one cannot use arbitrary axiomatizations in this approach. In particular, we give an example of a valid equation of Heyting algebras of rank 1 that cannot be derived, within the setting of distributive lattices, from other equations of rank 0-1 that are known to provide a full axiomatization of Heyting algebras. In addition, we use properties of Heyting algebras that are not directly equational, and thus our work does not yield a method that applies in general. Nevertheless, we expect that the approach, though it would have to be tailored, is likely to be successful in other instances as well.

The paper is organized as follows. In Section 2 we recall the so-called Birkhoff (discrete) duality for distributive lattices. We use this duality in Section 3 to build free weak Heyting algebras and in Section 4 to build free Heyting algebras. We conclude the paper by listing some future work.

2 Discrete Duality for Distributive Lattices

We recall that a non-zero element a of a distributive lattice D is called *join-irreducible* if for every $b, c \in D$ we have that $a \leq b \vee c$ implies $a \leq b$ or $a \leq c$. For each distributive lattice (DL for short) D let $J(D)$ denote the set of all join-irreducible elements of D. Let also $\leq$ be the restriction of the order of D to $J(D)$. Then $(J(D), \leq)$ is a poset. Recall also that for every poset X a subset $U \subseteq X$ is called a *downset* if $x \in U$ and $y \leq x$ imply $y \in U$. For each poset X we denote by $\mathcal{O}(X)$ the distributive lattice $(\mathcal{O}(X), \cap, \cup, \emptyset, X)$ of all downsets of X. Then every finite distributive lattice D is isomorphic to the lattice of all downsets of $(J(D), \leq)$ and vice versa, every poset X is isomorphic to the poset of join-irreducible elements of $\mathcal{O}(X)$. We call $(J(D), \leq)$ the *dual poset* of D and we call $\mathcal{O}(X)$ the *dual lattice* of X.

This duality can be extended to the duality of the category $\mathbf{DL}_{fin}$ of finite bounded distributive lattices and bounded lattice morphisms and the category $\mathbf{Pos}_{fin}$ of finite posets and order-preserving maps. In fact, if $h : D \to D'$ is a bounded lattice morphism, then the restriction of $h^\flat$, the lower adjoint of h, to $J(D')$ is an order-preserving map between $(J(D'), \leq')$ and $(J(D), \leq)$, and if $f : X \to X'$ is an order-preserving map between two posets X and X', then $f^\downarrow : \mathcal{O}(X) \to \mathcal{O}(X')$, $S \mapsto {\downarrow}f(S)$ is $\bigvee$-preserving and its upper adjoint $(f^\downarrow)^\sharp = f^{-1} : \mathcal{O}(X') \to \mathcal{O}(X)$ is a bounded lattice morphism. Moreover, injective bounded lattice morphisms (i.e. embeddings or, equivalently, regular monomorphisms) correspond to surjective order-preserving maps, and surjective lattice morphisms (homomorphic images) correspond to order embeddings that are in one-to-one correspondence with subsets of the corresponding poset.

We also recall that an element a, $a \neq 1$, of a distributive lattice D is called *meet-irreducible* if for every $b, c \in D$ we have that $b \wedge c \leq a$ implies $b \leq a$ or $c \leq a$. We let $M(D)$ denote the set of all meet-irreducible elements of D.

Proposition 2.1. *Let D be a finite distributive lattice. Then for every $p \in J(D)$, there exists $\kappa(p) \in M(D)$ such that $p \not\leq \kappa(p)$ and for every $a \in D$ we have*

$$p \leq a \ or \ a \leq \kappa(p).$$

Proof. For $p \in J(D)$, let $\kappa(p) = \bigvee\{a \in D \mid p \not\leq a\}$. Then it is clear that the condition involving all $a \in D$ holds. Note that if $p \leq \kappa(p) = \bigvee\{a \in D \mid p \not\leq a\}$, then, applying the join-irreducibility of p, we get $a \in D$ with $p \not\leq a$ but $p \leq a$, which is clearly a contradiction. So it is true that $p \not\leq \kappa(p)$. Now we show that $\kappa(p)$ is meet irreducible. First note that since p is not below $\kappa(p)$, the latter cannot be equal to 1. Also, if $a, b \not\leq \kappa(p)$ then $p \leq a, b$ and thus $p \leq a \wedge b$. Thus it follows that $a \wedge b \not\leq \kappa(p)$. This concludes the proof of the proposition.

Proposition 2.2. *Let X be a finite set and $F_{DL}(X)$ the free distributive lattice over X. Then the poset $(J(F_{DL}(X)), \leq)$ of join-irreducible elements of $F_{DL}(X)$ is isomorphic to $(\mathcal{P}(X), \supseteq)$, where $\mathcal{P}(X)$ is the power set of X and each subset $S \subseteq X$ corresponds to the conjunction $\bigwedge S \in F_{DL}(X)$. Moreover, for $x \in X$ and $S \subseteq X$ we have*

$$\bigwedge S \leq x \ iff \ x \in S.$$

Proof. This is equivalent to the disjunctive normal form representation for elements of $F_{DL}(X)$.

3 Weak Heyting Algebras

3.1 Freely Adding Weak Implications

Definition 3.1. *[8] A pair $(A, \rightarrow)$ is called a* weak Heyting algebra[1] *if A is a bounded distributive lattice and $\rightarrow: A^2 \rightarrow A$ a weak implication, that is, a binary operation satisfying the following axioms for all $a, b, c \in A$:*

(1) $a \rightarrow a = 1$,
(2) $a \rightarrow (b \wedge c) = (a \rightarrow b) \wedge (a \rightarrow c)$.
(3) $(a \vee b) \rightarrow c = (a \rightarrow c) \wedge (b \rightarrow c)$.
(4) $(a \rightarrow b) \wedge (b \rightarrow c) \leq a \rightarrow c$.

Let D and D' be distributive lattices. We let $\rightarrow (D \times D')$ denote the set $\{a \rightarrow b : a \in D$ and $b \in D'\}$. We stress that this is just a set bijective with $D \times D'$. The implication symbol is just a formal notation. For every distributive lattice D we also let $F_{DL}(\rightarrow (D \times D))$ denote the free distributive lattice over $\rightarrow (D \times D)$. Moreover, we let

$$H(D) = F_{DL}(\rightarrow (D \times D))/_{\approx}$$

where $\approx$ is the DL congruence generated by the axioms (1)–(4). We want to stress that we are not thinking of the axioms as a basis for an equational theory for a binary operation $\rightarrow$ here. The point of view is that of describing a bounded distributive lattice by generators and relations. That is, we want to find the quotient of the free bounded distributive lattice over the set $\rightarrow (D \times D)$ with respect to the lattice congruence generated by the pairs of elements of $F_{DL}(\rightarrow (D \times D))$ in (1)–(4) with a, b, c ranging over D. For an element $a \rightarrow b \in F_{DL}(\rightarrow (D \times D))$ we denote by $[a \rightarrow b]_{\approx}$ the $\approx$ equivalence class of $a \rightarrow b$.

The rest of the section will be devoted to showing that for each finite distributive lattice D the poset $(J(H(D)), \leq)$ is isomorphic to $(\mathcal{P}(J(D)), \subseteq)$. Below we give a dual proof of this fact. The dual proof, which relies on the fact that identifying two elements of an algebra simply corresponds to throwing out those points of the dual that are below one and not the other, is produced in a simple, modular, and systematic way that doesn't require any prior insight.

We start with a finite distributive lattice D and the free DL generated by the set

$$\rightarrow (D \times D) = \{a \rightarrow b \mid a, b \in D\}$$

of all formal arrows over D. As follows from Proposition 2.2, $J(F_{DL}(\rightarrow (D \times D)))$ is isomorphic to the power set of $\rightarrow (D \times D)$, ordered by reverse inclusion. Each subset of $\rightarrow (D \times D)$ corresponds to the conjunction of the elements in that subset; the empty set of course corresponds to 1. Now we want to take quotients of this free distributive lattice wrt various lattice congruences, namely the ones generated by the set of instances of the axioms of weak Heyting algebras.
The axiom $x \rightarrow x = 1$.

Here we want to take the quotient of $F_{DL}(\rightarrow (D \times D))$ with respect to the lattice congruence of $F_{DL}(\rightarrow (D \times D))$ generated by the set $\{(a \rightarrow a, 1) \mid a \in D\}$. By duality this quotient is given dually by the *subset*, call it P_1, of our initial poset

---

[1] In [8] weak Heyting algebras are called 'weakly Heyting algebras'.

$P_0 = J(F_{DL}(\to (D \times D)))$, consisting of those join-irreducibles of $F_{DL}(\to (D \times D))$ that do not violate this axiom. Thus, for $S \in J(F_{DL}(\to (D \times D)))$, S is admissible provided

$$\forall a \in D \quad (\bigwedge S \leq 1 \quad \Longleftrightarrow \quad \bigwedge S \leq a \to a).$$

Since all join-irreducibles are less than or equal to 1, it follows that the only join-irreducibles that are admissible are the ones that are below $a \to a$ for all $a \in D$. That is, viewed as subsets of $\to (D \times D)$, only the ones that contain $a \to a$ for each $a \in D$:

$$P_1 = \{S \in P_0 \mid a \to a \in S \text{ for each } a \in D\}.$$

The axiom $x \to (y \wedge z) = (x \to y) \wedge (x \to z)$.

We now want to take a further quotient and thus we want to keep only those join-irreducibles from P_1 that do not violate this second axiom. That is, $S \in P_1$ is admissible provided

$$\forall a, b, c \quad (\bigwedge S \leq a \to (b \wedge c) \quad \Longleftrightarrow \quad \bigwedge S \leq a \to b \text{ and } \bigwedge S \leq a \to c).$$

which means

$$\forall a, b, c \quad (a \to (b \wedge c) \in S \quad \Longleftrightarrow \quad a \to b \in S \text{ and } a \to c \in S).$$

Proposition 3.2. *The poset P_2 of admissible join-irreducibles at this stage is order isomorphic to the set*

$$Q_2 = \{f : D \to D \mid \forall a \in D \quad f(a) \leq a\}$$

ordered pointwise.

Proof. An admissible S from
P_2 corresponds to the function $f_S : D \to D$ given by

$$f_S(a) = \bigwedge \{b \in D \mid a \to b \in S\}.$$

In the reverse direction a function in P_2 corresponds to the admissible set

$$S_f = \{a \to b \mid f(a) \leq b\}.$$

The proof that this establishes an order isomorphism is a straightforward verification.

The axiom $(x \vee y) \to z = (x \to z) \wedge (y \to z)$.

We want the subposet of P_2 consisting of those f's such that

$$\forall a, b, c \quad ((a \vee b) \to c \in S_f \quad \Longleftrightarrow \quad a \to c \in S_f \text{ and } b \to c \in S_f).$$

To this end notice that

$$\forall a, b, c \quad ((a \vee b) \to c \in S_f \iff (a \to c \in S_f \text{ and } b \to c \in S_f))$$
$$\iff \forall a, b, c \quad (f(a \vee b) \leq c \iff (f(a) \leq c \text{ and } f(b) \leq c))$$
$$\iff \forall a, b \quad f(a \vee b) = f(a) \vee f(b).$$

That is, the poset, P_3, of admissible join-irreducibles left at this stage is isomorphic to the set

$$Q_3 = \{f : D \to D \mid f \text{ is join preserving and } \forall a \in D \quad f(a) \le a\}.$$

The axiom $(x \to y) \wedge (y \to z) \le x \to z$.

It is not hard to see that this yields, in terms of join-preserving functions $f : D \to D$,

$$\begin{aligned}
Q_4 &= \{f \in Q_3 \mid \forall a \in D \; f(a) \le f(f(a))\} \\
&= \{f : D \to D \mid f \text{ is join-preserving and } \forall a \in D \; f(a) \le f(f(a)) \le f(a) \le a\} \\
&= \{f : D \to D \mid f \text{ is join-preserving and } \forall a \in D \; f(f(a)) = f(a) \le a\}.
\end{aligned}$$

We note that the elements of Q_4 are nuclei [13] on the order-dual lattice of D. Since the f's in Q_4 are join and 0 preserving, they are completely given by their action on $J(D)$. The additional property shows that these functions have lots of fixpoints. In fact, we can show that they are completely described by their join-irreducible fixpoints.

Lemma 3.3. *Let* $f \in Q_4$, *then for each* $a \in D$ *we have*

$$f(a) = \bigvee \{r \in J(D) \mid f(r) = r \le a\}.$$

Proof. Clearly $\bigvee \{r \in J(D) \mid f(r) = r \le a\} \le f(a)$. For the converse, let r be maximal in $J(D)$ wrt the property that $r \le f(a)$. Now it follows that

$$r \le f(a) = f(f(a)) = \bigvee \{f(q) \mid J(D) \ni q \le f(a)\}.$$

Since r is join-irreducible, there is $q \in J(D)$ with $q \le f(a)$ and $r \le f(q)$. Thus $r \le f(q) \le q \le f(a)$ and by maximality of r we conclude that $q = r$. Now $r \le f(q)$ and $q = r$ yields $r \le f(r)$. However, $f(r) \le r$ as this holds for any element of D and thus $f(r) = r$. Since any element in a finite lattice is the join of the maximal join-irreducibles below it, we obtain

$$\begin{aligned}
f(a) &= \bigvee \{r \in J(D) \mid r \text{ is maximal in } J(D) \text{ wrt } r \le f(a)\} \\
&\le \bigvee \{r \in J(D) \mid f(r) = r \le f(a)\} \le f(a).
\end{aligned}$$

Finally, notice that if $f(r) = r \le f(a)$ then as $f(a) \le a$, we have $f(r) = r \le a$. Conversely, if $f(r) = r \le a$ then $r = f(r) = f(f(r)) \le f(a)$ and we have proved the lemma.

Proposition 3.4. *The set of functions in* Q_4, *ordered pointwise, is order isomorphic to the powerset of* $J(D)$ *in the usual inclusion order.*

Proof. The order isomorphism is given by the following one-to-one correspondence

$$\begin{aligned}
Q_4 &\leftrightarrows \mathcal{P}(J(D)) \\
f &\mapsto \{p \in J(D) \mid f(p) = p\} \\
f_T &\leftarrowtail T
\end{aligned}$$

where $f_T : D \to D$ is given by $f_T(a) = \bigvee\{p \in J(D) \mid T \ni p \leq a\}$. Using the lemma, it is straightforward to see that these two assignments are inverse to each other. Checking that f_T is join preserving and satisfies $f^2 = f \leq id_D$ is also straightforward. Finally, it is clear that $f_T \leq f_S$ if and only if $T \subseteq S$.

Theorem 3.5. *Let D be a finite distributive lattice and $X = (J(D), \leq)$ its dual poset. Then*

1. *The poset $(J(H(D)), \leq)$ is isomorphic to the poset $(\mathcal{P}(X), \subseteq)$ of all subsets of X ordered by inclusion.*
2. *$J(H(D)) = \{[\bigwedge_{q \notin T}(q \to \kappa(q))]_\approx \mid T \subseteq J(D)\}$, (where $\kappa(q)$ is the element defined in Proposition 2.1).*

Proof. As shown above, the poset $J(H(D))$, obtained from $J(F_{DL}(\to(D \times D)))$ by removing the elements that violate the congruence schemes (1)–(4), is isomorphic to the poset Q_4, and Q_4 is in turn isomorphic to $\mathcal{P}(J(D))$ ordered by inclusion, see Proposition 3.4.

In order to prove the second statement, let $q \in J(D)$, and consider $q \to \kappa(q) \in F_{DL}(\to (D \times D))$. If we represent $H(D)$ as the lattice of downsets $\mathcal{O}(J(H(D)))$, then the action of the quotient map on this element is given by

$$
\begin{aligned}
F_{DL}(\to (D \times D)) \ &\to \ H(D) \\
q \to \kappa(q) \ &\mapsto \ \{T' \in \mathcal{P}(J(D)) \mid q \to \kappa(q) \in S_{T'}\}.
\end{aligned}
$$

Now

$$
\begin{aligned}
q \to \kappa(q) \in S_{T'} \ &\Longleftrightarrow \ f_{T'}(q) \leq \kappa(q) \\
&\Longleftrightarrow \ \bigvee(\downarrow q \cap T') \leq \kappa(q) \\
&\Longleftrightarrow \ q \notin T'.
\end{aligned}
$$

The last equivalence follows from the fact that $a \leq \kappa(q)$ if and only if $q \not\leq a$ and the only element of $\downarrow q$ that violates this is q itself. We now can see that for any $T \subseteq J(D)$ we have

$$
\begin{aligned}
F_{DL}(\to (D \times D)) \ &\to \ H(D) \\
[\bigwedge_{q \notin T}(q \to \kappa(q))]_\approx \ &\mapsto \ \{T' \in \mathcal{P}(J(D)) \mid \forall q \ (q \notin T \Rightarrow q \to \kappa(q) \in S_{T'}\} \\
&= \{T' \in \mathcal{P}(J(D)) \mid \forall q \ (q \notin T \Rightarrow q \notin T'\} \\
&= \{T' \in \mathcal{P}(J(D)) \mid \forall q \ (q \in T' \Rightarrow q \in T\} \\
&= \{T' \in \mathcal{P}(J(D)) \mid T' \subseteq T\}.
\end{aligned}
$$

That is, under the quotient map $F_{DL}(\to (D \times D)) \to H(D)$, the elements $\bigwedge_{q \notin T}(q \to \kappa(q))$ are mapped to the principal downsets $\downarrow T$, for each $T \in \mathcal{P}(J(D)) = J(H(D))$. Since these principal downsets are exactly the join-irreducibles of $\mathcal{O}(J(H(D))) = H(D)$, we have that $\{[\bigwedge_{q \notin T}(q \to \kappa(q))]_\approx \mid T \subseteq J(D)\} = J(H(D))$.

3.2 Free Weak Heyting Algebras

In the coalgebraic approach to generating the free algebra, it is a fact of central importance that H as described here is actually a functor. That is, for a DL homomorphism $h : D \to E$ one can define a DL homomorphism $H(h) : H(D) \to H(E)$ so that H becomes a functor on the category of DLs. To see this, we only need to note that H is defined by rank 1 axioms, which the given axioms (1)-(4) for weak Heyting algebras clearly are. Therefore, H gives rise to a functor $H : \mathbf{DL} \to \mathbf{DL}$ [2,15]. Moreover, the category of weak Heyting algebras is isomorphic to the category $Alg(H)$ of the algebras for the functor H. For the details of such correspondences we refer to [2,1,11,5,15]. We would like to give a concrete description of how H applies to DL homomorphisms. We describe this in algebraic terms here and give the dual construction via Birkhoff duality.

Let $h : D \to E$ be a DL homomorphism. Recall that the dual map from $J(E)$ to $J(D)$ is just the lower adjoint $h^\flat$ with domain and codomain properly restricted. By abuse of notation we will just denote this map by $h^\flat$, leaving it to the reader to decide what the proper domain and codomain is. Now $H(D) = F_{DL}(\to (D \times D))/<Ax(D)>$, where $<Ax(D)>$ is the DL congruence generated by $Ax(D)$ and $Ax(D)$ is the set of all instances of the axioms (1)-(4) with $a, b, c \in D$. Also let q_D be the quotient map corresponding to mod'ing out by $<Ax(D)>$. The map $h : D \to E$ yields a map $h \times h : D \times D \longrightarrow E \times E$ and this of course yields a lattice homomorphism $F_{DL}(h \times h) : F_{DL}(\to (D \times D)) \longrightarrow F_{DL}(\to (E \times E))$. Now the point is that $F_{DL}(h \times h)$ carries elements of $Ax(D)$ to elements of $Ax(E)$ and thus in particular to elements of $<Ax(E)>$ (it is an easy verification and only requires h to be a homomorphism for axiom schemes (2) and (3)). This is equivalent to saying that $Ax(D) \subseteq Ker(q_E \circ F_{DL}(h \times h))$ and thus $<Ax(D)> \subseteq Ker(q_E \circ F_{DL}(h \times h))$, or equivalently that there is a unique map $H(h) : H(D) \to H(E)$ that makes the following diagram commute

$$
\begin{array}{ccc}
F_{DL}(\to (D \times D)) & \xrightarrow{F_{DL}(h \times h)} & F_{DL}(\to (E \times E)) \\
\downarrow{\scriptstyle q_D} & & \downarrow{\scriptstyle q_E} \\
H(D) & \xdashrightarrow{H(h)} & H(E).
\end{array}
$$

The dual diagram is

$$
\begin{array}{ccc}
\mathcal{P}(D \times D) & \xleftarrow{(h \times h)^{-1}} & \mathcal{P}(E \times E) \\
\uparrow{\scriptstyle e_D} & & \uparrow{\scriptstyle e_E} \\
\mathcal{P}(J(D)) & \xdashleftarrow{\mathcal{P}(h^\flat)} & \mathcal{P}(J(E))
\end{array}
$$

The map $e_D : \mathcal{P}(D) \hookrightarrow \mathcal{P}(D \times D)$ is the embedding, via Q_4 and so on into P_0 as obtained above. That is, $e_D(T) = \{a \to b \mid \forall p \in T \ (p \le a \Rightarrow p \le b\}$. Now in this dual setting, the fact that there is a map $\mathcal{P}(h^\flat)$ is equivalent to the fact that $(h \times h)^{-1} \circ e_E$ maps into the image of the embedding e_D. This is easily verified:

$$
\begin{aligned}
(h \times h)^{-1}(e_E(T)) &= \{a \rightarrow b \mid \forall q \in T \ (q \leq h(a) \Rightarrow q \leq h(b)\} \\
&= \{a \rightarrow b \mid \forall q \in T \ (h^\flat(q) \leq a \Rightarrow h^\flat(q) \leq b\} \\
&= \{a \rightarrow b \mid \forall p \in h^\flat(T) \ (p \leq a \Rightarrow p \leq b\} \\
&= e_D(h^\flat(T)).
\end{aligned}
$$

Thus we can read off directly what the map $\mathcal{P}(h^\flat)$ is: it is just forward image under $h^\flat$. That is, if we call the dual of $h : D \rightarrow E$ by the name $f : J(E) \rightarrow J(D)$, then $\mathcal{P}(f) = f[\,]$ where $f[\,]$ is the lifted forward image mapping subsets of $J(E)$ to subsets of $J(D)$. Finally, we note that $\mathcal{P}$ satisfies $\mathcal{P}(f)$ is an embedding if and only if f is injective, and $\mathcal{P}(f)$ is surjective if and only if f is surjective.

Since weak Heyting algebras are the algebras for the functor H, we can make use of coalgebraic methods for constructing free weak Heyting algebras. Similarly to [5], where free modal algebras and free distributive modal algebras were constructed, we construct finitely generated free weak Heyting algebras as initial algebras of $Alg(H)$. That is, we have a sequence of bounded distributive lattices, each embedded in the next:

$$
\begin{aligned}
n \quad &\longrightarrow \quad F_{DL}(n), \text{ the free bounded distributive lattice on } n \text{ generators} \\
D_0 \quad &= \quad F_{DL}(n) \\
D_{k+1} &= \quad D_0 + H(D_k), \text{ where } + \text{ is the coproduct in } \mathbf{DL} \\
i_0 : D_0 &\rightarrow \ D_0 + H(D_0) \ = D_1 \text{ the embedding given by coproduct} \\
i_k : D_k &\rightarrow \ D_{k+1} \text{ where } i_k = id_{D_0} + H(i_{k-1})
\end{aligned}
$$

For $a, b \in D_k$, we denote by $a \rightarrow_k b$ the equivalence class $[a \rightarrow b]_\approx \in H(D_k) \subseteq D_{k+1}$. Now, by applying the technique of [2], [1], [11], [5] to weak Heyting algebras, we arrive at the following theorem.

Theorem 3.6. *The direct limit $(D_\omega, (D_k \rightarrow D_\omega)_k)$ in $\mathbf{DL}$ of the system $(D_k, i_k : D_k \rightarrow D_{k+1})_k$ with the binary operation $\rightarrow_\omega : D_\omega \times D_\omega \rightarrow D_\omega$ defined by $a \rightarrow_\omega b = a \rightarrow_k b$, for $a, b \in D_k$ is the free n-generated weak Heyting algebra when we embed n in D_ω via $n \rightarrow D_0 \rightarrow D_\omega$.*

Now we will look at the dual of $(D_\omega, \rightarrow_\omega)$. Let $X_0 = \mathcal{P}(n)$ be the dual of D_0 and let

$$
X_{k+1} = X_0 \times \mathcal{P}(X_k)
$$

be the dual of D_{k+1}.

Theorem 3.7. *The sequence $(X_k)_{k < \omega}$ with maps $\pi_k : X_0 \times \mathcal{P}(X_k) \rightarrow X_k$ defined by*

$$
\pi_k(x, A) = (x, \pi_{k-1}[A])
$$

is dual to the sequence $(D_k)_{k < \omega}$ with maps $i_k : D_k \rightarrow D_{k+1}$. In particular, the π_k's are surjective.

Proof. The dual of D_0 is $X_0 = \mathcal{P}(n)$, and since $D_{k+1} = D_0 + H(D_k)$, it follows that $X_{k+1} = X_0 \times \mathcal{P}(X_k)$ as sums go to products and as H is dual to $\mathcal{P}$. For the maps, $\pi_0 : X_0 \times \mathcal{P}(X_0) \to X_0$ is just the projection onto the first coordinate since i_0 is the injection given by the sum construction. We note that π_0 is surjective. Now the dual $\pi_k : X_{k+1} = X_0 \times \mathcal{P}(X_k) \to X_k = X_0 \times \mathcal{P}(X_{k-1})$ of $i_k = id_{D_0} + H(i_{k-1})$ is $id_{X_0} \times \mathcal{P}(\pi_{k-1})$ which is exactly the map given in the statement of the theorem. Note that a map of the form $X \times Y \to X \times Z$ given by $(x, y) \mapsto (x, f(y))$ where $f : Y \to Z$ is surjective if and only the map f is. Also, as we saw above $\mathcal{P}(\pi_k)$ is surjective if and only if π_k is. Thus by induction, all the π_k's are surjective.

4   Heyting Algebras

4.1   Freely Adding Heyting Implications

Definition 4.1. *[13] A weak Heyting algebra* $(A, \to)$ *is called a* Heyting algebra, HA *for short, if the following two axioms are satisfied for all* $a, b \in A$:

(5) $b \leq a \to b$,
(6) $a \wedge (a \to b) \leq b$.

Since both D and $H(D)$ are embedded in $D + H(D)$ (where $+$ is the coproduct in the category of distributive lattices) we will not distinguish between the elements of D and $H(D)$ and their images in $D + H(D)$. It is a well-known consequence of duality that the dual of the coproduct $D + H(D)$ is the product $J(D) \times J(H(D))$, where $(p, T) \leq a \in D$ if and only if $p \leq a$ and $(p, T) \leq \alpha \in H(D)$ if and only if $T \leq \alpha$. The latter implies in particular that $(p, T) \leq a \to b$ if and only if $a \to b \in S_T$ if and only if $f_T(a) \leq b$ if and only if, for each $q \in T$ we have $q \leq a$ implies $q \leq b$. Let $\equiv$ be a distributive lattice congruence of the lattice $D + H(D)$ generated by the axioms (5)–(6) viewed as congruence schemes. We denote $(D + H(D))/_{\equiv}$ by $V(D)$. For a poset P, call $T \subseteq P$ *rooted* provided there is a $p \in P$ with $p \in T \subseteq {\downarrow}p$, see [10]. Though a rooted subset T is completely determined just by T, we often write (p, T) to identify the root. We denote the set of all rooted subsets of P by P^r.

Theorem 4.2. *Let D be a distributive lattice and $X = (J(D), \leq)$ its dual poset. Then*

1. *The poset* $(J(V(D)), \leq)$ *is isomorphic to the poset* $(X^r, \subseteq)$ *of all rooted subsets of X ordered by inclusion.*
2. $J(V(D)) = \{p \wedge \bigwedge_{q \notin T} q \to \kappa(q) : J(D) \supseteq T$ *is rooted with root p*$\}$.

Proof. We start from the coproduct $D + H(D)$, or dually speaking from the poset $P = J(D) \times J(H(D)) = J(D) \times \mathcal{P}(J(D))$ and we impose the axiom scheme (5), which means dually that we obtain a subset $P_5 \subseteq P$ of all join-irreducible elements that are admissible wrt the axiom scheme (5). That is, $(p, T) \in P_5$ if and only if

$$\forall a, b \in D \quad ((p, T) \leq b \quad \Rightarrow \quad (p, T) \leq a \to b)$$
$$\Longleftrightarrow \quad \forall a, b \in D \quad (p \leq b \quad \Rightarrow \quad f_T(a) \leq b)$$
$$\Longleftrightarrow \quad \forall a \in D \quad (f_T(a) \leq p)$$
$$\Longleftrightarrow \quad \forall q \in T \quad (q \leq p).$$

That is, the poset dual to the lattice obtained by mod'ing out by the axiom scheme (5) is

$$P_5 = \{(p, T) \mid T \subseteq \downarrow p\}.$$

Now further imposing the axiom scheme (6), we retain those elements of $(p, T) \in P_5$ satisfying

$$
\begin{aligned}
&\quad \forall a, b \in D \quad \big(((p, T) \le a \text{ and } (p, T) \le a \to b) \quad \Rightarrow \quad (p, T) \le b\big) \\
&\Longleftrightarrow \quad \forall a, b \in D \quad (p \le a \text{ and } f_T(a) \le b) \quad \Rightarrow \quad p \le b) \\
&\Longleftrightarrow \quad \forall b \quad (f_T(p) \le b) \quad \Rightarrow \quad p \le b) \\
&\Longleftrightarrow \quad p \le f_T(p) = \bigvee\{q \in T \mid q \le p\} \\
&\Longleftrightarrow \quad p \in T.
\end{aligned}
$$

That is, $P_6 = \{(p, T) \mid p \in T \subseteq \downarrow p\}$, which corresponds exactly to the set of all rooted subsets of $J(D)$ ordered by inclusion. This proves the first statement. The second statement is now an easy consequence of this and Theorem 3.5.

Let D be a finite distributive lattice and X its dual poset. Then $D + H(D)$ is dual to $X \times \mathcal{P}(X)$. Consequently, the canonical embedding $i : D \hookrightarrow D + H(D)$ corresponds to the first projection $\pi^1 : X \times \mathcal{P}(X) \to X$ mapping a pair (x, T) for $x \in X$ and $T \subseteq X$ to x. Let $h : D + H(D) \twoheadrightarrow V(D)$ be the quotient map. Then it follows from Theorem 4.2 that h corresponds to an embedding $e : X^r \to X \times \mathcal{P}(X)$ mapping each rooted subset T to $(root(T), T)$. Now we define $j : D \to V(D)$ as the composition $j = h \circ i$. Then, by duality, the dual of j is the map $\pi : X^r \to X$ such that $\pi(T) = root(T)$, or denoting T by (x, T) we have $\pi(x, T) = x$. This implies that π is surjective and therefore, by duality, $j : D \to V(D)$ is an embedding.

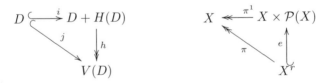

4.2   Free Heyting Algebras

In this section we relate our method to that of Ghilardi [10]. Consider the sequence

$$D_0 \xrightarrow{\;j_0\;} D_1 \xrightarrow{\;j_1\;} D_2 \ldots$$

where D_0 is the free distributive lattice on n generators, $D_{k+1} = V(D_k)$, and $j_k : D_k \to D_{k+1}$ is the embedding of D_k into $V(D_k)$ discussed in the previous section. Now let H be any n-generated Heyting algebra. Let $H_0 = <n>$, $H_{k+1} =< H_k \cup \{a \to_H b \mid a, b \in H_k\} >$ where all these are DL subalgebras of H generated by the given sets. Then we have a sequence

$$H_0 \xrightarrow{\;g_0\;} H_1 \xrightarrow{\;g_1\;} H_2 \ldots$$

as well as maps $\to_k$: $H_k \times H_k \to H_{k+1}$ given by $a \to_k b = a \to_H b$ whenever $a, b \in H_k$. By freeness of D_0, there is a quotient map $q_0 : D_0 \twoheadrightarrow H_0$, and since H_1 is generated by $H_0 \cup \to_H (H_0 \times H_0)$ and H satisfies (1)-(6), it follows that $ker(g_0 \circ q_0) \supseteq ker(j_0)$ and thus $g_0 \circ q_0$ factors through j_0. By induction on this argument we get a sequence of quotient maps q_k making a commutative diagram

$$
\begin{array}{ccccccc}
D_0 & \xhookrightarrow{\ j_0\ } & D_1 & \xhookrightarrow{\ j_1\ } & D_2 & \xhookrightarrow{\ j_2\ } & \cdots \\
\downarrow{\scriptstyle q_0} & & \downarrow{\scriptstyle q_1} & & \downarrow{\scriptstyle q_2} & & \\
H_0 & \xhookrightarrow{\ g_0\ } & H_1 & \xhookrightarrow{\ g_1\ } & H_2 & \xhookrightarrow{\ g_2\ } & \cdots
\end{array}
$$

On the lower sequence we have that each map is a 'partial' homomorphism in the sense that for each $k \geq 0$ we have $\to_k$: $H_k \times H_k \to H_{k+1}$ and for $k \geq 1$ in the sequence $H_{k-1} \xrightarrow{g_{k-1}} H_k \xrightarrow{g_k} H_{k+1}$ we have $g_k(a \to_{k-1} b) = g_{k-1}(a) \to_k g_{k-1}(b)$. Now because this is a HA implication and each finite DL is a HA we have in addition that $a \to_k b = g_k(a) \to_{H_{k+1}} g_k(b)$ for each k. This is of course very special to HAs. As was applied in [10], this property is equivalent to saying that the dual sequence

$$ Q_0 \xleftarrow{\ \pi_0\ } Q_1 \xleftarrow{\ \pi_1\ } Q_2 \cdots $$

of maps are 'partial p-morphisms', i.e., for each $k \geq 1$

$$ \forall \tau \in Q_{k+1}\, \forall S \in Q_k\, (S \leq \pi_k(\tau) \Rightarrow \exists \tau' \in Q_{k+1}\, (\tau' \subseteq \tau \,\&\, \pi_{k-1}\pi_k(\tau') = \pi_{k-1}(S))). $$

Note that the commutative diagram between the D_k and the H_k sequences translates to a dual diagram

$$
\begin{array}{ccccccc}
P_0 & \xleftarrow{\ r_0\ } & P_1 & \xleftarrow{\ r_1\ } & P_2 & \xleftarrow{\ r_2\ } & \cdots \\
\uparrow{\scriptstyle i_0} & & \uparrow{\scriptstyle i_1} & & \uparrow{\scriptstyle i_2} & & \\
Q_0 & \xleftarrow{\ \pi_0\ } & Q_1 & \xleftarrow{\ \pi_1\ } & Q_2 & \xleftarrow{\ \pi_2\ } & \cdots
\end{array}
$$

which tells us that $Q_{k+1} \subseteq Q_k^r$, the set of rooted subsets of Q_k, and that the action of the π_k's is to take the root. Now, a second fact that is very special to HAs is that not only is $Q_{k-1} \twoheadleftarrow Q_k \twoheadleftarrow Q_{k+1}$ a partial p-morphism diagram, but so is $Q_{k-1} \twoheadleftarrow Q_k \hookleftarrow \tau$ for any $\tau \in Q_{k+1}$ viewed as a subset of Q_k (and thus as an embedding). The ensuing property on rooted subsets $\tau \in Q_k^r$ for them to be admissible in a sequence of Q_k's for a Heyting algebra H is easily derivable in the same manner as our earlier calculations. This was done by Ghilardi in [10] and results in

$$ \forall T \in \tau\, \forall S \in Q_k\, (S \leq T \Rightarrow \exists T' \in \tau\, (T' \leq T \,\&\, root(T') = root(S))) \qquad \text{(G)} $$

The point is now that since, in each step and for each n-generated HA, H, the admissible rooted subsets can at most be those satisfying (G), if we start from the largest initial poset namely $R_0 = P_0 = J(F_{DL}(n))$ and proceed with $R_1 = R_0^r$, and $R_{k+1} = \{\tau \in R_k^r \mid \tau$ satisfies (G)$\}$ then, for any Heyting algebra with dual sequence $\{Q_k\}$ we have

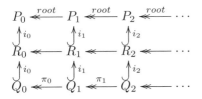

and thus H factors through $\varinjlim \mathcal{O}(R_k)$ and this latter algebra thus has the universal mapping property for HA. By the same argument, the same is true for any algebra given by any sequence R'_k between the P_k's and the R_k's for which the local operations $\to_k$ glue together correctly. However there is of course no guarantee that any of these algebras are themselves Heyting algebras (at most one is, as it is then the free n-generated HA).

Theorem 4.3. *The limit $\varinjlim \mathcal{O}(R_k)$ of the sequence $\{\mathcal{O}(R_k)\}_{k \in \omega}$ in the category* **DL** *of distributive lattices is the free Heyting algebra on n generators.*

Proof. (Sketch) It follows from the discussion above that $\varinjlim \mathcal{O}(R_k)$ satisfies the required universal properties of the free algebra. Therefore, we only need to show that $\varinjlim \mathcal{O}(R_k)$ is a Heyting algebra, or dually that $\varprojlim R_k = R_\omega$, with the standard topology of the inverse limit, is an Esakia space (see e.g., [4, Section 2.3.3], for details of this duality). The crucial part is that $\uparrow U$ must be clopen for U clopen in $\varprojlim R_k$. For each $k \in \omega$ let $\pi_k^\omega : R_\omega \to R_k$ be the projection map. A subset U is clopen in the limit R_ω provided $U = (\pi_{k-1}^\omega)^{-1}(V)$ for some $k-1$ and $V \subseteq R_{k-1}$ and then $U = (\pi_k^\omega)^{-1}(V_k)$ with $V_k = \pi_k^{-1}(V)$ is also true. Now, clearly $\uparrow U \subseteq (\pi_k^\omega)^{-1}(\uparrow V_k)$. The crux of this proof is that (G) implies that the reverse inclusion also holds. To see this, let $x \in R_\omega$ with $\pi_k^\omega(x) = x_k \geq y_k$ for some $y_k \in V_k$. Applying (G) with $\tau = \pi_{k+1}^\omega(x)$ and $T = root(\tau) = x_k$ and $S = y_k$, there is $T' \leq T, T' \in \tau$ with $root(S) = root(T')$. Now note that $root(T') = root(S) = root(y_k) \in \pi_{k-1}(V_k) = V$. Thus T' is in V_k. Also, we now take $y_{k+1} = \tau' = \tau \cap {\downarrow} T'$. As mentioned by Ghilardi this is easily seen to be admissible again, and since $root(\tau') = T' \in V_k$, then $y_{k+1} = \tau' \in (\pi_{k+1})^{-1}(V_k) = V_{k+1}$. Also $y_{k+1} = \tau' \leq \tau = x_{k+1}$ and in this way we build a sequence $y = (y_k) \in (\pi_{k-1}^\omega)^{-1}(V)$ with $y \leq x$. This proves $x \in {\uparrow} U$ and we are done.

We conclude this section with a few points on where this leaves us in the quest for a systematic approach to the generative description of free finitely generated algebras in DL based varieties. First an example concerning the choice of axiomatization.

Example 4.4. Let D be a finite distributive lattice and let $H'(D)$ denote $F_{DL}(\to (D \times D))$ modulo axioms (1),(2), (3) of Definition 3.1. This means that the dual of $H'(D)$ is isomorphic to the set $Q_3 = \{f : D \to D \mid f$ is join-preserving and $\forall a \in D\ f(a) \leq a\}$. Since the f's are join preserving we may consider them as order preserving functions $f : J(D) \to \mathcal{O}(J(D)) \cong D$ as this restriction uniquely determines f. We also let $V'(D)$ denote $D + H'(D)$ modulo axioms (5),(6) of Definition 4.1. The dual of $D + H'(D)$ is isomorphic to $J(D) \times P_3$ and by imposing axiom schemes (5) and (6), we get

$$P' = \{(p, f) \mid f(q) \subseteq {\downarrow} q \cap {\downarrow} p, f(p) = {\downarrow} p\}.$$

Then we can show that in general, $V'(D)$ is not isomorphic to $V(D)$.

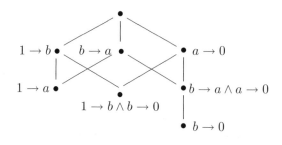

In fact, the inequality $(a \rightarrow b) \wedge (b \rightarrow c) \leq a \rightarrow c$ will not be valid on $V'(D)$ for all $a, b, c \in D$, whereas on $V(D)$ it is valid by definition. To see this consider the four element chain $0 < a < b < 1$. The poset P', where we write each S_f as the conjunction of the $q \rightarrow f(q)$ for which $f(q) < q$ is depicted above.

This poset is larger than $J(V(D))$ as the point $(b \rightarrow a) \wedge (a \rightarrow 0)$ is not in $J(V(D))$. We recall that axioms (1),(2) of Definition 3.1 and (5),(6) of Definition 4.1 are already sufficient to axiomatize Heyting algebras; see e.g., [13, Lemma 1.10] or [4, Theorem 2.2.6]. In logical terms the above observation means that the inequality $(a \rightarrow b) \wedge (b \rightarrow c) \leq (a \rightarrow c)$ is an example of a valid rank 1 inequality of the theory of Heyting algebras whose derivation is not a DL derivation on the basis of the axioms.

Finally a remark on mod'ing out to make the partial operations into operations.

Remark 4.5. Notice that our first sequence for the HA case with $D_{k+1} = V(D_k)$ is definitely 'too free' and we can make a first cut on general principles. For any $k \geq 1$ and for $a, b \in D_{k-1}$, we may take $j_k(a \rightarrow_{k-1} b)$ or $j_{k-1}(a) \rightarrow_k j_{k-1}(b)$, and if $\{H_k\}$ is any sequence obtained from an algebra it factors through the DL congruence sequence generated by

$$\forall a, b \in D'_{k-1} \quad j_k(a \rightarrow_{k-1} b) \approx j_{k-1}(a) \rightarrow_k j_{k-1}(b)$$

with $D'_0 = D_0$ and $D'_1 = D_1$. Let $\{P'_k\}$ be the corresponding dual sequence. In the case of HA one can prove the following

Claim: $\tau \in (P'_k)^r$ is admissible if and only if for all $p \in root(\tau)$ there is $T \in \tau$ with $root(T) = p$.

Mod'ing out this congruence ensures that the corresponding sequence of algebras yields an $\rightarrow$-algebra in the limit. However mod'ing out must have destroyed the truth of (1)-(4) as one can show that this is not quite Ghilardi's sequence. For example, for the four element chain as in the above example, with P its dual, $P_1 = P^r$, we have that $\tau = \{P, \{1, b\}, \{a, b\}, \{a\}\}$ satisfies the above admissibility condition but does not satisfy (G) as can be seen from taking $T = \{1, b\}$ and $P_1 \ni S = \{b\} \leq T$.

5   Conclusions and Future Work

In this paper we described finitely generated free (weak) Heyting algebras using an initial algebra-like construction. The main idea is to split the axiomatization of Heyting algebras into its rank 1 and non-rank 1 parts. The rank 1 reducts of Heyting algebras are weak Heyting algebras. For weak Heyting algebras we applied the standard initial algebra construction and then adjusted it for Heyting algebras. We used Birkhoff duality for finite distributive lattices and finite posets to obtain the dual characterization of the finite posets that approximate the duals of free algebras. As a result we obtained Ghilardi's representation of these posets in a more systematic and transparent way.

There are a few possible directions for further research. As we mentioned in the introduction, although we considered Heyting algebras (intuitionistic logic), this method could be applied to other non-classical logics. More precisely, the method is available if a signature of the algebras for this logic can be obtained by adding an extra operator to a locally finite variety. Thus, various non-rank 1 modal logics such as $S4$, $K4$ and other more complicated modal logics, as well as distributive modal logics, are the obvious candidates. On the other hand, one cannot always expect to have such a simple representation of free algebras. The algebras corresponding to other many-valued logics such as MV-algebras, l-groups, BCK-algebras and so on, are other examples where this method could lead to interesting representations. The recent work [6] that connects ontologies with free distributive algebras with operators shows that such representations of free algebras are not only interesting from a theoretical point of view, but could have very concrete applications.

Acknowledgements. The authors are grateful to Mamuka Jibladze for reading and commenting on early versions of the paper.

References

1. Abramsky, S.: A Cook's tour of the finitary non-well-founded sets. In: Artemov, S., et al. (eds.) We Will Show Them: Essays in honour of Dov Gabbay, pp. 1–18. College Publications (2005)
2. Adámek, J., Trnková, V.: Automata and algebras in categories. Mathematics and its Applications (East European Series), vol. 37. Kluwer, Dordrecht (1990)
3. Aguzzoli, S., Gerla, B., Marra, V.: Gödel algebras free over finite distributive lattices. Annals of Pure and Applied Logic 155(3), 183–193 (2008)
4. Bezhanishvili, N.: Lattices of Intermediate and Cylindric Modal Logics. PhD thesis, University of Amsterdam (2006)
5. Bezhanishvili, N., Kurz, A.: Free modal algebras: A coalgebraic perspective. In: Mossakowski, T., Montanari, U., Haveraaen, M. (eds.) CALCO 2007. LNCS, vol. 4624, pp. 143–157. Springer, Heidelberg (2007)
6. Bruun, H., Coumans, D., Gehrke, M.: Distributive lattice-structured ontologies. In: Kurz, A., Lenisa, M., Tarlecki, A. (eds.) CALCO 2009. LNCS, vol. 5728, pp. 267–283. Springer, Heidelberg (2009)
7. Butz, C.: Finitely presented Heyting algebras. Technical report, BRICS, Arhus (1998)
8. Celani, S., Jansana, R.: Bounded distributive lattices with strict implication. Math. Log. Q. 51(3), 219–246 (2005)
9. Chagrov, A., Zakharyaschev, M.: Modal Logic. The Clarendon Press (1997)
10. Ghilardi, S.: Free Heyting algebras as bi-Heyting algebras. Math. Rep. Acad. Sci. Canada XVI. 6, 240–244 (1992)
11. Ghilardi, S.: An algebraic theory of normal forms. Ann. Pure Appl. Logic 71, 189–245 (1995)

12. Ghilardi, S., Zawadowski, M.: Sheaves, Games and Model Completions. Kluwer Academic Publishers, Dordrecht (2002)
13. Johnstone, P.T.: Stone spaces. Cambridge University Press, Cambridge (1982)
14. Kurz, A., Pattinson, D.: Coalgebraic modal logic of finite rank. MSCS 15, 453–473 (2005)
15. Kurz, A., Rosický, J.: The Goldblatt-Thomason theorem for coalgebras. In: Mossakowski, T., Montanari, U., Haveraaen, M. (eds.) CALCO 2007. LNCS, vol. 4624, pp. 342–355. Springer, Heidelberg (2007)
16. Pataraia, D.: High order cylindric algebras (in preparation, 2009)
17. Pattinson, D., Schröder, L.: Beyond rank 1: Algebraic semantics and finite models for coalgebraic logics. In: Amadio, R.M. (ed.) FOSSACS 2008. LNCS, vol. 4962, pp. 66–80. Springer, Heidelberg (2008)
18. Schröder, L., Pattinson, D.: PSPACE bounds for rank-1 modal logics. In: Proc. 21st IEEE Symposium on Logic in Computer Science (LICS 2006), pp. 231–242 (2006)
19. Venema, Y.: Algebras and coalgebras. In: Blackburn, P., van Benthem, J., Wolter, F. (eds.) Handbook of Modal Logic, pp. 331–426. Elsevier, Amsterdam (2007)

Distributive Lattice-Structured Ontologies

Hans Bruun[1], Dion Coumans[2], and Mai Gehrke[2],[*]

[1] Technical University of Denmark, Denmark
[2] Radboud University Nijmegen, The Netherlands

Abstract. In this paper we describe a language and method for deriving ontologies and ordering databases. The ontological structures arrived at are distributive lattices with attribution operations that preserve $\vee$, $\wedge$ and $\bot$. The preservation of $\wedge$ allows the attributes to model the natural join operation in databases. We start by introducing ontological frameworks and knowledge bases and define the notion of a solution of a knowledge base. The import of this definition is that it specifies under what condition all information relevant to the domain of interest is present and it allows us to prove that a knowledge base always has a smallest, or terminal, solution. Though universal or initial solutions almost always are infinite in this setting with attributes, the terminal solution is finite in many cases.

We describe a method for computing terminal solutions and give some conditions for termination and non-termination. The approach is predominantly coalgebraic, using Priestley duality, and calculations are made in the terminal coalgebra for the category of bounded distributive lattices with attribution operations.

1 Introduction

In traditional relation database models used in commercial applications, information is stored in tuples. Research in knowledge representation however seeks models of information that are able to accommodate more complex information forms. A key issue in this respect is classification structures in the form of formal ontologies. While lattices are just what is called for to model the subsumption relation of concepts captured by ontologies, they fall short of providing tuples supported in the relational database model. In [9] Fischer Nilsson showed that lattices with additional (unary) operations preserving not only $\bot$ and $\vee$, but also $\wedge$ provide a common framework generalising ontologies and traditional databases. The tuples in relational databases may be represented as meets of basic attribution terms $a_i(c_i)$ and the database relations are then represented as lattice joins of such meets of basic attribution terms. This representation together with preservation of $\wedge$ and $\bot$ by the attribution operations provide database natural join simply as lattice meet. What makes this framework interesting from the application perspective is thus that it combines and enriches relational databases with lattice classifications. A preliminary algorithmization was implemented for

---

[*] The authors would like to thank the referees for their very useful input.

A. Kurz, M. Lenisa, and A. Tarlecki (Eds.): CALCO 2009, LNCS 5728, pp. 267–283, 2009.
© Springer-Verlag Berlin Heidelberg 2009

the OntoQuery project [11] which was concerned with content-based querying of large text databases. A description of an updated version of the algorithm is available in [4]. Here we describe the setting and the mathematical ideas behind the algorithm and analyse it in mathematical terms.

In algebraic terms, we want to solve a given generators and relations problem (modelling the dependencies between the basic concepts of the expert domain in question) that imposes the appropriate order and algebra structure on a given set of terms (which correspond to the current entries of the database/ontology). This approach to generative knowledge representation generalises the work of Oles which was used in IBM's medical knowledge representation tool [10]. Oles' work differs from ours in two aspects. It doesn't allow additional operations in the type and the only solution that is identified is the universal solution. Even in the plain distributive lattice setting this solution is typically much too big and involves many irrelevant concepts.

In this paper we introduce what we believe to be a key concept, namely that of classification of a term w.r.t. a solution of an ontological framework, and show that, given a set of terms, a generators and relations problem has a minimal solution that classifies all the given terms. We call this the terminal solution and it is the smallest meaningful solution - also in Oles' setting. In the setting with attributes, the universal solution of a generators and relations problem is typically infinite, and thus the generative approach to knowledge representation is impossible without the key notion of a terminal solution. While the terminal solution is typically finite, this is not always the case. We give sufficient conditions for it to be finite and infinite, respectively. We conjecture that our condition for nontermination is essentially sharp, but this remains an open problem.

The algebraic logics treated here are related to description logic [2] in the sense that both formalisms concern generator and relations problems for distributive lattices with additional operations, see e.g. [12,8,1]. The operations in description logic are more general as they may just preserve $\wedge$ or $\vee$ while ours preserve both. However, the main difference lies in the problems treated: the main issue in description logic is, given two compound concept descriptions (i.e., terms in the free distributive lattice with attribution operations over the set of basic concepts), to determine whether the equivalence class of one of them subsumes the equivalence class of the other in the universal solution of the generators and relations problem. This is a very local piece of information that concerns only the two terms in question. By contrast, we give the decomposition of a term into a join of join-irreducibles relative to the universal solution (that is, when such decomposition exists). This yields global information about that term, and when we do this for each term of interest, we obtain the full terminal solution. This is a global solution. In addition, finding the normal form for a term identifies the irreducible building blocks of the ontology which may not be readily identifiable from the generators and relation problem. This is the sense in which these ontologies are generative. Furthermore, our methods should readily generalise to the setting of operations preserving only $\wedge$ or $\vee$ using the finitely generated final coalgebras for modal logic as developed in [3].

In Section 2 we introduce the key concepts of this work, including ontological frameworks, knowledge bases, classification of a term and terminal solutions. In Section 3 we recall a few facts from duality theory and canonical extensions, and in Section 4 we apply these facts to obtain the existence as well as a useful description of terminal solutions. In Section 5 we describe the final coalgebra for the variety we work in and in the final section we use this to give a method for computing terminal solutions. In Section 6 we also supply some results on termination and non-termination.

2   Ontological Frameworks and Knowledge Bases

An *ontological framework*, $\mathcal{O} = (C, A, \Pi)$, consists of three finite sets. The first, C, is a set of basic concept names. We think of these as generating elements. The second, A, is a set of attribution operation, or attribute, symbols. We assume these operations to be unary. We write $T_{L_\perp A}(C)$ for the collection of terms built up from concepts from C using attributes in A and the lower bounded lattice operation symbols in $L_\perp = \{\vee, \wedge, \perp\}$. The third set, Π, is a set of so-called terminological axioms. The elements of Π are pairs (r, s) of terms from $T_{L_\perp A}(C)$. The idea is that the terminological axioms specify pairs of expressions that for ontological reasons must be identified.

As a simple example one may consider an ontological framework on real estate. The set C then consists of concepts relevant to real estate, like 'flat', 'villa', names of geographical regions, different sizes, etc. We may express the fact that the geographical region r_2 is a subregion of r_1 by the terminological axiom $(r_1, r_1 \vee r_2)$. The set A may contain for instance an attribute L for 'located in', which maps a region like r_1 to the concept $L(r_1)$ which (extensionally) should designate all real estates located in region r_1.

There are various reasons for which we only allow the use of $\perp$ in the specification of ontological frameworks. First of all, it is quite common in ontology to include $\perp$ as the inconsistent concept. However, a universal concept is less meaningful, especially in the database setting. Secondly, for our attribution operations, $\top$ and $\perp$ play very different roles. We will elaborate on this further on.

The fundamental idea is that an ontological framework specifies a class of ontological structures that are the solutions for the framework. In this work we restrict ourselves to considering solutions that have a lower bounded distributive lattice structure with attribution operations, one for each $a \in A$. We require the attribution operations to preserve $\vee$, $\wedge$, and $\perp$. Requiring the preservation of $\wedge$ allows the attributes to model the so called natural join operation in databases. In the mathematical work below we add $\top$ to the type as well as the laws $a(\top) = \top$ for $a \in A$ and call the corresponding algebras DLAs. This facilitates our construction and does not interact with our solutions: removing the top from a solution will always yield a subalgebra of type $L_\perp A$. Thus the named element $\top$ is not a part of the ontologically meaningful solution and is just added to make the mathematics smoother. It is important that in ontologies (such as ontologically ordered databases) the top need not be preserved by the attribution operations. The top we add here is just a mathematical gadget and it will always remain an unreachable join as the terminological axioms never include top. It does not influence the action of the attributes on the elements of the actual ontology and in particular, in case it exists, the action of the attributes on the top of the ontology.

Every DLA generated by C is a homomorphic image of $F_{DLA}(C)$, the free bounded distributive lattice with attribution operations generated by C. Thus if we want solutions of the ontological framework $\mathcal{O} = (C, A, \Pi)$ to be DLAs that are generated by C we only have to look among homomorphic images of $F_{DLA}(C)$. As said, the terminological axioms, that is, the elements of Π, are pairs of $DLA_\perp$ terms in the basic concept names. Each such axiom may be seen as an identification of particular elements in $F_{DLA}(C)$. This leads to the following definition:

Definition 1. *Let $\mathcal{O} = (C, A, \Pi)$ be an ontological framework. A solution of $\mathcal{O}$ is a homomorphic image of $F_{DLA}(C)$ in which r is identified with s for each pair (r, s) in Π.*

Since homomorphic images correspond to quotients of the domain by congruences, solutions are those quotients that are given by congruences containing Π. One of the most fundamental facts from general algebra about congruences is that there is a least congruence containing all the pairs in Π. This is the join of the principal congruences $\theta_{(r,s)}$ for $(r, s) \in \Pi$. The corresponding quotient of $F_{DLA}(C)$ is the least-collapsed algebra satisfying the terminological axioms and all other quotients satisfying all the terminological axioms are quotients of it. This yields the following theorem.

Theorem 1. *Let $\mathcal{O} = (C, A, \Pi)$ be an ontological framework. There exists a quotient $h_{\mathcal{O}} : F_{DLA}(C) \to F_{\mathcal{O}}$ of $F_{DLA}(C)$ satisfying the following conditions:*

1. *$h_{\mathcal{O}} : F_{DLA}(C) \to F_{\mathcal{O}}$ is a solution of $\mathcal{O} = (C, A, \Pi)$;*
2. *If $h : F_{DLA}(C) \to D$ is any solution of $\mathcal{O} = (C, A, \Pi)$, then there is a unique homomorphism $h_D : F_{\mathcal{O}} \to D$ so that the diagram*

$$F_{DLA}(C) \longrightarrow D$$
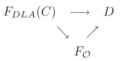
$$F_{\mathcal{O}}$$

commutes.

The solutions of $\mathcal{O}$ are exactly the quotients of $F_{\mathcal{O}}$.

Definition 2. *Let $\mathcal{O} = (C, A, \Pi)$ be an ontological framework. We call the quotient $h_{\mathcal{O}} : F_{DLA}(C) \to F_{\mathcal{O}}$ the universal solution of $\mathcal{O}$.*

This universal solution may also be described by:

$$F_{\mathcal{O}} = F_{DLA}(C)/\theta_\Pi$$

$$\text{with } \theta_\Pi = \bigvee \{\theta_{(r,s)} : (r, s) \in \Pi\} = \bigcap \{\theta : \Pi \subseteq \theta\}.$$

The universal solution is typically much too large in the sense that there will be many points that are not relevant to the underlying domain or database for which the ontology is made. Accordingly, we give a definition of knowledge base that also specifies a set of domain terms.

Definition 3. *A* Knowledge Base *(KB), $\mathcal{B} = (C, A, \Pi, I)$, is an ontological framework $\mathcal{O} = (C, A, \Pi)$ together with a specified finite set I of DLA terms. We call the elements of I inserted or inhabited terms and $\mathcal{O}$ the associated ontological framework.*

The idea of this definition is that the inserted terms are the ones that actually correspond to concepts of interest, or, in a more database oriented view, to terms for which data is available and has to be classified by the ontology. In the setting of our real estate example, the inserted terms could be the real estates some agent has for sale. The significance of this definition lies in the companion definition of a solution of a KB. First we define the notion of a classification of a term w.r.t. a solution of an ontological framework.

Definition 4. *Let $\mathcal{O} = (C, A, \Pi)$ be an ontological framework, $h : F_{DLA}(C) \rightarrow D$ a solution of $\mathcal{O}$, and $t \in F_{DLA}(C)$. Then $t_1, \ldots, t_n \in F_{DLA}(C)$ is a* classification *of t with respect to $h : F_{DLA}(C) \rightarrow D$ provided for each i we have $t_i \leq t$ and*

$$h(t) = h(t_1) \vee \ldots \vee h(t_n).$$

This formal definition generalises the idea of classifications as one has in taxonomies: e.g., the animal kingdom is divided into a disjunction of subclasses (mammals, ...).

Notice that if $t_1, \ldots, t_n \in F_{DLA}(C)$ is a classification of t, then $t_1 \vee \ldots \vee t_n \leq t$ in $F_{DLA}(C)$. If we actually have equality in $F_{DLA}(C)$, then it is of course a classification w.r.t. any ontological framework, but these classifications are trivial. The interesting ones are the ones forced by Π. Notice also that any classification of a term t w.r.t. the universal solution is also a classification of t w.r.t. any other solution of $\mathcal{O}$. However, the implication does not hold in the other direction. For the trivial quotient, $F_{DLA}(C) \rightarrow \mathbf{1}$, the term $\bot$ by itself is a classification of any term t. This of course tells us that $F_{DLA}(C) \rightarrow \mathbf{1}$ is not a very useful solution. A solution of a knowledge base is a solution of the associated ontological framework that is faithful to the universal solution in terms of classifying inserted terms. We are now ready to give the formal definition.

Definition 5. *Let $\mathcal{B} = (C, A, \Pi, I)$ be a KB. A* solution *of $\mathcal{B}$ is a solution $h : F_{DLA}(C) \rightarrow D$ of the associated ontological framework $\mathcal{O} = (C, A, \Pi)$ with the additional property that, for each inserted term $t \in I$, every classification $t_1, \ldots, t_n$ of t w.r.t. $h : F_{DLA}(C) \rightarrow D$ is also a classification of t w.r.t. $h_{\mathcal{O}} : F_{DLA}(C) \rightarrow F_{\mathcal{O}}$.*

Note that, for a knowledge base $\mathcal{B}$, a solution $h : F_{DLA}(C) \rightarrow D$ of the associated ontological framework is a solution of $\mathcal{B}$ if and only if, for each inserted term t, we have that $h(t) \leq h(s)$ implies $h_{\mathcal{O}}(t) \leq h_{\mathcal{O}}(s)$ for all terms s.

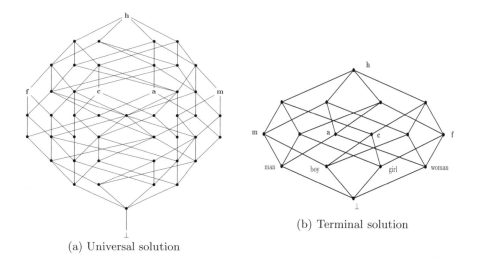

(a) Universal solution

(b) Terminal solution

Fig. 1. The human example

Clearly, the universal solution of the ontological framework associated to a knowledge base is a solution of the knowledge base. The import of the definition is that it allows us to work with smaller solutions than the universal one which still have all information relevant to the domain that has to be searched or classified.

Consider the concept human. Let $C = \{h, a, c, m, f\}$, where we think of these five concepts as human, adult, child, male and female. The free distributive lattice generated by these five unrelated concepts has more than 7000 elements. Introducing a terminological axiom to identify human with the disjuction of adult and child $(h, a \vee c)$ and one to identify human with the disjuction of male and female $(h, m \vee f)$, gives us an (attribute free) ontological framework whose universal solution $F_\mathcal{O}$ has 49 elements and is depicted in Figure 1(a). The boldface $\mathbf{h}, \mathbf{a}, \mathbf{c}, \mathbf{m}$ and $\mathbf{f}$ indicate the map $C \to F_\mathcal{O}$. Inserting just the concept h we obtain a knowledge base for which the least solution (which is what we will define to be the terminal solution) is the 16 element Boolean lattice depicted in Figure 1(b). Four new concepts: man $(m \wedge a)$, boy $(m \wedge c)$, girl $(f \wedge c)$, and woman $(f \wedge a)$ have been identified as the four concepts pertinent to the classification of human in this ontological framework and the terminal solution is the lattice generated by these four as irreducibles.

We will now use duality to obtain a more transparent description of the terminal solution. Most importantly, we will see that a knowledge base always has a terminal solution and, for a broad class of knowledge bases, this terminal solution is finite. Proving these facts as well as giving a description of an algorithm for finding the terminal solution of a knowledge base is best done coalgebraically or using duality theory.

3   Duality for Distributive Lattices

We will solve the problem of finding a solution of a knowledge base dually using Priestley duality for distributive lattices extended to incorporate the representation of attribution operations. The (Priestley) dual of a distributive lattice D is an ordered topological frame $(P_D, \tau, \leq)$. The set P_D may be taken to be the set of prime filters of D equipped with the reversed inclusion order. We define for each $d \in D$ a set $\hat{d} = \{\wp \in P_D : d \in \wp\}$. The topology τ on P_D is the topology generated by $\{\hat{d}, (\hat{d})^c : d \in D\}$, where $(\hat{d})^c$ denotes the complement of $\hat{d}$ in P_D. The dual lattice of this topological frame is the lattice of its clopen downsets, which is isomorphic to the original lattice D.

We also want to view the additional operations dually. The notion of canonical extension is particularly well-suited for this. The canonical extension encodes the topological dual frame in algebraic terms. A thorough explanation of canonical extensions may be found in [7]. One may obtain an incarnation of the canonical extension D^σ of a bounded distributive lattice D as the lattice of all downsets of the frame $(P_D, \tau, \leq)$ which is denoted by $\mathcal{D}(P_D)$. The original lattice, D, embeds in D^σ via:

$$D \to D^\sigma$$

$$d \mapsto \hat{d} = \{\wp \in P_D : d \in \wp\}.$$

On the other hand, the points of the dual space P_D may be obtained as the *completely join irreducible* elements of the canonical extension of the lattice. These are the points satisfying the infinitary version of join irreducibility: p is completely join irreducible provided $p = \bigvee_I u_i$ implies $p = u_i$ for some $i \in I$. We denote this set by $J^\infty(D^\sigma)$.

The power of canonical extensions lies in the fact that the assignment is categorical, that is, any bounded lattice homomorphism $h : D \to E$ extends to a complete homomorphism between the canonical extensions, $h^\sigma : D^\sigma \to E^\sigma$. The extension is effected by using the fact that the canonical extension of D is generated from the image of D by means of arbitrary meets and joins, see Theorem 3.25 in [7].

For a lattice homomorphism $h : D \to E$, the dual of h is given by the lower adjoint of h^σ restricted to the dual space, $P_E = J^\infty(E^\sigma)$, of E, i.e. by the function $(h^\sigma)^\flat \upharpoonright J^\infty(E^\sigma) : P_E \to P_D$ which is continuous and order preserving. This function is well-defined as h^σ sends completely join-irreducible elements of E^σ to completely join irreducible elements of D^σ. Every attribution operation on D may be viewed as a lattice homomorphism $D \to D$. Hence, every attribute a on D dually is captured by a continuous, order-preserving function $f_a : P_D \to P_D$.

One may show that a DLA-quotient $h : D \to E$ corresponds dually to a topologically closed subset $P_E \subseteq P_D$, closed under the actions of the maps f_a, dual to the attributes a on D. The extension $h^\sigma : D^\sigma \to E^\sigma$ may in that case be seen as the map

$$h^\sigma : \mathcal{D}(P_D) \to \mathcal{D}(P_E)$$

$$U \mapsto U \cap P_E.$$

It is clear that this map is a complete homomorphism and its lower adjoint is the map

$$(h^\sigma)^\flat : \mathcal{D}(P_E) \to \mathcal{D}(P_D)$$
$$V \mapsto {\downarrow}V$$

where ${\downarrow}V$ denotes the downset of V in P_D.

4   The Terminal Solution of a Knowledge Base

In the approach to ontological frameworks presented here, as in [10], our solutions are presented dually. However, the dual point of view is also particularly useful for understanding our newly introduced notion of a solution of a knowledge base.

To see this, recall that solutions of an ontological framework correspond to quotients of its universal solution. That is, every solution of $\mathcal{O}$ is of the form $h = g \circ h_\mathcal{O} : F_{DLA}(C) \to F_\mathcal{O} \to D$. The maps $h_\mathcal{O}$ and g are DLA morphisms and thus in particular bounded lattice morphisms. As a consequence their canonical extensions are complete lattice homomorphisms, which have lower adjoints. We obtain the following characterization of solutions of knowledge bases.

Theorem 2. *Let $\mathcal{B} = (C, A, \Pi, I)$ be a knowledge base, $\mathcal{O} = (C, A, \Pi)$ the associated ontological framework, and $h = g \circ h_\mathcal{O} : F_{DLA}(C) \to F_\mathcal{O} \to D$ a solution of $\mathcal{O}$. The following conditions are equivalent:*

1. *h is a solution of the knowledge base $\mathcal{B}$;*
2. *The retraction $F_\mathcal{O}^\sigma \xrightarrow{g^\sigma} D^\sigma \xrightarrow{(g^\sigma)^\flat} F_\mathcal{O}^\sigma$ fixes $h_\mathcal{O}(t)$ for each $t \in I$;*
3. *For each $t \in I$, $\max(\widehat{h_\mathcal{O}(t)})$ is contained in P_D;*

where $\widehat{h_\mathcal{O}(t)}$ denotes the clopen downset of $P_\mathcal{O}$ (the frame dual to $F_\mathcal{O}$), corresponding to $h_\mathcal{O}(t)$, $\max(\widehat{h_\mathcal{O}(t)})$ is the set of its maximal elements and P_D is the closed subspace of $P_\mathcal{O}$ dual to the quotient $g : F_\mathcal{O} \to D$.

Proof. As explained in the previous section, the composition in condition *2* may be viewed as

$$\mathcal{D}(P_\mathcal{O}) \xrightarrow{g^\sigma} \mathcal{D}(P_D) \xrightarrow{(g^\sigma)^\flat} \mathcal{D}(P_\mathcal{O})$$
$$S \mapsto S \cap P_D \mapsto {\downarrow}(S \cap P_D).$$

It is a basic fact of duality theory that every clopen S has enough maximal points in the sense that every element of S is below a maximal element of S. Thus it is clear that a clopen S is fixed by the above composition if and only if the maximal points of S are elements of P_D. This shows that conditions *2* and *3* of the theorem are equivalent. Their equivalence with *1* is an exercise in duality theory.

For any knowledge base $\mathcal{B} = (C, A, \Pi, I)$, the universal solution $h_\mathcal{O} : F_{DLA}(C) \to F_\mathcal{O}$ of the associated ontological framework $\mathcal{O} = (C, A, \Pi)$ is the greatest solution of $\mathcal{B}$ in the sense that any other solution factors through it. Every KB also has a least solution in the sense of the following theorem.

Theorem 3. *Let $\mathcal{B} = (C, A, \Pi, I)$ be a knowledge base. Then there exists a quotient $h_\mathcal{B} : F_{DLA}(C) \to D_\mathcal{B}$ of $F_{DLA}(C)$ satisfying the following conditions:*

1. $h_\mathcal{B} : F_{DLA}(C) \to D_\mathcal{B}$ *is a solution of* $\mathcal{B} = (C, A, \Pi, I)$;
2. *If* $h : F_{DLA}(C) \to D$ *is any solution of* $\mathcal{B} = (C, A, \Pi, I)$, *then there is a unique homomorphism* $h_D : D \to D_\mathcal{B}$ *so that the diagram*

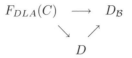

commutes.

Proof. By Theorem 2 and the general duality facts stated in Section 3, a solution of $\mathcal{B} = (C, A, \Pi, I)$ corresponds dually to a subspace Q of $P_\mathcal{O}$ that is topologically closed, is closed under the map f_a for each $a \in A$, and for which $\max(\widehat{h_\mathcal{O}(t)}) \subseteq Q$ for each $t \in I$. Since each of these three requirements on Q is preserved under arbitrary intersection, it follows that $\mathcal{B}$ has a least solution.

Note that this argument goes through unscathed in the setting of arbitrary modalities: We just need Q closed under relational image w.r.t. the dual Kripke relations R_a instead of under the f_as. In fact, using the canonical extension perspective, one can see that the proof goes through even for monotone operations (not necessarily join or meet preserving).

Definition 6. *Let* $\mathcal{B} = (C, A, \Pi, I)$ *be a knowledge base. We call the quotient* $h_\mathcal{B} : F_{DLA}(C) \to D_\mathcal{B}$ *the* terminal solution *of* $\mathcal{B}$.

The terminal solution may also be described dually in a generative way.

Theorem 4. *Let* $\mathcal{B} = (C, A, \Pi, I)$ *be a knowledge base. The terminal solution of* $\mathcal{B}$ *is described dually by*

$$P_\mathcal{B} = \overline{\left(\bigcup \{ f_w(\max(\widehat{h_\mathcal{O}(t)})) \mid w \in A^*, t \in I \} \right)},$$

where, for every word $w = a_0 \ldots a_n$ *in* A^*, $f_w = f_{a_n} \circ \ldots \circ f_{a_0}$, *the map dual to* $a_0 \circ \ldots \circ a_n$; *and for a subset* $P \subseteq P_\mathcal{O}$, $\overline{P}$ *denotes the topological closure of* P.

5 Description of $F_{DLA}(C)$ and Its Dual Frame

Since every solution is a quotient of the free DLA, we study the structure of $F_{DLA}(C)$ and its dual frame. These are particularly simple to understand on the basis of DLs. Let $A^*(C)$ denote the free algebra over C of type A. That is, $A^*(C) = \{w(c) : w \in A^* \text{ and } c \in C\}$ where we equate $\lambda(c)$ with c for each $c \in C$ (λ being the empty word) and we have operations $a^{A^*(C)} : A^*(C) \to A^*(C), w(c) \mapsto aw(c)$ for each $a \in A$. Generate the free bounded distributive lattice over $A^*(C)$, $F_{DL}(A^*(C))$, and define, for each $a \in A$, a unary operation $a^F : F_{DL}(A^*(C)) \to F_{DL}(A^*(C))$ to be the unique bounded lattice homomorphism given by $x \mapsto a^{A^*(C)}(x)$ for $x \in A^*(C)$. Then $(F_{DL}(A^*(C)), (a^F)_{a \in A})$ is the free DLA over C.

The dual space of $F_{DL}(A^*(C))$ is the ordered Cantor space $2^{A^*(C)}$ (with the order inherited from 2). In particular, the prime filters of $F_{DL}(A^*(C))$ are in one-to-one correspondence with the subsets of $A^*(C)$. The principal filters, i.e. the ones with a minimum element, are given by $\uparrow d$, where d is a join-irreducible element of $F_{DL}(A^*(C))$. These correspond to the finite subsets of, or finite conjunctions over, $A^*(C)$. The remaining subsets of $A^*(C)$ (an uncountable number of them) correspond to the non-principal filters or infinite conjunctions over $A^*(C)$. Hence, we may view the underlying set of the dual frame of $F_{DL}(A^*(C))$ as the collection of all conjunctions over $A^*(C)$. This is in fact precisely what it is in canonical extension terms. We will denote both the dual frame and its underlying set by $P(A^*(C))$. Dual to each attribution operation, we have a continuous order preserving function $f_a : P(A^*(C)) \to P(A^*(C))$. It is not hard to see that a conjunction x is sent to $a\backslash x$ which is the conjunction of all those $w(c)$ such that $aw(c)$ is one of the conjuncts in x.

Definition 7. *Let x be a conjunction over $A^*(C)$. The A-depth of x, notation $d_A(x)$, is n provided n is the smallest natural number such that $x \in P(A^{\leq n}(C))$ where $A^{\leq n}(C) = \bigcup_{k=0}^{n} A^k(C)$, and, for each set X, $P(X)$ denotes the set of all conjunctions over X, otherwise the A-depth of x is infinite. We will call conjunctions of finite depth* basic conjunctions *and call this part of the dual space the* finite part *of the dual space.*

Organising the basic conjunctions according to their A-depth, the poset $P(A^*(C))$ may be depicted as shown in figure 2. We omit the meet operation, writing for example $ca(c)$ for $c \wedge a(c)$.

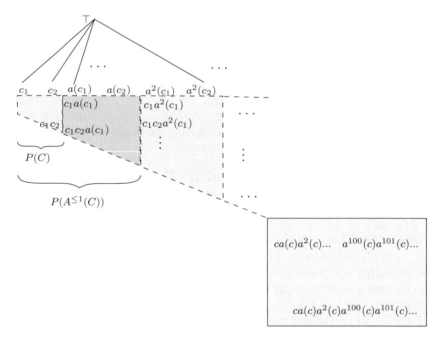

Fig. 2. $P(A^*(C))$

This graded view of the dual space is closely related to the coalgebraic construction that takes advantage of the fact that DLA is $Alg(H)$ for a functor on DL since the axioms for the attribution operations are all of rank 1, see e.g. [3]. In the coalgebraic setting $P(A^*(C))$ is obtained as the inverse limit of the posets

$$P(C) \leftarrow P(A^{\leq 1}(C)) \leftarrow P(A^{\leq 2}(C)) \cdots$$

where the projection $P(A^{\leq k-1}(C)) \leftarrow P(A^{\leq k}(C))$ is the map that, from a conjunction, removes all conjuncts of depth k. Notice that we also have the embeddings going forward, making these into retractions. The inverse limit of the projections is the dual space, and the direct limit of the embeddings is what we call the finite part of the dual space. Finally, for each $a \in A$, the map $f_a : P(A^*(C)) \to P(A^*(C))$ dual to the attribute a, is given by the maps $f_a^k : P(A^{\leq k}(C)) \to P(A^{\leq k-1}(C))$ which have the same action as f_a as described above.

6   Computing the Terminal Solution

In this section we will discuss how to find the terminal solution of a given knowledge base. We start by considering knowledge bases without attributes. In this case the construction of solutions is considerably simpler as the free algebra and its dual frame are finite and the topology on the dual frame is the trivial topology.

6.1   The Attribute-Free Case

In the attribute-free case we work over the finite poset $P(C)$ as described above, and we want to calculate

$$P_\mathcal{B} = \left(\bigcup \{ max(\widehat{h_\mathcal{O}(t)}) \mid t \in I \} \right).$$

That is, we need to calculate $max(\widehat{h_\mathcal{O}(t)})$ for each $t \in I$ where $h_\mathcal{O}$ is the quotient map corresponding to the congruence $\theta_\Pi = \bigvee \{ \theta_{(r,s)} \mid (r,s) \in \Pi \}$. Dually, this last equality tells us $P_\mathcal{O} = \bigcap \{ P_{(r,s)} \mid (r,s) \in \Pi \}$ where $P_{(r,s)}$ is the dual of $F_{DL}(C)/\theta_{(r,s)}$. This will allow us to devise a method to find $P_\mathcal{B}$ in which we compute with the posets $P_{(r,s)}$. Since we want to compute maximal antichains it makes more sense to work with a slightly different representation of the duality where the lattice dual to a poset is given in a way similar to the Hoare powerdomain construction.

Definition 8. *For any poset P, let $H(P)$ be the set of all anti-chains of P with the Hoare order:*

$$T \leq T' \quad \text{if and only if} \quad \forall p \in T \; \exists p' \in T' \quad p \leq p'.$$

It is not hard to see that this makes $H(P)$ into a lattice with the operations given by, for $T, T' \in H(P)$,

$$T \vee T' = max(T \cup T'),$$
$$T \wedge T' = max\{p \in P : \exists q \in T \; \exists q' \in T' \; (p \leq q \text{ and } p \leq q')\}.$$

For a given finite poset P, $H(P)$ is isomorphic to the collection $\mathcal{D}(P)$ of downsets of P by $T \mapsto {\downarrow}T$ and with inverse $S \mapsto max(S)$. For a finite lattice D, writing $J(D)$ for the set of join-irreducible elements of D, $D \cong \mathcal{D}(J(D)) \cong H(J(D))$ with $d \mapsto \hat{d} = {\downarrow}d \cap J(D) \mapsto max({\downarrow}d \cap J(D))$ the latter of which we will denote by $\tilde{d}$. Note that in the case of a term $t \in F_{DL}(C)$, the anti-chain $\tilde{t}$ consists of the pure conjunctions in the disjunctive normal form of t. That is, computing $\tilde{t}$ is just computing the disjunctive normal form of t.

Remark 1. For a poset P and a subposet $P' \subseteq P$, the corresponding quotient map $h : H(P) \to H(P')$ is given by $T \mapsto max[({\downarrow}T) \cap P']$. A nice property of this representation of the lattice dual to a finite poset is that the lower adjoint of h, $h^{\flat} : H(P') \to H(P)$, is just the identity map, and in particular, since adjoints satisfy $h^{\flat}h(T) \leq T$, we have $h(T) \leq T$ for all $T \in H(P)$. Further, if $h : H(P) \to H(P')$ factors through $h' : H(P) \to H(P'')$, i.e. if $P' \subseteq P'' \subseteq P$, then $h(T) \leq h'(T) \leq T$ for all $T \in H(P)$.

For a pair $(r, s) \in \Pi$, the quotient $F_{DL}(C)/\theta_{(r,s)}$ corresponds to the subset $P_{(r,s)}$ of all $p \in P(C)$ satisfying $p \leq r \iff p \leq s$, or equivalently $p \leq r \vee s \Rightarrow p \leq r \wedge s$. Thus we may assume our pairs (r, s) are given with $r \geq s$ and then

$$P_{(r,s)} = P(C) - Q_{(r,s)} \quad \text{where} \quad Q_{(r,s)} = \{p \in P(C) \mid p \leq r \text{ but } p \nleq s\}.$$

The following proposition provides the basic step of the method for computing $P_{\mathcal{B}}$ for an attribute-free knowledge base $\mathcal{B}$.

Proposition 1. *Let $(r, s) \in F_{DL}(C)^2$ and $t \in F_{DL}(C)$. Then the quotient map $h_{(r,s)}$ corresponding to $\theta_{(r,s)}$ is given by*

$$h_{(r,s)} : H(P(C)) \quad \to \quad H(P_{(r,s)})$$
$$\tilde{t} \quad \mapsto \quad max\Big(\bigcup_{p \in \tilde{t}} h_{(r,s)}(p)\Big)$$

where

$$h_{(r,s)}(p) = \begin{cases} p & \text{if } p \nleq \tilde{r} \\ max(\{pp' : p' \in \tilde{s}\}) & \text{if } p \leq \tilde{r} \end{cases}$$

Note that $p \leq \tilde{r}$ provided there is $p' \in \tilde{r}$ with $p \leq p'$ and the latter holds if and only if $p' \subseteq p$ when these are viewed as subsets of C. Also, the conjunction of pure conjunctions $p, p' \in P(C)$ is just the union of these viewed as subsets of C.

Proof. First of all $h_{(r,s)}$ is join preserving, so $h_{(r,s)}(\tilde{t}) = \bigvee_{p \in \tilde{t}} h_{(r,s)}(p))$ which is $max(\bigcup_{p \in \tilde{t}} h_{(r,s)}(p))$. Also $P_{(r,s)} = P(C) - \{p \in P(C) \mid p \leq r \text{ but } p \nleq s\}$. Thus it is clear that if $p \nleq \tilde{r}$ then $h_{(r,s)}(p) = p$. If on the other hand $p \leq \tilde{r}$, then $h_{(r,s)}(p) = h^{\flat}_{(r,s)}(h_{(r,s)}(p)) = p \wedge \tilde{s}$. Also $p \wedge \tilde{s} = max(\{pp' : p' \in \tilde{s}\})$ since $P(C)$ is closed under $\wedge$.

Now we are able to compute $h_{\mathcal{O}}(\tilde{t})$. We start by computing $\tilde{t}$, i.e. the disjunctive normal form of t. Next we generate a sequence $T_0, T_1, T_2, \ldots$ by applying the subroutine described above repeatedly for varying choices of $(r, s) \in \Pi$.

Let $(r_0, s_0), \ldots, (r_k, s_k)$ be a list of the elements in Π and define an infinite sequence $\{(u_n, v_n)\}$ by repeating this list time after time. That is, $\{(u_n, v_n)\}$ is the sequence $(r_0, s_0), \ldots, (r_k, s_k), (r_0, s_0), \ldots, (r_k, s_k), \ldots$. We then define

$$T_0 = \tilde{t}$$
$$T_n = h_{(u_n, v_n)}(T_{n-1}).$$

By Remark 1 it follows that

$$\tilde{t} = T_0 \geq T_1 \geq T_2 \geq \ldots \qquad \geq h_{\mathcal{O}}(\tilde{t}).$$

Since $H(P(C))$ is a finite set, this sequence must eventually be constant. In fact, one may show that once it has stayed constant through a whole cycle of choices of $(r, s) \in \Pi$, we have reached $h_{\mathcal{O}}(\tilde{t})$.

Having computed $h_{\mathcal{O}}(\tilde{t})$ for all inserted terms t, one finds $P_{\mathcal{B}}$ by taking the union of all their maximal elements.

6.2   The General Case

Adding attribution operations clearly complicates matters. We have to deal with an infinite lattice and with its dual space, which carries a non-trivial topology. However, some things remain the same: all clopen downsets of $P(A^*(C))$ are in fact downsets of finite anti-chains of the finite part of $P(A^*(C))$, and, with the graded or coalgebraic perspective on $P(A^*(C))$, we can work by approximating in finite DLs of the same type as in the attribute-free case (the generating set is just $A^{\leq n}(C)$ for some n rather than C). In the sequel we will not make a notational difference, for DLA terms t, between the term and the corresponding anti-chain $\tilde{t}$, nor between $h_{\mathcal{O}}(t)$, $\widehat{h_{\mathcal{O}}(t)}$ and $max(\widehat{h_{\mathcal{O}}(t)})$.

We first study the structure of $P_{\mathcal{O}}$. Recall, $P_{\mathcal{O}} = \bigcap_{(r,s) \in \Pi} P_{(r,s)}$, where $P_{(r,s)}$ is the subspace of $P(A^*(C))$ corresponding dually to the quotient $F_{DLA}(C) \rightarrow F_{DLA}(C)/\theta_{(r,s)}$. In the attribute-free setting $P_{(r,s)}$ is obtained by removing all points below r that are not below s (assuming, as above, without loss of generality that $r > s$). However, with a set of attributes involved, $P_{(r,s)}$ is not given by $P(A^*(C)) - (\downarrow r - \downarrow s)$, as this set is not closed under the actions of the maps f_a. Dually, this corresponds to the fact that requiring $r = s$ induces $w(r) = w(s)$ for every composition of attributes $w \in A^*$. However, we may understand DLA-quotients of $F_{DLA}(C)$ via DL-quotients of $F_{DL}(A^*(C))$. We will use superscripts DLA and DL to indicate which setting we are working in.

As $Q^{DL}_{(r,s)} = \downarrow r - \downarrow s$ is open in $P(A^*(C))$ and f_w is continuous, it follows that $P(A^*(C)) - \bigcup_{w \in A^*} f_w^{-1}(Q^{DL}_{(r,s)})$ is a closed subspace of $P(A^*(C))$. One easily proves that it is the largest closed subspace of $P(A^*(C))$ closed under the f'_as for which r and s get identified under the dual quotient map. By the adjunction property it follows that $f_w(x) \in Q^{DL}_{(r,s)}$ iff $x \in Q^{DL}_{(w(r), w(s))}$, hence $f_w^{-1}(Q^{DL}_{(r,s)}) = Q^{DL}_{(w(r), w(s))}$ and thus

$$P_{\mathcal{O}}^{DLA} = P(A^*(C)) - \bigcup_{\substack{w \in A^* \\ (r,s) \in \Pi}} Q^{DL}_{(w(r), w(s))}.$$

This means we can attempt to find $h_{\mathcal{O}}(t)$ by applying the same subroutine as in the attribute-free case: Starting from a term t (which is necessarily an anti-chain in the finite part) we repeatedly apply quotient maps $h_{(x,y)}^{DL}$ (as described in Proposition 1) for $(x,y) \in \Pi^* = \{(w(r), w(s)) : (r,s) \in \Pi, w \in A^*\}$. Note that for any term t' we reach in this way, $t \geq t' \geq h_{\mathcal{O}}(t)$. If at some point we reach a term t' which satisfies $h_{(x,y)}^{DL}(t') = t'$ for all $(x,y) \in \Pi^*$, then $t' = h_{\mathcal{O}}(t)$.

To check whether or not $h_{(x,y)}^{DL}(t') = t'$ for *all* $(x,y) \in \Pi^*$, we only have to consider finitely many pairs in Π^*. To explain why this is the case, we start with two definitions.

Definition 9. *Let t be a term and $(x,y) \in F_{DL_\perp A}(C)^2$. We say t is rejected by (x,y) if $h_{(x,y)}^{DL}(t) \neq t$.*

We generalize the notion of A-depth given in Definition 7 for basic conjunctions.

Definition 10. *Let $t \in F_{DLA}(C)$, $t \neq \top, t \neq \bot$. We have identified t with a non-empty finite anti-chain of non-empty basic conjunctions (namely, the ones that occur in its disjunctive normal form). We define the A-depth of t, $d_A(t)$, to be the maximum of $d_A(p)$ for p a basic conjunction in t.*

For two basic conjunctions p and p', $p \leq p'$ implies $d_A(p) \geq d_A(p')$. Hence, a term t' being rejected by $(x,y) \in \Pi^*$, implies that $d_A(t') \geq d_A(p)$ for some p in x. So to check whether t' is rejected, we only have to consider the finite collection of elements of the form $(w(r), w(s))$ with $(r,s) \in \Pi$ and $w \in A^{\leq d_A(t)}$.

There do however exist knowledge bases with inserted terms t for which $h_{\mathcal{O}}(t)$ is not in the finite part of the space $P(A^*(C))$. Consider for example $\mathcal{B} = (\{c\}, \{a\}, \{(c, ca(c))\}, \{c, a(c)\})$. Starting from the inserted term c we get the following chain:

$$c \xrightarrow{(c,ca(c))} ca(c) \xrightarrow{(a(c),a(c)a^2(c))} ca(c)a^2(c) \xrightarrow{(a^2(c),a^2(c)a^3(c))} ca(c)a^2(c)a^3(c)\dots$$

Above each arrow we have indicated which element of Π^* was used. In this case, $h_{\mathcal{O}}(c)$ is the conjunction $p = \bigwedge_{n \geq 0} a^n(c)$, an infinite point of the dual frame. The dual of the terminal solution is the two element chain, $\{p, q = \bigwedge_{n \geq 1} a^n(c)\}$ for which $f_a(p) = p$ and $f_a(q) = p$. Obtaining solutions in which f_a has fixpoints or maps downwards is impossible in the finite part of $P(A^*(C))$.

In the above example, the solution is finite but requires using points of the infinite part of $P(A^*(C))$. There are also KBs for which the terminal solution is infinite, e.g. $\mathcal{B} = (\{c,d\}, \{a\}, \{(c, ca(c) \vee cd)\}, \{c\})$. We call a solution of a KB *totally finite* if it lies within the finite part of $P(A^*(C))$. This implies in particular that the solution is finite. The algorithm for finding the terminal solution described above terminates if and only if the terminal solution is totally finite. We would like to be able to determine whether or not a given KB has a totally finite terminal solution. This problem has not been solved as far as we are aware. We do give two sufficient conditions for termination and a condition that implies non-termination, but these are not exhaustive.

Definition 11. *Let $\mathcal{B} = (C, A, \Pi, I)$ be a KB, p be a basic conjunction and $c \in C$. We define the A_c-depth of p to be the smallest $n \in \mathbb{N} \cup \{-1\}$ such that*

$p \in P(A^*(C - \{c\}) \cup A^{\leq n}(\{c\}))$. *In general, for a term* $t \in F_{DLA}(C) - \{\bot, \top\}$, *and a concept* $c \in C$, $d_{A_c}(t)$ *is defined to be the maximum of* $d_{A_c}(p)$ *for* p *a basic conjunction in* t.

From now on, we will omit A and just write $d(t)$ and $d_c(t)$.

Theorem 5. *(Termination condition I) Let* $\mathcal{B} = (C, A, \Pi, I)$ *be a KB. If, for every terminological axiom* $(r, s) \in \Pi$, *there is a concept* c *that occurs in* r, *i.e.* $d_c(r) \geq 0$, *with, for all* $(u, v) \in \Pi$, $d_c(v) \leq d_c(u)$, *then the terminal solution of* $\mathcal{B}$ *is totally finite.*

Proof. We will show that, under the condition stated in the theorem, $h_\mathcal{O}(t)$ is finite, i.e. it is a finite set of basic conjunctions, for every term t. It thus follows that the terminal solution is totally finite.

Let t be a term and define $\Pi' = \{(w(r), w(s)) : (r, s) \in \Pi, w \in A^{\leq d(t)}(C)\}$. This is a finite set of terminological axioms and therefore we may apply the algorithm for the attribute-free case and find a term t' that is not rejected by any of the terminological axioms in Π'. We claim that t' is not rejected by any element of Π^*.

Let $(x, y) = (w(r), w(s))$ be an element of Π^* that is not in Π'. By assumption there exists a concept c that occurs in r such that for all $(u, v) \in \Pi$, $d_c(v) \leq d_c(u)$. Using the fact that t' is the result of finitely many applications of the subroutine of Proposition 1, one easily shows that $d_c(t') = d_c(t)$. By assumption $(x, y) \notin \Pi'$ and hence $d_c(x) > d(t) \geq d_c(t) = d_c(t')$ and thus $t' \not\leq x$ and t' is not rejected by (x, y).

Theorem 6. *(Termination condition II) Let* $\mathcal{B} = (C, A, \Pi, I)$ *be a KB such that for every terminological axiom* $(r, s) \in \Pi$, $d(r) = d(s)$. *Then the terminal solution of* $\mathcal{B}$ *is totally finite.*

Proof. Again it suffices, for a given term t, to consider the set $\Pi' = \{(w(r), w(s)) : (r, s) \in \Pi, w \in A^{\leq d(t)}\}$ as, under the given condition, the depth of a term does not change when applying the subroutine.

It is easy to find KBs with totally finite terminal solutions that do not satisfy either of the above conditions. Finding a sharp syntactic termination condition does not seem probable to us but the above conditions cover most actual applications as concepts and their attribution translates typically are disjoint (e.g., in an ontology for real estate, $c =$ a geographical region and the concept $a(c)$ of being a piece of real estate located in region c, only have the inconsistent concept as common subsumer and a^2 of anything is inconsistent). Nevertheless, finding a sharp condition for total finiteness seems worthwhile and we expect that an algorithmic approach may be more fruitful. A first step in this direction is made by the non-termination condition below. For simplicity, we formulate it only for basic conjunctions, but it is readily extended to incorporate disjunctions of basic conjunctions. We start with a definition.

Definition 12. *For two words* w, w', *we say* w' *is an* extension *of* w, *notation* $w \sqsubseteq w'$, *if there exists a word* w'' *such that* $w' = ww''$.

Theorem 7. *(Non-termination condition) Let $\mathcal{B} = (C, A, \Pi, I)$ be a KB such that all terminological axioms are pairs of basic conjunctions and let $t \in I$ be a basic conjunction. If there exists a chain of rejections:*

$$t = t_0 \xrightarrow{w_0(r_0, s_0)} t_1 \xrightarrow{w_1(r_1, s_1)} \cdots \xrightarrow{w_{L-1}(r_{L-1}, s_{L-1})} t_L$$

such that there exist $n < L$ and $w_L \in A^*$ with $f_{w_n}(t_n) = f_{w_L}(t_L)$ and for every k with $n \leq k \leq L - 1$, $w_k \sqsubseteq w_{k+1}$, then $h_\mathcal{O}(t)$ is an infinite point of the dual frame.

Proof. It suffices to show that we may extend the chain of rejections given above ad infinitum. We may assume $n = 0$. As, by assumption, $f_{w_L}(t_L) = f_{w_0}(t_0) \leq r_0$, by adjunction $t_L \leq w_L(r_0)$. Similarly one shows $t_L \not\leq w_L(s_0)$. Hence, t_L is rejected by $w_L(r_0, s_0)$ and we define $t_{L+1} = t_L \wedge w_L(s_0)$. As $w_0 \sqsubseteq w_1$, there exists a word w_1' such that $w_1 = w_0 w_1'$. One may show that t_{L+1} is rejected by $(w_L w_1')(r_1, s_1)$ and we define $t_{L+2} = t_{L+1} \wedge (w_L w_1')(s_1)$. It is readily seen that one may continue this way, thereby forming an infinite chain of rejections.

This non-termination condition is not sharp. Consider for example the knowledge base $\mathcal{B} = (\{c, d, e\}, \{a\}, \{z_1, z_2, z_3\}, \{pa(p)\})$, where $p = cde$ and

$$z_1 = (cde, cdea(c)), \quad z_2 = (a(c)a(d), a(c)a(d)a^2(e)), \quad z_3 = (a^2(c), a^2(c)a^2(d)).$$

There exists no chain of rejections satisfying the non-termination condition given above. Nevertheless, $h_\mathcal{O}(pa(p))$ is an infinite conjunction as we have the following unending chain of rejections,

$$pa(p) \xrightarrow{z_2} pa(p)a^2(e) \xrightarrow{az_1} pa(p)a^2(c)a^2(e) \xrightarrow{z_3} pa(p)a^2(p) \xrightarrow{az_2} pa(p)a^2(p)a^3(e) \xrightarrow{a^2 z_1} \cdots$$

However, when we add the pair az_1 to our set Π, we do recognize the chain above as one satisfying our non-termination condition. So we may broaden the class of KB's that are recognized as having an infinite terminal solution by adding some of the elements of Π^* to the terminological axioms Π. This raises the question whether there is a general way to extend, for a given knowledge base $\mathcal{B} = (C, A, \Pi, I)$, the set of terminological axioms Π to a finite set $\Pi' \subseteq \Pi^*$, such that $\mathcal{B}$ has a totally finite terminal solution if and only if $\mathcal{B}' = (C, A, \Pi', I)$ does not satisfy the above non-termination condition (extended to disjunctions). We have no counterexample to this, the question is still open.

7   Related and Future Work

The approach to generative knowledge representation presented in this paper is a generalisation of the work of Oles [10]. However, Oles only identifies the universal solution of a given ontological framework thus making work with additional operators impossible. Algorithmically, however, Oles also only computes (max of) $h_\mathcal{O}(t)$ for the terms t that are of interest. For this reason we expect that in the applications to medical expert systems that he mentions, it is in fact the terminal solution that one works with. By introducing the concept of a classification of a term w.r.t. a solution of an ontological framework we are able to identify this solution as the least one with a certain property, thereby giving its mathematical significance. This in turn makes it feasible to work with additional operations. While we consider the identification of the concept of terminal solution a significant advancement, the fact that these need not be finite raises new questions. Our termination and non-termination conditions cover most practical applications, but finding a sharp termination condition seems an interesting and worthwhile problem. We feel that a coalgebraic approach to this problem is most promising. Since the dual of a DL is up-to exponentially bigger than the generating set, this is the worst case complexity, but in practice the algorithm performs better and this should be studied experimentally. Further, we have restricted ourselves to deterministic modalities, but the concept and existence of a terminal solution does not depend on this. We expect that our results readily can be extended to non-deterministic modalities. Finally, thanks to helpful input from the referees, we have made first steps in understanding the relation of this work to description logic. It requires further study and discussion and we hope the publication of this paper will help to achieve this goal.

References

1. Baader, F., Lutz, C., Sturm, H., Wolter, F.: Fusions of Description Logics and Abstract Description Systems. J. Artif. Intell. Res (JAIR) 16, 1–58 (2002)
2. Baader, F., Nutt, W.: Basic description logics. In: Baader, F., et al. (eds.) The description logic handbook: theory, implementation, and applications, pp. 43–95. Cambridge University Press, Cambridge (2003)
3. Bezhanishvili, N., Kurz, A.: Free modal algebras: a coalgebraic perspective. In: Mossakowski, T., Montanari, U., Haveraaen, M. (eds.) CALCO 2007. LNCS, vol. 4624, pp. 143–157. Springer, Heidelberg (2007)
4. Bruun, H.: The developement of a lattice structured database. IMM Technical report, Technical University of Denmark, DTU (2006),
   http://www2.imm.dtu.dk/pubdb/views/publication_details.php?id=4967
5. Bruun, H., Fischer Nilsson, J., Gehrke, M.: Lattice ordered ontologies (preprint)
6. Davey, B.A., Priestley, H.A.: Introduction to Lattices and Order, 2nd edn. Cambridge University Press, Cambridge (2002)
7. Gehrke, M., Jónsson, B.: Bounded distributive lattice expansions. Mathematica Scandinavica 94(1), 13–45 (2004)
8. Ghilardi, S., Santocanale, L.: Algebraic and Model-Theoretic Techniques for Fusion Decidability in Modal Logic. In: Vardi, M., Voronkov, A. (eds.) Logic for Programming, Artificial Intelligence and Reasoning (LPAR 2003). LNCS (LNAI), vol. 2850, pp. 152–166. Springer, Heidelberg (2003)
9. Nilsson, J.F.: Relational data base model simplified and generalized as algebraic lattice with attribution. In: FQAS 1994, pp. 101–104 (1994)
10. Oles, F.J.: An application of lattice theory to knowledge representation. Theoretical Computer Science 249, 163–196 (2000)
11. OntoQuery project net site, http://www.ontoquery.dk
12. Sofronie-Stokkermans, V.: Automated theorem proving by resolution in nonclassical logics. Annals of Mathematics and Artificial Intelligence 49(1-4), 221–252 (2007)

A Duality Theorem for Real C^* Algebras

M. Andrew Moshier[1] and Daniela Petrişan[2,*]

[1] Department of Mathematics and Computer Science, Chapman University, USA
[2] Department of Computer Science, University of Leicester, UK

Abstract. The full subcategory of proximity lattices equipped with some additional structure (a certain form of negation) is equivalent to the category of compact Hausdorff spaces. Using the Stone-Gelfand-Naimark duality, we know that the category of proximity lattices with negation is dually equivalent to the category of real C^* algebras. The aim of this paper is to give a new proof for this duality, avoiding the construction of spaces. We prove that the category of C^* algebras is equivalent to the category of skew frames with negation, which appears in the work of Moshier and Jung on the bitopological nature of Stone duality.

1 Introduction

Real C^* algebras have come under attention in theoretical computer science in the work of [9] as an abstraction of the notion of "test" or "observation." A standard motivating example is the ring of measurable maps on a measure space. In the work of Kozen [8], and Saheb-Djahromi [11], measure spaces are used to model programs with probabilistic choice. A measurable map can be thought of as providing the result of making an observation on the points of the measure space.

Another way to think about observations has a longer history in domain theory. There, the open sets of a Scott topology are regarded as forming a logic of observable properties [12,13,15]. For many decades, the two approaches have been known to be related via the two Stone duality theorems: First, the category of C^* algebras is dually equivalent to the category of compact Hausdorff spaces; second, the category of Boolean algebras is dually equivalent to the category of completely disconnected, compact Hausdorff spaces. Domain theory has traditionally taken its cue from the latter theorem, see [1]. Hence the emphasis on the "logical" view as opposed to the "real ring" view.

In this note, we investigate the connections between these two approaches by showing that the category of C^* algebras is dually equivalent to the category of proximity lattices with negation. Importantly, we do this without going through the construction of spaces, so we can isolate the results that make use of the axiom of choice.

A second motivating example appears in [9] : the bisimulation equivalence classes of labelled Markov processes are gathered into a 'universal' LMP, which

---

[*] Thanks to Nick Bezhanishvili for valuable comments and suggestions.

A. Kurz, M. Lenisa, and A. Tarlecki (Eds.): CALCO 2009, LNCS 5728, pp. 284–299, 2009.

is presented as the spectrum of a C^* algebra of formal linear combinations of labelled trees. The same object can be regarded as the solution of a domain equation in the category of coherent domains, see [4,14].

2 Preliminaries: The Categories at Issue

2.1 Commutative Real C^* Algebras

Definition 1. *A partially ordered ring is a commutative ring $(A, +, \cdot, 0, 1)$ endowed with a partial order $\leqslant$ compatible with the ring structure (i.e. $a \leqslant b$ implies $a + c \leqslant b + c$ for all c, and $ac \leqslant bc$ for all $c \geqslant 0$) and such that $a^2 \geqslant 0$ for all $a \in A$.*

Let A be a partially ordered commutative ring satisfying the following three conditions:

(a) The additive group $(A, +)$ is divisible and torsion-free.
(b) For every $a \in A$ there exists an integer n such that $n \cdot 1 \geqslant a$.
(c) If for all integers $n \geqslant 0$ we have $n \cdot a \leqslant 1$, then $a \leqslant 0$.

From (a) we obtain that A has a $\mathbb{Q}$-algebra structure. Condition (b) implies that the real number:

$$\|a\| = \inf\{q \in \mathbb{Q}| - q \leqslant a \leqslant q\}$$

is well defined. It is easy to see that $\| - \|$ is a semi-norm, and moreover because of (c) one can actually prove that $\| - \|$ is a norm.

Definition 2. *A partially ordered commutative ring $(A, +, \cdot, 0, 1, \leqslant)$, satisfying (a)-(c) above, is called a C^* algebra if additionally it is complete in the norm $\| - \|$.*

The completeness condition implies that the $\mathbb{Q}$-algebra structure on a C^* algebra can be uniquely extended to a $\mathbb{R}$-algebra structure. In what follows, for any $r \in \mathbb{R}$ we will denote the element $r \cdot 1$ of a C^* algebra simply by r.

The next useful observations regarding C^* algebras are proved in [5]. In a C^* algebra the order relation is determined by the ring structure. One can show that any $a \geqslant 0$ has a unique non-negative square root. So we have $a \geqslant 0$ if and only if a is a square. Therefore any ring homomorphism between C^* algebras is order preserving. Moreover it is also continuous wrt the norm induced topology. We can define the *absolute value* of an element a to be the unique non-negative square root of a^2, and we will denote it by $|a|$. A C^* algebra A is lattice ordered if we define the join and the meet of a and b by $a \vee b = \frac{a+b+|a-b|}{2}$ and $a \wedge b = \frac{a+b-|a-b|}{2}$, respectively.

Lemma 1. *Let A be a C^* algebra and $a \in A$. Then a can be written uniquely as a difference of nonnegative elements $a = a^+ - a^-$, such that $a^+ a^- = 0$. Moreover:*

$$|a| = a^+ + a^- \qquad (a \wedge b)^+ = a^+ \wedge b^+ \qquad (a \vee b)^+ = a^+ \vee b^+$$
$$(a \wedge b)^- = a^- \vee b^- \qquad (a \vee b)^- = a^- \wedge b^-$$

Proof. For the existence, take $a^+ = a \vee 0$ and $a^- = -(a \wedge 0)$. Now assume that $b, c \geqslant 0$, such that $a = b - c$ and $bc = 0$. It follows that $ab = b^2$. We can easily check that $(2b - a)^2 = a^2$. But $2b - a \geqslant 0$, so we have that $2b - a = |a|$, or equivalently $b = a^+$. This also implies $c = a^-$, so the uniqueness is proved. The remaining equalities are easily verified. □

Definition 3. *An element a of a C^* algebra is called strictly positive, and we write $a > 0$, iff there exists some strictly positive $q \in \mathbb{Q}$ such that $a \geqslant q$.*

Definition 4. *We will denote by C^*-Alg the category of C^* algebras with ring homomorphisms.*

Lemma 2. *Let A be a C^* algebra, and $a, b, c \in A$. Then:*
   a) If $a, b, c \geqslant 0$ we have that $a + b > 0$ and $a + c > 0$ is equivalent to $a + b \wedge c > 0$.
   b) If $a, b \geqslant 0$ we have that $a + b > 0 \Leftrightarrow a \vee b > 0$
   c) If $a \geqslant 0$ and $a + b^+ > 0$ and $a + b^- > 0$ for some $b \in A$ then $a > 0$.

Proof. a) Since $a + b \geqslant a + b \wedge c$ and $a + c \geqslant a + b \wedge c$ the direct implication is trivial. Conversely, there exists a positive rational number q and $x, y \geqslant 0$ such that $a + b = q + x$ and $a + c = q + y$. One can check that $a + b \wedge c = q + x \wedge y$. Since $x \wedge y \geqslant 0$ we have that $a + b \wedge c \geqslant q > 0$.

b) Assume $a + b > 0$. There exists $q \in \mathbb{Q}, q > 0$ such that $a + b \geqslant q$. But then $a \vee b = \frac{1}{2}(a + b + |a - b|) \geqslant \frac{q}{2} + \frac{|a-b|}{2} \geqslant \frac{q}{2} > 0$. Conversely assume that $a \vee b > 0$. We know that $a \wedge b \geqslant 0$ and that $a + b = a \vee b + a \wedge b$, hence $a + b$ is strictly positive.

c) There exists $q \in \mathbb{Q}, q > 0$ such that $a + b^+ \geqslant q$ and $a + b^- \geqslant q$. This implies that $a^2 + a|b| \geqslant q^2$, or equivalently $a(a + |b|) \geqslant q^2$. Since there exists a positive integer n such that $q \leqslant a + |b| \leqslant n$, we have that $a + |b|$ is invertible, and $(a + |b|)^{-1} \geqslant \frac{1}{n}$. Therefore $a \geqslant \frac{q^2}{n} > 0$. □

2.2   Proximity Lattices and Skew Frames with Negation

Proximity lattices have been introduced in [7], as a way to represent stably compact spaces. It is well known that the category StKTop of stably compact spaces is isomorphic to the category of ordered compact Hausdorff spaces. For the purpose of this paper we are interested in structures which correspond to compact Hausdorff spaces, so the order must be trivial. On the algebraic side this corresponds to proximity lattices equipped with an additional operation, a certain form of negation, as introduced in [10]. We would like to prove the duality between C^*-Alg and the category of proximity lattices with negation, formally defined below.

Definition 5. *A tuple $(L, \vee, \wedge, tt, ff, \prec)$ is called a strong proximity lattice iff $(L, \vee, \wedge, tt, ff)$ is a distributive lattice and $\prec$ is a binary relation satisfying the following axioms for all $x, x', a, y, y' \in L$:*
   • *$x \prec tt$ and $ff \prec x$*
   • *$x \prec y$ and $x \prec y' \Leftrightarrow x \prec y \wedge y'$; $x \prec y$ and $x' \prec y \Leftrightarrow x \vee x' \prec y$*
   • *$a \wedge x \prec y$ implies that there exists $a' \in L$ such that $a \prec a', a' \wedge x \prec y$*
   • *$x \prec y \vee a$ implies that there exists $a' \in L$ such that $a' \prec a, x \prec y \vee a'$*

Definition 6. *A proximity lattice with negation is a strong proximity lattice* $(L, \prec)$ *equipped with an additional unary operation* $\neg$ *satisfying:*

$$(\neg \text{ - } \neg) \qquad \neg \circ \neg = id$$
$$(\neg \text{ - } \prec) \qquad x \wedge y \prec z \Leftrightarrow y \prec z \vee \neg x$$

The relation $\prec$ is called a *strong entailment relation*, and it can be deduced from the above axioms that it is also transitive, interpolative and satisfies a cut-rule.

One can think of proximity lattices as abstract bases of ordered compact Hausdorff spaces in which tokens are names for pairs (u^+, u^-), where u^+ is an upper open and u^- is a lower open and the two are disjoint. Because a proximity lattice does not have a token corresponding to every such pair, the morphisms, which need to encode continuous monotonic functions, will be rather relations than functions between the carrier sets.

A *consequence relation* from a proximity lattice $\mathcal{L} = (L, \prec_{\mathcal{L}})$ to $\mathcal{M} = (M, \prec_{\mathcal{M}})$ is a binary relation $\vdash \subseteq L \times M$ so that $\prec_L; \vdash \,= \,\vdash$ and $\vdash; \prec_M \,= \,\vdash$.

Definition 7. *A morphism from* $(L, \prec_L)$ *to* $(M, \prec_M)$ *is a pair* $(\vdash^*, \vdash_*)$ *where* $\vdash^* \subseteq L \times M$ *and* $\vdash_* \subseteq M \times L$ *are consequence relations and furthermore they are adjoint, i.e.* $\prec_L \,\subseteq\, \vdash_*; \vdash^*$ *and* $\vdash^*; \vdash_* \,\subseteq\, \prec_M$.

Theorem 1. *The category* Prox *whose objects are strong proximity lattices and morphisms as defined above, is equivalent to the category of ordered compact Hausdorff spaces. Furthermore this equivalence cuts down to an equivalence between the full subcategory of proximity lattices with negation* Prox$_\neg$ *and the category of compact Hausdorff spaces.*

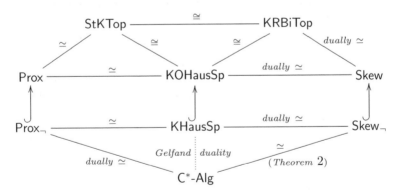

However if the proximity lattice is complete in a certain sense, then we can replace consequence relations with functions (contravariantly). This is where the skew frames come into picture. These structures first appeared in the work of the first author with Achim Jung on the bitopological nature of Stone duality. The category of compact regular skew frames, which will be defined below is proved to be dually equivalent to the category of compact regular bitopological spaces, which in turn is isomorphic to the category of stably compact spaces and compact ordered Hausdorff spaces. This equivalence restricts to an equivalence between a full subcategory of compact regular skew frames with negation and compact Hausdorff spaces. However, the duality between the category of skew frames and the category of proximity lattices can be proved directly. The relation between these categories is summarized in the above diagram. We list below the formal definitions of these notions and refer the reader to [6] for more details and motivation.

Definition 8. *A skew frame is a tuple:* $\mathcal{L} = (L, \sqcap, \bigsqcup^{\uparrow}, \bot, \wedge, \vee, tt, ff, \prec)$ *which satisfies:*

1. $(L, \wedge, \vee, tt, ff)$ *is a distributive lattice.*
2. $\sqcap$ *is a binary operation; the partial order* $\sqsubseteq$ *defined by* $a \sqsubseteq b$ *iff* $a \sqcap b = a$ *(which will be referred as the* information order*) has a least element* $\bot$ *and directed suprema* $\bigsqcup^{\uparrow}$.
3. $\prec$ *is contained in* $\leqslant$ *and it is an entailment relation, i.e. it satisfies the following four conditions:*

$$(ff \text{ - } \prec) \; ff \prec \gamma$$
$$(\prec \text{ - } tt) \; \gamma \prec tt$$
$$(\vee \text{ - } \prec) \; \gamma \prec \delta, \, \gamma' \prec \delta \Leftrightarrow \gamma \vee \gamma' \prec \delta$$
$$(\prec \text{ - } \wedge) \; \gamma \prec \delta, \, \gamma \prec \delta' \Leftrightarrow \gamma \prec \delta \wedge \delta'$$

4. *the following mixed laws hold:*

$$(\sqcap \text{ - } =) \;\; \gamma \sqcap \delta = (\gamma \wedge \bot) \vee (\delta \wedge \bot) \vee (\gamma \wedge \delta)$$
$$(\prec \text{ - } \sqsubseteq) \; \gamma \sqsubseteq \gamma', \, \delta \sqsubseteq \delta', \, \gamma \prec \delta \text{ implies } \gamma' \prec \delta'$$

Definition 9. *A skew frame* $\mathcal{L}$ *is called compact if for any directed sets in the information order* C *and* D, *we have that* $\bigsqcup^{\uparrow} C \prec \bigsqcup^{\uparrow} D$ *implies the existence of* $\delta \in D$ *and* $\gamma \in C$ *such that* $\gamma \prec \delta$.

Definition 10. *A skew frame* $\mathcal{L}$ *is regular iff for any element* $\beta \in L$ *it holds that* $\beta \sqsubseteq \bigsqcup^{\uparrow} \{\alpha \sqcap \gamma \mid \alpha \prec \beta \prec \gamma\}$. *(It always holds in a skew frame that the set on the right hand side of the previous equality is directed in the information order.)*

Definition 11. *A compact regular skew frame with negation is a compact regular skew frame, endowed additionally with a unary operation* $\neg$, *which satisfies the axioms* $(\neg \text{ - } \neg)$ *and* $(\neg \text{ - } \prec)$ *of Definition 6.*

Definition 12. *We denote by* Skew$_{\neg}$ *the category whose objects are compact regular skew frames with negation and morphisms are functions between the carrier sets preserving all the operations and* $\prec$.

The results gathered in the next Proposition are proved in Sect. 7 and 9 of [6].

Proposition 1. *In a compact regular skew frame with negation* $\mathcal{L}$ *the following hold:*

1. *If* $\gamma \prec \delta$, $\gamma' \leqslant \gamma$ *and* $\delta \leqslant \delta'$ *then* $\gamma' \prec \delta'$.

2. $\sqcap$, $\wedge$ and $\vee$ distribute over each other.
3. If γ, δ are bounded in the information order then there exists $\sup_{\sqsubseteq}\{\gamma, \delta\}$.
4. $(\mathcal{L}, \sqsubseteq)$ is bounded complete.
5. In the frame $\downarrow_{\sqsubseteq} tt = \{x \in \mathcal{L} \mid x \sqsubseteq tt\}$ the restriction of the information order coincides with the restriction of $\leqslant$. In the frame $\downarrow_{\sqsubseteq} ff = \{x \in \mathcal{L} \mid x \sqsubseteq ff\}$, the restriction of the information order coincides with the restriction of $\geqslant$.
6. We have $\gamma \sqcap tt = \gamma \vee \bot$ and $\gamma \sqcap ff = \gamma \wedge \bot$. Moreover $\gamma = \sup_{\sqsubseteq}\{\gamma \sqcap tt, \gamma \sqcap ff\}$ and $\bot = tt \sqcap ff$.
7. The negation satisfies the axioms:
   $(\neg - \sqsubseteq)$ $\gamma \sqsubseteq \delta \Rightarrow \neg\gamma \sqsubseteq \neg\delta$ and
   $(\neg - \sqcap)$ $\gamma \sqcap \neg\delta = \bot \Leftrightarrow \{\gamma, \delta\}$ is bounded wrt $\sqsubseteq$
8. If $a \leqslant b$ then $a \sqcap b = \sup_{\sqsubseteq}\{a \sqcap tt, b \sqcap ff\}$
9. For all $a, b, c \in \mathcal{L}$ we have that $a \prec c$, $b \prec c \Rightarrow a \sqcap b \prec c$ and $a \prec b$, $a \prec c \Rightarrow a \prec b \sqcap c$

Lemma 3. *Let $\mathcal{L}$ be a skew frame. The following hold:*

1. *If $\alpha_1, \ldots, \alpha_n, \beta_1 \ldots \beta_n \in \mathcal{L}$ such that $\alpha_i \leqslant \beta_j$ for all $1 \leqslant i, j \leqslant n$ then the set $\{\alpha_1 \sqcap \beta_1, \ldots, \alpha_n \sqcap \beta_n\}$ has a supremum wrt $\sqsubseteq$, namely $(\alpha_1 \vee \cdots \vee \alpha_n) \sqcap (\beta_1 \wedge \cdots \wedge \beta_n)$.*
2. *If $\mathcal{L}$ is a compact regular skew frame with negation then if $\gamma \prec \delta$ then $\neg\delta \prec \neg\gamma$.*

Proof. 1. follows from Proposition 1.8 applied for $\bigvee_{i=1}^{n} \alpha_i$ and $\bigwedge_{i=1}^{n} \beta_i$, and Proposition 1.5. A proof for 2. can be found in [6]. $\qquad\square$

3   From C^* Algebras to Skew Frames with Negation

Assume A is a C^* algebra. We define:

1. The binary operations $\wedge$ and $\vee$ as in Sect. 2.1;
2. The entailment relation $\prec$ as follows: for all a, $b \in A$, $a \prec b$ if and only if $a^- + b^+ > 0$ in the sense of Definition 3;
3. The binary operation $\sqcap$ by: $a \sqcap b = (a \wedge 0) \vee (b \wedge 0) \vee (a \wedge b)$;
4. The negation $\neg : A \to A$ is defined by $\neg a = -a$.

One can check without too much difficulty the next result:

Lemma 4. *A is lattice ordered and the operations $\wedge$ and $\vee$ distribute over each other.*

Starting with the C^* algebra A, we can obtain almost trivially a proximity lattice with negation $([-1, 1], \wedge, \vee, 1, -1, \prec, -)$. Here $[-1, 1] = \{x \in A \mid -1 \leqslant x \leqslant 1\}$. However it seems easier to obtain C^* algebras from skew frames with negation, so we will focus on the equivalence between $\mathsf{C}^*\text{-}\mathsf{Alg}$ and $\mathsf{Skew}_\neg$. We want to construct a skew frame $\mathcal{L}_A$. As it is for now, the entailment relation $\prec$ defined on A is not contained in the logical order $\leqslant$ and moreover A is not a bounded lattice. In order to achieve this, we define the carrier set of $\mathcal{L}_A$ to be the quotient of A by an equivalence relation $\sim$, defined by $a \sim b$ iff $\forall c \in A$ $a \prec c \Leftrightarrow b \prec c$ and $\forall c \in A$ $c \prec a \Leftrightarrow c \prec b$. We will denote by $[a]$ the $\sim$-equivalence class of a.

Lemma 5. *For any $\lambda, \gamma, \delta \in A$ we have that $\lambda \wedge \gamma \prec \delta$ is equivalent to $\lambda \prec \delta \vee (\neg \gamma)$*

Proof. $\lambda \wedge \gamma \prec \delta$ is equivalent to $(\lambda \wedge \gamma)^- + \delta^+ > 0$, or $\lambda^- \vee \gamma^- + \delta^+ > 0$. Using Lemma 2, this is equivalent to $\lambda^- \vee \gamma^- \vee \delta^+ > 0$ and $\lambda^- + (-\gamma)^+ \vee \delta^+ > 0$. The conclusion of the lemma follows from the fact that $(-\gamma)^+ \vee \delta^+ = ((-\gamma) \vee \delta)^+ = (\delta \vee (\neg \gamma))^+$. $\qquad\square$

Lemma 6. *$\sim$ is a congruence relation for $\wedge, \vee, \neg$ and $\mathcal{L}_A = A/\sim$ is a De Morgan lattice, having $ff = [-1]$ and $tt = [1]$ as the bottom and top elements respectively.*

Proof. Suppose $a \sim b$ and $c \in A$. We have to prove that $a \wedge c \prec x \Rightarrow b \wedge c \prec x$. But $a \wedge c \prec x \Leftrightarrow a \prec x \vee \neg c \Leftrightarrow b \prec x \vee \neg c \Leftrightarrow b \wedge c \prec x$. Now suppose $x \prec a \wedge c$. By ($\prec$ - $\wedge$), this implies that $x \prec a$ and $x \prec c$. Since $a \sim b$, we obtain that $x \prec b$. Using again ($\prec$ - $\wedge$), we obtain that $x \prec b \wedge c$. The proof for $\vee$ works in the same lines. Notice that $x \prec -a \Leftrightarrow x^- + a^- > 0 \Leftrightarrow a \prec -x \Leftrightarrow b \prec -x \Leftrightarrow x \prec -b$, so $\sim$ is a congruence for $-$ as well.

In order to show that $[1]$ is a top element, we can prove that for each $a \in A$ and each natural number $n > 0$ we have that $a \sim na$. Since for each a there exists some n such that $a \leqslant n$, we conclude that $a \sim \frac{1}{n}a \leqslant 1$. Therefore $[a] \leqslant [1]$. Similarly $[-1]$ is the least element wrt $\leqslant$. $\qquad\square$

We can define an entailment relation on $\mathcal{L}_A$, denoted also by $\prec$, (by abuse of notation), as follows: $[a] \prec [b]$ iff $a \prec b$. It follows from the definition of $\sim$ that $\prec$ is a well-defined relation on $\mathcal{L}_A$.

Lemma 7. *$\prec$ is a transitive entailment relation on $\mathcal{L}_A$ contained in $\leqslant$.*

Proof. In order to prove that $\prec$ is an entailment relation, by Lemma 6, it is enough to check the defining conditions for arbitrary elements of A, rather than $\sim$-equivalence classes: ($\prec -tt$) and ($ff- \prec$) are satisfied since $1^+ = (-1)^- 1 = 1$. In order to prove ($\prec -\wedge$) and ($\vee- \prec$) apply Lemma 2.a) for $\gamma^-, \delta^+, \delta'^+$ and for $\delta^+, \gamma^-, \gamma'^-$, respectively. Also use the fact that $\delta^+ \wedge \delta'^+ = (\delta \wedge \delta')^+$ and $(\gamma \vee \gamma')^- = \gamma^- \wedge \gamma'^-$, respectively.

Let us prove now that $\prec$ is transitive. Assume that $\gamma \prec \delta$ and $\delta \prec \epsilon$. This means that $\gamma^- + \delta^+$ and $\delta^- + \epsilon^+$ are strictly positive. The fact that $\gamma^- + \epsilon^+ > 0$, (and consequently that $\gamma \prec \delta$), follows from Lemma 2.c) applied for $a = \gamma^- + \epsilon^+$ and $b = \delta$. Using Lemma 2.b) and c) one can show that the entailment relation $\prec$ also satisfies a cut rule:

$$\theta \wedge \phi \prec \psi \text{ and } \phi \prec \psi \vee \theta \text{ implies } \phi \prec \psi \tag{1}$$

Now let us show that $\prec \subseteq \leqslant$. We have to prove that $a \prec b$ implies $a \wedge b \sim a$.

• If $a \prec c$ then $a \wedge b \prec c$ by Proposition 1.1. Conversely, suppose that $a \wedge b \prec c$. Since $a \prec b \leqslant c \vee b$, using Proposition 1.1, we have that $a \prec b \vee c$. Now from the cut rule (1) we can conclude that $a \prec c$.

• If $c \prec a$ then by the transitivity of $\prec$ we also have that $c \prec b$. Now $c \prec a \wedge b$ follows from ($\prec$ - $\wedge$). Conversely, if $c \prec a \wedge b$, then $c \prec a$ follows from ($\prec$ - $\wedge$). $\qquad\square$

Using Lemma 6, we can define $\sqcap$ an $\mathcal{L}_A$ by $[a] \sqcap [b] = [a \sqcap b]$.

Lemma 8. *1. The partial order $\sqsubseteq$ defined by $[a] \sqsubseteq [b]$ iff $[a] \sqcap [b] = [a]$ has a least element, namely $\bot = [0]$ and directed suprema $\bigsqcup^{\uparrow}$.*
2. $(\prec \text{ - } \sqsubseteq)$ holds in $\mathcal{L}_A$.

Proof. 1. The fact that $[0]$ is the least element wrt $\sqsubseteq$ follows from the definition of $\sqcap$. The directed completeness follows from the completeness of A wrt to $\| \ \|$.

2. Notice that $\gamma \sqsubseteq \delta$ if and only if $\gamma^- \leqslant \delta^-$ and $\gamma^+ \leqslant \delta^+$. This is a consequence of the fact that $\wedge$ and $\vee$ distribute over $\sqcap$ and of the fact that for $\alpha, \beta \geqslant 0$ we have that $\alpha \sqcap \beta = \alpha \wedge \beta$.

Assume $\gamma \sqsubseteq \gamma', \delta \sqsubseteq \delta'$ and $\gamma \prec \delta$. Using the above observation we know that $\gamma^- \leqslant \gamma'^-$ and $\delta^+ \leqslant \delta'^+$. This shows that $\gamma'^- + \delta'^+ \geqslant \gamma^- + \delta^+$, therefore $\gamma'^- + \delta'^+$ is strictly positive, or equivalently $\gamma' \prec \delta'$. $\qquad\square$

Proposition 2. $\mathcal{L}_A$ *is a compact regular skew frame with negation.*

Proof. From the lemma above we can already see that $\mathcal{L}_A$ is a skew frame. Also notice that the negation axioms are satisfied. The first one is trivially checked, while the second follows from Lemma 5. It remains to check that $\mathcal{L}_A$ is compact and regular. We will find useful the next remark: for any element $\alpha \in A$ we have that $[\alpha] \sqcap [1] = [\alpha^+]$ and $[\alpha] \sqcap [-1] = [\alpha^-]$.

Let us prove the regularity of $\mathcal{L}_A$. Assume $[\beta] \in \mathcal{L}_A$. For any positive integer n we have that $\beta - \frac{1}{n} \prec \beta \prec \beta + \frac{1}{n}$. Notice that since $\beta - \frac{1}{n} \leqslant \beta + \frac{1}{n}$ it follows that $(\beta - \frac{1}{n}) \sqcap (\beta + \frac{1}{n}) = \sup_{\sqsubseteq}\{(\beta - \frac{1}{n})^+, (\beta + \frac{1}{n})^-\}$.

Using the fact that $\sup_{n \in \mathbb{N}}(\beta - \frac{1}{n})^+ = \beta^+$ and similarly $\sup_{n \in \mathbb{N}}(\beta + \frac{1}{n})^- = \beta^-$, we can conclude that $[\beta] = \bigsqcup^{\uparrow}_{n \in \mathbb{N}}[\beta - \frac{1}{n}] \sqcap [\beta + \frac{1}{n}] \sqsubseteq \bigsqcup^{\uparrow}\{\alpha \sqcap \gamma \mid \alpha \prec [\beta] \prec \gamma\}$.

Now let's prove compactness. Assume $([d_i])_{i \in I}$ is a $\sqsubseteq$-directed subset of $\mathcal{L}$ and that $[\gamma] \prec \bigsqcup^{\uparrow}[d_i]$. This means that $\gamma^- + \sup_{\leqslant} d_i^+ \geqslant q$ for some positive rational q, thus there exists some i such that $\gamma^- + d_i^+ \geqslant q$, or $\gamma \prec d_i$. Similarly we can prove that $\bigsqcup^{\uparrow}[d_i] \prec [\gamma]$ implies the existence of some i such that $[d_i] \prec [\gamma]$. $\qquad\square$

Let $f : A \to B$ be a C^* algebra morphism. We know that f preserves order, therefore for any $a \in A$ we have that $f(a^+) = f(a)^+$ and $f(a^-) = f(a)^-$. Also for any rational number r, $f(r) = r$.

Lemma 9. *If $a, b \in A$ then the following hold:*

1. If $a \prec_A b$ then $f(a) \prec_B f(b)$.
2. If $a \sim_A b$ then $f(a) \sim_B f(B)$.

Proof. 1. Since $a \prec_A b$ we know that $a^- + b^+ \geqslant q$ for some positive rational number q. Using the fact that f preserves order we obtain that $f(a^-) + f(b^+) \geqslant q$, or equivalently $f(a)^- + f(b)^+ \geqslant q$. This means that $f(a) \prec_B f(b)$.

2. Assume $a \sim_A b$ and $c \in B$ such that $c \prec_B f(a)$. We will prove the existence of some a' such that $a' \prec_A a$ and $c \prec_B f(a')$. Because $c \prec_B f(a)$ there exists a positive rational number q such that $c^- + f(a)^+ \geqslant 2q$. Let $a' := a - q$.

One can prove that $a' \prec_A a$. On the other hand $f(a') = f(a) - q$, therefore
$c^- + f(a')^+ = c^- + \frac{f(a)-q+|f(a)-q|}{2} \geqslant c^- + \frac{f(a)+|f(a)|}{2} - q \geqslant q$ which proves that
$c \prec_B f(a')$. Since $a \sim_A b$, we also have that $a' \prec_A b$, hence $f(a') \prec_B f(b)$.
Using the transitivity of $\prec_B$ we get that $c \prec_B f(b)$. Similarly one can show that
$f(a) \prec_B c$ is equivalent to $f(b) \prec_B c$, so $f(a) \sim_B f(b)$. □

Now we can define a functor $S : \mathsf{C^*\text{-}Alg} \to \mathsf{Skew}_\neg$ by $S(A) = \mathcal{L}_A$. If $f : A \to B$
is a C^* algebra morphism, define $S(f) : \mathcal{L}_A \to \mathcal{L}_B$ by $S(f)([a]) = [f(a)]$. This is
well defined by the above lemma.

4   From Skew Frames with Negation to C^* Algebras

Definition 13. *Let $\mathcal{L}$ be a skew frame. A Urysohn map is a $\mathbb{Q}$-indexed family
of elements $\phi = \{\phi_q\}_{q \in \mathbb{Q}}$ such that*

- *If $p > q$ then $\phi_p \prec \phi_q$;*
- *For all $q \in \mathbb{Q}$, $\phi_q = \bigsqcup^\uparrow \{\phi_p \sqcap \phi_r \mid p > q > r\}$;*
- *$\bigsqcup^\uparrow_{q \in \mathbb{Q}}(f\!f \sqcap \phi_q) = f\!f$;*
- *$\bigsqcup^\uparrow_{q \in \mathbb{Q}}(tt \sqcap \phi_q) = tt$;*

Definition 14. *A Urysohn map $\phi = \{\phi_q\}_{q \in \mathbb{Q}}$ is called bounded if there exist
$p, q \in \mathbb{Q}$ such that $\phi_p = f\!f$ and $\phi_q = tt$.*

Let us denote by $\mathsf{Ury}(\mathcal{L})$ the set of bounded Urysohn maps on $\mathcal{L}$. The reason
why we call these families of elements Urysohn maps is that, intuitively, such a
family corresponds to the construction of a real-valued function, as in the proof
of Urysohn Lemma. Let us see an example, which may shed some light on the
intuition behind Urysohn maps.

Example 1. Assume (X, τ) is a compact Hausdorff space and $f : X \to \mathbb{R}$ is a
continuous function. Let $\mathcal{L}_X = \{(U, V) \mid U, V \in \tau; \ U \cap V = \emptyset\}$. If we define

$$
\begin{aligned}
(U,V) \wedge (U',V') &= (U \cap U', V \cup V') & tt &= (X, \emptyset) \\
(U,V) \vee (U',V') &= (U \cup U', V \cap V') & f\!f &= (\emptyset, X) \\
(U,V) \sqcap (U',V') &= (U \cap U', V \cap V') & \bot &= (\emptyset, \emptyset) \\
(U,V) \prec (U',V') &\text{ iff } U' \cup V = X & \neg(U,V) &= (V, U)
\end{aligned}
$$

then $(\mathcal{L}_X, \wedge, \vee, tt, f\!f, \bot, \sqcap, \neg, \prec)$ is a compact regular skew frame with negation.
Now, if we consider $U_p = \{x \in X \mid f(x) > p\}$ and $V_p = \{x \in X \mid f(x) < p\}$,
then $\phi_f = \{(U_p, V_p)\}_{p \in \mathbb{Q}}$ is a Urysohn map which contains enough information
to recover the function f.

In what follows we will endow $\mathsf{Ury}(\mathcal{L})$ with a C^* algebra structure. But first let
us introduce what we will call the 'constant' Urysohn maps. If $\mathcal{L}$ is a skew frame
and $p \in \mathbb{Q}$ then the family $\hat{p} = \{\hat{p}_q\}_{q \in \mathbb{Q}}$ defined by $\hat{p}_q = f\!f$ for $q > p$, $\hat{p}_p = \bot$ and
$\hat{p}_q = tt$ for $q < p$ is a bounded Urysohn map.

Let ϕ and ψ be Urysohn maps and $r \in \mathbb{Q}$. We define the set:

$$[\phi + \psi, r] = \{(\phi_p \wedge \psi_q) \sqcap (\phi_s \vee \psi_t) \mid p + q > r > s + t\}$$

Lemma 10. *If $p+q > r > s+t$ then $\phi_p \wedge \psi_q \leqslant \phi_s \vee \psi_t$ and $(\phi_p \wedge \psi_q) \sqcap (\phi_s \vee \psi_t) = \sup_{\sqsubseteq} \{\phi_p \sqcap \psi_q \sqcap tt, \phi_s \sqcap \psi_t \sqcap ff\}$*

Proof. We have that $p > s$ or $q > t$. In the former case we obtain that $\phi_p \prec \phi_s$. But $\prec$ is contained in $\leqslant$, therefore $\phi_p \leqslant \phi_s$, which implies that

$$\phi_p \wedge \psi_q \leqslant \phi_s \vee \psi_t \tag{2}$$

If $q > t$, then (2) can be obtained similarly. The second part of the lemma follows from Proposition 1.8. $\qquad\square$

Lemma 11. *For any $r \in \mathbb{Q}$ and any two Urysohn maps ϕ, ψ the set $[\phi+\psi, r]$ has a supremum wrt $\sqsubseteq$. If we define $(\phi+\psi)_r = \bigsqcup[\phi+\psi, r]$, the family $((\phi+\psi)_r)_{r \in \mathbb{Q}}$ is a Urysohn map. In particular for any rational numbers p, q we have that $\hat{p} + \hat{q} = \widehat{p+q}$.*

Proof. First we have to show that $\phi + \psi$ is well defined, that is that for each rational number r, the set $[\phi + \psi, r]$ has a supremum with respect to $\sqsubseteq$. Any finite subset $X \subseteq [\phi + \psi, r]$ has a supremum. This follows from Lemma 10 and Lemma 3.1. The set $\{\bigsqcup X \mid X \subseteq [\phi+\psi, r], X \text{ is finite}\}$ is $\sqsubseteq$ - directed and it has a supremum, that is exactly the supremum of $[\phi + \psi, r]$.

Next we have to check that $\phi + \psi$ is indeed a Urysohn map. Assume $p > q$. There exists rational numbers $a > c$ and $b > d$ such that $p > a + b > c + d > q$ and $\phi_a \vee \psi_b \prec \phi_c \wedge \psi_d$. It follows from Lemma 10 that $(\phi_a \vee \psi_b) \sqcap ff \sqsubseteq (\phi+\psi)_p$ and $(\phi_c \wedge \psi_d) \sqcap tt \sqsubseteq (\phi+\psi)_q$. From Proposition 1.9. it follows that $(\phi_a \vee \psi_b) \sqcap ff \prec (\phi_c \wedge \psi_d) \sqcap tt$. Using $(\prec - \sqsubseteq)$ we obtain that $(\phi + \psi)_p \prec (\phi + \psi)_q$.

The proof for the fact that $(\phi + \psi)$ satisfies the remaining conditions of a Urysohn map should impose no difficulty. $\qquad\square$

Lemma 12. $(\mathrm{Ury}(\mathcal{L}), +, \hat{0})$ *is a commutative, divisible and torsion-free monoid.*

Proof. Commutativity is trivial by the definition of $+$. It is easy to see that $\hat{0}$ is neutral element and that the monoid is torsion free. The key to prove that $\mathrm{Ury}(\mathcal{L})$ is divisible is to show that for positive integers n and any Urysohn map ϕ, the family $(\phi/n)_q = \phi_{nq}$ is a well defined Urysohn map. In order to prove associativity we will consider the set:

$$\{(\phi_p \wedge \psi_q \wedge \theta_r) \sqcap (\phi_t \vee \psi_u \vee \theta_v) \mid p+q+r > s > t+u+v\}$$

Notice that if $p+q+r > s > t+u+v$ then $\phi_p \wedge \psi_q \wedge \theta_r \leqslant \phi_t \vee \psi_u \vee \theta_v$. We can apply Lemma 3.1. in order to prove that the set above has a $\sqsubseteq$-supremum. Moreover one can check that this supremum coincides with both $(\phi + (\psi + \theta))_s$ and $((\phi + \psi) + \theta)_s$. $\qquad\square$

Lemma 13. *For any Urysohn map ϕ define $\phi_p^{\neg} = \neg(\phi_{-p})$. The family $\phi^{\neg} = (\phi_p^{\neg})_{p \in \mathbb{Q}}$ is a Urysohn map which additionally satisfies $\phi + \phi^{\neg} = \hat{0}$.*

Proof. The fact that $\phi^\neg$ is a Urysohn map is not difficult to prove. One has to combine Lemma 3.2, ($\neg- \sqsubseteq$) and the fact that $\neg f\!\!f = tt$.

We will prove that $\phi^\neg + \phi = \hat{0}$. Assume $r > 0$. We have to show that $\bigsqcup_{p+q>r>s+t} (\phi_p \wedge \phi_q^\neg) \sqcap (\phi_s \vee \phi_t^\neg) = f\!\!f$. Recall that $(\phi_p \wedge \phi_q^\neg) \sqcap (\phi_s \vee \phi_t^\neg) = \sup_\sqsubseteq\{(\phi_p \wedge \phi_q^\neg) \sqcap tt, (\phi_s \vee \phi_t^\neg) \sqcap f\!\!f\}$. First let us notice that

$$y \prec x \quad \Rightarrow \quad \neg x \wedge y = f\!\!f \tag{3}$$

Indeed we have $y \prec \neg\neg x \vee f\!\!f$ and from the definition of the negation on a skew frame it follows that $\neg x \wedge y \prec f\!\!f$. But $\prec$ is contained in $\leqslant$, therefore $\neg x \wedge y = f\!\!f$.

Because $p + q > r > 0$ we have that $p > -q$, so $\phi_p \prec \phi_{-q}$. Using (3) we obtain that $\phi_p \wedge \neg\phi_{-q} = f\!\!f$. Hence $(\phi_p \wedge \phi_q^\neg) \sqcap tt = f\!\!f \sqcap tt = \bot$. This shows that in fact $(\phi_p \wedge \phi_q^\neg) \sqcap (\phi_s \vee \phi_t^\neg) = (\phi_s \vee \phi_t^\neg) \sqcap f\!\!f$. It remains to prove that $\bigsqcup_{s+t<r}(\phi_s \vee \phi_t^\neg) \sqcap f\!\!f = f\!\!f$.

Let ϵ be a positive rational number such that $2\epsilon < r$. Notice that $\bigsqcup_{s\in\mathbb{Q}}(\phi_{s+\epsilon} \vee \phi_{\epsilon-s}^\neg) \sqcap f\!\!f \sqsubseteq \bigsqcup_{s+t<r}(\phi_s \vee \phi_t^\neg) \sqcap f\!\!f \sqsubseteq f\!\!f$. Next we will prove that $\bigsqcup_{s\in\mathbb{Q}}(\phi_{s+\epsilon} \vee \neg\phi_{s-\epsilon}) \sqcap f\!\!f = f\!\!f$.

Claim. If $a > b > c > d$ are rational numbers, then $\sup_\sqsubseteq\{\phi_a \sqcap \neg\phi_c \sqcap f\!\!f, \phi_b \sqcap \neg\phi_d \sqcap f\!\!f\} = \phi_a \sqcap \neg\phi_d \sqcap f\!\!f$.

This follows without much difficulty from (3). Using this claim we can now see that $\sup_\sqsubseteq((\phi_{s+\epsilon} \vee \neg\phi_{s-\epsilon}) \sqcap f\!\!f, (\phi_s \vee \neg\phi_{s-2\epsilon}) \sqcap f\!\!f) = (\phi_{s+\epsilon} \vee \neg\phi_{s-2\epsilon}) \sqcap f\!\!f$. Applying the claim inductively we obtain that $\sup_\sqsubseteq\{(\phi_{s+(i+2)\epsilon} \vee \neg\phi_{s+i\epsilon}) \sqcap f\!\!f \mid -m \leqslant i \leqslant n-2\} = (\phi_{s+n\epsilon} \vee \neg\phi_{s-m\epsilon}) \sqcap f\!\!f$. For all natural numbers $n \geqslant m + 2$ we have $(\phi_{s+n\epsilon} \vee \neg\phi_{s-m\epsilon}) \sqcap f\!\!f \sqsubseteq \bigsqcup_{s\in\mathbb{Q}}(\phi_{s+\epsilon} \vee \neg\phi_{s-\epsilon}) \sqcap f\!\!f$. But ϕ and $\phi^\neg$ are bounded Urysohn maps, therefore for n and m large enough we have that $\phi_{s+n\epsilon} = f\!\!f$ and $\neg\phi_{s-m\epsilon} = \phi_{m\epsilon-s}^\neg = f\!\!f$. We can conclude that $\bigsqcup_{s\in\mathbb{Q}}(\phi_{s+\epsilon} \vee \neg\phi_{s-\epsilon}) \sqcap f\!\!f = f\!\!f$, and finally that $\bigsqcup_{p+q>r>s+t} (\phi_p \wedge \phi_q^\neg) \sqcap (\phi_s \vee \phi_t^\neg) = f\!\!f$.

For $r < 0$ we can prove analogously that $(\phi+\phi^\neg)_r = tt$. Since $\phi^\neg$ is a Urysohn map it follows that $(\phi^\neg + \phi)_0 = tt \sqcap f\!\!f = \bot$. So $\phi + \phi^\neg = \hat{0}$.   □

Lemma 14. *If ϕ and ψ are Urysohn maps then $\phi \wedge \psi$ defined by $(\phi \wedge \psi)_p = \phi_p \wedge \psi_p$ is also a Urysohn map. Similarly one can define the Urysohn map $\phi \vee \psi$.*

The lemma above can be easily checked; it introduces an order on $\mathsf{Ury}(\mathcal{L})$ and we will call a Urysohn map non-negative if $\phi \geqslant \hat{0}$ in this order. First we will define multiplication only for non-negative Urysohn maps and then we will extend this operation to the entire $\mathsf{Ury}(\mathcal{L})$.

Lemma 15. *Let $\mathcal{L}$ be a skew frame with negation. Then $\phi^+ = \phi \vee \hat{0}$ and $\phi^- = (\phi \wedge \hat{0})^\neg$ are non-negative Urysohn maps, such that $\phi = \phi^+ - \phi^-$.*

Proof. It is obvious that ϕ^+ is a positive Urysohn map. As for ϕ^-, we will show that $\phi^- = \phi^\neg \vee \hat{0}$. Indeed $\phi_p^- = \neg(\phi \wedge \hat{0})_{-p} = \neg(\phi_{-p} \wedge \hat{0}_{-p}) = \neg\phi_{-p} \vee \neg\hat{0}_{-p} = \phi_p^\neg \vee \hat{0}_p^\neg = (\phi^\neg \vee \hat{0})_p$. The last equality holds because $\hat{0}^\neg = \hat{0}$. This shows that ϕ^- is also a positive Urysohn map. Moreover, $\phi^+ - \phi^- = \phi^+ + (\phi^-)^\neg = \phi \vee \hat{0} + \phi \wedge \hat{0}$.

We have that $(\phi \vee \hat{0})_r = \phi_r$ for $r > 0$, $(\phi \vee \hat{0})_0 = \phi_0 \vee \perp$ and $(\phi \vee \hat{0})_r = tt$ for $r < 0$. Similarly $(\phi \wedge \hat{0})_r = \phi_r$ for $r < 0$, $(\phi \wedge \hat{0})_0 = \phi_0 \wedge \perp$ and $(\phi \wedge \hat{0})_r = ff$ for $r > 0$. Let us prove that $\phi \vee \hat{0} + \phi \wedge \hat{0} = \phi$. Let r be a positive rational number and p, q, s, t rational numbers such that $p + q > r > s + t$. Using the fact that $tt \sqcap ff = \perp$ we can easily see that

$$(\phi \vee \hat{0})_p \sqcap (\phi \wedge \hat{0})_q \sqcap tt = \begin{cases} \perp & q > 0 \\ \phi_p \sqcap \phi_q \sqcap tt & q < 0 \end{cases}$$

But if $p > 0 > q$ then $\phi_p \prec \phi_q$, therefore $\phi_p \sqcap tt \sqsubseteq \phi_q \sqcap tt$, or equivalently $\phi_p \sqcap \phi_q \sqcap tt = \phi_p \sqcap tt$. So, we have that $\sup_{\sqsubseteq}\{(\phi \vee \hat{0})_p \sqcap (\phi \wedge \hat{0})_q \sqcap tt \mid p + q > r\} = \sup_{\sqsubseteq}\{\phi_p \sqcap tt \mid p > r\}$. Similarly, one can prove that $\sup_{\sqsubseteq}\{(\phi \vee \hat{0})_s \sqcap (\phi \wedge \hat{0})_t \sqcap ff \mid s + t < r\} = \sup_{\sqsubseteq}\{\phi_t \sqcap ff \mid t < r\}$. This shows that $(\phi \vee \hat{0} + \phi \wedge \hat{0})_r = \bigsqcup_{p > r > t} \phi_p \sqcap \phi_t = \phi_r$. For $r \leqslant 0$ the proof is similar. $\square$

Lemma 16. *Assume ϕ and ψ are non-negative Urysohn maps and $r > 0$ is a rational number. Then the set*

$$[\phi \circ \psi, r] = \{(\phi_p \wedge \psi_q) \sqcap (\phi_s \vee \psi_t) \mid p, q, s, t > 0;\ pq > r > st\}$$

has a supremum wrt $\sqsubseteq$. The map $\phi \circ \psi$ defined by

$$(\phi \circ \psi)_r = \begin{cases} \sqcup[\phi \circ \psi, r] & r > 0 \\ tt & r < 0 \\ \bigsqcup^{\uparrow}_{p > 0}(\phi \circ \psi)_p \sqcap tt & r = 0 \end{cases}$$

is a Urysohn map.

Now we can extend the definition of the multiplication to the entire $\mathsf{Ury}(\mathcal{L})$. If ϕ, ψ are arbitrary Urysohn maps, we define $\phi \circ \psi$ to be $\phi^+ \circ \psi^+ - \phi^- \circ \psi^+ - \phi^+ \circ \psi^- + \phi^- \circ \psi^-$.

Lemma 17. $(\mathsf{Ury}(\mathcal{L}), +, \circ, \hat{0}, \hat{1})$ *is a commutative ring and the order induced on $\mathsf{Ury}(\mathcal{L})$ by $\wedge$ defined in Lemma 14 is compatible with the ring structure.*

Proof. The fact that the multiplication is a commutative operation follows easily from its definition. Let us prove that $\hat{1}$ is a neutral element. It is enough to show that for a non-negative Urysohn map ϕ and for a positive rational number r we have that $(\phi \circ \hat{1})_r = \phi_r$. But $(\phi \circ \hat{1})_r \sqcap tt = \bigsqcup_{pq > r;\, p, q > 0} \phi_p \sqcap \hat{1}_q \sqcap tt$. We know that $\phi_p \sqcap \hat{1}_q \sqcap tt = \perp$ if $q \geqslant 1$ and $\phi_p \sqcap \hat{1}_q \sqcap tt = \phi_p \sqcap tt$ if $q < 1$. This shows that $(\phi \circ \hat{1})_r \sqcap tt = \phi_r \sqcap tt$. Analogously one can check that $(\phi \circ \hat{1})_r \sqcap ff = \phi_r \sqcap ff$. It only remains to show that multiplication distributes over addition. It is not difficult to see that it is enough to prove distributivity only for non-negative Urysohn maps, and this is a standard verification.

We need to show that if $\phi \leqslant \psi$ then $\phi + \theta \leqslant \psi + \theta$. It is enough to show that for all rational numbers p we have $(\phi + \theta)_p \sqcap tt \sqsubseteq (\psi + \theta)_p \sqcap tt$ and $(\psi + \theta)_p \sqcap ff \sqsubseteq (\phi + \theta)_p \sqcap ff$. Since $\phi \leqslant \psi$ we have that for all rational numbers r

$\phi_r \sqcap tt \sqsubseteq \psi_r \sqcap tt$. Consequently for all rational numbers r, s such that $r + s > p$ we have $\phi_r \sqcap \theta_s \sqcap tt \sqsubseteq \psi_r \sqcap \theta_s \sqcap tt$, so $(\phi + \theta)_p \sqcap tt \sqsubseteq (\psi + \theta)_p \sqcap tt$.

Suppose $\theta \geqslant \hat{0}$. We have that $\phi \leqslant \psi$ implies that $\phi^+ \leqslant \psi^+$ and $\phi^- \geqslant \psi^-$. The proof for the fact that $\phi^+ \circ \theta \leqslant \psi^+ \circ \theta$ and $\psi^- \circ \theta \leqslant \phi^- \circ \theta$ is similar to the one for the compatibility with addition, and we can just add the two inequalities to get the desired result.   □

Lemma 18. *If $\hat{1} \geqslant n\phi$ for all $n \in \mathbb{N}$ then $\phi \leqslant \hat{0}$. Moreover for every Urysohn map ϕ there exists a natural number n such that $\phi \leqslant n\hat{1}$.*

Proof. We can prove that $(n\phi)_p = \phi_{p/n}$. Since $\hat{1} \geqslant n\phi$ we can deduce that $\phi_{p/n} = ff$ for every $p > 1$ and for every positive integer n. Thus $\phi_p = ff$ for all $p > 0$. It follows that $\phi_0 \in \downarrow \sqsubseteq ff$, so $\phi_0 \leqslant \perp$. It is obvious that for all $p < 0$ we have $\phi_p \leqslant tt$, therefore $\phi \leqslant \hat{0}$. For the second part of the lemma, we use that ϕ is bounded, so there exists $q \in \mathbb{Q}$ such that $\phi_q = ff$. This implies that $\phi_p = ff$ for all $p > q$. Now if we choose $n > q$ it is easy to see that $\phi \leqslant \hat{n}$. But $\hat{n} = n\hat{1}$, so we are done.   □

So far we have proved that $\mathsf{Ury}(\mathcal{L})$ is a partially ordered commutative ring satisfying properties (a)-(c) from the definition of a C^* algebra. It follows from a general argument described in Sect. 2.1 that $\|\cdot\|$ defined by $\|\phi\| = \inf\{q \in \mathbb{Q} | -\hat{q} \leqslant \phi \leqslant \hat{q}\}$ for all $\phi \in \mathcal{L}$ is a norm. It remains to check:

Lemma 19. $\mathsf{Ury}(\mathcal{L})$ *is complete in the metric induced by $\|\cdot\|$.*

Proof. Suppose $\phi^{(n)}$ is a Cauchy sequence of Urysohn maps. For all positive ϵ there exists a $n_\epsilon \in \mathbb{N}$ such that for all $n, m \geqslant n_\epsilon$ we have $\|\phi^{(n)} - \phi^{(m)}\| < \epsilon$. This is equivalent to the fact that for all $\epsilon \in \mathbb{Q}_+^*$ there exists $n_\epsilon \in \mathbb{N}$ such that for all $n, m \geqslant n_\epsilon$ we have $-\hat{\epsilon} \leqslant \phi^{(n)} - \phi^{(m)} \leqslant \hat{\epsilon}$, or equivalently, using the fact that $(\phi + \hat{\epsilon})_p = \phi_{p-\epsilon}$:

$$\forall \epsilon \in \mathbb{Q}, \epsilon > 0 \; \exists n_\epsilon \in \mathbb{N} \; such \; that \; \forall n, m \geqslant n_\epsilon \forall p \in \mathbb{Q} : \; \phi^{(m)}_{p+\epsilon} \leqslant \phi^{(n)}_p \leqslant \phi^{(m)}_{p-\epsilon} \quad (4)$$

We want to construct a Urysohn map which will be the limit of this sequence. We will define it by $\phi_p = \bigsqcup_{\epsilon \in \mathbb{Q}_+} \bigsqcup_{n \geqslant n_\epsilon} \phi^{(n)}_{p+\epsilon} \sqcap \phi^{(n)}_{p-\epsilon}$. We should check that the set $\mathcal{X} = \{\phi^{(n)}_{p+\epsilon} \sqcap \phi^{(n)}_{p-\epsilon} | \epsilon \in \mathbb{Q}_+; n \geqslant n_\epsilon\}$ has a supremum wrt $\sqsubseteq$. First let's see that if ϵ_1, ϵ_2 are positive rationals and $n_1 \geqslant n_{\epsilon_1}$ and $n_2 \geqslant n_{\epsilon_2}$ then $\phi^{(n_1)}_{p+\epsilon_1} \leqslant \phi^{(n_2)}_{p-\epsilon_2}$. Indeed if $n_3 \in \mathbb{N}$ is such that $n_3 \geqslant n_{\epsilon_1}$ and $n_3 \geqslant n_{\epsilon_2}$, then using (4) twice, we get $\phi^{(n_1)}_{p+\epsilon_1} \leqslant \phi^{(n_3)} \leqslant \phi^{(n_2)}_{p-\epsilon_2}$. By Lemma 3 we obtain that each finite subset of $\mathcal{X}$ has a supremum, hence $\mathcal{X}$ has a supremum. Next we prove that $\phi = (\phi_p)_p \in \mathbb{Q}$ is a Urysohn map. Suppose $p > q$. Let ϵ be a positive rational such that $p - \epsilon > q + \epsilon$ and a natural number $n \geqslant n_\epsilon$. We have that $\phi^{(n)}_{p\pm\epsilon} \prec \phi^{(n)}_{q\pm\epsilon}$. Using Proposition 1.9 we obtain that $\phi^{(n)}_{p+\epsilon} \sqcap \phi^{(n)}_{p-\epsilon} \prec \phi^{(n)}_{q+\epsilon} \sqcap \phi^{(n)}_{q-\epsilon}$. Applying $(\prec - \sqsubseteq)$ we obtain that $\phi_p \prec \phi_q$. Let us prove that $\phi_r = \bigsqcup_{p>r>q} \phi_p \sqcap \phi_q$. The fact that $\bigsqcup_{p>r>q} \phi_p \sqcap \phi_q \sqsubseteq \phi_r$ is trivial. The other way around, we will prove that for arbitrary positive rational ϵ and $n \geqslant n_\epsilon$, we have that $\phi^{(n)}_{r+\epsilon} \sqcap \phi^{(n)}_{r-\epsilon} \sqsubseteq \bigsqcup_{p>r>q} \phi_p \sqcap \phi_q$. We have that $\phi_{r+\epsilon}^{(n)} \sqcap tt = \bigsqcup_{\delta>0} \phi_{r+\epsilon+\delta}^{(n)} \sqcap tt = \bigsqcup_{p>r} \phi_{p+\epsilon}^{(n)} \sqcap tt \sqsubseteq \bigsqcup_{p>r} \phi_p \sqcap tt$. Similarly, $\phi_{r-\epsilon}^{(n)} \sqcap ff \sqsubseteq \bigsqcup_{q<r} \phi_q \sqcap ff$. From these we conclude that $\phi_{r+\epsilon}^{(n)} \sqcap \phi_{r-\epsilon}^{(n)} = \sup_{\sqsubseteq} \{\phi_{r+\epsilon}^{(n)} \sqcap tt, \phi_{r-\epsilon}^{(n)} \sqcap ff\} \sqsubseteq \sup_{\sqsubseteq} \{\bigsqcup_{p>r} \phi_p \sqcap tt, \bigsqcup_{q<r} \phi_q \sqcap ff\} = \bigsqcup_{p>r>q} \phi_p \sqcap \phi_q$. The proof for the fact that $\bigsqcup_{p \in \mathbb{Q}} \phi_p \sqcap tt = tt$ and $\bigsqcup_{p \in \mathbb{Q}} \phi_p \sqcap ff = ff$ is easier.

It remains to check that the Urysohn map ϕ is indeed the limit of the sequence $\phi^{(n)}$. For each positive ϵ we choose n_ϵ as in (4) and we claim that for each $n \geqslant n_\epsilon$ we have that $\|\phi - \phi^{(n)}\| \leqslant \epsilon$. We have to check that for each rational number p and for each $n \geqslant n_\epsilon$ we have that $\phi_{p+\epsilon}^{(n)} \leqslant \phi_p \leqslant \phi_{p-\epsilon}^{(n)}$. By the definition of ϕ_p we already have that $\phi_{p+\epsilon}^{(n)} \sqcap tt \sqsubseteq \phi_p \sqcap tt$. We also need to see that $\phi_p \sqcap ff \sqsubseteq \phi_{p+\epsilon}^{(n)} \sqcap ff$. We have that $\phi_p \sqcap ff = \bigsqcup_{\delta>0} \bigsqcup_{m \geqslant n_\delta} \phi_{p-\delta}^{(m)} \sqcap ff$. As we have seen above $\phi_{p+\epsilon}^{(n)} \leqslant \phi_{p-\delta}^{(m)}$ implies that $\phi_{p-\delta}^{(m)} \sqcap ff \sqsubseteq \phi_{p+\epsilon}^{(n)} \sqcap ff$. Therefore we have that $\phi_p \sqcap ff \sqsubseteq \phi_{p+\epsilon}^{(n)} \sqcap ff$. Similarly one can prove that $\phi_p \leqslant \phi_{p-\epsilon}^{(n)}$ for all rationals p, and for all $n \geqslant n_\epsilon$. $\square$

5   The Equivalence Theorem

In this section we prove that C*-Alg is equivalent to Skew$_\neg$.

Lemma 20. *Let A be a C^* algebra and let M denote a maximal ideal. The following hold:*

1. *$x \in M$ if and only if $x^+ \in M$ and $x^- \in M$.*
2. *If a, $b \in A$ are such that $a \sim b$, then $a \in M$ if and only if $b \in M$.*
3. *If a, $b \in A$ are such that $a + p \sim b + p$ for all $p \in \mathbb{Q}$, then $a = b$.*

Proof. 1. We prove only the nontrivial implication. Suppose $x \in M$. Because $x^+ x^- = 0 \in M$ and M is also a prime ideal it follows that $x^+ \in M$ or $x^- \in M$. But $x = x^+ - x^-$, so both of x^+ and x^- belong to M.

2. Assume $a \in M$. Then there exists a real number p such that $b - p \in M$ (p is actually $\|b\|_M$ or $-\|b\|_M$ as described in [5]). If $p > 0$ then one can check that $b - p \prec b$, hence $b - p \prec a$. This means that $(b-p)^- + a^+ \geqslant q$ for some $q \in \mathbb{Q}_+^*$. Using point 1. above we obtain that both $(b-p)^-$ and a^+ are elements of M. Hence $(b-p)^- + a^+ \in M$ but this is a contradiction because strictly positive elements in a C^* algebra are invertible. The case $p < 0$ leads to a contradiction as well. Therefore $p = 0$, so $b \in M$.

3. It is not difficult to see that $a + p \sim b + p$ for all $p \in \mathbb{R}$. Assume by contradiction that $a \neq b$. Since the Jacobson radical of a C^* algebra is null, we know that there exists a maximal ideal M_0 such that $a - b \notin M_0$. But there exists $p \in \mathbb{R}$ such that $b - p \in M_0$. By the point 2. above we also have that $a - p \in M_0$, so $(a-p) - (b-p) = a - b \in M_0$. Contradiction. $\square$

Theorem 2. C*-Alg *and* Skew$_\neg$ *are equivalent categories.*

Proof (Sketch). We will prove that the functor $S : $ C*-Alg $\to$ Skew$_\neg$ defined at the end of Sect. 3 is faithful, full and surjective on objects. To this end, let $f : A \to B$ and $g : A \to B$ be C^* algebras morphisms such that $S(f) = S(g)$. This means that for all $a \in A$ we have $f(a) \sim g(a)$. Using the fact that f, g are ring morphisms and that $f(p) = g(p) = p$ for all $p \in \mathbb{Q}$ we can derive that for all rational numbers p and for all $a \in A$ we have that $f(a) + p \sim g(a) + p$. Now we can apply Lemma 20.3 to conclude that $f(a) = g(a)$ for all $a \in A$. So S is faithful. In order to prove that S is full, assume $g : SA \to SB$ is a morphism in $\mathsf{Skew}_\neg$. We want to find $f : A \to B$ such that $Sf = g$. If $a \in A$ we have that $([a - p])_{p \in \mathbb{Q}}$ is a Urysohn map in SA. The key is to show that there exists a unique $b \in B$ such that $g([a - p]) = [b - p]$ for all p. Then we can define $f(a) = b$. If $\mathcal{L}$ is a skew frame with negation then we will prove that $S(\mathsf{Ury}(\mathcal{L})) \simeq \mathcal{L}$. We construct a map $\alpha : S(\mathsf{Ury}(\mathcal{L})) \to \mathcal{L}$ defined by $\alpha([\phi]) = \phi_0$ for all Urysohn maps ϕ and prove that α is a well defined isomorphism of skew frames with negation.   □

6   Conclusions and Further Work

In the introduction, we mention that there are two notions of 'test'. One is motivated by Stone's representation of Boolean algebras as clopen sets of a topology. On this view (dominant in computer science), a topology forms a system of observable properties on one's system. The other is motivated by Stone-Gelfand-Naimark's representation of commutative C^* algebras as rings of real-valued continuous functions on compact Hausdorff spaces. On this view, the real-valued functions form a system of tests. The results of this paper establish directly the connection between these two views.

It would be interesting to compare our result to other approaches to constructive Gelfand duality, developed by Banaschewski and Mulvey [2], or more recently by Coquand and Spitters [3].

The completeness axiom for C^* algebras plays approximately the same role as the directed completeness of skew frames. In both settings, this completeness seems to be the only source of impredicativity (roughly, the need for axioms of the form $\forall A \subseteq X \exists x \in X.\phi$). Absent completeness, C^* algebras are even first-order structures. Likewise, proximity lattices with negation are first-order structures. It is well known that completeness is a rather innocuous assumption because any "incomplete C^* algebra" can be completed uniquely (impredicatively, of course). This suggests that our results could be used to develop a purely predicative proof that proximity lattice with negation are equivalent to a suitable category of rings that are essentially C^* algebras without completeness. Obviously, this would require a change in the morphisms of the ring category. We would no longer be able to take ring homomorphisms, but would instead need a suitable notion of "approximable map," adapted to ordered rings. With this, one could then move on to formulating other notions from analysis (e.g., Stone-Weierstraß) in predicative terms.

References

1. Abramsky, S.: Domain theory in logical form. Ann. Pure Appl. Logic 51 (1991)
2. Banaschewski, B., Mulvey, C.J.: A globalisation of the Gelfand duality theorem. Ann. Pure Appl. Logic 137(1-3), 62–103 (2006)

3. Coquand, T., Spitters, B.: Constructive Gelfand duality for C*-algebras. In: Mathematical Proceedings of the Cambridge Philosophical Society. Cambridge University Press, Cambridge (2009)
4. Desharnais, J., Gupta, V., Jagadeesan, R., Panangaden, P.: Approximating labelled Markov processes. Inf. Comput. 184(1), 160–200 (2003)
5. Johnstone, P.: Stone Spaces. Cambridge University Press, Cambridge (1982)
6. Jung, A., Moshier, M.A.: On the bitopological nature of Stone duality. Technical Report CSR-06-13, School of Computer Science, University of Birmingham (2006)
7. Jung, A., Sünderhauf, P.: On the duality of compact vs. open. In: Andima, S., Flagg, R.C., Itzkowitz, G., Misra, P., Kong, Y., Kopperman, R. (eds.) Papers on General Topology and Applications: Eleventh Summer Conference at the University of Southern Maine. Annals of the New York Academy of Sciences, vol. 806, pp. 214–230 (1996)
8. Dexter, K.: Semantics of probabilistic programs. In: SFCS 1979: Proceedings of the 20th Annual Symposium on Foundations of Computer Science (sfcs 1979), Washington, DC, USA, pp. 101–114. IEEE Computer Society Press, Los Alamitos (1979)
9. Mislove, M.W., Ouaknine, J., Pavlovic, D., Worrell, J.B.: Duality for labelled markov processes. In: Walukiewicz, I. (ed.) FOSSACS 2004. LNCS, vol. 2987, pp. 393–407. Springer, Heidelberg (2004)
10. Moshier, M.A.: On the relationship between compact regularity and Gentzen's cut rule. Theoretical Comput. Sci. 316, 113–136 (2004)
11. Saheb-Djahromi, N.: Cpo's of measures for nondeterminism. Theor. Comput. Sci. 12, 19–37 (1980)
12. Smyth, M.: Power domains and predicate transformers: a topological view. In: Díaz, J. (ed.) ICALP 1983. LNCS, vol. 154. Springer, Heidelberg (1983)
13. Smyth, M.: Topology. In: Handbook of Logic in Computer Science. OUP (1993)
14. van Breugel, F., Mislove, M.W., Ouaknine, J., Worrell, J.B.: An intrinsic characterization of approximate probabilistic bisimilarity. In: Gordon, A.D. (ed.) FOSSACS 2003. LNCS, vol. 2620, pp. 200–215. Springer, Heidelberg (2003)
15. Vickers, S.J.: Topology Via Logic. CUP (1989)

Conway Games, Coalgebraically*

Furio Honsell and Marina Lenisa

Dipartimento di Matematica e Informatica, Università di Udine
via delle Scienze 206, 33100 Udine, Italy
furio.honsell@comune.udine.it, lenisa@dimi.uniud.it

Abstract. Using *coalgebraic methods*, we extend *Conway's original theory of games* to include *infinite games* (*hypergames*). We take the view that a play which goes on forever is a *draw*, and hence rather than focussing on winning strategies, we focus on *non-losing strategies*. Infinite games are a fruitful metaphor for non-terminating processes, *Conway's sum* of games being similar to *shuffling*. Hypergames have a rather interesting theory, already in the case of generalized *Nim*. The theory of hypergames generalizes Conway's theory rather smoothly, but significantly. We indicate a number of intriguing directions for future work. We briefly compare infinite games with other notions of games used in computer science.

Keywords: Conway games, coalgebraic games, non-losing strategies.

1 Introduction

We focus on *combinatorial games*, that is no chance 2-player games, the two players being conventionally called *Left* (L) and *Right* (R). Such games have *positions*, and in any position there are rules which restrict L to move to any of certain positions, called the *Left positions*, while R may similarly move only to certain positions, called the *Right positions*. L and R move in turn, and the game is of *perfect knowledge*, *i.e.* all positions are public to both players. The game ends when one of the players has no move, the other player being the winner. Many games played on boards are combinatorial games, *e.g. Nim, Domineering, Go, Chess*. Games, like Nim, where for every position both players have the same set of moves, are called *impartial*. More general games, like Domineering, Go, Chess, where L and R may have different sets of moves are called *partizan*.

Combinatorial Game Theory started at the beginning of 1900 with the study of the famous impartial game Nim. In the 1930s, Sprague and Grundy generalized the results on Nim to all impartial finite (*i.e.* terminating) games, [Gru39,Spra35]. In the 1960s, Berlekamp, Conway, Guy introduced the theory of partizan games, which first appeared in the book "On Numbers and Games" [Con01]. In [Con01], the theory of games is connected to the theory of *surreal numbers*.

---

* Work supported by ART PRIN Project prot. 2005015824, by the FIRB Project RBIN04M8S8 (both funded by MIUR), and by the ESF Research Networking Programme GAMES.

A. Kurz, M. Lenisa, and A. Tarlecki (Eds.): CALCO 2009, LNCS 5728, pp. 300–316, 2009.

However, in [Con01], the author focusses only on *finite, i.e. terminating games.* Infinite games are neglected as ill-formed or trivial games, not interesting for "busy men", and their discussion is essentially confined to a single chapter. Infinity (or *loopy*) games have been later considered in [BCG82], Chapters 11-12. However, in Chapter 12 the authors focus on well-behaved classes of impartial games, which can be dealt with a generalization of the Grundy-Sprague theory, due to Smith [Smi66]. In Chapter 11, a theory for the special class of partizan *fixed loopy games* is developed; a game is *fixed* if infinite plays are winning either for L or R player. On the contrary, in the present paper we develop a general *coalgebraic account* of infinite games, taking the different (but sometimes more natural) view that an infinite play is a *draw.* We call such games *hypergames.*

Infinite games are extremely useful in various fields, such as Mathematical Logic and Computer Science. The importance of games for Computer Science comes from the fact that they capture in a natural way the notion of *interaction.* Infinite games model in a faithful way *reactive processes* (operating systems, controllers, communication protocols, etc.), that are characterised by their *non-terminating* behaviour and perpetual *interaction* with their environment.

The coalgebraic account of games developed in this paper is very natural and it paves the way to a smooth and insightful treatment of infinite games. It allows us to consider games up-to bisimilarity, and to generalize operations and relations on them as congruences up-to bisimilarities. Moreover, the coalgebraic setting makes explicit the common nature between processes and games. For hypergames the notion of *winning strategy* has to be replaced by that of *non-losing strategy,* since we take non terminating plays to be *draws.* Hypergames can be naturally defined as a *final coalgebra* of *non-wellfounded sets* (*hypersets*), which are the sets of a universe of Zermelo-Fraenkel satisfying an Antifoundation Axiom, see [FH83,Acz88]. Our theory of hypergames generalizes the original theory on finite games of [Con01] rather smoothly, but significantly. Our main results amount to a Determinacy and a Characterization Theorem of non-losing strategies on hypergames. The latter requires (a non-trivial) generalization of Conway's partial order relation on games to hypergames. Once hypergames are defined as a final coalgebra, operations on games, such as *disjunctive sum,* can be naturally extended to hypergames, by defining them as *final morphisms* into the coalgebra of hypergames. We will also discuss the class of impartial hypergames. In particular, we will extend the theory of Grundy-Sprague and Smith, based on the canonical *Nim games,* by introducing suitable *canonical ∞-hypergames.*

Finally, we will briefly compare our hypergames with other games arising in Combinatorial Game Theory and in Computer Science.

Summary. In Section 2, we recall Conway's theory of finite games and winning strategies. In Secton 3, we introduce hypergames as a final coalgebra, and we develop the theory of hypergames and non-losing strategies, which extends Conway's theory. In Section 4, we study in particular the theory of impartial hypergames. Comparison with related games and directions for future work appear in Section 5.

2   The Theory of Conway Games

We recall that Conway games are 2-player games, the two players are called *Left* (L) and *Right* (R). Such games have *positions*, and in any position p there are rules which restrict Left to move to any of certain positions, called the *Left positions* of p, while Right may similarly move only to certain positions, called the *Right positions* of p. Since we are interested only in the abstract structure of games, we can regard any position p as being completely determined by its Left and Right options, and we shall use the notation $p = (P^L, P^R)$, where P^L, P^R denote sets of positions. Games are identified with their initial positions. Left and Right move in turn, and the game ends when one of the two players does not have any option. Conway considers only terminating (inductively defined) games. These can be viewed as an *initial algebra* of a suitable functor, which we define *e.g.* on the category Class* of classes of (possibly non-wellfounded) sets and functional classes.

Definition 1 (Conway Games). *The set of* Conway Games $\mathcal{G}$ *is inductively defined by*

- *the empty game* $(\{\}, \{\}) \in \mathcal{G}$;
- *if* $P, P' \subseteq \mathcal{G}$, *then* $(P, P') \in \mathcal{G}$.

Equivalently, $\mathcal{G}$ *is the carrier of the* initial algebra $(\mathcal{G}, id)$ *of the functor* $F :$ Class* $\to$ Class*, *defined by* $F(X) = \mathcal{P}(X) \times \mathcal{P}(X)$ *(with usual definition on morphisms).*
Games will be denoted by small letters, e.g. p, *with* $p = (P^L, P^R)$ *and* p^L, p^R *generic elements of* P^L, P^R. *We denote by* Pos_p *the set of* positions *hereditarily reachable from* p.

Some simple games. The simplest game is the empty one, *i.e.* $(\{\}, \{\})$, which will be denoted by 0. Then we define the games $1 = (\{0\}, \{\})$, $-1 = (\{\}, \{0\})$, $* = (\{0\}, \{0\})$.

Winning strategies. In the game 0, the player who starts will lose (independently whether he plays L or R), since there are no options. Thus the second player (II) has a winning strategy. In the game 1 there is a winning strategy for L, since, if L plays first, then L has a move to 0, and R has no further move; otherwise, if R plays first, then he loses, since he has no moves. Symmetrically, -1 has a winning strategy for R. Finally, the game $*$ has a winning strategy for the first player (I), since he has a move to 0, which is losing for the next player.

   Formally, we first define *(finite) plays* over a game p as alternating sequences of moves on the game, starting from the initial position. One might think that the following definitions are a little involved, but this is necessarily so if we want to "dot all our i's and cross all our t's".

Definition 2 (Finite Plays). *Let* $p = (P^L, P^R)$ *be a game. The set of* finite plays *over* p, $FPlay_p$, *is defined by:*

$$\pi = p_1^{K_1} \ldots p_n^{K_n} \in FPlay_p, \quad for \ n \geq 0, \quad iff$$

- $K_1, \ldots, K_n \in \{L, R\}$;
- $(K_1 = L \ \wedge \ p_1^{K_1} \in P^L) \ \vee \ (K_1 = R \ \wedge \ p_1^{K_1} \in P^R)$

- $\forall i. \ 1 \leq i < n. \ (p_i^{K_i} = (P_i^L, P_i^R) \wedge p_{i+1}^{K_{i+1}} \in P_i^{\overline{K_i}}), \ \text{where } \overline{K} = \begin{cases} L & \text{if } K = R \\ R & \text{if } K = L. \end{cases}$

We denote by $FPlay_p^{LI}$ the set of plays starting with a Left move and ending with a Right move, i.e. $\{p_1^{K_1} \ldots p_n^{K_n} \in FPlay_p \mid K_1 = L \ \wedge \ K_n = R, \ n \geq 0\}$, and by $FPlay_p^{LII}$ the set of plays starting with a Right move and ending with a Right move, i.e. $\{p_1^{K_1} \ldots p_n^{K_n} \in FPlay_p \mid K_1 = R \ \wedge \ K_n = R, \ n \geq 1\}$. Similarly, we define $FPlay_p^{RI}$ and $FPlay_p^{RII}$.

Only *finite* plays can arise on a Conway game.

Winning strategies for a given player can be formalized as functions on plays ending with a move of the opponent player, telling which is the next move of the given player:

Definition 3 (Winning Strategies). Let $p = (P^L, P^R)$ be a game.
- A winning strategy on p for LI, i.e. *for Left as I player*, is a partial function $f : FPlay_p^{LI} \to Pos_p$, such that:

- $\forall \pi \in FPlay_p^{LI}. \ f(\pi) = p_{n+1}^L \implies \pi p_{n+1}^L \in FPlay_p$;
- f is defined on the empty play ϵ, denoted by $f(\epsilon) \downarrow$;
- $\forall \pi \in FPlay_p^{LI}. \ (f(\pi) = p_{n+1}^L \ \wedge \ p_{n+1}^L = (P_{n+1}^L, P_{n+1}^R) \implies$
$$\forall p_{n+2}^R \in P_{n+1}^R. \ f(\pi p_{n+1}^L p_{n+2}^R) \downarrow).$$

- A winning strategy on p for LII, i.e. *for Left as II player*, is a partial function $f : FPlay_p^{LII} \to Pos_p$, such that:

- $\forall \pi \in FPlay_p^{LII}. \ f(\pi) = p_{n+1}^L \implies \pi p_{n+1}^L \in FPlay_p$;
- for all $p^R \in P^R, \ f(p^R) \downarrow$;
- $\forall \pi \in FPlay_p^{LII}. \ (f(\pi) = p_{n+1}^L \ \wedge \ p_{n+1}^L = (P_{n+1}^L, P_{n+1}^R) \implies$
$$\forall p_{n+2}^R \in P_{n+1}^R. \ f(\pi p_{n+1}^L p_{n+2}^R) \downarrow).$$

- Winning strategies on p for RI and RII *are defined similarly, as partial functions* $f : FPlay_p^{RI} \to Pos_p$ and $f : FPlay_p^{RII} \to Pos_p$, respectively.
- A winning strategy on p for L *is a partial function* $f_L : FPlay_p^{LI} \cup FPlay_p^{LII} \to Pos_p$ such that $f_L = f_{LI} \cup f_{LII}$, for f_{LI}, f_{LII} winning strategies for LI and LII.
- A winning strategy on p for R *is a partial function* $f_R : FPlay_p^{RI} \cup FPlay_p^{RII} \to Pos_p$ such that $f_R = f_{RI} \cup f_{RII}$, for f_{RI}, f_{RII} winning strategies for RI and RII.
- A winning strategy on p for I *is a partial function* $f_I : FPlay_p^{LI} \cup FPlay_p^{RI} \to Pos_p$ such that $f_I = f_{LI} \cup f_{RI}$, for f_{LI}, f_{RI} winning strategies for LI and RI.
- A winning strategy on p for II *is a partial function* $f_{II} : FPlay_p^{LII} \cup FPlay_p^{RII} \to Pos_p$ such that $f_{II} = f_{LII} \cup f_{RII}$, for f_{LII}, f_{RII} winning strategies for LII and RII.

By induction on games, one can prove that any game has exactly one winner (see [Con01] for more details):

Theorem 1 (Determinacy, [Con01]). *Any game has a winning strategy either for L or for R or for I or for II.*

In [Con01], a relation $\gtrsim$ on games is introduced, inducing a partial order (which is a total order on the subclass of games corresponding to numbers). Such relation allows to characterize games with a winning strategy for L, R, I or II (see Theorem 2 below).

Definition 4. *Let* $x = (X^L, X^R)$, $y = (Y^L, Y^R)$ *be games. We define, by induction on games:*

$$x \gtrsim y \ \ \text{iff} \ \ \forall x^R \in X^R.\ (y \not\gtrsim x^R) \ \wedge \ \forall y^L \in Y^L.\ (y^L \not\gtrsim x) \ .$$

Furthermore, we define:
- $x > y$ *iff* $x \gtrsim y \ \wedge \ y \not\gtrsim x$
- $x \sim y$ *iff* $x \gtrsim y \ \wedge \ y \gtrsim x$
- $x \| y$ *(x fuzzy y)* *iff* $x \not\gtrsim y \ \wedge \ y \not\gtrsim x$

Notice that $\not\gtrsim$ does not coincide with $<$, *e.g.* $* = (\{0\}, \{0\})$ is such that $* \not\gtrsim 0$ holds, but $* \gtrsim 0$ does not hold.

As one may expect, $1 > 0 > -1$, while for the game $*$ (which is not a number), we have $*\|0$.

The following important theorem gives the connection between games and numbers, and it allows to characterize games according to winning strategies:

Theorem 2 (Characterization, [Con01]). *Let* x *be a game. Then*

$x > 0$ *(x is positive)*	iff	x has a winning strategy for L.
$x < 0$ *(x is negative)*	iff	x has a winning strategy for R.
$x \sim 0$ *(x is zero)*	iff	x has a winning strategy for II.
$x\|0$ *(x is fuzzy)*	iff	x has a winning strategy for I.

Generalizations of Theorems 1 and 2 to infinite games will be discussed in Section 3.

3   The Theory of Hypergames

Here we extend the class of games originally considered by Conway, by introducing *hypergames*, where plays can be unlimited. For such games the notion of winning strategy has to be replaced by that of *non-losing strategy*, since we take non terminating plays to be draws. In this section, we develop the theory of hypergames, which generalizes the one for finite games. Special care requires the generalization of the Characterization Theorem 2.

Hypergames can be naturally defined as a *final coalgebra* on non-wellfounded sets:

Definition 5 (Hypergames). *The set of* Hypergames $\mathcal{H}$ *is the carrier of the final coalgebra* $(\mathcal{H}, id)$ *of the functor* $F : Class^* \to Class^*$, *defined by* $F(X) = \mathcal{P}(X) \times \mathcal{P}(X)$ *(with usual definition on morphisms).*

Defining hypergames as a final coalgebra, we immediately get a *Coinduction Principle* for reasoning on infinite games:

Lemma 1. *A F-bisimulation on the coalgebra* $(\mathcal{H}, id)$ *is a symmetric relation* $\mathcal{R}$ *on hypergames such that, for any* $x = (X^L, X^R)$, $y = (Y^L, Y^R)$,

$$x\mathcal{R}y \implies (\forall x^L \in X^L.\exists y^L \in Y^L.x^L\mathcal{R}y^L) \wedge (\forall x^R \in X^R.\exists y^R \in Y^R.x^R\mathcal{R}y^R) .$$

Coinduction Principle. *Let us call a F-bisimulation on* $(\mathcal{H}, id)$ *a hyperbisimulation. The following principle holds:*

$$\frac{\mathcal{R} \ hyperbisimulation \quad x\mathcal{R}y}{x = y}$$

All important notions and constructions on games turn out to be invariant w.r.t. hyperbisimilarity, in particular hyperbisimilar games will have the same outcome. Moreover, the coalgebraic representation of games naturally induces a minimal representative for each bisimilarity equivalence class.

Some simple hypergames. Let us consider the following pair of simple hypergames: $a = (\{b\}, \{\})$ and $b = (\{\}, \{a\})$. If L plays as II on a, then she immediately wins since R has no move. If L plays as I, then she moves to b, then R moves to a and so on, an infinite play is generated. This is a draw. Hence L has a non-losing strategy on a. Simmetrically, b has a non-losing strategy for R.

Now let us consider the hypergame $c = (\{c\}, \{c\})$. On this game, any player (L, R, I, II) has a non-losing strategy; namely there is only the non-terminating play consisting of infinite c's.

It is remarkable that the formal definition of non-losing strategy is precisely the same as that of winning strategy, *i.e.* a function on finite plays (see Definition 3). This shows that the definition of non-losing strategy is the natural generalization to hypergames of the notion of winning strategy.

The main difference in the theory of hypergames with respect to the theory of games is that on a hypergame we can have non-losing strategies for various players at the same time, as in the case of the game c above.

To prove Theorem 3 below, which is the counterpart of Theorem 1 of Section 2, we use the following lemma, that follows from the definition of non-losing strategy:

Lemma 2. *Let p be a hypergame.*

- *If L as I player does not have a non-losing strategy on p, then R as II player has a non-losing strategy on p.*
- *If L as II player does not have a non-losing strategy on p, then R as I player has a non-losing strategy on p.*
- *Symmetrically for R.*

Theorem 3 (Determinacy). *Any hypergame has a non-losing strategy at least for one of the players L, R, I, II.*

Proof. Assume by contradiction that p has no non-losing strategies for L, R, I, II. Then in particular p has no non-losing strategy for LI or for LII. Assume the first case holds (the latter can be dealt with similarly). Then, by Lemma 2, p has a non-losing strategy for RII. Hence, by hypothesis there is no non-losing strategy for RI. But then, by Lemma 2, there is a non-losing strategy for LII. Therefore, by definition, there is a non-losing strategy for II. Contradiction.   □

Theorem 3 above can be sharpened, by considering when the non-losing strategy f is in particular a winning strategy, *i.e.* it only generates terminating plays. First, we state the following lemma:

Lemma 3. *Let p be a hypergame.*
- *If L as I player has a winning strategy on p, then R as II player does not have a non-losing strategy on p.*
- *If L as II player has a winning strategy on p, then R as I player does not have a non-losing strategy on p.*
- *Symmetrically for R.*
- *If L as I player has a non-losing strategy on p, but no winning strategies, then R as II player has a non-losing strategy on p.*
- *If L as II player has a non-losing strategy on p, but no winning strategies, then R as I player has a non-losing strategy on p.*
- *Symmetrically for R.*

Theorem 4. *Let p be a hypergame. Then either case 1 or case 2 arises:*
1. *There exists a winning strategy for exactly one of the players L, R, I, II, and there are no non-losing strategies for the other players.*
2. *There are no winning strategies, but there is a non-losing strategy for L or R or I or II. Furthermore:*
   (a) *if there is a non-losing strategy for L or R, then there is a non-losing strategy for at least one of the players I or II;*
   (b) *if there is a non-losing strategy for I or II, then there is a non-losing strategy for at least one of the players L or R;*
   (c) *if there are non-losing strategies for both L and R, then there are non-losing strategies also for both I and II;*
   (d) *if there are non-losing strategies for both I and II, then there are non-losing strategies also for both L and R.*

Proof. 1) If L has a winning strategy, then by Lemma 3 both RI and RII have no non-losing strategies. Hence neither R nor I nor II have non-losing strategies.
2a) Assume that there is a non-losing strategy for L, but no winning strategies. Then there is a non-losing strategy but no winning strategies for LI or for LII. Then, assume *w.l.o.g.* that there is a non-losing strategy but no winning strategies for LI, by Lemma 3 there is a non-losing strategy also for RII. Therefore, since there are non-losing strategies for LII and RII, then there is a non-losing strategy for II.
2c) If there are non-losing strategies both for L and R, then we there are non-losing strategies for LI, LII, RI, RII, thus there are non-losing strategies also for I and II.
The remaining items are proved similarly.   □

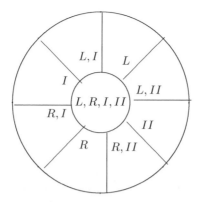

Fig. 1. The space of hypergames

According to Theorem 4 above, the space of hypergames can be decomposed as in Figure 1. For example, the game $c = (\{c\}, \{c\})$ belongs to the center of the space, while the games $a = (\{b\}, \{\})$ and $b = (\{\}, \{a\})$ belong to the sectors marked with L, II and R, II, respectively.

3.1   Characterization Theorem for Non-losing Strategies

The generalization to hypergames of Theorem 2 of Section 2 is quite subtle, because it requires to extend the relation $\gtrsim$ to hypergames, and this needs particular care. We would like to define such relation *by coinduction*, as the greatest fixpoint of a monotone operator on relations, however the operator which is naturally induced by the definition of $\gtrsim$ on games (see Definition 4) is *not* monotone. This problem can be overcome as follows.

Observe that the relation $\gtrsim$ on games is defined in terms of the relation $\not\gtrsim$. Vice versa $\not\gtrsim$ is defined in terms of $\gtrsim$. Therefore, on hypergames the idea is to define both relations at the same time, as the greatest fixpoint of the following operator on pairs of relations:

Definition 6. *Let $\Phi : \mathcal{P}(\mathcal{H} \times \mathcal{H}) \times \mathcal{P}(\mathcal{H} \times \mathcal{H}) \longrightarrow \mathcal{P}(\mathcal{H} \times \mathcal{H}) \times \mathcal{P}(\mathcal{H} \times \mathcal{H})$ be the operator defined by:*

$$\Phi(\mathcal{R}_1, \mathcal{R}_2) = (\{(x,y) \mid \forall x^R.y\mathcal{R}_2 x^R \ \wedge \ \forall y^L.y^L \mathcal{R}_2 x\},$$
$$\{(x,y) \mid \exists x^R.y\mathcal{R}_1 x^R \ \vee \ \exists y^L.y^L \mathcal{R}_1 x\})$$

The above operator turns out to be monotone componentwise. Thus we can define:

Definition 7. *Let the pair ($\gtrsim$, $\not\gtrsim$) be the greatest fixpoint of Φ. Furthermore, we define:*
- $x > y$ *iff* $x \gtrsim y \ \wedge \ y \not\gtrsim x$
- $x \sim y$ *iff* $x \gtrsim y \ \wedge \ y \gtrsim x$
- $x \| y$ *iff* $x \not\gtrsim y \ \wedge \ y \not\gtrsim x$

As an immediate consequence of Tarski's Theorem, the above definition of the pair of relations ($\gtrsim$, $\not\gtrsim$) as the greatest fixpoint of Φ gives us *Coinduction Principles*, which will be extremely useful:

Coinduction Principles. *We call Φ-bisimulation a pair of relations* $(\mathcal{R}_1, \mathcal{R}_2)$ *such that* $(\mathcal{R}_1, \mathcal{R}_2) \subseteq \Phi(\mathcal{R}_1, \mathcal{R}_2)$. *The following principles hold:*

$$\frac{(\mathcal{R}_1, \mathcal{R}_2) \ \textit{Φ-bisimulation} \quad x\mathcal{R}_1 y}{x \gtrsim y} \qquad \frac{(\mathcal{R}_1, \mathcal{R}_2) \ \textit{Φ-bisimulation} \quad x\mathcal{R}_2 y}{x \not\gtrsim y}$$

Notice that the pair of relations ($\gtrsim$, $\not\gtrsim$) on hypergames extends the corresponding pair on games, the latter being the least fixpoint of Φ.

Moreover, somewhat surprisingly at a first sight, notice that the relations $\gtrsim$ and $\not\gtrsim$ are *not* disjoint. *E.g.* the game $c = (\{c\}, \{c\})$ is such that both $c \gtrsim 0$ and $c \not\gtrsim 0$ (and also $0 \gtrsim c$ and $0 \not\gtrsim c$) hold. However, this is perfectly consistent in the hypergame scenario, since it is in accordance with the fact that some hypergames have non-losing strategies for more than one player. Namely, we have:

Theorem 5 (Characterization). *Let x be a hypergame. Then*

$x > 0$ *(x is positive)*	*iff*	*x has a non-losing strategy for L.*
$x < 0$ *(x is negative)*	*iff*	*x has a non-losing strategy for R.*
$x \sim 0$ *(x is zero)*	*iff*	*x has a non-losing strategy for II.*
$x \| 0$ *(x is fuzzy)*	*iff*	*x has a non-losing strategy for I.*

Proof. ($\Rightarrow$) Assume $x > 0$, *i.e.* $x \gtrsim 0$ and $0 \not\gtrsim x$. We show how to build a non-losing strategy for L. We have to build non-losing strategies both for LI and LII. For LII: since $x \gtrsim 0$, then, by definition, $\forall x^R . 0 \not\gtrsim x^R$, *i.e.*, for any R move x^R, $0 \not\gtrsim x^R$. Let $x^R = (X^{RL}, X^{RR})$, then $\exists x^{RL} \in X^{RL} . x^{RL} \gtrsim 0$, that is there exists a L move x^{RL} such that $x^{RL} \gtrsim 0$. Thus we can apply again the two steps above, going on forever or stopping when R cannot move. For LI: since $0 \not\gtrsim x$, then $\exists x^L . x^L \gtrsim 0$. Thus by the previous case there is a non-losing strategy for LII on x^L.

The other cases are dealt with similarly.

($\Leftarrow$) We proceed by coinduction, by showing all the four cases at the same time. Let
$\mathcal{R}_1 = \{(x, 0) \mid x$ has a non-losing strategy for $LII\} \cup$
$\qquad\qquad\qquad \{(0, x) \mid x$ has a non-losing strategy for $RII\}$,
$\mathcal{R}_2 = \{(x, 0) \mid x$ has a non-losing strategy for $RI\} \cup$
$\qquad\qquad\qquad \{(0, x) \mid x$ has a non-losing strategy for $LI\}$.
We prove that $(\mathcal{R}_1, \mathcal{R}_2)$ is a Φ-bisimulation. There are various cases to discuss. We only show one case, the others being similar. We prove that, if $x\mathcal{R}_1 0$ and x has a non-losing strategy for LII, then $\forall x^R . 0\mathcal{R}_2 x^R$. If LII has a non-losing strategy on x, then, by definition, for all x^R there is a non-losing strategy for LI, hence $\forall x^R . 0\mathcal{R}_2 x^R$. $\qquad\square$

The following table summarizes the Characterization Theorem:

Non-losing strategies	Relations w.r.t. 0	
L	$x > 0$	$x \gtrsim 0 \ \wedge \ 0 \not\gtrsim x$
R	$x < 0$	$x \not\gtrsim 0 \ \wedge \ 0 \gtrsim x$
II	$x \sim 0$	$x \gtrsim 0 \ \wedge \ 0 \gtrsim x$
I	$x \| 0$	$x \not\gtrsim 0 \ \wedge \ 0 \not\gtrsim x$

Properties of $\gtrsim$. The following proposition, which can be proved by coinduction, generalizes to hypergames the corresponding results of [Con01]:

Proposition 1. *For all hypergames x, we have*

$$x \not\gtrsim x^R \ \wedge \ x^L \not\gtrsim x \ \wedge \ x \gtrsim x \ \wedge \ x \sim x \ .$$

However, contrary to what happens on games, the relation $\gtrsim$ is *not* a partial order on hypergames (and $\sim$ is *not* an equivalence), since $\gtrsim$ fails to be transitive.

Counterexample. Let $b = (\{b\}, \{b\})$ and $a = (\{a\}, \{0\})$. Then $b \gtrsim 0$, since b has non-losing strategies for all the players. Moreover, one can show that $a \gtrsim b$, by coinduction, by considering the relations $\mathcal{R}_1 = \{(a, b)\} \cup \{(0, b)\}$ and $\mathcal{R}_2 = \{(b, a)\} \cup \{(b, 0)\}$. Thus we have $a \gtrsim b \ \wedge \ b \gtrsim 0$. However, one can easily check that $a \gtrsim 0$ does not hold.

The problem is that the "pivot" b in the above counterexample allows unlimited plays. Namely, if we restrict ourselves to "well-behaved" pivots, then we recover transitivity, *i.e.*:

Lemma 4. *Let x, y, z be hypergames such that y is "well-behaved", i.e. y has no unlimited plays. If $x \gtrsim y \ \wedge \ y \gtrsim z$, then $x \gtrsim z$.*

Proof. (Sketch) One can proceed by coinduction, by showing that the relations $\mathcal{R}_1 = \{(x, z) | \exists y \text{ well-behaved. } x \gtrsim y \ \wedge \ y \gtrsim z\}$ and $\mathcal{R}_2 = \{(z, x) | \exists y \text{ well-behaved.}$ $x \gtrsim y \ \wedge \ z \not\gtrsim y\} \cup \{(z, x) | \exists y \text{ well-behaved. } y \not\gtrsim x \ \wedge \ y \gtrsim z\}$ form a Φ-bisimulation. The difficult part is to prove that $\mathcal{R}_2$ is included in the second component of $\Phi(\mathcal{R}_1, \mathcal{R}_2)$. Here is where we need the hypothesis that b is well-behaved. □

3.2 Sum and Negation of Hypergames

There are various ways in which we can play several different (hyper)games at once. One way consists, at each step, in allowing the next player to select any of the component games and making any legal move on that game, the other games remaining unchanged. The following player can either choose to move in the same component or in a different one. This kind of compound games can be formalized through the *(disjunctive) sum*, [Con01]. The following coinductive definition extends to hypergames the definition of Conway sum:

Definition 8 (Hypergame Sum). *The* sum *on hypergames is given by the the final morphism* $+ : (\mathcal{H} \times \mathcal{H}, \alpha_+) \longrightarrow (\mathcal{H}, id)$, *where the coalgebra morphism* $\alpha_+ : \mathcal{H} \times \mathcal{H} \longrightarrow F(\mathcal{H} \times \mathcal{H})$ *is defined by*

$\alpha_+(x, y) = (\{(x^L, y) \mid x^L \in X^L\} \cup \{(x, y^L) \mid y^L \in Y^L\},$
$$\{(x^R, y) \mid x^R \in X^R\} \cup \{(x, y^R) \mid y^R \in Y^R\}) \ .$$

That is $+$ is such that:
$$x + y = (\{x^L + y \mid x^L \in X^L\} \cup \{x + y^L \mid y^L \in Y^L\},$$
$$\{x^R + y \mid x^R \in X^R\} \cup \{x + y^R \mid y^R \in Y^R\}) \ .$$

The definition of hypergame sum resembles that of *shuffling* on processes. In fact it coincides with interleaving, when impartial games are considered.

For concrete examples of sum games see Section 4, where generalized Nim and Traffic Jam games are discussed.

Another operation on games, which admits an immediate coinductive extension to hypergames, is *negation*, where the roles of L and R are exchanged:

Definition 9 (Hypergame Negation). *The* negation *of a hypergame is given by the final morphism* $- : (\mathcal{H}, \alpha_-) \longrightarrow (\mathcal{H}, id)$, *where the coalgebra morphism* $\alpha_- : \mathcal{H} \longrightarrow F(\mathcal{H})$ *is defined by*
$$\alpha_-(x) = (\{-x^R \mid x^R \in X^R\}, \{-x^L \mid x^L \in X^L\}) \ .$$

That is $-$ is such that: $-x = (\{-x^R \mid x^R \in X^R\}, \{-x^L \mid x^L \in X^L\}) \ .$

In particular, if x has a non-losing strategy for LI (LII), then $-x$ has a non-losing strategy for RI (RII), and symmetrically. Taking seriously L and R players and not fixing a priori L or R to play first, makes the definition of $-$ very natural.

In the following propositions, we summarize some interesting results on sum and negation, that can be extended to hypergames:

Proposition 2
i) $x - x \sim 0$.
ii) $x \gtrsim 0 \wedge y \gtrsim 0 \implies x + y \gtrsim 0$.
iii) If $y \sim 0$, *then* $x + y$ *has the same outcome as* x.
iv) If $y - z \sim 0$, *then the games* $x + y$ *and* $x + z$ *have the same outcome.*

The proofs of the items in the above proposition are similar to the ones provided in [Con01], pag.76, based on the construction of winning/non-losing strategies.

Proposition 3
i) $x > y$ *iff* $x - y$ *has a non-losing strategy for L.*
ii) $x < y$ *iff* $x - y$ *has a non-losing strategy for R.*
iii) $x \sim y$ *iff* $y - x$ *has a non-losing strategy for II.*
iv) $x \| y$ *iff* $y - x$ *has a non-losing strategy for I.*

The implications ($\Rightarrow$) in the above proposition are proved by building non-losing strategies, using the definitions of $\gtrsim$ and $\lesssim$, while the converse implications are proved using the Φ-coinduction principle.

4   The Theory of Impartial Hypergames

In this section, we focus on *impartial hypergames*, where, at each position, L and R have the same moves. Such hypergames can be simply represented by $x = X$, where X is the set of moves (for L or R). Coalgebraically, this amounts to say that impartial hypergames are a final coalgebra of the powerset functor.

In this section, we first recall the Grundy-Sprague theory for dealing with finite impartial games, then we discuss the theory of impartial hypergames, using Smith generalization of Grundy-Sprague results. In particular, we show how to provide a more complete account of such a theory, by introducing a class of *canonical hypergames*, extending the *Nim numbers*. This can only be given in the hypergame setting. We illustrate our results on an example.

4.1   The Grundy-Sprague Theory

Central to the theory of Grundy-Sprague, [Gru39,Spra35], is *Nim*, a classical impartial game, which is played with a number of heaps of matchsticks. The legal move is to strictly decrease the number of matchsticks in any heap (and throw away the removed sticks). A player unable to move because no sticks remain is the loser.

The Nim game with one heap of size n can be represented as the Conway game $*n$, defined (inductively) by

$$*n = \{*0, *1, \ldots, *(n-1)\} .$$

Namely, with a heap of size n, the options of the next player consist in moving to a heap of size $0, 1, \ldots, n-1$. The number n is called the *Grundy number* of the game. Clearly, if $n = 0$, the II player wins, otherwise player I has a winning strategy, moving to $*0$.

Nim games $*n$ are called *nimbers*, to distinguish them from the games n representing numbers, which have a different definition, see [Con01] for more details. Nimbers correspond to von Neumann finite numerals in Set Theory.

Nim games are central in game theory, since there is a classical result (by Grundy and Sprague, independently, [Gru39,Spra35]) showing that any impartial game "behaves" as a Nim game, or, using Conway terminology, is $\sim$-equivalent to a single-heap Nim game (see [Con01], Chapter 11). The algorithm for discovering the Nim game (or the Grundy number) corresponding to a given impartial game x proceeds inductively as follows. Assume that the Grundy numbers of the options of x are $n_0, n_1, \ldots$, then the Grundy number of x is the *minimal excludent (mex)* of $n_0, n_1, \ldots$ The mex of a list of numbers $n_0, n_1, \ldots$ is the least natural number which does not appear among $n_0, n_1, \ldots$ Then, having the Grundy number of (the positions of) a game, we know the winning strategy for that game.

Sums of impartial games. Here we explain how, using the above theory and the sum on Nim numbers, one can easily deal with compound impartial hypergames.

An example of a compound impartial game is the Nim game with more heaps. Using sum, the Nim game with two heaps of sizes n_1, n_2 can be represented as the Conway game $*n_1 + *n_2$. By the general result by Grundy-Sprague on impartial games, such game is also equivalent to a Nim game with a single heap, and thus there is a Grundy number n such that $*n \sim *n_1 + *n_2$. The sum of Nim numbers is particularly easy to compute and, as we will see, it is useful for analyzing the sum of generic impartial games. Thus, it deserves a special definition; following [Con01], we define the *Nim sum* $+_2$ by: $n_1 +_2 n_2 = n$, where n is the Nim number corresponding to the sum game $*n_1 + *n_2$. The Nim sum is quite easy to calculate, since one can show that it amounts to binary sum without carries. *E.g.* $1 +_2 3 = 2$, since $01 \oplus 11 = 10$, where $\oplus$ is binary sum.

In general, in order to analyze the sum of impartial games, one can proceed as follows. Using the Grundy-Sprague algorithm, one can compute the Nim numbers corresponding to the compound games. Then, Nim-summing such numbers one gets the Nim number corresponding to the starting game. If the result is 0, there is a winning strategy for the II player, otherwise there is a winning strategy for the I player, who can move to a position of Nim sum 0.

4.2   The Smith Theory in the Hypergame Setting

In [Con01], Chapter 11, the author briefly analyzes infinite impartial games, even if they escape his inductive definition. These games are represented as finite or infinite, cyclic graphs, having a node for each position of the game, and a direct edge from p to q when it is legal to move from p to q. Thus they exactly correspond to non-wellfounded sets, or impartial hypergames, in our setting. Theorem 3 specializes to impartial hypergames as follows:

Theorem 6
Any impartial hypergame has non-losing strategies either for I player or for II player or for both.

Smith [Smi66] extended the Grundy-Sprague theory on impartial games to cover infinite games (see [Con01], pag. 133-135). In particular, Smith provides an algorithm (which works for a large class of cyclic graphs) for marking the nodes of the game graph with naturals (ordinals if the graph is infinite) plus some infinity symbols. This generalizes the Grundy-Sprague inductive algorithm, based on the *mex*, for computing the Grundy number of an impartial game. From Smith's marking one can then immediately discover whether a given position is winning for I, for II or it is a draw.

Smith's Marking of the Game Graph, [Smi66]. A position p in the graph will be marked with the number n if the following conditions hold. Firstly, n must be the *mex* (minimal excludent) of all numbers that already appear as marks of any of the options of p. Secondly, each of the positions immediately following p which has not been marked with some number less than n must already have an option marked by n. We continue in this way until it is impossible to mark any further node with any ordinal number, and then attach the symbol ∞ to any remaining node (which we call *unmarked*). Finally, the label of a position marked as n is n, while the label of an unmarked position is the symbol ∞ followed by the labels of all marked options as subscripts, see for example the graph of Fig. 2.

Now, the following result holds:

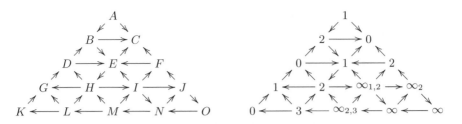

Fig. 2. The graph of an impartial hypergame, and Smith's marking

Theorem 7 (see [Con01], pag. 134). *A position marked as n is a II player win if and only if n is 0, otherwise it is a I player win. A position marked by ∞_K, where K is a set of naturals, is a I player win if and only if $0 \in K$, otherwise it is a draw.*

The above theorem can be proved by induction on n. The idea underlying such technique is that a node marked by n behaves as the Nim game $*n$. This can be viewed as a "canonical game" $\sim$-corresponding to the given node. However, the theory, as it is presented in the literature, is not completely satisfactory for ∞ nodes, since ∞ symbols do not correspond to "canonical infinite games". In the sequel, we show how to do this in our setting of hypergames.

Let us consider a position p marked with ∞_K. We claim that such node behaves as the (canonical) hypergame

$$*\infty_K = \{*\infty\} \cup \{*k \mid k \in K\} \,,$$

where $*\infty = \{*\infty\}$.

Namely, one can show that:

Theorem 8. *If x is the canonical hypergame associated to a position p in a graph, then*

$x\|0$ *iff x has a non-losing strategy for I.*
$x \sim 0$ *iff x has a non-losing strategy for II.*

Proof. For positions marked by n, the thesis follows immediately from Theorem 7. Then let p be a position marked by ∞_K. Using Theorem 7, we only need to prove that:

(a) the hypergame $*\infty_K$ has subscript 0 iff $*\infty_K\|0$ but not $*\infty_K \sim 0$;

(b) the hypergame $*\infty_K$ has no subscript 0 iff $*\infty_K\|0$ and $*\infty_K \sim 0$.

$(a \Rightarrow)$ First of all, notice that $*\infty$ is such that $*\infty \gtrsim 0$ and $0 \gtrsim *\infty$. Assume $0 \in K$. Then $*\infty_K \not\gtrsim 0$, since $0 \gtrsim *\infty$. Similarly $0 \not\gtrsim *\infty_K$, since $*\infty \gtrsim 0$. Hence $*\infty_K\|0$. Moreover, neither $*\infty_K \gtrsim 0$ nor $0 \gtrsim *\infty_K$ hold, since $0 \not\gtrsim 0$ does not hold.

$(a \Leftarrow)$ Assume $*\infty_K\|0$, but not $*\infty_K \sim 0$. Assume by contradiction that 0 is not subscript of ∞_K. Then, since for all $k \in K$. $*k \not\gtrsim 0 \wedge 0 \not\gtrsim *k$, we have $*\infty_K \sim 0$. Contradiction.

$(b \Rightarrow)$ Assume $0 \notin K$. Then $*\infty_K \| 0$, since $*\infty \gtrsim 0$ and $0 \gtrsim *\infty$. Moreover $*\infty_K \sim 0$, since, for all elements $x \in *\infty_K$, $x \not\gtrsim 0$ and $0 \not\gtrsim x$.

$(b \Rightarrow)$ Assume $*\infty_K \| 0$ and $*\infty_K \sim 0$. If by contradiction $0 \in K$, then $*\infty_K \sim 0$ does not hold, since $0 \not\gtrsim 0$ does not hold.   □

Thus, generalizing to impartial hypergames, Grundy-Sprague result on impartial games, we have:

Theorem 9. *Any impartial hypergame behaves either like a Nim game or like a hypergame of the shape $*\infty_K$.*

In the following, we show that our canonical hypergames are well-behaved also w.r.t. sum.

Traffic Jams and Generalized Sums. Following [Con01], we consider a concrete hypergame to illustrate how compound impartial hypergames are handled using Smith's generalized marking algorithm and the extension to ∞-nodes of the Nim sum. Let us consider the following concrete game, corresponding to the game graph in Fig. 2. We can think of the graph as the map of a fictitious country, where nodes correspond to towns, and edges represent motorways between them. The initial position of the game corresponds to the town where a vehicle is initially placed. Each player has to move such vehicle to a next town along the motorway. If this is not possible, then the player loses. Theorem 7 tells us which player has a non-losing strategy in any position. Now there is a natural generalization of the above traffic game, where more than one vehicle is considered. We assume that each town is big enough to accommodate all vehicles at once, if needed. At each step, the current player chooses a vehicle to move. Such game corresponds to the sum of the hypergames with single vehicles. In order to compute non-losing strategies for the sum game, one can use the *generalized Nim sum*, which amounts to the Nim sum extended to ∞-nodes as follows:

$$n +_2 \infty_K = \infty_K +_2 n = \infty_{\{k +_2 n \ | \ k \in K\}} \qquad \infty_K +_2 \infty_H = \infty .$$

Thus for example, if we have vehicles at positions H and I in Fig. 2, then the game is winning for I player, since $2 +_2 \infty_{1,2} = \infty_{2 +_2 1, 2 +_2 2} = \infty_{3,0}$. While a game with vehicles in I and J is a draw, since $\infty_{1,2} +_2 \infty_2 = \infty$.

On the other hand, having assigned canonical hypergames to the nodes of the graph, one could use hypergame sum (as defined in Definition 8) for summing them. Hence the question naturally arises whether canonical hypergames are well-behaved w.r.t. sum. The answer is positive, since one can prove that the hypergame sum behaves as the canonical hypergame corresponding to the result of the extended Nim sum, *i.e.*:

Proposition 4. *Let $*\infty_K, *\infty_H$ be hypergames. Then*
i) the hypergame $*\infty_K + *n$ behaves as the hypergame $*\infty_{\{k +_2 n \ | \ k \in K\}}$;
ii) the hypergame $*\infty_K + *\infty_H$ behaves as the hypergame $*\infty$.

Proof. (Sketch) Both items i) and ii) are proved using Theorem 5, by showing that the sum game has a non-losing strategy for L, R, I, II iff the corresponding game has one.   □

5   Comparison with Related Work and Directions for Future Work

Loopy games. The theory of general loopy games, where infinite plays can be either winning for L,R or draws is very difficult. For instance, already for the case of fixed games (where no draws are admitted), determinacy fails if the Axiom of Choice is assumed. In [BCG82], Chapter 11, fixed loopy games are studied. A $\geq$ relation is introduced, which is proved to be transitive and it allows to approximate the behavior of a loopy game, possibly with finite games. But this technique works only if certain fixpoints exist. Such theory has been later further developed and revisited in other works, see *e.g.* [San02,San02a].

On the contrary, our theory allows to deal with the class of games where infinite plays are draws in a quite general and comprehensive way. We plan to investigate more general classes of (possibly mixed) loopy games, where infinite plays can be considered as winning or draws.

Games and automata. The notion of hypergame that we have investigated in this paper is related to the notion of infinite game considered in the automata theoretic approach, originating in work of Church, Büchi, McNaughton and Rabin (see *e.g.* [Tho02]). In this approach, games are defined by the graphs of positions. L and R have different options, in general, but L is always taken as first player. Only games with infinite plays are considered. These games are fixed, according to the above definition. Winning strategies are connected with automata, and also the problem of a (efficient) computation of such strategies is considered. However, recently, non-losing strategies have been considered also in this setting, *e.g.* in the context of model checking for the μ-calculus, see [GLLS07].

Games for semantics of logics and programming languages. Game Semantics was introduced in the early 90's in the construction of the first fully complete model of Classical Multiplicative Linear Logic [AJ94], and of the first syntax-independent fully abstract model of PCF, by Abramsky-Jagadeesan-Malacaria, Hyland-Ong, and Nickau, independently. Game Semantics has been used for modeling a variety of programming languages and logical systems, and more recently for applications in computer-assisted verification and program analysis, [AGMO03]. In Game Semantics, 2-players games are considered, which are in some way related to Conway games, despite the rather different presentation. For more details see [AJ94]. The main difference between the Game Semantics approach and our approach lies in the fact that, in Game Semantics, infinite plays are always considered as winning for one of the two players, as in the case of Conway's fixed games.

Traced categories of games. In [Joy77], Joyal showed how Conway (finite) games and winning strategies can be endowed with a structure of a traced category. This work admits an extension to loopy games, when these are fixed. However, when draws are considered, Joyal's categorical construction apparently does not work, since we lose closure under composition (this is related to the fact that our relation $\gtrsim$ is not transitive). As future work, we plan to study traced categories for general infinite games, and to investigate the trace operation in a coalgebraic setting.

Games and coalgebras. In [BM96], a simple coalgebraic notion of game is introduced and utilized. It is folklore that bisimilarity can be defined as a 2-player game, where one player tries to prove bisimilarity, while the other tries to disprove it. This game turns out to be a fixed game in the sense of [BCG82], where infinite plays are winning for the player who tries to prove bisimilarity.

Conumbers. Conway's numbers [Con01] amount to Conway's games x such that no member of X^L is $\gtrsim$ any member of X^R, and all positions of a number are numbers. Thus, once we have defined hypergames and the relations $\gtrsim$, $\not\gtrsim$, we can define the subclass of *conumbers*, together with suitable operations extending those on numbers. It would be interesting to investigate the properties of such a class of hypergames. An intriguing point is whether it is possible to define a partial order, since, as seen in this paper, the relation $\gtrsim$ is not transitive on hypergames.

Compound games. In this paper, we have considered the (disjunctive) sum for building compound games. However, there are several different ways of combining games, which are analyzed in [Con01], Chapter 14, for the case of finite games. It would be interesting to extend to hypergames such theory on compound games.

References

AGMO03.   Abramsky, S., Ghica, D.R., Murawski, A.S., Ong, C.-H.L.: Applying Game Semantics to Compositional Software Modeling and Verifications. In: Jensen, K., Podelski, A. (eds.) TACAS 2004. LNCS, vol. 2988, pp. 421–435. Springer, Heidelberg (2004)

AJ94.   Abramsky, S., Jagadesaan, R.: Games and Full Completeness for Multiplicative Linear logic. Journal of Symbolic Logic 59, 543–574 (1994)

Acz88.   Aczel, P.: Non-wellfounded sets, CSLI Lecture Notes 14, Stanford (1988)

BM96.   Barwise, J., Moss, L.: Vicious Circles, CSLI Lecture Notes 60, Stanford (1996)

BCG82.   Berlekamp, E., Conway, J., Guy, R.: Winning Ways. Academic Press, London (1982)

Con01.   Conway, J.H.: On Numbers and Games, 2nd edn. A K Peters Ltd. (2001) (first edition by Academic Press, 1976)

FH83.   Forti, M., Honsell, F.: Set-theory with free construction principles. Ann. Scuola Norm. Sup. Pisa 10(4), 493–522 (1983)

GLLS07.   Grumberg, O., Lange, M., Leucker, M., Shoham, S.: When Not Losing Is Better than Winning: Abstraction and Refinement for the Full μ-calculus. Information and Computation 205(8), 1130–1148 (2007)

Gru39.   Grundy, P.M.: Mathematics and games. Eureka 2, 6–8 (1939)

Joy77.   Joyal, A.: Remarques sur la Theorie des Jeux a deux personnes. Gazette des sciences mathematiques du Quebec 1(4) (1977); English translation by Houston, R. (2003)

San02.   Santocanale, L.: Free μ-lattices. J. Pure Appl. Algebra 168, 227–264 (2002)

San02a.   Santocanale, L.: μ-bicomplete categories and parity games. Theor. Inform. Appl. 36(2), 195–227 (2002)

Smi66.   Smith, C.A.B.: Graphs and composite games. J. Combin. Th. 1, 51–81 (1966)

Spra35.   Sprague, R.P.: Über mathematische Kampfspiele. Tohoku Math. J. 41, 438–444 (1935-1936)

Tho02.   Thomas, W.: Infinite games and verification. In: Brinksma, E., Larsen, K.G. (eds.) CAV 2002. LNCS, vol. 2404, pp. 58–64. Springer, Heidelberg (2002)

A Semantical Approach
to Equilibria and Rationality

Dusko Pavlovic[*]

Kestrel Institute and Oxford University
dusko@{kestrel.edu,comlab.ox.ac.uk}

"An equilibrium does not appear because agents are rational, but rather agents appear to be rational because an equilibrium has been reached.[...] The task for game theory is to formulate a notion of rationality."

Larry Samuelson [20, p. 3]

Abstract. Game theoretic equilibria are mathematical expressions of rationality. Rational agents are used to model not only humans and their software representatives, but also organisms, populations, species and genes, interacting with each other and with the environment. Rational behaviors are achieved not only through conscious reasoning, but also through spontaneous stabilization at equilibrium points.

Formal theories of rationality are usually guided by informal intuitions, which are acquired by observing some concrete economic, biological, or network processes. Treating such processes as instances of computation, we reconstruct and refine some basic notions of equilibrium and rationality from some basic structures of computation.

It is, of course, well known that equilibria arise as fixed points; the point is that semantics of computation of fixed points seems to be providing novel methods, algebraic and coalgebraic, for reasoning about them.

1 Introduction

Game theory studies distributed processes where the resources are shared among the agents with different, inconsistent, and often adversarial goals. Originally devised as a tool of economics, politics, and warfare, game theory recently became an indispensable tool of algorithmics, especially as the processes and the problems of computation spread from computers to networks. The other way around, the algorithmic aspects of game theory have attracted a lot of attention on their own, leading to fruitful interactions between economics and algorithmics [7,15,22].

[*] Supported by ONR and EPSRC.

A. Kurz, M. Lenisa, and A. Tarlecki (Eds.): CALCO 2009, LNCS 5728, pp. 317–334, 2009.

In semantics of computation, often viewed as the stylistic dual of algorithmics, the paradigm of game also played a crucial role, and led to the solutions of some deep and long standing problems [1,11]; yet the resulting toolkit of *game semantics* [2] remained largely disjoint from the game theoretic methods, and concerns. While this may very well be justified by the different, and perhaps even disjoint goals of game semantics and game theory, the growing importance of the computational aspects of game theory continues to spur the explorations of a different conceptual link: *If gaming is computation, which semantical and programming methodologies apply to it?*

The present paper provides a belated account of some initial explorations in this direction, going back to a joint project with Samson Abramsky. The upshot is that the basic models of computation readily extend to capture the basic notions of game theory: the tools for reasoning about *choice*, be it possibilistic or probabilistic, and the tools to compute fixed points of possibilistic and probabilistic processes, turn out to be readily applicable to designing and programming strategic behavior. The approach seems promising in both directions: on one hand, the semantical view of games provides a convenient formal framework for conceptual analyses and concrete computations of response profiles and equilibria; the other way around, the game theoretic view of the computational processes opens an alley towards modeling a wide range of network interactions of increasing practical interest.

As a running example, we use what may be the smallest and the deepest problem of game theory: Prisoners' Dilemma. In its standard solution, traditional game theory recommends selfishness as the only rational strategy here, although "staying the course" of this strategy leads to an ostensive loss for everyone. Can our semantical tools dispel the irrationality of this standard solution, and provide a better model of rationality? We propose and analyze several refinements of the basic model of strategic reasoning, and show how the implementations of the optimization task of gaming can be refined, and their rationality improved. Some familiar semantical tools turn out to allow computing more informative equilibria, e.g. where players' preferences are quantified, rather than just partially ordered, and where the payoffs can be used dynamically (e.g. reinvested, or discounted), and not just accrued. This seems to suggest that applying semantical methodologies to program strategies may offer some new solutions, besides being fun.

Outline of the paper. Section 2 sketches a bird's eye view of program and process semantics, and points to the place of games in that landscape. In section 3, we reconstruct the familiar notions of Nash equilibrium and evolutionary stable strategy, as they could be obtained by running relational (nondeterministic) programs. We also discuss some nonstandard equilibrium concepts, which can be easily designed and implemented in this framework. In section 4, we lift these equilibrium concepts from the relational to a stochastic framework, where they can be obtained as stationary distributions of Markov chains. In order to remain close to the usual game-theoretic models, in both these sections games are viewed as stateless processes. In section 5, we discuss the role of state, i.e. position, in semantics of gaming. Section 6 summarizes the paper.

2   Program and Process Semantics of Games

Semantics of a natural language evolves through speech and through use of the language. Semantics of a programming language requires moreover a design effort, because it concerns not only communication between people, but also programming computers, and they need to be designed before they are built. However, as the notion of a computer is changing from a machine in a box to a world wide network, the simple notion of a program diversifies. Some programs acquire strategic, i.e. game theoretic aspects. We sketch a way to capture these aspects in a well studied framework of fixed point semantics, where coalgebras are always present in one way or another.

2.1   Program Semantics

In categorical semantics, a program is denoted by an arrow $A \xrightarrow{f} B$ in a category $\mathbb{D}$, where the objects A and B denote some data types, of the inputs and of the outputs of f, respectively. It is assumed that the category $\mathbb{D}$ has cartesian products , so that we can also represent a program $A \times C \xrightarrow{g} B \times D \times X$ with multiple inputs and multiple outputs.

But running a program does not just map data to data; it also causes a whole range of other observable effects. E.g., a computation may not terminate, or it may terminate with several possible outputs for the same input; or it may change the state of the computer, or of another resource. Such computational effects can be captured by computational *monads* [13]. Originally proposed as a tool of semantics, monads have been widely endorsed as a convenient programming tool [23]. In the meantime, an alternative presentation of essentially equivalent semantical structure has been proposed, in terms of *premonoidal categories* [19]. The category $\mathbb{D}$ of data types, with cartesian products and simple *deterministic* maps between them, is extended to a category $\mathbb{C}$ with the same data types as objects, but with the computations with nontrivial effects as its morphisms. Along the inclusion $\mathbb{D} \hookrightarrow \mathbb{C}$, the cartesian products of $\mathbb{D}$ are mapped into the *premonoidal* tensor products in $\mathbb{C}$. The relevant semantical structure is sometimes called *Freyd category* [18].

While the models of games that involve some of the well-studied computational effects seem quite interesting for future research, in the present paper we only consider the simplest effects of the *choice* operations, and in particular of the possibilistic (relational) and probabilisitic (randomized) choice. The reason is that these choice operations already come about in game theory, so that we can display some familiar ideas from a slightly different angle.

From the rich tool chest of program semantics, we shall thus consider only two simple but fundamental categories of computations:

- FRel of finite sets and relations, and
- SRel of finite sets and stochastic relations.

A morphism in either of these categories will be denoted by a crossed arrow $\longrightarrow$. While a binary relation $A \overset{R}{\longrightarrow} B$ in FRel can be viewed as a matrix $B \times A \overset{R}{\longrightarrow} \{0,1\}$, a stochastic relation $A \overset{P}{\longrightarrow} B$ in SRel is a matrix $B \times A \overset{P}{\longrightarrow} [0,1]$, i.e. $P = (p_{ji})_{B \times A}$, where $p_{ji} \in [0,1]$ and $\sum_{j \in B} p_{ji} = 1$ holds for all $i \in A$. Intuitively, the entry p_{ji} can thus be viewed as the probability that the input $i \in A$ will result in the output $j \in B$. The composition in SRel is the matrix composition. Both categories of computations $\mathbb{C} = $ FRel, SRel have the same cartesian subcategory of deterministic maps $\mathbb{D} = $ Set. They both happen to be monoidal, rather than premonoidal.

2.2   Processes and Controls

We model processes simply as programs that depend on a state, and may change it. If the state space is represented by an object X in a category of computations $\mathbb{C}$, then a process is thus a morphism $A \times X \overset{R}{\longrightarrow} B \times X$. Since every category of computations inherits along the inclusion $\mathbb{D} \hookrightarrow \mathbb{C}$ the cartesian diagonals $A \overset{\delta}{\longrightarrow} A \times A$ and the projections $A \times B \overset{\pi_1}{\longrightarrow} A$ and $A \times B \overset{\pi_2}{\longrightarrow} B$, we can separate the data part and the state part of a process as

- R_B : $A \times X \overset{R}{\longrightarrow} B \times X \overset{\pi_1}{\longrightarrow} B$ and
- R_X : $A \times X \overset{R}{\longrightarrow} B \times X \overset{\pi_2}{\longrightarrow} X$.

A process can be ongoing, and its outputs may be used to determine the inputs to be fed back into it. This is expressed through *feedback* $B \times X \overset{\phi}{\longrightarrow} A$. To stabilize a process, a *control* $X \overset{\gamma}{\longrightarrow} A$ can be extracted as a fixed point

$$
\frac{A \times X \overset{R}{\longrightarrow} B \times X \qquad\qquad B \times X \overset{\phi}{\longrightarrow} A}{A \times X \overset{R}{\longrightarrow} B \times X \overset{\phi}{\longrightarrow} A}
$$

$$
\begin{array}{ccc}
X & \overset{\gamma = Fix_A(\phi \circ R)}{\longrightarrow} & A \\
{\scriptstyle\langle \gamma, \mathrm{id}\rangle} \downarrow & & \uparrow {\scriptstyle\phi} \\
A \times X & \underset{R}{\longrightarrow} & B \times X
\end{array}
$$

Such fixed point operations play a central role in modeling processes and controls. We shall see that they play a central role in modeling games. The fixed point operations in FRel and SRel are spelled out in the Appendix.

For a categorical insider, we add that any Freyd-category [18] with a family of Conway fixed point operators [6,21] should suffice for the (as yet putative) research in *abstract game theory*. Equivalently, a traced Freyd category will do as well [4].

Examples. The simplest example of a process is a *Mealy machine*. A deterministic one is simply a function $A \times X \longrightarrow B \times X$. A nondeterministic (possibilistic) one is a relation $A \times X \nrightarrow B \times X$. A probabilistic automaton can in principle be viewed as a stochastic relation of the same type, just in SRel rather than in FRel. Other examples of processes include Markov chains... and even games.

2.3   Games as Processes

An m-player game is a process $A \times X \xrightarrow{\varrho} B \times X$ where the inputs, the outputs and the state consist of m components, i.e.

$$A = \prod_{i \in m} A_i \qquad B = \prod_{i \in m} B_i \qquad X = \prod_{i \in m} X_i$$

where we represent ordinals following von Neumann, in the form $m = \{0, 1, \ldots, m-1\}$. The inputs A_i are thought of as the *moves* available to the i-th player; the outputs B_i are her *payoffs*; the states in X_i are the *positions* that she can observe. The payoff types B_i are usually ordered, and this ordering expresses player's preference.

Games can thus be viewed as a special case of controllable processes, described in the preceding section. The optimization task of control is, however, slightly different. First of all, it is distributed: instead of a *global control*, each player designs and implements an *individual strategy*. And secondly, these strategies are not designed using feedback, to respond to the outputs, but rather to respond to the inputs supplied by the other players:

$$\cfrac{\cfrac{\cfrac{A \times X \xrightarrow{\varrho} B \times X}{A_{-i} \times X_i \xrightarrow{RS_i} A_i} \; (\star)}{A \times X \xrightarrow{RS=\langle RS_i \circ \pi_i \rangle_{i \in m}} A} \; (\star\star)}{X \xrightarrow{RS^\bullet = Fix_A(RS)} A} \; (\diamondsuit)$$

At step $(\star)$, each player i implements her rationality through a *response relation* RS_i to all of the opponents' moves, which are chosen from

$$A_{-i} = \prod_{k \in m, \; k \neq i} A_k$$

At step $(\star\star)$, the individual response relations RS_i are gathered into the response profile RS, which is simply the m-tuple of the relations RS_i.

Finally, at step $(\diamondsuit)$, the equilibrium $RS^\bullet$ is computed, as the fixed point of the response profile RS. The equilibrium is an m-tuple of relations $X \xrightarrow{RS_i^\bullet} A_i$.

It tells, for each position from X, how will the game be played if each player responds by RS_i to everyone else's responses $RS_{-i} = \langle RS_k \circ \pi_k \rangle_{k \in m \setminus \{i\}}$. The equilibrium is thus the global (social) result of the local (individual) preferences and of distributed reasoning (programming) in pursuit of these preferences.

How does an equilibrium come about? The usual explanation is that each player i knows everyone's preferences, and can thus construct RS_k, for all $k \in m$, on her own, and thus compute the profile RS and the equilibrium $RS^\bullet = Fix(RS)$. But this explanation should be taken as a metaphor. In reality, equilibria are often reached e.g. in biological systems, and in other genuinely distributed processes, where the agents do not perform explicit local computations, or reason about each other. Moreover, even in the cases where all strategies and their fixed points could conceivably be computed at each node, the fact that there are usually many equilibria gives rise to the question how do the players coordinate to meet at one equilibrium. This is where game theory enters the conceptual realm of "invisible hand" and equilibrium selection [20]. Modeling such genuinely distributed processes is one of the most interesting challenges of the computational semantics of gaming.

In the rest of the paper, we explore more closely each of the steps in the above derivation of the response strategies and equilibria. We begin with step ($\star$), where each player "programs" her response to other players' possible moves.

3   Strategies as Nondeterministic Programs

To reconstruct the first concepts of standard game theory, we first consider one shot games, i.e. where $X = 1$. This is standard in traditional game theory, where it is assumed that each player chooses a strategy in advance, and plays it out no matter what. The notion of position, or state, is thus abstracted away. Another standard assumption is that the payoffs are uniquely determined for each player, by mapping each tuple of moves in A to a tuple of payoffs in B. The game thus boils down to a function $A \xrightarrow{\varrho} B$. If there are 2 players, called 0 and 1, and if their payoffs are in $B_0 = B_1 = \mathbb{R}$, this gives the usual bimatrix form $A_0 \times A_1 \xrightarrow{\varrho} \mathbb{R} \times \mathbb{R}$. If there are 3 players, the game can be viewed as 3 tri-matrices, etc.

Even if the game, represented by the functions that compute the payoffs, is completely deterministic, the fact that the players need to choose between the various possible moves makes their strategies into nondeterministic programs. In this section, we view them as relations; later we view them as stochastic relations.

3.1   Designing and Refining Relational Strategies

We assume that the payoff types B_i are ordered, and that the players prefer higher payoffs. Each of them thus programs a response strategy towards the goal of maximizing his payoffs. We first consider some simple implementations, and then show how they can be *refined*.

A. Best response simply maximizes the payoff

$$s_{-i} \; BR_i \; s_i \iff \forall t_i \in A_i. \; \varrho_i(t_i, s_{-i}) \leq \varrho_i(s_i, s_{-i})$$

B. Stable response refines the view by taking the possible opponents' responses into account:

$$
\begin{aligned}
s_{-i} \; SR_i \; s_i \iff & \forall t_i \in A_i. \; \varrho_i(t_i, s_{-i}) \leq \varrho_i(s_i, s_{-i}) \land \\
& (\varrho_i(t_i, s_{-i}) = \varrho_i(s_i, s_{-i}) \Rightarrow \\
& \forall t_{-i} \in A_{-i}. \; \varrho_i(t_i, t_{-i}) \leq \varrho_i(s_i, t_{-i}))
\end{aligned}
$$

The idea is that a stable response s_i to s_{-i} should remain optimal in some neighborhood of s_{-i}. When the payoff function ϱ_i is linear, it is easy to prove that the above definition captures this. Indeed, if the opponents deviate from s_{-i} and play $(1 - \varepsilon)s_{-i} + \varepsilon t_{-i}$ for a small $\varepsilon > 0$, then a stable response s_i will still be the best, because the validity of

$$(1 - \varepsilon)\varrho_i(s_i, s_{-i}) + \varepsilon\varrho_i(s_i, t_{-i}) \geq (1 - \varepsilon)\varrho_i(t_i, s_{-i}) + \varepsilon\varrho_i(t_i, t_{-i})$$

for all t_i, follows from the above definition of SR_i.

C. Uniform response goes a step further by taking the opponents' *best* response into account:

$$
\begin{aligned}
s_{-i} \; UR_i \; s_i \iff & s_{-i} \; BR_i \; s_i \land \\
& \forall t_{-i} \in A_{-i}. \; s_i \; BR_{-i} \; t_{-i} \Rightarrow t_{-i} \; BR_i \; s_i
\end{aligned}
$$

where $s_i \; BR_{-i} \; t_{-i}$ abbreviates $\forall k \in m. \; k \neq i \Rightarrow (s_i, t_{-i,k}) \; BR_k \; t_k$. The best response s_i is thus required to remain optimal not only with respect to s_{-i}, but also with respect to the opponents' best responses to the profiles that include s_i. This is a rational, but very strong requirement: the relation UR_i may be empty. We mention it as a first attempt to refine the response by anticipating opponents' responses to it. The next example proceeds in this direction, while assuring a nonempty set of responses.

D. Constructive response. While the uniform response captures the best responses to *all* of the opponent's responses, the constructive response relation also capture the responses that may not be the best responses to a fixed opponent's move, but are better than what the best response would turn into in the context of opponent's rational responses to it.

$$
\begin{aligned}
s_{-i} \; CR_i \; s_i \iff & \forall t_i \in A_i. \; \varrho_i(s_i, s_{-i}) < \varrho_i(t_i, s_{-i}) \Rightarrow \\
& \exists t_{-i} \in A_{-i}. \; \varrho_{-i}(t_i, t_{-i}) > \varrho_{-i}(t_i, s_{-i}) \land \\
& \varrho_i(t_i, t_{-i}) < \varrho_i(s_i, s_{-i})
\end{aligned}
$$

Example. Prisoners' Dilemma is a famous 2-player game, usually presented with a single state $X = 1$ and two moves, "*cooperate*" and "*defect*", thus $A_0 = A_1 = \{c, d\}$, and $B_0 = B_1 = \mathbb{R}$. Players' preferences are given by a payoff function $\varrho = \langle \varrho_0, \varrho_1 \rangle : \{c, d\} \times \{c, d\} \longrightarrow \mathbb{R} \times \mathbb{R}$ which can be presented by the bimatrix

telling that $\varrho_0(c, c) = \varrho_1 = 10$, $\varrho_0(c, d) = 0$, $\varrho_1(c, d) = 11$ etc. The point is that players' local reasoning leads to globally suboptimal outcome: each player seems forced to play d, because he wins more whether the opponent plays c or d; but if both players play d they both win less than if they both play c. The constructive response allows the players to keep the strategy c as a candidate solution. Although $\varrho_0(c, c)$ gives lower payoff than $\varrho_0(d, c)$, the player 0 knows that the profile (d, c) is unlikely to happen, because $\varrho_1(d, d) > \varrho_1(d, c)$. So he keeps (c, c) as better than (d, d).

Of course, this form of rationality does not offer the worst case protection, and may not seem rational at all, because there is no guarantee that (c, c) will happen either. Indeed, $\varrho_1(c, d) > \varrho_1(c, c)$ is likely to motivate player 1 to defect, which leads to the worst outcome for player 0, since $\varrho_0(c, d) < \varrho_0(c, c)$.

However, if player 1 follows the rationality of CR, and not BR, then he'll also consider cooperating, because of the threat that player 0 would retaliate in response to his defection, and $\varrho_1(d, d) < \varrho_1(c, c)$. So the possibility of the solution (c, c) depends on whether the players share the same rationality CR. This sharing cannot be coordinated in the relational model of one-round Prisoners' Dilemma. We shall later see how some more precise models do allow this.

3.2   Playing Out the Strategies: Computing the Equilibria

For each of the described notions of response, we now consider the corresponding notion of equilibrium, derived at step ($\spadesuit$) in section 2.3. The relational fixed point operators are described in the Appendix.

A. Rationalizability and the Nash equilibrium. The Nash best response relations yield the system of n relations BR, which we write as

$$s\ BR\ t \iff \forall i \in n.\ s_{-i}\ BR_i\ t_i$$

It is not hard to see that the strong fixed point (see appendix) yields the solutions of that system, i.e.

$$BR^{\bullet}s \iff s\ BR\ s$$
$$\iff \forall i \in n.\ s_{-i}\ BR_i\ s_i$$

This, of course, means that s is a *Nash equilibrium* [14]. On the other hand, the weak fixed point extracts the transitive closure of BR, i.e. the smallest relation BR^* satisfying

$$BR^*s \iff \exists t.\ BR^*t \wedge tBRs$$

In game theory, the strategies $\{s_i \mid BR_i^*s_i\}$ are said to be *rationalizable* [5,17]. An s_i is rationalizable if and only if there is a rationalizable counterstrategy t_{-i} for which s_i is the best response.

B. Stable Strategies. Given

$$s\ SR\ t \iff \forall i \in n.\ s_{-i}\ SR_i\ t_i$$

the fixed point

$$
\begin{aligned}
SR^\bullet s \iff & \forall i \in n.\ s_{-i}\ SR_i\ s_i \\
\iff & \forall t \in A. \forall i \in n.\ \varrho_i(t_i, s_{-i}) \le \varrho_i(s_i, s_{-i}) \wedge \\
& \varrho_i(s_i, t_{-i}) = \varrho_i(s_i, s_{-i}) \Rightarrow \\
& \forall t_{-i} \in A_{-i}.\ \varrho_i(t_i, t_{-i}) \le \varrho_i(s_i, t_{-i})
\end{aligned}
$$

is an *evolutionary stable strategy*, which is a straightforward generalization of the concept due to biologist John Maynard-Smith [12][1]. On the other hand, the weak fixed point

$$SR^*s \iff \exists t. SR^*t \wedge tSRs$$

yields the new class of *stably rationalizable strategies*. Unfolding the above equivalence tells that s is stably rationalizable iff every s_i is the best response for some stably rationalizable t_{-i}, and *moreover*, whenever t_i is another best response to t_{-i}, as good as s_i, then s_i is at least as good as t_i with respect to the other counterstrategies.

C. Uniform equilibria and profiles.

$$
\begin{aligned}
UR^\bullet\ (s_i, s_{-i}) \iff & \forall i \in n.\ s_{-i}\ UR_i\ s_i \\
\iff & \forall i \in n.\ s_{-i}\ BR_i\ s_i\ \wedge \\
& \forall t_{-i} \in A_{-i}.\ s_i\ BR_{-i}\ t_{-i} \Rightarrow t_{-i}\ BR_i\ s_i
\end{aligned}
$$

where BR_{-i} is like in 3.1C. A uniform equilibrium s is thus a Nash equilibrium such that each its components s_i is a uniform move, in the sense that it lies in the set

$$U_i = \{s_i \in A_i \mid \forall t_{-i} \in A_{-i}.\ s_i\ BR_{-i}\ t_{-i} \Rightarrow t_{-i} BR_i s_i\}$$

---

[1] He considered the symmetric case, where all players have the same preferences and the same choice of actions.

A Nash equilibrium thus fails to be uniform whenever some opponent has an alternative best response. The uniformity of a response i-th player assures that it is the best response also with respect to such alternatives. In a sense, the uniformity requirement only eliminates the unreliable Nash equilibria from the search space.

The weak fixed point

$$UR^* \, s \iff \exists t.UR^*t \wedge tURs$$

yields the new class of *uniformly rationalizable strategies*. Unfolding the above equivalence tells that s is uniformly rationalizable iff every s_i is a uniform best response for some uniformly rationalizable t_{-i}.

D. Constructive equilibrium

$$CR^\bullet \, s \iff \forall i.s_{-i}CR_i s_i$$

As it stands, this equilibrium includes the Nash equilibria, *and* the fixed points of CR, chosen because they yield better payoff than the equilibria. While CR itself does not guarantee the feasilibity of any CR_i-response of a particular player, the CR-equilibrium does guarantee that all players have the same CR-justification.

Remark. The above characterizations of equilibria guarantee provide no existence guarantees: e.g., the set $BR^\bullet$ of the Nash equilibria, of course, always exists, but it can be empty. The existence, of course, requires additional side conditions, such as the convexity of the set of strategies [14].

Example. For Prisoners' Dilemma, both (c, c) and (d, d) are constructive equilibria. The former is unstable, since each player can improve her immediate payoff by defecting. This gain can be offset by the loss from retaliation, and can be irrational, especially if the value of (c, c) is much larger than the value of (d, d).

But the relational view of the strategic choices cannot express these quantitative considerations. In the next section, we explore a refinement where they can be expressed.

4   Strategies as Randomized Programs

In this section, we consider the framework where the preferences are quantified: the strategic choices are expressed as probability distributions over the available moves. A strategy is thus a randomized program.[2] — Is it possible to improve the rationality of strategic behaviors by quantifying the preferences, and biasing them more towards the more favorable moves?

---

[2] The payoff functions can also be viewed as randomized programs, capturing games that involve some form of *gambling*. But this leads to an essentially different type of game theory[9].

In the standard game theoretic reasoning, the payoffs are only used as a convenient way to express players' preference ordering. Indeed, any affine transformation of a payoff matrix represents the same game. In the present section, this is not the case any more. *We assume that all payoffs are non-negative*, and normalize them into probability distributions.

A. Best response distribution is just a normalization of the payoff function:

$$s_{-i} \ BD_i \ s_i = \frac{\varrho_i(s_i, s_{-i})}{\sum_{t_i \in A_i} \varrho_i(t_i, s_{-i})}$$

where $A_i \times A_{-i} \xrightarrow{BD_i} [0, 1]$ is viewed as a fuzzy relation $A_{-i} \xrightarrow{BD_i} A_i$. The idea is that $s_{-i} \ BD_i \ s_i$ (which can be viewed as the matrix entry in the row s_i and the column s_{-i}) records not only that s_i is the best response to s_{-i}, like BR_i did, i.e. not just that s_i is preferred to the other responses t_i; but $s_{-i} \ BD_i \ s_i$ also quantifies how much better is s_i than t_i, in terms of the difference $(s_{-i} \ BD_i \ s_i) - (s_{-i} \ BD_i \ t_i)$.

B. Stable response distribution measures not only how good is s_i as a response to s_{-i}, but also how good is it, on the average, with respect to the other countermoves t_{-i}.

$$s_{-i} \ SD_i \ s_i = \frac{\varrho_i(s_i, s_{-i}) \cdot \sum_{t_{-i} \in A_{-i}} \varrho_i(s_i, t_{-i})}{\sum_{t_i \in A_i} \varrho_i(t_i, s_{-i}) \cdot \sum_{t_{-i} \in A_{-i}} \varrho_i(t_i, t_{-i})}$$

Like in the case of the relational stable response, if s_i and t_i yield are equally good as responses to s_{-i}, then s_i remains a stable response if it is at least as good as t_i with respect to all other countermoves t_{-i}. Moreover, s_i will now remain stable even if it is not as good as t_i with respect to each other t_{-i}, but just if it is as good *on the average*. In fact, if s_i is much better than t_i on the average, the probability $s_{-i} \ SD_i \ s_i$ may be greater than $s_{-i} \ SD_i \ t_i$, even if $\varrho_i(s_i, s_{-i}) < \varrho_i(t_i, s_{-i})$.

C. Uniform response distribution multiplies the probability of a response s_i to s_{-i} by the payoffs from the response s_i to all other countermoves t_{-i}, averaged by the likelihood that t_{-i} may occur as the countermove against s_i, which is taken to be proportional with $\varrho_{-i}(s_i, t_{-i})$.

$$s_{-i} \ UD_i \ s_i = \frac{\varrho_i(s_i, s_{-i}) \cdot \sum_{t_{-i} \in A_{-i}} \varrho_i(s_i, t_{-i}) \cdot \varrho_{-i}(s_i, t_{-i})}{\sum_{t_i \in A_i} \varrho_i(t_i, s_{-i}) \cdot \sum_{t_{-i} \in A_{-i}} \varrho_i(t_i, t_{-i}) \cdot \varrho_{-i}(t_i, t_{-i})}$$

If it happens that s_i is a good response across all the best countermoves t_{-i} against it, then s_i is assigned a high uniform response probability. This was expressed in the uniform response relation too. The preference is now not only quantified, but also *smoothened out*, as to have a high uniform response probability as soon as is a good response to the likely countermoves just *on the average*.

D. Constructive response distribution To simplify **notation** for a sequence of values $\langle f(s, y) \rangle_{s \in A}$ renormalized into a probability distribution over A, we shall henceforth write

$$\lfloor f(s, y) \rfloor_s = \frac{f(s, y)}{\sum_{t \in A} f(t, y)}$$

The upshot is that we get $\sum_{s \in A} \lfloor f(s, y) \rfloor_s = 1$. The subscript s, denoting the renormalized variable, will be omitted when clear from the context.

We also write

$$a_+ = \left\{ \begin{array}{ll} a & \text{if } a > 0 \\ 0 & \text{otherwise} \end{array} \right\} = \frac{|a| + a}{2}$$

Now define

$$s_{-i} \, CD_i \, s_i = \left\lfloor \varrho_i(s_i, s_{-i}) \right.$$
$$+ \bigvee_{t_i \in A_i} \left(\varrho_i(t_i, s_{-i}) - \varrho_i(s_i, s_{-i}) \right)_+ \cdot$$
$$\sum_{t_{-i} \in A_{-i}} \left(\varrho_{-i}(t_i, t_{-i}) - \varrho_{-i}(t_i, s_{-i}) \right)_+ \cdot$$
$$\left. \left(\varrho_i(s_i, s_{-i}) - \varrho_i(t_i, t_{-i}) \right)_+ \right\rfloor_{s_i}$$

The idea behind constructive distribution is that the probabilistic weight of s_i as a response to s_{-i} is now increased to equal the weight of a t_i that may be a better response to s_{-i} alone, but for which there is a threat of the countermoves t_{-i}, which are better for the opponent than s_{-i}, but worse for the player.

Examples. Response distributions for Prisoners' Dilemma are now

$$BD_i = \begin{pmatrix} \frac{10}{21} & 0 \\ \frac{11}{21} & 1 \end{pmatrix} \qquad SD_i = \begin{pmatrix} \frac{25}{58} & 0 \\ \frac{33}{58} & 1 \end{pmatrix}$$

$$UD_i = \begin{pmatrix} \frac{1000}{1011} & 0 \\ \frac{11}{1011} & 1 \end{pmatrix} \qquad CD_i = \begin{pmatrix} \frac{19}{30} & 0 \\ \frac{11}{30} & 1 \end{pmatrix}$$

where the columns represent the opponent's moves c and d, the rows the player's own responses, and the entries the suggested probability for each response.

Stochastic equilibria. Stochastic response profiles are Markov chains, and the induced equilibria are their stationary distributions. Playing out the randomized response strategies and computing stochastic equilibria is thus placed in a rich and well ploughed field [16].

A stochastic Nash equilibrium is a uniform fixed point $BD^{\bullet} = Fix(BD)$, which can be computed as in Appendix B. Since each player participating in the profile BD responds by a mixed strategy where the frequency of a move is proportional to the payoff that it yields, the condition $BD^{\bullet} = BD \circ BD^{\bullet}$ means that $BD^{\bullet}$ *maximizes everyone's average payoff*. Formally, this is a consequence of the fact that BD is a stochastic matrix, and that 1 is its greatest eigenvalue, so that the images of the vectors in the eigenspace of 1 are of maximal length.

The stochastic equilibria $SB^{\bullet}$, $UB^{\bullet}$ and $CB^{\bullet}$ maximize players' average payoffs in a similar manner, albeit for more refined notions of averaging, captured by their more refined response distributions.

Back to the Dilemma. The strategies UD and CD above recommend cooperation as a better response to peer's cooperation. One might thus hope that, by taking into account the *average* payoffs, the stochastic approach may overcome the myopic rationality of *defection* as the equilibrium in Prisoners' Dilemma. Unfortunately, it is easy to see the only fixed point of any response disrribution in the form $RD = \begin{pmatrix} p & 0 \\ 1-p & 1 \end{pmatrix}$ is the vector $\begin{pmatrix} 0 \\ 1 \end{pmatrix}$, as soon as $p < 1$. Defection is the only equilibrium.

Let us try to understand why. Suppose that it is assured that both players play a constructive strategy. If they both assume that the other one will cooperate, each of them will cooperate at the first step with a probability $\frac{19}{30}$, which seems favorable. However, under the same assumption, the probability that either of them will cooperate at both of the first *two* steps is $\left(\frac{19}{30}\right)^2 = \frac{361}{900}$, which is not so favorable. And it exponentially converges to 0. With a static strategy repeated over and over, any probability p that the opponent will cooperate in one step leads to the probability p^n that he will cooperate in n steps, which becomes 0 in the long run. Trust and cooperation require memory and adaptation, which can be implemented in position games.

5   Position and Memory

The positions in a position game $A \times X \xrightarrow{\varrho} B \times X$ are recorded in the state space $X = \prod_{i \in m} X_i$, where the projection X_i shows what is visible to the player i. In games of perfect information, all of X is visible to all players. Even if the payoff function $A \times X \xrightarrow{\varrho_B} B$, the players can use the positions to adapt their strategies, and $A \times X \xrightarrow{\varrho_X} X$ should update the position as each move is made.

For instance, in Iterated Prisoners' Dilemma, each player chooses a sequence of moves $\sigma_i = \langle s_i^0, s_i^1, \ldots, s_i^n \rangle$ and collects at each step the payoff $\varrho(s_0^{\ell}, s_1^{\ell})$. But the moves s_i^{ℓ} can be chosen adaptively, taking into account the previous ℓ moves. These moves can be recorded as the position. E.g., set $X = (\{c, d\} \times \{c, d\})^*$, and besides the payoff bimatrix $\{c, d\}^2 \xrightarrow{\varrho_B} \mathbb{R}^2$, declare the position update

$\{c, d\}^2 \xrightarrow{\varrho_X} (\{c, d\}^2)^*$ to be the list function $\varrho_X(s, x) = s :: x$. What are the rational strategies now?

Axelrod reports about the Iterated Prisoners' Dilemma tournaments in [3]. E.g., one of the simplest and most successful strategies was *tit-for-tat.* It uses a rudimentary notion of position, recording just the last move: i.e., $X = \{c, d\}^2$, and $\{c, d\}^2 \xrightarrow{\varrho_X} \{c, d\}^2$ is the identity function. The tit-for-tat strategy $X \overset{TT_i}{\twoheadrightarrow} A_i$ is simply to repeat opponent's last move:

$$\langle x_0, x_1 \rangle \; TT_0 \; x_1 \qquad\qquad \langle x_0, x_1 \rangle \; TT_1 \; x_0$$

If both players stick with this strategy, then they will

- either forever cooperate, or forever defect — if they agree initially,
- or forever alternate — if they initially disagree.

Within an n-round iterated game, this is clearly not an equilibrium strategy, since each player can win that game by switching any c to d. However, both players' total gains in the game will be higher if they cooperate. That is why a cooperative strategy may be rational when the cumulative gains within a tournament are taken into account, while it may not be a rational way to win a single party of the same game, or to assure a higher payoff from a single move.

The upshot is that a game, viewed in strategic form, may thus lead to three completely different games, depending on whether the payoffs are recorded per move, or per n rounds against the same opponent, or per tournament against many opponents playing different strategies. While the different situations arguably determine different rationalities, which can be captured by different normal forms, the process view of a game, with the various positions through which it may evolve, displays not only the semantical relations between the different views of the same game, but also a dynamical view of adaptive strategies.

As a final example, consider a version of Iterated Prisoners' Dilemma, where the positions $X = \mathbb{R} \times \mathbb{R}$ record the cumulative gains of both players. The cumulative payoff function and the position update function thus happen to be identical, $\{c, d\}^2 \times \mathbb{R}^2 \xrightarrow{\varrho} \mathbb{R}^2$. To give the game a sense of the moment, let us assume that the gains are subject to a galloping inflation rate of 50% per round, i.e. that the cumulative payoffs are given by the bimatrix

$10 + \frac{x_1}{2}$ $10 + \frac{x_0}{2}$	$11 + \frac{x_0}{2}$ $\frac{x_0}{2}$
$\frac{x_1}{2}$ $11 + \frac{x_0}{2}$	$1 + \frac{x_1}{2}$ $1 + \frac{x_0}{2}$

where $x = \langle x_0, x_1 \rangle \in \mathbb{R}^2$ is the position, i.e. the previous gains. Suppose that the player uses the position-sensitive form of the constructive rationality

$$(s_{-i}, x)\, CD_i\ s_i = \left\lfloor \varrho_i(s_i, s_{-i}, x) \right.$$
$$+ \bigvee_{t_i \in A_i} \Big(\varrho_i(t_i, s_{-i}, x) - \varrho_i(s_i, s_{-i}, x) \Big)_+ \cdot$$
$$\sum_{t_{-i} \in A_{-i}} \Big(\varrho_{-i}(t_i, t_{-i}, \xi) - \varrho_{-i}(t_i, s_{-i}, \xi) \Big)_+ \cdot$$
$$\left. \Big(\varrho_i(s_i, s_{-i}, \xi) - \varrho_i(t_i, t_{-i}, \xi) \Big)_+ \right\rfloor_{s_i}$$

where $\xi = \varrho_X(t_i, s_{-i}, x)$ is the position reached after the profile (t_i, s_{-i}) is played at the position x. The response distribution is then

$$CD_i(x) = \begin{pmatrix} \frac{38+x}{60+2x} & \frac{x}{2+2x} \\ \frac{22+x}{60+2x} & \frac{2+x}{2+2x} \end{pmatrix}$$

Since x changes at each step, the profile CD is not a Markov chain any more. Its fixed point is cumbersome to compute explicitly, although it converges fast numerically. In any case, it is intuitively clear that the high inflation rate motivates the players to cooperate. If they do, the cumulative payoff for each of them approaches $\$10 \cdot \sum_{k=0}^{\infty} \frac{1}{2^k} = \$20$. If they both defect, their cumulative payoffs are $\$1 \cdot \sum_{k=0}^{\infty} \frac{1}{2^k} = \$2$. If they begin to cooperate and accumulate $\$x$ each, and then one defects, he will acquire an advantage of $\$11$ for that move. But after 10 further moves, with both players defecting, his advantage will reduce to about 1 cent, and the cumulative payoff for both players will again boil down to $\$2$.

6   Conclusions and Future Work

We explored the semantical approaches to gaming from three directions: through relational programming of strategies in section 3, through quantifying preferences in terms of distributions (rather than preorders) in section 4, and finally by taking into account the positions and the process aspects of gaming, in section 5.

The advantage of viewing strategies as programs is that they can be refined, together with the notion of rationality that they express. To illustrate this point, we discussed in section 3 some simple refinements of the standard equilibrium concepts.

The advantage of viewing strategies as *randomized* programs is that the problem of equilibrium selection [20] can be attacked by the Markov chain methods. Mixed strategies are, of course, commonly used in game theory. They assure the existence of Nash equilibria. The mixture is interpreted either as a mixed population of players, each playing a single strategy, or as the probability distribution with which the single player chooses a single strategy [12]. However, when equilibria are computed as stationary distributions of Markov chains, the mixture provides additional information which can be used to coordinate equilibrium selection. The concrete methods to extract and use this information need to be worked out in future research.

The most interesting feature of the semantical view of games is the dynamics of gaming, as it evolves from position to position. This feature has only been touched upon in the the present paper. On one hand, it leads beyond the Markovian realm, and equilibria are harder to compute. But on the other hand, in practice, the important rational solutions are often attained through genuinely adaptive, position sensitive strategies. The toy example of Prisoners' Dilemma already shows that a widely studied science of rationality may miss even the basic forms of social rationality because of small technical shortcomings. Combining semantics of computation and game theory may help eliminate them.

Acknowledgement. Through years, I have benefited from many conversations with Samson Abramsky, on a wide range of ideas about games and semantics.

References

1. Abramsky, S., Malacaria, P., Jagadeesan, R.: Full completeness for pcf. Information and Computation 163, 409–470 (2000)
2. Abramsky, S., McCusker, G.: Game semantics. In: Schwichtenberg, H., Berger, U. (eds.) Computational Logic: Proceedings of the 1997 Marktoberdorf Summer School, pp. 1–56. Springer, Heidelberg (1999)
3. Axelrod, R.M.: The evolution of cooperation. Basic Books, New York (1984)
4. Benton, N., Hyland, M.: Traced premonoidal categories. Informatique Théorique et Applications 37(4), 273–299 (2003)
5. Bernheim, D.B.: Rationalizable strategic behavior. Econometrica 52, 1007–1028 (1984)
6. Bloom, S.L., Ésik, Z.: Iteration theories: the equational logic of iterative processes. Springer-Verlag New York, Inc, New York (1993)
7. Borodin, A., El-Yaniv, R.: Online computation and competitive analysis. Cambridge University Press, New York (1998)
8. Crole, R.L., Pitts, A.M.: New foundations for fixpoint computations: FIX-hyperdoctrines and the FIX-logic. Inf. Comput. 98(2), 171–210 (1992)
9. Dubins, L.E., Savage, L.J.: How to Gamble If You Must. McGraw-Hill, New York (1965)
10. Hasegawa, M.: The uniformity principle on traced monoidal categories. Elec. Notes in Theor. Comp. Sci. 69, 1014 (2003)
11. Hyland, J.M.E., Ong, C.-H.L.: On full abstraction for pcf: i, ii, and iii. Inf. Comput. 163(2), 285–408 (2000)
12. Smith, J.M.: Evolution and the Theory of Games. Cambridge University Press, Cambridge (1982)
13. Moggi, E.: Notions of computation and monads. Inf. Comput. 93(1), 55–92 (1991)
14. Nash, J.: Non-cooperative games. The Annals of Mathematics 54(2), 286–295 (1951)

15. Nisan, N., Roughgarden, T., Tardos, E., Vazirani, V.V.: Algorithmic Game Theory. Cambridge University Press, New York (2007)
16. Norris, J.R.: Markov Chains. Cambridge Series in Statistical and Probabilistic Mathematics. Cambridge University Press, Cambridge (1998)
17. Pearce, D.: Rationalizable strategic behavior and the problem of perfection. Econometrica 52, 1029–1050 (1984)
18. Power, A.J., Thielecke, H.: Closed Freyd- and κ-categories. In: Wiedermann, J., Van Emde Boas, P., Nielsen, M. (eds.) ICALP 1999. LNCS, vol. 1644, pp. 625–634. Springer, Heidelberg (1999)
19. Power, J., Robinson, E.: Premonoidal categories and notions of computation. Mathematical. Structures in Comp. Sci. 7(5), 453–468 (1997)
20. Samuelson, L.: Evolutionary Games and Equilibrium Selection. Series on Economic Learning and Social Evolution. MIT Press, Cambridge (1997)
21. Simpson, A.K., Plotkin, G.D.: Complete axioms for categorical fixed-point operators. In: Proceedings of the Fifteenth Annual IEEE Symposium on Logic in Computer Science (LICS 2000), pp. 30–41. IEEE Computer Society Press, Los Alamitos (2000)
22. Tardos, É.: Network games. In: Babai, L. (ed.) STOC, pp. 341–342. ACM, New York (2004)
23. Wadler, P.: Monads for functional programming. In: Jeuring, J., Meijer, E. (eds.) AFP 1995. LNCS, vol. 925, pp. 24–52. Springer, Heidelberg (1995)

A   Appendix: Fixed Points in FRel

For a relation $A \times X \xrightarrow{R} A$ in the monoidal category $(\mathsf{FRel}, \times, 1)$, the standard fixed point operator (induced by its simple trace structure) gives $X \xrightarrow{R^\bullet} A$, defined

$$x R^\bullet a \iff (x, a) R a$$

On the other hand, the order structure of FRel induces another fixed point operator, where $X \xrightarrow{R^*} A$ is defined as the smallest relation satisfying $\langle id, R^* \rangle R = R^*$, i.e.

$$x R^* a \iff \exists c \in A.\ x R^* c \wedge (x, c) R a$$

For each x, the set $x R^* = \{a \mid x R^* a\}$ is just the image of the transitive closure of $A \xrightarrow{xR} A$. It can be defined inductively, as

$$x R^* a \iff \forall n \in \mathbb{N}.\ x R^n a, \text{ where}$$
$$x R^0 a \iff \exists a' \in A.\ (x, a') R a$$
$$x R^{n+1} a \iff \exists c \in A.\ x R^n c \wedge (x, c) R a$$

or in terms of the image $x R(C) = \{a \mid \exists c \in C.\ (x, c) R a\}$ and

$$x R^* = \bigcap_{n=1}^{\infty} x R^n(A)$$

If the containment order on $\wp A$ represents information, so the singletons $\{a\}$ are maxima, and A is the minimum, then the above intersection is the least upper bound, and R^* is the least fixed point. Indeed, the containment $R^\bullet \subseteq R^*$ means that in the information order $R^\bullet \sqsupseteq R^*$.

B   Appendix: Fixed Points in SRel

A stochastic matrix $k \times m \xrightarrow{H} k$ can be viewed as an m-tuple of square stochastic matrices $k \xrightarrow{H_i} k$. By the Perron-Frobenius theorem, each H_i has 1 as the principal eigenvalue. This can also be directly derived from the fact that the rows of each $H_i - I$ must be linearly dependent, since the sum of the entries of each of its columns is 0. The fixed vectors of each H_i thus lie in its eigenspace of 1. But this space may be of a high dimension. Which m-tuple of vectors is the *uniform* fixed point of H [8,21]?

The uniform fixed points arise from the trace operations [10,4]. Let $k \times m \xrightarrow{H^*} k$ be formed from the projectors $k \xrightarrow{H_i^*} k$ to the principal eigenspaces of $k \xrightarrow{H_i} k$. The uniform fixed point $m \xrightarrow{H^\bullet} k$ of $k \times m \xrightarrow{H} k$ can be obtained by tracing out k in $\widehat{H} = \langle H^*, H^* \rangle : k \times m \rightarrowtail k \times k$, defined by

$$\widehat{h}_{\langle v,w \rangle \langle u,i \rangle} = h^*_{v \langle u,i \rangle} h^*_{w \langle u,i \rangle}$$

The uniform fixed point is thus $H^\bullet = (h^\bullet_{ui})_{k \times m}$ where

$$h^\bullet_{ui} = \left\lfloor \sum_{w \in k} \widehat{h}_{\langle w,u \rangle \langle w,i \rangle} \right\rfloor = \frac{\sum_{w \in k} h^*_{w \langle w,i \rangle} h^*_{u \langle w,i \rangle}}{\sum_{v,w \in k} h^*_{w \langle w,i \rangle} h^*_{v \langle w,i \rangle}}$$

To check that this is a fixed point of H, i.e. that $H \langle H^\bullet, I \rangle = H^\bullet : m \rightarrowtail k$, note that $\widetilde{H} = \langle H^\bullet, I \rangle : m \rightarrowtail k \times m$ is

$$\widetilde{h}_{uji} = \left\{ \begin{array}{l} h^\bullet_{ui} \text{ if } j = i \\ 0 \quad \text{otherwise} \end{array} \right\} = \left\{ \begin{array}{l} \frac{\sum_{w \in k} h^*_{w \langle w,i \rangle} h^*_{u \langle w,i \rangle}}{\sum_{v,w \in k} h^*_{w \langle w,i \rangle} h^*_{v \langle w,i \rangle}} \text{ if } j = i \\ 0 \qquad\qquad\qquad \text{otherwise} \end{array} \right.$$

Now $H \widetilde{H} = H^\bullet : m \rightarrowtail k$ is satisfied iff

$$h^\bullet_{ui} = \sum_{\langle v,j \rangle \in k \times m} h_{u \langle v,j \rangle} \left\lfloor \sum_{w \in k} h^*_{w \langle w,i \rangle} h^*_{u \langle w,i \rangle} \right\rfloor = \left\lfloor \sum_{w \in k} h^*_{w \langle w,i \rangle} \sum_{v \in k} h_{u \langle v,i \rangle} h^*_{v \langle w,i \rangle} \right\rfloor$$

holds for each $i \in m, u \in k$. But this is valid because $\sum_{v \in k} h_{u \langle v,i \rangle} h^*_{v \langle w,i \rangle} = h^*_{v \langle w,i \rangle}$, i.e. $H_i H_i^* = H_i^*$ holds by the definition of H^*.

Van Kampen Colimits as Bicolimits in Span[*]

Tobias Heindel[1] and Paweł Sobociński[2]

[1] Abt. für Informatik und angewandte KW, Universität Duisburg-Essen, Germany
[2] ECS, University of Southampton, United Kingdom

Abstract. The exactness properties of coproducts in extensive categories and pushouts along monos in adhesive categories have found various applications in theoretical computer science, e.g. in program semantics, data type theory and rewriting. We show that these properties can be understood as a single universal property in the associated bicategory of spans. To this end, we first provide a general notion of Van Kampen cocone that specialises to the above colimits. The main result states that Van Kampen cocones can be characterised as exactly those diagrams in $\mathbb{C}$ that induce bicolimit diagrams in the bicategory of spans $Span_{\mathbb{C}}$, provided that $\mathbb{C}$ has pullbacks and enough colimits.

Introduction

The interplay between limits and colimits is a research topic with several applications in theoretical computer science, including the solution of recursive domain equations, using the coincidence of limits and colimits. Research on this general topic has identified several classes of categories in which limits and colimits relate to each other in useful ways; extensive categories [5] and adhesive categories [21] are two examples of such classes.

Extensive categories [5] have coproducts that are "well-behaved" with respect to pullbacks; more concretely, they are disjoint and universal. Extensivity has been used by mathematicians [4] and computer scientists [25] alike. In the presence of products, extensive categories are distributive [5] and thus can be used, for instance, to model circuits [28] or to give models of specifications [11]. Sets and topological spaces inhabit extensive categories while quasitoposes are not, in general, extensive [15].

Adhesive categories [20,21] have pushouts along monos that are similarly "well-behaved" with respect to pullbacks – they are instances of Van Kampen squares. Adhesivity has been used as a categorical foundation for double-pushout graph transformation [20,7] and has found several related applications [8,27]. Toposes are adhesive [22] but quasitoposes, in general, are not [14].

Independently of our work, Cockett and Guo proposed *Van Kampen* (VK) *colimits* [6] as a generalisation of Van Kampen squares. The main examples of VK-colimits include coproducts in extensive categories and pushouts along monos in adhesive categories. Another example is a strict initial object; moreover, in a

---

[*] Research partially supported by EPSRC grant EP/D066565/1.

Barr-exact category, any regular epimorphism is a VK-coequaliser of its kernel pair [12, Theorem 3.7(d)].

The definition of VK-colimits relies only on elementary notions of category theory. This feature, while attractive, obscures their relationship with other categorical concepts. More abstract characterisations exist for extensive and adhesive categories. For instance, a category $\mathbb{C}$ is extensive if and only if the functor $+ : \mathbb{C} \downarrow A \times \mathbb{C} \downarrow B \to \mathbb{C} \downarrow A + B$ is an equivalence for any $A, B \in \mathbb{C}$ [23,5]; adhesive categories can be characterised in a similar manner [21]. Our definition of VK-cocone will be of the latter kind, i.e. in terms of an equivalence of categories. We also provide an elementary characterisation in the spirit of Cockett and Guo.

This paper contains one central result: VK-cocones are those diagrams that are bicolimit diagrams when embedded in the associated bicategory of spans. This characterises "being Van Kampen" as a universal property. We believe that this insight captures and explains the essence of the various aforementioned well-behaved colimits studied in the literature.

Spans are known to theoretical computer scientists through the work of Katis, Sabadini and Walters [16] who used them to model systems with boundary, see also [10]. The bicategory of spans over $\mathbb{C}$ contains $\mathbb{C}$ via a canonical embedding $\Gamma : \mathbb{C} \to Span_{\mathbb{C}}$, that is the identity on objects and takes each arrow $C -f\to D$ of $\mathbb{C}$ to its *graph* $C \leftarrow\text{id}- C -f\to D$. Spans generalise partial maps [26]: those spans that have a monomorphism as their left leg, as well as relations[1]. Bicolimits are the canonical notion of colimit in a bicategory.

There is some interesting related recent work. Milius [25] showed that coproducts are preserved (as a lax-adjoint-cooplimit) in the 2-category of relations over an extensive category $\mathbb{C}$. Cockett and Guo [6] have investigated the general conditions under which partial map categories are join-restriction categories: roughly, certain colimits in the underlying category are required to be VK-cocones.

Structure of the paper. In §1 we isolate the relevant class of bicategories and recall the related notions. We also describe the bicategory of spans $Span_{\mathbb{C}}$. In §2 we give a definition of VK-cocones together with an elementary characterisation and several examples. In §3 we recall the definition of bicolimits and prove several technical lemmas that allow us to pass between related concepts in $\mathbb{C}$ and $Span_{\mathbb{C}}$. Our main characterisation theorem is proved in §4.

1   Preliminaries

Here we introduce background on bicategories [3] and some notational conventions. For the basic notions of category, functor and natural transformation, the reader is referred to [24]. Our focus is the bicategory of spans over a category $\mathbb{C}$ with a choice of pullbacks (cf. Example 3). In order to avoid unnecessary bookkeeping, we only consider bicategories[2] that strictly satisfy the identity axioms.

---

[1] In the presence of a factorisation system in $\mathbb{C}$.
[2] We have found the bicategory of spans easier to work with than the biequivalent 2-category [19].

Definition 1 (Strictly unitary bicategories). A *strictly unitary* (SU) *bicategory $\mathscr{B}$* consists of:

- a collection ob $\mathscr{B}$ of *objects*;
- for $A, B \in$ ob $\mathscr{B}$ a category $\mathscr{B}(A, B)$, the objects and arrows of which are called, respectively, the *arrows* and the *2-cells* of $\mathscr{B}$. Composition is denoted by $\circ$ and referred to as *vertical composition*. Given $(f\colon A \to B) \in \mathscr{B}(A, B)$, its identity 2-cell will be denoted $\iota_f\colon f \to f$. Each $\mathscr{B}(A, A)$ contains a special object $\mathrm{id}_A\colon A \to A$, called the *identity arrow*;
- for $A, B, C \in$ ob $\mathscr{B}$, a functor $c_{A,B,C}\colon \mathscr{B}(A, B) \times \mathscr{B}(B, C) \to \mathscr{B}(A, C)$ called *horizontal composition*. On objects, $c_{A,B,C}\langle f, g \rangle$ is written $g \circ f$, while on arrows $c_{A,B,C}\langle \gamma, \delta \rangle$ it is $\delta * \gamma$. For any $f\colon A \to B$ we have $\mathrm{id}_B \circ f = f = f \circ \mathrm{id}_A$;
- for $A, B, C, D \in$ ob $\mathscr{B}$, arrows $f\colon A \to B$, $g\colon B \to C$ and $h\colon C \to D$ an *associativity natural isomorphism* $\alpha_{A,B,C,D}(f, g, h)\colon h \circ (g \circ f) \to (h \circ g) \circ f$. It satisfies the coherence axioms: for any composable f, g, h, k, we have $\alpha_{f,\mathrm{id},g} = \iota_{g \circ f}$ and also that the following 2-cells are equal:

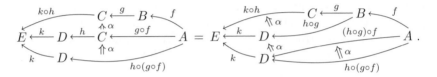

Example 2. Any (ordinary) category $\mathbb{C}$ is an (SU-)bicategory with trivial 2-cells.

Example 3 (Span bicategory [3]). Assume that $\mathbb{C}$ has a choice of pullbacks that preserves identities: for any cospan $X -f\to Z \leftarrow g- Y$ there exists an object $X \times_Z Y$ and span $X \leftarrow g'- X \times_Z Y -f'\to Y$ that together with f and g form a pullback square. Moreover if f is id_X then $X \times_Z Y = Y$ and $f' = \mathrm{id}_Y$.[3] *Span*$_{\mathbb{C}}$ has:

- as objects, the objects of $\mathbb{C}$, i.e. ob *Span*$_{\mathbb{C}} =$ ob $\mathbb{C}$;
- as arrows from A to B, the $\mathbb{C}$-spans $A \leftarrow l- U -r\to B$. We shall usually write $(l, r)\colon A \rightharpoonup B$ to denote such a span and refer to U as its *carrier*. When $l = \mathrm{id}$ and $r\colon A \to B$ we will write simply $(, r)\colon A \rightharpoonup B$.
  The composition with another span $B \leftarrow p- W -q\to C$ is obtained via the chosen pullback as illustrated to the right; however this composition is only associative up to canonical isomorphism. The identity on an object A is the span $A \leftarrow \mathrm{id}- A -\mathrm{id}\to A$.
- its 2-cells $\xi\colon (l, r) \to (l', r')$ are $\mathbb{C}$-arrows $\xi\colon U \to U'$ between the respective carriers such that $l' \circ \xi = l$ and $r' \circ \xi = r$.

For our purposes it suffices to consider *strict homomorphisms* between SU-bicategories.

---

[3] Note that this is a harmless assumption since the choice of the particular pullback diagram for any span is insignificant.

Definition 4 (Strict homomorphisms [3]). Let $\mathscr{A}$ and $\mathscr{B}$ be SU-bicategories. A *strict homomorphism* $\mathcal{F}: \mathscr{A} \to \mathscr{B}$ consists of a function $\mathcal{F}: \mathrm{ob}\,\mathscr{A} \to \mathrm{ob}\,\mathscr{B}$ and a family of functors $\mathcal{F}(A, B): \mathscr{A}(A, B) \to \mathscr{B}(\mathcal{F}A, \mathcal{F}B)$ such that:

(i) for all $A \in \mathscr{A}$, $\mathcal{F}(\mathrm{id}_A) = \mathrm{id}_{\mathcal{F}A}$;
(ii) for all $f: A \to B$, $g: B \to C$ in $\mathscr{A}$, $\mathcal{F}(g \circ f) = \mathcal{F}(g) \circ \mathcal{F}(f)$;
(iii) $\mathcal{F}\alpha_{A,B,C,D} = \alpha_{\mathcal{F}A,\mathcal{F}B,\mathcal{F}C,\mathcal{F}D}$.

Example 5. The following strict homomorphisms will be of interest to us:

– the covariant embedding $\Gamma: \mathbb{C} \to Span_{\mathbb{C}}$ which acts as the identity on objects and takes an arrow $f: C \to D$ to its graph $(\,, f): C \rightharpoonup D$;
– $\Gamma\mathcal{F}: \mathbb{J} \to Span_{\mathbb{C}}$ where $\mathcal{F}: \mathbb{J} \to \mathbb{C}$ is a functor;
– given an SU-bicategory $\mathscr{B}$ and $B \in \mathrm{ob}\,\mathscr{B}$, we shall abuse notation and denote the strict homomorphism from $\mathbb{J}$ to $\mathscr{B}$ which is constant at B by $\Delta_{\mathbb{J}}B$, often omitting the subscript $\mathbb{J}$. Note that in the case of $\mathscr{B} = Span_{\mathbb{C}}$, "$\Delta = \Gamma\Delta$".

Definition 6 (Lax transformations). Given strict homomorphisms $\mathcal{F}, \mathcal{G}: \mathscr{A} \to \mathscr{B}$ between SU-bicategories, a *(lax) transformation* consists of arrows $\kappa_A: \mathcal{F}A \to \mathcal{G}A$ for $A \in \mathscr{A}$ and 2-cells $\kappa_f: \mathcal{G}f \circ \kappa_A \Rightarrow \kappa_B \circ \mathcal{F}f$ for $f: A \to B$ in $\mathscr{A}$ such that:

$$
\begin{array}{ccc}
\mathcal{F}A & \xrightarrow{\mathcal{F}f} & \mathcal{F}B \\
{\scriptstyle\kappa_A}\downarrow & {\scriptstyle\kappa_f}\nearrow & \downarrow{\scriptstyle\kappa_B} \\
\mathcal{G}A & \xrightarrow[\mathcal{G}f]{} & \mathcal{G}B
\end{array}
$$

(i) $\kappa_{\mathrm{id}_A} = \iota_{\kappa_A}$ for each $A \in \mathscr{A}$;
(ii) for any $f: A \to B$, $g: B \to C$ in $\mathscr{A}$, the following 2-cells are equal:

A transformation is said to be *strong* when all the κ_f are invertible 2-cells. Given $B \in \mathscr{B}$ and a homomorphism $\mathcal{M}: \mathbb{J} \to \mathscr{B}$, a *pseudo-cocone* $\lambda: \mathcal{M} \to \Delta B$ is a synonym for a strong transformation $\lambda: \mathcal{M} \to \Delta B$.

Because bicategories have 2-cells, there are morphisms between transformations. They are called *modifications* and are defined as follows.

Definition 7 (Modifications [3,18]).
Given natural transformations κ, λ from $\mathcal{F}$ to $\mathcal{G}$, a modification $\Xi: \kappa \to \lambda$ consists of 2-cells $\Xi_A: \kappa_A \to \lambda_A$ for $A \in \mathscr{A}$ such that, for all $f: A \to B$ in $\mathscr{A}$, $\lambda_f \circ (\iota_{\mathcal{G}f} * \Xi_A) = (\Xi_B * \iota_{\mathcal{F}f}) \circ \kappa_f$.

$$
\kappa_A \left(\begin{array}{c}\mathcal{F}A \xrightarrow{\mathcal{F}f} \mathcal{F}B \\ {\scriptstyle\Xi_A}\Rightarrow \nearrow{\scriptstyle\lambda_A} \nearrow \\ \mathcal{G}A \xrightarrow[\mathcal{G}f]{} \mathcal{G}B\end{array}\right)_{\lambda_f}^{\lambda_B} = \kappa_A \left(\begin{array}{c}\mathcal{F}A \xrightarrow{\mathcal{F}f} \mathcal{F}B \\ {\scriptstyle\kappa_f}\nearrow \,\,\, \Rightarrow{\scriptstyle\Xi_B} \\ \mathcal{G}A \xrightarrow[\mathcal{G}f]{} \mathcal{G}B\end{array}\right)_{\kappa_B}^{\lambda_B}.
$$

Composition is componentwise, the identity modification on κ is $I_\kappa = \{\iota_{\kappa_A}\}_{A \in \mathscr{A}}$.

Given SU-bicategories $\mathscr{A}$ and $\mathscr{B}$, let $\mathrm{Hom}_l\,[\mathscr{A}, \mathscr{B}]$ denote the SU-bicategory of homomorphisms, lax transformations and modifications. Let $\mathrm{Hom}\,[\mathscr{A}, \mathscr{B}]$ denote the corresponding SU-bicategory with arrows the strong transformations.

2   Van Kampen Cocones

Here we give the definition of Van Kampen cocones together with an elementary characterisation. Let us consider coproducts as a motivating example. A coproduct diagram $A -i_1\to A + B \leftarrow i_2- B$ in a category $\mathbb{C}$ is a cocone of the two-object diagram $\langle A, B \rangle$. If $\mathbb{C}$ has pullbacks along coproduct injections then each $x: X \to A + B$ gives rise to $i_1{}^*x: i_1{}^*X \to A$ and $i_2{}^*x: i_2{}^*X \to B$ by pulling back along i_1 and i_2, respectively. Then $x \mapsto \langle i_1{}^*x, i_2{}^*x \rangle$ defines the functor $\langle i_1{}^*\_, i_2{}^*\_ \rangle: (\mathbb{C} \downarrow A + B) \to (\mathbb{C} \downarrow A \times \mathbb{C} \downarrow B)$ on objects. The coproduct satisfies the properties expected in an extensive category when this functor is an equivalence (see [5]).

The situation readily generalises as follows: replace $\langle i_1, i_2 \rangle$ by any cocone $\kappa: \mathcal{D} \to \Delta A$ from a functor $\mathcal{D}: \mathbb{J} \to \mathbb{C}$ to an object A in a category $\mathbb{C}$ with (enough) pullbacks. Any arrow $x: X \to A$ induces a natural transformation $\Delta x: \Delta X \to \Delta A$ and since also $\kappa: \mathcal{D} \to \Delta A$ is a natural transformation, the former can be pulled back along the latter in the functor category $[\mathbb{J}, \mathbb{C}]$ yielding a natural transformation $\kappa^*(\Delta x): \kappa^*(\Delta X) \to \mathcal{D}$.

$$
\begin{array}{ccc}
X & \quad & \Delta X \longleftarrow \kappa^*(\Delta X) \\
{\scriptstyle x}\downarrow & & {\scriptstyle \Delta x}\downarrow \quad\quad \downarrow{\scriptstyle \kappa^*(\Delta x)} \\
A & & \Delta A \longleftarrow_{\kappa} \mathcal{D}
\end{array}
$$

The described operation extends to a functor $\kappa^*(\Delta\_)$ from $\mathbb{C} \downarrow A$ to (a full subcategory of) $[\mathbb{J}, \mathbb{C}] \downarrow \mathcal{D}$ using the universal property of pullbacks; it maps morphisms with codomain A to *cartesian transformations* with codomain $\mathcal{D}$.

Definition 8 (Cartesian transformations). Let $\mathcal{E}, \mathcal{D} \in [\mathbb{J}, \mathbb{C}]$ be functors and let $\tau: \mathcal{E} \to \mathcal{D}$ be a natural transformation. Then τ is a *cartesian (natural) transformation* if all naturality squares are pullback squares, i.e. if the pair $\mathcal{E}_i \leftarrow\tau_i- \mathcal{D}_i -\mathcal{D}_u\to \mathcal{D}_j$ is a pullback of $\mathcal{E}_i -\mathcal{E}_u\to \mathcal{E}_j \leftarrow\tau_j- \mathcal{D}_j$ for all $u: i \to j$ in $\mathbb{J}$.

$$
\begin{array}{ccc}
\mathcal{E}_i & \xrightarrow{\mathcal{E}_u} & \mathcal{E}_j \\
{\scriptstyle \tau_i}\downarrow & & \downarrow{\scriptstyle \tau_j} \\
\mathcal{D}_i & \xrightarrow{\mathcal{D}_u} & \mathcal{D}_j
\end{array}
$$

Let $[\mathbb{J}, \mathbb{C}] \downarrow \mathcal{D}$ be the slice category over $\mathcal{D}$, which has natural transformations with codomain $\mathcal{D}$ as objects. Let $[\mathbb{J}, \mathbb{C}] \Downarrow \mathcal{D}$ denote the full subcategory of $[\mathbb{J}, \mathbb{C}] \downarrow \mathcal{D}$ with the cartesian transformations as objects.

Definition 9 (Van Kampen cocones). Let $\mathbb{C}$ be any category, let $\mathcal{D}: \mathbb{J} \to \mathbb{C}$ be a functor, and let $\kappa: \mathcal{D} \to \Delta_{\mathbb{J}} A$ be a cocone such that pullbacks along each coprojection κ_i exist $(i \in \mathbb{J})$. Then κ is *Van Kampen* (VK) if the functor $\kappa^*(\Delta_{\mathbb{J}}\_): \mathbb{C} \downarrow A \to [\mathbb{J}, \mathbb{C}] \Downarrow \mathcal{D}$ is an equivalence of categories.

Extensive and adhesive categories have elementary characterisations that are special cases of the following.

Proposition 10 (Elementary VK characterisation). *Suppose that $\mathbb{C}$ has pullbacks and $\mathbb{J}$-colimits, $\mathcal{D}: \mathbb{J} \to \mathbb{C}$ is a functor and $\kappa: \mathcal{D} \to \Delta_{\mathbb{J}} A$ a cocone such that $\mathbb{C}$ has pullbacks along κ_i $(i \in \mathbb{J})$. Then $\kappa: \mathcal{D} \to \Delta_{\mathbb{J}} A$ is Van Kampen iff for every cartesian transformation $\tau: \mathcal{E} \to \mathcal{D}$, arrow $x: X \to A$ and cocone $\beta: \mathcal{E} \to \Delta_{\mathbb{J}} X$ such that $\kappa \circ \tau = \Delta x \circ \beta$, the following are equivalent:*

(i) $\beta: \mathcal{E} \to \Delta_{\mathbb{J}} X$ is a $\mathbb{C}$-colimit;
(ii) $\mathcal{D}_i \leftarrow\tau_i- \mathcal{E}_i -\beta_i\to X$ is a pullback of $\mathcal{D}_i -\kappa_i\to A \leftarrow x- X$ for all $i \in \mathbb{J}$.   □

Cockett and Guo's [6] definition of Van Kampen colimits is the equivalence of (i) and (ii) in our Proposition 10. The two definitions are thus very close and coincide in the presence of the relevant pullbacks and colimits.

Remark 11. Under the assumptions of Proposition 10, any Van Kampen cocone $\kappa \colon \mathcal{D} \to \Delta A$ is a colimit diagram of $\mathcal{D}$ in $\mathbb{C}$ (take $\tau = \mathrm{id}_{\mathcal{D}}$ and $x = \mathrm{id}_A$).

Example 12. The following well-known concepts are examples of VK-cocones:

   (i) a strict initial object is a VK-cocone for the functor from the empty category;

   (ii) a coproduct diagram in an extensive category [5] is a VK-cocone for a functor from the discrete two object category;

A VK-cocone from a span is what has been called a *Van Kampen square* [21].

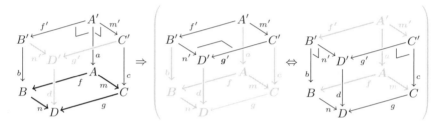

Fig. 1. Van Kampen square

Example 13 (Van Kampen square). A commutative square $B \xleftarrow{f} A \xrightarrow{m} C$, $B \xrightarrow{n} D \xleftarrow{g} C$ is *Van Kampen* when for each commutative cube as illustrated in Figure 1 on the left that has pullback squares as rear faces, its *top face is a pushout square* if and only if its *front faces are pullback squares* (cf. Figure 1).

In the left hand diagram in Figure 1, the two arrows $B \xleftarrow{f} A \xrightarrow{m} C$ describe a diagram from the three object category $\cdot \leftarrow \cdot \to \cdot$, and the cospan $B \xrightarrow{n} D \xleftarrow{g} C$ gives a cocone for this diagram. That the back faces are pullback squares means that we have a cartesian transformation from $B' \xleftarrow{f'} A' \xrightarrow{m'} C$ to $B \xleftarrow{f} A \xrightarrow{m} C$.

Adhesive categories are thus precisely categories with pullbacks in which pushouts along monomorphisms exist and are VK-cocones.

3   Van Kampen Cocones in a Span Bicategory

We begin the study of VK-cocones in the span bicategory by explaining roughly how Van Kampen squares induce bipushout squares in the bicategory of spans via the embedding Γ. An illustration of this is given in Figure 2.

At the base of Figure 2(a) is (the image of) a $\mathbb{C}$-span $B \xleftarrow{f} A \xrightarrow{m} C$ in $\mathit{Span}_{\mathbb{C}}$, i.e. a span of spans. Further, if two spans $(b, b') \colon B \to E$ and $(c, c') \colon C \to E$ are a pseudo-cocone for $B \xleftarrow{(,f)} A \xrightarrow{(,m)} C$ then taking pullbacks of b along f and

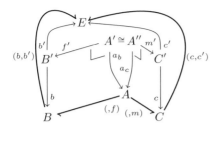
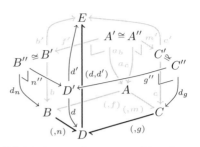

(a) $\mathcal{S}pan_{\mathbb{C}}$-cocone of the $\mathbb{C}$-span $B \leftarrow f- A -m\rightarrow C$

(b) Mediating $\mathcal{S}pan_{\mathbb{C}}$-morphism from the $\mathbb{C}$-cocone $B -n\rightarrow D \leftarrow g- C$

Fig. 2. Cocones and mediating morphisms consisting of spans

c along m (in $\mathbb{C}$) yields isomorphic objects over A, say a_b and a_c; as a result we obtain two pullback squares that will be the back faces of a commutative cube.

Next, let the bottom of Figure 2(b) be (the image of) a commuting $\mathbb{C}$-square, thus yielding another pseudo-cocone of $B \leftarrow(,f)- A -(,m)\rightarrow C$, namely $B -(,n)\rightarrow D \leftarrow(,g)- C$. If there is a mediating morphism $(d,d')\colon D \rightharpoonup E$ from the latter cocone to the one of Figure 2(a) $(B -(b,b')\rightharpoonup E \leftarrow(c,c')- E)$ then pulling back d along n and g results in morphisms which are isomorphic to b and c, respectively, say d_n and d_g; the resulting pullback squares provide the front faces of a cube.

Now, if $B -n\rightarrow D \leftarrow g-C$ is a VK-cocone of $B \leftarrow f- A -m\rightarrow C$ then such a mediating morphism can be constructed by taking D' as the pushout of B' and C' over either one of A' or A''. The morphisms $d\colon D' \rightarrow E$ and $d'\colon D' \rightarrow D$ arise from the universal property of pushouts, everything commutes and the front faces are pullback squares because of the VK-property. Further this mediating morphism is essentially unique, which means that given any other span $(e,e')\colon D \rightharpoonup E$ such that both $(b,b') \cong (e,e') \circ (,n)$ and $(c,c') \cong (e,e') \circ (,g)$ hold, the two spans (e,e') and (d,d') are isomorphic via a unique isomorphism.

Though this sketch lacks relevant technical details, it nevertheless may suffice to convey the flavor of the diagrams that are involved in the proof of the fact that Van Kampen squares in $\mathbb{C}$ induce bipushouts in $\mathcal{S}pan_{\mathbb{C}}$. Moreover, also the converse holds, i.e. if the image of a pushout is a bipushout in $\mathcal{S}pan_{\mathbb{C}}$ then it is a Van Kampen square.

3.1 Span Bicolimits

Clearly any diagram in $\mathcal{S}pan_{\mathbb{C}}$ can be "decomposed" into a diagram in $\mathbb{C}$: each arrow in $\mathcal{S}pan_{\mathbb{C}}$ gives two $\mathbb{C}$-arrows from a carrier object; further a 2-cell in $\mathcal{S}pan_{\mathbb{C}}$ is an $\mathbb{C}$-arrow between the carriers satisfying certain commutativity requirements.

We shall start with further observations along these lines. Roughly we are able to "drop a dimension" in the following sense. First, it is easy to see that $[\mathbb{J}, \mathbb{C}]$ inherits a choice of pullbacks from $\mathbb{C}$. In particular, it follows that $\mathcal{S}pan_{[\mathbb{J},\mathbb{C}]}$ is an SU-bicategory. Now, given $\mathcal{F}, \mathcal{G} \in [\mathbb{J}, \mathbb{C}]$ we note that:

- spans of natural transformations from $\mathcal{F}$ to $\mathcal{G}$ correspond to lax transformations from $\Gamma\mathcal{F}$ to $\Gamma\mathcal{G}$; and
- morphisms of such spans are the counterpart of modifications.

The following lemma makes this precise.

Lemma 14. *There is a strict homomorphism*

$$\Gamma\colon Span_{[\mathbb{J},\mathbb{C}]} \to \mathrm{Hom}_l\,[\mathbb{J}, Span_{\mathbb{C}}]$$

that takes $\mathcal{F} \in [\mathbb{J}, \mathbb{C}]$ to $\Gamma\mathcal{F}$ and is full and faithful on both arrows and 2-cells.

Proof (proof sketch). Here we only give the definition of Γ as checking the details involves tedious calculations.

A span of natural transformations $(\varphi, \psi)\colon \mathcal{F} \rightharpoonup \mathcal{G}$ with carrier $\mathcal{H}$ is mapped to a lax transformation $\Gamma_{\mathcal{F},\mathcal{G}}(\varphi, \psi)$ as follows: for each $i \in \mathbb{J}$, we put $\kappa_i := (\varphi_i, \psi_i)\colon \mathcal{F}_i \rightharpoonup \mathcal{G}_i$, and for each morphism $u\colon i \to j$ in $\mathbb{J}$, we define a 2-cell $\kappa_u\colon (,\mathcal{G}_u) \circ \kappa_i \to \kappa_j \circ (,\mathcal{F}_u)$ as sketched to the right. More explicitly, using that $\mathcal{F}_u \circ \varphi_i = \varphi_j \circ \mathcal{H}_u$ holds by naturality of φ, the arrow $\kappa_u\colon \mathcal{H}_i \to \mathcal{F}_i \times_{\mathcal{F}_j} \mathcal{H}_j$ is the unique one satisfying $\varphi_i = \pi_1 \circ \kappa_u$ and $\mathcal{H}_u = \pi_2 \circ \kappa_u$. To check that

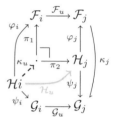

κ_u is a 2-cell it remains to check the equation $\psi_j \circ \pi_2 \circ \kappa_u = \psi_j \circ \mathcal{H}_u = \mathcal{G}_u \circ \psi_i$, which follows by the naturality of ψ.

Further, a 2-cell between spans $(\varphi, \psi), (\varphi', \psi')\colon \mathcal{F} \rightharpoonup \mathcal{G}$ with respective carriers $\mathcal{H}, \mathcal{H}'$ is a natural transformation $\xi\colon \mathcal{H} \to \mathcal{H}'$ satisfying both $\varphi' \circ \xi = \varphi$ and $\psi' \circ \xi = \psi$. This induces a modification $\{\xi_i\}_{i \in \mathbb{J}}\colon \Gamma(\varphi, \psi) \to \Gamma(\varphi', \psi')$. $\square$

Corollary 15. *For any functor $\mathcal{F} \in [\mathbb{J}, \mathbb{C}]$, the strict homomorphism Γ defines a natural isomorphism between the following two functors of type $[\mathbb{J}, \mathbb{C}] \to \mathbf{Cat}$:*

$$Span_{[\mathbb{J},\mathbb{C}]}(\mathcal{F}, \_) \cong \mathrm{Hom}_l\,[\mathbb{J}, Span_{\mathbb{C}}](\Gamma\mathcal{F}, \Gamma\_).$$

$\square$

The above lemma and corollary can be adapted to talk about strong transformations instead of lax ones (this will recur when we discuss bicolimits formally). This restriction to strong transformations has a counterpart on the other side of the isomorphism of Corollary 15: we need to restrict to those spans in $Span_{[\mathbb{J},\mathbb{C}]}(\mathcal{F}, \mathcal{G})$ that have a cartesian transformation from the carrier to $\mathcal{F}$.

Recall that a cartesian transformation between functors is a natural transformation with all naturality squares pullbacks (cf. Definition 8). It is an easy exercise to show that cartesian natural transformations include all natural isomorphisms and are closed under pullback. Hence – in a similar way as one can restrict the span bicategory to all partial map spans, i.e. those with the left leg monic – we let $Span_{[\mathbb{J},\mathbb{C}]}^{\leftharpoonup}$ be the (non-full) sub-bicategory of $Span_{[\mathbb{J},\mathbb{C}]}$ that has all those spans $(\varphi, \psi)\colon \mathcal{F} \rightharpoonup \mathcal{G}$ in $Span_{[\mathbb{J},\mathbb{C}]}(\mathcal{F}, \mathcal{G})$ as arrows of which the left leg φ is a cartesian transformation. Adapting the proof of Lemma 14, one obtains the following proposition.

Proposition 16. *There is a strict homomorphism* $\Gamma\colon Span^{\leftarrow}_{[\mathbb{J},\mathbb{C}]} \to \mathrm{Hom}\,[\mathbb{J}, Span_{\mathbb{C}}]$ *which is full and faithful on both arrows and 2-cells. For any functor* $\mathcal{F} \in [\mathbb{J}, \mathbb{C}]$, Γ *defines a natural isomorphism between the following functors* $[\mathbb{J}, \mathbb{C}] \to \mathbf{Cat}$:

$$Span^{\leftarrow}_{[\mathbb{J},\mathbb{C}]}(\mathcal{F}, \_) \;\cong\; \mathrm{Hom}\,[\mathbb{J}, Span_{\mathbb{C}}](\Gamma\mathcal{F}, \Gamma\_). \qquad\qquad \square$$

The above lets us pass between diagrams in $Span_{\mathbb{C}}$ and $\mathbb{C}$: for example the strong transformations of homomorphisms to $Span_{\mathbb{C}}$ are those spans of natural transformations of functors to $\mathbb{C}$ that have a cartesian first leg; the modifications of the former are the morphisms of spans of the latter. This observation will be useful when relating the notion of bicolimit in $Span_{\mathbb{C}}$ with the notion of VK-cocone in $\mathbb{C}$.

An elementary definition of bicolimits. For our purposes we need to recall only the definition of (conical) bicolimits [17] for functors with an (ordinary) small category $\mathbb{J}$ as domain. Given a homomorphism $\mathcal{M}\colon \mathbb{J} \to \mathscr{B}$, a bicolimit of $\mathcal{M}$ is an object $\mathrm{bicol}\,\mathcal{M} \in \mathscr{B}$ with a pseudo-cocone $\kappa\colon \mathcal{M} \to \Delta(\mathrm{bicol}\,\mathcal{M})$ such that "pre-composition" with κ gives an equivalence of categories

$$\mathscr{B}(\mathrm{bicol}\,\mathcal{M}, X) \simeq \mathrm{Hom}\,[\mathbb{J}, \mathscr{B}](\mathcal{M}, \Delta X) \qquad\qquad (1)$$

that is natural in X (i.e. the right hand side is essentially representable as a functor $\lambda X. \mathrm{Hom}\,[\mathbb{J}, \mathscr{B}](\mathcal{M}, \Delta X)\colon \mathscr{B} \to \mathbf{Cat}$); the pair $\langle \mathrm{bicol}\,\mathcal{M}, \kappa \rangle$ is referred to as *the bicolimit of* $\mathcal{M}$. We will often speak of $\kappa\colon \mathcal{M} \to \Delta\mathrm{bicol}\,\mathcal{M}$ as a bicolimit without mentioning the pair $\langle \mathrm{bicol}\,\mathcal{M}, \kappa \rangle$ explicitly.

To make the connection with the elementary characterisation of Van Kampen cocones in Proposition 10, we shall use the fact that equivalences of categories can be characterised as full, faithful functors that are essentially surjective on objects, and work with the following equivalent, elementary definition.

Definition 17 (Bicolimits). Given an su-bicategory $\mathscr{B}$, a category $\mathbb{J}$ and a strict homomorphism $\mathcal{M}\colon \mathbb{J} \to \mathscr{B}$, a *bicolimit* for $\mathcal{M}$ consists of:

- an object $\mathrm{bicol}\,\mathcal{M} \in \mathscr{B}$;
- a pseudo-cocone $\kappa\colon \mathcal{M} \to \Delta\mathrm{bicol}\,\mathcal{M}$: for each object $i \in \mathbb{J}$ an arrow $\kappa_i\colon \mathcal{M}_i \to \mathrm{bicol}\,\mathcal{M}$, and for each $u\colon i \to j$ in $\mathbb{J}$ an invertible 2-cell $\kappa_u\colon \kappa_i \to \kappa_j \circ \mathcal{M}_u$ satisfying the axioms required for κ to be a strong transformation.

The bicolimit satisfies the following universal properties.

(i) essential surjectivity:
  for any pseudo-cocone $\lambda\colon \mathcal{M} \to \Delta X$, there exists $h\colon \mathrm{bicol}\,\mathcal{M} \to X$ in $\mathscr{B}$ and an invertible modification $\Theta\colon \lambda \to \Delta h \circ \kappa$; and

(ii) fullness and faithfullness:
  for any $h, h'\colon \mathrm{bicol}\,\mathcal{M} \to X$ in $\mathscr{B}$ and each modification $\Xi\colon \Delta h \circ \kappa \to \Delta h' \circ \kappa$, there is a unique 2-cell $\xi\colon h \to h'$ satisfying $\Xi = \Delta\xi * I_\kappa$ (and hence ξ is invertible iff Ξ is).

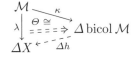

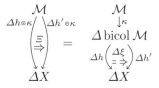

The pair $\langle h, \Theta \rangle$ of Condition (i) is called a *mediating cell* from κ to λ.

Condition (ii) of this definition implies that mediating cells from a bicolimit to a pseudo-cocone are *essentially unique*: any two such mediating cells $\langle h, \Theta \rangle$ and $\langle h', \Theta' \rangle$ are isomorphic since $\Theta' \circ \Theta^{-1} \colon \Delta h \circ \kappa \to \Delta h' \circ \kappa$ corresponds to a unique invertible 2-cell $\zeta \colon h \to h'$ such that $\Theta' \circ \Theta^{-1} = \Delta\zeta * I_\kappa$.

To relate the notion of bicolimit with the characterisation of VK-cocones of Proposition 10, we shall reformulate the above elementary definition. Given a pseudo-cocone $\kappa \colon \mathcal{M} \to \Delta C$, a morphism $h \colon C \to D$ will be called *universal for κ* or *κ-universal* if, given any other morphism $h' \colon C \to D$ with a modification $\Xi \colon \Delta h \circ \kappa \to \Delta h' \circ \kappa$, there exists a *unique* 2-cell $\xi \colon h \to h'$ satisfying $\Xi = \Delta\xi * I_\kappa$; further, a mediating cell $\langle h, \Theta \rangle$ is called universal, if the morphism h is universal. The motivation behind this terminology and the slightly redundant statement of the following proposition will become apparent in §4; its proof is straightforward.

Proposition 18. *A pseudo-cocone $\kappa \colon \mathcal{M} \to \Delta C$ from a diagram $\mathcal{M}$ to C is a bicolimit iff both of the following hold:*

(i) *for any pseudo cocone $\lambda \colon \mathcal{M} \to \Delta D$ there is a universal mediating cell $\langle h \colon C \to D, \Theta \rangle$ from κ to λ;*

(ii) *all arrows $h \colon C \to D$ are universal for κ.* □

We are interested in bicolimits of strict homomorphisms of the form $\Gamma\mathcal{F}$ where $\mathcal{F} \colon \mathbb{J} \to \mathbb{C}$ is a functor and $\Gamma \colon \mathbb{C} \to Span_\mathbb{C}$ is the covariant embedding of $\mathbb{C}$. The defining equivalence of bicolimits in (1) specialises as follows:

$$Span_\mathbb{C}(\text{bicol } \Gamma\mathcal{F}, X) \simeq \mathrm{Hom}\,[\mathbb{J}, Span_\mathbb{C}](\Gamma\mathcal{F}, \Delta X).$$

Using Proposition 16, this is equivalent to:

$$Span_\mathbb{C}(\text{bicol } \Gamma\mathcal{F}, X) \simeq Span_{[\mathbb{J},\mathbb{C}]}^{\Leftarrow}(\mathcal{F}, \Delta X).$$

We shall exploit working in $Span_{[\mathbb{J},\mathbb{C}]}^{\Leftarrow}$ in the following lemma which relates the concepts involved in the elementary definition of bicolimits with diagrams in $\mathbb{C}$. It will serve as the technical backbone of our main theorem.

Lemma 19 (Mediating cells and universality for spans).

Let $\kappa \colon \mathcal{F} \to \Delta C$ be a cocone in $\mathbb{C}$ of a diagram $\mathcal{F} \in [\mathbb{J}, \mathbb{C}]$, and let $\lambda \colon \Gamma\mathcal{F} \to \Delta D$ be a pseudo-cocone in $Span_\mathbb{C}$ where $\lambda_i = (\varphi_i, \psi_i)$ for all $i \in \mathbb{J}$:

(i) *to give a mediating cell*

$$\langle C \xleftarrow{h_1} H \xrightarrow{h_2} D, \Theta \colon \lambda \to \Delta(h_1, h_2) \circ \Gamma\kappa \rangle$$

from $\Gamma\kappa$ to λ is to give a cocone $\vartheta \colon \mathcal{H} \to \Delta H$ where $\mathcal{H}$ is the carrier functor of the image of λ in $Span_{[\mathbb{J},\mathbb{C}]}^{\Leftarrow}(\mathcal{F}, \Delta D)$ (cf. Proposition 16) such that the resulting three-dimensional diagram (†) in $\mathbb{C}$ (to the right) commutes and its lateral faces $\begin{smallmatrix}\mathcal{H}_i \\ \downarrow \\ \mathcal{F}_i\end{smallmatrix} \begin{smallmatrix}\to \\ \Box \\ \to\end{smallmatrix} \begin{smallmatrix}H \\ \downarrow \\ C\end{smallmatrix}$ are pullbacks;

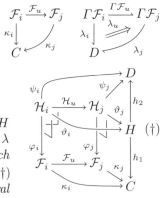

(ii) to give a modification $\Xi\colon \Delta(h_1, h_2)\circ\Gamma\kappa \to \Delta(h_1', h_2')\circ\Gamma\kappa$ for a pair of spans $(h_1, h_2), (h_1', h_2')\colon C \to D$ is to give a cartesian transformation $\Xi\colon \mathcal{F} \times_{\Delta C} \Delta H \to \mathcal{F} \times_{\Delta C} \Delta H'$ such that the two equations $\pi_1' \circ \Xi = \pi_1$ and $(\Delta h_2') \circ \pi_2' \circ \Xi = (\Delta h_2) \circ \pi_2$ hold.

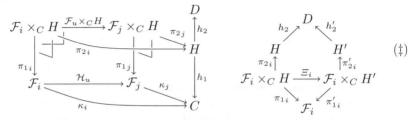

$(\ddagger)$

Here $\mathcal{F} \xleftarrow{\pi_1} \mathcal{F} \times_{\Delta C} \Delta H \xrightarrow{\pi_2} \Delta H$ is the pullback of $\mathcal{F} \xrightarrow{\kappa} \Delta C \xleftarrow{\Delta h_1} \Delta H$ as sketched in ($\ddagger$) above and similarly for $h'\colon H' \to C$.

Further, to give a cell $\xi\colon (h_1, h_2) \to (h_1', h_2')$ that satisfies $\Delta\xi * I_{\Gamma\kappa} = \Xi$ is to give a $\mathbb{C}$-arrow $\xi\colon H \to H'$ which satisfies the three equations $h_1' \circ \xi = h_1$, $h_2' \circ \xi = h_2$ and $\Delta\xi \circ \pi_2 = \pi_2' \circ \Xi$;

(iii) given a span $(h_1, h_2)\colon C \to D$, if the pullback of κ along h_1 is a colimit, i.e. if $\pi_2\colon \mathcal{F} \times_{\Delta C} \Delta H \to \Delta H$ is a colimit, then (h_1, h_2) is universal for $\Gamma\kappa$;

(iv) conversely, if (h_1, h_2) is universal for $\Gamma\kappa$, then $\pi_2\colon \mathcal{F} \times_{\Delta C} \Delta H \to \Delta H$ is a colimit – provided that some colimit of $\mathcal{F} \times_{\Delta C} \Delta H$ exists in $\mathbb{C}$.

Proof. Both (i) and (ii) are immediate consequences of Proposition 16.

As for (iii), we need to show that every modification $\Xi\colon \Delta(h_1, h_2) \circ \Gamma\kappa \to \Delta(h_1', h_2') \circ \Gamma\kappa$ is equal to $\Delta\xi * I_{\Gamma\kappa}$ for a unique $\xi\colon (h_1, h_2) \to (h_1', h_2')$. By (ii), Ξ is a natural transformation $\Xi\colon \mathcal{F} \times_{\Delta C} \Delta H \to \mathcal{F} \times_{\Delta C} \Delta H'$. Then, by naturality of Ξ, we have that $\pi_{2j}' \circ \Xi_j \circ (\mathcal{F}_u \times_{\Delta C} \Delta H) = \pi_{2i}' \circ \Xi_i$ holds for all $u\colon i \to j$ in $\mathbb{J}$, and since π_2 is a colimit we have a unique $\xi\colon H \to H'$ satisfying $\xi \circ \pi_{2i} = \pi_{2i}' \circ \Xi_i$ for all $i \in \mathbb{J}$. The equations $h_i = h_i' \circ \xi$ follow from the universal property of π_2 (and the properties of Ξ). To show uniqueness of ξ, let $\zeta\colon (h_1, h_2) \to (h_1', h_2')$ be a 2-cell such that $\Xi = \Delta\zeta * I_{\Gamma\kappa}$; then using the second statement of Lemma 19(ii), $\Delta\zeta \circ \pi_2 = \pi_2' \circ \Xi$; hence $\zeta = \xi$ follows since π_2 is a colimit. In summary, (h_1, h_2) is universal for $\Gamma\kappa$.

To show (iv), let $\langle H', \vartheta \rangle$ be a colimit of $\mathcal{F} \times_{\Delta C} \Delta H$. Now, it suffices to show that there is a $\mathbb{C}$-morphism $\xi\colon H \to H'$ such that $\vartheta = \Delta\xi \circ \pi_2$.[4] By the universal property of ϑ, we obtain unique $\mathbb{C}$-arrows $h_1'\colon H' \to C$ and $h_2'\colon H' \to D$ such that $\Delta h_1' \circ \vartheta = \kappa \circ \pi_1$ and $\Delta h_2' \circ \vartheta = \Delta h_2 \circ \pi_2$. It also follows that the two equations $h_1 \circ k = h_1'$ and $h_2 \circ k = h_2'$ hold. Pulling back κ along h_1' yields a span $\mathcal{F} \xleftarrow{\pi_1'} \mathcal{F} \times_{\Delta C} \Delta H' \xrightarrow{\pi_2'} \Delta H'$; we then obtain a natural transformation $\Xi\colon \mathcal{F} \times_{\Delta C} \Delta H \to \mathcal{F} \times_{\Delta C} \Delta H'$ which satisfies $\pi_1 = \pi_1' \circ \Xi$ and $\vartheta = \pi_2' \circ \Xi$, and hence also $\Delta h_2 \circ \pi_2 = \Delta h_2' \circ \vartheta = \Delta h_2' \circ \pi_2' \circ \Xi$. By (ii), this defines a modification

---

[4] The reason is that once such a ξ is provided, there is a unique $k\colon H' \to H$ satisfying $\Delta k \circ \vartheta = \pi_2$, and thus $\xi \circ k = \text{id}_{H'}$ by the universal property of colimits; moreover $k \circ \xi = \text{id}_H$ must hold since (h_1, h_2) is universal for $\Gamma\kappa$.

$\Xi\colon \Delta(h_1, h_2)\circ\Gamma\kappa \to \Delta(h'_1, h'_2)\circ\kappa$. Using universality, we get a unique $\xi\colon H \to H'$ such that $h'_1 \circ \xi = h_1$, $h'_2 \circ \xi = h_2$ and $\Delta\xi \circ \pi_2 = \pi'_2 \circ \Xi = \vartheta$. □

4 Van Kampen Cocones as Span Bicolimits

Here we prove the main result of this paper, Theorem 22. Roughly speaking, the conclusion is that (under natural assumptions – existence of pullbacks and enough colimits in $\mathbb{C}$) to be VK in $\mathbb{C}$ is to be a bicolimit in $\mathcal{Span}_{\mathbb{C}}$. The consequence is that "being VK" is a universal property; in $\mathcal{Span}_{\mathbb{C}}$ rather than in $\mathbb{C}$.

The proof relies on a correspondence between the elementary characterisation of Van Kampen cocones in $\mathbb{C}$ of Proposition 10 and the universal properties of pseudo-cocones in $\mathcal{Span}_{\mathbb{C}}$ of Proposition 18. More precisely, given a colimit $\kappa\colon \mathcal{M} \to \Delta C$ in $\mathbb{C}$, we shall show that:

- $\Gamma\kappa$-universality *of all* spans $(h_1, h_2)\colon C \nrightarrow D$ corresponds to the implication (ii) $\Rightarrow$ (i) of Proposition 10, which is also known as pullback-stability or *universality* of the colimit κ;
- *existence of some* universal mediating cell from $\Gamma\kappa$ to any $\lambda\colon \Gamma\kappa \to \Delta D$ is the counterpart of the implication (i) $\Rightarrow$ (ii) of Proposition 10, which – for want of a better name – we here refer to as "*converse universality*" of κ;
- thus, $\Gamma\kappa$ is a bicolimit in $\mathcal{Span}_{\mathbb{C}}$ if and only if the colimit κ is Van Kampen.

The first two points are made precise by the statements of the following two lemmas. The third point is the statement of the main theorem.

Lemma 20 (Converse universality). *Let $\mathcal{F} \in [\mathbb{J}, \mathbb{C}]$ where $\mathbb{C}$ has pullbacks and for all $(\tau\colon \mathcal{E}\to\mathcal{F}) \in [\mathbb{J}, \mathbb{C}]\Downarrow\mathcal{F}$ a colimit of $\mathcal{E}$ exists. Then $\kappa\colon \mathcal{F} \to \Delta_{\mathbb{J}}C$ satisfies "converse universality" iff given any pseudo-cocone $\lambda\colon \Gamma\mathcal{F} \to \Delta D$, there exists a universal mediating cell $\langle(h_1, h_2), \Theta\rangle$ from $\Gamma\kappa$ to λ in $\mathcal{Span}_{\mathbb{C}}$.*

Proof. ($\Rightarrow$) Suppose that $\lambda\colon \Gamma\mathcal{F} \to \Delta D$ is a pseudo-cocone in $\mathcal{Span}_{\mathbb{C}}$.

For $u\colon i \to j$ in $\mathbb{J}$, we obtain a commutative diagram, as illustrated (cf. Proposition 16). Let $\vartheta\colon \mathcal{H} \to \Delta H$ be the colimit of $\mathcal{H}$; thus we obtain $h_1\colon H \to C$ and $h_2\colon H \to D$ making diagram (†) commute. By converse universality, the side faces $^{\mathcal{H}_i}_{\mathcal{F}_i}\downarrow^{\to}_{\to}\downarrow^{H}_C$ are pullback squares; using Lemma 19(ii) we get an invertible modification $\Theta\colon \lambda \to \Delta(h_1, h_2) \circ \Gamma\kappa$. That (h_1, h_2) is universal follows from Lemma 19(iii) since ϑ is a colimit.

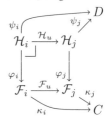

($\Leftarrow$) If in diagram (†) $D = \mathrm{col}\,\mathcal{H}$ and $\langle D, \psi\rangle$ is the corresponding colimit, we first use the assumption to obtain a universal mediating cell $\langle(h_1, h_2), \Theta\rangle$ from $\Gamma\kappa$ to $\lambda^{(\varphi, \psi)}$ where $\lambda^{(\varphi, \psi)}$ is the pseudo-cocone corresponding to the cartesian transformations $\varphi\colon \mathcal{H} \to \mathcal{F}$ and $\psi\colon \mathcal{H} \to \Delta D$ such that $\lambda^{(\varphi, \psi)}_i = (\varphi_i, \psi_i)$ as in Lemma 19(i); the latter also provides $\vartheta\colon \mathcal{H} \to \Delta H$ such that $h_2 \circ \vartheta_i = \psi_i$ and all $^{\mathcal{H}_i}_{\mathcal{F}_i}\downarrow^{\to}_{\to}\downarrow^{H}_C$ are pullback squares.

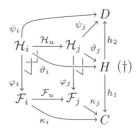

It suffices to show that $h_2 = \mathrm{id}_H$. However, by the universal property of the colimit $\langle D, \psi \rangle$, there is an arrow $k \colon D \to h$ such that $k \circ \psi_i = \vartheta_i$. The equation $h_2 \circ k = \mathrm{id}_D$ holds because $\langle D, \psi \rangle$ is a colimit in $\mathbb{C}$, and $k \circ h_2 = \mathrm{id}_H$ follows since (h_1, h_2) is universal for $\Gamma\kappa$. □

Lemma 21 (Universality). *Let $\mathcal{F} \in [\mathbb{J}, \mathbb{C}]$ where $\mathbb{C}$ has pullbacks such that for all $(\tau \colon \mathcal{E} \to \mathcal{F}) \in [\mathbb{J}, \mathbb{C}] \Downarrow \mathcal{F}$, a colimit of $\mathcal{E}$ exists. Then $\kappa \colon \mathcal{F} \to \Delta C$ satisfies universality iff every morphism $(h_1, h_2) \colon C \to D$ in $Span_{\mathbb{C}}$ is universal for $\Gamma\kappa$.*

Proof. ($\Rightarrow$) Any morphism (h_1, h_2) leads to a diagram (‡) where all the side-faces are pullbacks. By universality of κ, the cocone π_2 of the top face is a colimit; thus (h_1, h_2) is universal for $\Gamma\kappa$ by Lemma 19(iii).
($\Leftarrow$) Suppose that in diagram (‡) the side faces are all pullbacks. By assumption (h_1, h_2) is universal for $\Gamma\kappa$, thus $\langle H, \pi_2 \colon \mathcal{F} \times_{\Delta C} \Delta H \to \Delta H \rangle$ is a colimit by Lemma 19(iv). □

Finally, these two lemmas together with Proposition 18 imply our main result.

Theorem 22. *Let $\mathcal{F} \in [\mathbb{J}, \mathbb{C}]$ where $\mathbb{C}$ has pullbacks and for all cartesian transformations $\tau \colon \mathcal{E} \to \mathcal{F}$, a colimit of $\mathcal{E}$ exists. Then a cocone $\kappa \colon \mathcal{F} \to \Delta C$ is Van Kampen iff $\Gamma\kappa \colon \Gamma\mathcal{F} \to \Delta C$ is a bicolimit in $Span_{\mathbb{C}}$.* □

5 Conclusion, Related Work and Future Work

We gave a general definition of Van Kampen cocone that captures several previously studied notions in computer science, topology, and related areas, showing that they are instances of the same concept. Moreover, we have provided two alternative characterisations: the first one is elementary, and involves only basic category theoretic notions; the second one exhibits it as a *universal property*: Van Kampen cocones are just those colimits that are preserved by the canonical covariant embedding into the span bicategory.

Although this result is purely category theoretic, there exists closely related work in theoretical computer science. Apart from the references already given in the introduction, we mention the unfolding semantics of Petri nets in terms of coreflections. The latter has been generalised to graph grammars in [1]. Using ω-adhesive categories, i.e. adhesive categories in which colimits of ω-chains of monomorphisms are Van Kampen, it is possible to give such a coreflective unfolding semantics for grammars which rewrite objects of an ω-adhesive category [2]. Moreover, the morphisms between grammars in the latter work have a direct relation to the span-bicategory since they are essentially (a 1-categorical counterpart of) spans which preserve the structure of grammars.

Finally, the definition of Van Kampen cocone allows for several natural variations. For example, one may replace the slice category over the object at the "tip" of cocones by a (full) subcategory of it; this is exactly the step from global descent to $\mathbb{E}$-descent [13] and is closely related to the proposals in [7,9] for a weakening of the notion of adhesivity. Alternatively, one may start with cocones or diagrams of a particular form. In this way quasi-adhesive categories [21] arise as in the latter only pushouts along *regular* monos are required to be VK; another example is the work of Cockett and Guo [6], where Van Kampen cocones exist for a class of diagrams that naturally arises in their study of join restriction categories. Thus, possibly combining the latter two ideas, several new forms of Van Kampen cocones and diagrams arise as the subject for future research.

Acknowledgment. The authors thank the anonymous referees for their helpful suggestions that have helped to significantly improve this article.

References

1. Baldan, P.: Modelling Concurrent Computations: from Contextual Petri Nets to Graph Grammars. PhD thesis, Dipartimento di Informatica, Università di Pisa (2000)
2. Baldan, P., Corradini, A., Heindel, T., König, B., Sobociński, P.: Unfolding grammars in adhesive categories. In: Kurz, A., Lenisa, M., Tarlecki, A. (eds.) CALCO 2009. LNCS, vol. 5728, pp. 350–366. Springer, Heidelberg (2009)
3. Bénabou, J.: Introduction to bicategories. In: Midwest Category Seminar. Lect. Notes Math., vol. 42, pp. 1–77. Springer, Heidelberg (1967)
4. Brown, R., Janelidze, G.: Van Kampen theorems for categories of covering morphisms in lextensive categories. J. Pure Appl. Algebra 119, 255–263 (1997)
5. Carboni, A., Lack, S., Walters, R.F.C.: Introduction to extensive and distributive categories. J. Pure Appl. Algebra 84(2), 145–158 (1993)
6. R. Cockett and X. Guo. Join restriction categories and the importance of being adhesive. Unpublished manuscript, slides from CT 2007 (2007), http://pages.cpsc.ucalgary.ca/~robin/talks/jrCat.pdf
7. Ehrig, H., Habel, A., Padberg, J., Prange, U.: Adhesive high-level replacement categories and systems. In: Ehrig, H., Engels, G., Parisi-Presicce, F., Rozenberg, G. (eds.) ICGT 2004. LNCS, vol. 3256, pp. 144–160. Springer, Heidelberg (2004)
8. Ehrig, H., König, B.: Deriving bisimulation congruences in the DPO approach to graph rewriting. In: Walukiewicz, I. (ed.) FOSSACS 2004. LNCS, vol. 2987, pp. 151–166. Springer, Heidelberg (2004)
9. Ehrig, H., Prange, U.: Weak adhesive high-level replacement categories and systems: A unifying framework for graph and petri net transformations. In: Futatsugi, K., Jouannaud, J.-P., Meseguer, J. (eds.) Algebra, Meaning, and Computation. LNCS, vol. 4060, pp. 235–251. Springer, Heidelberg (2006)
10. Gadducci, F., Heckel, R., Llabrés, M.: A bi-categorical axiomatisation of concurrent graph rewriting. In: Category theory and computer science, CTCS 1999. Elect. Notes Theor. Comput. Sc., vol. 29 (1999)
11. Jacobs, B.: Parameters and parametrization in specification, using distributive categories. Fundamenta Informaticae 24(3), 209–250 (1995)
12. Janelidze, G., Sobral, M., Tholen, W.: Beyond barr exactness: Effective descent morphisms. In: Pedicchio, M.C., Tholen, W. (eds.) Categorical Foundations: Special Topics in Order, Topology, Algebra, and Sheaf Theory. Cambridge University Press, Cambridge (2004)
13. Janelidze, G., Tholen, W.: Facets of descent, I. Applied Categorical Structures 2(3), 245–281 (1994)

14. Johnstone, P.T., Lack, S., Sobociński, P.: Quasitoposes, quasiadhesive categories and artin glueing. In: Mossakowski, T., Montanari, U., Haveraaen, M. (eds.) CALCO 2007. LNCS, vol. 4624, pp. 312–326. Springer, Heidelberg (2007)

15. Johnstone, P.T.: Sketches of an Elephant: A topos theory compendium, vol. 1. Clarendon Press (2002)

16. Katis, P., Sabadini, N., Walters, R.F.C.: Bicategories of processes. J. Pure Appl. Algebra 115, 141–178 (1997)

17. Kelly, G.M.: Elementary observations on 2-categorical limits. Bull. Austral. Math. Soc. 39, 301–317 (1989)

18. Kelly, G.M., Street, R.H.: Review of the elements of 2-categories. Lect. Notes Math., vol. 420, pp. 75–103 (1974)

19. Lack, S.: A 2-categories companion [math.CT] (February 19, 2007) arXiv:math/0702535v1

20. Lack, S., Sobociński, P.: Adhesive categories. In: Walukiewicz, I. (ed.) FOSSACS 2004. LNCS, vol. 2987, pp. 273–288. Springer, Heidelberg (2004)

21. Lack, S., Sobociński, P.: Adhesive and quasiadhesive categories. RAIRO Theor. Inform. Appl. 39(3), 511–546 (2005)

22. Lack, S., Sobociński, P.: Toposes are adhesive. In: Corradini, A., Ehrig, H., Montanari, U., Ribeiro, L., Rozenberg, G. (eds.) ICGT 2006. LNCS, vol. 4178, pp. 184–198. Springer, Heidelberg (2006)

23. Lawvere, F.W.: Some thoughts on the future of category theory. In: category theory, Como. Lect. Notes Math., vol. 1488, pp. 1–13. Springer, Heidelberg (1991)

24. Mac Lane, S.: Categories for the Working Mathematician. Springer, Heidelberg (1998)

25. Milius, S.: On colimits in categories of relations. Appl. Categor. Str. 11(3), 287–312 (2003)

26. Robinson, E., Rosolini, G.: Categories of partial maps. Inf. Comput. 79(2), 95–130 (1988)

27. Sassone, V., Sobociński, P.: Reactive systems over cospans. In: Logic in Computer Science, LiCS 2005, pp. 311–320. IEEE Press, Los Alamitos (2005)

28. Walters, R.F.C.: Categories and Computer Science. Carslaw Publications (1991)

Unfolding Grammars in Adhesive Categories[*]

Paolo Baldan[1], Andrea Corradini[2], Tobias Heindel[3],
Barbara König[3], and Paweł Sobociński[4]

[1] Dipartimento di Matematica Pura e Applicata, Università di Padova, Italy
[2] Dipartimento di Informatica, Università di Pisa, Italy
[3] Abteilung für Informatik und Angewandte Kognitionswissenschaft, Universität
Duisburg-Essen, Germany
[4] ECS, University of Southampton, United Kingdom

Abstract. We generalize the unfolding semantics, previously developed
for concrete formalisms such as Petri nets and graph grammars, to the
abstract setting of (single pushout) rewriting over adhesive categories.
The unfolding construction is characterized as a coreflection, i.e. the
unfolding functor arises as the right adjoint to the embedding of the
category of occurrence grammars into the category of grammars.

As the unfolding represents potentially infinite computations, we need
to work in adhesive categories with "well-behaved" colimits of ω-chains
of monomorphisms. Compared to previous work on the unfolding of Petri
nets and graph grammars, our results apply to a wider class of systems,
which is due to the use of a refined notion of grammar morphism.

1   Introduction

When modelling systems one often needs a truly concurrent semantics providing
explicit information concerning causality, conflict and concurrency of events in
computations. This is clearly the case if one wants to understand and investigate
the inherent concurrency of a given system, but truly concurrent models are also
a cornerstone of verification techniques based on partial order methods [17]. In
fact, the latter avoid the enumeration of all possible interleavings of events, and,
in this way – especially in the case of highly concurrent systems – yield very
compact descriptions of the behaviour of a system.

One such partial order method is the unfolding approach: it "unravels" a
system and produces a structure which fully describes its concurrent behaviour,
including all reachable states and the mutual dependencies of all possible steps.

Unfolding techniques were first introduced for Petri nets and later extended
to several other formalisms, e.g. to graph transformation systems, which in turn
generalize (various extensions of) Petri nets. However, there are many types of
graph transformation formalisms – based on undirected and directed graphs,
hypergraphs, graphs with scopes, graphs with second-order edges, and so forth.

---

[*] Supported by DFG project SANDS and project AVIAMO of the University of
Padova.

A. Kurz, M. Lenisa, and A. Tarlecki (Eds.): CALCO 2009, LNCS 5728, pp. 350–366, 2009.

Hence a more abstract notion of unfoldings is called for, which exhibits the "essence" of the unfolding technique underlying all these special cases.

To this aim, we propose an *abstract* unfolding procedure which applies uniformly to all these rewriting mechanisms. Following the line of research of [9,14], we shall regard system states as objects of a category $\mathbb{C}$ satisfying suitable properties. Part of the properties of $\mathbb{C}$ ensure a meaningful notion of $\mathbb{C}$-object rewriting, while other additional properties are required to guarantee, first, that the unfolding procedure is feasible and, second, that the unfolding can be characterized as a co-reflection in the style of [20].

The approach to rewriting that we will use is the *single pushout approach* (SPO) [16]. This is one of the most commonly used *algebraic approaches* to rewriting, alternative to the *double pushout* (DPO) approach [10], where some subtle complications due to the inhibiting effects of DPO rewriting are avoided.

As a categorical framework we consider *adhesive categories* [14], which turn out to be appropriate for SPO rewriting as the needed pushouts in the partial map category Par($\mathbb{C}$) can be shown to exist. After having provided an algorithm to construct all finite prefixes of the unfolding, a crucial step consists in joining these parts into a single structure. To ensure that this is possible, we need that colimits of ω-chains of monomorphisms exist and satisfy suitable properties. Adhesive categories having sufficiently well-behaved colimits of ω-chains of monomorphisms will be called ω-*adhesive* (see also [12,8]).

The main result states that the unfolding construction induces a *coreflection*, i.e. it can be expressed as a functor that is right adjoint to the embedding of the category of occurrence grammars, the category where unfoldings live, into the category of all grammars. In order to define the category of grammars we introduce an original notion of grammar morphism which is similar to the graph grammar morphisms proposed in [4] but more concrete; as a consequence, we can treat uniformly the whole class of grammars, without the need to restrict to so-called *semi-weighted* grammars as it was done in several approaches for Petri nets (see, e.g. [18]) and for graph grammars [4].

Roadmap: In order to motivate at a more intuitive level the definitions and constructions which will follow, we first sketch the general ideas of our work. Note that we work in a setting of abstract objects (which could be sets, multisets, graphs, etc.) which are rewritten according to a rule by removing (the image of) its left-hand side and by gluing its right-hand side to the remaining object. According to the SPO approach, the left- and right-hand sides of a rule are related by a partial map, i.e. a span $L \hookleftarrow K \to R$ in the underlying category where the left leg is a mono. As it is usually done in unfolding approaches we restrict to linear rules where both legs are mono; for a schematic representation see Figure 1(a).

A rule essentially indicates what is deleted (▤), what is preserved (■) and what is created (▥). This can either be represented by a span as in Figure 1(a) or by a combined representation (see Figure 1(b)). Very roughly, given a (linear) rule $L \hookleftarrow K \rightarrowtail R$ (or $L \supseteq K \subseteq R$ in a more set-based notation), an object G that contains the left-hand side L is rewritten to $G \backslash (L \backslash K) \cup R$. This however is properly defined only if the complement exists.

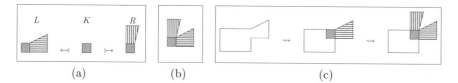

Fig. 1. (a) Rule as a partial map; (b) Combined rule representation; (c) Schematic representation of an unfolding step

In a general setting the so-called dangling condition issue arises. In the case of graphs, the dangling condition can be understood as follows: what happens if a node is to be removed, and such a node is attached to an edge which is not explicitly deleted? There are two ways to resolve this issue: the DPO solution which forbids the rewriting step, and the SPO solution which removes the edge. Since the inhibiting effects of the first solution lead to serious complications in the theory, we follow the latter path, which, as we will discuss, amounts to defining the term $G\backslash(L\backslash K)$ using the more general construction of relative pseudo-complements, known from lattice theory.

The unfolding of a fixed start object provides a partial order representation of the set of derivations in the grammar starting from such an object. Intuitively the construction works as follows: look for an occurrence of a left-hand side of a rule and, instead of replacing it, attach the right-hand side as in Figure 1(c) and record the occurrence of the rule. Doing this iteratively one obtains a growing object, called a type object, which is possibly infinite, and a set of rules that describe the dependencies on this type object.

Now, in order to characterize the unfolding construction abstractly and to show its universality, we will need the following concepts:

- *A notion of category which allows to define* (SPO) *rewriting properly:* For this we will use adhesive categories which can be used for defining abstractly a notion of rewriting which enjoys suitable Church-Rosser properties. (Section 2)
- *An analogue of occurrence nets:* As the unfolding of a Petri net is a special kind of net, the unfolding construction in this abstract setting will produce a special kind of grammar, which will be characterized as an occurrence grammar. In occurrence grammars suitable notions of causality, concurrency and conflict can be defined, allowing for a "static" characterization of reachable states as concurrent objects. (Section 4)
- *Well behaved ω-colimits:* In order to be able to construct potentially infinite unfoldings, we have to be able to glue together a countable chain of finite prefixes of the unfolding. To this aim we require that colimits of ω-chains exist and are well-behaved: adhesive categories enjoying this property are called ω-adhesive. The notion of ω-adhesivity is a natural extension of adhesivity that enjoys several closure properties. (Section 5)
- *A category of grammars and the coreflection result:* Finally we will present a coreflection result, i.e. we will show that the unfolding is in a sense the "best"

approximation of the original grammar in the realm of occurrence grammars. In order to do this we have to introduce a category of grammars, defining a suitable notion of grammar morphism. (Section 5)

2   Adhesive Categories for SPO Rewriting

We will use adhesive categories [14] as a basis for the rewriting framework. For this we fix an (adhesive) category $\mathbb{C}$ to which all objects and morphisms belong.

Definition 1 (Adhesive category). *A category is* adhesive *if*

1. *it has all pullbacks;*
2. *pushouts along monomorphisms exist, i.e. for each span $B \xleftarrow{f} A \xrightarrowtail{m} C$ with monic m, a pushout $B \xrightarrow{n} D \xleftarrow{g} C$ exists, yielding a* pushout square $\begin{smallmatrix} B & \leftarrow & A \\ \downarrow & & \downarrow \\ D & \leftarrow & C \end{smallmatrix}$;
3. *all pushouts along monos are Van Kampen squares, i.e. given a cube diagram as shown below with: (i) m monic, (ii) the bottom square a pushout and (iii) the back squares pullbacks, we have that the top square is a pushout iff the front squares are pullbacks.*

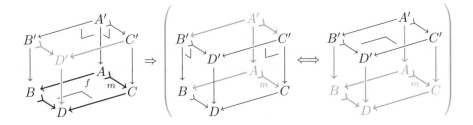

It is known that every topos is adhesive [15]. The subobjects of an object in an adhesive category form a distributive lattice, a fact which we will make more precise in the following.

Definition 2 (Subobject poset). *Let $T \in \mathbb{C}$ be an object. Two monomorphisms $a \colon A \rightarrowtail T$, $a' \colon A' \rightarrowtail T$ are* isomorphic *if there exists an isomorphism $i \colon A \to A'$ with $a = a' \circ i$. Such an equivalence class is called the subobject represented by a. Then the subobject poset $\langle \mathrm{Sub}(T), \sqsubseteq \rangle$ has isomorphism classes $[a \colon A \rightarrowtail T]$ of monomorphisms over T as elements. Further, given two monomorphisms $a \colon A \rightarrowtail T$ and $b \colon B \rightarrowtail T$, the inclusion $[a] \sqsubseteq [b]$ holds if there exists $j \colon A \rightarrowtail B$ such that $a = b \circ j$.*

Proposition 3 (Distributive subobject lattices [14]). *Any subobject poset in an adhesive category is a distributive lattice, where the meet $[a] \sqcap [b]$ of two subobjects $[a]$, $[b]$ is obtained by taking the pullback of their representatives and the join $[a] \sqcup [b]$ is obtained by taking a pullback, followed by a pushout (i.e. adhesive categories have effective unions).*

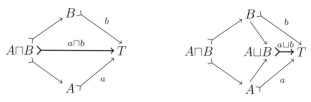

Another operation on subobjects that is directly connected to the SPO rewriting mechanism is *relative pseudo-complementation* [7] (cf. Proposition 6).

Definition 4 (Relative pseudo-complement). *Let $\langle L, \sqsubseteq \rangle$ be a lattice. The relative pseudo-complement* (RPC) *of a with respect to b, written $a \mapsto b$, is an element d satisfying $a \sqcap x \sqsubseteq b \iff x \sqsubseteq d$ for all $x \in L$. It is unique if it exists.*

The lattice L is relatively pseudo-complemented (RPC) *if the RPC $a \mapsto b$ exists for all pairs $a, b \in L$.*

In a finite distributive lattice the RPC of a w.r.t. b always exists and can be obtained as $a \mapsto b = \bigsqcup \{y \mid a \sqcap y \sqsubseteq b\}$. We consider the following two special cases:

- In the case of a powerset lattice, given two sets $B, A \in \wp(M)$, the RPC of A w.r.t. B, is the set $M \setminus (A \setminus B) = \{m \in M \mid m \notin A \text{ or } m \in B\}$.
- In the case of subobject lattices, if $[a], [b] \in \mathrm{Sub}(T)$, with $[a] \sqsupseteq [b]$, the RPC $[c] = [a] \mapsto [b]$ with $c \colon (A \mapsto B) \rightarrowtail T$ gives rise to a particular pullback square $\begin{smallmatrix} A & \twoheadleftarrow & B \\ \downarrow & & \downarrow \\ T & \twoheadleftarrow & A \mapsto B \end{smallmatrix}$.
  In particular, note that that the RPC $[a] \mapsto [a]$ is $[\mathrm{id}_T]$ (if the top arrow of the pullback is an iso so is the bottom arrow).
  As an example we consider the category of directed graphs and graph morphisms (i.e., the functor category $\mathbf{Set}^{\bullet \rightrightarrows \bullet}$) which is known to be a topos and thus adhesive. Given the two graph inclusions shown in (a) below, the RPC is given in (b) and yields the pullback square shown in (c). This construction corresponds to taking the *largest pullback complement.*

Having relatively pseudo-complemented subobject lattices will be important for SPO rewriting. In SPO rewriting a rule is essentially a partial map which specifies what is deleted, what is preserved and what is created. Given a category $\mathbb{C}$ with pullbacks along monomorphisms, its category of partial maps is defined as follows (see [19]).

Definition 5 (Partial maps). *The category $\mathrm{Par}(\mathbb{C})$ of partial maps (in $\mathbb{C}$) has the same objects as $\mathbb{C}$, i.e. $\mathrm{ob}(\mathrm{Par}(\mathbb{C})) = \mathrm{ob}(\mathbb{C})$. An arrow in $\mathrm{Par}(\mathbb{C})$ is a $\mathbb{C}$-span $A \leftarrowtail^m X \xrightarrow{f} B$ with monic m, taken up to isomorphisms at X (Fig. 2(a)). It is called a* partial map *and is written $(m_{(X)}f) \colon A \rightharpoonup B$ or just $(m, f) \colon A \rightharpoonup B$.*

If m is an isomorphism, then (m, f) is called a total map. *The identity on an object A is $(\mathrm{id}_A, \mathrm{id}_A) \colon A \rightharpoonup A$ (Fig. 2(b)); composition is defined via pullbacks (Fig. 2(c)).*

Note that a partial map $(m, f) \colon A \rightharpoonup B$ is monic in $\mathrm{Par}(\mathbb{C})$ if and only if it is total and f is monic in $\mathbb{C}$; hence we often write $C \rightarrowtail^g \rightarrow D$ instead of $C -(\mathrm{id}, g) \rightarrow D$.

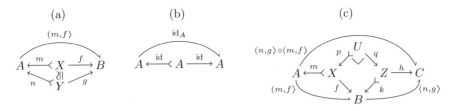

Fig. 2. Diagrams showing partial maps

Proposition 6 (Partial map pushouts). *Let* $\mathbb{C}$ *be an adhesive category. Then in* Par($\mathbb{C}$) *pushouts along monomorphisms exist if and only if subobject lattices in* $\mathbb{C}$ *are relatively pseudo-complemented.*

The pushout can be obtained as depicted below. Note that the vertical partial maps are monic and hence we omit the first legs which are isos. Starting from the diagram on the left, one first obtains D *as* RPC *of* m *w.r.t.* $m \circ \alpha$ *and then constructs* C *as pushout of* d *and* β *in* $\mathbb{C}$.

Besides assuming that $\mathbb{C}$ is adhesive we also require that the operation of taking RPCs is *stable under pullback* (in the sense of Lemma 1.4.13 in [13]), a fact needed later to show that SPO rewriting is preserved by grammar morphisms. All toposes and all adhesive categories known to the authors satisfy this requirement. Note that it would also be possible to develop the theory in another setting such as Heyting categories [13].

3   Grammars and Grammar Morphisms as Simulations

The basic entities of single-pushout rewriting are rules, which are partial maps, and grammars, which are collections of rules with a start object. As it is standard in unfolding approaches, we restrict to linear and consuming rules. Further, rewriting amounts to taking pushouts along the rules of a grammar.

Definition 7 (Rules and SPO rewriting). *A* $\mathbb{C}$-*rule is a partial map span* $q = L \leftarrow\!\!\alpha\!\!\prec K -\beta\rightarrow R$. *It is called* linear *if* β *is monic and* consuming *if* α *is not an isomorphism. We denote by* $\mathscr{R}_C$ *the class of consuming, linear* $\mathbb{C}$-*rules.*

Let $A \in \mathbb{C}$ *be an object. A (monic)* match *for a rule* $q = L \leftarrow\!\!\alpha\!\!\prec K \succ\!\!\beta\!\!\rightarrow R$ *into* A *is a monomorphism* $m \colon L \rightarrowtail A$ *in* Par($\mathbb{C}$). *Then* q rewrites A *(at* m*) to* B, *written* $A \models\langle q,m\rangle\Rightarrow B$ *or simply* $A \models q\Rightarrow B$, *if there is a pushout square* ${}^{L}_{A}\!\downarrow\overset{\longrightarrow}{\ulcorner}\downarrow{}^{R}_{B}$ *in* Par($\mathbb{C}$), *i.e. if a pushout* $A -b\rightarrow B \leftarrow n- R$ *of* $A \leftarrow\!\!m\!\!\prec L -(\alpha,\beta)\rightarrow R$ *exists; in this situation* $A \models\langle q,m\rangle\Rightarrow B$ *is also referred to as an* SPO *rewriting step.*

Example 8. The graph $S = $ ⌾↣○↣○ models a tiny network: vertices are network nodes, edges are directed network links, and looping edges represent stored messages. Further $q_1 = $ ⌾↣○ ↢ ○↣○ ↣ ○↣⌾ and $q_2 = $ ○ ↢ ∅ ↣ ∅ model message dispatching and failure of network nodes, respectively. Now q_1 can rewrite S, namely ⌾↣○↣○ ⊨q_1⇒ ○↣⌾↣○. In the latter state, the failure of the middle network node is captured by ○↣⌾↣○ ⊨q_2⇒ ○   ○, i.e. dangling edges are removed.

Definition 9 (Slice category). *For an object $T \in \mathbb{C}$, the* slice category over T, *denoted $\mathbb{C} \downarrow T$, has $\mathbb{C}$-morphisms A–a→T with codomain T as objects. A $\mathbb{C} \downarrow T$-morphism $\psi \colon (A$–a→$T) \rightarrow (B$–b→$T)$ is a $\mathbb{C}$-morphism $\psi \colon A \rightarrow B$ satisfying $a = b \circ \psi$. Further, we denote by $\lfloor \_ \rfloor_T \colon \mathbb{C} \downarrow T \rightarrow \mathbb{C}$ the obvious forgetful functor, mapping A–a→T to A, and acting as the identity on morphisms.*

A grammar will be defined as a set of rules with a start object. This is in analogy to Petri nets where we regard the latter as a set of transitions with an initial marking. More precisely, as described in detail in [4], the token game of a Petri net with place set P can be modelled by SPO rewriting in the slice category $\mathbb{S} \downarrow P$, where $\mathbb{S}$ is the category of (finite) sets and functions: multisets are encoded as functions with co-domain P. Abstracting away from sets, we work in the slice category $\mathbb{C} \downarrow T$ for a given "place" object T, also called *type object*.

Example 10. Fixing the *type graph* $T = $ ⌾, we give a typed version of Example 8. The double fins correspond to network links, i.e. the typed version of ⌾↣○↣○ is ⌾↠○↠○ where the morphisms into T is the unique one preserving the fins; the typed rules are given by ⌾↠○ ↢ ○↠○ ↣ ○↠⌾ and ○ ↢ ∅ ↣ ∅.

Notation: We introduce a convention for rules $q \in \mathscr{R}_{\mathbb{C} \downarrow T}$: we will always assume $q = l_q \leftarrow_{\alpha_q} \prec k_q \succ_{\beta_q} \rightarrow r_q$ and $|q|_T = L_q \leftarrow_{\alpha_q} \prec K_q \succ_{\beta_q} \rightarrow R_q \in \mathscr{R}_\mathbb{C}$, where the latter is the obvious untyped version of q.

If $\mathbb{C}$ is adhesive and RPCs are stable under pullback, then $\mathbb{C} \downarrow T$ has the same properties (cf. Proposition 29 and [14]). Now *typed* grammars are defined as follows.

Definition 11 (Typed grammar). *Let $T \in \mathbb{C}$ be an object called the type object. A T-typed grammar G is a pair $G = \langle Q, s \colon S \rightarrow T \rangle$ where $Q \subseteq \mathscr{R}_{\mathbb{C} \downarrow T}$ is a set of linear, consuming $\mathbb{C} \downarrow T$-rules, and S–s→$T \in \mathbb{C} \downarrow T$ is the start object.*

The rewriting relation over $\mathbb{C} \downarrow T$-objects associated with G is defined by $a \vdash_G \Rightarrow b$ if $a \vdash_q \Rightarrow b$ for some $q \in Q$; further an object $a \in \mathbb{C} \downarrow T$ is reachable in G if $s \vdash_G \Rightarrow^* a$, where $\vdash_G \Rightarrow^*$ is the transitive-reflexive closure of $\vdash_G \Rightarrow$.

In the rest of the paper we restrict ourselves to *finite* grammars.

Definition 12 (Finite grammar). *Let $G = \langle Q, s \colon S \rightarrow T \rangle$ be a grammar; then G is finite if the start object and all left and right hand sides are finite, i.e. $\mathrm{Sub}(S)$ is finite and $\mathrm{Sub}(L_q)$, $\mathrm{Sub}(R_q)$ are finite for all $q \in Q$. Moreover for each rule there are at most finitely many other rules with isomorphic left-hand sides, i.e. the set $\{q' \mid l_q \cong l_{q'} \& q' \in Q\}$ is finite for each $q \in Q$.*

Finiteness of a grammar ensures that every reachable object is finite. As a consequence, using Proposition 6 and existence of RPCs in finite lattices, it also guarantees that for every rule q and match m of q into a reachable object, the pushout of q and m exists in $\text{Par}(\mathbb{C})$: this implies that (as it is usual for the SPO approach) rewriting is possible at any match.

Retyping operations and grammar morphisms. Now we equip grammars with a notion of morphism, turning them into a category. Following the ideas in the literature on Petri nets and graph transformation, a morphism relating two systems should induce a simulation among them, in the sense that every computation in the source system is mapped to a computation of the target system. Another desirable property is that a notion of morphism defined in our abstract setting should "specialize" to the corresponding notions proposed for systems such as Petri nets and graph grammars. The morphisms we introduce below will satisfy the first requirement, i.e. a grammar morphism will describe how the target grammar can simulate the source grammar. Concerning the second property, the proposed notion of morphism is more concrete: for example, when $\mathbb{C}$ is the category of graphs, a graph grammar morphism of [4] might be induced by several different ones according to our definition. However, this greater explicitness allows to characterize the unfolding as a coreflection without restricting to the so-called *semi-weighted* grammars (cf. [18,4,1]).

In analogy to Petri nets, where morphisms are monoid homomorphisms preserving the net structure, a morphism between two grammars typed over T and T', respectively, will be a functor $\mathcal{F}: \mathbb{C}{\downarrow}T \to \mathbb{C}{\downarrow}T'$ that preserves the rules and the start object, and comes equipped with some additional information.

Definition 13 (Retyping operation). *A retyping operation* $\mathcal{F}: \mathbb{C}{\downarrow}T \to \mathbb{C}{\downarrow}T'$ *is a pair* $\mathcal{F} = \langle F, \varphi \rangle$ *where* $F: \mathbb{C}{\downarrow}T \to \mathbb{C}{\downarrow}T'$ *is a functor mapping each object* $A{-}a{\to}T$ *to* $\mathcal{F}(a) = F(A){-}F(a){\to}T'$, *and* $\varphi: (|\_|_{T'} \circ F) \overset{.}{\to} |\_|_T$ *is a cartesian natural transformation.*[1]

Every morphism $f: T \to T'$ *in* $\mathbb{C}$ *induces a* (canonical) *retyping operation* $\sharp f = \langle f \circ \_, \text{id}_\_ \rangle$ *where the functor* $f \circ \_ : \mathbb{C}{\downarrow}T \to \mathbb{C}{\downarrow}T'$ *post-composes any* $\mathbb{C}{\downarrow}T$-*object with* f, *and* $\text{id}_\_$ *is the family of identities* $\{\text{id}_A: A \to A\}_{(A\overset{a}{\to}T)\in\mathbb{C}{\downarrow}T}$.

This definition is closely related to the *pullback-retyping* used in [4]. In fact, as illustrated to the right, the action of a retyping operation $\langle \mathcal{F}, \varphi \rangle: \mathbb{C}{\downarrow}T \to \mathbb{C}{\downarrow}T'$ is pulling back along φ_{id_T} followed by composition with $\mathcal{F}(\text{id}_T)$, which is retyping along the span $T \leftarrow \mathcal{F}(T) \to T'$, according to [4].

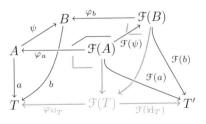

---

[1] This means that $\varphi = \{F(A) -\varphi_a\to A\}_{a\in\mathbb{C}{\downarrow}T}$ is a family of arrows such that for each arrow $\psi: a \to b$ in $\mathbb{C}{\downarrow}T$, the span $A \leftarrow\varphi_a- F(A) -|F(\psi)|\to F(B)$ is a $\mathbb{C}$-pullback of $A -|\psi|\to B \leftarrow\varphi_b- F(B)$, yielding a pullback square $\begin{smallmatrix} B & \leftarrow & F(B) \\ \uparrow & & \uparrow \\ A & \leftarrow & F(A) \end{smallmatrix}$.

The definition of grammar morphisms now is as follows.

Definition 14 (Typed grammar morphism). *Let* $G = \langle Q, s\colon S \to T \rangle$ *and* $G' = \langle Q', s'\colon S' \to T' \rangle$ *be typed grammars in* $\mathbb{C}$. *Then a grammar morphism* $\mathcal{F}\colon G \to G'$ *is a retyping operation* $\mathcal{F} = \langle F, \varphi \rangle\colon \mathbb{C}{\downarrow}T \to \mathbb{C}{\downarrow}T'$ *such that*

(i) the start object is preserved, i.e. $\mathcal{F}(s) = s'$, *and*
(ii) for any rule $q \in Q$, *the image* $\mathcal{F}(q) := \mathcal{F}(l_q) \leftarrow \mathcal{F}(\alpha_q) - \mathcal{F}(k_q) - \mathcal{F}(\beta_q) \to \mathcal{F}(r_q)$
    is either a rule in G', *i.e.* $\mathcal{F}(q) \in Q'$, *or an identity rule, i.e.* $\mathcal{F}(q) = \mathcal{F}(l_q) \xleftarrow{id} \mathcal{F}(l_q) \xrightarrow{id} \mathcal{F}(l_q)$.

We will now prove the *Simulation Lemma:* it shows that each grammar morphism maps rewriting steps in the domain to corresponding ones in the co-domain, which are either applications of rules in the target grammar or identity steps. Hence grammar morphisms preserve reachability.

Lemma 15 (Simulation Lemma). *Let* $\mathcal{F} = \langle F, \varphi \rangle\colon G \to G'$ *be a morphism between grammars* $G = \langle Q, s\colon S \to T \rangle$ *and* $G' = \langle Q', s'\colon S' \to T' \rangle$. *Then for any rewriting step* $a \models_{\langle q,m \rangle} \Rightarrow b$ *in* G *we have that* $\mathcal{F}(a) \models_{\langle \mathcal{F}(q), \mathcal{F}(m) \rangle} \Rightarrow \mathcal{F}(b)$.

Proof (Sketch). Suppose that $(A - a \to T) \models_{\langle q,m \rangle} \Rightarrow (B - b \to T)$, with match $m\colon l \rightarrowtail a$ for a rule $q = l \leftarrow \alpha - k - \beta \to r \in Q$. Using Proposition 6, there is a diagram of the form $_a^l \curlyvee \boxplus \boxminus \curlyvee_b^r$. As pushouts and pullbacks in $\mathbb{C}{\downarrow}T$ are constructed in $\mathbb{C}$, it is enough to consider the underlying $\mathbb{C}$-diagram $_A^L \curlyvee \boxplus \boxminus \curlyvee_B^R$. Now the morphism $\langle F, \varphi \rangle$ provides not only arrows $\varphi_a\colon \mathcal{F}(A) \to A$ and $\varphi_b\colon \mathcal{F}(B) \to B$ into the "tips" of the two squares, but actually a pair of "fitting" pullback cubes over $_A^L \curlyvee \boxplus \boxminus \curlyvee_B^R$. The top face of the resulting double cube is $_{\mathcal{F}(A)}^{\mathcal{F}(L)} \curlyvee \boxplus \boxminus \curlyvee_{\mathcal{F}(B)}^{\mathcal{F}(R)}$, which by Proposition 6 is a Par($\mathbb{C}$) pushout square because RPCs and pushouts of pairs of monomorphisms are stable under pullback.

4 Unfolding Grammars into Occurrence Grammars

Every typed grammar can be *unfolded* by recording all possible sequences of rewriting steps originating from the start object. In analogy to the constructions proposed for Petri nets and graph grammars, the structure that we will obtain is a *(non-deterministic) occurrence grammar*, which is a partial order representation of all possible computations. Finite initial parts of the (full) unfolding of the grammar – so-called *prefixes* – give a compact representation of the behaviour of the grammar up to a certain *causal depth*.

In this section we introduce the class of occurrence grammars and show that in such grammars reachable objects can be characterized statically, by means of suitably defined dependency relations (causality and asymmetric conflict) between rules. This allows to avoid "solving" reachability problems while constructing the truncations of the full unfolding, i.e. the algorithm presented at the end of this section builds the unfolding in a static manner.

4.1   Occurrence Grammars

To properly define occurrence grammars, we need to recall from [2] the corresponding relations between rules of typed grammars, which can be described in words as follows. Given two rules q and q', then q *causes* q' if q produces something needed by q' to be activated, and q *can be disabled* by q' if q' destroys something on which q depends.

Definition 16 (Causality, conflict). *Let $G = \langle Q, s\colon S \to T \rangle$ be a typed grammar; then G is* mono-typed *if s is monic and, for each rule $q \in Q$ all three of l_q, k_q and r_q are monic, yielding subobjects $[l_q], [k_q]$ and $[r_q] \in \mathrm{Sub}(T)$. If G is mono-typed, a pair of rules $q, q' \in Q$ may be related in any of the following ways.*

$<\ :\ q$ *directly causes* q',   *written* $q < q'$, *if* $r_q \sqcap l_{q'} \not\sqsubseteq k_q$
$\ll\ :\ q$ *can be disabled by* q',   *written* $q \ll q'$, *if* $l_q \sqcap l_{q'} \not\sqsubseteq k_{q'}$

Further, the asymmetric conflict *relation is* $\nearrow := <^+ \cup (\ll \setminus \mathrm{id}_Q)$ *where* $<^+$ *is the transitive closure of* $<$ *and* $\ll \setminus \mathrm{id}_Q$ *the irreflexive version of* $\ll$; *moreover*

– the direct causes *of q,*   *are given by* $\lfloor q \rfloor = \{q' \in Q \mid q' < q\}$, *and*
– the (complete) causes *of q,*   *are given by* $\lfloor q \rfloor = \{q' \in Q \mid q' <^* q\}$.

Any subobject $[a] \in \mathrm{Sub}(T)$ may be related to a rule $q \in Q$ in a similar way:

$<\ :\ q$ *directly causes* $[a]$,   *written* $q < a$, *if* $r_q \sqcap a \not\sqsubseteq k_q$, *and*
$<_{\mathrm{co}}\ :\ [a]$ *is (partly) consumed by* q',   *written* $a <_{\mathrm{co}} q'$, *if* $a \sqcap l_{q'} \not\sqsubseteq k_{q'}$,

and we also have the following sets:

– the consumers *of $[a]$,*   *are* $\lceil a \rceil = \{q' \in Q \mid a <_{\mathrm{co}} q'\}$ *and*
– the (complete) causes *of $[a]$,*   *are* $\lfloor a \rfloor = \{q' \in Q \mid \exists q \in Q.\, q' <^* q < a\}$.

Now we are ready to define *occurrence grammars*, which are a generalization of occurrence nets. We will see later that in an occurrence grammar with type object T, rule applications can be interpreted as consuming and producing subobjects of T (Proposition 19).

Definition 17 (Occurrence Grammar). *An occurrence grammar is a mono-typed grammar $O = \langle Q, s\colon S \rightarrowtail T \rangle$ with a countable set of rules Q such that*

1. *the type object is the union of all right hand sides, i.e.* $\mathrm{id}_T \cong s \sqcup \bigsqcup_{q \in Q} r_q$,
2. *the transitive-reflexive closure $<^*$ of causality $<$ is a partial order,*
3. *for each rule $q \in Q$, $\lfloor q \rfloor$ is finite, and $\nearrow|_{\lfloor q \rfloor} := \nearrow \cap (\lfloor q \rfloor \times \lfloor q \rfloor)$ is acyclic,*
4. *the start object has no causes, i.e. $\lfloor s \rfloor = \varnothing$,*
5. *there are no backward conflicts, i.e. $r_q \sqcap r_{q'} \sqsubseteq k_q \sqcup k_{q'}$ for all $q \neq q' \in Q$,*
6. *left-hand sides are properly produced, i.e. $l_q \sqsubseteq s \sqcup \bigsqcup_{p' \in \lfloor q \rfloor} r_{p'}$ for all $q \in Q$.*

Example 18. Consider the following occurrence grammar which exactly captures the two rewriting steps of Examples 8 and 10. The type graph is ⊗↠⊗↠⊕ where the dot and the plus discern the different nodes; the start object is ⊗↠○↠⊕, and we have the two rules ⊗↠○ ↢ ○↠○ ↣ ○↠⊗ and ○ ↢ ∅ ↣ ∅ corresponding to the dispatching of the message and the breakdown of the middle node. As it will become clear later, this grammar is part of the unfolding of the grammar in Examples 8 and 10.

Properties of occurrence grammars. In the theory of Petri nets, a characteristic property of occurrence nets is that they are safe, i.e. that every reachable marking is a set of places, rather than a proper multiset. The analogous result for occurrence grammars reads as follows.

Proposition 19 (Safety). *Let* $O = \langle Q, S \succ\!s\!\rightarrow T \rangle$ *be an occurrence grammar and let* $s \models_O \Rightarrow^* a \in \mathbb{C} \downarrow T$ *be a reachable object. Then* a *is monic.*

Hence reachable objects of occurrence grammars can be seen as subobjects of the type object. In the unfolding algorithm below, instead of considering reachable objects, we can concentrate on the statically characterized *concurrent subobjects*, as they are exactly the ones contained in reachable subobjects.

Definition 20 (Concurrent subobject). *Let* $O = \langle Q, S \succ\!s\!\rightarrow T \rangle$ *be an occurrence grammar. A subobject* $\lfloor a \rfloor \in \mathrm{Sub}(T)$ *is called a* concurrent subobject of O *if (i)* $\lfloor a \rfloor$ *is finite, (ii)* $\lfloor a \rfloor \cap \ulcorner a \urcorner = \varnothing$, *and (iii)* $\nearrow|_{\lfloor a \rfloor}$ *is acyclic.*

Intuitively, $\lfloor a \rfloor$ is concurrent when its set of causes is finite and conflict free (condition (i) and (iii), respectively) and there are no causal dependencies between subobjects of $\lfloor a \rfloor$ (condition (ii)).

Proposition 21 (Static coverability). *Let* $O = \langle Q, S \succ\!s\!\rightarrow T \rangle$ *be an occurrence grammar, and* $\lfloor a \rfloor \in \mathrm{Sub}(T)$ *be a subobject. Then* $\lfloor a \rfloor$ *is concurrent if and only if there is some reachable object* b *such that* $a \sqsubseteq b$.

4.2   The Unfolding Construction

The idea of the unfolding procedure for a given grammar G, is to construct a chain of growing occurrence grammars U_n. Each U_n represents all computations up to *causal depth* n where the depth of a concurrent computation is the length of a maximally parallel execution of the computation. Finally the full unfolding U_G will arise as the "union" of the chain $\{U_n \text{ "}\subseteq\text{" } U_{n+1}\}_{n \in \mathbb{N}}$. This is a concrete algorithmic description of the unfolding. As shown in the next section, the unfolding can be characterised in a succint and elegant way as the right adjoint functor to the inclusion of the category of occurrence grammars into the category of grammars.

Definition 22 (Unfolding algorithm). *Let* $G = \langle Q, S -s\!\rightarrow T \rangle$ *be a finite grammar.* We will construct a chain $U_0 \text{ "}\subseteq\text{" } U_1 \text{ "}\subseteq\text{" } \ldots \text{ "}\subseteq\text{" } U_n \ldots$ of occurrence grammars $U_n = \langle Q_n, S -s_n\!\rightarrow T_n \rangle$ that come equipped with *folding morphisms* $\mathfrak{F}_n \colon U_n \rightarrow G$ mapping rule occurrences in each n-th unfolding U_n to the original grammar G; further each $\mathfrak{F}_n$ will be induced by a *folding arrow* $T_n -\lambda_n\!\rightarrow T$, i.e. $\mathfrak{F}_n = \sharp\lambda_n$ (see Definition 13).

Base case. The 0-th unfolding U_0 contains the start object of G and no rules, i.e. $U_0 = \langle \varnothing, S \succ\!\text{id}\!\rightarrow S \rangle$. The folding arrow is $\lambda_0 \colon T_0 = S \rightarrow T$, which induces $\mathfrak{F}_0 = \sharp\lambda_0$.

Induction step. Going from U_n to U_{n+1} consists in adding the next level of causal depth. The central operation of this step can be described as the *non-consuming* application of all rules with all possible (new) matches to T_n "in parallel" – here the non-consuming rule application of a rule $q = L \leftarrow\alpha\prec K \succ\beta\rightarrow R$ at a match $m\colon L \rightarrowtail T$ is the application of $q^+ := K \leftarrow\text{id}\prec K \succ\beta\rightarrow R$ at $m \circ \alpha\colon K \rightarrowtail T$ (see Figure 1(c)).

A *new match* or a *new occurrence* of a rule $q \in Q$ in the n-th unfolding $\langle Q_n, S -s_n\rightarrow T_n\rangle$ via the folding $\mathfrak{F}_n$ is a monomorphism $\nu\colon L_q \rightarrowtail T_n$ such that the corresponding subobject of $\mathrm{Sub}(T_n)$ is concurrent, and that satisfies $\lambda_n \circ \nu = l_q$; additionally, ν must be *new*, which means that ν is not an occurrence of q that is already present in Q_n, i.e. there is no rule $q' \in Q_n$ such that $\nu = l_{q'}$ and q is the image of q' w.r.t. $\mathfrak{F}_n$.

Let $\{\nu_i\colon L_{q_i} \rightarrowtail T_n \mid i \in I_n\}$ be the set of all new matches where the index set $I_n = \{1, \ldots, m\}$ is finite as G is finite. Now consider the diagram below, consisting of the matches ν_i and the rule morphisms $\alpha_{q_i}, \beta_{q_i}$ for $i \in I_n$.

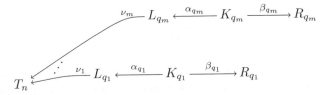

Take the colimit of the diagram above in $\mathbb{C}$, by a stepwise computation of pushouts of monos, obtaining the morphisms t_n, k_i, r_i into T_{n+1} for $i \in I_n$. Furthermore since every object in the diagram above is typed over T, we obtain the folding arrow $\lambda_{n+1}\colon T_{n+1} \to T$ as a mediating arrow.

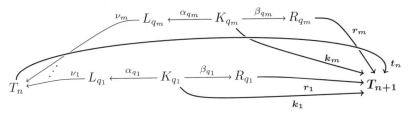

Now the new rule occurrences form the set $Q'_{n+1} := \{(t_n \circ \nu_i) \leftarrow\alpha_{q_i} - k_{q_i} -\beta_{q_i}\rightarrow r_{q_i} \mid i \in I_n\}$ and are at depth level $n+1$; further the complete set Q_{n+1} of rules of U_{n+1} is $Q_{n+1} = Q'_{n+1} \cup \sharp t_n(Q_n)$.

To complete the object part of the $(n+1)$-th unfolding, we just need to define $U_{n+1} := \langle Q_{n+1}, S\rightarrowtail t_n\circ s_n\rightarrow T_{n+1}\rangle$. Further the folding morphism $\mathfrak{F}_{n+1}\colon U_{n+1} \to G$ is given by $\mathfrak{F}_{n+1} := \sharp\lambda_{n+1}$, which is induced by the folding arrow $T_{n+1} -\lambda_{n+1}\rightarrow T$.

Summarizing, we have inductively defined an ω-chain of occurrence grammars $U_0 \text{``}\subseteq\text{''} U_1 \text{``}\subseteq\text{''} \cdots \text{``}\subseteq\text{''} U_n \ldots{}_\infty$, where each U_n has components $\langle Q_n, s_n\colon S\rightarrowtail T_n\rangle$, folding morphisms $\sharp\lambda_n\colon U_n \to G$, and "inclusion" morphisms $\sharp t_n\colon U_n \rightarrowtail U_{n+1}$.

Example 23. We sketch the unfolding of the grammar $\langle\{q'_1, q'_2\}, s\colon \bigcirc\!\!\rightarrowtail\!\!\circ\!\!\rightarrowtail\!\!\bullet \to \bigcirc\!\!\bullet\rangle$ where the rules q'_1 and q'_2 are $\bigcirc\!\!\rightarrowtail\!\!\circ \leftarrowtail \circ\!\!\rightarrowtail\!\!\circ \rightarrowtail \circ\!\!\rightarrowtail\!\!\bigcirc$ and $\circ \leftarrowtail \varnothing \rightarrowtail \varnothing$, respectively.

We start with the type graph $T_0 = \bigcirc\!\!\rightarrowtail\!\!\circ\!\!\rightarrowtail\!\!\bullet$. In the first unfolding step we find the single rule occurrence $\bigcirc\!\!\rightarrowtail\!\!\circ$ of q'_1 and the three occurrences $\circ$, $\circ$, and $\bullet$ of q'_2 and add the corresponding right-hand sides, yielding the type graph $T_1 = \bigcirc\!\!\bigcirc\!\!\rightarrowtail\!\!\bullet$.

In the second unfolding there is only one occurrence of q_1, namely ⚬⟶⊕ , and adding the right-hand side yields the type graph $T_2 = $ ⚬⟶⚬⟶⊕ . Now as there are no further new matches, the latter unfolding is actually the full unfolding.

5 ω-Adhesive Categories and the Coreflection Result

In this section we propose ω-adhesive categories as a framework in which the unfolding construction is feasible and can be characterized as the right adjoint to the inclusion functor from the full sub-category of occurrence grammars into the category of all finite grammars. Note however that once we have the extra structure of ω-adhesive categories, we could also relax the condition on grammars from finite to countable.

As we mentioned at the beginning of Section 4.2, the unfolding U_G of a grammar G will be a single occurrence grammar that represents the complete chain of truncations generated by the algorithm of Definition 22. The colimits that we will use to construct U_G and to prove the coreflection result are *Van Kampen* (VK) *fans*: they are the ω-chain counterpart of Van Kampen squares, the latter being the central concept in the definition of adhesive categories in [14].

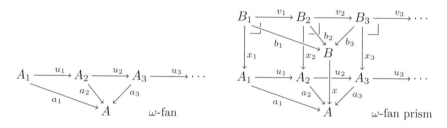

Definition 24 (ω-adhesive categories). *An ω-fan is an ω-chain diagram $\mathcal{A} = \{A_n -u_n\rightarrow A_{n+1}\}_{n\in\mathbb{N}}$ with a cocone $\alpha = \{A_n -a_n\rightarrow A\}_{n\in\mathbb{N}}$ (see the left one of the displayed diagrams); it is a colimit ω-fan if α is a colimit of $\mathcal{A}$, and it is a Van Kampen fan if in each ω-fan prism over it, as illustrated in the right one of the displayed diagrams, having pullback squares $\begin{smallmatrix}B_i & \rightarrow & B_{i+1}\\ \downarrow & & \downarrow\\ A_i & \rightarrow & A_{i+1}\end{smallmatrix}$ as back faces, the top face is a colimit ω-fan if and only if all lateral trapezia $\begin{smallmatrix}B_i & \rightarrow & B\\ & \searrow & \downarrow\\ & & A_i \rightarrow A\end{smallmatrix}$ are pullbacks.*

Now a category is ω-adhesive *if it is adhesive, and moreover*

– *it has colimits of monic ω-chains $\{A_n \succ^{u_n}\rightarrow A_{n+1}\}_{n\in\mathbb{N}}$, and*
– *colimits of monic ω-chains give rise to Van Kampen fans.*

From now on, we assume $\mathbb{C}$ to be ω-adhesive. To ensure soundness of the full unfolding construction in Definition 26, we need the following lemma, which can be shown in analogy to Lemma 2.3 of [14].

Lemma 25 (Monic VK-fans). *Let $\mathbb{C}$ be any category, let $\{A_n \succ^{u_n}\rightarrow A_{n+1}\}_{n\in\mathbb{N}}$ be a monic ω-chain paired with a cocone $\{A_n -a_n\rightarrow A\}_{n\in\mathbb{N}}$ such that they together form a Van Kampen fan. Then each $a_n\colon A_n \rightarrowtail A$ is monic.*

Definition 26 (Full unfolding). *Let* $G = \langle Q, s \colon S \to T \rangle$ *be a finite grammar, and let* $\{U_n \rightarrowtail_{\sharp t_n} U_{n+1}\}_{n \in \mathbb{N}}$ *be the chain constructed as in Definition 22, where* $U_n = \langle Q_n, s_n \colon S \rightarrowtail T_n \rangle$ *and* $t_n \colon T_n \rightarrowtail T_{n+1}$ *for each* $n \in \mathbb{N}$. *To define the* full *unfolding* U_G, *let* $\iota = \{i_n \colon T_n \rightarrowtail T^U\}_{n \in \mathbb{N}}$ *be the colimit of the* ω*-chain diagram* $\mathcal{T} = \{T_n \rightarrowtail_{t_n} T_{n+1}\}_{n \in \mathbb{N}}$, *and put* $U_G := \langle \bigcup_{n \in \mathbb{N}} \sharp i_n(Q_n), i_0 \colon S \rightarrowtail T^U \rangle$.

Finally, to define the folding morphism $\mathfrak{F} \colon U_G \to G$, *let the* $\lambda_n \colon T_n \to T$ *be as in Definition 22. By the universal property of the colimit* ι, *there is a unique folding arrow* $\lambda \colon T^U \to T$ *satisfying* $\lambda \circ i_n = \lambda_n$ *for all* $n \in \mathbb{N}$; *now put* $\mathfrak{F} := \sharp \lambda$.

Proposition 27 (Completeness of the unfolding). *Let* G *be a grammar and* $\sharp \lambda \colon U_G \to G$ *be the folding morphism from the full unfolding* U_G.

Then each derivation in G *has a unique counterpart in* U_G, *i.e. for each rewriting sequence* $s \models_{\langle q'_1, m'_1 \rangle} \Rightarrow a'_1 \cdots \models_{\langle q'_n, m'_n \rangle} \Rightarrow a'_n$ *in* G, *there is a unique sequence* $s^U \models_{\langle q_1, m_1 \rangle} \Rightarrow a_1 \cdots \models_{\langle q_n, m_n \rangle} \Rightarrow a_n$ *in the unfolding* U_G *such that* m'_i, q'_i, a'_i *are the images of* m_i, q_i, a_i *under the retyping with* λ.

This proposition is sufficient for many applications, but it does not rule out that U_G might contain superfluous information. The coreflection result ensures that the unfolding with the folding morphism $\mathfrak{F} \colon U_G \to G$ is the "minimal" or – more precisely – *universal* choice of an occurrence grammar O and a morphism $\mathcal{H} \colon O \to G$ that is complete in the sense of Proposition 27.

Theorem 28 (Coreflection). *Let* $\mathfrak{F} \colon U_G \to G$ *be the folding morphism from the unfolding* U_G *of a finite grammar* G. *Then for each occurrence grammar* O *and morphism* $\mathcal{H} \colon O \to G$ *there is a unique morphism* $\mathcal{V} \colon O \to U_G$ *such that* $\mathcal{H} = \mathfrak{F} \circ \mathcal{V}$.

Proof (idea). Existence and uniqueness of some morphism $\mathcal{V} = \langle V, \zeta \rangle$ follow from two facts: first, the morphism $\mathcal{H} = \langle \mathcal{H}, \vartheta \rangle$ determines ζ and $\lfloor \_ \rfloor_{T^U} \circ V$, and in fact the only information missing is the value $V(\mathrm{id}_{T'})$ – here $O = \langle Q', s' \colon S' \to T' \rangle$; second, the type object T' is the "tip" of a VK fan, the diagram of which is determined by the start object and all the right hand sides. Pulling back this fan along $\vartheta_{\mathrm{id}_{T'}}$ yields again a colimit fan, and $V(\mathrm{id}_{T'})$ arises as a uniquely determined mediating morphism. $\square$

This theorem directly implies that the unfolding construction extends to a functor from the category of finite (or even countable) grammars to that of occurrence grammars, which in turn means that the category of occurrence grammars is a coreflective subcategory of the category of finite grammars.

As for examples and counter examples of ω-adhesive categories: the category $\mathbb{S}$ of *finite* sets – the "primordial" elementary topos – is *not* ω-adhesive; as a fact, any elementary topos is ω-adhesive if and only if it has countable sums. Hence the category of sets is ω-adhesive, and more generally any Grothendieck topos is. Further examples arise via the following constructions.

Proposition 29 (Closure of ω-adhesivity). *Let* $\mathbb{C}$ *and* $\mathbb{D}$ *be* ω*-adhesive categories. Then the following categories are again* ω*-adhesive:*

- *the product category* $\mathbb{C} \times \mathbb{D}$*;*
- *the slice category* $\mathbb{C} {\downarrow} T$ *for any* $T \in \mathbb{C}$*;*
- *the co-slice category* $I {\downarrow} \mathbb{C}$ *for any* $I \in \mathbb{C}$*;*
- *the functor category* $[\mathbb{X}, \mathbb{C}]$ *for any category* $\mathbb{X}$*;*
- *the Artin-Wraith glueing* $\mathbb{C} {\downarrow} \mathcal{F}$*, i.e. the comma category* $\mathbb{C} {\downarrow} \mathcal{F}$ *for any functor* $\mathcal{F} \colon \mathbb{D} \to \mathbb{C}$ *that preserves pullbacks.*

Proof (Sketch). In each case pullbacks and the relevant colimits are constructed componentwise.

In addition we know that the slice construction preserves stability of pseudo-complementation. Note also that all examples of "graph categories" mentioned in the introduction (undirected and directed graphs, hypergraphs, graphs with scopes, graphs with second-order edges, etc.) are ω-adhesive and RPCs are stable under pullback.

6   Related Work and Conclusion

Our work is strongly related to earlier work on true concurrency in the setting of adhesive categories. For instance [14] shows parallel and sequential independence results for adhesive rewriting systems. As a next step, in [2] it has been shown how one can represent computations of a system as processes, i.e. as deterministic occurrence grammars. In the present paper we generalize this work to non-deterministic occurrence grammars (or branching processes) that record all events of a set of possible computations.

The central contribution is the generalization of the unfolding technique to the abstract setting of ω-adhesive categories and the theorem that the unfolding construction, "unravelling" a grammar into an occurrence grammar, can be characterized as a coreflection. As this result holds in any ω-adhesive category with some mild restrictions it applies to numerous application-relevant instances of graph-like structures, and hence it is unnecessary to prove it again and again.

Furthermore we have introduced a new notion of grammar morphisms where the retyping is given by a functor. This allows us to treat also non-semi-weighted grammars, i.e. grammars where the start graph or the right-hand sides might not be injectively typed. Otherwise technical complications arise because of the presence of "too much symmetry" in the structure which is being unfolded and hence the uniqueness of arrow $\underline{\mathcal{V}}$ in Theorem 27 cannot be guaranteed. Another solution to the symmetry problem has been proposed in [11], for the case of Petri nets and with a different notion of morphism.

Our definition of grammar morphism is not the most general one that can be conceived: it was inspired by previous works on graph transformation [1] and on unfolding constructions characterized as coreflections [20]. It is not obvious to what extent the coreflection result of Section 5 could be extended to more general definitions of morphisms: this is an interesting topic for future work.

The unfolding represents all computations as well as all reachable objects of the original grammar in a single acyclic branching structure. Hence, as observed in [17,3], it can serve as the basis for partial order verification techniques. For instance, we plan to generalize the notion of finite complete prefix to the abstract framework of the present paper. Another direction is to adapt the model-based diagnosis techniques of [6,5]; the latter depend on the preservation of products of grammars by the unfolding functor, which is ensured by the coreflection result. In future work we will further investigate products in our category of grammar morphisms.

References

1. Baldan, P.: Modelling Concurrent Computations: from Contextual Petri Nets to Graph Grammars. PhD thesis, Dipartimento di Informatica, Università di Pisa (2000)
2. Baldan, P., Corradini, A., Heindel, T., König, B., Sobociński, P.: Processes for adhesive rewriting systems. In: Aceto, L., Ingólfsdóttir, A. (eds.) FOSSACS 2006. LNCS, vol. 3921, pp. 202–216. Springer, Heidelberg (2006)
3. Baldan, P., Corradini, A., König, B.: A framework for the verification of infinite-state graph transformation systems. Information and Computation 206, 869–907 (2008)
4. Baldan, P., Corradini, A., Montanari, U., Ribeiro, L.: Unfolding Semantics of Graph Transformation. Information and Computation 205, 733–782 (2007)
5. Baldan, P., Chatain, T., Haar, S., König, B.: Unfolding-based diagnosis of systems with an evolving topology. In: van Breugel, F., Chechik, M. (eds.) CONCUR 2008. LNCS, vol. 5201, pp. 203–217. Springer, Heidelberg (2008)
6. Benveniste, A., Fabre, E., Haar, S., Jard, C.: Diagnosis of asynchronous discrete event systems, a net unfolding approach. IEEE Transactions on Automatic Control 48(5), 714–727 (2003)
7. Birkhoff, G.: Lattice Theory. American Mathematical Society (1967)
8. Cockett, R., Guo, X.: Join restriction categories and the importance of being adhesive. Unpublished manuscript, slides from CT 2007 (2007), http://pages.cpsc.ucalgary.ca/~robin/talks/jrCat.pdf
9. Ehrig, H., Habel, A., Kreowski, H.-J., Parisi-Presicce, F.: Parallelism and Concurrency in High-Level Replacement Systems. Mathematical Structures in Computer Science 1, 361–404 (1991)
10. Ehrig, H., Pfender, M., Schneider, H.J.: Graph-grammars: an algebraic approach. In: Proc. of IEEE Conf. on Automata and Switching Theory, pp. 167–180 (1973)
11. Hayman, J., Winskel, G.: The unfolding of general Petri nets. In: Proc. of FSTTCS 2008. Dagstuhl Seminar Proceedings, vol. (08004) (2008)
12. Heindel, T., Sobociński, P.: Van Kampen colimits as bicolimits in Span. In: Kurz, A., Lenisa, M., Tarlecki, A. (eds.) CALCO 2009. LNCS, vol. 5728, pp. 335–349. Springer, Heidelberg (2009)
13. Johnstone, P.T.: Sketches of an Elephant, vol. 1. Oxford Science Publications (2002)
14. Lack, S., Sobociński, P.: Adhesive and quasiadhesive categories. Theoretical Informatics and Applications 39(2), 511–546 (2005)

15. Lack, S., Sobociński, P.: Toposes are adhesive. In: Corradini, A., Ehrig, H., Montanari, U., Ribeiro, L., Rozenberg, G. (eds.) ICGT 2006. LNCS, vol. 4178, pp. 184–198. Springer, Heidelberg (2006)
16. Löwe, M.: Algebraic approach to single-pushout graph transformation. Theoretical Computer Science 109, 181–224 (1993)
17. McMillan, K.L.: Symbolic Model Checking. Kluwer Academic Publishers, Dordrecht (1993)
18. Meseguer, J., Montanari, U., Sassone, V.: On the semantics of Place/Transition Petri nets. Mathematical Structures in Computer Science 7(4), 359–397 (1997)
19. Robinson, E., Rosolini, G.: Categories of partial maps. Inf. Comput. 79(2), 95–130 (1988)
20. Winskel, G.: Event structures. In: Brauer, W., Reisig, W., Rozenberg, G. (eds.) APN 1986. LNCS, vol. 255, pp. 325–392. Springer, Heidelberg (1987)

Views on Behaviour Protocols and Their Semantic Foundation*

Sebastian S. Bauer and Rolf Hennicker

Institut für Informatik, Ludwig-Maximilians-Universität München, Germany
{bauerse,hennicker}@pst.ifi.lmu.de

Abstract. We consider UML protocol state machines where states are annotated with invariants and transitions are labelled with pre- and postconditions of operations. We claim that the meaning of a protocol strongly depends on the role in which the protocol is actually used. In our study we distinguish three different views on protocols: The user's and the implementor's perspective, which both are inspired by the contract principle, and an interaction perspective which focuses on collaborations. For each view we define a model-theoretic semantics based on labelled transition systems and simulation relations integrating control flow and the evolution of data states. Our approach is compositional in the sense that correct user and implementation models can be composed to a correct interaction model. Moreover, we define a refinement relation for protocols which retains our compositionality results.

1 Introduction

Protocols are an important means to specify externally visible behaviours and to regulate collaborations of software units. There is a long tradition to specify and apply protocols in software development. Historically, one can distinguish essentially two protocol styles considered in the literature. First, there are control-flow-based protocols which specify sequences of actions as legal execution paths in some given context. This type of protocol is supported by various formalisms, most prominently by process algebras, but also by specialised frameworks tailored to software components and their interfaces; cf. e.g. [10,6]. There is another type of protocol which focuses instead on the information stored in software units. These approaches consider the evolution of data states typically specified by invariants and pre- and postconditions of operations; cf. e.g. [7,12]. Some approaches also provide means to combine both aspects, CSP-OZ [13] and the protocol state machines of the UML [9]. Concerning the expressive power, the advantage of the latter is that control flow and (changing) data states are integrated in a mutually dependent way such that it is possible to specify the effect of an operation on data states depending on the current control state in which the operation is invoked. UML protocol state machines define protocols which can be attached to a classifier, typically to an interface or to a port.

---

* This research has been partially supported by the GLOWA-Danube project 01LW0602A2 sponsored by the German Federal Ministry of Education and Research.

A. Kurz, M. Lenisa, and A. Tarlecki (Eds.): CALCO 2009, LNCS 5728, pp. 367–382, 2009.

In this paper we are interested in a rigorous formal foundation of behavioural protocols which integrate control flow and data states in the sense of UML protocol state machines. To get an idea of the meaning of protocol state machines, let us look at the UML specification [9] where one can read, e.g., the following.

(a) "The protocol state machine defines all allowed transitions for each operation" and "...what result is expected from their use."
(b) "It specifies which operations of the classifier can be called in which state and under which condition, thus specifying the allowed call sequences on the classifier's operations." and "Protocol state machines help define the usage mode of the operations ...".
(c) The protocol of a port "describes valid interactions at this interaction point".

Though these informal descriptions do not really clarify what the precise meaning of a protocol state machine is, they give at least a hint that there are different views under which a protocol can be considered. Statement (a) can be understood as a requirement for the implementation of the operations offered, e.g., by a provided interface. Thus, (a) corresponds to the implementor's view on a protocol. Obviously, statement (b) refers to requirements that a user of the operations belonging, e.g., to a required interface must satisfy. Thus, (b) corresponds to the user's view on a protocol. Finally, statement (c) shows that there is still a third view where a protocol is seen as a specification of admissible interactions. As a consequence we claim that there is not a single interpretation for a given protocol but we must carefully distinguish between the different roles under which the protocol is used and express the correponding requirements in the semantics of each view.

For the formalisation of the different kinds of protocol semantics we follow a model-theoretic approach where we propose, for a given protocol P, an implementation semantics $[\![P]\!]_{impl}$, a user semantics $[\![P]\!]_{use}$, and an interaction semantics $[\![P]\!]_{iact}$. In each case, $[\![P]\!]_v$ (with $v \in \{impl, use, iact\}$) consists of a class of models, given by appropriate kinds of labelled transition systems, which are considered to be correct implementation, user, and interaction models with respect to the requirements expressed by P under the respective view. For our correctness notions we introduce different kinds of simulation relations between protocol states and model states (representing control and data), which respect invariants and pre- and postconditions. The use of simulation relations is quite usual in event based formalisms but there are less approaches which integrate data states, cf. e.g. [8]. We are not aware of any work on simulations between protocols on the specification level, with pre- and postconditions, and transition systems on the model level. Transition systems are also frequently used for the semantics of UML state machines (see [3] for an overview) but they do neither reflect the contract principle nor the different views described above. With regard to our previous work, the semantics elaborated for class specifications in [2] and for component interfaces in [1] can be considered as special cases of the current approach without taking into account control states and user models.[1]

On the basis of our semantic definitions we study the (synchronous) composition of user and implementation models and we show that any correct user and any correct implementation of a protocol P are composable to an interaction model which satisfies the interaction requirements expressed by P. Thus user and implementation models can be developed independently of each other. Then we focus on refinements of protocols which, according to the loose semantics approach, can be simply defined by model class inclusions; an idea which stems from the theory of algebraic specifications and their implementations; cf. e.g. [11]. We provide a syntactic criterion for protocol refinement in terms of protocol simulation. Finally, we introduce a composition operator for protocols which generalises the (synchronous) parallel composition operator for labelled transition systems to take into account state invariants and pre- and postconditions. As a final result, we show that under the assumption of protocol refinement the model-theoretic composition of user and implementation models is compatible with protocol composition.

Outline. After providing some technical preliminaries we introduce, in Sect. 2, protocol signatures and the syntax of behaviour protocols. Then we focus on the different kinds of protocol semantics, implementation semantics in Sect. 3, user semantics in Sect. 4, and interaction semantics in Sect. 5. In Sect. 5 we also provide our first compositionality result for user and implementation models. In Sect. 6, we discuss refinement and composition of protocols and in Sect. 7 we finish with some concluding remarks.

Preliminaries. Our approach is based on labelled transition systems which will occur in different variations. Generally, a *labelled transition system* (LTS) $M = (Q, q_0, L, \Delta)$ consists of a set Q of states, the initial state $q_0 \in Q$, a set L of labels, and a transition relation $\Delta \subseteq Q \times L \times Q$.[2] We also consider LTS *with* $\bot$ which have a transition relation $\Delta \subseteq Q \times L \times Q^\bot$ where $Q^\bot = Q \cup \{\bot\}$. A transition of the form $(q, l, \bot)$ expresses divergence. An LTS (with $\bot$) is *input-enabled* if for any state $q \in Q$ and label $l \in L$ there is at least one transition with source state q and label l. Reachability of states $q \in Q$ (and of $\bot$) is defined in the obvious way. An LTS *with state labelling* is a quintuple $M = (Q, q_0, L, \Delta, l)$ where (Q, q_0, L, Δ) is an LTS and $l : Q \to D$ is a function associating to each state $q \in Q$ an element of some given set D.

2   Behaviour Protocols

In this section we define the syntactical constructs used for building behaviour protocols which will be graphically represented by UML protocol state machines.

---

[1] In fact, glass box models in [2] and interface models in [1] can be considered as special cases of the implementation models introduced here where the control flow aspect is omitted; similarly the idea behind black box models in [2] is related to the current notion of an interaction model.

[2] A transition $(q, l, q') \in \Delta$ is frequently denoted by $q \xrightarrow{l} q'$.

For their abstract syntax we use mathematical notations which define precisely the necessary concepts in accordance with the UML notation. The essential difference to the UML is that we do not consider composite states; we will, however, be more precise in our assumptions on constraints expressing invariants, pre- and postconditions, and their interpretation.

We assume given a class $Sign_{\mathrm{Obs}}$ of *observer signatures*. Each observer signature $\Sigma_{\mathrm{Obs}} \in Sign_{\mathrm{Obs}}$ defines a set $Obs(\Sigma_{\mathrm{Obs}})$ of *observers* that can be used to observe the externally visible data of a state. For any observer signature Σ_{Obs} we assume given the following sets of Σ_{Obs}-*predicates* which both are closed under the usual logical connectives (like $\wedge$, $\vee$, etc.):

(1) a set $\mathcal{S}(\Sigma_{\mathrm{Obs}})$ of *state predicates* φ, and
(2) a set $\mathcal{T}(\Sigma_{\mathrm{Obs}})$ of *transition predicates* π.

State predicates will be used to express invariants on states and transition predicates will be used to relate pre- and poststates of transitions. We assume that to each state predicate φ and transition predicate π there is associated a set $\mathrm{var}(\varphi)$ ($\mathrm{var}(\pi)$ resp.) of (logical) variables. We do not fix a special form of observer signatures, observers, and predicates; for more details see, e.g., [1]. In our examples observers are just visible state variables of some primitive types. For the representation of predicates we will use an OCL-like notation; cf. [12].

Protocol signatures extend observer signatures by introducing operations. An *operation op* has the form $opname(X_{\mathrm{in}})$ where X_{in}, in the following denoted by $\mathrm{var}_{\mathrm{in}}(op)$, is a (possibly empty) sequence of input variables. For the sake of simplicity we do not consider output variables here which, however, could be easily integrated in our framework.

Definition 1 (Protocol signature). *A protocol signature* $\Sigma = (\Sigma_{\mathrm{Obs}}, Op)$ *consists of an observer signature* $\Sigma_{\mathrm{Obs}} \in Sign_{\mathrm{Obs}}$ *and a set* Op *of operations.*

Example 1. Our running example is taken from [5] and extended by state variables. We consider a turnstile located at the entrance of a subway. The observer signature of the turnstile consists of two visible state variables of integer type: *passed* for the number of persons that have passed the turnstile and *fare* for the actual costs of a trip. There are three operations: *coin*($x : int$) for dropping a coin with amount x into the turnstile's slot, *pass*() for passing through the turnstile and *ready*() for deactivating a possible alarm mode of the turnstile. Formally, the turnstile has the protocol signature $\Sigma_{\mathrm{Turnstile}} = (\{passed : int, fare : int\}, \{coin(x : int), pass(), ready()\})$.   ♣

A protocol expresses constraints on the observable behaviour of a given classifier like an interface or a port. Protocols specify the legal sequences of operation invocations, their invocation conditions, and their expected effect in terms of data changes. Moreover, a data constraint can be associated to any protocol state. Formally, a behaviour protocol over a given protocol signature $\Sigma = (\Sigma_{\mathrm{Obs}}, Op)$ is a labelled transition system whose transitions are labelled by triples of the form $(\varphi_{\mathrm{pre}}, op, \varphi_{\mathrm{post}})$. Thereby op represents the invocation of an operation $op \in Op$,

φ_{pre} is a state predicate over Σ_{Obs} representing a precondition under which the operation can be called in the actual protocol state and φ_{post} is a transition predicate over Σ_{Obs} expressing the postcondition that must hold after the execution of the operation when the next protocol state is reached. Thus protocols can constrain both the control flow and the evolution of data. In particular, a behaviour protocol can express that calling an operation op in a protocol state, say s_1, has a different effect on the data as if the same operation is called in another protocol state, say s_2. To formalise data constraints on states, we use a state labelling function which assigns to each protocol state a state predicate, called *state invariant*.[3]

Definition 2 (Σ-Protocol). *Let $\Sigma = (\Sigma_{\mathrm{Obs}}, Op)$ be a protocol signature. A Σ-protocol is an LTS $P = (S, s_0, PLabel_\Sigma, \Delta, l)$ with state labelling l, where*

a) S is a set of protocol states,
b) $s_0 \in S$ is the initial protocol state,
c) $PLabel_\Sigma = \{(\varphi_{\mathrm{pre}}, op, \varphi_{\mathrm{post}}) \mid op \in Op, \varphi_{\mathrm{pre}} \in \mathcal{S}(\Sigma_{\mathrm{Obs}}), \varphi_{\mathrm{post}} \in \mathcal{T}(\Sigma_{\mathrm{Obs}}),$ $\mathrm{var}(\varphi_{\mathrm{pre}}) \subseteq \mathrm{var}_{\mathrm{in}}(op), \mathrm{var}(\varphi_{\mathrm{post}}) \subseteq \mathrm{var}_{\mathrm{in}}(op)\}$ is the set of protocol labels,
d) $\Delta \subseteq S \times PLabel_\Sigma \times S$ is a protocol transition relation, and
e) $l : S \to \mathcal{S}(\Sigma_{\mathrm{Obs}})$ is a state labelling with $\mathrm{var}(l(s)) = \emptyset$ for all $s \in S$.

Since pre- and postconditions usually depend on the arguments of an operation they may contain (at most) the operation's input variables, cf. c), while state invariants must not contain any variable, cf. e).

Example 2. The protocol state machine in Fig. 1 describes a protocol $P_{\mathrm{Turnstile}}$ for the interface of the subway turnstile introduced in Ex. 1. The graphical UML notation can be translated in an obvious way to the abstract syntax defined above. For instance, concrete labels of the form $[\varphi_{\mathrm{pre}}]op/[\varphi_{\mathrm{post}}]$ represent protocol labels $(\varphi_{\mathrm{pre}}, op, \varphi_{\mathrm{post}})$. For the specification of state invariants and

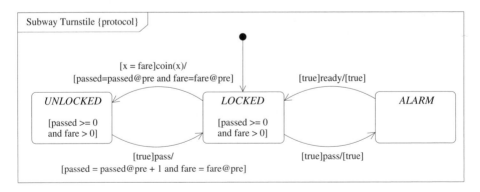

Fig. 1. UML Protocol State Machine of a Subway Turnstile

---

[3] State invariants can vary from state to state; if the same invariant is used for all protocol states then this would correspond to the classical concept of an invariant.

pre- and postconditions we use an OCL-like notation. The state $LOCKED$ is the initial state where the fare must be positive and the number of passed persons must not be negative. If in this state a coin is deposited whose value matches the fare the turnstile will become unlocked while the values of the state variables remain unchanged. In the state $UNLOCKED$ a person can pass through the turnstile with the effect that the number of passed persons is increased by one while the fare remains unchanged and the state $LOCKED$ is reached again. If a person tries to pass the turnstile without dropping a coin into its slot the turnstile switches into an alarm mode. The alarm can be shut off by invoking *ready*. The protocol is loose, e.g., it does not specify the effect on the data for transitions entering and leaving the $ALARM$ state. Hence the fare and the number of passed persons could be corrupted in this exceptional situations.   ♣

Since a protocol is a purely syntactic construct, it is an obvious issue to ask for its precise meaning. We claim that there is no single answer since there are different views under which a protocol can be considered. For instance, an interface to which a protocol is attached to can play the role of a provided or of a required interface. Hence, the protocol will express different requirements depending whether it is seen from the implementor's or from the user's point of view. In each case the control flow specified by a protocol together with the constraints on the data imposes obligations and guarantees for the supplier *and* for the client which will be reflected by the respective protocol semantics provided in Sect. 3 and Sect. 4. Moreover, there is even a third view on a protocol if it is used to describe the legal interactions between two components which will be addressed in Sect. 5.

3   Implementation Semantics of Protocols

In this section we consider protocols being attached to a *provided* interface. Then an implementor (supplier) is responsible for the implementation of the operations offered by the interface in accordance with the protocol. Before we can discuss implementation correctness, we first have to formalise the notion of an implementation. For this purpose we follow a model-theoretic approach. The idea is that for a given protocol signature $\Sigma = (\Sigma_{\mathrm{Obs}}, Op)$, a Σ-implementation model provides an abstract representation of a program which realises the operations of Op. Technically, we use labelled transition systems where labels represent operation calls (with actual input parameters) and a transition connects the state before the operation call and the state after execution of the called operation. To model divergence we use the distinguished element $\perp$ and a transition to $\perp$ models the fact that the execution of an operation does either not terminate or halts due to a runtime error. We assume that an operation can always be called as long as $\perp$ is not reached, i.e. that the transition system is input-enabled.

The states of a Σ-implementation model must carry information of both, the control flow and the actual data values observable by the state observers of $Obs(\Sigma_{\mathrm{Obs}})$. To discriminate both aspects we distinguish between control states and data states. A data state is determined by the (current) values of the observers. Hence, we associate to any observer signature Σ_{Obs} the class of data states $DState(\Sigma_{\text{Obs}}) = [Obs(\Sigma_{\text{Obs}}) \rightarrow \mathcal{D}]$ of total functions mapping observers to values in a predefined semantic domain $\mathcal{D}$. In the turnstile example a data state σ determines two integer values $\sigma(fare)$ and $\sigma(passed)$. The underlying state space of a Σ-implementation model is then formed by the cartesian product of a set of control states and a set of data states.

The labels of a Σ-implementation model are formally defined as follows: First, we define the set $Label_\Sigma = \{(op, \rho) \mid op \in Op, \rho : \text{var}_{\text{in}}(op) \rightarrow \mathcal{D}\}$ where each pair (op, ρ) represents an invocation of the operation op with actual input parameters determined by a valuation $\rho : \text{var}_{\text{in}}(op) \rightarrow \mathcal{D}$. The labels of a Σ-implementation model have then the form $in(op, \rho)$ where the symbol in expresses the fact that the operation invocation (op, ρ) is considered as an input for the implementation (on which the implementation reacts).

Definition 3 (Σ-Implementation Model). *Let $\Sigma = (\Sigma_{\text{Obs}}, Op)$ be a protocol signature. A Σ-implementation model (Σ-implementation for short) is an input-enabled LTS with $\bot$ of the form $M = (C \times Q, (c_0, \sigma_0), In(Label_\Sigma), \Delta)$ where*

  a) *$C \times Q$ is a set of implementation states over a set C of control states and a set $Q \subseteq DState(\Sigma_{\text{Obs}})$ of data states,*
  b) *$(c_0, \sigma_0) \in C \times Q$ is the initial state,*
  c) *$In(Label_\Sigma) = \{in(op, \rho) \mid (op, \rho) \in Label_\Sigma\}$ is the set of (input) labels, and*
  d) *$\Delta \subseteq (C \times Q) \times In(Label_\Sigma) \times (C \times Q)^\bot$ is a transition relation.*

The class of all Σ-implementation models is denoted by $Impl(\Sigma)$.

Let us now discuss implementation correctness w.r.t. a given protocol. From the implementor's point of view a protocol can be considered as a contract which guarantees (for the implementor) that each invocation of an operation occurs in accordance with the given protocol transitions. This means that an operation is only called in a state where the call is admissible and that the given precondition for this operation invocation is satisfied. If these assumptions are not met then the implementation can have any arbitrary behaviour including divergence. Under the assumption of the correct usage of the operations according to the protocol the implementor must ensure that the implementation does not diverge, that in each reachable state the specified state invariants are valid, and that after execution of a correctly invoked operation the specified postcondition is satisfied. To formalise these conditions we require that for a given Σ-protocol P and a Σ-implementation M there exists a simulation relation such that M simulates P in accordance with the protocol transitions and the specified data constraints. For the formalisation of this idea we must first be able to interpret data constraints, given in the form of state predicates and transition predicates, with respect to the data state space $DState(\Sigma_{\text{Obs}})$. For this purpose, we assume that, for any observer signature Σ_{Obs}, the Σ_{Obs}-predicates are equipped with a *satisfaction relation* $\models$ for Σ_{Obs}-states σ and valuations ρ such that the logical connectives are interpreted as usual:

(1) $\sigma; \rho \vDash \varphi$ for a state predicate φ, a data state $\sigma \in DState(\Sigma_{\mathrm{Obs}})$, and a valuation $\rho : \mathrm{var}(\varphi) \to \mathcal{D}$;

(2) $\sigma, \sigma'; \rho \vDash \pi$ for a transition predicate π, data states $\sigma, \sigma' \in DState(\Sigma_{\mathrm{Obs}})$, and a valuation $\rho : \mathrm{var}(\pi) \to \mathcal{D}$.

If $\mathrm{var}(\varphi) = \emptyset$ the satisfaction relation must be independent of any valuation ρ and we write briefly $\sigma \vDash \varphi$. Again, to be generic, we do not assume any particular satisfaction relation. For instance, if boolean OCL expressions are used for the predicates then the satisfaction relation would correspond to checking whether an expression evaluates to *true*; cf. [2]. We are now able to define our notion of implementation simulation in terms of a relation between protocol states and implementation states.

Definition 4 (Implementation Simulation). *Let* $P = (S, s_0, PLabel_\Sigma, \Delta_P, l)$ *be a Σ-protocol, and* $M = (C \times Q, (c_0, \sigma_0), In(Label_\Sigma), \Delta_M) \in Impl(\Sigma)$. *An implementation simulation between P and M is a relation* $\precsim_{impl} \subseteq S \times (C \times Q)$ *such that*

a) $s_0 \precsim_{impl} (c_0, \sigma_0)$ *and* $\sigma_0 \vDash l(s_0)$,

b) *for all* $s \precsim_{impl} (c, \sigma)$ *with* $\sigma \vDash l(s)$, *for all* $(\varphi_{\mathrm{pre}}, op, \varphi_{\mathrm{post}}) \in PLabel_\Sigma$, *and for all valuations* $\rho : \mathrm{var}_{\mathrm{in}}(op) \to \mathcal{D}$, *it holds:*

if $s \xrightarrow{(\varphi_{\mathrm{pre}}, op, \varphi_{\mathrm{post}})} s' \in \Delta_P$ *and* $\sigma; \rho \vDash \varphi_{\mathrm{pre}}$, *then* $(c, \sigma) \xrightarrow{in(op, \rho)} \bot \notin \Delta_M$,

and for all transitions $(c, \sigma) \xrightarrow{in(op, \rho)} (c', \sigma') \in \Delta_M$, $s' \precsim_{impl} (c', \sigma')$ *and* $\sigma, \sigma'; \rho \vDash \varphi_{\mathrm{post}}$ *and* $\sigma' \vDash l(s')$.

The idea of condition b) is that for any protocol transition, whenever (the data part of) the source state of an implementation transition satisfies the precondition of the invoked operation, then the implementation does not diverge and for all target states the postcondition and the required invariant are valid. M may arbitrarily react to operations calls which are not admissible according to P.

Definition 5 (Implementation Correctness). *Let P be a Σ-protocol, and let* $M \in Impl(\Sigma)$. *M is a correct Σ-implementation of P, denoted by* $M \vDash_{impl} P$, *if there is an implementation simulation between P and M. The class of all correct Σ-implementations of P is defined by* $[\![P]\!]_{impl} = \{M \in Impl(\Sigma) \mid M \vDash_{impl} P\}$.

4   User Semantics of Protocols

In this section we consider the user's (client's) point of view where a protocol is attached to a *required* interface. Then the user is responsible for the correct usage of the operations of the required interface in accordance with the protocol. For a given protocol signature $\Sigma = (\Sigma_{\mathrm{Obs}}, Op)$, a Σ-user model describes an abstraction of an algorithm where the actions are invocations of the required operations according to the needs of the user. The state space of a user model does only include control states and the transition relation determines the sequences of operation invocations. So far, we have not taken into account data states which belong to the states of implementation models. But, since data states are assumed to be visible, a user might decide which operation to invoke next in dependency of the (current) data state of an implementation; e.g. to guarantee preconditions. To formalise the possible dependency of an operation invocation on data states we include data states in the transition labels of user models which then have the form $(\sigma, out(op, \rho))$ with $\sigma \in DState(\Sigma_{Obs})$ and with the operation call (op, ρ) surrounded by the symbol out to indicate that the operation invocation is an output of the user.

Definition 6 (Σ-User Model). *Let $\Sigma = (\Sigma_{Obs}, Op)$ be a protocol signature. A Σ-user model (Σ-user for short) is an LTS of the form $N = (C, c_0, ULabel_\Sigma, \Delta)$ where*

a) *C is a set of control states,*
b) *$c_0 \in C$ is the initial state,*
c) *$ULabel_\Sigma = DState(\Sigma_{Obs}) \times Out(Label_\Sigma)$ is the set of labels where $Out(Label_\Sigma) = \{out(op, \rho) \mid (op, \rho) \in Label_\Sigma\}$, and*
d) *$\Delta \subseteq C \times ULabel_\Sigma \times C$ is a transition relation.*

The class of all Σ-users is denoted by $User(\Sigma)$.

Example 3. Let us consider a user of the turnstile who first inserts a coin whose value is equal to the required fare and then passes through the turnstile. A corresponding user model has three control states *start*, *paid*, and *passed*, and for all data states σ and σ' the transitions[4]

$$start \xrightarrow{(\sigma, out(coin(\sigma(fare))))} paid, \quad paid \xrightarrow{(\sigma', out(pass()))} passed.$$

This means that for each data state σ of a turnstile implementation the user first inserts a coin whose value meets exactly the required fare in state σ and then, independently of the implementation state, he/she proceeds by passing through the turnstile. An abstract program representing this situation could be formulated by `int f = getFare(); coin(f); pass();`. Another example would be a user who only pays the fare and passes if the fare does not exceed the value of 5 which could be represented by `int f = getFare(); if (f <= 5) {coin(f); pass();}`. Finally, a third user who is willing to pay exactly the amount of 5 and then will anyway try to pass could be represented by `coin(5); pass();`. Since, in a concrete turnstile implementation, the fare may be different from 5 the last user exhibits a wrong behaviour w.r.t. the turnstile protocol. ♣

We will now formalise what it means for a user to behave correctly w.r.t. a given protocol. From the user's point of view a protocol can again be seen as a contract which requires that the user invokes an operation only in a state where the call is admissible and where the given precondition is satisfied. On the other hand, the user can rely on the fact that in all (reachable) states the state invariants are valid and that after the execution of an operation the given postcondition

---

[4] $Coin(\sigma(fare)))$ is an abbreviation for $(coin(x : int), \rho)$ with $\rho(x) = \sigma(fare)$.

is satisfied. These conditions can again be formalised by means of a simulation relation. Given a protocol P and a user model N, the idea is that any transition of N should be in accordance with (i.e. simulated by) some transition of P. Though data states do not belong to the states of a user they play still a crucial role for the transitions of a user model and therefore must be taken into account in the simulation relation. Hence, a user simulation is a relation between control states of the user together with data states on the one side and protocol states on the other side.

Definition 7 (User Simulation). *Let* $P = (S, s_0, PLabel_\Sigma, \Delta_P, l)$ *be a* Σ-*protocol and* $N = (C, c_0, ULabel_\Sigma, \Delta_N)$ *be a* Σ-*user model. A user simulation between* N *and* P *is a relation* $\precsim_{use} \subseteq (C \times DState(\Sigma_{Obs})) \times S$ *such that*

a) $(c_0, \sigma) \precsim_{use} s_0$ *for all* $\sigma \in DState(\Sigma_{Obs})$ *with* $\sigma \vDash l(s_0)$,
b) *for all* $(c, \sigma) \precsim_{use} s$ *with* $\sigma \vDash l(s)$ *and for all* $(op, \rho) \in Label_\Sigma$, *it holds:*

    if $c \xrightarrow{(\sigma, out(op, \rho))} c' \in \Delta_N$, *then a transition* $s \xrightarrow{(\varphi_{pre}, op, \varphi_{post})} s' \in \Delta_P$ *exists such that*

    (1) $\sigma; \rho \vDash \varphi_{pre}$,
    (2) $(c', \sigma') \precsim_{use} s'$ *for all* $\sigma' \in DState(\Sigma_{Obs})$ *for which* $\sigma, \sigma'; \rho \vDash \varphi_{post}$ *and* $\sigma' \vDash l(s')$.

Condition a) says that any combination of the initial user control state c_0 with a data state σ that satisfies the invariant of the initial protocol state s_0 must be in relation to s_0. The idea behind condition b) is that when the user is in a control state c and invokes an operation under a data state σ, then the invocation must be permitted by some protocol transition whose precondition φ_{pre} is satisfied by σ; cf. (1). By (2) it is guaranteed that in the next related states the data part satisfies the postcondition of the protocol transition (and the invariant of the target protocol state) which expresses the protocol's guarantee for the user.

Definition 8 (User Correctness). *Let* P *be a* Σ-*protocol, and let* $N \in User(\Sigma)$. N *is a correct* Σ-*user of* P, *denoted by* $N \vDash_{use} P$, *if there is a user simulation between* N *and* P. *The class of all correct* Σ-*users of* P *is defined by* $\llbracket P \rrbracket_{use} = \{N \in User(\Sigma) \mid N \vDash_{use} P\}$.

For instance, the first two users of the turnstile described in Ex. 3 are correct w.r.t. the protocol $P_{Turnstile}$ while the third one is not.

5   Interaction Semantics of Protocols and Compositionality

As already pointed out there is a third view on protocols if they are used to specify the legal interactions between two connected components which synchronise on operation invocations. Before we discuss interaction correctness, we introduce interaction models which are again labelled transition systems. The idea is that for a given protocol signature $\Sigma = (\Sigma_{Obs}, Op)$, a Σ-interaction model describes the (runtime) interactions that may occur when a user and an implementation model communicate through a synchronous communication channel. Thus the state space of a Σ-interaction model must carry information on the control flow of the interaction *and* on the evolution of data states. The labels of an interaction model have the form $sync(op, \rho)$ to express synchronisation on an operation invocation (op, ρ). Similar to implementation models, a transition connects the state before the invocation happens and the state after the execution of the called operation. Of course, divergence must also be taken into account to model the case where an operation invoked during an interaction does not terminate or fails. An essential difference to implementation models is, that interaction models are not input-enabled because this would not make any sense for interactions.

Definition 9 (Σ-Interaction Model). *Let* $\Sigma = (\Sigma_{\mathrm{Obs}}, Op)$ *be a protocol signature. A Σ-interaction model is an LTS with $\perp$ of the form* $K = (C \times Q, (c_0, \sigma_0), Sync(Label_\Sigma), \Delta)$ *where* $Sync(Label_\Sigma) = \{ sync(op, \rho) \mid (op, \rho) \in Label_\Sigma \}$. *All other parts are defined in the same way as for Σ-implementation models. The class of all Σ-interaction models is denoted by $IAct(\Sigma)$.*

Given a protocol P, interaction correctness w.r.t. P means that the order of operation invocations occurring during an ongoing interaction as well as the change of data states must be in accordance with the protocol. For the formalisation we consider interaction simulations between interaction models and protocols. In contrast to the definitions of implementation simulation and user simulation, which where essentially governed by the contract principle, the definition of interaction simulation combines both views by requiring that any transition of a given interaction model K must be simulated by some transition of a given protocol P while respecting all the constraints on the data states, i.e. invariants, preconditions *and* postconditions; cf. condition b) below.

Definition 10 (Interaction Simulation). *Let* $P = (S, s_0, PLabel_\Sigma, \Delta_P, l)$ *be a Σ-protocol and* $K = (C \times Q, (c_0, \sigma_0), Sync(Label_\Sigma), \Delta_K) \in IAct(\Sigma)$. *An interaction simulation between K and P is a relation* $\precsim_{iact} \subseteq (C \times Q) \times S$ *such that*

a) $(c_0, \sigma_0) \precsim_{iact} s_0$ *and* $\sigma_0 \vDash l(s_0)$,

b) for all $(c, \sigma) \precsim_{iact} s$ *and for all* $(op, \rho) \in Label_\Sigma$, *it holds:*

$\quad$ *if* $(c, \sigma) \xrightarrow{sync(op,\rho)} (c', \sigma') \in \Delta_K$, *then there exists* $s \xrightarrow{(\varphi_{\mathrm{pre}}, op, \varphi_{\mathrm{post}})} s' \in \Delta_P$ *such that* $(c', \sigma') \precsim_{iact} s'$ *and*

$\qquad$ *1)* $\sigma \vDash l(s)$ *and* $\sigma; \rho \vDash \varphi_{\mathrm{pre}}$,

$\qquad$ *2)* $\sigma' \vDash l(s')$ *and* $\sigma, \sigma'; \rho \vDash \varphi_{\mathrm{post}}$,

c) for all $(c, \sigma) \precsim_{iact} s$ *and for all* $(op, \rho) \in Label_\Sigma$, $(c, \sigma) \xrightarrow{sync(op,\rho)} \perp \notin \Delta_K$.

Due to conditions a) and b) there exists for each reachable state $(c, \sigma) \in C \times Q$ of K a protocol state $s \in S$ such that $(c, \sigma) \precsim_{iact} s$. Hence, by condition c) of the simulation relation, $\perp$ is not reachable in K which is indeed an important requirement for any correct interaction.

Definition 11 (Interaction Correctness). *Let P be a Σ-protocol, and let $K \in IAct(\Sigma)$. K is a correct Σ-interaction model of P, denoted by $K \models_{iact} P$, if there is an interaction simulation between K and P. The class of all correct Σ-interaction models of P is defined by $[\![P]\!]_{iact} = \{K \in IAct(\Sigma) \mid K \models_{iact} P\}$.*

In the following we show that user and implementation models can be composed in a straightforward way to form an interaction model. Technically, the composition of a user model N and an implementation model M results from synchronising the output labels $out(op, \rho)$ of N with the input labels $in(op, \rho)$ of M by taking into account that the current data state of the implemenation must fit to the data state the user has assumed (in its label) when issuing the operation call. The data space of the resulting composition model $N \otimes M$ consists, as expected, of the product of the user (control) states and the implementation states. In fact, $N \otimes M$ represents the runtime behaviour of the two communicating user and implementation models.

Definition 12 (Model Composition). *Let $N = (C_N, c_{N,0}, ULabel_\Sigma, \Delta_N) \in User(\Sigma)$ and $M = (C_M \times Q, (c_{M,0}, \sigma_0), In(Label_\Sigma), \Delta_M) \in Impl(\Sigma)$. The model composition of N and M is the LTS with $\perp$*

$$N \otimes M = (C_N \times C_M \times Q, (c_{N,0}, c_{M,0}, \sigma_0), Sync(Label_\Sigma), \Delta_{N \otimes M})$$

where the transition relation $\Delta_{N \otimes M}$ is the least relation satisfying

a) if $c_N \xrightarrow{(\sigma, out(op,\rho,))} c'_N \in \Delta_N$ and $(c_M, \sigma) \xrightarrow{in(op,\rho)} (c'_M, \sigma') \in \Delta_M$,
   then $(c_N, c_M, \sigma) \xrightarrow{sync(op,\rho)} (c'_N, c'_M, \sigma') \in \Delta_{N \otimes M}$ and
b) if $c_N \xrightarrow{(\sigma, out(op,\rho))} c'_N \in \Delta_N$ and $(c_M, \sigma) \xrightarrow{in(op,\rho)} \perp \in \Delta_M$,
   then $(c_N, c_M, \sigma) \xrightarrow{sync(op,\rho)} \perp \in \Delta_{N \otimes M}$.

Note that for all states (c_N, c_M, σ) of the composition, if $(c_N, (\sigma, out(op, \rho)), c'_N)$ is a user transition in Δ_N then there actually exists an implementation transition with source state (c_M, σ) and label $in(op, \rho)$ in Δ_M since M is input-enabled. Obviously, $N \otimes M$ is a Σ-interaction model where $C_N \times C_M$ are the control states and Q the data states. Our first compositionality result says that, for a given protocol P, the composition of any correct user N of P with any correct implementation M of P exhibits a correct interaction behaviour w.r.t. P which, in particular, does not diverge. Thus user and implementation models can be developed independently of each other.

Theorem 1. *Let P be a Σ-protocol. If $N \in [\![P]\!]_{use}$ and $M \in [\![P]\!]_{impl}$, then $N \otimes M \in [\![P]\!]_{iact}$.*

Proof. By assumption there exists a user simulation $\lesssim_{use}$ between N (with states C_N) and P (with states S) and an implementation simulation $\lesssim_{impl}$ between P and M (with states $C_M \times Q$). In order to show that $N \otimes M \models_{iact} P$ we must find an interaction simulation $\lesssim_{iact} \subseteq (C_N \times C_M \times Q) \times S$ between $N \otimes M$ and P. We define $(c_N, c_M, \sigma) \lesssim_{iact} s :\Leftrightarrow (c_N, \sigma) \lesssim_{use} s \wedge s \lesssim_{impl} (c_M, \sigma) \wedge \sigma \models l(s)$. It is straightforward to show that $\lesssim_{iact}$ is indeed an interaction simulation between $N \otimes M$ and P. □

6   Protocol Refinement

The compositionality result of the last section considers the case where user and implementation models conform to the *same* protocol P. In general, however, the user may take into account another protocol than the implementation does. In this case, to ensure that user and implementation can still work properly together, it is essential that the two protocols are compatible, which we will express by protocol refinement. Let us assume that there is a protocol P considered by a user and a protocol P' considered by an implementor. To define what it means that P' is a refinement of P one must take into account the user's *and* the implementor's perspectives. Each user N, which behaves correctly w.r.t. P, expects that it will finally be bound to an arbitrary implementation which respects P. In order to guarantee that any implementation $M \in [\![P']\!]_{impl}$ also respects P, we must require $[\![P']\!]_{impl} \subseteq [\![P]\!]_{impl}$. Conversely, each implementation M, which is correct w.r.t. P', expects that it will finally be used by an arbitrary user which behaves correctly w.r.t. P'. In order to guarantee that any user $N \in [\![P]\!]_{use}$ is also correct w.r.t. P', we must require $[\![P]\!]_{use} \subseteq [\![P']\!]_{use}$. This leads to the following definition of protocol refinement (which is obviously transitive).[5]

Definition 13 (Protocol Refinement). *Let P and P' be Σ-protocols. P' is a refinement of P, written $P \rightsquigarrow P'$, if $[\![P']\!]_{impl} \subseteq [\![P]\!]_{impl}$ and $[\![P]\!]_{use} \subseteq [\![P']\!]_{use}$.*

The notion of refinement is based on protocol semantics. To be able to check protocol refinement in practice, we define a syntactic refinement simulation.

Definition 14 (Refinement Simulation). *Let $P = (S, s_0, PLabel_\Sigma, \Delta, l)$ and $P' = (S', s_0', PLabel_\Sigma, \Delta', l')$ be two Σ-protocols. $P \rightarrow_{ref} P'$ if there exists a refinement simulation between P and P' given by a relation $\sim_{ref} \subseteq S \times S'$ s.t.*

a) $s_0 \sim_{ref} s_0'$, and for all $\sigma \in DState(\Sigma_{Obs})$, $\sigma \vDash l'(s_0')$ implies $\sigma \vDash l(s_0)$,

b) for all $s \sim_{ref} s'$ and $(\varphi_{pre}, op, \varphi_{post}) \in PLabel_\Sigma$, if $(s, (\varphi_{pre}, op, \varphi_{post}), \underline{s}) \in \Delta$, then there exists $(s', (\varphi'_{pre}, op, \varphi'_{post}), \underline{s}') \in \Delta'$ such that $\underline{s} \sim_{ref} \underline{s}'$, and for all $\sigma, \sigma' \in DState(\Sigma_{Obs})$ and $\rho : var_{in}(op) \to \mathcal{D}$,

1) if $\sigma; \rho \vDash \varphi_{pre}$ and $\sigma \vDash l(s)$, then $\sigma; \rho \vDash \varphi'_{pre}$ and $\sigma \vDash l'(s')$,

2) if $\sigma; \rho \vDash \varphi_{pre}$ and $\sigma \vDash l(s)$ and $\sigma, \sigma'; \rho \vDash \varphi'_{post}$ and $\sigma' \vDash l'(s')$, then $\sigma, \sigma'; \rho \vDash \varphi_{post}$ and $\sigma' \vDash l(\underline{s})$.

The crucial idea behind condition b) is that any transition of the protocol P can be simulated by some transition of P' where the precondition may be weakened and the postcondition may be strengthened; an idea which stems from behavioural subtyping in object-oriented specifications; cf. e.g. [13].

---

[5] We assume that both protocols are formed over the same signature Σ. An extension to the case where the signatures of P and P' are matched, e.g. by a signature morphism, should be straightforward if the underlying observer signatures remain unchanged. Otherwise, one has to elaborate more on an appropriate refinement notion for data states w.r.t. some kind of observer signature morphism.

Theorem 2. *If $P \to_{ref} P'$, then $P \rightsquigarrow P'$.[6]*

Proof. Assume $P \to_{ref} P'$, i.e. there exists a refinement simulation $\sim_{ref} \subseteq S \times S'$ where S and S' are the states of P and P' resp.. For $M \in [\![P']\!]_{impl}$ with states $C \times Q$ there exists an implementation simulation $\lesssim'_{impl}$ between P' and M. We define a relation $\lesssim_{impl} \subseteq S \times (C \times Q)$ by $s \lesssim_{impl} (c, \sigma) :\Leftrightarrow \exists s' \in S'.s \sim_{ref} s' \land s' \lesssim'_{impl} (c, \sigma)$. It can be shown that $\lesssim_{impl}$ is an implementation simulation between P and M and hence $M \in [\![P]\!]_{impl}$. For $N \in [\![P]\!]_{use}$ with states C there exists a user simulation $\lesssim_{use}$ between N and P. We define a relation $\lesssim'_{use} \subseteq (C \times DState(\Sigma_{Obs})) \times S'$ by $(c, \sigma) \lesssim'_{use} s' :\Leftrightarrow \exists s \in S.(c, \sigma) \lesssim_{use} s \land s \sim_{ref} s' \land \sigma \vDash l(s')$. It can be shown that $\lesssim'_{use}$ is a user simulation between N and P' and hence $N \in [\![P']\!]_{use}$. □

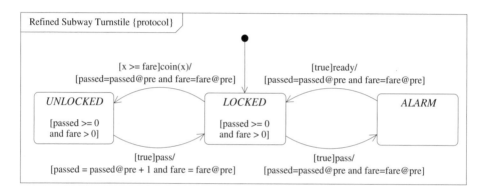

Fig. 2. Refined UML Protocol State Machine of a Subway Turnstile

Example 4. Fig. 2 defines a protocol $P'_{Turnstile}$ which is a refinement of $P_{Turnstile}$; cf. Ex. 2. The refined protocol strengthens the postcondition for entering and leaving the *ALARM* state by requiring that the number of passed persons and the fare remain unchanged. On the other hand, the precondition of inserting a coin into the turnstile's slot is weakened. ♣

We will now extend the compositionality result of Thm. 1 by taking into account refinement. Our first observation is that, given arbitrary Σ-protocols P and P', $N \in [\![P]\!]_{use}$, and $M \in [\![P']\!]_{impl}$, the model composition $N \otimes M$ can have an arbitrary, even divergent, behaviour. If, however, $P \rightsquigarrow P'$ is a protocol refinement, then $N \otimes M$ is a correct interaction model w.r.t. P *and* w.r.t. P'.

Corollary 1. *Let P and P' be two Σ-protocols, $N \in [\![P]\!]_{use}$, and $M \in [\![P']\!]_{impl}$. If $P \rightsquigarrow P'$, then $N \otimes M \in [\![P]\!]_{iact} \cap [\![P']\!]_{iact}$.*

Proof. By assumption, $N \in [\![P]\!]_{use}$ and, since $P \rightsquigarrow P'$, $M \in [\![P]\!]_{impl}$. Hence, by Thm. 1, $N \otimes M \in [\![P]\!]_{iact}$. Similarly, $M \in [\![P']\!]_{impl}$ and, since $P \rightsquigarrow P'$, $N \in [\![P']\!]_{use}$. Thus we obtain, by Thm. 1, $N \otimes M \in [\![P']\!]_{iact}$. □

---

[6] Whether the other direction holds is an open issue.

It is an obvious question whether the intersection $[\![P]\!]_{iact} \cap [\![P']\!]_{iact}$ can be characterised by the interaction semantics of a single protocol. For this purpose, we define an operator $\boxtimes$ for the composition of Σ-protocols P and P'.

Definition 15 (Protocol Composition). *Let $P = (S, s_0, PLabel_\Sigma, \Delta, l)$ and $P' = (S', s_0', PLabel_\Sigma, \Delta', l')$ be two Σ-protocols. Then the protocol composition of P and P' is the Σ-protocol $P \boxtimes P' = (S \times S', (s_0, s_0'), PLabel_\Sigma, \Delta_\boxtimes, l_\boxtimes)$ where*

a) the transition relation $\Delta_\boxtimes$ is the least relation satisfying: if $(s, s') \in S_\boxtimes$, $(s, (\varphi_{\mathrm{pre}}, op, \varphi_{\mathrm{post}}), \underline{s}) \in \Delta$ and $(s', (\varphi'_{\mathrm{pre}}, op, \varphi'_{\mathrm{post}}), \underline{s}') \in \Delta'$ then

$$(s, s') \xrightarrow{(\varphi_{\mathrm{pre}} \wedge \varphi'_{\mathrm{pre}}, op, \varphi_{\mathrm{post}} \wedge \varphi'_{\mathrm{post}})} (\underline{s}, \underline{s}') \in \Delta_\boxtimes,$$

b) the state labelling $l_\boxtimes : S \times S' \to \mathcal{S}(\Sigma_{\mathrm{Obs}})$ is defined by $l_\boxtimes(s, s') = l(s) \wedge l'(s')$ for all $(s, s') \in S \times S'$.

The composition operator for Σ-protocols synchronises transitions which share the same operation by building the conjunction of the respective preconditions and postconditions. In this way it extends the usual parallel composition of labelled transition systems which synchronises on shared actions.

Lemma 1. *Let P and P' be Σ-protocols. $[\![P \boxtimes P']\!]_{iact} = [\![P]\!]_{iact} \cap [\![P']\!]_{iact}$.*

Proof. $\subseteq$: Let $K \in [\![P \boxtimes P']\!]_{iact}$ be an interaction model with states $C \times Q$. Hence, there exists an interaction simulation $\lesssim_{iact}^\boxtimes \subseteq (C \times Q) \times (S \times S')$. W.l.o.g. we show that $K \in [\![P]\!]_{iact}$. For this purpose, we have to find an interaction simulation $\lesssim_{iact} \subseteq (C \times Q) \times S$ between K and P. We define $(c, \sigma) \lesssim_{iact} s :\Leftrightarrow \exists s' \in S'.(c, \sigma) \lesssim_{iact}^\boxtimes (s, s')$ which is an interaction simulation between K and P.
 $\supseteq$: Let $K \in [\![P]\!]_{iact} \cap [\![P']\!]_{iact}$. Therefore, we have interaction simulations $\lesssim_{iact}$ and $\lesssim'_{iact}$ between K and P and between K and P', respectively. We define the relation $(c, \sigma) \lesssim_{iact}^\boxtimes (s, s') :\Leftrightarrow (c, \sigma) \lesssim_{iact} s \wedge (c, \sigma) \lesssim_{iact} s'$ which is an interaction simulation between K and $P \boxtimes P'$. $\qquad\square$

We are now able to present our final result which shows that in the presence of protocol refinement semantic composition of user and implementation models is compatible with syntactic composition of protocols.

Theorem 3. *Let P and P' be Σ-protocols. If $N \in [\![P]\!]_{use}$, $M \in [\![P']\!]_{impl}$, and $P \rightsquigarrow P'$, then $N \otimes M \in [\![P \boxtimes P']\!]_{iact}$.*

Proof. By Corollary 1 and Lemma 1. $\qquad\square$

7   Conclusions

We have investigated a model-theoretic foundation of behaviour protocols in terms of labelled transition systems and simulation relations which integrate control flow and evolution of data states, thus formalising the implementor's view, the user's view and the interaction view on protocols. On this basis we have studied protocol refinement and we have provided compositionality results relating model composition and protocol composition. Important issues of future research concern the development of verification techniques, the extension to asynchronous communication, and the integration into the port-based Java/A component model [4]. Then we will consider ports with provided *and* required interfaces such that semantic models integrate both, the role of an implementation and of a user at the same time.

Acknowledgements. We thank the anonymous reviewers of the submitted version of this paper for many valuable remarks and we are grateful to Fernando Orejas for his useful comments to a draft of this paper.

References

1. Bidoit, M., Hennicker, R.: An algebraic semantics for contract-based software components. In: Meseguer, J., Roşu, G. (eds.) AMAST 2008. LNCS, vol. 5140, pp. 216–231. Springer, Heidelberg (2008)
2. Bidoit, M., Hennicker, R., Knapp, A., Baumeister, H.: Glass-box and black-box views on object-oriented specifications. In: Proc. 2nd IEEE Int. Conf. Software Engineering and Formal Methods (SEFM 2004), pp. 208–217. IEEE Computer Society Press, Los Alamitos (2004)
3. Eshuis, R.: Reconciling statechart semantics. Sci. Comput. Program. 74(3), 65–99 (2009)
4. Knapp, A., Janisch, S., Hennicker, R., Clark, A., Gilmore, S., Hacklinger, F., Baumeister, H., Wirsing, M.: Modelling the CoCoME with the Java/A component model. In: Rausch, A., Reussner, R., Mirandola, R., Plášil, F. (eds.) The Common Component Modeling Example. LNCS, vol. 5153, pp. 207–237. Springer, Heidelberg (2008)
5. Martin, R.C.: UML tutorial: Finite state machines. Published article on Object Mentor Webpage (January 1998),
   http://www.objectmentor.com/resources/articles/umlfsm.pdf (last accessed on 2009-02-05)
6. Mencl, V.: Specifying component behavior with port state machines. Electr. Notes Theor. Comput. Sci. 101, 129–153 (2004)
7. Meyer, B.: Object-Oriented Software Construction, 2nd edn. Prentice–Hall, Upper Saddle River (1997)
8. Mousavi, M.R., Reniers, M.A., Groote, J.F.: Notions of bisimulation and congruence formats for SOS with data. Inf. Comput. 200(1), 107–147 (2005)
9. Object Management Group (OMG). UML Superstructure Specification 2.1.2. Technical report, OMG (2007),
   http://www.omg.org/spec/UML/2.1.2/Superstructure/PDF/
   (last accessed on 2009-02-05)
10. Plasil, F., Visnovsky, S.: Behavior protocols for software components. IEEE Trans. Software Eng. 28(11), 1056–1076 (2002)
11. Sannella, D.T., Tarlecki, A.: Towards formal development of programs from algebraic specifications: Implementations revisited. Acta Informatica 25(3), 233–281 (1988)
12. Warmer, J., Kleppe, A.: The Object Constraint Language, 2nd edn. Addison Wesley, Reading (2003)
13. Wehrheim, H.: Behavioural Subtyping in Object-Oriented Specification Formalisms. Habilitation thesis, Universität Oldenburg (2002)

Correctness, Completeness and Termination of Pattern-Based Model-to-Model Transformation

Fernando Orejas[1], Esther Guerra[2], Juan de Lara[3], and Hartmut Ehrig[4]

[1] Universitat Politècnica de Catalunya, Spain
orejas@lsi.upc.edu
[2] Universidad Carlos III de Madrid, Spain
eguerra@inf.uc3m.es
[3] Universidad Autónoma de Madrid, Spain
jdelara@uam.es
[4] Technische Universität Berlin, Germany
ehrig@cs.tu-berlin.de

Abstract. Model-to-model (M2M) transformation consists in transforming models from a source to a target language. Many transformation languages exist, but few of them combine a declarative and relational style with a formal underpinning able to show properties of the transformation. Pattern-based transformation is an algebraic, bidirectional, and relational approach to M2M transformation. Specifications are made of patterns stating the allowed or forbidden relations between source and target models, and then compiled into low level operational mechanisms to perform source-to-target or target-to-source transformations. In this paper, we study the compilation into operational triple graph grammar rules and show: (i) correctness of the compilation of a specification without negative patterns; (ii) termination of the rules, and (iii) completeness, in the sense that every model considered relevant can be built by the rules.

1 Introduction

Model-to-model (M2M) transformation is an enabling technology for recent software development paradigms, like Model-Driven Development. It consists in transforming models from a source to a target language and is useful, e.g. to migrate between language versions, to transform a model into an analysis domain, and to refine a model. In some cases, after performing the transformation, the source and target models can be modified separately. Therefore, it is useful to be able to execute transformations both in the forward and backward directions to recover consistency. Thus, an interesting property of M2M transformation languages is to allow specifying transformations in a direction-neutral way, from which forward and backwards transformations can be automatically derived.

In recent years, many M2M specification approaches have been proposed [1,2,13,14,15,17,18] with either *operational* or *declarative* style. The former languages explicitly describe the operations needed to create elements in the target model from elements in the source, i.e they are unidirectional. Instead, in declarative approaches, a description of the mappings between source and target models is provided, from which *operational mechanisms* are generated to transform in the forward and backward directions.

In this paper, we are interested in declarative, bidirectional M2M transformation languages. Even though many language proposals exist, few have a formal basis enabling the analysis of specifications or the generated operational mechanisms [19]. In previous work [3], we proposed a new graphical, declarative, bidirectional and formal approach to M2M transformation based on triple patterns. Patterns specify the allowed or forbidden relations between two models and are similar to graph constraints [6], but for triple graphs. The latter are structures made of three graphs representing the source and target models, as well as the correspondence relations between their elements. Thus, in pattern-based transformation we define the set of valid pairs of source and target models by constraints, and not by rules. Then, patterns are compiled into operational rules working on triple graphs to perform forward and backward transformations.

In the present work, we prove certain properties of the compilation of pattern-based specifications into rules. First, we show that our compilation mechanism generates graph grammars that are terminating. This result is interesting as it means that we do not need to use external control mechanisms for rule application [11]. Second, we prove that the transformation rules are sound with respect to the positive fragment of the specification. This means that a triple graph satisfies all positive patterns in a specification if and only if it is terminal with respect to the generated rules. In other words, the operational mechanisms actually do their job, and this corresponds to the notion of *correctness* in [19]. Finally, we also prove completeness of the rules, i.e. that the rules are able to produce any model *generated* by the original M2M specification. These generated graphs are a meaningful subset of all the models satisfying the specification.

We think that this work paves the way to using formal methods in a key activities of Model-Driven Development: the specification and execution of M2M transformations. The paper is organized as follows. Section 2 provides an introduction to triple graphs and to the transformation rules used in this paper. Section 3 introduces M2M pattern specifications, their syntax and semantics. Section 4 is the core of the paper: we introduce the transformation rules associated to a pattern specification and we prove their termination, soundness and completeness. In Section 5 we compare our approach with some other approaches to M2M transformation. Finally, in Section 6 we draw some conclusions and we sketch some future work. In addition, along the paper we use a small running example describing the transformation of class diagrams into relational schemas [17]. The report [16] includes the full proofs for all the results.

2   Preliminaries

This section introduces the basic concepts that we use throughout the paper about triple graphs and triple graph transformation. Triple graphs [18] model the relation between two graphs called source and target through a correspondence graph and a span of graph morphisms. In this sense, if we consider that models are represented by graphs, triple graphs may be used to represent transformations, as well as transformation information through the connection graph.

Definition 1 (Triple Graph and Morphism). *A triple graph $G = (G_S \overset{c_S}{\leftarrow} G_C \overset{c_T}{\rightarrow} G_T)$ (or just $G = \langle G_S, G_C, G_T \rangle$ if c_S and c_T may be considered implicit) consists of three graphs G_S, G_C, and G_T, and two morphisms c_S and c_T. A triple graph morphism $m = (m_S, m_C, m_T) : G^1 \rightarrow G^2$ is made of three graph morphisms m_S, m_C, and m_T such that $m_S \circ c_S^1 = c_S^2 \circ m_C$ and $m_T \circ c_T^1 = c_T^2 \circ m_C$.*

Given a triple graph G, we write $G|_X$ for $X \in \{S, T\}$ to refer to a triple graph whose X-component coincides with G_X and the other two components are the empty graph, e.g. $G|_S = \langle G_S, \emptyset, \emptyset \rangle$. Similarly, given a triple graph morphism $h : G_1 \rightarrow G_2$ we also write $h|_X : G_1|_X \rightarrow G_2|_X$ to denote the morphism whose X-component coincides with h_X and whose other two components are the empty morphism between empty graphs. Finally, given G, we write i_G^X to denote the inclusion $i_G^X : G|_X \rightarrow G$, where the X-component is the identity and where the other two components are the (unique) morphism from the empty graph into the corresponding graph component.

Triple graphs form the category **TrG**, which can be formed as the functor category **Graph**$^{\cdot \leftarrow \cdot \rightarrow \cdot}$. In principle, we may consider that **Graph** is the standard category of graphs. However, the results in this paper still apply when **Graph** is a different category, as long as it is an adhesive-HLR category [12,6] and satisfies the additional property of n-factorization (see below). For instance, **Graph** could also be the category of typed graphs or the category of attributed (typed) graphs.

Definition 2 (Jointly surjective morphisms). *A family of graph morphisms $\{H_1 \overset{f_1}{\rightarrow} G, \ldots, H_n \overset{f_n}{\rightarrow} G\}$ is* jointly surjective *if for every element e (a node or an edge) in G there is an e' in H_k, with $1 \leq k \leq n$ such that $f_k(e') = e$.*

A property satisfied by graphs and by triple graphs, is n-factorization, a generalization (and also a consequence) of the property of pair factorization [6]:

Proposition 1 (n-factorization). *Given a family of graph morphisms $\{H_1 \overset{f_1}{\rightarrow} G, \ldots, H_n \overset{f_n}{\rightarrow} G\}$ with the same codomain G, there exists a graph H, a monomorphism m and a jointly surjective family of morphisms $\{H_1 \overset{g_1}{\rightarrow} H, \ldots, H_n \overset{g_n}{\rightarrow} H\}$ such that the diagram below commutes:*

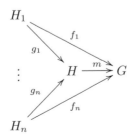

It may be noticed that if a category satisfies the n-factorization property and $\{f_1, \ldots, f_n\}$ in the above diagram are monomorphisms then so are $\{g_1, \ldots, g_n\}$.

In this paper, a graph transformation rule is a monomorphism $(L \xrightarrow{r} R)$, possibly equipped with some Negative Application Conditions (NACs) in its left-hand side for limiting its application. The reason is that in our approach we just need to use *non-deleting rules*. Hence, rule application is defined by a pushout. Moreover, in our case, when applying a rule to a given graph G, it is enough to consider the case where the morphism that matches L to G is a mono:

Definition 3 (Non-Deleting Triple Rule, Rule Application, Terminal Graph). *A (non-deleting) triple rule $p = \langle N, L \xrightarrow{r} R \rangle$, consists of a triple monomorphism r and a finite set of negative application conditions $N = \{L \xrightarrow{n_i} N_i\}_{i \in I}$, where each n_i is a triple monomorphism.*

A monomorphism $m : L \to G$ is a match for the rule $p = \langle N, L \xrightarrow{r} R \rangle$ if m satisfies all the NACs in N, i.e. for each NAC $L \xrightarrow{n_i} N_i$ there is no monomorphism $h : N_i \to G$ such that $m = h \circ n_i$. Given a match $m : L \to G$ for p, the application of p to G via m, denoted $G \Rightarrow_{p,m} H$, is defined by the pushout below:

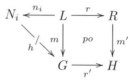

where m' is called the comatch of this rule application.

Given a set TR of transformation rules, a triple graph G is terminal for TR if no rule in TR can be applied to G.

3   Pattern-Based Model-to-Model Transformation

Triple patterns are similar to graph constraints [6,9]. We use them to describe the allowed and forbidden relationships between the source and target models in a M2M transformation. In particular, we consider two kinds of patterns. Negative patterns, which are denoted by just a triple graph, describe relationships that should not occur between the source and target models. This means that, from a formal point of view, negative patterns are just like negative graph constraints, i.e. a (triple) graph G satisfies the negative pattern N if N is not a subgraph of G (up to isomorphism). Positive patterns specify possible relationships between source and target models. Positive patterns consist of a set of negative premises and a conclusion. As we can see below, satisfaction of P-patterns does not coincide exactly with satisfaction of graph constraints.

Definition 4 (Patterns, M2M Specification). *An N-pattern, denoted $N(Q)$ consists of a triple graph Q. A P-pattern $S = \{N(Q \xrightarrow{n_j} C_j)\}_{j \in J} \Rightarrow Q$ consists of*

- *The conclusion, given by the triple graph Q.*
- *The negative premises $N(Q \xrightarrow{n_j} C_j)$, given by the inclusions $Q \xrightarrow{n_j} C_j$.*

An M2M specification SP is a finite set of P and N-patterns. Given a specification SP, we denote by SP^+ (respectively, SP^-) the positive fragment of SP, i.e. the set of all P-patterns in SP (respectively, the set of all N-patterns in SP).

A P-pattern is intended to describe a class of transformations denoted by triple graphs. P-patterns describe, simultaneously, source-to-target and target-to-source transformations. In this sense, there are two notions of satisfaction associated to P-patterns: forward satisfaction, associated to source-to-target transformations and backward satisfaction, associated to target-to-source transformations. Then, a triple graph G forward satisfies a pattern $\{N(Q \xrightarrow{n_j} C_j)\}_{j \in J} \Rightarrow Q$ if whenever Q_S is embedded in the source of G (and the premises are forward satisfied by the embedding), Q is embedded in G. And an embedding m of Q_S in G_S forward satisfies a premise $N(Q \xrightarrow{n_j} C_j)$ if there is no embedding m' of $(C_j)_S$ in G_S such that m' extends m. Backward satisfaction is the converse notion.

Definition 5 (Pattern Satisfaction)

- *A monomorphism $m\colon Q|_S \to G$ forward satisfies a negative premise $N(Q \xrightarrow{n_j} C_j)$, denoted $m \models_F N(Q \xrightarrow{n_j} C_j)$ if there does not exist a monomorphism $g\colon (C_j)|_S \to G$ such that $m = g \circ (n_j)|_S$. Similarly, $m\colon Q|_T \to G$ backward satisfies a negative premise $N(Q \xrightarrow{n_j} C_j)$, denoted $m \models_B N(Q \xrightarrow{n_j} C_j)$ if there does not exist a monomorphism $g\colon (C_j)|_T \to G$ such that $m = g \circ (n_j)|_T$.*
- *A triple graph G forward satisfies a P-pattern $\mathcal{S} = \{N(Q \xrightarrow{n_j} C_j)\}_{j \in J} \Rightarrow Q$, denoted $G \models_F \mathcal{S}$, if for every monomorphism $m\colon Q|_S \to G$, such that, for every j in J, $m \models_F N(Q \xrightarrow{n_j} C_j)$, there exists a monomorphism $m'\colon Q \to G$ such that $m = m' \circ i_Q^S$:*

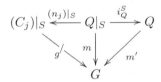

- *A triple graph G backward satisfies a P-pattern $\mathcal{S} = \{N(Q \xrightarrow{n_j} C_j)\}_{j \in J} \Rightarrow Q$, denoted $G \models_B \mathcal{S}$, if for every monomorphism $m\colon Q|_T \to G$, such that, for every j in J, $m \models_B N(Q \xrightarrow{n_j} C_j)$, there exists a monomorphism $m'\colon Q \to G$ such that $m = m' \circ i_Q^T$.*
- *G satisfies $\mathcal{S}$, denoted $G \models \mathcal{S}$, if $G \models_F \mathcal{S}$ and $G \models_B \mathcal{S}$.*
- *A triple graph G satisfies an N-pattern $N(Q)$, denoted $G \models N(Q)$ if there is no monomorphism $h\colon Q \to G$.*

Though the abuse of notation, given a monomorphism $m\colon Q \to G$, we may also say that m forward satisfies a negative premise if the monomorphism $i_G^S \circ m|_S\colon Q|_S \to G$ forward satisfies it.

Example 1. The left of Fig. 1 shows a specification describing a transformation between class diagrams and relational schemas [17]. When considered in the

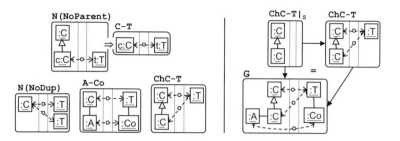

Fig. 1. Example patterns (left). Triple graph satisfying the specification (right).

forward direction, the transformation creates a database schema to store in tables the attributes of the classes, where classes of the same inheritance hierarchy are mapped to the same table. The first pattern C-T states that top-level classes (i.e., those without parents in the inheritance hierarchy) are mapped to tables. Note that we use a notation similar to UML object diagrams (i.e. c:C represents a node c of type Class). N(NoDup) is an N-pattern that forbids associating two tables to the same class. A-Co and ChC-T are P-patterns with an empty set of premises. A-Co says that attributes of a class are stored in columns of the associated table. Finally, ChC-T specifies that children and parent classes are mapped to the same table. The right of the same figure shows an example of satisfaction. Graph G satisfies all patterns in the specification and, in particular, the diagram shows how an occurrence of ChC-T$|_S$ is extended to ChC-T.

4   Correctness, Completeness, and Termination of Transformations

An M2M specification S can be used in different scenarios [11]. We can build a target model from a source model (or vice-versa), check whether two models can be mapped according to S, or synchronize two models that previously satisfied S but that were modified separately. Each scenario needs a specialized operational mechanism. Here we cover the first scenario. Starting from a specification S, we generate a triple graph grammar to perform forward transformations. Obviously, the same techniques could be applied for implementing backward transformations. The basic idea is to see the given P-patterns as tiles that have to "cover" a given source model, perhaps with some overlapping. The target model obtained by gluing the target parts of these patterns is the result of the transformation. In addition, the N-patterns allow us to discard some possible models.

Given a source model, a pattern specification will normally have many models. In particular, there may be several non-isomorphic triple graphs sharing the same source graph. This means that there may be several correct transformations for that source graph. Our technique will non-deterministically allow us to obtain all the transformations satisfying the specification. We think that this is the only reasonable approach, if a priori we cannot select any preferred model. It should be obvious that following this kind of approach it is impossible to build some models of the given specification. In particular, it would be impossible to generate models whose target and connection part cannot be generated using the given patterns as described above. For instance, models whose target part includes some nodes of a given type not mentioned in the patterns. We think that restricting our attention to this kind of *generated models* is reasonable in this context. This is similar to the "No Junk" condition in algebraic specification.

Our approach is based on associating to a given specification SP a set of forward transformation rules $TR(SP)$. These rules have, in the left-hand side, a graph including the source part of the conclusion of a positive pattern and part of the target and the connection part. In the right-hand side they have the whole conclusion of the pattern. The idea is that these rules may be used to build "a piece" of the target and the connection graphs, when we discover an occurrence of the source part of a pattern on the given source graph. Rules may include part of the target and connection part of the pattern because this part of the pattern may have been already built by another pattern (rule) application. In addition, the negative premises in the given positive patterns are transformed into NACs of the given rules. Moreover, if we want these rules to be terminating, then we may include some additional NACs that ensure the termination of the set of transformation rules associated to all the P-patterns of a given specification. It should be clear that we can define a set of backward rules in a similar way.

Definition 6 (Forward Transformation Rules for Patterns). *To every P-pattern $\mathcal{S} = \{N(Q \xrightarrow{n_j} C_j)\}_{j \in J} \Rightarrow Q$, we associate the set of forward transformation rules $TR(\mathcal{S})$ consisting of all the rules $r = \langle NAC(r), L_r \xrightarrow{i} Q \rangle$, where:*

- *L_r is a triple graph such that $Q|_S \subseteq L_r \subset Q$ and i is the monomorphism associated to the inclusion $L_r \subset Q$.*
- *$NAC(r)$ is the set that includes a NAC $n'_j : L_r \rightarrow C'_j$ for each premise $N(n_j : Q \rightarrow C_j)$ in $\mathcal{S}$, where n'_j and C'_j are defined up to isomorphism by the pushout depicted on the left below.*

The set of terminating transformation rules associated to $\mathcal{S}$, $TTR(\mathcal{S})$ is the set of all rules $\langle NAC(r) \cup TNAC(r), L_r \xrightarrow{i} Q \rangle$ such that $\langle NAC(r), L_r \xrightarrow{i} Q \rangle \in TR(\mathcal{S})$ and $TNAC(r)$ is the set of all the termination NACs for r, i.e. all the monomorphisms $n : L_r \rightarrow T$ where there is a monomorphism $f_2 : Q \rightarrow T$ such that n and f_2 are jointly surjective and the diagram on the right below commutes:

$$
\begin{array}{ccc}
Q|_S & \xrightarrow{(n_j)|_S} & (C_j)|_S \\
{\scriptstyle i^S_{L_r}} \downarrow & po & \downarrow \\
L_r & \xrightarrow{n'_j} & C'_j
\end{array}
\qquad\qquad
\begin{array}{ccc}
Q|_S & \xrightarrow{i^S_Q} & Q \\
{\scriptstyle i^S_{L_r}} \downarrow & & \downarrow {\scriptstyle f_2} \\
L_r & \xrightarrow{n} & T
\end{array}
$$

Example 2. Fig. 2 shows the two forward rules generated from pattern C-T presented in Example 1. The first one uses $L = Q|_S$, while the LHS of the second

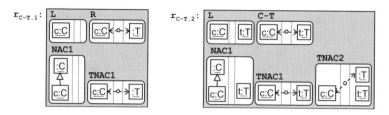

Fig. 2. Forward rules generated from pattern `C-T`

reuses an existing table. Both rules include a NAC (named NAC1), generated from the negative pre-condition `NoParent` of the pattern. The termination NACs ensure that each class is connected to at most one table.

In some related work (e.g., [6,4]) termination is ensured by just the termination NAC $L_r \rightarrow Q$. This NAC is enough to ensure finite termination if the set of possible matches of the given rule does not change after applying the transformation rules, i.e. if the possible matches of the rule are always *essential matches*, according to the terminology in [6,4]. However, this is not the case in our context. Our rules may have triple graphs in the left-hand side with non-empty target or connection part. As a consequence, at some point, there may exist non-essential matches, and this may cause, if we would only use that NAC, that the resulting transformation system is not terminating, as the example below shows.

Example 3. Suppose we have the rule shown to the left of Fig. 3, which is one of the backward rules that we would derive from pattern `ChC-T` if we want to apply our technique to implement backward model transformations. And suppose that we only add a termination NAC (labelled `TNAC`) equal to its RHS. The rule creates a new child of a class connected to a table. Then, the sequence of transformations that starts as shown to the right of the same figure does not terminate. This is so, because the rule adds a new match for the LHS in each derivation, thus being able to produce an inheritance hierarchy of any depth.

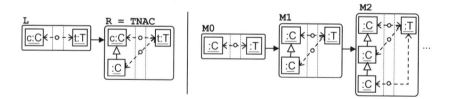

Fig. 3. Backward rule (left). Non-terminating sequence (right).

According to Definition 6, the set of termination NACs for the rule includes the three graphs depicted in Fig. 4. TNAC2 is isomorphic to `TNAC`, but it identifies the class in L with the child class. Then it is clear that TNAC2 avoids applying the rule to M_1, thus ensuring termination.

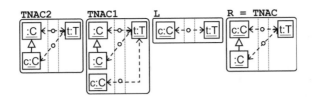

Fig. 4. Rule with termination NACs

Definition 7 (Forward Transformation Rules for Specifications). *Given a pattern specification SP, we define the set of forward transformation rules associated to SP:*

$$TR(SP) = \bigcup_{\mathcal{S} \in SP^+} TR(\mathcal{S})$$

Similarly, TTR(SP) is the set of terminating transformation rules associated to the patterns in SP^+.

Our first result shows that $TTR(SP)$ is terminating. To show this result, we first need to notice that the transformation rules never modify the source part of the given triple graphs. Then, the key to show this theorem is a specific property of our termination NACs ensuring that, if we have transformed the graph G_1 into the graph G_2 using a rule r with match $m_1 \colon L_r \to G_1$, then we cannot apply the same rule with match $m_2 \colon L_r \to G_2$ if the source parts of the domains of m_1 and m_2 coincide. The reason is that, if we have already applied r with match m_1 then the graph G_2 will already embed L_r and Q (via m_2 and m_1, respectively). Moreover if the source parts of the domains of m_1 and m_2 coincide, then we can ensure the existence of embeddings $n \colon L_r \to T$ and $h \colon T \to G_2$, where n is a termination NAC and $h \circ n = m_2$. Implicitly, this means that the termination NACs impose finite bounds on the number of times that an element of the given source graph can be part of a match.

Theorem 1 (Termination). *For any finite pattern specification SP, TTR(SP) is terminating.*

Our second main result shows that a triple graph is terminal for $TTR(SP)$ if and only if it forward satisfies the positive patterns in SP. Obviously, we cannot ensure that if G is terminal then G will also satisfy the negative patterns in SP, since they play no role in the construction of $TTR(SP)$.

Theorem 2 (Correctness). *For any finite pattern specification SP, $G \models_F SP^+$ if and only if G is terminal with respect to TTR(SP).*

To prove this theorem, first, we show that a morphism forward satisfies a premise of a pattern if and only if it satisfies the corresponding NACs in the associated rules. Then, we can see that if G is a forward model of SP^+ and h is a match for a rule r associated to a pattern $\mathcal{S}$ that forward satisfies all the negative premises in $\mathcal{S}$, then h will not satisfy a termination NAC in the rule. Conversely, we can prove that if $h: Q|_S \to G$ is a monomorphism that satisfies all the premises in S and h does not satisfy the termination NAC of the rule $Q|_S \to Q$ then there exists an $h': Q \to G$ that extends h.

With respect to completeness, as discussed above, we are only interested in the models whose elements in the target and connection part can be considered to be there because some pattern prescribes that they must be there. We call these graphs SP-generated.

Definition 8 (SP-Generated Graphs). *Given a pattern specification SP, a triple graph G is SP-generated if there is a finite family of P-patterns $\{\mathcal{S}_k\}_{k \in K}$, with $\mathcal{S}_k = \{N(Q_k \overset{n_j}{\to} C_{jk})\}_{j \in J} \Rightarrow Q_k$ in SP, and a family of monomorphisms $\{Q_k \overset{f_k}{\to} G\}_{k \in K}$ such that every f_k forward satisfies all the premises in $\mathcal{S}_k$, and $f_1, \ldots, f_n, i_G^S$ are jointly surjective. In this case, we also say that G is generated by the patterns $\mathcal{S}_1, \ldots \mathcal{S}_n$ and the morphisms $f_1, \ldots f_n$.*

Example 4. Fig. 5 presents three SP-generated graphs from the example specification of Fig. 1, together with the family of patterns that generates them. It must be noted that the same pattern may occur several times in the given family generating the graph. For instance, the pattern A-Co is used twice for generating G_3. Graph G_1 is generated by pattern C-T, but is not a model of the specification as the child class and its attribute need to be translated. On the contrary, graphs G_2 and G_3 are models of the specification.

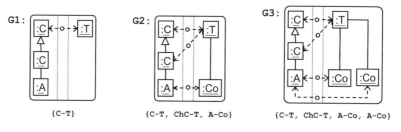

Fig. 5. SP-generated graphs

Our next result shows that, given a source graph G_S using the rules from $TR(SP)$, and starting from the graph $\langle G_S, \emptyset, \emptyset \rangle$, we can generate exactly all SP-generated graphs H such that $H_S = G_S$. Obviously not all SP-generated graphs need to be models of the specification (for instance $\langle G_S, \emptyset, \emptyset \rangle$ is generated by the empty family of patterns), but this result ensures that if H describes a valid model transformation of G_S and H only contains nodes and edges that the patterns prescribe that must be present, then H can be obtained by graph transformation.

Theorem 3 (Characterization of SP-Generated Graphs). *Given a pattern specification SP, G is an SP-generated graph if and only if $G|_S \Rightarrow^* G$ using rules from $TR(SP)$.*

The proof of this theorem goes as follows. First, we prove that if G is generated by $\mathcal{S}_1, \ldots \mathcal{S}_n$ and $f_1, \ldots f_n$ then there is a series of transformations:

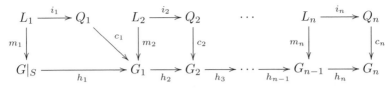

such that $G = G_n$ up to isomorphism, where the rules involved in these transformations are obtained by the pullback:

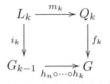

Conversely, we can prove that if G can be obtained by a series of transformations as the one above then G is generated by the patterns associated to these rules, together with the morphisms $f_1, \ldots f_n$, where $f_k = h_n \circ \cdots \circ h_{k+1} \circ c_k$.

As a direct consequence of this theorem we immediately get our first completeness result:

Corollary 1 (Completeness). *Given a pattern specification SP, if G is SP-generated and $G \models_F SP$ then $G|_S \Rightarrow^* G$ using rules from $TR(SP)$.*

There are two aspects in the previous completeness result which may be considered not fully satisfactory. On one hand, we have proved completeness of $TR(SP)$, i.e. a non-terminating transformation system. On the other hand, the notion of generated model may not completely follow our intuition. In particular, according to Def. 8, a given pattern may be used several times with the same match to generate several different parts of the target and connection graphs. Next, we provide a more restrictive notion of SP-generated graphs, namely strictly SP-generated graphs, and then we show that strictly SP-generated forward models are the terminal graphs of our terminating transformation systems.

Definition 9 (Strictly SP-Generated Graphs). *Given a pattern specification SP, a triple graph G is strictly SP-generated if G is an SP-generated graph and for every P-pattern $S = \{N(Q \xrightarrow{n_j} C_j)\}_{j \in J} \Rightarrow Q$, if $f_1 \colon Q \to G$ and $f_2 \colon Q \to G$ are two monomorphisms such that $(f_1)_S = (f_2)_S$ and both forward satisfy all the premises $n_j \colon Q \to C_j$, then $f_1 = f_2$.*

Notice that in the above definition it is enough to ask that either f_1 or f_2 forward satisfy all the premises of the pattern since this depends only on the source component of the morphisms. Therefore, since both morphisms coincide in their source component, if one of them forward satisfies a premise so will do the other morphism.

Example 5. In Fig. 5, graphs G_1 and G_2 are strictly generated, while G_3 is not, because both occurrences of A-Co share the same source.

Theorem 4 (Completeness for strictly SP-Generated Graphs). *Given a pattern specification SP, if G is strictly SP-generated and $G \models_F SP$ then $G|_S \Rightarrow^* G$ using rules from $TTR(SP)$ and, moreover, G is a terminal graph.*

The key to prove the above theorem is to show that if G is a strictly SP-generated graph and we assume that G is generated by a minimal family of patterns $S_1, \ldots S_n$ and monomorphisms $f_1, \ldots f_n$, then the minimality of the family ensures that all the matches in the above derivation satisfy the corresponding termination NACs, which means that the rules may be applied.

Finally, by Theorem 2 and Theorem 4, we have:

Corollary 2 (Soundness and Completeness). *Given a pattern specification SP consisting of positive patterns, and a strictly SP-generated graph G then $G \models_F SP$ if and only if $G|_S \Rightarrow^* G$ using rules from $TTR(SP)$ and G is a terminal graph.*

Remark 1. Corollary 2 tells us that, given a set of positive patterns SP and a source graph G_S, the set of all strictly SP-generated forward models of SP, whose S-component coincides with G_S, is included in the set of terminal graphs obtained from $\langle G_S, \emptyset, \emptyset \rangle$. However, if SP includes some negative patterns, then some (or perhaps all) of these terminal graphs may fail to satisfy these additional patterns. As said above, this is completely reasonable since negative patterns have not played any role in the construction of $TR(SP)$ or $TTR(SP)$. However, the negative patterns can be added as NACs into the transformation rules as described, for instance, in [6]. In this case, it will be impossible to transform a graph G_1 into G_2 if G_2 would violate some negative pattern. Then, the transformation system could be considered more efficient since the derivation tree associated to a given start graph would be pruned from all the graphs violating the negative patterns. However, in this case, our soundness and completeness results would slightly change. In particular, a terminal graph would not necessarily be a model of the given positive patterns anymore. More precisely, a graph would be terminal if it is a model of the given positive patterns or if all its possible transformations violate a negative pattern.

Example 6. Fig. 6 shows some derivations starting from a given graph G_0 using the generated terminating forward rules. All graphs in the derivations are strictly SP-generated. Hence all graphs in Fig. 5 are reachable. Notice that G_3 in Fig. 5 is not reachable using the terminating rules, as the rule generated from A-Co is not applicable to G_2. Graphs G_2 and G_5 are terminal w.r.t. $TTR(SP)$: the former is a forward model of SP, and the latter is only a forward model of SP^+ because the N-pattern N(NoDup) is not satisfied. As stated in the remark, we could add additional NACs to the generated rules to forbid applying a rule

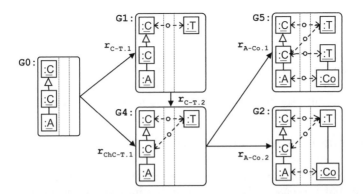

Fig. 6. Some derivations using the generated terminating forward rules

creating an occurrence of N-patterns. In that case, rule $r_{A-Co.1}$ would not be applied to G_4 and therefore graph G_5 would not be reached.

5   Related Work

Some declarative approaches to M2M transformation are unidirectional, e.g. PMT [20] or Tefkat [13], while we generate both forward and backward transformations. Among the visual, declarative and bidirectional approaches, a prominent example is the OMG's standard language QVT-relational [17]. Relations in QVT contain *when* and *where* clauses to guide the execution of the operational mechanisms. In our case, it is not necessary because the generated rules are terminating, correct and complete. Moreover, QVT lacks a formal semantics, complicating the analysis. On the contrary, our patterns have a formal semantics, which makes them amenable to verification.

TGGs [18] formalize the synchronized evolution of two graphs through declarative rules. From this specification, low level operational TGG rules are derived to perform forward and backward transformations, similar to our case. However, whereas in declarative TGG rules dependencies must be made explicit (i.e. TGG rules must declare which elements should exist and which ones are created), in our patterns this information is derived when generating the rules.

Completeness and correctness of forward and backward transformations was proved for TGGs in [7]. Termination was not studied because TGGs need a control mechanism to guide the execution of the operational rules, such as priorities [10] or their coupling to editing rules [5]. This is not necessary with our patterns, but we need to ensure finite termination of the operational mechanisms. The conditions for information preservation of forward and backward transformations was studied in [5]. Moreover, in [8] the results in [5] are extended to the case of triple graph grammars with NACs. A similar result can be adapted for pattern-based specifications.

In our initial presentation of pattern-based transformation [3], we introduced some deduction operations able to generate new patterns from existing ones.

In the present paper, we have simplified the framework by eliminating such deduction operations, but enriching the process of generating operational rules. The new generation process ensures completeness as it generates each possible LHS for the rules, which however could not be guaranteed with the deduction operations. However, such operations can be used as heuristics to generate less rules, or to reduce the non-confluent behaviour of the transformations.

6   Conclusions and Future Work

In this paper we have demonstrated three properties of the compilation mechanisms of pattern-based M2M specifications into graph grammar rules: finite termination, correctness with respect to the positive fragment of the specification and completeness. The first result allows using the generated rules without any external control mechanism for rule execution, as a difference from current approaches [11,17,18]. The correctness result ensures soundness of the operational mechanisms, and as remarked in the article, it can be easily extended to correctness of specifications including negative patterns. Finally, completeness guarantees that if the M2M specification has a model, then it can be found by the operational mechanism.

Our results provide a formal foundation for our approach to pattern-based M2M transformations. However, from a practical point of view, we believe that the techniques presented in this paper have to be complemented with other techniques that ensure some good performance. More precisely, our results guarantee that using our approach we can obtain all the (generated) models of given specification, which means all the possible transformations that are correct according to the given specification. However, on one hand, this is in general an exponential number of models, which implies that computing all these models would be not feasible. On the other hand, typically, there may be some preferred kind of model (for instance, models that are minimal in some sense) and, as a consequence, we would not be interested in computing all the models, but only the preferred ones. In addition, in order to make pattern-based transformation useful for Model-Driven Development, we are currently addressing further challenges: handling attributes in M2M specifications, supporting advanced meta-modelling concepts like inheritance and integrity constraints, and tool support. Moreover, we believe that pattern-based transformation can be used as a formal basis for other transformation languages, like QVT.

Acknowledgments. Work supported by the Spanish Ministry of Science and Innovation, projects METEORIC (TIN2008-02081), MODUWEB (TIN2006-09678) and FORMALISM (TIN2007-66523). Moreover, part of this work was done during a sabbatical leave of the first author at TU Berlin, with financial support from the Spanish Ministry of Science and Innovation (grant ref. PR2008-0185). We thank the referees for their useful comments.

References

1. Akehurst, D.H., Kent, S.: A relational approach to defining transformations in a metamodel. In: Jézéquel, J.-M., Hussmann, H., Cook, S. (eds.) UML 2002. LNCS, vol. 2460, pp. 243–258. Springer, Heidelberg (2002)
2. Braun, P., Marschall, F.: Transforming object oriented models with BOTL. ENTCS 72(3) (2003)
3. de Lara, J., Guerra, E.: Pattern-based model-to-model transformation. In: Ehrig, H., Heckel, R., Rozenberg, G., Taentzer, G. (eds.) ICGT 2008. LNCS, vol. 5214, pp. 426–441. Springer, Heidelberg (2008)
4. Ehrig, H., Ehrig, K., de Lara, J., Taentzer, G., Varró, D., Varró-Gyapay, S.: Termination criteria for model transformation. In: Cerioli, M. (ed.) FASE 2005. LNCS, vol. 3442, pp. 49–63. Springer, Heidelberg (2005)
5. Ehrig, H., Ehrig, K., Ermel, C., Hermann, F., Taentzer, G.: Information preserving bidirectional model transformations. In: Dwyer, M.B., Lopes, A. (eds.) FASE 2007. LNCS, vol. 4422, pp. 72–86. Springer, Heidelberg (2007)
6. Ehrig, H., Ehrig, K., Prange, U., Taentzer, G.: Fundamentals of algebraic graph transformation. Springer, Heidelberg (2006)
7. Ehrig, H., Ermel, C., Hermann, F.: On the relationship of model transformations based on triple and plain graph grammars. In: Proc. GRAMOT 2008 (2008)
8. Ehrig, H., Hermann, F., Sartorius, C.: Completeness and correctness of model transformations based on triple graph grammars with negative application conditions. In: GT-VMT 2009. Electronic Communications of the EASST (to appear, 2009)
9. Heckel, R., Wagner, A.: Ensuring consistency of conditional graph rewriting - a constructive approach. ENTCS 2 (1995)
10. Königs, A.: Model transformation with triple graph grammars. In: Proc. MTiP 2005 (2005)
11. Königs, A., Schürr, A.: Tool integration with triple graph grammars - a survey. ENTCS 148(1), 113–150 (2006)
12. Lack, S., Sobocinski, P.: Adhesive categories. In: Walukiewicz, I. (ed.) FOSSACS 2004. LNCS, vol. 2987, pp. 273–288. Springer, Heidelberg (2004)
13. Lawley, M., Steel, J.: Practical declarative model transformation with Tefkat. In: Bruel, J.-M. (ed.) MoDELS 2005. LNCS, vol. 3844, pp. 139–150. Springer, Heidelberg (2006)
14. MOLA. MOdel transformation LAnguage, http://mola.mii.lu.lv/
15. MTF. Model Transformation Framework , http://www.alphaworks.ibm.com/tech/mtf
16. Orejas, F., Guerra, E., de Lara, J., Ehrig, H.: Correctness, completeness and termination of pattern-based model-to-model transformation (long version). Technical Report 2009/09, TU Berlin, Fak. IV (2009)
17. QVT (2005), http://www.omg.org/docs/ptc/05-11-01.pdf
18. Schürr, A.: Specification of graph translators with triple graph grammars. In: Mayr, E.W., Schmidt, G., Tinhofer, G. (eds.) WG 1994. LNCS, vol. 903, pp. 151–163. Springer, Heidelberg (1995)
19. Stevens, P.: Bidirectional model transformations in QVT: Semantic issues and open questions. In: Engels, G., Opdyke, B., Schmidt, D.C., Weil, F. (eds.) MODELS 2007. LNCS, vol. 4735, pp. 1–15. Springer, Heidelberg (2007)
20. Tratt, L.: A change propagating model transformation language. JOT 7(3), 107–126 (2008)

Constructor-Based Institutions

Daniel Găină, Kokichi Futatsugi, and Kazuhiro Ogata

Graduate School of Information Science
Japan Advanced Institute of Science and Technology (JAIST)
{daniel,kokichi,ogata}@jaist.ac.jp

Abstract. Many computer science applications concern properties which are true of a restricted class of models. We present a couple of constructor-based institutions defined on top of some base institutions by restricting the class of models. We define the proof rules for these logics formalized as institutions, and prove their completeness in the abstract framework of institutions.

1 Introduction

Equational specification and programming constitute the bases of the modern algebraic specification languages (like CafeOBJ [9], CASL [2] and Maude [6]), the other features being somehow built on top of it. In 1935 Birkhoff first proved a completeness result for equational logic, in the unsorted case. Goguen and Meseguer extended the result to the many-sorted case, providing a full algebraization of finitary many-sorted equational deduction. In [7] the result is cast in the framework of institutions capturing both the finitary and the infinitary cases. Here we present an institution-independent completeness result for constructor-based logics [4] obtained from a base logic, basically by restricting the class of models.

Applications concern properties which are true of a restricted class of models. In most of the cases the models of interest include the initial model(s) of a set of axioms. Some approaches consider the initial semantics and reason about properties which are true of initial model [15]. Our work takes into account the generation aspects of software systems by considering the constructor-based institutions. The signatures of these institutions consist of a signature in the base institution and a distinguished set of operation symbols called *constructors*. The result sorts of the constructors are called *constrained* and a sort which is not constrained it is *loose*. The constructors determine the class of *reachable* models which are of interest from the user point of view. Intuitively the carrier sets of such models consist of constructor-generated elements. The sentences and the satisfaction condition are preserved from the base institution. In order to obtain a constructor-based institution the signature morphisms of the base institution are restricted such that the *reducts* of the reachable models along the signature morphisms are also reachable. In the examples presented here it is simply required that constructors are preserved by signature morphisms, and no

A. Kurz, M. Lenisa, and A. Tarlecki (Eds.): CALCO 2009, LNCS 5728, pp. 398–412, 2009.

"new" constructors are introduced for "old" constrained sorts (for sorts being in the image of some constrained sorts of the source signature).

We provide proof rules for the constructor-based institutions and we prove a completeness result using institution independent techniques. However the completeness is relative to a family of *sufficient complete* basic specifications (Σ, Γ) with signature Σ and set Γ of sentences. Intuitively (Σ, Γ) is sufficient complete when any term constructed with symbols from Σ and variables of sort loose can be reduced to a term formed with constructors and variables of sort loose using the equations from Γ.

For lack of space, we omit most of the proofs of our theorems, propositions and lemmas. However, interested readers can find these proofs in the extended version of this paper [10].

In Section 2 we present the notions of institution and entailment system; these will constitute the base for expressing the soundness and completeness properties for a logic, i.e. the semantic deduction coincides with the syntactic provability. After presenting some well known examples of institutions we give the definitions of some important concepts that are used in this paper. Section 3 introduces the abstract concept of universal institution and reachable universal weak entailment system which is proved sound and complete with respect to a class of reachable models, under conditions which are also investigated. Section 4 contains the main result. The entailment system developed in the previous section is borrowed by constructor-based institutions through institution morphisms. Soundness is preserved, and completeness is relative to a family of sets of sentences. Section 5 concludes the paper and discusses the future work.

2 Institutions

Institutions were introduced in [12] with the original goal of providing an abstract framework for algebraic specifications of computer science systems.

Definition 1. *An institution consists of*
  - a category $\mathbb{S}ig$, *whose objects are called* signatures.
  - a functor $\mathbb{S}en : \mathbb{S}ig \to \mathbb{S}et$, *providing for each signature a set whose elements are called* (Σ-)sentences.
  - a functor $\mathbb{M}od : \mathbb{S}ig^{op} \to \mathbb{C}at$, *providing for each signature* Σ *a category whose objects are called* (Σ-)models *and whose arrows are called* (Σ-)morphisms.
  - a relation $\models_\Sigma \subseteq |\mathbb{M}od(\Sigma)| \times \mathbb{S}en(\Sigma)$ *for each* $\Sigma \in |\mathbb{S}ig|$, *called* ($\Sigma$-)satisfaction, *such that for each morphism* $\varphi : \Sigma \to \Sigma'$ *in* $\mathbb{S}ig$, *the following* satisfaction condition *holds:* $M' \models_{\Sigma'} \mathbb{S}en(\varphi)(e)$ *iff* $\mathbb{M}od(\varphi)(M') \models_\Sigma e$, *for all* $M' \in |\mathbb{M}od(\Sigma')|$ *and* $e \in \mathbb{S}en(\Sigma)$.

Following the usual notational conventions, we sometimes let $\_ \upharpoonright_\phi$ denote the reduct functor $\mathbb{M}od(\varphi)$ and let φ denote the sentence translation $\mathbb{S}en(\varphi)$. When $M = M' \upharpoonright_\varphi$ we say that M' is a φ-*expansion* of M, and that M is the φ-*reduct* of M'.

*Example 1 (First-order logic (**FOL**) [12]).* Signatures are first-order many-sorted signatures (with sort names, operation names and predicate names); sentences are the usual closed formulae of first-order logic built over atomic formulae given either as equalities or atomic predicate formulae; models are the usual first-order structures; satisfaction of a formula in a structure is defined in the standard way.

*Example 2 (Constructor-based first-order logic (**CFOL**)).* The signatures (S, F , F^c, P) consist of a first-order signature (S, F, P), and a distinguished set of *constructors* $F^c \subseteq F$. The constructors determine the set of *constrained* sorts $S^c \subseteq S$: $s \in S^c$ iff there exists a constructor $\sigma \in F^c_{w \to s}$. We call the sorts in $S^l = S - S^c$ *loose*. The (S, F, F^c, P)-sentences are the *universal constrained first-order sentences* of the form $(\forall X)\rho$, where X is a finite set of constrained variables[1], and ρ is a first-order formula formed over the atoms by applying Boolean connectives and quantifications over finite sets of loose variables[2]. The (S, F, F^c, P)-models are the usual first-order structures M with the carrier sets for the constrained sorts consisting of interpretations of terms formed with constructors and elements of loose sorts, i.e. there exists a set Y of variables of loose sorts, and a function $f : Y \to M$ such that for every constrained sort $s \in S^c$ the function $f^\#_s : (T_{F^c}(Y))_s \to M_s$ is a surjection, where $f^\#$ is the unique extension of f to a (S, F^c, P)-morphism. A constructor-based first-order signature morphisms $\varphi : (S, F, F^c, P) \to (S_1, F_1, F^c_1, P_1)$ is a first-order signature morphism $\varphi : (S, F, P) \to (S_1, F_1, P_1)$ such that the constructors are preserved along the signature morphisms, and no "new" constructors are introduced for "old" constrained sorts: if $\sigma \in F^c$ then $\varphi(\sigma) \in F^c_1$, and if $\sigma_1 \in (F^c_1)_{w_1 \to s_1}$ and $s_1 \in \varphi(S^c)$ then there exists $\sigma \in F^c$ such that $\varphi(\sigma) = \sigma_1$. Variants of constructer-based first-order logic were studied in [4] and [3].

*Example 3 (Constructor-based Horn clause logic (**CHCL**)).* This institution is obtained from **CFOL** by restricting the sentences to *universal Horn sentences* of the form $(\forall X)(\forall Y) \wedge H \Rightarrow C$, where X is a finite set of constrained variables, Y is a finite set of loose variables, H is a finite set of atoms, and C is an atom. **CHCL**$_\infty$ is the infinitary extension of **CHCL** obtained by allowing *infinitary universal Horn sentences* $(\forall X)(\forall Y) \wedge H \Rightarrow C$ where the sets X, Y and H may be infinite.

*Example 4 (Order-sorted algebra (**OSA**) [14]).* An order-sorted signature $(S, \leq , F)$ consists of an algebraic signature (S, F), with a partial ordering $(S, \leq)$ such that the following *monotonicity condition* is satisfied $\sigma \in F_{w_1 \to s_1} \cap F_{w_2 \to s_2}$ and $w_1 \leq w_2$ imply $s_1 \leq s_2$. A morphism of **OSA** signatures $\varphi : (S, \leq , F) \to (S', \leq', F')$ is just a morphism of algebraic signatures $(S, F) \to (S', F')$ such that the ordering is preserved, i.e. $\varphi(s_1) \leq' \varphi(s_2)$ whenever $s_1 \leq s_2$. Given an order-sorted signature $(S, \leq, F)$, an order-sorted $(S, \leq, F)$-algebra is a (S, F)-algebra M such that $s_1 \leq s_2$ implies $M_{s_1} \subseteq M_{s_2}$, and $\sigma \in F_{w_1 \to s_1} \cup F_{w_2 \to s_2}$ and $w_1 \leq w_2$ imply $M^{w_1, s_1}_\sigma = M^{w_2, s_2}_\sigma$ on M_{w_1}. Given order-sorted $(S, \leq, F)$-algebras

---

[1] $X = (X_s)_{s \in S}$ is a set of constrained variables if $X_s = \emptyset$ for all $s \in S^l$.

[2] $Y = (Y_s)_{s \in S}$ is a set of loose variables if $Y_s = \emptyset$ for all $s \in S^c$.

M and N, an order-sorted $(S, \leq, F)$-homomorphism $h : M \to N$ is a (S, F)-homomorphism such that $s_1 \leq s_2$ implies $h_{s_1} = h_{s_2}$ on M_{s_1}.

Let $(S, \leq, F)$ be an order-sorted signature. We say that the sorts s_1 and s_2 are in the same *connected component* of S iff $s_1 \equiv s_2$, where $\equiv$ is the least equivalence on S that contains $\leq$. An **OSA** signature $(S, \leq, F)$ is *regular* iff for each $\sigma \in F_{w_1 \to s_1}$ and each $w_0 \leq w_1$ there is a unique least element in the set $\{(w, s) \mid \sigma \in F_{w \to s} \text{ and } w_0 \leq w\}$. For regular signatures $(S, \leq, F)$, any $(S, \leq, F)$-term has a least sort and the initial $(S, \leq, F)$-algebra can be defined as a term algebra, cf. [14]. A partial ordering $(S, \leq)$ is *filtered* iff for all $s_1, s_2 \in S$, there is some $s \in S$ such that $s_1 \leq s$ and $s_2 \leq s$. A partial ordering is *locally filtered* iff every connected component of it is filtered. An order-sorted signature $(S, \leq, F)$ is *locally filtered* iff $(S, \leq)$ is locally filtered, and it is *coherent* iff it is both locally filtered and regular. Hereafter we assume that all **OSA** signatures are coherent.

The atoms of the signature $(S, \leq, F)$ are equations of the form $t_1 = t_2$ such that the least sort of the terms t_1 and t_2 are in the same connected component. The sentences are closed formulas built by application of boolean connectives and quantification to the equational atoms. Order-sorted algebras were extensively studied in [13] and [14].

Example 5 (Constructor-based order-sorted logic (COSA)). This institution is defined on top of **OSA** similarly as **CFOL** is defined on top of **FOL**. The constructor-based order-sorted signatures $(S, \leq, F, F^c)$ consists of an order-sorted signature $(S, \leq, F)$, and a distinguished set of operational symbols $F^c \subseteq F$, called *constructors*, such that $(S, \leq, F^c)$ is an order-sorted signature (the monotonicity and coherence conditions are satisfied). As in the first-order case the constructors determine the set of *constrained* sorts $S^c \subseteq S$: $s \in S^c$ iff there exists a constructor $\sigma \in F_{w \to s}^c$ with the result sort s. We call the sorts in $S^l = S - S^c$ *loose*. The $(S, \leq, F, F^c)$-sentences are the *universal constrained order-sorted sentences* of the form $(\forall X)\rho$, where X is finite set of constrained variables, and ρ is a formula formed over the atoms by applying Boolean connectives and quantifications over finite sets of loose variables. The $(S, \leq, F, F^c)$-models are the usual order-sorted $(S, \leq, F)$-algebras with the carrier sets for the constrained sorts consisting of interpretations of terms formed with constructors and elements of loose sorts, i.e. there exists a set of loose variables Y, and a function $f : Y \to M$ such that for every constrained sort $s \in S^c$ the function $f_s^{\#} : (T_{F^c}(Y))_s \to M_s$ is a surjection, where $f^{\#}$ is the unique extension of f to a $(S, \leq, F^c)$-morphism. A signature morphism $\varphi : (S, \leq, F, F^c) \to (S_1, \leq_1, F_1, F_1^c)$ is an order-sorted signature morphism such that the constructors are preserved along the signature morphisms, no "new" constructors are introduced for "old" constrained sorts, and if $s_1' \leq_1 s_1''$ and there exists $s'' \in S^c$ such that $s_1'' = \varphi(s'')$ then there exists $s' \in S^c$ such that $s_1' = \varphi(s')$.

Constructor-based Horn order-sorted algebra (**CHOSA**) is obtained by restricting the sentences of **COSA** to universal Horn sentences. Its infinitary variant **CHOSA**$_\infty$ is obtained by allowing infinitary universal Horn sentences.

We introduce the following institutions for technical reasons.

*Example 6 (Generalized first-order logic (**GFOL**)).* Its signatures (S, S^c, F, P) consist of a first-order signature (S, F, P) and a distinguished set of constrained sorts $S^c \subseteq S$. A sort which is not constrained is loose. A *generalized first-order signature morphism* between (S, S^c, F, P) and (S_1, S_1^c, F_1, P_1) is a simple signature morphism between (S, F, P) and $(S_1, F_1 + T_{F_1}, P_1)$, i.e. constants can be mapped to terms. The sentences are the universal constrained first-order sentences and the models are the usual first-order structures.

Generalized Horn clause logic (**GHCL**) is the restriction of **GFOL** to universal Horn sentences, and **GHCL$_\infty$** is extending **GHCL** by allowing infinitary universal Horn sentences.

*Example 7 (Generalized order-sorted algebra (**GOSA**)).* This institution is a variation of **OSA** similarly as **GFOL** is a variation of **FOL**. Its signatures distinguish a subset of constrained sorts and the signature morphisms allow mappings of constants to terms. The sentences are the universal constrained order-sorted sentences. **GHOSA** and **GHOSA$_\infty$** are defined in the obvious way.

Entailment systems. A *sentence system* $(\mathbb{S}ig, \mathbb{S}en)$ consists of a category of signatures $\mathbb{S}ig$ and a sentence functor $\mathbb{S}en : \mathbb{S}ig \to \mathbb{S}et$. An *entailment system* $\mathcal{E} = (\mathbb{S}ig, \mathbb{S}en, \vdash)$ consists of a sentence system $(\mathbb{S}ig, \mathbb{S}en)$ and a family of entailment relations $\vdash = \{\vdash_\Sigma\}_{\Sigma \in |\mathbb{S}ig|}$ between sets of sentences with the following properties:

(*Anti-monotonicity*) $E_1 \vdash_\Sigma E_2$ if $E_2 \subseteq E_1$,

(*Transitivity*) $E_1 \vdash_\Sigma E_3$ if $E_1 \vdash_\Sigma E_2$ and $E_2 \vdash_\Sigma E_3$, and

(*Unions*) $E_1 \vdash_\Sigma E_2 \cup E_3$ if $E_1 \vdash_\Sigma E_2$ and $E_1 \vdash_\Sigma E_3$.

(*Translation*) $E \vdash_\Sigma E'$ implies $\varphi(E) \vdash_{\Sigma'} \varphi(E')$ for all $\varphi : \Sigma \to \Sigma'$

When we allow infinite unions, i.e. $E \vdash_\Sigma \bigcup_{i \in J} E_i$ if $E \vdash_\Sigma E_i$ for all $i \in J$, we call the entailment system *infinitary*. We say that $\mathcal{E} = (\mathbb{S}ig, \mathbb{S}en, \vdash)$ is a *weak entailment system* (abbreviated *WES*) when it does not satisfy the *Translation* property. *The semantic entailment system* of an institution $\mathcal{I} = (\mathbb{S}ig, \mathbb{S}en, \mathbb{M}od, \models)$ consists of $(\mathbb{S}ig, \mathbb{S}en, \models)$. When there is no danger of confusion we may omit the subscript Σ from $\vdash_\Sigma$ and for every signature morphism $\varphi \in \mathbb{S}ig$, we sometimes let φ denote the sentence translation $\mathbb{S}en(\varphi)$.

Remark 1. The weak entailment system of an institution is an entailment system whenever is sound and complete.

Let $\mathcal{I} = (\mathbb{S}ig, \mathbb{S}en, \mathbb{M}od, \models)$ be an institution and $\mathcal{M} = \{\mathcal{M}_\Sigma\}_{\Sigma \in |\mathbb{S}ig|}$ a family of classes of models, where $\mathcal{M}_\Sigma \subseteq \mathbb{M}od(\Sigma)$ for all signatures $\Sigma \in |\mathbb{S}ig|$. A WES $\mathcal{E} = (\mathbb{S}ig, \mathbb{S}en, \vdash)$ of the institution $\mathcal{I}$ is *sound* (resp. *complete*) *with respect to* $\mathcal{M}$ when $E \vdash e$ implies $M \models (E \Rightarrow e)$[3] (resp. $M \models (E \Rightarrow e)$ implies $E \vdash e$) for all sets of sentences $E \subseteq \mathbb{S}en(\Sigma)$, sentences $e \in \mathbb{S}en(\Sigma)$ and models $M \in \mathcal{M}_\Sigma$. We say that $\mathcal{E}$ is *sound* (resp. *complete*) when $\mathcal{M}_\Sigma = |\mathbb{M}od(\Sigma)|$ for all signatures Σ.

---

[3] $M \models (E \Rightarrow e)$ iff $M \models E$ implies $M \models e$.

We call the WES $\mathcal{E} = (\mathbb{S}ig, \mathbb{S}en, \vdash)$ *compact* whenever: if $\Gamma \vdash E_f$ and $E_f \subseteq \mathbb{S}en(\Sigma)$ is finite then there exists $\Gamma_f \subset \Gamma$ finite such that $\Gamma_f \vdash E_f$. For each WES $\mathcal{E} = (\mathbb{S}ig, \mathbb{S}en, \vdash)$ one can easily construct the *compact sub-WES* $\mathcal{E}^c = (\mathbb{S}ig, \mathbb{S}en, \vdash^c)$ by defining the entailment relation $\vdash^c$ as follows: $\Gamma \vdash^c E$ iff for each $E_f \subseteq E$ finite there exists $\Gamma_f \subseteq \Gamma$ finite such that $\Gamma_f \vdash E_f$.

Remark 2. $(\mathbb{S}ig, \mathbb{S}en, \vdash^c)$ is a WES.

Basic sentences. A set of sentences $E \subseteq \mathbb{S}en(\Sigma)$ is called *basic* [8] if there exists a Σ-model M_E such that, for all Σ-models M, $M \models E$ iff there exists a morphism $M_E \to M$.

Lemma 1. *Any set of atomic sentences in* **GHCL** *and* **GHOSA** *is basic.*

Internal logic. The following institutional notions dealing with logical connectives and quantifiers were defined in [16].

Let Σ be a signature of an institution,

- a Σ-sentence $\neg e$ is a *(semantic) negation* of the Σ-sentence e when for every Σ-model M we have $M \models_\Sigma \neg e$ iff $M \not\models_\Sigma e$,

- a Σ-sentence $e_1 \wedge e_2$ is a *(semantic) conjunction* of the Σ-sentences e_1 and e_2 when for every Σ-model M we have $M \models_\Sigma e_1 \wedge e_2$ iff $M \models_\Sigma e_1$ and $M \models_\Sigma e_2$, and

- a Σ-sentence $(\forall \chi)e'$, where $\Sigma \xrightarrow{\chi} \Sigma' \in \mathbb{S}ig$ and $e' \in \mathbb{S}en(\Sigma')$, is a *(semantic) universal χ-quantification* of e' when for every Σ-model M we have $M \models_\Sigma (\forall \chi)e'$ iff $M' \models_{\Sigma'} e'$ for all χ-expansions M' of M.

Very often quantification is considered only for a restricted class of signature morphisms. For example, quantification in **FOL** considers only the finitary signature extensions with constants. For a class $\mathcal{D} \subseteq \mathbb{S}ig$ of signature morphisms, we say that the institution has universal $\mathcal{D}$-quantifications when for each $\chi : \Sigma \to \Sigma'$ in $\mathcal{D}$, each Σ'-sentence has a universal χ-quantification.

Reachable models. Consider two signature morphisms $\chi_1 : \Sigma \to \Sigma_1$ and $\chi_2 : \Sigma \to \Sigma_2$ of an institution. A signature morphisms $\theta : \Sigma_1 \to \Sigma_2$ such that $\chi_1; \theta = \chi_2$ is called a *substitution* between χ_1 and χ_2.

Definition 2. *Let $\mathcal{D}$ be a broad subcategory of signature morphisms of an institution. We say that a Σ-model M is $\mathcal{D}$-reachable if for each span of signature morphisms $\Sigma_1 \xleftarrow{\chi_1} \Sigma_0 \xrightarrow{\chi} \Sigma$ in $\mathcal{D}$, each χ_1-expansion M_1 of $M \restriction_\chi$ determines a substitution $\theta : \chi_1 \to \chi$ such that $M \restriction_\theta = M_1$.*

Proposition 1. *In* **GHCL**, *assume that $\mathcal{D}$ is the class of signature extensions with (possibly infinite number of) constants. A model M is $\mathcal{D}$-reachable iff its elements are exactly the interpretations of terms.*

One can replicate the above proposition for **GHOSA** too. Note that for each set E of atomic sentences in **GHCL** or **GHOSA**, the model M_E defining E as basic set of sentences is $\mathcal{D}$-reachable, where $\mathcal{D}$ is the class of signature extensions with constants.

Definition 3. *We say that a signature morphism* $\varphi : \Sigma \to \Sigma'$ *is finitary if it is finitely presented* [4] *in the comma category* $\Sigma/\mathbb{S}ig$ [5].

In typical institutions the extensions of signatures with finite number of symbols are finitary.

Definition 4. *Let* $\mathcal{D}^c$ *and* $\mathcal{D}^l$ *be two broad sub-categories of signature morphisms. We say that a* Σ-*model* M *is* $(\mathcal{D}^c, \mathcal{D}^l)$-*reachable if for every signature morphism* $\chi : \Sigma \to \Sigma'$ *in* $\mathcal{D}^c$ *and each* χ-*expansion* M' *of* M *there exists a signature morphism* $\varphi : \Sigma \to \Sigma''$ *in* $\mathcal{D}^l$, *a substitution* $\theta : \chi \to \varphi$, *and a* Σ''-*model* M'' *such that* $M'' \restriction_\theta = M'$.

The two notions of reachability, apparently different, are closely related.

Proposition 2. *Let* $\mathcal{D}^c$, $\mathcal{D}^l$ *and* $\mathcal{D}$ *be three broad sub-categories of signature morphisms such that* $\mathcal{D}^c, \mathcal{D}^l \subseteq \mathcal{D}$. *A* Σ-*model* M *is* $(\mathcal{D}^c, \mathcal{D}^l)$-*reachable if there exists a signature morphism* $\Sigma \xrightarrow{\varphi} \Sigma' \in \mathcal{D}$ *and a* φ-*expansion* M' *of* M *such that*

1. M' *is* $\mathcal{D}$-*reachable, and*
2. *whether*
   (a) $\varphi \in \mathcal{D}^l$, *or*
   (b) *every signature morphism in* $\mathcal{D}^c$ *is finitary and* φ *is the vertex of a directed co-limit* $(\varphi_i \xrightarrow{u_i} \varphi)_{i \in J}$ *of a directed diagram* $(\varphi_i \xrightarrow{u_{i,j}} \varphi_j)_{(i \leq j) \in (J, \leq)}$ *in* $\Sigma/\mathbb{S}ig$, *and* $\varphi_i \in \mathcal{D}^l$ *for all* $i \in J$.

Note that the above proposition comes in two variants: infinitary and finitary. The infinitary variant corresponds to the first condition ($\varphi \in \mathcal{D}^l$) and is applicable to infinitary institutions, such as **GHCL$_\infty$** and **GHOSA$_\infty$** while the finitary variant is applicable to **GHCL** and **GHOSA**. Throughout this paper we implicitly assume that in **GHCL, GHOSA, GHCL$_\infty$** and **GHOSA$_\infty$**, $\mathcal{D}$ represents the subcategory of signature morphisms which consists of signature extensions with constants; $\mathcal{D}^c$ represents the subcategory of signature morphisms which consists of signature extensions with constants of constrained sorts; $\mathcal{D}^l$ represents the subcategory of signature morphisms which consists of signature extensions with constants of loose sorts. In the finitary cases, such as **GHCL** and **GHOSA**, we assume that the signature morphisms in $\mathcal{D}^c$ and $\mathcal{D}^l$ are finitary.

The following is a corollary of Proposition 2.

---

[4] An object A in a category $\mathcal{C}$ is called *finitely presented* ([1]) if
  - for each directed diagram $D : (J, \leq) \to \mathcal{C}$ with co-limit $\{Di \xrightarrow{\mu_i} B\}_{i \in J}$, and for each morphism $A \xrightarrow{g} B$, there exists $i \in J$ and $A \xrightarrow{g_i} Di$ such that $g_i; \mu_i = g$,
  - for any two arrows g_i and g_j as above, there exists $i \leq k, j \leq k \in J$ such that $g_i; D(i \leq k) = g_j; D(j \leq k) = g$.

[5] The objects of $\Sigma/\mathbb{S}ig$ are signature morphisms $\Sigma \xrightarrow{\chi} \Sigma'$ with the source Σ, and the arrows between objects $\Sigma \xrightarrow{\chi_1} \Sigma_1$ and $\Sigma \xrightarrow{\chi_2} \Sigma_2$ are also signature morphisms $\varphi : \Sigma_1 \to \Sigma_2$ such that $\chi_1; \varphi = \chi_2$.

Corollary 1. *In* **GHCL** *and* **GHCL**$_\infty$, *a* Σ-*model* M, *where* $\Sigma = (S, S^c, F, P)$, *is* $(\mathcal{D}^c, \mathcal{D}^l)$-*reachable iff there exists a set of loose variables* Y *and a function* $f : Y \rightarrow M$ *such that for every constrained sort* $s \in S^c$ *the function* $f_s^\# :$ $(T_F(Y))_s \rightarrow M_s$ *is surjective, where* $f^\#$ *is the unique extension of* f *to a* (S, F, P)-*morphism.*

One can replicate the above corollary for **GHOSA** and **GHOSA**$_\infty$ too. Since the three sub-categories of signature morphisms $\mathcal{D}$, $\mathcal{D}^c$ and $\mathcal{D}$ are fixed in concrete institutions, we will refer to $\mathcal{D}$-reachable model(s) as ground reachable model(s), and to $(\mathcal{D}^c, \mathcal{D}^l)$-reachable model(s) as reachable model(s) [3].

3   Universal Institutions

Let $I = (\mathbb{S}ig, \mathbb{S}en, \mathbb{M}od, \models)$ be an institution, $\mathcal{D} \subseteq \mathbb{S}ig$ be a broad subcategory of signature morphisms, and $\mathbb{S}en^\bullet$ be a sub-functor of $\mathbb{S}en$ (i.e. $\mathbb{S}en^\bullet : \mathbb{S}ig \rightarrow \mathbb{S}et$ such that $\mathbb{S}en^\bullet(\Sigma) \subseteq \mathbb{S}en(\Sigma)$ and $\varphi(\mathbb{S}en^\bullet(\Sigma)) \subseteq \mathbb{S}en^\bullet(\Sigma')$, for each signature morphism $\varphi : \Sigma \rightarrow \Sigma'$). We denote by $\mathcal{I}^\bullet$ the institution $(\mathbb{S}ig, \mathbb{S}en^\bullet, \mathbb{M}od, \models)$. We say that $\mathcal{I}$ is a $\mathcal{D}$-*universal institution over* $\mathcal{I}^\bullet$ [7] when

- $(\forall\chi)\rho \in \mathbb{S}en(\Sigma)$ for all signature morphisms $\Sigma \xrightarrow{\chi} \Sigma' \in \mathcal{D}$ and sentences $\rho \in \mathbb{S}en^\bullet(\Sigma')$, and

- any sentence of $\mathcal{I}$ is of the form $(\forall\chi)\rho$ as above.

The completeness results below comes both in a finite and an infinite variant, the finite one being obtained by adding (to the hypotheses of the infinite one) all the finiteness hypotheses marked in the brackets.

The *reachable universal weak entailment system* (*RUWES*) developed in this section consists of four layers: the "atomic" layer which in abstract settings is assumed but is developed in concrete examples, the layer of the *weak entailment system with implications* (*IWES*), the layer of the *generic universal weak entailment system* (*GUWES*) and the upmost layer of the RUWES of $\mathcal{I}$. The soundness and the completeness at each layer is obtained relatively to the soundness and completeness of the layer immediately below.

Reachable universal weak entailment systems (RUWES). Let us assume a $\mathcal{D}^c$-universal institution $I = (\mathbb{S}ig, \mathbb{S}en, \mathbb{M}od, \models)$ over $I_2 = (\mathbb{S}ig, \mathbb{S}en_2, \mathbb{M}od, \models)$ such that I_2 has $\mathcal{D}^l$-quantifications for a broad subcategory $\mathcal{D}^l \subseteq \mathbb{S}ig$ of signature morphisms. We define the following properties, i.e. proof rules, for the WES of $\mathcal{I}$. (*Case splitting*) $\Gamma \vdash_\Sigma (\forall\chi)e_1$ if $\Gamma \vdash_\Sigma (\forall\varphi)\theta(e_1)$ for all sentences $(\forall\varphi)\theta(e_1)$ such that $\varphi \in \mathcal{D}^l$ and $\theta : \chi \rightarrow \varphi$ is a substitution, where $\Gamma \subseteq \mathbb{S}en(\Sigma)$ and $(\forall\chi)e_1 \in \mathbb{S}en(\Sigma)$. (*Substitutivity*) $(\forall\chi)\rho \vdash_\Sigma (\forall\varphi)\theta(\rho)$, for every sentence $(\forall\chi)\rho \in \mathbb{S}en(\Sigma)$ and any substitution $\theta : \chi \rightarrow \varphi$.

In **GHCL**, assume a set Γ of Σ-sentences and a Σ-sentence $(\forall x)\rho$ such that x is a constrained variable. In this case, *Case splitting* says that if for any term t formed with loose variables and operation symbols from Σ, we have $\Gamma \vdash (\forall Y)\rho(x \leftarrow t)$ [6], where Y are all (loose) variables which occur in t, then

---

[6] $\rho(x \leftarrow t)$ is the formula obtained from ρ by substituting t for x.

we have proved $\Gamma \vdash (\forall x)\rho$. In most of the cases the set of terms t formed with loose variables and operation symbols from a given signature [7] is infinite which implies that the premises of *Case splitting* are infinite, and thus, the corresponding entailment system is not compact.

Given a compact WES $\mathcal{E}_2 = (\mathbb{S}ig, \mathbb{S}en_2, \vdash^2)$ for I_2, the RUWES of $\mathcal{I}$ consists of the least WES over $\mathcal{E}_2$ closed under *Substitutivity* and *Case splitting*. This is the finitary version of the RUWES, and is applicable to **GHCL** and **GHOSA**. Note that the resulting entailment system is not compact (even if $\mathcal{E}_2$ is compact) since *Case splitting* is an infinitary rule. The infinitary variant is obtained by dropping the compactness condition, and by considering the infinitary WES for $\mathcal{I}$, and is applicable to **GHCL$_\infty$** and **GHOSA$_\infty$**.

Proposition 3. *The RUWES of $\mathcal{I}$ is sound with respect to all $(\mathcal{D}^c, \mathcal{D}^l)$-reachable models if the WES of $\mathcal{I}_2$ is sound with respect to all $(\mathcal{D}^c, \mathcal{D}^l)$-reachable models.*

Note that *Case splitting* is sound with respect to all $(\mathcal{D}^c, \mathcal{D}^l)$-reachable models while *Substitutivity* is sound to all models.

Theorem 1 (Reachable universal completeness). *The RUWES of $\mathcal{I}$ is complete with respect to all $(\mathcal{D}^c, \mathcal{D}^l)$-reachable models if*

1. *the WES of $\mathcal{I}_2$ is complete with respect to all $(\mathcal{D}^c, \mathcal{D}^l)$-reachable models (and compact), and*
2. *for each set of sentences $E \subseteq \mathbb{S}en_2(\Sigma)$ and each sentence $e \in \mathbb{S}en_2(\Sigma)$, we have $E \models e$ iff $M \models (E \Rightarrow e)$ for all $(\mathcal{D}^c, \mathcal{D}^l)$-reachable models M.*

Generic universal weak entailment systems (GUWES). Let us assume a $\mathcal{D}^l$-universal institution $I = (\mathbb{S}ig, \mathbb{S}en, \mathbb{M}od, \models)$ over $\mathcal{I}_1$ with $\mathbb{S}en_1$ the sub-functor of $\mathbb{S}en$. We define the following property, for the WES of $\mathcal{I}$.
(Generalization) $\Gamma \vdash_\Sigma (\forall \chi)e'$ iff $\chi(\Gamma) \vdash_{\Sigma'} e'$, for every set $\Gamma \subseteq \mathbb{S}en(\Sigma)$ and any sentence $(\forall \chi)e' \in \mathbb{S}en(\Sigma)$, where $\chi : \Sigma \to \Sigma'$.

Given a compact WES $\mathcal{E}_1 = (\mathbb{S}ig, \mathbb{S}en_1, \vdash^1)$ for $\mathcal{I}_1$, the GUWES of $\mathcal{I}$ consists of the least WES over $\mathcal{E}_1$, closed under *Substitutivity* and *Generalization*. This is the finitary version of the GUWES, and is applicable to the restriction of **GHCL** and **GHOSA** to the sentences quantified over finite sets of variables of loose sorts. Its infinitary variant is obtained by dropping the compactness condition, and by considering the infinitary WES of $\mathcal{I}$; it is applicable to the restriction of **GHCL$_\infty$** and **GHOSA$_\infty$** to the sentences quantified over sets (possible infinite) of loose variables.

Proposition 4. *The GUWES of $\mathcal{I}$ is sound whenever the WES of $\mathcal{I}_1$ is sound.*

Theorem 2 (Generic universal completeness). *Let $\mathcal{D}$ be a broad subcategory of signature morphisms such that $\mathcal{D}^l \subseteq \mathcal{D}$. Assume that*

1. *the WES of $\mathcal{I}_1$ is complete (and compact), and*
2. *for each set of sentences $E \subseteq \mathbb{S}en_1(\Sigma)$ and each sentence $e \in \mathbb{S}en_1(\Sigma)$, we have $E \models_\Sigma e$ iff $M \models_\Sigma (E \Rightarrow e)$ for all $\mathcal{D}$-reachable models M.*

---

[7] We consider terms modulo renaming variables.

Then we have

1. *the GUWES of $\mathcal{I}$ is complete (and compact), and*
2. $\Gamma \models_{\Sigma} (\forall\varphi)e'$, *where* $(\Sigma \xrightarrow{\varphi} \Sigma') \in \mathcal{D}^l$, *iff* $M' \models_{\Sigma'} (\varphi(\Gamma) \Rightarrow e')$ *for all $\mathcal{D}$-reachable models M'.*

We have proved that the GUWES of $\mathcal{I}$ is complete which implies that it is complete with respect to all $(\mathcal{D}^c, \mathcal{D}^l)$-models, for any subcategory $\mathcal{D}^c \subseteq \mathcal{D}$ of signature morphisms. Therefore, the first condition of Theorem 1 is fulfilled. The following remark addresses the second condition of Theorem 1.

Remark 3. Under the assumption of Theorem 2, for any subcategory $\mathcal{D}^c \subseteq \mathcal{D}$ of signature morphisms, we have $\Gamma \models_{\Sigma} (\forall\varphi)e'$ iff $M \models_{\Sigma} (\Gamma \Rightarrow (\forall\varphi)e')$ for all $(\mathcal{D}^c, \mathcal{D}^l)$-reachable models M.

Weak entailment systems with implications (IWES). Assume an institution $\mathcal{I} = (\mathbb{S}ig, \mathbb{S}en, \mathbb{M}od, \models)$, a sub-functor $\mathbb{S}en_0 : \mathbb{S}ig \to \mathbb{S}et$ of $\mathbb{S}en$ such that

- $(\bigwedge H \Rightarrow C) \in \mathbb{S}en(\Sigma)$, for all (finite) sets of sentences $H \subseteq \mathbb{S}en_0(\Sigma)$ and any sentence $C \in \mathbb{S}en_0(\Sigma)$, and
- any sentence in $\mathcal{I}$ is of the form $(\bigwedge H \Rightarrow C)$ as above.

We denote the institution $(\mathbb{S}ig, \mathbb{S}en_0, \mathbb{M}od, \models)$ by $\mathcal{I}_0$. We define the following proof rules for the IWES of $\mathcal{I}$.

(Implications) $\Gamma \vdash_{\Sigma} (\bigwedge H \Rightarrow C)$ iff $\Gamma \cup H \vdash_{\Sigma} C$, for every set $\Gamma \subseteq \mathbb{S}en(\Sigma)$ and any sentence $\bigwedge H \Rightarrow C \in \mathbb{S}en(\Sigma)$.

Given a compact WES $\mathcal{E}_0 = (\mathbb{S}ig, \mathbb{S}en_0, \vdash^0)$ for $\mathcal{I}_0$, the IWES of $\mathcal{I}$ consists of the least WES over $\mathcal{E}_0$, closed under the rules of *Implications*. This is the finitary version of the IWES for $\mathcal{I}$, and is applicable to the restriction of **GHCL** and **GHOSA** to the sentences formed without quantifiers. Its infinitary variant is obtained by dropping the compactness condition and by considering the infinitary WES for $\mathcal{I}$; it is applicable to the restriction of **GHCL**$_\infty$ and **GHOSA**$_\infty$ to the quantifier-free sentences.

Proposition 5. *[7] Let us assume that*

1. *the WES of $\mathcal{I}_0$ is sound, complete (and compact),*
2. *every set of sentences in $\mathcal{I}_0$ is basic, and*
3. *there exits a broad subcategory $\mathcal{D} \subseteq \mathbb{S}ig$ such that for each set $B \subseteq \mathbb{S}en_0(\Sigma)$ there is a $\mathcal{D}$-reachable model M_B defining B as basic set of sentences.*

Then we have

1. *the IWES of $\mathcal{I}$ is sound, complete (and compact), and*
2. $E \models e$ *iff* $M \models (E \Rightarrow e)$ *for all $\mathcal{D}$-reachable models M.*

Atomic weak entailment systems (AWES). In order to develop concrete sound and complete universal WES we need to define sound and complete WES for the "atomic" layer of each institution.

Proposition 6. *[7] Let* $\mathbf{GHCL}_0$ *be the restriction of* $\mathbf{GHCL}$ *to the atomic sentences. The WES of* $\mathbf{GHCL}_0$ *generated by the rules below is sound, complete and compact.*

(Reflexivity) $\emptyset \vdash t = t$, *where* t *is a term.*

(Symmetry) $t = t' \vdash t' = t$, *where* t, t' *are terms.*

(Transitivity) $\{t = t', t' = t''\} \vdash t = t''$, *where* t, t', t'' *are terms.*

(Congruence) $\{t_i = t'_i | 1 \leq i \leq n\} \vdash \sigma(t_1, ..., t_n) = \sigma(t'_1, ..., t'_n)$, *where* $t_i, t'_i \in T_F$ *are terms and* σ *is an operation symbol.*

(PCongruence) $\{t_i = t'_i | 1 \leq i \leq n\} \cup \{\pi(t_1, ..., t_n)\} \vdash \pi(t'_1, ..., t'_n)$, *where* t_i, t'_i *are terms and* π *is a predicate symbol.*

One can define a sound, complete and compact AWES for the atomic part of $\mathbf{GHOSA}$ by considering all the proof rules from Proposition 6, except the last one which deals with predicates. The following is a corollary of Theorem 1.

Corollary 2. *[Completeness of the* $\mathbf{GHCL}$*] The RUWES of* $\mathbf{GHCL}$ *generated by the rules of Case splitting, Substitutivity, Generalization, Implications, Reflexivity, Symmetry, Transitivity, Congruence and PCongruence is sound and complete with respect to all reachable models.*

4   Borrowing Completeness

Let $\mathcal{I}' = (\mathbb{S}ig', \mathbb{S}en', \mathbb{M}od', \models')$ and $\mathcal{I} = (\mathbb{S}ig, \mathbb{S}en, \mathbb{M}od, \models)$ be two institutions. An *institution morphism* $(\phi, \alpha, \beta) : \mathcal{I}' \to \mathcal{I}$ consists of

  - a functor $\phi : \mathbb{S}ig' \to \mathbb{S}ig$, and
  - two natural transformations $\alpha : \phi; \mathbb{S}en \Rightarrow \mathbb{S}en'$ and $\beta : \mathbb{M}od' \Rightarrow \phi^{op}; \mathbb{M}od$

such that the following satisfaction condition for institution morphisms holds: $M' \models'_{\Sigma'} \alpha_{\Sigma'}(e)$ iff $\beta_{\Sigma'}(M') \models_{\phi(\Sigma')} e$, for every signature $\Sigma' \in \mathbb{S}ig'$, each Σ'-model M', and any $\phi(\Sigma')$-sentence e.

Definition 5. *We say that a WES* $\mathcal{E} = (\mathbb{S}ig, \mathbb{S}en, \vdash)$ *of an institution* $\mathcal{I} = (\mathbb{S}ig, \mathbb{S}en, \mathbb{M}od, \models)$ *is* Ω*-complete, where* $\Omega = (\Omega_{\Sigma})_{\Sigma \in |\mathbb{S}ig|}$ *is a family of sets of sentences (* $\Omega_{\Sigma} \subseteq \mathcal{P}(\mathbb{S}en(\Sigma))$ *for all signatures* Σ*) iff* $\Gamma \models_{\Sigma} E$ *implies* $\Gamma \vdash_{\Sigma} E$ *for all* $\Gamma \in \Omega_{\Sigma}$.

Remark 4. Let $(\phi, \alpha, \beta) : \mathcal{I}' \to \mathcal{I}$ be an institution morphism (where $\mathcal{I}' = (\mathbb{S}ig', \mathbb{S}en', \mathbb{M}od', \models')$ and $\mathcal{I} = (\mathbb{S}ig, \mathbb{S}en, \mathbb{M}od, \models)$). Every WES $\mathcal{E} = (\mathbb{S}ig, \mathbb{S}en, \vdash)$ for $\mathcal{I}$ generates freely a WES $\mathcal{E}' = (\mathbb{S}ig', \mathbb{S}en', \vdash')$ for $\mathcal{I}'$, where $\mathcal{E}'$ is the least WES closed under the rules $\alpha_{\Sigma'}(\Gamma) \vdash'_{\Sigma'} \alpha_{\Sigma'}(E)$, where Σ' is a signature in $\mathcal{I}'$ and $\Gamma \vdash_{\phi(\Sigma')} E$ is a deduction in $\mathcal{E}$.

Theorem 3. *Consider*

1. *an institution morphism* $(\phi, \alpha, \beta) : \mathcal{I}' \to \mathcal{I}$ *(where* $\mathcal{I}' = (\mathbb{S}ig', \mathbb{S}en', \mathbb{M}od', \models')$ *and* $\mathcal{I} = (\mathbb{S}ig, \mathbb{S}en, \mathbb{M}od, \models)$*) such that* $\alpha_{\Sigma'}$ *is surjective for all* $\Sigma' \in |\mathbb{S}ig'|$,
2. *a class of models* $\mathcal{M} = (\mathcal{M}_{\Sigma})_{\Sigma \in |\mathbb{S}ig|}$ *(in* $\mathcal{I}$*) such that* $\beta_{\Sigma'}(|\mathbb{M}od'(\Sigma')|) \subseteq \mathcal{M}_{\phi(\Sigma')}$ *for all signatures* $\Sigma' \in |\mathbb{S}ig|$, *and*

3. a WES $\mathcal{E} = (\mathbb{S}ig, \mathbb{S}en, \vdash)$ for $\mathcal{I}$ which is sound and complete with respect to $\mathcal{M}$.

Then the entailment system $\mathcal{E}' = (\mathbb{S}ig', \mathbb{S}en', \vdash')$ of $\mathcal{I}'$ determined by $\mathcal{E}$ is sound and Ω-complete, where for every signature $\Sigma' \in |\mathbb{S}ig'|$ we have $\Gamma' \in \Omega_{\Sigma'}$ iff $\Gamma = \alpha_{\Sigma'}^{-1}(\Gamma')$ has the following property: $M \models_{\phi(\Sigma')} \Gamma$ implies $M \in \beta_{\Sigma'}(|\mathbb{M}od'(\Sigma')|)$, for any $M \in \mathcal{M}_{\phi(\Sigma')}$.

Proof. Since $\alpha_{\Sigma'}$ is surjective, for all signatures $\Sigma' \in |\mathbb{S}ig'|$, $\mathcal{E}' = (\mathbb{S}ig', \mathbb{S}en', \vdash')$ with $\vdash'_{\Sigma'} = \alpha_{\Sigma'}(\vdash_{\phi(\Sigma')})$, for all signatures $\Sigma' \in |\mathbb{S}ig'|$, is the WES of $\mathcal{I}'$ determined by the institution morphism (ϕ, α, β).

1. Suppose that $\Gamma' \vdash'_{\Sigma'} E'$ and let M' be a Σ'-model such that $M' \models' \Gamma'$. By the definition of $\mathcal{E}'$ there exists $\Gamma \vdash_{\phi(\Sigma')} E$ such that $\alpha_{\Sigma'}(\Gamma) = \Gamma'$ and $\alpha_{\Sigma'}(E) = E'$. By the satisfaction condition for the institution morphisms we have $\beta_{\Sigma'}(M') \models_{\phi(\Sigma')} \Gamma$. Since $\mathcal{E}$ is sound with respect to $\mathcal{M}$ we have $M \models_{\phi(\Sigma')} (\Gamma \Rightarrow E)$ for all models $M \in \mathcal{M}_{\phi(\Sigma')}$. Because $\beta_{\Sigma'}(M') \in \mathcal{M}_{\phi(\Sigma')}$ we have that $\beta_{\Sigma'}(M') \models_{\phi(\Sigma')} (\Gamma \Rightarrow E)$ which implies $\beta_{\Sigma'}(M') \models_{\phi(\Sigma')} E$. By the satisfaction condition for institution morphisms we get $M' \models'_{\Sigma'} \alpha_{\Sigma'}(E)$. Hence $M' \models'_{\Sigma'} E'$.

2. Assume $\Gamma' \models_{\Sigma'} E'$, where $\Gamma' \in \Omega$, and let $\Gamma = \alpha_{\Sigma'}^{-1}(\Gamma')$ and $E = \alpha_{\Sigma'}^{-1}(E')$. Note that $M \models (\Gamma \Rightarrow E)$ for all $M \in \mathcal{M}_\Sigma$. Indeed for any $M \in \mathcal{M}_\Sigma$ we have: $M \models_{\phi(\Sigma')} \Gamma$ implies $M \in \beta_{\Sigma'}(|\mathbb{M}od'(\Sigma')|)$; so, there exists a Σ'-model M' such that $\beta_{\Sigma'}(M') = M$ and by satisfaction condition for institution morphisms $M' \models' \Gamma'$ which implies $M' \models' E'$; applying again satisfaction condition we obtain $M \models E$. Since $\mathcal{I}$ is complete with respect to $\mathcal{M}$ we have $\Gamma \vdash E$ which implies $\Gamma' \vdash' E'$.

In order to develop sound and complete WES for the constructor-based institutions we need to set the parameters of Theorem 3. We define the institution morphism $\Delta_{\mathbf{HCL}} = (\phi, \alpha, \beta) : \mathbf{CHCL} \to \mathbf{GHCL}$ such that

1. the functor ϕ maps
   - every **CHCL** signature (S, F, F^c, P) to a **GHCL** signature (S, S^c, F, P), where S^c is the set of constrained sorts determined by F^c, and
   - every **CHCL** signature morphism $(\varphi^{sort}, \varphi^{op}, \varphi^{pred})$ to the **GHCL** signature morphism $(\varphi^{sort}, \varphi^{op}, \varphi^{pred})$;
2. α is the identity natural transformation (recall that $\mathbb{S}en(S, F, F^c, P) = \mathbb{S}en(S, S^c, F, P)$, where S^c is a the set of constrained sorts determined by the constructors in F^c), for every **CHCL** signature (S, F, F^c, P) we have $\alpha_{(S,F,F^c,P)} = 1_{\mathbb{S}en(S,F,F^c,P)}$;
3. β is the inclusion natural transformation (note that every (S, F, F^c, P)-model M is also a (S, S^c, F, P)-model; indeed if there exists a set of loose variables Y and a function $f : Y \to M$ such that for every constrained sort $s \in S^c$ the function $f_s^\# : (T_{F^c}(Y))_s \to M_s$ is a surjection, where $f^\#$ is the unique extension of f to a (S, F^c, P)-morphism, then for every constrained sort $s \in S^c$ the function $\overline{f}_s : (T_F(Y))_s \to M_s$ is a surjection too, where $\overline{f}$ is the unique extension of f to a (S, F, P)-morphism), for every **CHCL** signature (S, F, F^c, P) the functor $\beta_{(S,F,F^c,P)} : \mathbb{M}od(S, F, F^c, P) \to \mathbb{M}od(S, S^c, F, P)$

is defined by $\beta_{(S,F,F^c,P)}(M) = M$ for all models $M \in |\mathrm{Mod}(S,F,F^c,P)|$ and $\beta_{(S,F,F^c,P)}(h) = h$ for all morphism $h \in \mathrm{Mod}(S,F,F^c,P)$.

Remark 5. A (S, S^c, F, P)-model M in **GHCL** is reachable iff there exists a set of loose variables Y and a function $f : Y \to M$ such that for every constrained sort $s \in S^c$ the function $f_s^\# : (T_{F^{cons}}(Y))_s \to M_s$ is surjective, where F^{cons} is the set operations with constrained resulting sorts, and $f^\#$ is the unique extension of f to a (S, F^{cons}, P)-morphism.

Definition 6. *A basic specification (Σ, Γ) in* **CHCL** *is sufficient complete, where $\Sigma = (S, F, F^c, P)$, if for every term t formed with operation symbols from F^{cons} (the set of operations with constrained resulting sorts) and loose variables from Y there exists a term t' formed with constructors and loose variables from Y such that $\Gamma \vdash_{(S,S^c,F,P)} (\forall Y)t = t'$ in* **GHCL**.

Since $\models_{(S,S^c,F,P)} \subseteq \vdash_{(S,S^c,F,P)}$ for all **GHCL**-signatures (S, S^c, F, P) (see Corollary 2 for the definition of $\vdash_{(S,S^c,F,P)}$), the condition $\Gamma \vdash_{(S,S^c,F,P)} (\forall Y)t = t'$ in Definition 6 is more general than if we assumed $\Gamma \models_{(S,S^c,F,P)} (\forall Y)t = t'$.

The following is a corollary of Theorem 3.

Corollary 3. *The entailment system of* **CHCL** *generated by the proof rules for* **GHCL** *is sound and Ω-complete, where $\Gamma \in \Omega_{(S,F,F^c,P)}$ iff $((S, S^c, F, P), \Gamma)$ is a sufficient complete* **CHCL**-*specification.*

Proof. We set the parameters of Theorem 3. The institution $\mathcal{I}'$ is **CHCL** and the institution $\mathcal{I}$ is **GHCL**. The institution morphism is $\Delta_{\mathbf{HCL}}$ and the entailment system $\mathcal{E}$ of **GHCL** is the least entailment system closed under the rules enumerated in Corollary 2. $\mathcal{M}$ is the class of all reachable models. We need to prove that for every sufficient complete specification $((S, F, F^c, P), \Gamma)$ and any reachable (S, S^c, F, P)-model M (where S^c is the set constrained sorts determined by F^c) we have: $M \models \Gamma$ implies $M \in |\mathrm{Mod}(S, F, F^c, P)|$. Because M is reachable by Remark 5 there exists a set Y of loose variables and a function $f : Y \to M$ such that for every constrained sort $s \in S^c$ the function $f_s^\# : (T_{F^{cons}}(Y))_s \to M_s$ is a surjection, where $f^\#$ is the unique extension of f to a (S, F^{cons}, P)-morphism. Because $((S, F, F^c, P), \Gamma)$ is sufficient complete, for every constrained sort $s \in S^c$ the function $\overline{f}_s : (T_{F^c}(Y))_s \to M_s$ is a surjection too, where $\overline{f}$ is the unique extension of f to a (S, F^c, P)-morphism.

Similar results as Corollary 3 can be formulated for **GHOSA**, **GHCL**$_\infty$, and **GHOSA**$_\infty$.

In general, the proof rules given here for the constructor-based institutions are not complete. Consider the signature (S, F, F^c, P) in **CHCL**, where $S = \{s\}$, $F_{\to s} = \{a, b\}$, $F^c = \{a\}$ and $P = \emptyset$. It is easy to notice that $\models a = b$ but there is no way to prove $\emptyset \vdash a = b$.

Assume that $S = Nat$, $F_{\to Nat} = \{0\}$, $F_{Nat \to Nat} = \{s\}$, $F_{NatNat \to Nat} = \{+\}$, $F^c_{\to Nat} = \{0\}$, $F^c_{Nat \to Nat} = \{s\}$ and $P = \emptyset$. Consider the following equations $\rho_1 = (\forall x : Nat)x + 0 = x$ and $\rho_2 = (\forall x : Nat)(\forall x' : Nat)x + (s\,x') = s(x + x')$. Then $((S, S^c, F), \{\rho_1, \rho_2\})$ is a sufficient complete specification. Intuitively, if $\Gamma \in \mathbb{S}en(S, F, F^c, P)$ specify that non-constructor operators are inductively defined with respect to the constructors then $((S, F, F^c, P), \Gamma)$ is a sufficient complete specification.

Structural Induction. In the constructor-based institutions presented here the carrier sets of the models consist of interpretations of terms formed with constructors and elements of loose sort. Thus, *Case Splitting* can be rephrased as follows:

(*Case Splitting*) $\Gamma \vdash (\forall x)\rho$ if $\Gamma \vdash (\forall Y)\rho(x \leftarrow t)$ for all terms t formed with constructors and variables of sort loose, where Γ is a set of sentences, and $(\forall x)\rho$ a sentence such that x is a constrained variable.

In order to prove the premises of *Case splitting*, in many cases, we use induction on the structure of terms. For any t formed with constructors in F^c and loose variables we have

(*Structural induction*) $\Gamma \vdash_{(S,F,F^c)} (\forall V)\rho(x \leftarrow t)$ if

1. (*Induction base*) for all $cons \in F^c_{\to s}$, $\Gamma \vdash_{(S,F,F^c)} \rho(x \leftarrow cons)$,
2. (*Induction step*) for all $\sigma \in F^c_{s_1 \ldots s_n \to s}$, $\Gamma \cup \{\rho(x \leftarrow x') \mid x' \in X\} \vdash_{(S,F \cup C,F^c)}$
   $\rho(x \leftarrow \sigma(c_1, \ldots, c_n))$, where
   - $C = \{c_1, \ldots, c_n\}$ is a set of new variables such that c_i has the sort s_i, for all $i \in \{1, \ldots, n\}$, and
   - $X \subseteq C$ is the set of variables with the sort s.

where V are all (loose) variables in t.

5 Conclusions and Future Work

We define the infinitary rules of *Case splitting* and show that the WES of **CHCL** is sound and complete with respect to all sufficient complete specifications. We define the rules of *Structural induction* to deal with the infinitary premises of *Case spliting* but the infinitary rules can not be replaced with the finitary ones in order to obtain a complete and compact WES because the class of sentences true of a class of models for a given constructor-based specification is not in general recursively enumerable. Gödel's famous incompleteness theorem show that this holds even for the specification of natural numbers.

Due the abstract definition of reachable model given here, one can easily define a constructor-based institution on top of some base institution by defining the constructor-based signatures as signature morphisms in the base institution. This construction may be useful when lifting the interpolation and amalgamation properties (necessary for modularization) from the base institution.

The area of applications provided by the general framework of the present work is much wider. For example we may consider partial algebras [5] , preorder algebras [9], or variations of these institutions, such as order-sorted algebra with transitions. The present work is much general than [7]. If $\mathcal{D}^c$ is the broad subcategory consisting of identity morphisms then all models are constrained and we may obtain the result in [7] concerning Horn institutions. It is to investigate the applicability of Theorem 1 to **GFOL** by adapting the completeness of first-order institutions developed in [11]. Then it is straightforward to construct an institution morphism **CFOL** → **GFOL** and obtain an entailment system sound and complete (relatively to a family of sufficient complete basic specifications) for **CFOL**.

References

1. Adámek, J., Rosický, J.: Locally Presentable and Accessible Categories. London Mathematical Society Lecture Notes, vol. 189. Cambridge University Press, Cambridge (1994)
2. Astesiano, E., Bidoit, M., Kirchner, H., Krieg-Brückner, B., Mosses, P.D., Sannella, D., Tarlecki, A.: Casl: the common algebraic specification language. Theor. Comput. Sci. 286(2), 153–196 (2002)
3. Bidoit, M., Hennicker, R.: Constructor-based observational logic. J. Log. Algebr. Program. 67(1-2), 3–51 (2006)
4. Bidoit, M., Hennicker, R., Kurz, A.: Observational logic, constructor-based logic, and their duality. Theor. Comput. Sci. 3(298), 471–510 (2003)
5. Burmeister, P.: A Model Theoretic Oriented Approach to Partial Algebras. Akademie-Verlag, Berlin (1986)
6. Clavel, M., Durán, F., Eker, S., Lincoln, P., Martí-Oliet, N., Meseguer, J., Talcott, C. (eds.): All About Maude - A High-Performance Logical Framework. LNCS, vol. 4350. Springer, Heidelberg (2007)
7. Codescu, M., Gaina, D.: Birkhoff completeness in institutions. Logica Universalis 2(2), 277–309 (2008)
8. Diaconescu, R.: Institution-independent ultraproducts. Fundamenta Informaticae 55(3-4), 321–348 (2003)
9. Diaconescu, R., Futatsugi, K.: Logical foundations of CafeOBJ. Theor. Comput. Sci. 285(2), 289–318 (2002)
10. Găină, D., Futatsugi, K., Ogata, K.: Constructor-based Institutions. Technical report, Japan Advanced Institute of Science and Technology, vol. IS-RR-2009-002, pp. 1–19 (May 29, 2009), http://hdl.handle.net/10119/8177
11. Găină, D., Petria, M.: Completeness by forcing (submitted)
12. Goguen, J.A., Burstall, R.: Institutions: Abstract model theory for specification and programming. Journal of the Association for Computing Machinery 39(1), 95–146 (1992)
13. Goguen, J.A., Diaconescu, R.: An Oxford survey of order sorted algebra. Mathematical Structures in Computer Science 4(3), 363–392 (1994)
14. Goguen, J.A., Meseguer, J.: Order-sorted algebra I: Equational deduction for multiple inheritance, overloading, exceptions and partial operations. Theoretical Computer Science 105(2), 217–273 (1992)
15. Goguen, J.A., Thatcher, J.W.: Initial algebra semantics. In: Annual Symposium on Switching and Automata Theory, pp. 63–77 (1974)
16. Tarlecki, A.: Bits and pieces of the theory of institutions. In: Poigné, A., Pitt, D.H., Rydeheard, D.E., Abramsky, S. (eds.) Category Theory and Computer Programming. LNCS, vol. 240, pp. 334–360. Springer, Heidelberg (1986)

DBtk: A Toolkit for Directed Bigraphs

Giorgio Bacci, Davide Grohmann, and Marino Miculan

Department of Mathematics and Computer Science, University of Udine, Italy
{giorgio.bacci,grohmann,miculan}@dimi.uniud.it

Abstract. We present DBtk, a toolkit for Directed Bigraphs. DBtk supports a textual language for directed bigraphs, the graphical visualization of bigraphs, the calculation of IPO labels, and the calculation of redex matchings. Therefore, this toolkit provides the main functions needed to implement simulators and verification tools.

1 Introduction

Bigraphical Reactive Systems (BRSs) [7] are a promising meta-model for ubiquitous (i.e., concurrent, communicating, mobile) systems. A calculus can be modeled as a BRS by encoding its terms (or states) as *bigraphs*, semi-structured data capable to represent at once both the location and the connections of the components of a system. The reduction semantics of the calculus is represented by a set of rewrite rules on this semi-structured data. Many calculi and models have been successfully represented as BRSs, such as CCS, Petri Nets, Mobile Ambients and, in the "directed" variant of [2], also Fusion calculus [5,6,3].

BRSs offer many general and powerful results. Particularly important for verification purposes is the possibility to derive systematically labelled transition systems via the so-called *IPO construction* [5], where the labels for a given agent are the *minimal* contexts which trigger a transition. Interestingly, the strong bisimilarity induced by this LTS is always a congruence.

Moreover, a "bigraphical simulation engine" would allow to obtain immediately a simulator for any calculus/model formalized as a BRS. The core of this engine will be the implementation of *redex matching*, that is, to determine when and where the left-hand side of a bigraphical reaction rule matches a bigraph; then this redex is replaced with the right-hand side of the same rule.

These are the main features offered by the *Directed Bigraphs Toolkit* (DBtk), which we describe in this paper. The architecture of DBtk is shown in Fig. 1. First, the toolkit defines data structures and operations for representing and manipulating bigraphs (Section 2). For more conveniently interacting with the user, bigraphs can be described using a language called DBL (Section 3), and also graphically visualized by means of a SVG representation (Section 4). Then, the toolkit provides the functions for calculating matchings of a redex within a bigraph (Section 5) and the RPOs and IPOs of directed bigraphs (Section 6).

Comparison with related work and directions for future work are in Section 7. DBtk can be found, with examples and more detailed descriptions, at http://www.dimi.uniud.it/grohmann/dbtk.

A. Kurz, M. Lenisa, and A. Tarlecki (Eds.): CALCO 2009, LNCS 5728, pp. 413–422, 2009.
© Springer-Verlag Berlin Heidelberg 2009

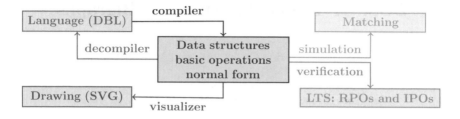

Fig. 1. Architecture of DBtk

2   Basic Data Structures and Operations

The Main module contains the basic data structures for representing directed bigraphs, and the two basic categorical operations: composition and tensor product. Two bigraphs can be composed by putting the roots of the "lower" one inside the sites (or holes) of the "upper" one, and by pasting the links having the same names in the common interface, whereas the tensor product juxtaposes two bigraphs if they do not share names on their inner and outer interfaces, respectively. For user convenience, the derived operations (such as sharing products and merge product) are also provided. Moreover, the module provides the support for the bigraph algebra, i.e., functions for constructing the elementary bigraphs (see Table 1), and the *normalization procedure*, which takes a generic bigraph and yields a triple of bigraphs (more precisely two wirings and a discrete bigraph) that represent its directed normal form (see [4]).

3   DBL, the Directed Bigraphical Language

The *directed bigraphical language* (DBL) is a (term) language for bigraphs, which follows the algebra defined in [4]. An expression in DBL can be compiled in the internal representation of bigraphs. Also a decompiler is provided, translating a bigraph represented in the internal data structures back to a term expression.

A bigraph definition begins with a signature definition, whose syntax is:

```
Signature [ACTIVITY NAME:#PORTS,...] ;
```

where `ACTIVITY` is picked from the set {active, passive, atomic}, `NAME` is the name of the control, and `#PORTS` is the number of the node ports. For example:

```
Signature [passive n1:2, atomic node_AT:0, active tr:1] ;
```

Then, the bigraph is described in a functional-style language, which allows to compose elementary bigraphs (whose syntax is in Table 1) and other expressions with the various operators of the algebra. Composition and tensor product are denoted by ° and *, respectively. As a shortcut, also the sharing operators are also provided: outer sharing product (/\); inner sharing product (\/); sharing product (||); prime outer product (/^\); and prime sharing product (|).

As an example, the following specification

Table 1. Syntax expressions for elementary directed bigraphs

Barren root (1) **merge 0** (special case of **merge**)	Closure ($\mathbf{Y}_y^x$) **x X y** (generalized to **[x,w] X [y,z]**)	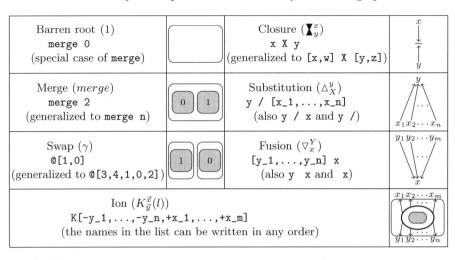
Merge (*merge*) **merge 2** (generalized to **merge n**)	Substitution ($\triangle_X^y$) **y / [x_1,...,x_n]** (also **y / x** and **y /**)	
Swap (γ) **@[1,0]** (generalized to **@[3,4,1,0,2]**)	Fusion ($\triangledown_x^Y$) **[y_1,...,y_n] x** (also **y x** and **x**)	
Ion ($K_{\vec{y}}^{\vec{x}}(l)$) **K[-y_1,...,-y_n,+x_1,...,+x_m]** (the names in the list can be written in any order)		

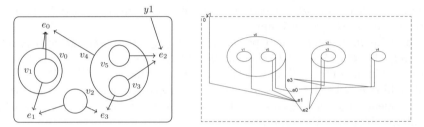

Fig. 2. A directed bigraph (left) and its SVG representation (right)

```
Signature [active t4:1,atomic t5:1,atomic t2:2];
let a = merge 1|X[x,z]|(y1X[a,y]*t4 [-w])°
                        (w\w*merge 1|t5 [+a]*y/y) in
  let b = (((wX[b,c]*merge 1)°
          (b/b*t4 [+c])°t2 [+b,-d]|t2 [+x,-e])°([d,e]X)) in
    let c = t2 [+z,+y] in a°(b*c)
```

corresponds to the directed bigraph shown in Fig. 2.

4   Graphical Visualization

In addition to the bigraphical language, DBtk allows to represent bigraphs in the XML-based SVG format, which is the W3C open standard for vector graphics; many web browsers have free, native support for SVG rendering. The translation function **dbg2svg** takes a directed bigraph data structure and returns a string containing its SVG representation. This function is fruitful in combination with the compiler module: bigraphs are defined in DBL, compiled to the DBG-data structure and then written to an SVG file (see Fig. 2, right).

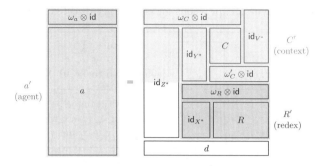

Fig. 3. Schema of a valid matching sentence

5   Matching

The main challenge for implementing the dynamics of bigraphical reactive systems is the *maching problem*, that is, to determine whether a reaction rule can be applied to rewrite a given bigraphical agent. To this end, we have to find when, and where, a redex appears in a bigraph; then this redex can be replaced with the corresponding reactum.

The matching problem can be defined formally as follows: given a ground bigraphical agent a' and a (parametric) redex R', there is a match of R' in a' if there exist a context C' and parameters d such that $a' = C' \circ (R' \otimes id) \circ d$. A matching can be decomposed in a set of simpler components, as in Fig. 3. It is evident that solving this problem is necessary to yield a mechanism for simulating bigraphical computations.

To solve this equation we have defined a set of inference rules that, by induction on the structure of a' and R', find the matching context C' and parameters d, preserving the equality at each inference step. So, we give an inductive characterization of a matching algorithm for directed bigraphs, along the lines of [1]. The main difference between the characterization presented in [1] is that we consider *directed* bigraphs (without binders). For this reason, both the structure of a matching sequence and many rules have to be changed and generalized.

We base our presentation on discrete decomposition of directed bigraphs:

Proposition 1 (Discrete decomposition). *Any directed bigraph G can be decomposed uniquely (up to iso) into a composition of the following form:*

$$G = (\omega \otimes \mathsf{id}) \circ D \circ (\omega' \otimes \mathsf{id})$$

where D is discrete and ω, ω' are two wirings satisfying the following conditions:
1. *in ω if two outer downward names are peer, then their target is an edge;*
2. *in ω' there are no edges, and no two inner upward names are peer (i.e. on inner upward names is a renaming, but outer downward names can be peer).*

For a formal proof see the discrete normal form in [4]. The two conditions ensure that all edges are in ω and that the two wirings are of the following forms:

$$\omega = \delta \, \mathbf{Y}_{\vec{y}}^{\vec{x}} \, \sigma \otimes \alpha \qquad\qquad \omega' = \delta' \otimes \beta$$

for some downward renaming α, upward renaming β, fusions δ, δ' and substitution σ. Note that for a ground bigraph g the wiring ω' is simply id_ε.

Using the same approach of [1] we now define *matching sentences* and a set of rules for deriving valid matching sentences.

Definition 1 (Matching Sentence). *A* matching sentence *is a 7-uple over wirings and directed bigraphs, written*

$$\omega_a, \omega_R, \omega'_C, \omega_C \vdash a, R \hookrightarrow C, d$$

where a, R, C, d are discrete directed bigraphs, R and C have not renamings (i.e. are just composed by discrete molecules), ω_a, ω_R, ω_C are wirings with no inner downward names, and ω'_C is a generic fusion.

A matching sentence is valid *if $a' = C'(\mathsf{id}_{Z+} \otimes R')d$ holds, where*

$$a' = (\omega_a \otimes \mathsf{id})a \qquad\qquad\qquad\qquad\qquad\qquad\qquad \textit{(agent)}$$
$$R' = (\omega_R \otimes \mathsf{id})(\mathsf{id}_{X+} \otimes R) \qquad\qquad\qquad\qquad \textit{(redex)}$$
$$C' = (\omega_C \otimes \mathsf{id} \otimes \mathsf{id}_{V-})\big(\mathsf{id}_{(Z \uplus Y)+} \otimes (\mathsf{id}_{V-} \otimes C)(\omega'_C \otimes \mathsf{id})\big) \qquad \textit{(context)}$$

Note that the notion of valid sentence precisely capture the abstract definition of matching: for a general match $a' = C'(\mathsf{id}_{Z+} \otimes R')d$, by Proposition 1 and some simplifications due to the groundness of a and d, we can decompose a', R', C' obtaining a corresponding valid matching sentence; conversely, if $\omega_a, \omega_R, \omega'_C, \omega_C \vdash a, R \hookrightarrow C, d$ is valid, by definition, there exist a', R' C' such that $a' = C'(\mathsf{id}_{Z+} \otimes R')d$.

The discrete decomposition of the agent, context and redex in a valid matching sentence is schematically shown in the picture in Figure 3. Composition and tensor product on bigraphs are depicted as vertical and horizontal juxtaposition.

We infer valid matching sentences using the inference system given in Fig. 4.

A detailed discussion of this system is out of the scope of this paper; we refer to [1] for explanation of a similar system. We just notice that we can prove that this matching inference system for directed bigraphs is sound and complete, that is, all valid matching sequences can be derived using these rules.

A redex can occur many times inside an agent, hence the resolving procedure is highly non-deterministic. This is implemented by a back-tracking technique based on two stacks: one for storing all the applicable rules, and the other for tracing all the different ways to apply a single rule. To solve a matching problem we create a `match` object for the pair (a', R'), whose constructor initializes the two stacks; then invoking the `next` method the next match (i.e. the pair (C', d)) found exploring the searching tree is returned, until no matches are found anymore.

An example of matching is given in Fig. 5. Consider the agent a in the picture above, and the redex R below. A matching of R into a split the agent itself in three parts: a context, the redex and some parameters. In this case, there exist two possible splitting: one identifies the redex's round node with the agent's round node inside the left rectangle, and the rectangle node with the one on the right in the agent. So, the context has the left rectangle and the below round

$$\text{PRIME-AXIOM} \quad \frac{p\colon \langle W^+\rangle \qquad \sigma\colon Z^+\to \qquad \alpha\colon W^+\to Z^+}{\sigma\alpha,\mathsf{id}_\varepsilon,\mathsf{id}_\varepsilon,\sigma\vdash p,\mathsf{id}_1\hookrightarrow \mathsf{id}_1,(\alpha\otimes\mathsf{id}_1)p}$$

$$\text{SWITCH}\quad \frac{\omega_a,\mathsf{id}_\varepsilon,\mathsf{id}_\varepsilon,\omega_C(\mathsf{id}_{Z^+}\otimes\sigma_R)\vdash p,\mathsf{id}_1\hookrightarrow P,d \qquad d\colon \langle m,Z^+\rangle}{\omega_a,\sigma_R,\mathsf{id}_\varepsilon,\omega_C\vdash p,P\hookrightarrow \mathsf{id}_1,d}$$

$$\text{ION}\quad \frac{\omega_a,\omega_R,\omega_C',\omega_C\vdash p,R\hookrightarrow P,d \qquad \sigma\colon \{\vec{y}\}^+\to}{\sigma\,\curlywedge\,\omega_a,\omega_R,\omega_C',\sigma\Delta_{\vec{x}}^{\vec{y}}\curlywedge\omega_C\vdash (K_\epsilon^{\vec{y}}\otimes\mathsf{id}_{U^+})p,R\hookrightarrow (K_\epsilon^{\vec{x}}\otimes\mathsf{id}_W)P,d}$$

$$\text{PAR}\quad \frac{\omega_a,\omega_R,\omega_C',\omega_C\curlywedge\sigma\vdash a,R\hookrightarrow C,d \qquad \omega_b,\omega_S,\omega_D',\omega_D\curlywedge\sigma\vdash b,S\hookrightarrow D,e}{\begin{array}{c}\omega_a\curlywedge\omega_b\ ,\ \omega_R\curlywedge\omega_S,\\ \omega_C'\otimes\omega_D'\ ,\ \omega_C\curlywedge\omega_D\curlywedge\sigma\end{array}\vdash a\otimes b,R\otimes S\hookrightarrow C\otimes D,d\otimes e}$$

$$\text{PERM}\quad \frac{\omega_a,\omega_R,\omega_C',\omega_C\vdash a,\bigotimes_i^m P_{\pi^{-1}(i)}\hookrightarrow C,d}{\omega_a,\omega_R,\omega_C',\omega_C\vdash a,\bigotimes_i^m P_i\hookrightarrow C\pi,d}$$

$$\text{MERGE}\quad \frac{\omega_a,\omega_R,\omega_C',\omega_C\vdash a,R\hookrightarrow C,d}{\omega_a,\omega_R,\omega_C',\omega_C\vdash (merge\otimes\mathsf{id}_Y)a,R\hookrightarrow (merge\otimes\mathsf{id}_X)C,d}$$

$$\text{EDGE-LIFT}\quad \frac{\begin{array}{c}\delta_C\colon T^-\to V^- \qquad \zeta_C\colon S^-\to W^- \qquad C\colon \langle n,W^-\rangle\to\\ \omega_a\ ,\ \sigma_R\otimes\tau_R\ ,\ \mathsf{id}_\varepsilon,\\ (\delta_C\curlyvee\triangledown_S)\delta_R\,\text{⅄}_{\vec{y}}^{\vec{x}}\,(\mathsf{id}_{\{\vec{y}\}^+}\curlywedge\Delta_{\vec{x}}^{\vec{y}}\widehat{\delta}_R(\Delta^T\curlywedge\widehat{\zeta}_C))\otimes\omega_C\end{array}\vdash a,R\hookrightarrow \widehat{C},d}{\omega_a,\delta_R\,\text{⅄}_{\vec{y}}^{\vec{x}}\,\sigma_R\otimes\tau_R,\delta_C\curlyvee\zeta_C,\omega_C\vdash a,R\hookrightarrow C,d}$$

$$\text{WIRING-AXIOM}\quad \frac{}{\Delta^y,\Delta^{\vec{x}},\mathsf{id}_\varepsilon,\Delta_{\vec{x}}^y\vdash \mathsf{id}_\varepsilon,\mathsf{id}_\varepsilon\hookrightarrow \mathsf{id}_\varepsilon,\mathsf{id}_\varepsilon}$$

$$\text{CLOSE}\quad \frac{\sigma_a,\omega_R,\omega_C',\sigma_C\vdash a,R\hookrightarrow C,d}{\begin{array}{c}(\delta\otimes\mathsf{id}_{U^+})(\text{⅄}_{\vec{y}}^{\vec{x}}\otimes\mathsf{id}_{U^+})\sigma_a\ ,\ \omega_R\\ \omega_C'\ ,\ (\delta\otimes\mathsf{id}_{W^+})(\text{⅄}_{\vec{w}}^{\vec{x}}\otimes\mathsf{id}_{W^+})\sigma_C\end{array}\vdash a,R\hookrightarrow C,d}$$

Fig. 4. Matching inference system

node of the agent, whilst the parameter contains both nodes contained by the agent's rectangle node on the right. In the other case, the matching identifies both the redex's nodes with the nodes inside the agent's right rectangle. Hence, the context contains all the remaining nodes, and the parameter is empty.

6   RPO and IPO

The RPO/IPO module provides the functions for constructing relative pushouts and idem pushouts (see Fig. 6). The implementation of the RPO construction follows faithfully the algorithm given in [2]; it takes four bigraphs (f_0,f_1,g_0,g_1) as depicted in Fig. 6(1), and yields the RPO triple (h_0,h_1,h) as in Fig. 6(2).

An example of an RPO construction is in Fig. 7. Intuitively, the construction works as follows: the common parts of the span f_0,f_1 are removed from the bound g_0,g_1 and placed in h; the nodes/edges of f_0 not present in f_1 are added to h_1, and analogously for h_0; finally the middle interface is computed and the parent and link maps are computed, preserving compositions.

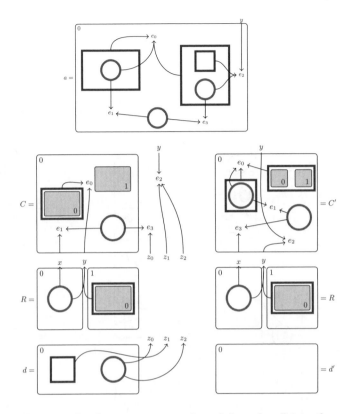

Fig. 5. An example: there are two matches of the redex R into the agent a

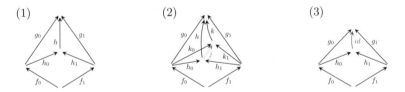

Fig. 6. A candidate RPO (1), an RPO (2) and an IPO (3)

The procedure calculating IPOs (see [2]) is split into two steps:

1. a check if the span of bigraphs has bounds;
2. the effective computation of the IPO.

Since the IPO construction is non-deterministic, an implementation must calculate all possible solutions. To do this, we define an object IPO, whose constructor takes two bigraphs (satisfying the consistency conditions, i.e., they must admit bounds) and internally computes all the possible sets (L^+, Q^+) and auxiliary functions $(\theta, \phi, \xi, \eta)$ needed in the construction. After the initialization, the IPO computation proceeds as follows: when the object method next is called, an

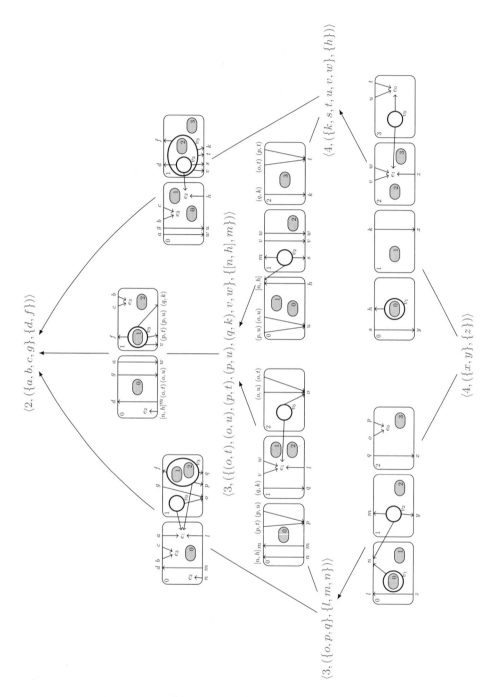

Fig. 7. An example of an RPO construction

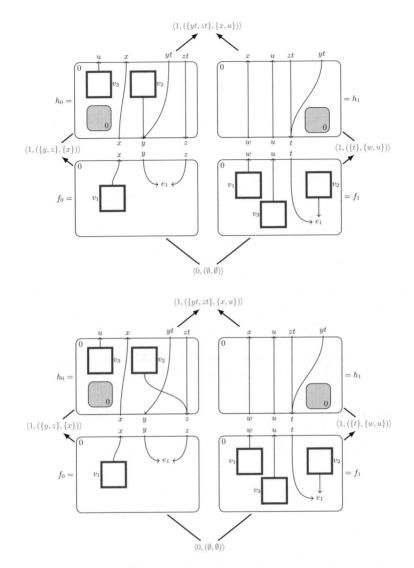

Fig. 8. An example: two IPOs are possible for the span f_0, f_1, one connects v_2 to y and the other v_2 to z

IPO is calculated by using the current internal state of the object, which is then updated for the next call to next. If no IPO is found, an exception is raised.

An example of an IPO calculation is shown in Fig. 8. Where, after checking that the span f_0, f_1 has bounds, the computation of IPO yields out to different IPOs. The first (shown above) have an arrow connecting the node v_2 to y, whilst the second an arrow connecting v_2 to z. Both cases are suitable, for the fact that v_2 is linked to e_1 in f_1, and v_2 is not present in f_0, but e_1 belongs to f_0 and it is accessible by two names y and z. So, to get a bound for our span, we must add v_2 to h_0 and connect its port to the edge e_1 in f_0, to do so, there are two possibility to target e_1: trough the name y or trough the name z.

7   Related and Future Work

Related work. The closest work is Birkedal et al. [1], where a language and a matching algoritm for *binding* bigraphs are presented. That work has been of inspiration for our matching inference system, although we had to change quite many details both in the rules and in the implementation due to the fact that we consider directed bigraphs and not binding bigraphs. Also, as far as we know, DBtk provides the first constructions of RPOs and IPOs for bigraphs.

Future work. The present system is still a prototype, and needs some refinements to mature to a ready-to-use tool. Among the features that can be added, we mention a graphical user interface (GUI), where the user can "draw" interactively a bigraph, a reaction rule and ultimately a whole BRSs. Another interesting development will be to consider *stochastic bigraphic reactive systems*, where rules are endowed with *rates*. We think that the matching procedure given in this paper can be successfully used within a Gillespie algorithm.

Acknowledgements. We thank our students Raul Pellarini, Patrik Osgnach, Federico Chiabai for their help in developing the code.

References

1. Birkedal, L., Damgaard, T.C., Glenstrup, A.J., Milner, R.: Matching of bigraphs. ENTCS 175(4), 3–19 (2007)
2. Grohmann, D., Miculan, M.: Directed bigraphs. In: Proc. XXIII MFPS. ENTCS, vol. 173, pp. 121–137. Elsevier, Amsterdam (2007)
3. Grohmann, D., Miculan, M.: Reactive systems over directed bigraphs. In: Caires, L., Vasconcelos, V.T. (eds.) CONCUR 2007. LNCS, vol. 4703, pp. 380–394. Springer, Heidelberg (2007)
4. Grohmann, D., Miculan, M.: An algebra for directed bigraphs. In: Proceedings of TERMGRAPH 2007. ENTCS, vol. 203(1), pp. 49–63. Elsevier, Amsterdam (2008)
5. Jensen, O.H., Milner, R.: Bigraphs and transitions. In: Proc. POPL, pp. 38–49 (2003)
6. Leifer, J.J., Milner, R.: Transition systems, link graphs and petri nets. Mathematical Structures in Computer Science 16(6), 989–1047 (2006)
7. Milner, R.: Bigraphical reactive systems. In: Larsen, K.G., Nielsen, M. (eds.) CONCUR 2001. LNCS, vol. 2154, pp. 16–35. Springer, Heidelberg (2001)

Constraint-Muse: A Soft-Constraint Based System for Music Therapy⋆

Matthias Hölzl[1], Grit Denker[2], Max Meier[1], and Martin Wirsing[1]

[1] Ludwig-Maximilians-Universität München
[2] SRI International, Menlo Park

Abstract. Monoidal soft constraints are a versatile formalism for specifying and solving multi-criteria optimization problems with dynamically
changing user preferences. We have developed a prototype tool for interactive music creation, called Constraint Muse, that uses monoidal soft
constraints to ensure that a dynamically generated melody harmonizes
with input from other sources. Constraint Muse provides an easy to use
interface based on Nintendo Wii controllers and is intended to be used in
music therapy for people with Parkinson's disease and for children with
high-functioning autism or Asperger's syndrome.

1 Introduction

Soft constraints are a versatile formalism for specifying and solving optimization
problems; they have been applied in many different areas such as assembly-line
sequencing [1] and Service-Level-Agreement negotiation [2]. In previous work we
have developed Monoidal Soft Constraints (MSCs), a soft-constraint formalism
particularly well-suited to multi-criteria optimization problems with dynamically
changing user preferences [3]. MSCs have successfully been applied to problems
such as optimizing software-defined radios [4], orchestrating services [5], and
allocating shelf space to products in supermarkets [6].

In this paper we show a novel application of soft constraints in the area of
music therapy. Various active and receptive approaches to music therapy have
successfully been used as part of the treatment or aftercare for conditions such as
chronic pain, eating disorders, developmental and learning disabilities, or brain
injuries [7]. Our current work was developed in the context of therapy for patients with Parkinson's disease; another planned application is in music therapy
for autistic children diagnosed with high-functioning autism (HFA), Asperger's
syndrome (AS), or Pervasive Development Disorder–Not Otherwise Specified
(PDD-NOS).

Our application is inspired by the recent availability of inexpensive novel input
devices for computers and game consoles. For example, the controller of the Wii
game console, called Wiimote, contains not only the elements commonly found
on game pads, such as buttons and direction sticks, but also a three dimensional
accelerometer and a camera so that it can be used as a pointing device. Several

---

⋆ This work has been partially sponsored by the project Sensoria IST-2005-016004.

A. Kurz, M. Lenisa, and A. Tarlecki (Eds.): CALCO 2009, LNCS 5728, pp. 423–432, 2009.
© Springer-Verlag Berlin Heidelberg 2009

applications for Wiimotes in the treatment of patients with Parkinson's disease have been proposed, e.g., for assessing movement and cognitive skills.

In our system we combine the facilities of the Wiimote with the algorithmic capabilities of soft-constraint systems. We have developed a prototype tool for interactive music creation, called Constraint Muse, that uses MSCs to ensure that a dynamically generated melody harmonizes with input from other sources. Constraint Muse provides an easy to use interface based on Wiimotes and is intended to be used in music therapy for people with Parkinson's disease and for autistic children. It allows users to influence the generation of melodies by performing various gestures with Wiimotes. This input, together with other factors, is processed by a solver for MSCs.

The structure of the paper is as follows: in the next section we present the therapeutical background that inspired the development of Constraint Muse, then we introduce the theory of monoidal soft constraints, Sect. 4 explains the application of soft constraints in Constraint Muse. The final section concludes and presents avenues for further work.

2   Music Therapy

According to the definition of the American Music Therapy Association, "Music Therapy is the clinical and evidence-based use of music interventions to accomplish individualized goals within a therapeutic relationship by a credentialed professional who has completed an approved music therapy program." [7]. The Constraint Muse system is part of a larger effort at SRI to develop technological support for novel therapeutic approaches. The initial work on Constraint Muse has been performed in the context of diagnosis and therapy for patients with Parkinson's disease, and we also plan to employ the system in an after-school program for autistic children with high-functioning autism, AS, or PDD-NOS.

2.1   Music Therapy for Patients with Parkinson's Disease

Work by Pacchetti et al. [8] has shown that music therapy significantly improves the motor skills and emotional experience of patients with Parkinson's disease. However, as the disease significantly impairs the motor skills it is difficult for patients to play a regular musical instrument such as the piano. One of our goals with the Constraint Muse system is to provide an "instrument" that allows patients at different stages of the disease and with various musical backgrounds to participate in therapeutic exercises.

2.2   Music Therapy for Autistic Children

There are several reasons why computer supported music therapy is well-suited for children diagnosed with HFA, PDD-NOS, or Asperger's Syndrome. The children are members of the "digital generation" [9] who are accustomed to using digital devices for fun, education, or for connecting with others. Research has suggested that computer-assisted teaching can result in a noticeable decrease in agitation and self-stimulatory behaviors [10], and also that "the computer places the child in control, allowing the child to become an independent learner" [11]. The incorporation of various types of technology into the children's education and therapy can improve the functional capabilities of children with autism [12].

The straightforward interaction with the Wiimote in the Constraint Muse system allows children to easily contribute to the creation of musical harmonies, tonalities and dynamics, and to perceive this creation process by their peers. The possibility to attach multiple Wiimotes to the Constraint Muse system enables new forms of interactions between children ranging from generating music as a collaborative game to expressing interpersonal relationships via musical dialogue.

3 Monoidal Soft Constraints

Constraints are restrictions on values that several variables may take simultaneously; more formally a constraint is a predicate over one or more variables. A constraint system is a set of constraints; a solution of a constraint system is an assignment of values to variables such that the conjunction of all constraints in the system is satisfied. For a general introduction see the books [13,14].

Soft constraints generalize the notion of constraint: a soft constraint is a function mapping its arguments into some domain D, e.g., the integers. Soft constraints are commonly used to describe problems where the "quality" or "cost" of solutions is important; we then optimize the overall value. Some well-known soft constraint formalisms are Valued Constraint Satisfaction Problems [15] and Semiring-based Soft Constraints [16].

In Constraint Muse the creation of melodies is controlled by *monoidal soft constraints (MSCs)* [3], a formalism which is targeted at the specification of multi-criteria optimization problems where changing contexts may influence the evaluation of constraints, and where the preferences between different constraints may vary dynamically. Originally we used the Maude soft constraint solver as described in [3], but it turned out to be unsuitable for interactive use in a soft real-time environment. Therefore, we have reimplemented an improved version of this solver in C#.

The basic idea behind monoidal soft constraints is to specify the computation of the result value in a two-step process: first the soft constraints compute measures for several different criteria, then the criteria are combined by a so-called preference relation. More precisely: let Var be a set of variables and D_{Var} the domain in which variables take their values. A *valuation* $v : Var \rightharpoonup D_{Var}$ is a partial map with finite support. Let $\mathcal{G} = \langle G_1, \dots, G_n \rangle$ be a family of ordered commutative monoids; each G_i corresponds to one of the criteria mentioned above. A (monoidal) soft constraint is a triple $\langle V; cst; i \rangle$ where $V \subseteq_{\mathsf{fin}} Var$ is a finite subset of the variables, and $cst : (V \rightarrow D_{Var}) \rightarrow G_i$ maps valuations defined for all variables in V to elements of G_i. The results of several soft constraints defined on the same grade G_i are multiplied using the monoid multiplication of G_i, therefore multiple constraints may contribute to a single criterion.

The final outcome of the optimization process, obtained by taking into account all grades, is taken from the ordered set of *ranks E*. The computation of the rank based on the grades is performed by the *preference relation* $p = \langle p_1, \ldots, p_n \rangle$ where $p_i : G_i \to (E \to E)$ is a left-compatible operation of G_i on E, i.e., an operation of G_i on E such that $\forall \alpha, \beta \in G_i : \forall x \in E : \alpha \leq \beta \implies \alpha x \leq \beta x$. Furthermore the p_i have to commute with each other, i.e.,

$$\forall i, j \leq n : \forall \alpha_i \in G_i, \alpha_j \in G_j : \forall x \in E : \alpha_i(\alpha_j x) = \alpha_j(\alpha_i x).$$

The resulting constraint theory is very expressive: grades only have to be monoids, and the set of ranks can be any ordered set, even one without internal composition operation. Solutions can be obtained, e.g., by branch-and-bound search; various optimizations are possible if the grades or ranks exhibit stronger algebraic structures.

4   The "Constraint Muse" System

The use of algorithms in music composition goes back for centuries. With the advent of electronic computers automated composition of music has become an active field of academic research and artistic endeavor. An overview of the field is given in [17,18]. Most approaches to algorithmic composition are non-interactive and therefore not suitable for an application such as Constraint Muse. Some interactive music programming languages exist, e.g., Cycling 74's Max system [19], but the Constraint Muse system is unique among these by using a sophisticated general-purpose mechanism for multi-criteria optimizations.

As mentioned in the introduction, the main user input to Constraint Muse is derived from Wii controllers, but it is also possible to use other inputs to influence the generated melodies, such as pre-recorded music or keyboard controllers. The inputs, as well as other information, e.g., the desired range of pitches for the current voice, are then used by a soft constraint system to create melodies.

4.1   Constraint-Based Music Creation

The influence of the controller on the generated pitches is modeled by a soft constraint

$$noteGrades_{Controller} : Pitch \to \mathbb{Z}$$

The controller input appears as a context parameter, not in the argument list of the constraint, since it is a value obtained from the constraint solver's environment and not a variable controlled by the solver. The value of *noteGrades* for a pitch ν represents the desirability of ν given user input *Controller*.

But user input is not the only influence on the pitch of a note, it should also fit into the harmonic structure of the other notes currently playing. Consonance between two pitches is computed by the following function, based on music theory from composer Howard Hanson [20]

$$consonance : Pitch \times Pitch \to \mathbb{Z}$$

Consonant intervals, such as perfect fifths, evaluate to positive values whereas dissonant intervals, e.g. the tritone, evaluate to negative values. Since our goal is to find a melody that is consonant with the other notes currently played by the system, the constraint $harmony_F$ sums the values of *consonance* over the set of all other simultaneously sounding pitches F:

$$harmony_F : Pitch \rightarrow \mathbb{Z}$$
$$harmony_F(\nu) = \sum_{f \in F} consonance(\nu, f)$$

Again, F is a context parameter as it is obtained from the environment. The range of possible notes for a given voice can be described by the constraint

$$possibleNotes : Pitch \rightarrow Bool$$

This constraint returns true if and only if a pitch is in the range for the current melody. A similar constraint can be used to model adherence to a scale. Note that while these constraints map into the Booleans, whether or not they are guaranteed to be satisfied in all solutions depends on the specified preference relation.

We use the lexicographically ordered set $E = Bool \times \mathbb{Z}$ as domain for the ranks, i.e., we treat the Boolean constraints for possible notes as hard constraints. The results of *noteGrades* and *harmony* are additively injected into the second component of E. The constraint solver, conceptually, computes the value of the constraint system for all possible pitches; the pitch with the highest value in E is the note actually played.

The constraint-solving algorithm as described in [3] has exponential worst-case time complexity. To satisfy the soft real-time requirements of Constraint Muse we have chosen a suitable heuristic function that avoids the search of exponentially many nodes for the average case. Furthermore, the constraint solver contains built-in measures to minimize the size of the search tree. For example, our search algorithm evaluates the *possibleNotes* function as early as possible in order to minimize the number of notes in the search space. Further optimizations are described in [6].

4.2   User Interaction

The Constraint Muse uses novel interaction modalities based on Nintendo's Wii remote (Wiimote, see Fig. 1). The end user is presented with a simple, game-like application that is designed with very low cognitive load for interaction. This is achieved by capitalizing on the Wiimote's natural interaction models which can be mapped to the structure of the generated music in intuitively obvious ways. The Wiimote has two unique features: First, it has a built-in tri-axial linear accelerometer that allows reading acceleration data of a user's hand movements. Second, the Wiimote has an infrared camera, built into the front face that can track infrared beacons in the IR sensor bar mounted on top of the display. This

Fig. 1. Wii Remote Controller (Wiimote)

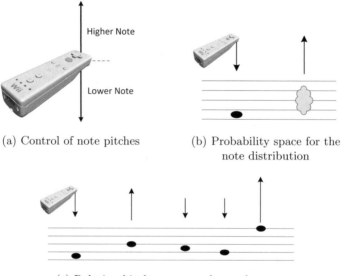

(a) Control of note pitches

(b) Probability space for the note distribution

(c) Relationship between strokes and notes

Fig. 2. First variant of the Constraint Muse system

capability is used to project a pointer on the display and manipulate that pointer based on the user's hand movements in 3D space. In addition, the Wiimote has a small built-in speaker that allows customized sound per device to be played back through the handheld for more realistic positional sound, and that can be used to inform each player about the contribution that their input is having on the overall composition. It also has a small built-in vibrator that allows haptic feedback to the user, similar to force feedback in traditional joysticks. In addition, the Wiimote has seven programmable buttons.

We have built several versions of Constraint Muse with different trade-offs between user control and accessibility; the choice of an appropriate version for a particular user depends on factors such as his knowledge of music theory, his feel for rhythm, and the degree to which his motor skills are impaired.

One version of Constraint Muse allows users to control the pitch of the played notes by up- and down-strokes of the Wiimote, see Fig. 2(a). The pitch change is relative to the previous note's pitch, with larger strokes leading to correspondingly higher variations in pitch; see Fig 2(c) for an illustration. In order to allow

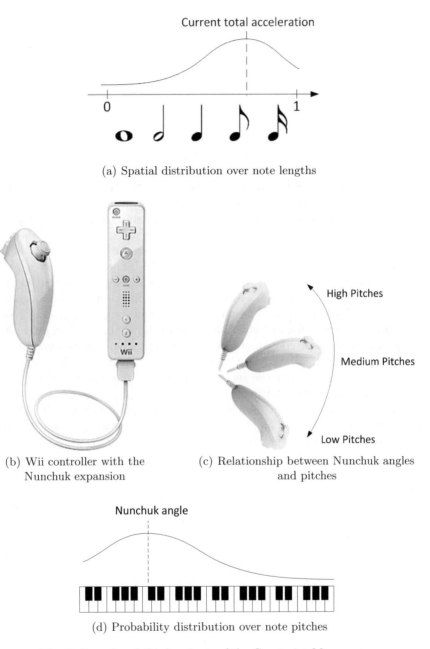

(a) Spatial distribution over note lengths

(b) Wii controller with the Nunchuk expansion

(c) Relationship between Nunchuk angles and pitches

(d) Probability distribution over note pitches

Fig. 3. Second and third variants of the Constraint Muse system

the system to harmonize the generated melody, there is no deterministic mapping between strokes and pitches, rather each stroke represents a probability density around the desired target pitch; see Fig. 2(b). This variant provides a great deal of influence and is particularly well suited for users who desire a large amount of control and who have good motor skills and "feel" for the rhythm. For some users, down-strokes are easier to execute than up-strokes; in these cases the melody tends to use only use the lowest register.

A second variant forgoes the direct connection between Wiimote movements and note pitch. Neither direction nor timing of the Wiimote is directly used to influence the generated melody. Instead, the total acceleration of the Wiimote influences the number of generated notes and their duration, with higher acceleration leading to larger numbers of shorter notes. The precise distribution is obtained by uniformly dividing the unit interval between the note durations, and computing a Gaussian distribution centered around the current normalized total acceleration, see Fig. 3(a). A second controller (the so-called Nunchuk, see Fig. 3(b)) is used to compute the average pitch. Similarly to the first variant, the angle in which the Nunchuk controller is held influences the pitch of the generated notes, see Fig. 3(c); in order to vary the melody without constant movement of the Nunchuk the angle of the Nunchuk controls the mean of a Gaussian distribution which is computed over all note pitches (Fig. 3(d)). With this variant, the generated melody is always perfectly in time and on pitch, since there is no deterministic control of these parameters. Therefore this version may be particularly suited for people with large motor deficits, but it does not offer very precise control over the generated music for more advanced players or patients with less severe impairments.

Therefore, a third version combines these two control methods in that the pitch is controlled by the Nunchuk controller and movement of the Wiimote directly triggers notes (but has no influence on the pitch of the generated notes). In contrast to the first version, the direction of the stroke with the Wiimote has no influence on the pitch of the note which eliminates the problem with up-strokes of the first version.

So far the Constraint Muse system has been tested by several users and has received very positive evaluations. Currently the generated melodies and harmonic structures fit well with open music styles, such as meditative music or Jazz. In certain musical styles, such as classical music or pop songs, the listener is used to compositions following a rather strict structure. Since Constraint Muse currently does not remember previously generated harmonies and melodies, it cannot create music in these styles.

5   Conclusions and Further Work

The Constraint Muse system hints at new possibilities for music therapy: even without being capable of playing a conventional musical instrument, patients can express themselves by creating music in real-time. By attaching multiple Wiimotes or other controllers it is also possible to create a shared experience for several users. This may offer novel therapeutic possibilities for diseases such as autism, and it may also prove useful in the treatment of other illnesses.

A systematic study about the use of Constraint Muse in an after-school program for autistic children is currently in preparation. Children participating in this program will use the system to collaboratively create music and thereby develop their emotional competencies.

Constraint Muse will allow students to intuitively control the composition of music in various tonalities, timbres, dynamics and moods. By playing, creating, and hearing sound effects and music, students will be better able to connect to the seven basic human emotions, as defined by Ekman [21], which are fear, sadness, anger, joy, surprise, disgust, and contempt, in relation to the generated harmonies and melodies.

The study will determine to what extent the consequential thinking of students can be strengthened by experiencing how each team member's sound and music affect the musical output of the whole group. Moreover, we will investigate how listening to the effects appearing when a student acts without regard for the rest of his group helps children to better understand the impact of their behavior on others.

Another line of research is the improvement of the interaction models and the musical capabilities of Constraint Muse. None of the current variants of the system allows users to adjust the volume or timbre of the music. Future versions will explore additional parameters to be controlled by the user. One of the important issues will be to find a good balance between providing detailed control of various parameters, predictability of the musical output, and ease of use of the system.

Currently Constraint Muse is only able to generate music with a very free form, since both harmonic structure and melodies are determined by the system without taking into account the previously generated output. As mentioned above, there are many styles of music where the listener expects repetitions, with or without variation, of certain themes. A promising approach in this direction consists of extending Constraint Muse with a memory component that stores previously generated melodies and harmonies. Additional soft constraints can then control how closely the freshly generated parts have to resemble the stored music.

References

1. Bergen, M.E., van Beek, P., Carchrae, T.: Constraint-based vehicle assembly line sequencing. In: AI 2001: Proceedings of the 14th Biennial Conference of the Canadian Society on Computational Studies of Intelligence, London, UK, pp. 88–99. Springer, Heidelberg (2001)
2. Bistarelli, S., Santini, F.: A nonmonotonic soft concurrent constraint language for sla negotiation. Electron. Notes Theor. Comput. Sci. 236, 147–162 (2009)
3. Hölzl, M., Meier, M., Wirsing, M.: Which soft constraints do you prefer? In: Rosu, G. (ed.) WRLA 2008. ENTCS (2008) (to appear)

4. Wirsing, M., Denker, G., Talcott, C., Poggio, A., Briesemeister, L.: A Rewriting Logic Framework for Soft Constraints. In: Proc. 6th Int. Workshop on Rewriting Logic and its Applications (WRLA 2006), Vienna, Austria, April 2006. ENTCS (2006)
5. Wirsing, M., Clark, A., Gilmore, S., Hölzl, M., Knapp, A., Koch, N., Schroeder, A.: Semantic-Based Development of Service-Oriented Systems. In: Najm, E., Pradat-Peyre, J.-F., Donzeau-Gouge, V.V. (eds.) FORTE 2006. LNCS, vol. 4229, pp. 24–45. Springer, Heidelberg (2006)
6. Meier, M.: Life is Soft. Master's thesis, LMU München and Univ. Augsburg (2008)
7. American Music Therapy Association: Web site of the American Music Therapy Association, http://www.musictherapy.org/ (last accessed 2009-06-01)
8. Pacchetti, C., Mancini, F., Aglieri, R., Fundarò, C., Martignoni, E., Nappi, G.: Active Music Therapy in Parkinson's Disease: An Integrative Method for Motor and Emotional Rehabilitation. Psychosom Med. 62 (2000)
9. Carlson, G.: Digital Media in the Classroom. CMP Books, San Francisco (2004)
10. Hileman, C.: Computer technology with autistic children. In: Autism Society of America National Conference, Milwaukee, Wisconsin (July 1996)
11. Stokes, S.: Assistive technology for children with autism. Report published for Wisconsin Department of Public Instruction (2001), http://www.csun.edu/cod/2001/proceedings/0155stokes.htm (last accessed 2009-04-16)
12. National Association of Special Education Teachers: Assistive technology for students with autism spectrum disorders (retrieved), http://www.naset.org/2709.0.html
13. Marriott, K., Stuckey, P.: Programming With Constraints: An Introduction. MIT Press, Cambridge (1998)
14. Dechter, R.: Constraint Processing. Morgan Kaufmann, San Francisco (2003)
15. Fargier, H., Lang, J., Martin-Clouaire, R., Schiex, T.: A constraint satisfaction framework for decision under uncertainty. In: Proc. of the 11th Int. Conf. on Uncertainty in Artificial Intelligence, pp. 175–180 (1996)
16. Bistarelli, S., Montanari, U., Rossi, F.: Semiring-based constraint satisfaction and optimization. J. ACM 44(2), 201–236 (1997)
17. Essl, K.: Algorithmic composition. In: Collins, N., d'Escrivan, J. (eds.) Cambridge Companion to Electronic Music. Cambridge University Press, Cambridge (2007)
18. Nierhaus, G.: Algorithmic Composition - Paradigms of Automated Music Generation. Springer, Heidelberg (2008)
19. Cycling 74: Cycling 74 Web Site, http://www.cycling74.com/ (last accessed 2009-06-02)
20. Hanson, H.: Harmonic Materials of Modern Music. Irvington (June 1960)
21. Ekman, P.: Emotions Revealed: Recognizing faces and feelings to improve communication and emotional life. Henry Holt, New York (2003)

CIRC: A Behavioral Verification Tool Based on Circular Coinduction

Dorel Lucanu[1], Eugen-Ioan Goriac[1], Georgiana Caltais[1], and Grigore Roşu[2]

[1] Faculty of Computer Science
Alexandru Ioan Cuza University, Iaşi, Romania
{dlucanu,egoriac,gcaltais}@info.uaic.ro
[2] Department of Computer Science
University of Illinois at Urbana-Champaign, USA
grosu@cs.uiuc.edu

Abstract. CIRC is a tool for automated inductive and coinductive theorem proving. It includes an engine based on circular coinduction, which makes CIRC particularly well-suited for proving behavioral properties of infinite data-structures. This paper presents the current status of the coinductive features of the CIRC prover, focusing on new features added over the last two years. The presentation is by examples, showing how CIRC can automatically prove behavioral properties.

1 Introduction

The first version of the coinductive engine of CIRC was presented in [7]. Meanwhile, important contributions have been made to the tool. One of the contributions is the implementation of the deterministic language ROC! [3] used for specifying proof strategies. This language provides a flexible way to combine basic proof actions, such as *reduction* and *coinduction expansion,* into coinductive proof schemata. ROC! was designed so that the combination of different proving techniques is possible.

Another contribution was the extension of the tool with *special contexts* [6], as explained below. An important technical aspect of our previous implementation of circular coinduction in CIRC [7] (see also [8]) was a freezing operator $\lceil\cdot\rceil$ used to add frozen versions $\lceil e \rceil$ of dynamically discovered coinductive hypotheses e to the behavioral specification. The role of the freezing operator is to enforce the use of coinductive hypotheses in a sound manner, forbidding their use in contextual reasoning. However, many experiments with CIRC have shown that this constraint to completely forbid the use of coinductive hypotheses in all contexts is too strong, in that some contexts are actually safe. The novel special context extension of CIRC allows the user to specify contexts in which the frozen coinductive hypotheses can be used. CIRC can be used to check that those contexts are indeed safe. Moreover, the current version of CIRC includes an algorithm for automatically computing a set of special contexts, which are then, also automatically, used in circular coinductive proofs.

A. Kurz, M. Lenisa, and A. Tarlecki (Eds.): CALCO 2009, LNCS 5728, pp. 433–442, 2009.
© Springer-Verlag Berlin Heidelberg 2009

The CIRC tool, user manual and many examples can be downloaded from `http://fsl.cs.uiuc.edu/CIRC`. CIRC can also be used online through its web interface at `http://fsl.cs.uiuc.edu/index.php/Special:CircOnline`.

The theoretical foundation for the coinductive engine included in CIRC is presented in [8]. So, many concepts and notations used in this paper are presented in detail there. In this paper we present a set of examples aiming to illustrate the main features of CIRC. These include non-trivial properties of streams, bisimulation of automata, how to handle goals when a possible infinite proof is detected, the use of special contexts.

2   Behavioral Specifications and CIRC

Behavioral specifications are pairs of the form $(\mathcal{B}, \Delta)$, where $\mathcal{B} = (S, \Sigma, E)$ is a many sorted algebraic specification and Δ is a set of Σ-contexts, called *derivatives*. A derivative in Δ is written as $\delta[*{:}h]$, where $*{:}h$ is a special variable of sort h designating the place-holder in the context δ. The sorts S are split in two classes: *hidden sorts*, $H = \{h \mid \delta[*{:}h] \in \Delta\}$, and *visible sorts*, $V = S \setminus H$. A Δ-*context* is inductively defined as follows: 1) each $\delta[*{:}h] \in \Delta$ is a context for h; and 2) if $C[*{:}h']$ is a a context for h' and $\delta[*{:}h] \in \Delta_{h'}$, then $C[\delta[*{:}h]]$ is a context for h, where $\Delta_{h'}$ is the subset of derivatives of sort h'. A Δ-*experiment* is a Δ-context of visible sort. If e is an equation of the form $(\forall X)t = t'$ and C a Δ-context appropriate for t and t', then $C[e]$ denotes the equation $(\forall X)C[t] = C[t']$. If $\delta \in \Delta$, then $\delta[e]$ is called a *derivative* of e. Given an entailment relation $\vdash$ over $\mathcal{B}$, the *behavioral entailment* relation is defined as follows: $\mathcal{B} \Vdash e$ iff $\mathcal{B} \vdash C[e]$ for each Δ-context C appropriate for the equation e. In this case, we say that $\mathcal{B}$ *behaviorally satisfies* the equation e. The reader is referred to [8] for more rigorous definitions, properties and other technical details.

Several of the examples we present in this paper are based on infinite streams. To specify the streams, we consider two sorts: a hidden sort *Stream* for the streams and a visible sort *Data* for the stream elements. The streams are defined in terms of head and tail, i.e., $hd(*{:}Stream)$ and $tl(*{:}Stream)$ are derivatives. For instance, the stream $(1{:}0{:}1)^\infty = 1{:}0{:}1{:}1{:}0{:}1\ldots$, which is mathematically defined by the equation $ozo = 1{:}0{:}1{:}ozo$, is behaviorally specified by the following equations written in terms of head and tail: $hd(ozo) = 1$, $hd(tl(ozo)) = 0$, $hd(tl^2(ozo)) = 1$ and $tl^3(ozo) = ozo$. The specifications of other hidden structures are obtained in a similar manner, by defining their behavior in terms of the derivatives.

CIRC is developed as an extension of the Full Maude language. The underlying entailment relation used in CIRC is $\mathcal{E} \vdash_{\rightarrow\leftarrow} (\forall X)t = t'$ if $\wedge_{i \in I}(u_i = v_i)$ iff $\mathrm{nf}(t) = \mathrm{nf}(t')$, where $\mathrm{nf}(t)$ is computed as follows:

- the variables X of the equations are turned into fresh constants;
- the condition equalities $u_i = v_i$ are added as equations to the specification;
- the equations in the specification are oriented and used as rewrite rules.

The circular coinduction engine implements the proof system given in [8] by a set of reduction rules $(\mathcal{B}, \mathcal{F}, \mathcal{G}) \Rightarrow (\mathcal{B}, \mathcal{F}', \mathcal{G}')$, where $\mathcal{B}$ represents the (original) algebraic specification, $\mathcal{F}$ is the set of frozen axioms and $\mathcal{G}$ is the current set of proof obligations. Here is a brief description of the reduction rules:

[Done]: $(\mathcal{B}, \mathcal{F}, \emptyset) \Rightarrow \cdot$
This rule is applied whenever the set of proof obligations is empty and indicates the termination of the reduction process.

[Reduce]: $(\mathcal{B}, \mathcal{F}, \mathcal{G} \cup \{\boxed{e}\}) \Rightarrow (\mathcal{B}, \mathcal{F}, \mathcal{G})$ if $\mathcal{B} \cup \mathcal{F} \vdash_{\rightarrow\leftarrow} e$
This rule is applied whenever the current goal is a $\vdash_{\rightarrow\leftarrow}$-consequence of $\mathcal{B} \cup \mathcal{F}$ and operates by removing $\boxed{e}$ from the set of goals.

[Derive]: $(\mathcal{B}, \mathcal{F}, \mathcal{G} \cup \{\boxed{e}\}) \Rightarrow (\mathcal{B}, \mathcal{F} \cup \{\boxed{e}\}, \mathcal{G} \cup \{\boxed{\Delta(e)}\})$ if $\mathcal{B} \cup \mathcal{F} \not\vdash_{\rightarrow\leftarrow} e$
This rule is applied when the current goal e is hidden and it is not a $\vdash_{\rightarrow\leftarrow}$-consequence. The current goal is added to the specification and its derivatives to the set of goals. $\boxed{\Delta(e)}$ denotes the set $\{\delta[e] \mid \delta \in \Delta\}$.

[Normalize]: $(\mathcal{B}, \mathcal{F}, \mathcal{G} \cup \{\boxed{e}\}) \Rightarrow (\mathcal{B}, \mathcal{F}, \mathcal{G} \cup \{\boxed{\mathrm{nf}(e)}\})$
This rule removes the current goal from the set of proof obligations and adds its normal form as a new goal. The normal form $\mathrm{nf}(e)$ of an equation e of the form $(\forall X)t = t'$ if $\wedge_{i \in I}(u_i = v_i)$ is $(\forall X)\mathrm{nf}(t) = \mathrm{nf}(t')$ if $\wedge_{i \in I}(u_i = v_i)$, where the constants from the normal forms are turned back into the corresponding variables.

[Fail]: $(\mathcal{B}, \mathcal{F}, \mathcal{G} \cup \{\boxed{e}\}) \Rightarrow (\mathcal{B} \cup \mathcal{F} \not\vdash_{\rightarrow\leftarrow} e$ and e is visible
This rule stops the reduction process with failure whenever the current goal e is visible and the corresponding normal forms are different.

It is easy to see that the reduction rules [Done], [Reduce], and [Derive] implement the proof rules with the same names given in [8]. The reduction rules [Normalize] and [Fail] have no correspondent in the proof system. [Normalize] is directly related to the particular definition for the basic entailment relation used in CIRC. [Fail] signals a failing stop of the reduction process and such a case needs (human) analysis in order to know the source of the failure. The soundness of CIRC follows by showing that the proof system in [8] is equivalent to its extension with the rule:

$$\frac{\mathcal{B} \cup \mathcal{F} \vdash \mathcal{G} \cup \boxed{\mathrm{nf}(e)}}{\mathcal{B} \cup \mathcal{F} \vdash \mathcal{G} \cup \boxed{e}} \ \text{[Normalize]}$$

The use of CIRC is very simple. First the user must define (or load if already defined) the equational specification $\mathcal{B}$ using a Full Maude-like syntax. For instance, the equational description of streams can be given as follows:

```
(theory STREAM-EQ is
  sorts Data Stream .                op hd : Stream -> Data .
  ops 0 1 : -> Data .                op tl : Stream -> Stream .
  op not : Data -> Data .
  eq not(0) = 1 .                    eq not(1) = 0 .
  op ozo : -> Stream   .
  eq hd(ozo) = 0 .                   eq hd(tl(tl(ozo))) = 0 .
  eq hd(tl(ozo)) = 1 .               eq tl(tl(tl(ozo))) = ozo .
  ...
endtheory)
```

Since Full Maude has no support for behavioral specifications, CIRC uses a new kind of modules, called *c-theories*, where the specific syntactic constructs are included. Here is a c-theory specifying the derivatives Δ for streams:

```
(ctheory STREAM is including STREAM-EQ .
  derivative hd(*:Stream) .
  derivative tl(*:Stream) .
endctheory)
```

As c-theories extend Full-Maude theories, the whole specification $(\mathcal{B}, \Delta)$ may be included into a single c-theory.

The user continues by loading several goals, expressed as (conditional) equations, using the command (add goal _ .) and then launches the coinductive proving engine using the command (coinduction .). The coinductive engine of CIRC has three ways to terminate:

- *successful termination*: the initial goals are behavioral consequences of the specification;
- *failing termination*: the system fails to prove a visible equation (in this case we do not know if the initial goals hold or not);
- *the maximum number of steps was exceeded*: either the execution does not terminate or the maximum number of steps is set to a too small value and should be increased.

However, the termination of CIRC is conditioned by the terminating property of the equational specification $\mathcal{B}$. For instance, CIRC does not terminate if Maude falls into a infinite rewriting when it computes a certain normal form.

3   CIRC at Work

In this section we present several scenarios on how to interact with the tool in order to prove behavioral properties. The definitions of the stream operations are given by mathematical specifications. The behavioral variants, expressed in terms of head and tail derivatives, are obtained in a similar way to that used for the definition of the stream *ozo* in Section 2.

Example 1. This example illustrates how CIRC implements the proof system given in [8]. The operations *even* and *odd* over streams are defined as follows: $odd(a:s) = a:even(s)$ and $even(a:s) = odd(s)$. Supposing that the file including the c-theory STREAM has the name stream.maude, we present how the user can verify that $zip(odd(S), even(S)) = S$:

```
Maude> in stream.maude
Maude> (add goal zip(odd(S:Stream), even(S:Stream)) = S:Stream .)
Maude> (coinduction .)
Proof succeeded.
  Number of derived goals: 2
  Number of proving steps performed: 13
  Maximum number of proving steps is set to: 256
Proved properties:
  zip(odd(S:Stream),odd(tl(S:Stream))) = S:Stream
```

From the output we conclude that CIRC needs to prove 2 extra derived subgoals in order to prove the initial property and that it performs 13 basic steps ([Reduce], [Derive], etc). Note that the superior limit for the number of basic steps is set to 256. The command (set max no steps _ .) is used to change this number. Exceeding this limit is a good clue for possible infinite computations.

The command (show proof .) can be used in order to visualize the applied rules from the proof system:

```
Maude> (show proof .)
|-  [* tl(zip(odd(S),odd(tl(S)))) *] = [* tl(S) *]
---------------------------------------------------------------- [Reduce]
|||-  [* tl(zip(odd(S),odd(tl(S)))) *] = [* tl(S) *]


|-  [* hd zip(odd(S),odd(tl(S))) *] = [* hd S *]
---------------------------------------------------------------- [Reduce]
|||-  [* hd zip(odd(S),odd(tl(S))) *] = [* hd S *]


1. |||-  [* hd zip(odd(S),odd(tl(S))) *] = [* hd S *]
2. |||-  [* tl(zip(odd(S),odd(tl(S)))) *] = [* tl(S) *]
---------------------------------------------------------------- [Derive]
|||-  [* zip(odd(S),odd(tl(S))) *] = [* S *]


|-  [* zip(odd(S),odd(tl(S))) *] = [* S *]
---------------------------------------------------------------- [Normalize]
|-  [* zip(odd(S),even(S)) *] = [* S *]
```

Comparing with the proof tree given in [8], we see that the [Derive] step is accompanied by a [Normalize] step.

Example 2. This is a non-trivial example of coinductive property over streams automatically proved with the circular coinduction. Let s denote the First Feigenbaum sequence: $(1:0:1:1:1:0:1:0:1:0:1:1)^\infty$ and *31zip* an operation defined by the equation $31zip(a_0:a_1:\ldots,b_0:b_1:\ldots) = a_0:a_1:a_2:b_0:a_3:a_4:a_5:b_1:\ldots$. We present below how the user can verify that $31zip(ozo, ozo) = s$:

```
Maude> (add goal 31zip(ozo, ozo) = s .)
Maude> (coinduction .)
Proof succeeded.
  Number of derived goals: 24
  Number of proving steps performed: 102
  Maximum number of proving steps is set to: 256
Proved properties:
  tl(tl(tl(31zip(ozo,tl(tl(ozo)))))) =
    tl(tl(tl(tl(tl(tl(tl(tl(tl(tl(s)))))))))))
  tl(tl(31zip(ozo,tl(tl(ozo))))) =
    tl(tl(tl(tl(tl(tl(tl(tl(tl(s))))))))))
  tl(31zip(ozo,tl(tl(ozo)))) = tl(tl(tl(tl(tl(tl(tl(tl(s)))))))))
  31zip(ozo,tl(tl(ozo))) = tl(tl(tl(tl(tl(tl(tl(s))))))))
  tl(tl(tl(31zip(ozo,tl(ozo))))) = tl(tl(tl(tl(tl(tl(tl(s)))))))
  tl(tl(31zip(ozo,tl(ozo)))) = tl(tl(tl(tl(tl(tl(s))))))
  tl(31zip(ozo,tl(ozo))) = tl(tl(tl(tl(tl(s)))))
  31zip(ozo,tl(ozo)) = tl(tl(tl(tl(s))))
  tl(tl(tl(31zip(ozo,ozo)))) = tl(tl(tl(s)))
  tl(tl(31zip(ozo,ozo))) = tl(tl(s))
  tl(31zip(ozo,ozo)) = tl(s)
  31zip(ozo,ozo) = s
```

The "proved properties" constitute the set F of the intermediate lemmas the circular coinduction discovered during the derivation process. The set $\boxed{F}$ of their frozen forms satisfies the Circularity Principle [8]: $\text{STREAM} \cup \boxed{F} \vdash \boxed{\Delta[F]}$, which implies $\text{STREAM} \Vdash F$.

Example 3. In this example we show how CIRC can be used for proving simultaneous several properties related to the famous Thue-Morse sequence. We first introduce the operations *not*, *zip*, and *f*, mathematically described by the following equations:

$$not(a\!:\!s) = not(a)\!:\!not(s)$$
$$zip(a\!:\!s, s') = a\!:\!zip(s', s)$$
$$f(a\!:\!s) = a\!:\!not(a)\!:\!f(s)$$

Note that the operation *not* is overloaded over data and streams, the right meaning resulting from the type of the argument. The Thue-Morse sequence is $morse = 0\!:\!zip(not(morse), tl(morse))$. It is known that *morse* and its complement are the only fixed points of the function f. If we try to prove this property with CIRC, then we obtain the following output:

```
Maude> (add goal f(morse) = morse .)
Maude> (coinduction .)
Stopped: the number of prover steps was exceeded.
```

Often, this message indicates that the execution of the circular coinductive engine does not terminate for the given goal(s). Analyzing the derived goals (these are obtained either by calling the command (set show details on .) before starting the coinduction engine or the command (show proof .) at the end), we observe that the fourth derived goal is normalized to $f(tl(morse)) = zip(tl(morse), not(tl(morse)))$. If we replace the subterm $tl(morse)$ with a free variable S, then we obtain a more general lemma: $f(S) = zip(S, not(S))$. CIRC can prove by circular coinduction simultaneously a set of goals. So, we try to simultaneously prove the two properties and we see that this time CIRC successfully terminates:

```
Maude> in stream.maude
Maude> (add goal f(morse) = morse .)
Maude> (add goal f(S:Stream) = zip(S:Stream, not(S:Stream)) .)
Maude> (coinduction .)
Proof succeeded.
  Number of derived goals: 8
  Number of proving steps performed: 41
  Maximum number of proving steps is set to: 256
Proved properties:
  tl(f(morse)) = tl(morse)
  tl(f(S:Stream)) = zip(not(S:Stream),tl(S:Stream))
  f(morse) = morse
  f(S:Stream) = zip(S:Stream,not(S:Stream))
```

A similar reasoning is used for proving the following properties:

$$31zip(ozo, g(altMorse)) = g(altMorse) \text{ and } 31zip(ozo, g(S)) = g(f(f(S)))$$

where *altMorse* is Thue-Morse sequence but defined using the alternative definition $altMorse = f(0\!:\!tl(altMorse))$ and g is the function over streams defined by $g(a_0\!:\!a_1\!:\!a_2\!:\!\ldots) = (a_0 + a_1)\!:\!g(a_1\!:\!a_2\!:\!\ldots)$. Here is the dialog with CIRC:

```
Maude> (add goal 31zip(ozo, g(altMorse)) = g(altMorse) .)
Maude> (add goal 31zip(ozo, g(S:Stream)) = g(f(f(S:Stream))) .)
Maude> (coinduction .)
Proof succeeded.
  Number of derived goals: 16
  Number of proving steps performed: 79
  Maximum number of proving steps is set to: 256
Proof properties:
  g(tl(f(tl(f(altMorse))))) = g(tl(f(tl(altMorse))))
  tl(tl(tl(31zip(ozo,g(S:Stream))))) = g(tl(f(tl(f(S:Stream)))))
  g(f(tl(f(altMorse)))) = g(f(tl(altMorse)))
  tl(tl(31zip(ozo,g(S:Stream)))) = g(f(tl(f(S:Stream))))
  g(tl(f(f(altMorse)))) = g(tl(altMorse))
  tl(31zip(ozo,g(S:Stream))) = g(tl(f(f(S:Stream))))
  g(f(f(altMorse))) = g(altMorse)
  31zip(ozo,g(S:Stream)) = g(f(f(S:Stream)))
```

The stream $g(altMorse)$ is studied in [9] and its equivalent definition with $31zip$ and g is given in [1].

Example 4. This example illustrates CIRC capabilities of proving fairness properties. Let g_1, g_2 and g_3 be three infinite streams defined in a mutually-recursive manner by the following equations: $g_1 = 0 : not(g_3)$, $g_2 = 0 : not(g_1)$ and $g_3 = 0 : not(g_2)$. The three streams model a ring of three gates, where the output of each gate is the negation of its current input, except for the initial output which is 0. We introduce the infinite stream *ones* representing the sequence 1^∞, defined by the equation *ones* $= 1 : ones$. We also consider the operator *extractOnes* that filters all the occurrences of the element 1 when provided a certain stream:

$$extractOnes(b_0 : b_1 : b_2 \ldots) = \begin{cases} 1 : extractOnes(b_1 : b_2 \ldots), & \text{if } b_0 = 1 \\ extractOnes(b_1 : b_2 \ldots), & otherwise \end{cases}$$

We want to use CIRC in order to prove the property that g_1 includes an infinite number of occurrences of the element 1. For proving this property the tool needs the lemma $not(not(a_0 : a_1 \ldots)) = a_0 : a_1 \ldots$ that can be also proved with CIRC in one derivation step and several equational reductions. Finally, the goal we want to prove is $extractOnes(g_1) = ones$:

```
Maude> (add goal extractOnes(g1) = ones .)
Maude> (coinduction .)
Proof succeeded.
  Number of derived goals: 6
  Number of proving steps performed: 30
  Maximum number of proving steps is set to: 256
Proof properties:
  extractOnes(g3) = ones
  extractOnes(g2) = ones
  extractOnes(g1) = ones
```

From the provided output, one can see that CIRC needed also to automatically prove two other similar fairness properties for the streams g_2 and g_3. However, the specification of *extractOnes* should be used with care because is not always terminating.

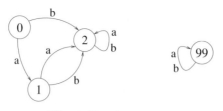

Fig. 1. Two bisimilar automata

Example 5. CIRC is also used for proving automata bisimilarity. In our example, we consider the two deterministic automata presented in Fig. 1. The transition function *delta* is defined over states modeled as elements from the hidden sort *State*, and respectively over the alphabet elements (*a* and *b*) modeled as elements from the visible sort *Alph*. The function is defined in terms of equations such as: $delta(0,a) = 1$, $delta(0,b) = 2$, $delta(99,a) = 99$, $delta(99,b) = 99$ and so on. In this case *delta* is the only observer, so the derivative set consists of *delta*-derivatives defined for every element from the alphabet: `delta(*:State, a)` and `delta(*:State, b)`. CIRC manages to prove that the state 0 of the first automaton is bisimilar to state 99 of the second automaton:

```
Maude> (add goal 0 = 99 .)
Maude> (coinduction .)
Proof succeeded.
  Number of derived goals: 6
  Number of proving steps performed: 30
  Maximum number of proving steps is set to: 256
Proved properties:
  2 = 99
  1 = 99
  0 = 99
```

The tool needs in this case to automatically prove two other bisimilarity relations between the states 2 and 99 and respectively 1 and 99.

Example 6. In this scenario we exhibit the use of the special contexts. A particular definition for special contexts was given for the first time in [4]. A more detailed presentation of special contexts and of how they are used to extend circular coinduction is given in [6].

We assume that we want to prove simultaneously the following goals: $morse = altMorse$ and $f(S) = zip(S, not(S))$. Here is dialog with CIRC:

```
Maude> in streams.maude
Maude> (add goal morse = altMorse .)
Maude> (add goal f(S:Stream) = zip(S:Stream, not(S:Stream)) .)
Maude> (coinduction .)
Stopped: the number of prover steps was exceeded.
```

At this moment, the command (show proof .) lists all the steps it managed to perform before reaching the limit. Two of the frozen hypotheses added to the specification are: $\boxed{zip(S, not(S))} = \boxed{f(S)}$ and $\boxed{tl(morse)} = \boxed{tl(altMorse)}$. In order to see what is happening, let us consider the complete definitions for *morse* and *altMorse*:

$$hd(morse) = 0 \qquad\qquad\qquad hd(altMorse) = 0$$

$$hd(tl(morse)) = 1 \qquad\qquad\qquad hd(tl(altMorse)) = 1$$

$$tl^2(morse) = zip(tl(morse), not(tl(morse))) \qquad tl^2(altMorse) = f(tl(altMorse))$$

CIRC cannot reduce $\boxed{tl^2(morse)} = \boxed{tl^2(altMorse)}$ because the above frozen hypotheses cannot be applied under zip and not. However, for this case, the deductions

$$\frac{\vdash \boxed{tl(morse)} = \boxed{tl(altMorse)}}{\vdash \boxed{not(tl(morse))} = \boxed{not(tl(altMorse))}}$$

$$\frac{\vdash \boxed{tl(morse)} = \boxed{tl(altMorse)}, \ \boxed{not(tl(morse))} = \boxed{not(tl(altMorse))}}{\vdash \boxed{zip(tl(morse), not(tl(morse)))} = \boxed{zip(tl(altMorse), not(tl(altMorse)))}}$$

are sound. We say that the contexts $not(*{:}Stream)$, $zip(*{:}Stream, S{:}Stream)$, and $zip(S{:}Stream, *{:}Stream)$ are *special*, because they allow to use frozen hypotheses under them in the deduction process.

The current version of CIRC includes an algorithm for computing special contexts for a given specification. To activate this algorithm, we have to set on the switch `auto contexts` before loading the c-theory, using the command (`set auto contexts on .`). After the c-theory is loaded, CIRC outputs the computed special contexts:

```
The special contexts are:
not *:Stream
zip(V#1:Stream,*:Stream)
zip(*:Stream,V#2:Stream)
```

Having the special contexts computed, CIRC successfully terminates the proof computation for the above property:

```
Maude> (add goal morse = altMorse .)
Maude> (add goal f(S:Stream) = zip(S:Stream, not(S:Stream)) .)
Maude> (coinduction .)
Proof succeeded.
  Number of derived goals: 8
  Number of proving steps performed: 41
  Maximum number of proving steps is set to: 256
Proved properties:
  tl(morse) = tl(altMorse)
  tl(f(S:Stream)) = zip(not(S:Stream),tl(S:Stream))
  morse = altMorse
  f(S:Stream) = zip(S:Stream,not(S:Stream))
```

We see above that CIRC does not report the context $f(*{:}Stream)$ as being special. The problem of computing the special contexts is not decidable, so the algorithm implemented in CIRC is not able to always find all the special contexts. CIRC includes the facility to declare a context as special, provided that the user guarantees for that.

Other examples of properties which can be proved using the special contexts are: $S_1 \times (S_2 + S_3) = S_1 \times S_2 + S_1 \times S_3$ and $(S_1 + S_2) \times S_3 = S_1 \times S_3 + S_2 \times S_3$ for streams, similar properties for infinite binary trees [5], the equivalence for basic process algebra, and well-definedness of stream definitions [10]. All these examples are available on the web site of the tool.

4   Conclusions

The development of CIRC started about three years ago. The main focus up to now was to implement the circular coinduction engine and to use it on many examples. The experience we gained in this period helped us better understand how circular coinduction and its extension with special contexts can be recast as a formal proof system [8,6].

The special contexts were useful for proving many properties of streams and infinite binary trees. They were also useful for proving the well-definedness of several stream definitions inspired from [10].

The potential of the ideas included in [4] is not exhausted. For instance, the case analysis is not included in the current version. A former version of CIRC included some facilities in this respect, but it needed manual assistance. One of our future aim is to extend CIRC with automated case analysis.

Acknowledgment.   The paper is supported in part by NSF grants CCF-0448501, CNS-0509321 and CNS-0720512, by NASA contract NNL08AA23C, by the Microsoft/Intel funded Universal Parallel Computing Research Center at UIUC, and by CNCSIS grant PN-II-ID-393.

References

1. Allouche, J.-P., Arnold, A., Berstel, J., Brlek, S., Jockusch, W., Plouffe, S., Sagan, B.E.: A sequence related to that of Thue-Morse. Discrete Mathematics 139, 455–461 (1995)
2. Allouche, J.-P., Shallit, J.: The ubiquitous prouhet-thue-morse sequence. In: Ding, C., Helleseth, T., Niederreiter, H. (eds.) Sequences and Their applications (Proc. SETA 1998), pp. 1–16. Springer, Heidelberg (1999)
3. Caltais, G., Goriac, E.-I., Lucanu, D., Grigoraş, G.: A Rewrite Stack Machine for ROC! Technical Report TR 08-02, "Al.I.Cuza" University of Iaşi, Faculty of Computer Science (2008), http://www.infoiasi.ro/~tr/tr.pl.cgi
4. Goguen, J., Lin, K., Roşu, G.: Conditional circular coinductive rewriting with case analysis. In: Wirsing, M., Pattinson, D., Hennicker, R. (eds.) WADT 2003. LNCS, vol. 2755, pp. 216–232. Springer, Heidelberg (2003)
5. Grigoraş, G., Caltais, D.L.G., Goriac, E.: Automated proving of the behavioral attributes. Accepted for the 4th Balkan Conference in Informatics, BCI 2009 (2009)
6. Lucanu, D., Roşu, G.: Circular Coinduction with Special Contexts. Technical Report UIUCDCS-R-2009-3039, University of Illinois at Urbana-Champaign (submitted, 2009)
7. Lucanu, D., Roşu, G.: Circ: A circular coinductive prover. In: Mossakowski, T., Montanari, U., Haveraaen, M. (eds.) CALCO 2007. LNCS, vol. 4624, pp. 372–378. Springer, Heidelberg (2007)
8. Roşu, G., Lucanu, D.: Circular Coinduction – A Proof Theoretical Foundation. In: Kurz, A., Lenisa, M., Tarlecki, A. (eds.) CALCO 2009. LNCS, vol. 5728, pp. 127–144. Springer, Heidelberg (2009)
9. Schroeder, M.R.: Fractals, Chaos, Power Laws. W.H. Freeman, New York (1991)
10. Zantema, H.: Well-definedness of streams by termination (submitted, 2009)

The Priced-Timed Maude Tool

Leon Bendiksen and Peter Csaba Ölveczky

Department of Informatics, University of Oslo
lobendik@ifi.uio.no, peterol@ifi.uio.no

Abstract. Priced-Timed Maude is a new rewriting-logic-based tool that extends Real-Time Maude to support the formal specification and analysis of systems in which the cost of performing actions and of idling plays a significant role. Our tool emphasizes expressiveness and ease of specification, and is particularly suitable for modeling distributed priced systems in an object-oriented style.

1 Introduction

Reasoning about the accumulated *cost* (say, *price* or *energy usage*) during behaviors is crucial in embedded systems and wireless sensor networks where minimizing overall energy consumption is critical. *Priced* (or *weighted*) *timed automata* (PTA) [1,2] extend timed automata to priced systems, and the PTA tool UPPAAL CORA [3] has been successfully applied to a range of applications [4]. While the restrictive PTA formalism ensures that many systems properties are decidable [5], such finite-control PTA do not support well the modeling of larger and advanced systems where different communication models, unbounded data structures, advanced functions, and/or object-oriented features such as class inheritance and dynamic object creation and deletion are needed.

This paper introduces the Priced-Timed Maude tool. Our tool extends the rewriting-logic-based Maude [6] and Real-Time Maude [7,8] tools to formally model and analyze priced and priced real-time systems. Priced-Timed Maude emphasizes expressiveness and ease of specification; the price to pay is that many system properties are no longer decidable. Being an extension of (Real-Time) Maude, our tool inherits its characteristic features:

- a general and expressive, yet simple and intuitive, specification formalism;
- suitability for specifying distributed systems in an object-oriented style;
- supporting a range of formal analysis methods, including symbolic simulation, cost/time-bounded reachability analysis, LTL model checking, and search for cost/time-optimal solutions.

Real-Time Maude has been successfully applied to a wide array of advanced state-of-the-art applications that are beyond the pale of timed automata, including the CASH scheduling algorithm [9] (where unbounded data structures are needed) and density and topology control algorithms for wireless sensor networks [10,11] (with novel forms of communication and where complex functions are needed for computing, e.g., the overlaps of different sensing areas).

A. Kurz, M. Lenisa, and A. Tarlecki (Eds.): CALCO 2009, LNCS 5728, pp. 443–448, 2009.

Priced-Timed Maude has been applied to benchmarks such as energy task graph scheduling, the airplane landing problem, and to the larger problem of efficiently routing passengers within a subway network while minimizing power consumption of the trains [12]. Given the experiences with Maude and Real-Time Maude, there is ample reason to believe that Priced-Timed Maude can be successfully applied to "real" state-of-the-art priced (timed) systems.

Price-Timed Maude is available free of charge at `http://www.ifi.uio.no/RealTimeMaude/PricedTimedMaude` and is described in detail in [12].

2   Specification and Analysis in Priced-Timed Maude

Recall that a Real-Time Maude specification is a tuple (Σ, E, IR, TR), where

- (Σ, E) is a theory in *membership equational logic* [13], with Σ a signature and E a terminating and confluent set of conditional equations and membership axioms. (Σ, E) specifies the system's state space as an algebraic data type.
- *IR* is a set of *conditional instantaneous rewrite rules* specifying the system's *instantaneous* (i.e., zero-time) local transition patterns.
- *TR* is a set of *tick rewrite rules* which model the time elapse in the system and have the form `{t} => {t'}` in time τ if *cond*, where τ is a term denoting the *duration* of the rule, t and t' are terms, and `{_}` is an operator encapsulating the global state.

All large Real-Time Maude applications have been specified in an object-oriented style, where the state of the system is represented by a term `{t}`, with t a term of sort `Configuration` that has the structure of a *multiset* of objects and messages.

Priced-Timed Maude extends Real-Time Maude by supporting the definition of the following kinds of *priced* rewrite rules:

- "global instantaneous priced rules" `{t} => {t'}` with cost u if *cond*
- "local object-oriented priced rules" c `=>` c' with cost u if *cond*, where c and c' are terms of sort `Configuration`
- priced tick rules of the form `{t} => {t'}` in time τ with cost u if *cond*,

where u denotes a term of sort `Cost` that may contain variables. A main technical difference w.r.t. Real-Time Maude is that *tick* rules apply to the *whole* state, whereas *priced* rules may also apply to *local* actions in object-oriented systems.

Priced-Timed Maude modules are executable under reasonable conditions. Priced-Timed Maude extends Real-Time Maude's analysis methods to the priced setting by supporting the following analysis methods:

- *priced rewriting* that simulates *one* behavior of the system up to a given time and/or cost limit;
- explicit-state breadth-first *search* that analyzes whether a state matching the search pattern can be reached from the initial state within a given time and/or cost limit;
- finding *optimal* solutions, such as the cheapest cost needed to reach a desired state; and
- explicit-state ("un-priced") *linear temporal logic model checking*.

3   Example: A Multiset of Simple Thermostats

The following Priced-Timed Maude module models a collection of fairly simple thermostats. Each ThStat object has an attribute temp denoting the current temperature, attributes low and high denoting the desired temperature range, and an attribute state that is on when the associated heater is turned on and is off otherwise. In our simple model, the temperature in a room increases by two degrees per time unit when its heater is turned on, otherwise the temperature decreases by one degree per time unit. The goal is to keep the temperature in each room in the desired range by turning the heater off when the temperature is high (rule turnOff) and by turning the heater on when it is low (rule turnOn). Furthermore, the rule turnOn can be nondeterministically applied at *any* time (modeling a user pushing a 'heater-on' button). The energy *cost* of turning on a heater depends on the current temperature (and is defined to be $(100 - x)/10 + 2$ for current temperature x), and each heater consumes 3 units of energy per time unit when it is turned on. (See [6] for an explanation of the Maude syntax.)

```
(ptomod HEATER is  including POSRAT-TIME-DOMAIN-WITH-INF .
                   including POSRAT-COST-DOMAIN .  including STRING .
  sort OnOff .   ops on off : -> OnOff .   subsort String < Oid .

  class ThStat | state : OnOff, low : PosRat, high : PosRat, temp : PosRat .

  vars T T' : PosRat .        var O : Oid .
  var C : Configuration .     var τ : Time .

  rl [turnOn] :
     < O : ThStat | state : off, temp : T >   =>
     < O : ThStat | state : on >  with cost (100 - T) quo 10 + 2 .

  rl [turnOff] :
     < O : ThStat | state : on, high : T, temp : T >   =>
     < O : ThStat | state : off > .

  crl [tick] :
     {C} => {delta(C,τ)} in time τ with cost (rate(C) * τ) if τ <= mte(C) .

  eq delta(< O : ThStat | state : on, temp : T >, τ) =
          < O : ThStat | temp : T + (2 * τ) > .
  eq delta(< O : ThStat | state : off, temp : T >, τ) =
          < O : ThStat | temp : T - τ > .

  eq mte(< O : ThStat | state : on, high : T, temp : T' >) = (T - T') / 2 .
  eq mte(< O : ThStat | state : off, low : T, temp : T' >) = T' - T .

  eq rate(< O : ThStat | state : off >) = 0 .
  eq rate(< O : ThStat | state : on >) = 3 .
endptom)
```

The tick rule is typical for object-oriented priced-timed systems, where `delta` defines the effect of time elapse on a system, `mte` defines the maximal time that can elapse until some action must be taken, and `rate` defines the cost per time unit when "idling." The skeletons of these functions are predefined; the user must only define them for single objects, as in the equations above.

The following command checks how much it costs to reach a state where the bathroom temperature is 78 degrees, from a state with three thermostat objects:

```
Maude> (find cheapest
  {< "BedRoom" : ThStat | state : off, low : 62, high : 70, temp : 66 >
   < "BathRoom" : ThStat | state : off, low : 74, high : 78, temp : 76 >
   < "Kitchen" : ThStat | state : on, low : 70, high : 76, temp : 72 >}
  =>*
  {< "BathRoom" : ThStat | temp : 78 >  C:Configuration}
  with no time limit.)

Solution C:Configuration --> ... ;
TIME_ELAPSED:Time --> 1 ;    TOTAL_COST_INCURRED:Cost --> 10
```

The minimum cost needed to reach a desired state is 10 (4 to turn on the bathroom heater, and 3 each to let the kitchen and bathroom heater burn for one time unit).

4   Comparison with UPPAAL CORA

Priced-Timed Maude is strictly more expressive than UPPAAL CORA, the only other formal tool for priced timed systems we know, since any 1PTA (a PTA with one cost variable) can easily be defined as a Priced-Timed Maude specification [12]. On the other hand, not only is rewriting Turing-complete, but the cost of applying a transition or of idling may be a function of the current state, which is not the case for PTA. The thermostat system in Section 3 cannot be modeled as a multi-priced timed automaton (because the "clock" rates are 3 and -1, and because the cost of turning on the heater is a function of the state). The state-of-the-art OGDC algorithm for wireless sensor networks [14] is another example of a priced real-time system—where energy consumption is a critical cost—that can be modeled in Price-Timed Maude (by modifying the Real-Time Maude model of OGDC (see [10]) that was developed before Price-Timed Maude existed), but that cannot be modeled as a PTA due to the need to model advanced data types and functions for computing, e.g., angles and overlaps between sensing areas.

Because of our tool's expressiveness, not only are most system properties undecidable in general, but it is also hard or impossible to apply techniques such as the use of priced zones. Instead, our tool performs explicit-state analysis by internally translating commands into Maude commands.

We have compared the performance of UPPAAL CORA and Priced-Timed Maude on the energy task graph scheduling benchmark, and found that Priced-Time Maude was significantly slower (by factors larger than 100 for large states;

see [12] for details). Nevertheless, the underlying Real-Time Maude tool has proved useful to exhaustively analyze very sophisticated state-of-the-art systems that *cannot* be modeled by timed automata. There is therefore reason to believe that Priced-Timed Maude can be successfully applied to advanced priced systems. Furthermore, since functions are defined by equations, which do not increase the state space, there may well be examples where Priced-Timed Maude outperforms UPPAAL CORA.

5   Future Work

This work should continue in different directions. First, we should further analyze the relationship between Priced-Timed Maude and UPPAAL CORA by investigating whether there exist priced systems that can be modeled as PTAs but whose Priced-Timed Maude analysis is more efficient than the corresponding UPPAAL CORA analysis. Second, the current implementation of Priced-Timed Maude only supports *local* priced rules in "flat" object-oriented specifications. Although all large Real-Time Maude applications have been specified as flat object-oriented systems, future versions of the Priced-Timed Maude tool should support *local* priced rules in "nested" object-oriented systems and in non-object-oriented systems. Third, we should define an appropriate "priced temporal logic" and support model checking of formulas in such a logic. Finally, we should extend Priced-Timed Maude to *multi-priced* systems, that is, to systems with more than one kind of cost, such as one where actions have, say, both a *pollution* and an *energy* cost.

References

1. Behrmann, G., Fehnker, A., Hune, T., Larsen, K.G., Pettersson, P., Romijn, J.M.T., Vaandrager, F.W.: Minimum-cost reachability for priced timed automata. In: Di Benedetto, M.D., Sangiovanni-Vincentelli, A.L. (eds.) HSCC 2001. LNCS, vol. 2034, pp. 147–161. Springer, Heidelberg (2001)
2. Alur, R., La Torre, S., Pappas, G.J.: Optimal paths in weighted timed automata. In: Di Benedetto, M.D., Sangiovanni-Vincentelli, A.L. (eds.) HSCC 2001. LNCS, vol. 2034, pp. 49–62. Springer, Heidelberg (2001)
3. Behrmann, G.: Uppaal Cora web page, http://www.cs.aau.dk/~behrmann/cora/ (accessed February 2009)
4. Behrmann, G., Larsen, K.G., Rasmussen, J.I.: Priced timed automata: Algorithms and applications. In: de Boer, F.S., Bonsangue, M.M., Graf, S., de Roever, W.-P. (eds.) FMCO 2004. LNCS, vol. 3657, pp. 162–182. Springer, Heidelberg (2005)
5. Bouyer, P., Larsen, K.G., Markey, N.: Model-checking one-clock priced timed automata. In: Seidl, H. (ed.) FOSSACS 2007. LNCS, vol. 4423, pp. 108–122. Springer, Heidelberg (2007)
6. Clavel, M., Durán, F., Eker, S., Lincoln, P., Martí-Oliet, N., Meseguer, J., Talcott, C.: All About Maude - A High-Performance Logical Framework. LNCS, vol. 4350. Springer, Heidelberg (2007)
7. Ölveczky, P.C., Meseguer, J.: The Real-Time Maude Tool. In: Ramakrishnan, C.R., Rehof, J. (eds.) TACAS 2008. LNCS, vol. 4963, pp. 332–336. Springer, Heidelberg (2008)

8. Ölveczky, P.C., Meseguer, J.: Semantics and pragmatics of Real-Time Maude. Higher-Order and Symbolic Computation 20(1-2), 161–196 (2007)
9. Ölveczky, P.C., Caccamo, M.: Formal simulation and analysis of the CASH scheduling algorithm in Real-Time Maude. In: Baresi, L., Heckel, R. (eds.) FASE 2006. LNCS, vol. 3922, pp. 357–372. Springer, Heidelberg (2006)
10. Ölveczky, P.C., Thorvaldsen, S.: Formal modeling, performance estimation, and model checking of wireless sensor network algorithms in Real-Time Maude. Theoretical Computer Science 410(2-3), 254–280 (2009)
11. Katelman, M., Meseguer, J., Hou, J.: Redesign of the LMST wireless sensor protocol through formal modeling and statistical model checking. In: Barthe, G., de Boer, F.S. (eds.) FMOODS 2008. LNCS, vol. 5051, pp. 150–169. Springer, Heidelberg (2008)
12. Bendiksen, L.: Specification and analysis of priced systems in Priced-Timed Maude. Master's thesis, University of Oslo (2008)
13. Meseguer, J.: Membership algebra as a logical framework for equational specification. In: Parisi-Presicce, F. (ed.) WADT 1997. LNCS, vol. 1376. Springer, Heidelberg (1998)
14. Zhang, H., Hou, J.C.: Maintaining sensing coverage and connectivity in large sensor networks. Wireless Ad Hoc and Sensor Networks: An International Journal 1 (2005)

A Tool Proving Well-Definedness of Streams Using Termination Tools

Hans Zantema

Department of Computer Science, TU Eindhoven, P.O. Box 513,
5600 MB Eindhoven, The Netherlands
H.Zantema@tue.nl
and
Institute for Computing and Information Sciences, Radboud University
Nijmegen, P.O. Box 9010, 6500 GL Nijmegen, The Netherlands

Abstract. A stream specification is a set of equations intended to define a stream, that is, an infinite sequence over a given data type. In [5] a transformation from such a stream specification to a TRS is defined in such a way that termination of the resulting TRS implies that the stream specification admits a unique solution. In this tool description we present how proving such well-definedness of several interesting boolean stream specifications can be done fully automatically using present powerful tools for proving TRS termination.

1 Introduction

Streams are among the simplest data types in which the objects are infinite. We consider streams to be maps from the natural numbers to some data type D. The basic constructor for streams is the operator ':' mapping a data element d and a stream s to a new stream $d : s$ by putting d in front of s. Using this operator we can define streams by equations. For instance, the stream zeros only consisting of 0's can be defined by the single equation $\mathsf{zeros} = 0 : \mathsf{zeros}$.

More complicated streams are defined using stream functions. For instance, a boolean steam c together with stream functions f and g is uniquely defined by

$$c = 1 : f(c) \qquad\qquad g(0, xs) = 0 : 1 : f(xs)$$
$$f(x : xs) = g(x, xs) \qquad\qquad g(1, xs) = 0 : f(xs),$$

in fact, c is the tail of the well-known Fibonacci stream. However, the following modification of this stream specification

$$c = 1 : f(c) \qquad\qquad g(0, xs) = 0 : 1 : f(xs)$$
$$f(x : xs) = g(x, xs) \qquad\qquad g(1, xs) = f(xs)$$

does **not** uniquely define a stream c.

The main issue we address is the following: given a stream specification like the two examples above, can we automatically check that it uniquely defines the streams and stream functions involved? If so, we call the stream specification

A. Kurz, M. Lenisa, and A. Tarlecki (Eds.): CALCO 2009, LNCS 5728, pp. 449–456, 2009.

well-defined. For this question in [5] a technique has been proposed in which the
stream specification S is transformed to its *observational variant* $\mathsf{Obs}(S)$, being a
term rewrite system (TRS). The main result of [5] states that if the TRS $\mathsf{Obs}(S)$
is terminating, then the stream specification S admits a unique solution. As for
proving termination of TRSs automatically during the last ten years strong tools
have been developed like AProVE [2] and TTT2 [3], our approach to prove well-
definedness of a stream specification S simply consists of proving termination of
$\mathsf{Obs}(S)$ by a tool like AProVE or TTT2. It is not even required to download such
a termination tool: both tools admit a web interface in which any TRS is easily
entered. Our tool transforms any boolean stream specification S in the format as
given here to $\mathsf{Obs}(S)$ in the TRS format as is accepted by the termination tools.
A full version with graphical user interface runs under Windows; a command
line version runs under Linux. Both versions are downloaded from

`www.win.tue.nl/~hzantema/str.zip`

together with a number of examples.

This paper is structured as follows. In Section 2 we present the basic definitions
and theorems. However, playing around with the tool and trying termination
tools on the resulting systems can be done without mastering all details. In
Section 3 we describe the features of the tool. In Section 4 we show a number of
examples. We conclude in Section 5.

2   Definitions and Theory

In stream specifications we have two sorts: s (stream) and d (data). We write
D for the set of data elements; here we restrict to the boolean case where $D =
\{0, 1\}$. We assume a particular symbol : having type $d \times s \to s$. For giving the
actual stream specification we need a set Σ_s of stream symbols, each being of
type $d^n \times s^m \to s$ for $n, m \geq 0$. Terms of sort s are defined in the obvious way.
As a notational convention we will use

- x, y, z for variables of sort d,
- u, u_i for terms of sort d,
- xs, ys, zs for variables of sort s,
- t, t_i for terms of sort s.

Definition 1. *A* stream specification (Σ_s, S), *or shortly* S, *consists of* Σ_s *as
given before, and a set* S *of equations of the shape* $f(u_1, \ldots, u_n, t_1, \ldots, t_m) = t$,
where

- $f \in \Sigma_s$ *is of type* $d^n \times s^m \to s$,
- $u_1, \ldots, u_n$ *are terms of sort* d,
- *for every* $i = 1, \ldots, m$ *the term* t_i *is either a variable of sort* s, *or* $t_i = x : xs$
  where x *is a variable of sort* d *and* xs *is a variable of sort* s,
- t *is any term of sort* s,
- *no variable occurs more than once in* $f(u_1, \ldots, u_n, t_1, \ldots, t_m)$,

- *for every term of the shape* $f(u_1, \ldots, u_n, u_{n+1} : t_1, \ldots, u_{n+m} : t_m)$ *where* $u_i \in D$ *for* $i = 1, \ldots, m + n$, *and* $t_1, \ldots, t_m$ *are variables, exactly one rule from* S *is applicable.*

Example 1. The Thue-Morse sequence morse can be specified by the following stream specification:

$$\begin{aligned}
\text{morse} &= 0 : \text{zip}(\text{inv}(\text{morse}), t(\text{morse})) & g(1, xs) &= 0 : \text{inv}(xs) \\
\text{inv}(x : xs) &= g(x, xs) & t(x : xs) &= xs \\
g(0, xs) &= 1 : \text{inv}(xs) & \text{zip}(x : xs, ys) &= x : \text{zip}(ys, xs).
\end{aligned}$$

Here g is introduced to meet our format: $\text{inv}(0 : xs)$ is not allowed as a left hand side. The operation t represents the tail function; the symbol tail is reserved for later use.

Stream specifications are intended to specify streams for the constants in Σ_s, and stream functions for the other elements of Σ_s. The combination of these streams and stream functions is what we will call a *stream model*. As streams over D are maps from the natural numbers to D we write D^ω for the set of all streams over D.

Definition 2. *A stream model is defined to consist of a set* $M \subseteq D^\omega$ *and a set of functions* $[f]$ *for every* $f \in \Sigma_s$, *where* $[f] : D^n \times M^m \to M$ *if the type of* f *is* $d^n \times s^m \to s$.

We write $\mathcal{T}_s$ for the set of ground terms of sort s. For $t \in \mathcal{T}_s$ the interpretation $[t]$ in a stream model is defined inductively by:

$$\begin{aligned}
[f(u_1, \ldots, u_n, t_1, \ldots, t_m)] &= [f]([u_1], \ldots, [u_n], [t_1], \ldots, [t_m]) \quad \text{for } f \in \Sigma_s \\
[u : t](0) &= [u] \\
[u : t](i) &= [t](i - 1) \quad\quad\quad\quad\quad\quad \text{for } i > 0
\end{aligned}$$

for $u, u_i \in D$ and $t, t_i \in \mathcal{T}_s$.

So in a stream model a user defined operator f is interpreted by the given function $[f]$, and the operator : applied on a data element d and a stream s is interpreted by putting d on the first position and shifting every stream element of s to its next position.

A stream model $(M, ([f])_{f \in \Sigma_s})$ is said to *satisfy* a stream specification (Σ_s, S) if $[\ell \rho] = [r \rho]$ for every equation $\ell = r$ in S and every ground substitution ρ. We also say that the specification *admits* the model.

Now we define a transformation Obs transforming the original stream specification S to its *observational variant* Obs(S). The basic idea is that the streams are observed by two auxiliary operator head and tail, of which head picks the first element of the stream and tail removes the first element from the stream, and that for every $t \in \mathcal{T}_s$ of type stream both head(t) and tail(t) can be rewritten by Obs(S).

First we define P(S) obtained from S by modifying the equations as follows. By definition every equation of S is of the shape $f(u_1, \ldots, u_n, t_1, \ldots, t_m) = t$ where for every $i = 1, \ldots, m$ the term t_i is either a variable of sort s, or $t_i = x : xs$ where x is a variable of sort d and xs is a variable of sort s. In case $t_i = x : xs$ then in the left hand side of the rule the subterm t_i is replaced by xs, while in the right hand side of the rule every occurrence of x is replaced by head(xs) and every occurrence of xs is replaced by tail(xs).

For example, the zip rule in Example 1 will be replaced by

$$\mathsf{zip}(xs, ys) = \mathsf{head}(xs) : \mathsf{zip}(ys, \mathsf{tail}(xs)).$$

Now we are ready to define Obs.

Definition 3. *Let* (Σ_s, S) *be a stream specification. Let* P(S) *be defined as above. Then* Obs(S) *is the TRS consisting of*

- *the two rules* head($x : xs$) $\to x$, tail($x : xs$) $\to xs$,
- *for every equation in* P(S) *of the shape* $\ell = u : t$ *the two rules*

$$\mathsf{head}(\ell) \to u, \quad \mathsf{tail}(\ell) \to t,$$

- *for every equation in* P(S) *of the shape* $\ell = r$ *with* root(r) $\neq :$ *the two rules*

$$\mathsf{head}(\ell) \to \mathsf{head}(r), \quad \mathsf{tail}(\ell) \to \mathsf{tail}(r).$$

From [5] we recall the two variants of the main theorem.

Theorem 1. *Let* (Σ_s, S) *be a stream specification for which the TRS* Obs(S) *is terminating. Then the stream specification admits a unique model* $(M, ([f])_{f \in \Sigma_s})$ *satisfying* $M = \{[t] \mid t \in \mathcal{T}_s\}$.

Theorem 2. *Let* (Σ_s, S) *be a stream specification for which the TRS* Obs(S) *is terminating and the only subterms of left hand sides of* S *of sort* d *are variables. Then the stream specification admits a unique model* $(M, ([f])_{f \in \Sigma_s})$ *satisfying* $M = D^\omega$.

In [5] it is shown by an example that this distinction is essential: uniqueness of stream functions does not hold if the stream specification is data dependent, that is, left hand sides contain symbols 0 and 1. Moreover, in [5] it has been shown that the technique is not complete: some fix-point specifications are well-defined while the observational variant is non-terminating.

3   The Tool

The full version of the tool with graphical user interface under Windows provides the following features:

- Boolean stream specifications can be entered, loaded, edited and stored. The format is the same as given here, with the only difference that for the operator ':' a prefix notation is chosen, in order to be consistent with the user defined symbols.

- By clicking a button the observational variant of the current stream specification is tried to be created. In doing so, all requirements of the definition of stream specification are checked. If they are not fulfilled, an appropriate error message is shown.
- If all requirements hold, then the resulting observational variant is shown on the screen by which it can be entered by cut and paste in a termination tool. Alternatively, it can be stored.
- An initial part of a stream can be computed by applying outermost rewriting based on the stream specification starting in a given term.
- Several stream specifications, including Fibonacci stream, Thue-Morse stream and Kolakoski stream are predefined. For all of these examples termination of the observational variant can be proved fully automatically both by AProVE [2] and TTT2 [3], proving well-definedness of the given stream specification.

The main facility, that is, generating the observational variant in case the syntax is correct, and generating an appropriate error message otherwise, is also provided by the command line verison under Linux.

None of the actions require substantial computation: for all features the result shows up instantaneously. On the other hand, proving termination of a resulting observational variant by a tool like AProVE or TTT2 may take some computation time, although never more than a few seconds for the given examples. This was one of the objectives of the project: the transformation itself should be simple and direct, while the real work to be done makes use of the power of current termination provers.

4 Examples

In this section we give a few examples to give an impression of how the input and output of our tool looks like. The first example is the Fibonacci stream as presented in the introduction. The input is

```
f(:(x,xs)) = g(x,xs)
g(0,xs) = :(0,:(1,f(xs)))
g(1,xs) = :(0,f(xs))
c = :(1,f(c))
fib = :(0,c)
```

Within a fraction of a second our tool transforms this to the observational variant

```
(VAR xs ys x)
(RULES
hd(f(xs)) -> hd(xs)
tl(f(xs)) -> zip(f(odd(tl(xs))),f(even(tl(xs))))
hd(zip(xs,ys)) -> hd(xs)
tl(zip(xs,ys)) -> zip(ys,tl(xs))
hd(odd(xs)) -> hd(xs)
tl(odd(xs)) -> even(tl(xs))
```

```
hd(even(xs)) -> hd(odd(tl(xs)))
tl(even(xs)) -> tl(odd(tl(xs)))
hd(c) -> 0
tl(c) -> :(1,c)
hd(:(x,xs)) -> x
tl(:(x,xs)) -> xs
)
```

As a second example we consider the Thue-Morse sequence as specified in Example 1:

```
morse = :(0,zip(inv(morse),t(morse)))
inv(:(x,xs)) = g(x,xs)
g(0,xs) = :(1,inv(xs))
g(1,xs) = :(0,inv(xs))
t(:(x,xs)) = xs
zip(:(x,xs),ys) = :(x,zip(ys,xs))
```

yielding as output:

```
(VAR xs ys x)
(RULES
hd(morse) -> 0
tl(morse) -> zip(inv(morse),t(morse))
hd(inv(xs)) -> hd(g(hd(xs),tl(xs)))
tl(inv(xs)) -> tl(g(hd(xs),tl(xs)))
hd(g(0,xs)) -> 1
tl(g(0,xs)) -> inv(xs)
hd(g(1,xs)) -> 0
tl(g(1,xs)) -> inv(xs)
hd(t(xs)) -> hd(tl(xs))
tl(t(xs)) -> tl(tl(xs))
hd(zip(xs,ys)) -> hd(xs)
tl(zip(xs,ys)) -> zip(ys,tl(xs))
hd(:(x,xs)) -> x
tl(:(x,xs)) -> xs
)
```

Finally we consider the stream specification

```
f(:(x,xs)) = g(x,xs)
g(0,xs) = :(1,f(xs))
g(1,xs) = :(0,f(f(xs)))
c = :(1,c)
conc = f(c)
```

yielding as output:

```
(VAR xs x)
(RULES
hd(f(xs)) -> hd(g(hd(xs),tl(xs)))
tl(f(xs)) -> tl(g(hd(xs),tl(xs)))
hd(g(0,xs)) -> 1
tl(g(0,xs)) -> f(xs)
hd(g(1,xs)) -> 0
tl(g(1,xs)) -> f(f(xs))
hd(c) -> 1
tl(c) -> c
hd(conc) -> hd(f(c))
tl(conc) -> tl(f(c))
hd(:(x,xs)) -> x
tl(:(x,xs)) -> xs
)
```

Both AProVE [2] and TTT2 [3] are able to prove termination each of the resulting TRSs within a few seconds, in this way proving well-definedness of all three original specifications. Unfortunately the generated proofs of both tools are quite long and not easily readable.

5   Conclusions, Related Work

We presented a technique and described a tool by which well-definedness of stream specifications can be proved fully automatically. This is done by proving termination of a transformed term rewrite system $\mathsf{Obs}(S)$ using any termination provers. In this way the power of present termination provers is exploited.

We mention two other approaches to this problem. The first is based on the well-known property that productivity implies well-definedness. An approach to prove productivity is given in [1], together with a corresponding tool. The second s by proving equality of two copies of the same specification by the tool Circ [4]. In the latter it is essential that the specification is deterministic.

However, for proving well-definedness of $f(c)$ in

$$f(0:xs) = 1:f(xs) \qquad c = 1:c$$
$$f(1:xs) = 0:f(f(xs))$$

both other approaches fail, while our approach succeeds. This specification in our format (introducing g) and its observational variant were presented in Section 4 as the third example.

Conversely, there are examples where the other approaches succeed and our approach fails. For instance, productivity of M of the specification

$$M = 1:\mathsf{even}(\mathsf{zip}(M,M))$$
$$\mathsf{zip}(a:s,t) = a:\mathsf{zip}(t,s)$$
$$\mathsf{even}(a:s) = a:\mathsf{odd}(s)$$
$$\mathsf{odd}(a:s) = \mathsf{even}(s)$$

is easily proved by the tool of [1], while our approach fails to prove well-definedness.

References

1. Endrullis, J., Grabmayer, C., Hendriks, D.: Data-oblivious stream productivity. In: Cervesato, I., Veith, H., Voronkov, A. (eds.) LPAR 2008. LNCS (LNAI), vol. 5330, pp. 79–96. Springer, Heidelberg (2008)
2. Giesl, J., et al.: Automated program verification environment (AProVE), http://aprove.informatik.rwth-aachen.de/
3. Korp, M., Sternagel, C., Zankl, H., Middeldorp, A.: Tyrolean termination tool 2. In: Treinen, R. (ed.) RTA 2009. LNCS, vol. 5595, pp. 295–304. Springer, Heidelberg (2009), http://colo6-c703.uibk.ac.at/ttt2/
4. Lucanu, D., Rosu, G.: CIRC: A circular coinductive prover. In: Mossakowski, T., Montanari, U., Haveraaen, M. (eds.) CALCO 2007. LNCS, vol. 4624, pp. 372–378. Springer, Heidelberg (2007)
5. Zantema, H.: Well-definedness of Streams by Termination. In: Treinen, R. (ed.) RTA 2009. LNCS, vol. 5595, pp. 164–178. Springer, Heidelberg (2009), www.win.tue.nl/~hzantema/str.pdf

Author Index